Macmillan Encyclopedia of Physics

Editorial Board

MACMILLAN
ENCYCLOPEDIA OF
PHYSICS

John S. Rigden

Editor in Chief

Volume 3

MACMILLAN REFERENCE USA

Simon & Schuster Macmillan

NEW YORK

Simon & Schuster and Prentice Hall International

LONDON MEXICO CITY NEW DELHI SINGAPORE SYDNEY TORONTO

Simon & Schuster Macmillan
1633 Broadway
New York, NY 10019

Library of Congress Catalog Card Number: 96-30977

PRINTED IN THE UNITED STATES OF AMERICA

Printing Number

1 2 3 4 5 6 7 8 9 10

LIBRARY OF CONGRESS CATALOGING-IN-PUBLICATION DATA

Macmillan Encyclopedia of Physics / John S. Rigden, editor in chief.
    p.   cm.
    Includes bibliographical references and index.
    ISBN 0-02-897359-3 (set).
    1. Physics—Encyclopedias.   I. Rigden, John S.
    QC5.M15   1996
    530′.03—dc20   96-30977
                                                      CIP

This paper meets the requirements of ANSI-NISO Z39.48-1992
(Permanence of Paper).

COMMON ABBREVIATIONS AND MATHEMATICAL SYMBOLS

=	equals; double bond	| |	absolute value of
≠	not equal to	+	plus
≡	identically equal to; equivalent to; triple bond	−	minus
		/	divided by
~	asymptotically equal to; of the order of magnitude of; approximately	×	multiplied by
		⊕	direct sum
		⊗	direct product
≈, ≃	approximately equal to	±	plus or minus
≅	congruent to; approximately equal to	∓	minus or plus
		√	radical
∝	proportional to	∫	integral
<	less than	∮	contour integral
>	greater than	Σ	summation
≮	not less than	Π	product
≯	not greater than	∂	partial derivative
≪	much less than	°	degree
≫	much greater than	°B	degrees Baumé
≤	less than or equal to	°C	degrees Celsius (centigrade)
≥	greater than or equal to	°F	degrees Fahrenheit
≰	not less than or equal to	!	factorial
≱	not greater than or equal to	′	minute
∪	union of	″	second
∩	intersection of	∇	curl
⊂	subset of; included in		
⊃	contains as a subset	ϵ_0	electric constant
∈	an element of	μ	micro-
∋	contains as an element	μ_0	magnetic constant
→	approaches, tends to; yeilds; is replaced by	μA	microampere
		μA h	microampere hour
⇒	implies; is replaced by	μC	microcoulomb
⇐	is implied by	μF	microfarad
↓	mutually implies	μg	microgram
⇔	if and only if	μK	microkelvin
⊥	perpendicular to	μm	micrometer
‖	parallel to	μm	micron

μm Hg	microns of mercury
μmol	micromole
μs, μsec	microsecond
μu	microunit
$\mu\Omega$	microhm
σ	Stefan–Boltzmann constant
Ω	ohm
Ω cm	ohm centimeter
Ω cm/(cm/cm^3)	ohm centimeter per centimeter per cubic centimeter
A	ampere
Å	angstrom
a	atto-
A$_s$	atmosphere, standard
abbr.	abbreviate; abbreviation
abr.	abridged; abridgment
Ac	Actinium
ac	alternating-current
aF	attofarad
af	audio-frequency
Ag	silver
A h	ampere hour
AIP	American Institute of Physics
Al	aluminum
alt	altitude
Am	americium
AM	amplitude-modulation
A.M.	ante meridiem
amend.	amended; amendment
annot.	annotated; annotation
antilog	antilogarithm
app.	appendix
approx	approximate (in subscript)
Ar	argon
arccos	arccosine
arccot	arccotangent
arccsc	arccosecant
arc min	arc minute
arcsec	arcsecant
arcsin	arcsine
arg	argument
As	arsenic
At	astatine
At/m	ampere turns per meter
atm	atmosphere
at. ppm	atomic parts per million
at. %	atomic percent
atu	atomic time unit
AU	astronomical unit
a.u.	atomic unit
Au	gold

av	average (in subscript)
b	barn
b.	born
B	boron
Ba	barium
bcc	body-centered-cubic
B.C.E.	before the common era
Be	beryllium
Bi	biot
Bi	bismuth
Bk	berkelium
bp	boiling point
Bq	becquerel
Br	bromine
Btu, BTU	British thermal unit
C	carbon
c	centi-
c.	circa, about, approximately
C	coulomb
c	speed of light
Ca	calcium
cal	calorie
calc	calculated (in subscript)
c.c.	complex conjugate
CCD	charge-coupled device
Cd	cadmium
cd	candela
CD	compact disc
Ce	cerium
C.E.	common era
CERN	European Center for Nuclear Research
Cf	californium
cf.	confer, compare
cgs, CGS	centimeter-gram-second (system)
Ci	curie
Cl	chlorine
C.L.	confidence limits
c.m.	center of mass
cm	centimeter
Cm	curium
cm^3	cubic centimeter
Co	cobalt
Co.	Company
coeff	coefficient (in subscript)
colog	cologarithm
const	constant
Corp.	Corporation
cos	cosine
cosh	hyperbolic cosine
cot	cotangent
coth	hyperbolic cotangent

cp	candlepower		e.u.	electron unit
cP	centipoise		eu	entropy unit
cp	chemically pure		Eu	europium
cpd	contact potential difference		eV	electron volt
cpm	counts per minute		expt	experimental (in subscript)
cps	cycles per second		F	farad
Cr	chromium		F	Faraday constant
cS	centistoke		f	femto-
Cs	cesium		F	fermi
csc	cosecant		F	fluorine
csch	hyperbolic cosecant		fc	foot-candle
Cu	copper		fcc	face-centered-cubic
cu	cubic		Fe	iron
cw	continuous-wave		fF	femtofarad
D	Debye		Fig. (pl., Figs.)	figure
d	deci-		fL	foot-lambert
$d.$	died		fm	femtometer
da	deka-		Fm	fermium
dB, dBm	decibel		FM	frequency-modulation
dc	direct-current		f. (pl., ff.)	following
deg	degree		fpm	fissions per minute
det	determinant		Fr	francium
dev	deviation		Fr	franklin
diam	diameter		fs	femtosecond
dis/min	disintegrations per minute		ft	foot
dis/s	disintegrations per second		ft lb	foot-pound
div	divergence		ft lbf	foot-pound-force
DNA	deoxyribose nucleic acid		f.u.	formula units
Dy	dysprosium		g	acceleration of free fall
dyn	dyne		G	gauss
E	east		G	giga-
e	electronic charge		g	gram
E	exa-		G	gravitational constant
e, exp	exponential		Ga	gallium
e/at.	electrons per atom		Gal	gal (unit of gravitational force)
e b	electron barn		gal	gallon
e/cm3	electrons per cubic centimeter		g-at.	gram-atom
ed. (pl., eds.)	editor		g.at. wt	gram-atomic-weight
e.g.	exempli gratia, for example		Gc/s	gigacycles per second
el	elastic (in subscript)		Gd	gadolinium
emf, EMF	electromotive force		Ge	germanium
emu	electromagnetic unit		GeV	giga-electron-volt
Eng.	England		GHz	gigahertz
Eq. (pl., Eqs.)	equation		Gi	gilbert
Er	erbium		grad	gradient
erf	error function		GV	gigavolt
erfc	error function (complement of)		Gy	gray
Es	einsteinium		h	hecto-
e.s.d.	estimated standard deviation		H	henry
esu	electrostatic unit		h	hour
et al.	et alii, and others		H	hydrogen
etc.	et cetera, and so forth		h	Planck constant

H.c.	Hermitian conjugate	ks, ksec	kilosecond
hcp	hexagonal-close-packed	kt	kiloton
He	helium	kV	kilovolt
Hf	hafnium	kV A	kilovolt ampere
hf	high-frequency	kW	kilowatt
hfs	hyperfine structure	kW h	kilowatt hour
hg	hectogram	$k\Omega$	kilohm
Hg	mercury	L	lambert
Ho	holmium	L	langmuir
hp	horsepower	l, L	liter
Hz	hertz	La	lanthanum
I	iodine	LA	longitudinal-acoustic
ICT	International Critical Tables	lab	laboratory (in subscript)
i.d.	inside diameter	lat	latitude
i.e.	id est, that is	lb	pound
IEEE	Institute of Electrical and Electronics Engineers	lbf	pound-force
		lbm	pound-mass
if	intermediate frequency	LED	light emitting diode
Im	imaginary part	Li	lithium
in.	inch	lim	limit
In	indium	lm	lumen
Inc.	Incorporated	lm/W	lumens per watt
inel	inelastic (in subscript)	ln	natural logarithm (base e)
ir, IR	infrared	LO	longitudinal-optic
Ir	iridium	log	logarithm
J	joule	Lr	lawrencium
Jy	jansky	LU	Lorentz unit
k, k_B	Boltzmann's constant	Lu	lutetium
K	degrees Kelvin	lx	lux
K	kayser	ly, lyr	light-year
k	kilo-	M	Mach
K	potassium	M	mega-
kA	kiloamperes	m	meter
kbar	kilobar	m	milli-
kbyte	kilobyte	m	molal (concentration)
kcal	kilocalorie	M	molar (concentration)
kc/s	kilocycles per second	m_e	electronic rest mass
kdyn	kilodyne	m_n	neutron rest mass
keV	kilo-electron-volt	m_p	proton rest mass
kG	kilogauss	$M_\odot$	solar mass (2×10^{33} g)
kg	kilogram	MA	megaamperes
kgf	kilogram force	mA	milliampere
kg m	kilogram meter	ma	maximum
kHz	kilohertz	mb	millibarn
kJ	kilojoule	mCi	millicurie
kK	kilodegrees Kelvin	Mc/s	megacycles per second
km	kilometer	Md	mendlelvium
kMc/s	kilomegacycles per second	MeV	mega-electron-volt; million electron volt
kn	knot		
kOe	kilo-oersted	Mg	magnesium
kpc	kiloparsec	mg	milligram
Kr	krypton	mH	millihenry

mho	reciprocal ohm	No.	number
MHz	megahertz	Np	neper
min	minimum	Np	neptunium
min	minute	ns, nsec	nanosecond
mK	millidegrees Kelvin; millikelvin	n/s	neutrons per second
mks, MKS	meter-kilogram-second (system)	n/s cm^2	neutrons per second per square centimeter
mksa	meter-kilogram-second ampere	ns/m	nanoseconds per meter
mksc	meter-kilogram-second coulomb	O	oxygen
ml	milliliter	$o()$	of order less than
mm	millimeter	$O()$	of the order of
mmf	magnetomotive force	obs	observed (in subscript)
mm Hg	millimeters of mercury	o.d.	outside diameter
Mn	manganese	Oe	oersted
MO	molecular orbital	ohm^{-1}	mho
Mo	molybdenum	Os	osmium
MOE	magneto-optic effect	oz	ounce
mol	mole	P	peta-
mol %, mole %	mole percent	P	phosphorus
mp	melting point	p	pico-
Mpc	megaparsec	P	poise
mph	miles per hour	Pa	pascal
MPM	mole percent metal	Pa	protactinium
Mrad	megarad	Pb	lead
ms, msec	millisecond	pc	parsec
mu	milliunit	Pd	palladium
MV	megavolt; million volt	PD	potential difference
mV	millivolt	pe	probable error
MW	megawatt	pF	picofarad
mwe, m (w.e.)	meter of water equivalent	pl.	plural
Mx	maxwell	P.M.	post meridiem
mμm	millimicron	Pm	promethium
MΩ	megaohm	Po	polonium
n	nano-	ppb	parts per billion
N	newton	p. (pl., pp.)	page
N	nitrogen	ppm	parts per million
N	normal (concentration)	Pr	praseodymium
N	north	psi	pounds per square inch
N, N_A	Avogadro constant	psi (absolute)	pounds per square inch absolute
Na	sodium	psi (gauge)	pounds per square inch gauge
NASA	National Aeronautics and Space Administration	Pt	platinum
		Pu	plutonium
nb	nanobarn	R (ital)	gas constant
Nb	niobium	R	roentgen
Nd	neodymium	Ra	radium
N.D.	not determined	rad	radian
NDT	nondestructive testing	Rb	rubidium
Ne	neon	Re	real part
n/f	neutrons per fission	Re	rhenium
Ni	nickel	rev.	revised
N_L	Loschmidt's constant	rf	radio frequency
nm	nanometer	Rh	rhodium
No	nobelium		

r.l.	radiation length	tanh	hyperbolic tangent
rms	root-mean-square	Tb	terbium
Rn	radon	Tc	technetium
RNA	ribonucleic acid	Td	townsend
RPA	random-phase approximation	Te	tellurium
rpm	revolutions per minute	TE	transverse-electric
rps, rev/s	revolutions per second	TEM	transverse-electromagnetic
Ru	ruthenium	TeV	tera-electron-volt
Ry	rydberg	Th	thorium
s, sec	second	theor	theory, theoretical (in subscript)
S	siemens	THz	tetrahertz
S	south	Ti	titanium
S	stoke	Tl	thallium
S	sulfur	Tm	thulium
Sb	antimony	TM	transverse-magnetic
Sc	scandium	TO	transverse-optic
sccm	standard cubic centimeter per minute	tot	total (in subscript)
		TP	temperature-pressure
Se	selenium	tr, Tr	trace
sec	secant	trans.	translator, translators; translated by; translation
sech	hyperbolic secant		
sgn	signum function	u	atomic mass unit
Si	silicon	U	uranium
SI	*Système International* (International System of Measurement)	uhf	ultrahigh-frequency
		uv, UV	ultraviolet
sin	sine	V	vanadium
sinh	hyperbolic sine	V	volt
SLAC	Stanford Linear Accelerator Center	VB	valence band
		vol. (pl., vols.)	volume
Sm	samarium	vol %	volume percent
Sn	tin	vs.	versus
sq	square	W	tungsten
sr	steradian	W	watt
Sr	strontium	W	West
STP	standard temperature and pressure	Wb	weber
		Wb/m^2	webers per square meter
Suppl.	Supplement	wt %	weight percent
Sv	sievert	W.u.	Weisskopf unit
T	tera-	Xe	xenon
T	tesla	Y	yttrium
t	tonne	Yb	ytterbium
Ta	tantalum	yr	year
TA	transverse-acoustic	Zn	zinc
tan	tangent	Zr	zirconium

JOURNAL ABBREVIATIONS

Acc. Chem. Res.
Accounts of Chemical Research
Acta Chem. Scand.
Acta Chemica Scandinavica
Acta Crystallogr.
Acta Crystallographica
Acta Crystallogr. Sec. A
Acta Crystallographica, Section A: Crystal
Physics, Diffraction, Theoretical, and General Crystallography
Acta Crystallogr. Sec. B
Acta Crystallographica, Section B: Structural
Crystallography and Crystal Chemistry
Acta Math. Acad. Sci. Hung.
Acta Mathematica Academiae Scientiarum
Hungaricae
Acta Metall.
Acta Metallurgica
Acta Oto-Laryngol.
Acta Oto-Laryngologica
Acta Phys.
Acta Physica
Acta Phys. Austriaca
Acta Physica Austriaca
Acta Phys. Pol.
Acta Physica Polonica
Adv. Appl. Mech.
Advances in Applied Mechanics
Adv. At. Mol. Opt. Phys.
Advances in Atomic, Molecular, and Optical
Physics
Adv. Chem. Phys.
Advances in Chemical Physics
Adv. Magn. Reson.
Advances in Magnetic Resonance
Adv. Phys.
Advances in Physics

Adv. Quantum Chem.
Advances in Quantum Chemistry
AIAA J.
AIAA Journal
AIChE J.
AIChE Journal
AIP Conf. Pro.
AIP Conference Proceedings
Am. J. Phys.
American Journal of Physics
Am. J. Sci.
American Journal of Science
Am. Sci.
American Scientist
Anal. Chem.
Analytical Chemistry
Ann. Chim. Phys.
Annales de Chimie et de Physique
Ann. Fluid Dyn.
Annals of Fluid Dynamics
Ann. Geophys.
Annales de Geophysique
Ann. Inst. Henri Poincaré
Annales de l'Institut Henri Poincaré
Ann. Inst. Henri Poincaré, A
Annales de l'Institut Henri Poincaré,
Section A: Physique Theorique
Ann. Inst. Henri Poincaré, B
Annales de l'Institut Henri Poincaré,
Section B: Calcul des Probabilites et
Statistique
Ann. Math.
Annals of Mathematics
Ann. Otol. Rhinol. Laryngol.
Annals of Otology, Rhinology, & Laryngology
Ann. Phys. (Leipzig)
Annalen der Physik (Leipzig)

Ann. Phys. (N.Y.)
Annals of Physics (New York)

Ann. Phys. (Paris)
Annales de Physique (Paris)

Ann. Rev. Mat. Sci.
Annual Reviews of Materials Science

Ann. Rev. Nucl. Part. Sci.
Annual Review of Nuclear and Particle
Science

Ann. Sci.
Annals of Science

Annu. Rev. Astron. Astrophys.
Annual Reviews of Astronomy and Astrophysics

Annu. Rev. Nucl. Part. Sci.
Annual Reviews of Nuclear and Particle
Science

Annu. Rev. Nucl. Sci.
Annual Review of Nuclear Science

Appl. Opt.
Applied Optics

Appl. Phys. Lett.
Applied Physics Letters

Appl. Spectrosc.
Applied Spectroscopy

Ark. Fys.
Arkiv foer Fysik

Astron. Astrophys.
Astronomy and Astrophysics

Astron. J.
Astronomical Journal

Astron. Nachr.
Astronomische Nachrichten

Astrophys. J.
Astrophysical Journal

Astrophys. J. Lett.
Astrophysical Journal, Letters to the Editor

Astrophys. J. Suppl. Ser.
Astrophysical Journal, Supplement Series

Astrophys. Lett.
Astrophysical Letters

Aust. J. Phys.
Australian Journal of Physics

Bell Syst. Tech. J.
Bell System Technical Journal

Ber. Bunsenges. Phys. Chem.
Berichte der Bunsengesellschaft für
Physikalische Chemie

Br. J. Appl. Phys.
British Journal of Applied Physics

Bull. Acad. Sci. USSR, Phys. Ser.
Bulletin of the Academy of Sciences of the
USSR, Physical Series

Bull. Am. Astron. Soc.
Bulletin of the American Astronomical Society

Bull. Am. Phys. Soc.
Bulletin of the American Physical Society

Bull. Astron. Instit. Neth.
Bulletin of the Astronomical Institutes of the
Netherlands

Bull. Chem. Soc. Jpn.
Bulletin of the Chemical Society of Japan

Bull. Seismol. Soc. Am.
Bulletin of the Seismological Society of
America

C. R. Acad. Sci.
Comptes Rendus Hebdomadaires des Seances
de l'Academie des Sciences

C. R. Acad. Ser. A
Comptes Rendus Hebdomadaires des Seances
de l'Academie des Sciences, Serie A:
Sciences Mathematiques

C. R. Acad. Ser. B
Comptes Rendus Hebdomadaires des Seances
de l'Academie des Sciences, Serie B: Sciences
Physiques

Can. J. Chem.
Canadian Journal of Chemistry

Can. J. Phys.
Canadian Journal of Physics

Can. J. Res.
Canadian Journal of Research

Chem. Phys.
Chemical Physics

Chem. Phys. Lett.
Chemical Physics Letters

Chem. Rev.
Chemical Reviews

Chin. J. Phys.
Chinese Journal of Physics

Class. Quantum Grav.
Classical and Quantum Gravity

Comments Nucl. Part. Phys.
Comments on Nuclear and Particle Physics

Commun. Math. Phys.
Communications in Mathematical Physics

Commun. Pure Appl. Math.
Communications on Pure and Applied
Mathematics

Comput. Phys.
Computers in Physics

Czech. J. Phys.
Czechoslovak Journal of Physics

Discuss. Faraday Soc.
Discussions of the Faraday Society

Earth Planet. Sci. Lett.
 Earth and Planetary Science Letters
Electron. Lett.
 Electronics Letters
Fields Quanta
 Fields and Quanta
Fortschr. Phys.
 Fortschritte der Physik
Found. Phys.
 Foundations of Physics
Gen. Relativ. Gravit.
 General Relativity and Gravitation
Geochim. Cosmochim. Acta
 Geochimica et Cosmochimica Acta
Geophys. Res. Lett.
 Geophysical Research Letters
Handb. Phys.
 Handbuch der Physik
Helv. Chim. Acta
 Helvetica Chimica Acta
Helv. Phys. Acta
 Helvetica Physica Acta
High Temp. (USSR)
 High Temperature (USSR)
IBM J. Res. Dev.
 IBM Journal of Research and Development
Icarus.
 Icarus. International Journal of the Solar System
IEEE J. Quantum Electron.
 IEEE Journal of Quantum Electronics
IEEE Trans. Antennas Propag.
 IEEE Transactions on Antennas and
  Propagation
IEEE Trans. Electron Devices
 IEEE Transactions on Electron Devices
IEEE Trans. Inf. Meas.
 IEEE Transactions on Instrumentation and
  Measurement
IEEE Trans. Inf. Theory
 IEEE Transactions on Information Theory
IEEE Trans. Magn.
 IEEE Transactions on Magnetics
IEEE Trans. Microwave Theory Tech.
 IEEE Transactions on Microwave Theory and
  Techniques
IEEE Trans. Nucl. Sci.
 IEEE Transactions on Nuclear Science
IEEE Trans. Sonics Ultrason. Ind. Eng. Chem.
 IEEE Transactions on Sonics Ultrasonics
  Industrial and Engineering Chemistry
Infrared Phys.
 Infrared Physics

Inorg. Chem.
 Inorganic Chemistry
Inorg. Mater. (USSR)
 Inorganic Materials (USSR)
Instrum. Exp. Tech. (USSR)
 Instruments and Experimental Techniques
  (USSR)
Int. J. Magn.
 International Journal of Magnetism
Int. J. Mod. Phys. A
 International Journal of Modern Physics A
Int. J. Quantum Chem.
 International Journal of Quantum Chemistry
Int. J. Quantum Chem. 1
 International Journal of Quantum Chemistry,
  Part 1
Int. J. Quantum Chem. 2
 International Journal of Quantum Chemistry,
  Part 2
Int. J. Theor. Phys.
 International Journal of Theoretical Physics
Izv. Acad. Sci. USSR, Atmos. Oceanic Phys.
 Izvestiya, Academy of Sciences, USSR,
  Atmospheric and Oceanic Physics
Izv. Acad. Sci. USSR, Phys. Solid Earth
 Izvestiya, Academy of Sciences, USSR, Physics
  of the Solid Earth
J. Acoust. Soc. Am.
 Journal of the Acoustical Society of America
J. Am. Ceram. Soc.
 Journal of the American Ceramic Society
J. Am. Chem. Soc.
 Journal of the American Chemical Society
J. Am. Inst. Electr. Eng.
 Journal of the American Institute of Electrical
  Engineers
J. Appl. Crystallogr.
 Journal of Applied Crystallography
J. Appl. Phys.
 Journal of Applied Physics
J. Appl. Spectrosc. (USSR)
 Journal of Applied Spectroscopy (USSR)
J. Atmos. Sci.
 Journal of Atmospheric Sciences
J. Atmos. Terr. Phys.
 Journal of Atmospheric and Terrestrial Physics
J. Audio Engin. Soc.
 Journal of the Audio Engineering Society
J. Chem. Phys.
 Journal of Chemical Physics
J. Chem. Soc.
 Journal of the Chemical Society

J. Chim. Phys.
Journal de Chemie Physique

J. Comput. Phys.
Journal of Computational Physics

J. Cryst. Growth
Journal of Crystal Growth

J. Electrochem. Soc.
Journal of Electrochemical Society

J. Fluid Mech.
Journal of Fluid Mechanics

J. Gen. Rel. Grav.
Journal of General Relativity and Gravitation

J. Geophys. Res.
Journal of Geophysical Research

J. Inorg. Nucl. Chem.
Journal of Inorganic and Nuclear Chemistry

J. Lightwave Technol.
Journal of Lightwave Technology

J. Low Temp. Phys.
Journal of Low-Temperature Physics

J. Lumin.
Journal of Luminescence

J. Macromol. Sci. Phys.
Journal of Macromolecular Science, [Part B] Physics

J. Mater. Res.
Journal of Materials Research

J. Math. Phys. (Cambridge, Mass.)
Journal of Mathematics and Physics (Cambridge, Mass.)

J. Math. Phys. (N.Y.)
Journal of Mathematical Physics (New York)

J. Mech. Phys. Solids
Journal of the Mechanics and Physics of Solids

J. Mol. Spectrosc.
Journal of Molecular Spectroscopy

J. Non-Cryst. Solids
Journal of Non-Crystalline Solids

J. Nucl. Energy
Journal of Nuclear Energy

J. Nucl. Energy, Part C.
Journal of Nuclear Energy, Part C: Plasma Physics, Accelerators, Themonuclear Research

J. Nucl. Mater.
Journal of Nuclear Materials

J. Opt. Soc. Am.
Journal of the Optical Society of America

J. Opt. Soc. Am. A
Journal of the Optical Society of America A

J. Opt. Soc. Am. B
Journal of the Optical Society of America B

J. Phys. (Moscow)
Journal of Physics (Moscow)

J. Phys. (Paris)
Journal de Physique (Paris)

J. Phys. A
Journal of Physics A: Mathematical and General

J. Phys. B
Journal of Physics B: Atomic, Molecular, and Optical Physics

J. Phys. C
Journal of Physics C: Solid State Physics

J. Phys. D
Journal of Physics D: Applied Physics

J. Phys. E
Journal of Physics E: Scientific Instruments

J. Phys. F
Journal of Physics F: Metal Physics

J. Phys. G
Journal of Physics G: Nuclear and Particle Physics

J. Phys. Chem.
Journal of Physical Chemistry

J. Phys. Chem. Ref. Data
Journal of Physical and Chemical Reference Data

J. Phys. Chem. Solids
Journal of Physics and Chemistry of Solids

J. Phys. Radium
Journal de Physique et le Radium

J. Phys. Soc. Jpn.
Journal of the Physical Society of Japan

J. Plasma Phys.
Journal of Plasma Physics

J. Polym. Sci.
Journal of Polymer Science

J. Polym. Sci., Polym. Lett. Ed.
Journal of Polymer Science, Polymer Letters Edition

J. Polym. Sci., Polym. Phys. Ed.
Journal of Polymer Science, Polymer Physics Edition

J. Quant. Spectros. Radiat. Transfer
Journal of Quantitative Spectroscopy & Radiative Transfer

J. Res. Natl. Bur. Stand.
Journal of Research of the National Bureau of Standards

J. Res. Natl. Bur. Stand. Sec. A
Journal of Research of the National Bureau of Standards, Section A: Physics and Chemistry

J. Res. Natl. Bur. Stand. Sec. B
　　Journal of Research of the National Bureau
　　　of Standards, Section B: Mathematical
　　　Sciences
J. Res. Natl. Bur. Stand. Sec. C
　　Journal of Research of the National Bureau
　　　of Standards, Section C: Engineering and
　　　Instrumentation
J. Rheol.
　　Journal of Rheology
J. Sound Vib.
　　Journal of Sound and Vibration
J. Speech Hear. Disord.
　　Journal of Speech and Hearing Disorders
J. Speech Hear. Res.
　　Journal of Speech and Hearing Research
J. Stat. Phys.
　　Journal of Statistical Physics
J. Vac. Sci. Technol.
　　Journal of Vacuum Science and Technology
J. Vac. Sci. Technol. A
　　Journal of Vacuum Science and Technology A
J. Vac. Sci. Technol. B
　　Journal of Vacuum Science and Technology B
JETP Lett.
　　JETP Letters
Jpn. J. Appl. Phys.
　　Japanese Journal of Applied Physics
Jpn. J. Phys.
　　Japanese Journal of Physics
K. Dan. Vidensk. Selsk. Mat. Fys. Medd.
　　Kongelig Danske Videnskabernes Selskab,
　　　Matematsik-Fysiske Meddelelser
Kolloid Z. Z. Polym.
　　Kolloid Zeitschrift & Zeitschrift für Polymere
Lett. Nuovo Cimento
　　Lettere al Nuovo Cimento
Lick Obs. Bull.
　　Lick Observatory Bulletin
Mater. Res. Bull.
　　Materials Research Bulletin
Med. Phys.
　　Medical Physics
Mem. R. Astron. Soc.
　　Memoirs of the Royal Astronomical Society
Mol. Cryst. Liq. Cryst.
　　Molecular Crystals and Liquid Crystals
Mol. Phys.
　　Molecular Physics
Mon. Not. R. Astron. Soc.
　　Monthly Notices of the Royal Astronomical
　　　Society

Natl. Bur. Stand. (U.S.), Circ.
　　National Bureau of Standards (U.S.),
　　　Circular
Natl. Bur. Stand. (U.S.), Misc. Publ.
　　National Bureau of Standards (U.S.),
　　　Miscellaneous Publications
Natl. Bur. Stand. (U.S.), Spec. Publ.
　　National Bureau of Standards (U.S.),
　　　Special Publications
Nucl. Data, Sect. A
　　Nuclear Data, Section A
Nucl. Fusion
　　Nuclear Fusion
Nucl. Instrum.
　　Nuclear Instruments
Nucl. Instrum. Methods
　　Nuclear Instruments & Methods
Nucl. Phys.
　　Nuclear Physics
Nucl. Phys. A
　　Nuclear Physics A
Nucl. Phys. B
　　Nuclear Physics B
Nucl. Sci. Eng.
　　Nuclear Science and Engineering
Opt. Acta
　　Optica Acta
Opt. Commun.
　　Optics Communications
Opt. Lett.
　　Optics Letters
Opt. News
　　Optics News
Opt. Photon. News
　　Optics and Photonics News
Opt. Spectrosc. (USSR)
　　Optics and Spectroscopy (USSR)
Percept. Psychophys.
　　Perception and Psychophysics
Philips Res. Rep.
　　Philips Research Reports
Philos. Mag.
　　Philosophical Magazine
Philos. Trans. R. Soc. London
　　Philosophical Transactions of the Royal Society
　　　of London
Philos. Trans. R. Soc. London, Ser. A
　　Philosophical Transactions of the Royal Society
　　　of London, Series A: Mathematical and
　　　Physical Sciences
Phys. (N.Y.)
　　Physics (New York)

Phys. Fluids
Physics of Fluids

Phys. Fluids A
Physics of Fluids A

Phys. Fluids B
Physics of Fluids B

Phys. Konden. Mater.
Physik der Kondensierten Materie

Phys. Lett.
Physics Letters

Phys. Lett. A
Physics Letters A

Phys. Lett. B
Physics Letters B

Phys. Med. Bio.
Physics in Medicine and Biology

Phys. Met. Metallogr. (USSR)
Physics of Metals and Metallography (USSR)

Phys. Rev.
Physical Review

Phys. Rev. A
Physical Review A

Phys. Rev. B
Physical Review B: Condensed Matter

Phys. Rev. C
Physical Review C: Nuclear Physics

Phys. Rev. D
Physical Review D: Particles and Fields

Phys. Rev. Lett.
Physical Review Letters

Phys. Status Solidi
Physica Status Solidi

Phys. Status Solidi A
Physica Status Solidi A: Applied Research

Phys. Status Solidi B
Physica Status Solidi B: Basic Research

Phys. Teach.
Physics Teacher

Phys. Today
Physics Today

Phys. Z.
Physikalische Zeitschrift

Phys. Z. Sowjetunion
Physikalische Zeitschrift der Sowjetunion

Planet. Space Sci.
Planetary and Space Science

Plasma Phys.
Plasma Physics

Proc. Cambridge Philos. Soc.
Proceedings of the Cambridge Philosophical Society

Proc. IEEE
Proceedings of the IEEE

Proc. IRE
Proceedings of the IRE

Proc. Natl. Acad. Sci. U.S.A.
Proceedings of the National Academy of Sciences of the United States of America

Proc. Phys. Soc. London
Proceedings of the Physical Society, London

Proc. Phys. Soc. London, Sect. A
Proceedings of the Physical Society, London, Section A

Proc. Phys. Soc. London, Sect. B
Proceedings of the Physical Society, London, Section B

Proc. R. Soc. London
Proceedings of the Royal Society of London

Proc. R. Soc. London, Ser. A
Proceedings of the Royal Society of London, Series A: Mathematical and Physical Sciences

Prog. Theor. Phys.
Progress of Theoretical Physics

Publ. Astron. Soc. Pac.
Publications of the Astronomical Society of the Pacific

Radiat. Eff.
Radiation Effects

Radio Eng. Electron. (USSR)
Radio Engineering and Electronics (USSR)

Radio Eng. Electron. Phys. (USSR)
Radio Engineering and Electronic Physics (USSR)

Radio Sci.
Radio Science

RCA Rev.
RCA Review

Rep. Prog. Phys.
Reports on Progress in Physics

Rev. Geophys.
Reviews of Geophysics

Rev. Mod. Phys.
Reviews of Modern Physics

Rev. Opt. Theor. Instrum.
Revue d'Optique, Theorique et Instrumentale

Rev. Sci. Instrum.
Review of the Scientific Instruments

Russ. J. Phys. Chem.
Russian Journal of Physical Chemistry

Sci. Am.
Scientific American

Sol. Phys.
Solar Physics

Solid State Commun.
   Solid State Communications
Solid State Electron.
   Solid State Electronics
Solid State Phys.
   Solid State Physics
Sov. Astron.
   Soviet Astronomy
Sov. Astron. Lett.
   Soviet Astronomy Letters
Sov. J. At. Energy
   Soviet Journal of Atomic Energy
Sov. J. Low-Temp. Phys.
   Soviet Journal of Low-Temperature
      Physics
Sov. J. Nucl. Phys.
   Soviet Journal of Nuclear Physics
Sov. J. Opt. Technol.
   Soviet Journal of Optical Technology
Sov. J. Part. Nucl.
   Soviet Journal of Particles and Nuclei
Sov. J. Plasma Phys.
   Soviet Journal of Plasma Physics
Sov. J. Quantum Electron.
   Soviet Journal of Quantum Electronics
Sov. Phys. Acoust.
   Soviet Physics: Acoustics
Sov. Phys. Crystallogr.
   Soviet Physics: Crystallography
Sov. Phys. Dokl.
   Soviet Physics: Doklady
Sov. Phys. J.
   Soviet Physics Journal
Sov. Phys. JETP
   Soviet Physics: JETP
Sov. Phys. Semicond.
   Soviet Physics: Semiconductors
Sov. Phys. Solid State
   Soviet Physics: Solid State
Sov. Phys. Tech. Phys.
   Soviet Physics: Technical Physics
Sov. Phys. Usp.
   Soviet Physics: Uspekhi
Sov. Radiophys.
   Soviet Radiophysics
Sov. Tech. Phys. Lett.
   Soviet Technical Physics Letters
Spectrochim. Acta
   Spectrochimica Acta
Spectrochim. Acta, Part A
   Spectrochimica Acta, Part A: Molecular
      Spectroscopy

Spectrochim. Acta, Part B
   Spectrochimica Acta, Part B: Atomic
      Spectroscopy
Supercon. Sci. Technol.
   Superconductor Science and Technology
Surf. Sci.
   Surface Science
Theor. Chim. Acta
   Theoretica Chimica Acta
Trans. Am. Cryst. Soc.
   Transactions of the American Crystallographic
      Society
Trans. Am. Geophys. Union
   Transactions of the American Geophysical
      Union
Trans. Am. Inst. Min. Metall. Pet. Eng.
   Transactions of the Amercian Institute of
      Mining, Metallurgical and Petroleum
      Engineers
Trans. Am. Nucl. Soc.
   Transactions of the American Nuclear Society
Trans. Am. Soc. Mech. Eng.
   Transactions of the American Society of
      Mechanical Engineers
Trans. Am. Soc. Met.
   Transactions of the American Society for
      Metals
Trans. Br. Ceramic Society
   Transactions of the British Ceramic Society
Trans. Faraday Society
   Transactions of the Faraday Society
Trans. Metall. Soc. AIME
   Transactions of the Metallurgical Society of
      AIME
Trans. Soc. Rheol.
   Transactions of the Society of Rheology
Ukr. Phys. J.
   Ukrainian Physics Journal
Z. Anal. Chem.
   Zeitschrift für Analytische Chemie
Z. Angew. Phys.
   Zeitschrift für Angewandte Physik
Z. Anorg. Allg. Chem.
   Zeitschrift für Anorganische und Allgemeine
      Chemie
Z. Astrophys.
   Zeitschrift für Astrophysik
Z. Elektrochem.
   Zeitschrift für Elektrochemie
Z. Kristallogr. Kristallgeom. Krystallphys. Kristallchem.
   Zeitschrift für Kristallographis, Kristallgeome-
      trie, Krystallphysik, Kristallchemie

Z. Metallk.
    Zeitschrift für Metallkunde

Z. Naturforsch.
    Zeitschrift für Naturforschung

Z. Naturforsch. Teil A
    Zeitschrift für Naturforschung, Teil A Physik, Physikalische Chemie, Kosmophysik

Z. Phys.
    Zeitschrift für Physik

Z. Phys. Chem. (Frankfurt am Main)
    Zeitschrift für Physikalische Chemie (Frankfurt am Main)

Z. Phys. Chem. (Leipzig)
    Zeitschrift für Physikalische Chemie (Leipzig)

M

MACH, ERNST

b. Chirlitz near Brno, Moravia, Austro-Hungarian Empire, February 18, 1838; d. Vaterstetten near Munich, Germany, February 19, 1916; *mechanics, acoustics, optics, aerodynamics, history and epistemology of physics.*

Mach was the eldest of three children of Johann Mach, a well-educated man who worked as tutor and farmer and deserved his reputation for silkworm cultivation and researches in some fields of biology. Mach was educated at home before entering the public Piarist Gymnasium in Kremsier (Kromeriz) in 1853. Three years later, when he had finished his studies at the gymnasium, Mach moved on to study physics at the University of Vienna. In 1860 Mach received his doctorate for a dissertation on electrical discharge and induction and began to teach physics. In 1864 he became professor of mathematics (in 1866 of physics) at the University of Graz. He left Graz in 1867 to accept a professorship and to direct the Institute of Physics at Prague University, remaining there for the next twenty-eight years. During that time he held several official positions, being named rector in 1879–1880 and 1883–1884, and was involved in the conflict of nations in Bohemia. In 1895 Mach moved to the University of Vienna as professor of the history and theory of the inductive sciences, but he suffered a stroke two years later and retired from his position at the university in 1901. In 1913, three years before his death, Mach moved to the home of his eldest son and closest colleague, Ludwig.

Mach's scientific work can be divided into three main categories: (1) physics, (2) physiology of the senses (psychology), and (3) philosophy and scientific methodology. These research topics should not be seen as isolated from each other; each was pursued in all of Mach's creative periods. However, the starting point and basis for his work was and always remained physics. During the 1860s he carried out a series of useful experiments to test the correctness of the Doppler effect and formula. Starting in the early 1880s he undertook experiments on phenomena associated with fast-flying projectiles. In this work, not only did he take the first split-second photographs of supersonic projectiles and the compression cones in the air surrounding them (now called Mach cones in his honor), his experiments also provided information about the flow relations for flying projectiles. This work represented an important contribution to the development of modern high-speed photography and ballistic measuring techniques. It also established his reputation as one of the pioneers of modern aerodynamics.

Mach tried hard to work out a general theory of sense perception, maintaining that all empirical knowledge and scientific theories were in the end reducible to statements about sensations. His strong interest in the relationship between physics and the psyche was a central motivation for his critical-historical analyses of the development of physical thinking. The connection between the two lines of research is represented in the three monographs *Mechanics* (1883), *Heat* (1896), and *Optics* (1921). Their titles must be regarded as programmatic. As a rigorous sensualist, he rejected such metaphysical concepts as absolute time and space, criticized Isaac Newton's physics, and showed the limits of a mechanistic description of physical phenomena. Albert Einstein, who read Mach's books as a young student, was strongly influenced by this critique and took it up directly in elaborating his theory of relativity—especially in forming the Mach principle that inertia results from the interaction between one body and all of the other bodies in the universe.

During the course of his physiological studies during the 1860s, Mach discovered the phenomenon of contrast increase (Mach bands), and he made important investigations into kinesthetic sensation, which led to the discovery that humans only have a sense organ for detecting accelerated movements (but not uniform movements) and that this organ is located in the semicircular canals of the inner ear. Mach's phenomenalistic and positivistic approach strongly influenced later scientists such as Einstein, Werner Heisenberg, and Wolfgang Pauli, while many impressionistic writers found in Mach's subject-object reflections a confirmation of their literary ambitions. Mach's ideas were one central source of ideas for the Vienna circle of philosophers. Mach's views were even felt in contemporary politics, as reflected in the writings of the Russian Bolsheviks.

See also: AERODYNAMICS; DOPPLER EFFECT; MACH'S PRINCIPLE

Bibliography

BLACKMORE, J. T. *Ernst Mach: His Work, Life, and Influence* (University of California Press, Berkeley, 1972).

BLACKMORE, J. T., ed. *Ernst Mach: A Deeper Look* (Kluwer, Dordrecht, 1992).

COHEN, R. S., and SEEGER, R. J., eds. *Ernst Mach: Physicist and Philosopher* (Reidel, Dordrecht, 1970).

HOFFMAN, D., and LAITKO, H. *Ernst Mach: Studien und Dokumente zu Leben und Werk* (Verlag, Berlin, 1991).

DIETER HOFFMANN

MACH'S PRINCIPLE

A woman is walking at constant speed toward the back of a train that is moving eastward at high speed. This simple situation illustrates relative motion. The woman is walking west relative to the train; the train is moving east with respect to the ground; the ground is moving west with respect to the train. The motion of any object here is described relative to another object. Equivalently, the motions are specified with the use of specific reference frames. The train is a reference system through which the woman is walking. The ground is a reference system with respect to which the train is moving.

All of these reference systems are inertial systems. In all such frames the simple laws of motion apply. If an object has no force acting on it, then the object will stay at rest if it is initially at rest. If it is initially in motion, it will continue its motion with constant velocity (unchanging speed and direction). If forces act on the object, its velocity will change; the object will accelerate. (Acceleration is the rate of change of velocity.) The acceleration will be proportional to the force acting and inversely proportional to the mass of the object. This tendency to continue the original motion, to have velocity changed only by forces, and for large-mass objects to be difficult to accelerate is precisely what is meant by inertia.

If the laws of motion, the laws governing acceleration and inertia, work in one reference frame it can be shown that they work equally well in any other frame of reference that is moving by it in a straight line at constant speed. All such frames are inertial frames. This equivalence of all inertial frames means that there is no single primary absolute frame for physics. For any motion of any object there will be some inertial reference frame relative to which that object is—at least momentarily—not moving. The velocity of an object therefore has no absolute meaning; only relative velocity has meaning (relative to another object, relative to a reference system).

There is no single best reference frame, but that does not mean that all reference frames are equally good, that all frames are inertial frames. As a reference frame, consider a powerful car when it is starting from a stoplight. A coin on the slippery dashboard slides toward the back of the car when the car starts and toward the front when the brakes are applied. There are no forward or backward forces on the coin. Rather there is acceleration relative to the car because the car is not an inertial frame. It is not moving with constant speed with respect to the ground, which is (at least approximately) an inertial frame. If the car goes around a corner, even at constant speed, it will again fail to be an inertial system, and the coin will demonstrate this by sliding to the side of the dashboard toward the outside of the turn. The coin will be exhibiting inertial behavior with respect to the ground but noninertial behavior (centrifugal acceleration) with respect to the car. Only when the car is moving across the ground in a straight line with constant speed does it qualify as an inertial reference system.

Velocity has no absolute meaning, but it appears that acceleration does have absolute meaning. The acceleration is the same in one inertial reference system as in another. When accelerations are different, it is an indication that a noninertial reference system (e.g., the braking car) is being used.

What determines if a reference frame is inertial? In the classical mechanics (physics of motion) of Isaac Newton the answer is that inertia is simply an inherent property of space. This answer was very unsatisfactory to many. The most effective arguments against this viewpoint were given by Ernst Mach, who argued that in an empty universe there could be no meaning whatever to motion. If there were no reference points in the universe, could we say that the car was starting forward, slowing down, cornering, or, for that matter, rolling and twisting? There could be no resistance to acceleration (since there could be no meaning to acceleration), so there could be no inertia. If this viewpoint is valid, then inertia and the determination of what is and what is not acceleration must be generated by the matter content of the universe. The tendency of the speeding car to resist cornering and braking, here on Earth, must be generated by interactions of the car with distant stars.

This viewpoint, which has become known as Mach's principle, is clearly in contradiction with Newton's classical theory of motion. Whether it agrees with the modern theory, based on Albert Einstein's work, is much less clear. Einstein was very much influenced by Mach's ideas, and his theory, general relativity, has predictions that are in apparent agreement with Mach's principle, in particular the dragging of inertial frames. A clear example of this is that the inertial (nonrotating) reference frames near the earth are slightly affected by the motion of the massive earth and therefore rotate slightly with respect to the distant stars. The effect is so small that it takes around 10 million years for the inertial frames to be dragged one revolution with respect to the distant stars, yet an experiment to measure this effect with satellite-borne gyroscopes is expected to be able to measure this effect before the end of the century. Though Machian in some ways, Einstein's theory has aspects that clearly contradict Mach's principle. In particular, it says that there are inertial frames even in an empty universe.

Mach's principle was put forth in rather vague form: Inertia should be due to interactions of material bodies. Because of this vagueness, it is likely that there will always be disagreement about what the principle exactly means and whether it is contained in a specific theory, like Einstein's general relativity.

See also: FRAME OF REFERENCE; FRAME OF REFERENCE, INERTIAL; INERTIAL MASS; MACH, ERNST; MASS; NEWTONIAN MECHANICS; NEWTON'S LAWS; RELATIVITY, GENERAL THEORY OF

Bibliography

FEUER, L. S. *Einstein and the Generations of Science* (Basic Books, New York, 1974).

GRAVES, J. C. *The Conceptual Foundations of Contemporary Relativity Theory* (MIT Press, Cambridge, MA, 1971).

MACH, E. *The Science of Mechanics: A Critical and Historical Account of Its Development,* trans. by K. Menger (Open Court Publishing, LaSalle, IL, 1960).

RICHARD H. PRICE

MAGNET

Popularly speaking, a magnet is an object that produces a magnetic field, which can be used to attract some metals. Magnets have an ancient history, appearing in Greek writings from as early as 800 B.C.E. Records show that the mineral magnetite, a magnetic oxide of iron, was mined in the Greek district of Magnesia. The mineral was known as loadstone in English. Some believe that it was used as a compass in China as early as the twenty-sixth century B.C.E., although others suggest that it was introduced there from Arab origins in the twelfth century C.E. By 1600 there was enough information for William Gilbert, an English scientist, to publish the book *De Magnte.*

Since then much work has been done in the field because magnets have many practical uses in motors, generators, beam handling equipment, and information recording devices of an astonishing variety. The most striking aspect of any magnet is that it has two different regions called poles. One is called a north seeking, or north pole (N) and the other is called a south seeking, or south pole (S). Just as in the case of the two kinds of electric charge, like poles repel and unlike poles attract. One very interesting feature of a magnet is that if you cut the magnet in half, trying to separate the N pole from the S pole, you will find that the resulting two magnets each have an N pole and an S pole. Nobody has yet managed to find or create a magnet with only one pole (i.e., a monopole). Magnets can be divided into two main classes—electromagnets, which require an external electric current and are discussed elsewhere, and permanent magnets, which require no external power to produce a magnetic field. This entry will deal exclusively with permanent magnets.

Permanent magnets have two properties that are particularly useful in classifying them. One is called remanence (r in Fig. 1). This property describes the amount of magnetization that remains after the initial magnetizing field has been removed. The larger this value, the stronger the magnet. The other property is called the coercivity (c in Fig. 1) and is a measure of the degree to which the magnet resists

demagnetization. The larger this value, the more permanent the magnet.

Most practical uses require the largest remanence consistent with the other desired qualities. Some uses require a large coercivity. These magnetic materials are called hard. Other uses require a small coercivity. These magnetic materials are called soft. Some magnetic materials are so soft that the thermal agitation of their atoms are enough to demagnetize the magnet.

The first practical use of a magnet was as a compass. Lodestone was used directly for this purpose. It was discovered that iron could be magnetized by stroking it with a permanent magnet or by tapping a bar which was held in a N-S direction. However, the coercivity was so small that time or rough handling would make the bar useless as a magnet. Alloys such as steel are much harder to magnetize but also are more difficult to demagnetize.

After the physics of magnetism was better understood, special alloys with high remanence were created. One of these, alnico—a combination of iron, nickel, cobalt, and aluminum—resulted in a great advance in the strength of permanent magnets. Rare earth elements have permitted further advances in strength. Samarium cobalt alloys were the first to be developed, but a neodymium-iron-boron combination produces the strongest magnets. These latter magnets are so strong that it is almost impossible to remove one the size of a golf ball from the side of a filing cabinet. In fact, they can even attract a dollar bill because of the small amount of iron oxide found in the printing ink. One difficulty with many of the modern alloys is that they are quite brittle and break easily if dropped or hit.

A good example of a practical use of permanent magnets is in small portable power generators. Magnets are used in place of field coils to provide a magnetic field through which coils of wire are forced so that a current can be generated. These magnets should be strong or have a large remanence; they should resist demagnetization, or have a large coercivity, and be light. If the generator is to be light, not subjected to rough handling, and does not have to be particularly inexpensive, the latest neodymium magnets could be used. If cost and damage resistance are important parameters, alnico magnets might be used.

Permanent magnets are used in practically all loudspeakers, large or small, and contribute significantly to the mass of the device. In a loudspeaker a solenoid (coil) is attached to a cone that vibrates to

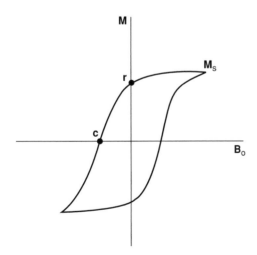

Figure 1 A magnetic hysteresis curve (illustrating the magnetization curve for the entire cycle of external field variation), where **M** is the magnetization, $\mathbf{B}_0$ is the applied magnetic field, **r** is the remanence, and **c** is the coercivity.

generate the sound. This solenoid is positioned in a permanent magnetic field produced by the magnet. Current variation in the winding of the solenoid provided by a power amplifier induces mechanical force to move the solenoid in the presence of the magnetic field and thus drive the loudspeaker.

The most widespread use of magnetic materials is in the information recording industry. Many people do not realize that an audio tape is a ribbon of plastic upon which a very large number of small magnets are glued. Recording surfaces for different purposes require different characteristics from the magnetic material.

Audiotapes have two major formulations: standard and metal oxide. Floppy disks for computers also come in two major formulations. Video recorders use magnetic tapes. Even computer hard disks depend on a thin coating of little magnetic particles. In all these cases, a delicate balance has to be maintained between the density of information placed on the tape, the tendency for the recorded signals to demagnetize themselves, and the complexity of the electronics used to organize the magnets on the tape.

See also: FIELD, MAGNETIC; MAGNETIC BEHAVIOR; MAGNETIC DOMAIN; MAGNETIC MATERIAL; MAGNETIC MOMENT; MAGNETIC MONOPOLE; MAGNETIC POLE; MAGNETIZATION

Bibliography

ABELE, M. G. *Structures of Permanent Magnets: Generation of Uniform Fields* (Wiley, New York, 1993).

HECHT, E. *Physics* (Brooks/Cole, Pacific Grove, CA, 1996).

PARKER, R. J. *Advances in Permanent Magnetism* (Wiley, New York, 1990).

A. F. BURR

MAGNETIC BEHAVIOR

Magnetic field refers to the stress system in the space surrounding a steady electric current. The electric current may flow in a wire, or it may be the electric current arising from the intrinsic spin of a charged elemental particle such as the electron or proton. A magnetic field also arises in the presence of a time varying electric field or in any frame of reference moving perpendicular to an electric field. A magnetic field **B** represents an isotropic pressure $B^2/8\pi$ (B in gauss) and a tension $B^2/4\pi$ in the direction of the field. In static conditions the field takes up a configuration in which the tension balances and confines the pressure. In electromagnetic radiation the pressure is responsible for the rapid time rate of change of the local electromagnetic momentum.

We should be aware that magnetic fields are a common element of our environment, from the magnetic latch on the cupboard and refrigerator doors, to the solenoid that opens and closes the water valves in the washing machine, to the field excitation in the electric generators and motors that run the fans, blowers, heat pumps, streetcars, and railroad locomotives that make up the technological world in which we spend our lives. Magnetic fields are equal partners with electric fields in the visible light by which these words are being read and in the radio waves that direct our television sets. Magnetic fields are the essential ingredient in the activity of stars like the Sun and in the space weather in the terrestrial magnetosphere in which we reside. The properties of magnetic fields are well known and relatively simple.

The intensity of a magnetic field is measured in gauss or in tesla, in association with electric currents measured in electrostatic units or in amperes, respectively. A current of 1 A represents 3×10^9 electrostatic units of current. Note, then, that a field of 1 T has an intensity of 10^4 G, comparable to the magnetic field produced by a powerful electromagnet with an iron core. By comparison, the magnetic field of Earth, in which we reside, has an intensity of about 0.3 G at the equator and 0.6 G at the poles. It is sufficient to orient a freely swinging bar of iron (e.g., an iron bolt suspended by a thread or the magnetic needle of a compass), but it is too weak to feel through handling iron objects. On the other hand, a magnetic field of 10 G would be quite noticeable, because the magnetic stresses increase as the square of the field strength.

Magnetic fields are produced in the laboratory and in the electrical machinery of the home and factory by permanent magnets and by electric currents flowing in wires usually wound into compact coils to provide an intense field. The magnetic field produced by an electric current can be enhanced by a factor of 10^3 or more by filling the space around the current with a ferromagnetic substance, such as annealed low-carbon steel. This is common practice in electric motors and transformers. It is customary to denote the magnetic field produced by the electric currents in the wires by **H**. The effect of **H** is to align the magnetic moments of some of the electrons in the atoms of the ferromagnetic material, providing a magnetic dipole moment **M** per unit volume. The total magnetic field **B** is then equal to the sum **H** + 4π**M**, which can be large compared to **H**. The relation between **H** and **B** is sometimes written **B** = μ**H**, where μ is called the magnetic permeability and is itself a function of **H**, with 4π**M** generally saturating and limited to values of the order of 10^4 G or less. Internal friction in the orientation of atoms provides hysteresis effects in the relation between **H** and **B.**

The behavior of matter in the presence of a magnetic field falls into three categories. For instance, some molecules have a weak magnetic dipole moment of their own; that is, the individual molecule is a tiny permanent magnetic. Consequently, when placed in a magnetic field there is a tendency for the magnetic moments of the molecules to align themselves with the applied magnetic field, thereby slightly enhancing the field. Such substances are said to be paramagnetic. The paramagnetic effect is proportional to $1/T$, because the thermal agitation by the temperature T tends to misalign the molecules.

The strongest and, therefore, the most familiar and useful effect is ferromagnetism arising from the organized alignment of electron spins. Each individual electron acts as a microscopic gyroscope, carrying its negative electric charge around with the spin to produce a magnetic dipole moment, so each electron is a permanent magnet. In metallic iron (as well as nickel, cobalt, etc.) the electron clouds of each atom overlap with the nearest neighbors, creating a situation in which two or more electrons in unfilled electron shells are pushed into alignment. This alignment extends through macroscopic regions of the metal crystal, producing magnetic domains with dimensions on the general order of 0.1 mm in which one or more electrons of each atom are aligned to produce a net magnetic dipole moment. Then, when an external magnetic field is applied, the magnetic moments of the many domains become aligned, much as the molecules are aligned in paramagnetic substances, except that the net effect is immensely stronger, with μ as large as 3 $\times$ 10^3 in soft iron. In some iron compounds (e.g., magnetite) and iron alloys (e.g., high-carbon steel, alnico) the alignment of the domains is "frozen" into the metal, providing the familiar and exceed-

ingly useful permanent magnet. As with paramagnetism, thermal agitation diminishes the cooperative alignment so that ferromagnetism declines to zero at the Curie temperature appropriate for each particular substance (1,043 K for ^{26}F).

The third magnetic category is diamagnetism, in which the orbits of the electrons in the atoms and molecules are free enough to respond to the applied magnetic field. The effect is that some of the centrifugal force of the orbiting electrons is transferred to the magnetic field, tending to push the field out of the material, so that μ is slightly less than one.

Note, then, that in any substance there are both paramagnetic and diamagnetic effects occurring simultaneously. The stronger effect determines the magnetic classification of the substance.

In view of the magnetic saturation in strong fields, ferromagnetic effects are limited, and the strongest fields are now produced using coils of superconducting wires to prevent resistive dissipation of the electric current. An electromotive force (voltage) V is applied to the superconducting coil of wire, initiating a current I. The energy input is VI. The energy goes into creating the magnetic field, which has an energy density $B^2/8\pi$ (B in G). The magnetic field strength B increases linearly with I so that the magnetic energy $\mathscr{E}$ is proportional to I^2, from which it follows that I and B increase linearly with time for a fixed applied potential V.

The magnetic field stresses tend to burst the coil of wire so that careful mechanical design is essential. The maximum fields achieved are limited by the critical field for the superconducting state in the wires, on the general order of 1–2 $\times$ 10^5 G. Stronger magnetic fields ($\sim 10^6$ G) can be achieved only under transient conditions by compressing the magnetic field within a metal tube imploded by detonation of a surrounding block of high explosive.

In this connection it is interesting to note that the mean magnetic field on the order of 5 G at the surface of the Sun is spontaneously compressed by the electrically conducting gas into widely separated bundles of 1–2 $\times$ 10^3 G, reaching 3 $\times$ 10^3 G in the formation of sunspots. Most stars have overall mean magnetic fields of 1–10^2 G, with fields in excess of 10^8 G in some white dwarf stars and 10^{11}–10^{12} G in some neutron stars.

Consider, then, the basic physical properties of magnetic fields. The presence of a magnetic field **B** at any point in space can (in the absence of electric fields) be demonstrated by its unique orientation of a freely suspended iron needle (a compass) in the direction of the vector **B**. There are no known magnetic charges in the universe (magnetic monopoles) to act as sources of magnetic field, in contrast with the general existence of elemental negative and positive electric charges (i.e., electrons and protons) that are sources of electric fields. On the other hand, the open end of a finely wound, long, thin current-carrying solenoid is a useful simulation of a magnetic charge g providing a radial magnetic field **B** with intensity locally falling off inversely with the square of the radial distance r from the open end.

To provide a quantitative expression for the equivalent charge g, consider a long thin solenoid of radius R, made up of n turns of wire per centimeter (with $nR \gg 1$) with a current I carried by the wire. The length L of the solenoid is very large compared to R. The magnetic field **B** contained within the solenoid is uniform across R in this limit, extending along the solenoid with an intensity $4\pi nI/c$ G, where c represents the speed of light in centimeters per second if I is in electrostatic units. The total magnetic flux Φ extending along the solenoid is $\Phi = \pi R^2 B$, issuing from the open end, from which it spreads out radially ($B \sim 1/r^2$ for $R \ll r \ll L$) to simulate a magnetic charge g such that $4\pi g = \Phi$. Thus, the equivalent magnetic charge of the open end is $g = \pi R^2 nI/c$.

A magnetic field **B** at any point is defined by $\mathbf{F}/g$, where **F** is the force exerted by the field on the equivalent magnetic charge g (open end) placed at that point, so that **B** is the force per unit charge.

Alternative means for experimental determination of **B** are the Lorentz force $\mathbf{I} \times \mathbf{B}/c$ exerted on an electric current **I** and the torque $\mathbf{m} \times \mathbf{B}$ exerted on a magnetic dipole moment **m**. These same forces exerted on the electronic structure of an atom produce modifications of the electromagnetic spectrum of the atom, which provide spectroscopic determination of **B** through the Zeeman effect.

With **B** defined in terms of the force **F**, it is evident that **B**, like the force **F**, is a vector described by its three orthogonal components (B_x, B_y, B_z) in Cartesian coordinates (x, y, z), or by the magnitude and direction of **B**. The magnetic field **B** is conveniently illustrated by field lines, mapping the direction of local compass pointings. Formally the field lines represent the instantaneous family of solutions of the two differential equations

$$\frac{dx}{B_x} = \frac{dy}{B_y} = \frac{dz}{B_z}.$$

The magnetic field associated with an electric current **I** flowing along a straight wire circles the wire with an intensity $2I/c\varpi$, declining inversely with distance ϖ from the wire. The direction of the field is conveniently determined by the right-hand rule, placing the thumb along the direction of **I** with the fingers of the right hand indicating the direction of the azimuthal field. A circular loop of wire of radius R carrying a current I has a magnetic field $2\pi I/cR$ at its center, with the field pointing along the axis of the loop. The magnetic dipole moment of the loop is $m = \pi R^2 I/c$, pointing along the field on the axis of the loop.

Formally, the electric current density **j** is related to **B** by Ampère's law $4\pi\mathbf{j} = c\boldsymbol{\nabla} \times \mathbf{B}$ under static conditions. The magnetic field is expressible in terms of **j** through the Biot–Savart integral

$$\mathbf{B}(\mathbf{r}) = \frac{1}{c} \int \frac{d^3\mathbf{r}'\mathbf{j}(\mathbf{r}') \times (\mathbf{r} - \mathbf{r}')}{\left|\mathbf{r} - \mathbf{r}'\right|^3}.$$

To achieve static conditions, the electric current is driven by an externally applied electromotive force so that the current is the cause of the magnetic field. Under time dependent conditions, the situation can be more complicated, for it must be understood that a magnetic field represents a real presence in space. The magnetic field has an energy density $B^2/8\pi$ (B in gauss) besides representing a stress system composed of an isotropic pressure $B^2/8\pi$ and a tension $B^2/4\pi$ along the field, as already noted. If there is no externally applied electromotive force, then the complete Maxwell equation

$$\partial\mathbf{E}/\partial t = c\boldsymbol{\nabla} \times \mathbf{B} - 4\pi j$$

must be employed. Insofar as **j** is inadequate to satisfy Ampère's law, magnetic energy is converted into electric field **E** striving to create the electric current for Ampère's law. In the presence of an electrically conducting medium with conductivity σ, then the magnetic field causes the current. The resistive heating j^2/σ ergs·cm^{-3}·s^{-1} is supplied by the declining magnetic energy density $B^2/8\pi$.

In the absence of a conducting medium it follows that $\mathbf{j} = 0$, and the magnetic field quickly converts itself into an electric field, with $\partial\mathbf{E}/\partial t = +c\boldsymbol{\nabla} \times \mathbf{B}$. The growing electric field has vanishing divergence and converts itself back into a magnetic field in the manner described by Faraday's law of induction $\partial\mathbf{B}/\partial t = -c\mathbf{\nabla} \times \mathbf{E}$, and so on. The result is an electromagnetic wave with phase velocity c, in which $\mathbf{E}$ and $\mathbf{B}$ are equal partners. Unfortunately, the symmetric roles of electric and magnetic fields in electromagnetic radiation are obscured by the commonly used SI system of units, in which $\mathbf{E}$ and $\mathbf{B}$ are given different dimensions ($MLT^{-2}Q^{-1}$ and $MT^{-1}Q^{-1}$, respectively, with Q representing electric charge). Consequently, the electrostatic system of units (esu), with $\mathbf{B}$ measured in gauss and $\mathbf{E}$ in stativolts per centimeter, has been employed here, wherein $\mathbf{B}$ and $\mathbf{E}$ have the same dimension ($M^{1/2}L^{-1/2}T^{-1}$), and their equality implies the same energy density and stress.

It is important to note that most of the universe is occupied by partially or wholly ionized gas so that electric fields in the reference frame of the gas are quickly canceled by the motion of free electrons. Hence, the magnetic field cannot move significantly relative to the gas; to do so would induce electric fields in the gas. The result is a peculiar elastic gas exhibiting the combined pressures of the gas and field and the magnetic tension $B^2/4\pi$ already mentioned. It is an observed fact that magnetic fields fill the universe, with few regions in which the magnetic field is not strong enough to have detectable consequences, and, in many situations, overwhelming consequences. The magnetic stresses provide spontaneous discontinuities (current sheets) in the magnetic fields, which provide rapid, and sometimes explosive, dissipation of magnetic energy. Thus, the frequent occurrence of gas at local temperatures way above ambient values (e.g., the solar corona, and the magnetosphere of Earth and other planets) is generally to be associated with the activity of the local magnetic fields. The combination of hydrodynamic and magnetic stresses provide such novelties as sunspots, the x-ray coronas of most stars, the galactic confinement of cosmic rays, and probably the cosmic rays themselves. The magnetic fields appear to be generated in the convecting liquid metal cores of planets, in the convective zones of stars, and evidently in the churning gaseous disks of galaxies. The free existence of magnetic fields in the universe is direct evidence of the scarcity, if not the total absence, of magnetic monopoles. If monopoles were present in any significant number, they would short circuit the general magnetic fields, just as the presence of free electrons cancels out any large-scale electric fields in the frame of reference of the ionized gases.

See also: AMPÈRE'S LAW; ANTIFERROMAGNETISM; COSMIC RAY; DIAMAGNETISM; ELECTROMAGNETIC INDUCTION, FARADAY'S LAW OF; ELECTROMAGNETISM; FERRIMAGNETISM; FERROMAGNETISM; FIELD, MAGNETIC; FIELD LINES; MAGNETIC MOMENT; MAGNETIC MONOPOLE; MAGNETOHYDRODYNAMICS; PARAMAGNETISM; ZEEMAN EFFECT

Bibliography

BLEANEY, B. I., and BLEANEY, B. *Electricity and Magnetism,* 4th ed. (Oxford University Press, Oxford, Eng., 1994).

JACKSON, J. D. *Classical Electrodynamics,* 2nd ed. (Wiley, New York, 1975).

PARKER, E. N. *Cosmical Magnetic Fields* (Clarendon Press, Oxford, Eng., 1979).

E. N. PARKER

MAGNETIC COOLING

Magnetic cooling, or adiabatic demagnetization, is a cryogenic technique whereby a paramagnetic substance is first isothermally magnetized and then adiabatically demagnetized to attain a lower temperature. This method had been proposed independently in 1926 by Peter Debye and William Giauque. Experimental verification was first accomplished in 1933.

In a paramagnetic system such as paramagnetic salt, application of a magnetic field aligns its spin system into a state of higher order. Entropy is a measure of randomness. Thus, at a given temperature, entropy of the system is lower when it is in a magnetic field. This is represented in Fig. 1, which shows entropy S as a function of temperature T for a paramagnetic salt with and without a magnetic field H. Application of the magnetic field H, with the temperature kept at T_1, is shown as path A to B; path B to C represents an adiabatic process that produces a temperature drop from T_1 to T_2. There is a limit to this method of cooling since below a certain temperature the zero-field entropy curve drops sharply and the steep slope renders further cooling by demagnetization ineffective. The slope and temperature of this drop-off varies from salt to salt and is due to the mutual interaction among its spins. Based on this characteristic, paramagnetic salts of choice for the pioneering experiments were gadolinium sulfate, cerium fluoride, and iron ammonium alum, all of which have now been replaced by cerium magnesium nitrate (CMN).

To achieve low temperature by adiabatic demagnetization, the paramagnetic salt is initially surrounded with liquid helium. The container is then surrounded by liquid helium, which in turn is surrounded by another container filled with liquid nitrogen. As a first step, a strong magnetic field is applied, producing a rise in temperature. The heat vaporizes the liquid helium in contact with the salt. Because of the high thermal conductivity of the helium gas, the heat is then removed efficiently and conducted to the surrounding liquid helium that is continually being pumped to maintain the lowest possible temperature until an equilibrium is reached. The volume containing the salt is evacuated at this point to isolate this space from its surrounding. In this state, the magnetic moments are aligned in the direction of the external field and the temperature is at the lowest possible attained by pumping on the liquid helium. At this point, the external field is suddenly removed and the salt attains its lowest temperature. (Even if the magnetic field is not reduced to zero and the final temperature does not quite reach that which corresponds to the sharp drop-off portion of the zero-field entropy-temperature plot, the magnetic dipole moment, which varies as the ratio H/T, is conserved. Some consequences of this are found to play a significant role in a similar process involving nuclear magnetism.)

When the field is removed, the system of aligned magnetic moments becomes slowly randomized to reach an equilibrium with the crystal lattice, while maintaining the entropy of the entire salt (combined spin and lattice system) unaltered (i.e., adiabatic). It is the coupling between the system of magnetic moments at extremely low spin temperature and the lattice at higher lattice temperature that produces the final low temperature. The lowest attainable temperature with this method is about 10^{-3} K. This level of low temperature can be reached by dilution refrigeration as well.

To achieve even lower temperatures, one resorts to nuclear adiabatic demagnetization of highly conducting metals such as copper. Dilution refrigeration and superconducting solenoids must be used to reach starting temperatures below 15 mK and magnetic fields greater than 5 T. The magnetic ordering temperatures for nuclear cooling (at which the interaction energy equals the thermal energy) are three orders of magnitude smaller than for the best paramagnetic salt. As the magnetic field is reduced, the nuclear magnetic moment is preserved, making temperatures below 1 μK assessable.

At very low temperatures, there are two limiting temperatures that can be approached, the nuclear spin temperature T_n, whereby nuclei approach thermal equilibrium, and the conduction electron temperature T_e, indicating interaction between nuclear spins and conduction electrons. These equilibrium temperatures are reached in characteristic times: τ_n, the spin-spin relaxation time, and τ_e, the spin-lattice relaxation time. The latter is longer at low temperatures. The spin-lattice relaxation time is proportional to the number of contributing electrons. Since only electrons near the Fermi surface contribute, their number is proportional to $1/T_e$, as is the relaxation time. The high conduction properties of copper, silver, and indium have made them the metals of choice for nuclear refrigeration. The lowest temperature achieved was 600 pK, obtained by cascade nuclear magnetic cooling of silver at the Helsinki University of Technology. Warm-up to 3 nK took three hours.

Cascade nuclear magnetic cooling operates on two stages. The first stage is 1 kg of copper magnetized to 8 T and cooled to 15 mK. A superconducting heat switch is opened, the field is reduced to 0.1 T, and the copper cools to about 200 μK. In the second stage, 2 g of copper or silver are then magnetized to 7 T. When equilibrium is reached, demagnetization to 20 mT produces a conduction electron

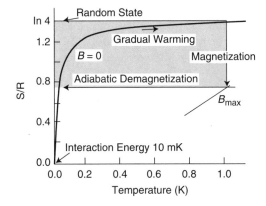

Figure 1 Entropy S as a function of temperature T for a paramagnetic salt, with and without a magnetic field H.

temperature of 50 μK in copper (200 μK in silver). When demagnetized, the nuclear spins are cooled to about 30 nK in copper.

New techniques for magnetic cooling continue to be developed. By quickly reversing the external field at the end of demagnetization, population inversions of the nuclear energy levels have been achieved, producing a negative temperature of -4 mK in silver.

See also: CRYOGENICS; ENTROPY; FERMI SURFACE; PARAMAGNETISM

Bibliography

LOUNASMAA, O. V. *Experimental Principles and Methods Below 1 K* (Academic Press, London, 1988).

ALICE L. NEWMAN

MAGNETIC DOMAIN

Below its ordering temperature, a ferromagnetic or ferrimagnetic sample exhibits a spontaneous magnetization. However, under many circumstances the magnetization of such a sample is broken up into discrete regions in each of which the magnetic moment of every unit cell is parallel to that of its neighbors. The net magnetization of the whole sample often approximates zero. These regions are called magnetic domains and the thin transitions between adjacent regions are domain walls. The character of domains in a given sample depends on intrinsic magnetic properties (magnetization, magnetocrystalline anisotropy, and magnetostriction) of the material, as well as the details of its physical microstructure including state of strain, overall shape, and the magnetic fields acting on the sample. It is in terms of domains that we understand the response of a specimen to magnetic fields. This domain theory represents a remarkable insight into the physics of magnetic behavior in applied magnetic fields

Two microscopic and one macroscopic effects underlie magnetic domains. These are a quantum mechanical effect called exchange interaction and a magnetocrystalline anisotropy energy both acting at the level of the atomic cells, and demagnetizing effects of the gross sample.

Simple Illustrative Example

Consider a thin single crystal sheet of a ferromagnetic material that has an easy axis normal to its surfaces as explained below. The exchange interaction impels the magnetization of each magnetic ion to be parallel to that of its neighbors. Anisotropy energy means that the energy of the magnetization **M** varies with direction in the crystal lattice. In this case the lowest energy corresponds to **M** parallel or antiparallel to the normal. It would require energy to turn it to any other direction. Since in the absence of external torques **M** tends to lie along an axis, the normal, we say that this crystal has a uniaxial magnetocrystalline anisotropy with an easy axis. It requires work, anisotropy energy, to turn **M** away from the normal.

If we first apply only the exchange interaction, the magnetization of every unit cell in the single crystal sample would be parallel to that of every other, and might point in any direction. The exchange energy for the whole sample E_e would be at a minimum. If we then introduce a uniaxial anisotropy with an easy axis, the total magnetization of the sample would be along that axis, and the total anisotropy energy E_a would also be at a minimum. For the simple case of a thin sheet with the easy axis normal, the **M** of the whole sample would point either out or in along one of the surface normals.

We know from electromagnetic theory that the discontinuity in the magnetization in going from inside the sample to outside the sample (often discussed in terms of magnetic poles) gives rise to a magnetic field $\mathbf{H}_d$ within the sheet. This demagnetizing field is in a direction opposite to **M** and so this is a high energy state. If the whole sheet were magnetized along one normal, a single domain state, the demagnetizing energy E_d of the sample would be a maximum.

We can reduce the total magnetic energy $E_{tot} = E_e + E_a + E_d$ by breaking the magnetization up into smaller domains. One possible configuration consists of parallel up and down domains of equal width as sketched in Fig. 1a. The spacing is determined by the fundamental magnetic properties of the material and by the thickness of the sheet. In going through the domain walls between the up and down domains, the magnetization of each unit cell changes cell by cell from up to down in the plane of the wall. Note that within the wall there is exchange energy because neighboring cells do not have magnetization in exactly the same direction, and there is anisotropy energy because the magnetization vector is not along an easy direction. Therefore each square centimeter of the wall has an energy. Introducing more walls lowers demagnetizing energy, but it increases wall energy, and so there is an equilibrium wall spacing.

If we apply a positive (+) external magnetic field H_{ap}, the up domains grow at the expense of the down domains. As H_{ap} increases, this continues until the down domains disappear at a saturation field H_{sat}. Fields above saturation have little effect on the magnetization. If the field is cycled from $+H_{sat}$ to $-H_{sat}$ to $+H_{sat}$, $M(H)$ does not exactly follow the same path when H is increasing as when it is decreasing, but encloses a finite area. This area represents the dissipation of energy, magnetic losses. The phenomenon of not retracing the same path is called hysteresis, the closed curve $M(H)$ is a hysteresis loop and we encounter its analog in a wide variety of physical phenomena.

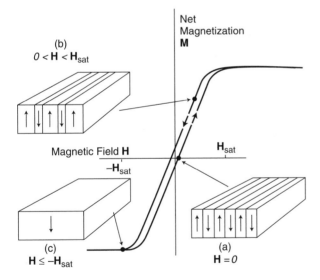

Figure 1 Possible domain configurations in a single crystal of magnetic material with an easy axis of magnetization. (a) At zero field there are ribbon domains of equal width. (b) As a finite field is applied, one set of domains grows at the expense of the other. (c) At fields $H \geq H_{sat}$ the magnetization is in a single domain. Note that as the field is cycled $M(H)$ encloses a finite area, a hysteresis loop.

More General Cases

Although the example given is simple, it is actually encountered. Many important magnetic materials are not uniaxial, but cubic and have 3 or 4 easy axes. Often samples are polycrystalline, and the easy axis direction varies from grain to grain. Real samples often contain nonmagnetic inclusions or grain boundaries. Furthermore, real samples often contain strain distributions that couple to the magnetization through magnetostriction. Finally, the detailed domain configuration depends on the past magnetic and thermal history of the sample. In all of these cases the magnetization process is more complicated than in our example.

Observation of Domains

Fortunately, a variety of experimental techniques are available by which we can visualize real domain structures. One of the most striking is applicable to magnetic materials that will transmit light such as the mixed oxide $Y_3Fe_5O_{12}$ called YIG. Linearly polarized light transmitted by a magnetic material undergoes a rotation of its axis of polarization that is proportional to the projection of the magnetization on the light ray (magneto-optical or Faraday rotation). If in our example above, light transmitted through the up domains exits with a rotation Φ then that through down domains would have a rotation $-\Phi$. By setting an analyzer appropriately, we can make up domains dark, and down domains light. Thus with the sample on the stage of a polarizing microscope we can see the stripe domains in bold contrast, and see how they respond to all sorts of applied fields, changes in temperature or strain, and how the walls interact with imperfections.

Phase Transition

From a thermodynamic point of view, the domain state in Fig. 1 is an example of two distinct states coexisting at a first-order transition. In positive field the magnetization up state is stable, in negative field the magnetization down state is stable. At zero field, both are stable as is any mixture all the way from an infinitesimal fraction of up state in predominant down state to an infinitesimal down state in a sea of up state. It is exactly analogous to the stable coexistence of ice and water at $0°C$.

Applications

Materials in which domain walls move easily so the magnetization responds readily to applied fields are called soft magnetic materials. Among other applications, these are used in the transformers and motors on which developed society depends. Materials with high anisotropy in which domain walls are locked in place, and cannot move are called hard magnetic materials, and in the extreme case permanent magnets. They are used in some motors and in loudspeakers. Magnetic recording depends on magnetic material so finely divided that each particle is too small to support a domain wall, and so the magnetization does not readily respond to small fields, and thus behaves like a tiny permanent magnet. The insight inherent in the domain theory is crucial to the design of a tremendous range of the magnetic materials used in these huge technological industries.

See also: ANTIFERROMAGNETISM; FERROMAGNETISM; FIELD, MAGNETIC; HYSTERESIS; MAGNET; MAGNETIC MATERIAL; MAGNETIZATION; MAGNETO-OPTICAL EFFECTS

Bibliography

CAREY, R., and ISAAC, E. D. *Magnetic Domains and Techniques for Their Observation* (Academic Press, New York, 1966).
CHIKAZUMI, S. *Physics of Magnetism* (Wiley, New York, 1964).
CRAIK, D. J. *Magnetism: Principles and Applications* (Wiley, New York, 1995).
MORRISH, A. H. *The Physics Principles of Magnetism* (Krieger, New York, 1980).

JOSEPH F. DILLON JR.

MAGNETIC FIELD

See FIELD, MAGNETIC

MAGNETIC FLUX

Magnetic flux is a simple well-defined quantity used in general physics when discussing Faraday's Law. Mathematically it can be expressed as

$$\Phi = \mathbf{B} \cdot \mathbf{A}, \tag{1}$$

where Φ is the magnetic flux, a scalar. In this equation $\mathbf{B}$ is a vector that stands for the magnetic field at some point in space. $\mathbf{B}$ is sometimes called the magnetic flux density because of this relationship. The symbol $\mathbf{A}$ is the area over which the flux is to be calculated. Remember that area is a vector with a direction perpendicular to the plane of the area. Note that the flux is the scalar product of the two vectors. In other words, the flux is the product of the area times that component of the magnetic field $\mathbf{B}$ that is perpendicular to the plane of the area. Note particularly that the flux will change if you change the area, the magnitude of $\mathbf{B}$, or the angle between $\mathbf{A}$ and $\mathbf{B}$. Note also that there is a slight ambiguity as to which of two directions the vector $\mathbf{A}$ is pointing. The convention is to have the vector point outward from a closed surface. If the surface is not closed, you can choose either direction. But you must be careful that you do not change your choice in the middle of a problem and that any other conventions (such as the direction of a current) are compatible with your choice. [The SI unit for measuring magnetic flux is the weber (Wb), where 1 Wb corresponds to 1 V·s.]

Equation (1) works fine if the magnitude and relative direction of $\mathbf{B}$ are the same over all the area involved. If they are not, then one must divide up the area into many little areas small enough so that $\mathbf{B}$ and the angle that $\mathbf{B}$ makes with $\mathbf{A}$ do not change. When this is done, Eq. (1) can be applied to each of these little areas. This procedure is the same as writing the integral expression

$$\Phi = \int_{s} \mathbf{B} \cdot d\mathbf{A}, \tag{2}$$

where the integral $\int_{s}$ means to perform the sum over the surface S.

In many problems, it will not be difficult to do the integration, but sometimes it takes a bit of imagination to spot the relevant area. A typical problem is to find the electromotive force (emf) generated when a wire moves through a magnetic field that is perpendicular to the plane traced out by the wire, that is, to find the potential difference between the ends of the wire of length ℓ as it moves with a velocity v through the field B that is pointing up, perpendicular to the paper. To solve this problem, one uses Faraday's law, which states that the rate of charge in the magnetic flux is equal to the emf. The wire will travel a distance vt in a time t. In this case the area involved is that traced by the wire, ℓvt in size. The charge in flux is then $\mathbf{B}\,\ell vt$. It took t amount of time to cut that amount of flux so the emf is $\mathbf{B}\,\ell v$ volts. Note that if the wire were moving up from the paper, the component of $\mathbf{B}$ in the plane of the area would be zero so that no emf would be generated.

See also: ELECTRIC FLUX; ELECTROMAGNETIC INDUCTION, FARADAY'S LAW OF; ELECTROMOTIVE FORCE; FARADAY EFFECT; FIELD, MAGNETIC

Bibliography

GRIFFITHS, D. J. *Introduction to Electrodynamics,* 2nd ed. (Prentice Hall, Englewood Cliffs, NJ, 1989).

HECHT, E. *Physics* (Brooks/Cole, Pacific Grove, CA 1996).

A. F. BURR

MAGNETIC LEVITATION

See LEVITATION, MAGNETIC

MAGNETIC MATERIAL

When placed in a magnetic field **H,** all materials develop a magnetization **M. M** is defined to be the vector sum of the atomic or electronic magnetic moments per unit volume of material. It is customary to divide materials into two broad classes: strong and weak. Strong magnetic materials are often called simply magnetic materials because they exhibit large values of magnetization below a temperature T_0 called an ordering temperature. Below T_0 the atomic or electronic moments line up in various configurations to create special classes of magnetic materials such as ferromagnets, antiferromagnets, ferrimagnets, speromagnets, spin glasses, and others. These types of (strong) magnetic materials form the basis for many engineering devices.

Weak magnetic materials are further classified by the sign of the magnetic susceptibility, which is defined as $\chi = M/H$. Materials for which $\chi > 0$ are called paramagnetic, and those for which $\chi < 0$ are called diamagnetic. Paramagnetism arises from the alignment by an applied field of permanent atomic magnetic moments in the direction of the field. Diamagnetism is the result of the rearrangement of electronic currents by the field, in such a way that **M** is in the opposite direction to **H.** In general, the strong magnetic materials become paramagnetic above their ordering temperatures where thermal fluctuations randomize their spin directions. For this reason the weakly magnetic materials are often called nonmagnetic.

Strong magnetic materials are further categorized as soft, hard, or semihard materials. To explain these terms it is necessary to consider the hysteresis loop shown in Fig. 1. The magnetization is measured as a function of applied field H as it traverses the path shown. M_s is defined as the saturation magnetization where the moments are fully aligned, M_r is the remanent magnetization when $H = 0$, and H_c is the coercive field, defined as the reversed field required to drive M to zero. Soft, hard, and semihard materials are rather arbitrarily defined to have H_c values, expressed in units of oersteds, in the ranges of 0.1–100 Oe, 5,000–20,000 Oe, and 500–3,000 Oe, respectively. Thus hard materials are difficult to demagnetize meaning they are good permanent magnets, soft materials are very easy to demagnetize so that they are useful as cores of alternating current devices, and semihard materials are in the middle, which renders their properties useful in devices where switching the direction of **M** is re-

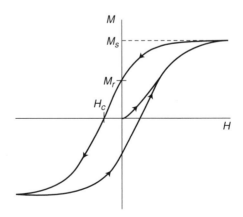

Figure 1   Hysteresis loop.

quired but not too easily. The most common such application is in recording devices such as magnetic disks and tapes. Each of these types of material will be considered separately below.

Soft Magnetic Materials

Soft magnetic materials are used mainly in electrical circuits in which the soft material is used to collect and amplify the flux generated by the electric currents. Typical alternating current (ac) applications include cores of inductors and transformers, and direct current (dc) applications include electromagnets and relays.

The hysteresis loops of soft magnetic materials have the important characteristics of very low coercivity and very high permeability, which is defined as the ratio $\mu = B/H$, where B is the magnetic induction $B = H + 4\pi M$ (Gaussian units). In order to achieve high μ values, up to 100,000 for low H values, it is necessary that the saturation magnetization M_s be very high and the coercivity H_c be very low. These properties ensure that the area enclosed by the hysteresis loop, which equals the energy loss in traversing the loop, is very small. In addition, the high M_s value means that as much magnetic flux as possible is carried by the magnetic circuit as is required in inductor and transformer cores.

Additional material properties required of soft magnets include (1) high homogeneity and freedom from second phases, impurities, stress, and crystallographic defects. This ensures that magnetization reversal through domain wall motion will be easy and thus H_c small; (2) high ordering (Curie) temperature; (3) high electrical resistivity to minimize energy losses through eddy currents; (4) good corrosion resistance and temperature stability; (5) easy machinability; and (6) inexpensive production.

In power transformer applications the direction of magnetization changes sixty times per second. This causes a similar number of mechanical stresses per second and leads often to an audible hum. The basic cause of this effect, a relationship between magnetization and mechanical stress called magnetostriction, has been an important problem for materials scientists to minimize.

Typically used alloys for ac applications include iron-silicon (with about 3 weight % Si), iron-aluminum, and nickel-iron (Permalloy). Certain compositions of the Ni-Fe alloys, called Mumetal and Supermalloy, have μ values up to 3×10^5 and H_c values as low as 0.01 Oe. Amorphous alloys produced by rapid quenching from the melt, at rates of about 10^6 K/s, also have good soft magnetic properties. Examples include $Fe_{80}B_{20}$ and $Fe_{40}Ni_{40}P_{14}B_6$ in which the metalloids (B and P) are present to promote easy glass formation.

For very high frequency applications it is necessary to use insulating materials because of the severe eddy current problems. Cubic or soft ferrites have the general chemical formula $MO \cdot Fe_2O_3$ where M is a transition metal such as Fe, Ni, Mn, or Zn. The soft ferrites can be classified as nonmicrowave ferrites for frequencies up to 500 MHz and microwave ferrites for use up to 500 GHz. Soft ferrites are used in electronic equipment such as telephone signal transmitters and receivers, and radio receiver antennas. Microwave ferrites such as yttrium-iron garnet are used as waveguide for electromagnetic-radiation and phase-shifter devices.

Hard (Permanent Magnet) Materials

Permanent magnets have a wide variety of uses where a constant magnetic field is needed without power usage via electric currents. In regions of space where the field is changing a force is exerted on another magnetized object. This leads to many devices such as holding devices, latches, actuators, loudspeakers, and motors. Fields generated by permanent magnets also are useful for bending the paths of charged particles in mass spectrometers or in magnetic resonance imaging devices.

The figure of merit for permanent magnets is called the energy product $(BH)_{max}$, which is the maximum value of the BH product in the demagnetizing part of a B(H) hysteresis loop. The units of the energy product are million gauss oersted (MGOe) in Gaussian units. This figure of merit is inversely proportional to the volume of magnetic material needed to produce a given field in a unit volume of space.

The materials and magnetic properties required for excellent hard magnetic materials include (1) high M_s since the maximum value of energy product obtainable is $4\pi^2 M_s^2$; (2) high remanence M_r; (3) a nearly rectangular hysteresis loop with $M_r \approx M_s$ and high H_c value ($> 2\pi M_s$); (4) high Curie temperature and thus high temperature stability at operating temperatures; (5) good corrosion resistance; (6) good mechanical properties; and (7) inexpensive production.

Given the large worldwide market for permanent magnet materials, of order two billion dollars, it is easy to understand the large research effort that has been devoted to understanding the physics of such materials so that better and less expensive compounds can be produced. Since about 1880 there has been a continual increase in the energy product, ranging from about 0.25 MGOe in 1880 for carbon steel to about 50 MGOe in 1990 for $Nd_2Fe_{14}B$.

Some commonly used hard materials include Alnicos, which are Al-Ni-Co alloys that consist of finely dispersed particles formed by heat treatment in an applied field. $(BH)_{max}$ for these materials is about 10 MGOe. Hard ferrite magnets such as $BaFe_{12}O_{19}$ are relatively inexpensive and derive their coercivity from high magnetocrystalline anisotropy. In recent years, great progress has been achieved by studying rare-earth transition metal compounds such as $SmCo_5$ [$H_c \approx 20$ kOe and $(BH)_{max} \approx 25$ MGOe], $Sm_2(Co_{0.85}Fe_{0.11}Mn_{0.04})_{17}$, and $Nd_2Fe_{14}B$.

Since nature might be expected to provide theoretical values of $(BH)_{max}$ up to about 150 MGOe, there is still great interest and activity in searching for new materials.

Semihard (Recording) Materials

Magnetic materials used for data storage on rigid or floppy disks or magnetic tape are classified as semihard materials. Small domains are written onto the film by application of fields from heads which produce sufficiently large fields are written onto the film by application of fields from heads which produce sufficiently large fields to reverse the magnetization locally to produce a "bit" of information corresponding to a domain magnetized in a certain direction. Since the information must be permanently stored, the magnetization must be stable and resistant to change; that is, the material must have a high magnetization, relatively high coercivity, and a fairly rectangular hysteresis loop. H_c values must not be so high, however, that the bits cannot be switched by a reasonably small field (2,000–3,000 Oe) available with present write heads. Reading of the information is achieved by a "read head" based either on sensing the field through flux changes in a coil or change in resistance of a soft magnetic material, as the disk is rotated past the read head.

The amount of data stored per unit area of disk (areal density) has been increasing rapidly. The rate is about a factor of 10 every decade. At the present time the areal density in commercial products is about 100 megabits per square inch (100 Mb/in²). The materials being used are typically Co-based al-

loys such as CoCrTa or CoCrPt, which have H_c values of about 2,000 Oe and magnetization directions in the plane of the film (longitudinal recording). Vigorous research activity is underway to increase the areal density to 10,000 Mb/in² by the year 2005. Further increases in density may require a different scheme in which the bits are magnetized perpendicular to the film plane (perpendicular recording).

Other types of data storage such as rewritable magneto-optical systems are under research and development for future high-density data-storage systems.

See also: ANTIFERROMAGNETISM; DIAMAGNETISM; ELECTROMAGNETISM; FERRIMAGNETISM; FERROMAGNETISM; FIELD, MAGNETIC; MAGNET; MAGNETIC DOMAIN; MAGNETIZATION; PARAMAGNETISM; SPIN GLASS

Bibliography

BERTRAM, H. N. *Theory of Magnetic Recording* (Cambridge University Press, Cambridge, Eng., 1994).

CRANGLE, J. *Solid-State Magnetism* (Von Nostrand Reinhold, New York, 1991).

HADJIPANAYIS, G. C., and PRINZ, G. A., eds. *Science and Technology of Nanostructured Magnetic Materials* (Plenum, New York, 1991).

JILES, D. *Introduction to Magnetism and Magnetic Materials* (Chapman and Hall, London, 1991).

MALLINSON, J. C. *The Foundations of Magnetic Recording*, 2nd ed. (Academic Press, New York, 1993).

McCURRIE, R. A. *Ferromagnetic Materials, Structure and Properties* (Academic Press, London, 1994).

PARKER, R. J. *Advances in Permanent Magnetism* (Wiley, New York, 1990).

DAVID J. SELLMYER

MAGNETIC MOMENT

Magnetic moment refers to the (typically) dipole magnetic field associated with current loops, the field of a permanent magnet, or the field owing to the intrinsic spin that many fundamental particles such as the electron possess. The common classical description is derived from considering the field of a single loop of current. At distances much larger than the loop the shape of the field is the same as if there were two opposite magnetic charges or poles separated by a small distance perpendicular to the plane of the real loop. This is a direct analogy to the electric dipole moment, which is the field due to two oppositely charged particles separated by a small distance. Often the term "magnetic moment" is replaced by "magnetic dipole moment."

Just as an external electric field $\mathbf{E}$ will create a torque τ on an electric dipole $\mathbf{p}$ given by $\tau = \mathbf{p} \times \mathbf{E}$, an external induction magnetic field will apply a torque on a current loop with the tendency to align the plane of the loop in the direction of the applied field. This torque is given by $\tau = \boldsymbol{\mu} \times \mathbf{B}$, where $\mathbf{B}$ is the magnetic induction field and $\boldsymbol{\mu}$ is called the magnetic moment. This vector expression for torque can be written more illustratively as $\tau = \mu B \sin \boldsymbol{\theta}$, where $\boldsymbol{\theta}$ is the angle between the axis of the loop of current carrying wire and the magnetic field. This expression can be compared to a calculation of the torque on a single current loop which is $\tau = iAB \sin \boldsymbol{\theta}$, where i is the current and A is the area of the loop. Comparing the two expressions for torque, the definition for magnetic moment is seen to be $\mu = iA$. If there are N turns in the current loop then the definition is $\mu = NiA$. This definition allows a measurement of magnetic moment using torque techniques.

At the atomic level the above definition is still useful if one considers the current to be that of a single electron with charge $-e$ circulating in an orbit of radius r with an angular frequency ω. This current is $i = -e\omega/2\pi$ so that the magnetic moment becomes $\mu = -e\boldsymbol{\omega}\, r^2/2$ for a circular loop. In terms of the orbital angular momentum, $\ell = m\boldsymbol{\omega}\, r^2$, the magnetic moment is $\mu = -(e/2m)\ell$, where m is the mass of the electron. This expression must be quantized which introduces Plank's constant $\hbar$ in such a manner that the magnetic moment becomes $\boldsymbol{\mu} = -(e\hbar/2m)\mathbf{I}$, where $\mathbf{I}$ is the orbital angular momentum quantum number. The expression $\mu = e\hbar/2m$ ($= \mu_B$) is the basic quantum of magnetic moment called the Bohr magneton. The energy of an orbital electron in a magnetic induction field then is $U = -\boldsymbol{\mu} \cdot \mathbf{B}$, where $\boldsymbol{\mu}$ is an integral multiple ℓ of the Bohr magneton.

Fundamental particles such as the electron, proton, neutron, and most others also possess an intrinsic magnetic moment called spin. The magnitude of the moment for the electron is given by $\mu = \mu_B(1/2)$, where the 1/2 represents the half-integral spin of the electron. In atomic cases where more than one electron may couple together (in this case in a manner called Russell–Saunders coupling), an atom or molecule may have both an total orbital angular momentum $\mathbf{L}$ ($=\mathbf{I}_1 + \mathbf{I}_2 + \ldots$) and a spin angular momentum $\mathbf{S}$ ($= \mathbf{s}_1 + \mathbf{s}_2 + \ldots$). The total moment is calculated by using the following expression: $\mu = \mu_B(\mathbf{L} + 2\mathbf{S})$, which precesses around the total angular momentum $\mathbf{J}$ ($= \mathbf{L} + \mathbf{S}$). The magnetic moment becomes $\mu = \mu_B\, g\, J$, where g is called the spectroscopic g value and is given as

$$g = 1 + \frac{J(J + 1) + S(S + 1) - L(L + 1)}{2J(J + 1)}.$$

The intrinsic moment of the proton is given by $\mu = e\hbar/2M$, where M is the mass of the proton. This number is called the nuclear magneton and is considerably smaller than the Bohr magneton owing to the relatively large mass in the denominator.

There is a further source of a magnetic moment that occurs on the application of a magnetic induction field to an orbital electron. Faraday's law of induction states that when a magnetic field is applied to a current loop, in this case the electron in its orbit, a magnetic field (moment) in the opposite direction will develop. This is the source of diamagnetism and occurs for any electron with a nonzero orbital angular momentum. It is a small effect and is masked in any material in which unpaired electrons exist.

Experiments to search for the existence of magnetic monopoles (magnetic charges) as the source of the intrinsic spin moment have failed and further suggest that these intrinsic moments are also due to the motion of a charge. For electrons, muons, and nucleons and their antiparticles, the evidence appears convincing. However, for the more elementary particles called hadrons the evidence is not yet conclusive.

See also: BOHR MAGNETON; DIPOLE MOMENT; ELECTRIC MOMENT; ELECTROMAGNETIC INDUCTION; ELECTROMAGNETIC INDUCTION, FARADAY'S LAW OF; ELEMENTARY PARTICLES; FIELD, MAGNETIC; NUCLEAR MOMENT

Bibliography

CONDON, E. U., and ODISHAW, H. *Handbook of Physics* (McGraw-Hill, New York, 1958).

JILES, D. *Introduction to Magetism and Magnetic Materials* (Chapman and Hall, London, 1991).

TIPLER, P. A. *Physics* (Worth, New York, 1982).

JOHN E. DRUMHELLER

MAGNETIC MONOPOLE

At least since the scholar-crusader Pierre of Maricourt wrote a letter about the properties of magnets in 1269, people have contemplated the idea of an isolated magnetic pole, even though no one has ever successfully cut a magnet into separate north and south poles. André-Marie Ampère in 1820, building on the work of Hans Christian Oersted, proposed that all magnetism is nothing but forces produced by and acting on electric currents. In the hands of Michael Faraday, this became the notion that magnetic fields consist of lines of magnetic flux circling around electric currents, with the flux having no beginning and no end. Positive and negative electric charges are sources and sinks of electric flux, respectively. Therefore, if magnetic flux lines have no beginning or end, then magnetic charges, or monopoles, cannot exist. The term "monopole" is meant as a contrast to such terms as "dipole" and "quadrupole," which refer to configurations that have no net charge. Thus, the Ampère theory implies that there is no such thing as a magnetic monopole and explains why none so far has been found in nature.

Why then should this hypothetical object continue to fascinate physicists? Current pictures of fundamental physics do not tell us merely that magnetic poles could exist, they also say a great deal about the properties poles must have if they really are there, giving guidance for ingenious experimental searches to find them. If poles turned up, surely one would learn important things about physics at very short-distance or high-energy scales. Even thinking about monopoles casts a powerful intellectual searchlight on physics theories, and also provides soluble model problems with practical applications.

Experimental Searches

Many searches have put stringent limits on the density of monopoles in the universe, using a whole variety of monopole properties. For example, a monopole passing through a superconducting loop of wire will change the current in the wire by an appreciable amount. In principle, no other process could duplicate this effect, so long as the loop remains unbroken: By Faraday's law, if the flux through the loop changes there must be an electromotive force around the loop, but that is not allowed in a superconductor. To prevent the total flux from changing, the current in the wire must change to produce enough magnetic flux to compensate for the flux lines pulled through the loop by the monopole. Thus, the large pole strength and the very special properties of superconductors provide a unique detection method for poles.

A very fast monopole would ionize matter in the same way as a large atomic nucleus. There have been searches using this, even though the ionization would not give a clear signal that the object carries magnetic rather than electric charge. A slow monopole can excite an atom or molecule when it comes very close to the atomic nucleus, and so a long trail of such excitations could indicate the passage of a pole. If monopoles were exceedingly massive, then just shaking or heating ordinary matter could pry loose the monopoles, which then would fall into a detector below the material. The large pole strength implies that monopoles could be ripped out of matter by applying a strong magnetic field; then one could look for the energy given up by the pole as it plows into a detector. Similarly, monopoles in interstellar space could absorb energy from galactic or intergalactic magnetic fields, thereby leaching the field strength. Eugene Parker has argued that this gives an upper bound on monopole densities, as appreciable magnetic fields are observed in the universe.

If a proton came close to a monopole, the pole might catalyze decay of the proton to lighter particles, something not yet seen under normal circumstances. Large underground detectors might reveal a trail of energy bursts coming from a monopole that passed straight through, catalyzing a whole string of proton decays on the way. If the rate of catalysis were great enough, then in large and dense objects ranging from major planets like Jupiter to neutron stars, the catalysis would produce enough heat to generate copious x rays. All of these phenomena have figured in experiments and astronomical observations, including studies of moon rocks, Antarctic ice, and magnetic ores. So far there are no confirmed positive results. Meanwhile, cosmologists have thought about how monopoles might have been generated during the evolution of the universe. It seems quite likely that they could arise, but, at least in the popular inflation cosmology, as quickly as they were produced the poles would move so far apart that the whole universe observable today might have only one or two left.

Uses of Magnetic Poles

The concept of an isolated pole has practical uses for measurement and calculation, even if in fact that pole is one end of a long thin magnet. Charles Augustin Coulomb used this idea to measure the force between magnetic poles, and determined that it has the same Coulomb's law form as the force between electric charges. That is, the force is directed along the line determined by the poles, inversely proportional to the square of the distance between them, and proportional to the product of pole strengths (repulsive if they are alike and attractive if unlike). Henri Poincare analyzed the scattering of electrons passing near one end of a long magnet, using the same monopole approximation to describe the magnetic field. His complete and exact solution exhibits the beautiful spiral motion of the electrons, and exploits the fact that there is a conserved total angular momentum, which includes besides the obvious contributions from the two particles an "electromagnetic spin" directed from charge to pole, equal in magnitude to the product of the charge with the pole strength.

Description of Magnets

Ampère, the monopole debunker himself, introduced another use for the monopole idea. In response to criticism by Faraday, he observed that outside a magnet one cannot tell whether it contains loops of electric current or pairs of opposite magnetic poles. All external phenomena work the same either way, but often they are easier to calculate using poles. Even up to quite recent times physicists have been confused about this ambiguity, and there have been suggestions, for example, that measurements of very slow atomic oscillations could determine whether or not there are monopoles inside the proton accounting for its magnetism. Ampère's principle assures that this is not possible.

Induced Magnetism

Consider an electrically charged particle moving freely in a spherical surface centered on a magnetic monopole. The particle moves in a circle, smaller than a great circle unless the pole strength is zero. Now imagine a gyroscope with its axis fixed at one end. If the other end is free to move, and moves much more slowly than the gyroscope is spinning, then the tip of the axis will move in a circle smaller than a great circle unless the spin is zero. Thus the tip of the gyroscope axis moves in exactly the same way as an electric charge in the field of a monopole, just as Poincare observed. The gyroscope is the original example of *induced magnetism,* a concept which Michael Berry has shown is quite general for systems with fast degrees of freedom like the gyroscope spin and slow degrees of freedom like the axis precession: Motion in one system without any magnetic fields imitates motion in another system with explicit magnetic forces.

Quantum Condition and Internal Pole Structure

With the advent of quantum mechanics, P. A. M. Dirac found that there is an important quantum constraint on possible magnetic poles: the product of the strength of any free magnetic pole and any free electric charge must be equal to an integer multiple of a smallest unit. One may find the value of Dirac's unit by insisting that Poincaré's electromagnetic spin be quantized, like any other angular momentum in quantum mechanics. This implies that the strength of a magnetic pole must be approximately equivalent to the electric charge of a nucleus with sixty-nine protons (i.e., an isotope of thulium). This charge is absolutely enormous for an elementary particle, and it implies that just like the thulium nucleus the monopole must have a large size on the scale of its Compton wavelength. The latter is the smallest distance in which one could confine a particle, consistently with Werner Heisenberg's uncertainty principle, without giving the object a kinetic energy far greater than its rest energy. The large monopole radius allows one to describe its internal structure by classical physics. However, that physics must go beyond ordinary electrodynamics, and indeed beyond the standard model, since these theories have room at best for point monopoles. As a result, the dramatic success of the standard model in describing observed phenomena up to quite high energies implies that the monopole radius must be so small, and the mass of the pole so large, that even the most powerful particle accelerators now contemplated would lack the energy to create a monopole-antimonopole pair. To find real monopoles, we shall have to observe them rather than create them.

Spin and Statistics

The electromagnetic spin for a system of charge and pole may have half-integer values. In conventional dynamics without monopoles, half-integer spin is associated with Fermi–Dirac statistics, while integer spin comes with Bose–Einstein statistics. If, for example, two spin-0 particles, one with electric charge and one with magnetic charge, form a composite with spin 1/2, it turns out that this "dyon" (particle with both electric and magnetic charge) has Fermi–Dirac statistics, so that the usual connection between spin and statistics is maintained. This is one more illustration of how snugly monopoles fit constraints already seen in other contexts, and therefore how thinking about them can deepen our understanding of fundamental dynamics.

Confinement of Magnetic or Electric Charge

Monopoles offer still another aid to understanding physics; imagine a north and a south pole inside a large piece of superconductor. By the Meissner effect (suppression of magnetic fields in superconductors) the two poles must feel an attractive force whose strength is independent of the distance between them. Thus no matter how much kinetic energy is supplied to set the poles flying apart, eventually that energy will be exhausted and they will be drawn back together. Consequently, only objects with net magnetic charge zero may be isolated from each other inside the superconductor. This phenomenon is called confinement of magnetic charge. In the accepted picture (quantum chromodynamics, or QCD) of the forces governing protons, neutrons, and other strongly interacting particles, these objects are regarded as composites of quarks, which carry an analogue of electric charge called color-electric charge. Quarks are never seen in isolation, and thus provide an example of color-electric confinement. In the mid-1970s Stanley Mandelstam and others suggested empty space might behave like a superconductor of color-magnetic poles, and therefore confine quarks, since this would represent a dual version (electric and magnetic phenomena interchanged) of the magnetic confinement expected in an ordinary electric superconductor. Nathan Seiberg, Edward Witten, and others have used versions of QCD exhibiting supersymmetry to strengthen the case for quark confinement through color-magnetic super-

conductivity. Even if regular magnetic poles are never observed, conclusive evidence that space is filled with color-magnetic poles may be found. If so, at least a type of monopole will be seen as a ubiquitous and essential part of our world.

See also: AMPÈRE, ANDRÉ-MARIE; AMPÈRE'S LAW; COLOR CHARGE; COULOMB, CHARLES AUGUSTIN; COULOMB'S LAW; DIRAC, PAUL ADRIEN MAURICE; ELECTROMOTIVE FORCE; FARADAY, MICHAEL; MAGNETIC FLUX; QUANTUM CHROMODYNAMICS; QUARK; QUARK CONFINEMENT; SPIN AND STATISTICS

Bibliography

CARRIGAN, R. A., JR. "Quest for the Magnetic Monopole." *Phys. Teach.* **13**, 391–398 (1977).

CARRIGAN, R. A., JR., and TROWER, W. P. "Superheavy Magnetic Monopoles." *Sci. Am.* **246** (4), 106–118 (1982).

FORD, K. W. "Magnetic Monopoles." *Sci. Am.* **209** (6), 122–131 (1963).

GOLDHABER, A. S., and TROWER, W. P., eds. *Magnetic Monopoles* (American Association of Physics Teachers, College Park, MD, 1990).

ALFRED S. GOLDHABER

MAGNETIC PERMEABILITY

When the magnetic field H (ampere per meter) is applied to a long thin rod specimen made of ferromagnetic material, the rod acquires magnetization I (tesla). Magnetization is defined as the magnetic moment per unit volume. In most cases for low magnetic fields, the magnetization is proportional to the strength of the field (i.e., $I = \chi H$, where χ is the magnetic susceptibility, which is measured by a unit of henry per meter). Suppose that a secondary coil of thin wire is wound around the specimen rod and the intensity of the field is changed at a rate of dH/dt. According to the law of electromagnetic induction, the voltage can be observed across the coil, which is partly proportional to dH/dt and partly proportional to dI/dt. Exactly speaking the voltage is proportional to dB/dt, where B is called the magnetic flux density or magnetic induction. This quantity is related to H and I by the relation $B = \mu_0 H + I$, where μ_0 is the permeability of vacuum ($\mu_0 = 4 \times$

Table 1 Relative Permeabilities of Typical Soft Magnetic Materials

Material	Composition	Initial Permeability	Maximum Permeability
Si-steel	3%Si-Fe	1,500	40,000
Amorphous	B-Si-C-Fe	15,000	450,000
Supermalloy	5Mo79Ni-Fe	100,000	1,000,000
Mn-Zn ferrite	50Mn50Zn	2,000	
Ni-Zn ferrite	30Ni70Zn	80	

10^{-7} H/m). The first term, $\mu_0 H$, is called the magnetization of vacuum, because even without a ferromagnetic specimen, the secondary coil produces the voltage proportional to $\mu_0 dH/dt$. Magnetic permeability μ is defined as the ratio B/H. Therefore, $B = \mu H$. Since $B = \mu H$ and $I = \chi H$, the relation $B = \mu_0 H + I$ becomes $\mu H = \mu_0 H + \chi H$, which simplifies to $\mu = \mu_0 + \chi$. Sometimes the relative susceptibility $\bar{\chi} = \chi/\mu_0$ (i.e., $\chi = \bar{\chi}\mu_0$) and the relative permeability $\bar{\mu} = \mu/\mu_0$ (i.e., $\mu = \bar{\mu}\mu_0$) are used. In this case, $\mu = \mu_0 + \chi$ becomes $\bar{\mu}\mu_0 = \mu_0 + \bar{\chi}\mu_0$, which simplifies to $\bar{\mu} = 1 + \bar{\chi}$. For soft magnetic materials, which exhibit very large magnetic susceptibility, the magnetic permeability is nearly equal to the magnetic susceptibility.

A change in B, symbolized by ΔB, can be measured by integrating the voltage in the secondary coil during a change with time in H, symbolized by ΔH, by means of a fluxmeter or some electronic integrator. The permeability is calculated by $\mu = \Delta B/\Delta H$. More conveniently, the ac voltage can be measured during application of an ac magnetic field, and the amplitudes of B and H can be compared.

Various magnetically soft materials are being used as the magnetic core materials. For a power transformer, the average slope of the B-H curve, which is called the maximum permeability μmax, is an important factor for characterizing performance. For a small transformer used for electronic devices, the more important factor is the initial slope (i.e., slope at the origin) of the B-H curve, which is called the initial permeability μ_a. Table 1 lists typical magnetically soft materials with their compositions, as well as their initial and maximum permeabilities (relative values).

Permeability is a structure-sensitive quantity. If magnetic sheets that are to be used as a core material are bent or cold-worked, the permeability is easily reduced by a factor of ten or so. To optimize performance, heat treatments are needed. In the case of ferrites or the magnetic oxides, the atmosphere during heat treatment has a great influence on their characteristic performance.

See also: ELECTROMAGNET; ELECTROMAGNETIC INDUCTION; ELECTROMAGNETIC INDUCTION, FARADAY'S LAW OF; FIELD, MAGNETIC; FERROMAGNETISM; MAGNETIC FLUX; MAGNETIC MOMENT; MAGNETIC SUSCEPTIBILITY; TRANSFORMER

Bibliography

CHIKAZUMI, S. *Physics of Magnetism* (Wiley, New York, 1964).

SOSHIN CHIKAZUMI

MAGNETIC POLE

A magnetic pole is the region at each end of a permanent magnet where the external field is the strongest. Each magnet has a north pole (so called because it tends to point north) and a south pole. All attempts, however, to precisely locate the pole are bound to fail. For example, if you cut a long bar magnet in half in an attempt to isolate the north pole from the south pole, you will find that each half still has both a north and a south pole. Nevertheless you can show that magnetic poles act much like electric charges in that like poles repel and unlike poles attract. The force between poles is even an $1/r^2$ force, just like Coulomb's law for charges.

There are theoretical reasons to hope for the existence of single magnetic poles, called magnetic monopoles. Maxwell's equations would be symmetric between electric and magnetic fields if monopoles existed.

For example, one of Maxwell's equations can be written as

$$\oint \mathbf{E} \cdot d\mathbf{A} = \frac{q_e}{\varepsilon_0},$$

where $\mathbf{E}$ is the electric field, $\mathbf{A}$ is the area to be integrated over, ε_0 is the permittivity of free space, and q_e is the electric charge. Unfortunately, the equivalent magnetic field equation must be written as

$$\oint \mathbf{B} \cdot d\mathbf{A} = 0,$$

where $\mathbf{B}$ is the magnetic field, because discrete magnetic poles have not been found.

Physicists highly value symmetry in their equations and theories, so they would prefer to write the magnetic equations as

$$\oint \mathbf{B} \cdot d\mathbf{A} = \mu_0 q_m,$$

where μ_0, the permeability, is the magnetic equivalent of ε_0 and q_m is the magnetic equivalent of the electric charge. Symmetry arguments have proven useful in the past so physicists like P. A. M. Dirac, as well as others, have spent much time studying the theoretical properties of discrete magnetic poles. However, magnetic change is not required; one can have a sensible theory without it. For example, the special theory of relativity is able to explain the existence of magnetic fields on the basis of moving charges or of charges in a moving frame of reference.

Nevertheless the impact that the discovery of a magnetic monopole would have on physics is such that many people have spent much time looking for one. Several announcements have been made of possible candidates for its discovery. At one time it was thought that certain tracks made in plastic which had been in space long enough to be damaged by cosmic rays could best be explained as tracks left by monopoles. Later a certain signal from a closed loop of wire was thought to be caused by a monopole passing through the loop, but in both cases further study showed that alternative explanations were more probable. However, all efforts to

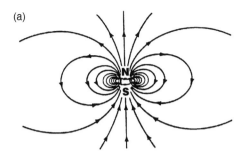

(a)

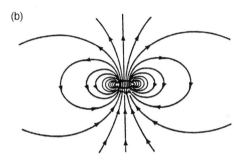

(b)

Figure 1 (a) A magnetic field from a magnetic dipole and (b) a magnetic field from a loop.

show the existence of magnetic monopoles have failed.

You can show that at distances that are large compared to the separation of the poles or the size of the loop, the field of a magnetic dipole is indistinguishable from the magnetic field generated by a current flowing in a small loop. Fig. 1a shows the external field of a north pole placed close to a south pole (a short bar magnet or magnetic dipole) and Fig. 1b shows the magnetic field produced by a small loop of wire oriented perpendicular to the plane of the page. Both have the same l/r^3 dependence that one would expect from a dipole field.

One can consider a permanent magnet to be a collection of aligned current loops, but this idea is just a convenient mental picture. A picture more closely connected with reality is to ascribe the permanent magnetic field to the spin of electrons in the outer shell of the atoms that make up the magnetic material. This moving charge is all that is needed to explain the existence of a magnetic field.

See also: CHARGE; COULOMB'S LAW; DIRAC, PAUL ADRIEN MAURICE; FIELD, ELECTRIC; FIELD, MAGNETIC; MAGNET; MAGNETIC BEHAVIOR; MAGNETIC MATERIAL; MAGNETIC MONOPOLE; MAXWELL'S EQUATIONS; SYMMETRY

Bibliography

HECHT, E. *Physics* (Brooks/Cole, Pacific Grove, CA, 1996).

A. F. BURR

MAGNETIC RESONANCE IMAGING

Nuclear magnetic resonance (NMR) is an excitation and detection of nuclear spins usually embedded in a multiatomic environment. In the presence of a strong external magnetic field $\mathbf{B}_0$, say in the z direction, each magnetic moment of the nucleus in the system (simply referred to as spins hereafter) precesses with a frequency ν characteristic of the nucleus. This is known as Larmor precession. Larmor frequency is given by the relation

$$\omega = \gamma \mathbf{B}, \qquad (1)$$

where ω is the angular frequency ($\omega = 2\pi\nu$) and γ is the gyromagnetic ratio. For a proton, γ proton is 42.6 MHz/T. $\mathbf{B}$ is the magnetic field as "seen" by the nucleus. For magnetic fields commonly employed in NMR, ω is in the radio-frequency (rf) range, usually in the megahertz range. When we apply an external rf pulse that has the same frequency as the Larmor frequency of the precessing spin, the spin absorbs the rf energy (nuclear magnetic resonance) and rotates away from the direction of the external field. Careful choice of amplitude, orientation, and duration of the pulse can direct the spin to flip in the direction $90°$ away from the external field $\mathbf{B}_0$ or in the $180°$ direction. The net magnetization $\mathbf{M}$ of the assembly of aligned spins prior to the application of the pulse is in the z direction, that is, $\mathbf{M}_z = \mathbf{M}$. When the rf pulse, such as a $90°$ pulse, is applied, an xy component $\mathbf{M}_{xy}$ appears. This can be detected by a receiving antenna that is usually the same as the pulse transmission coil. The rf pulse does not last sufficiently long to flip all the spins (saturation). Thus this step can be seen as attaining partial saturation.

When the pulse is turned off, the transverse magnetization $\mathbf{M}_{xy}$ begins to decrease as the spins flip back to the positive z direction to reach an equilibrium state. This relaxation process is called the free induction decay (FID). This relaxation process (or decay of transverse magnetization) is detected by the receiving coil. In NMR, signals such as these are recorded as plots of signal intensity versus time. These time domain data are then transformed into frequency domain by a computer routine called Fast Fourier Transform (FFT). Fourier transformed data show signal intensity as a function of frequency and are called the NMR spectrum.

NMR Signals

The intensity of the NMR signal is determined by, among other factors, the number of spins that respond to the rf signal. For hydrogen (proton spin), which is amply present in the human body, NMR signal is sensitive to the environment of hydrogen nuclei. Bonding strength of hydrogen within a molecule also determines the signal intensity. It is known that a tightly bound hydrogen nuclei, such as those in bones, do not contribute significantly to the signal. Thus the density of the loosely bound hydrogen, called spin density, is one of the principal parameters for NMR. When the external field is turned off suddenly, the spin system becomes randomized with a time constant T_1 (spin-lattice relaxation time). This is an exponential decay process much like radioactive decay, and T_1 is the time constant for the decay. This is another significant parameter for NMR. It was discovered by Raymond Damadian in 1971 that T_1 of proton spins associated with diseased tissue is longer than that for corresponding healthy tissue. The transverse magnetization $\mathbf{M}_{xy}$ also decays exponentially with yet another time constant T_2. This is the third significant parameter. T_2 can be detected and measured by observing the free induction decay. For clinical observation, however, the apparent T_2 (i.e., T_2^*) measured by FID is smaller due to the large area from which the signals arrive as the signals lose some phase coherence. Spin-lattice relaxation time T_1 and spin-spin relaxation time T_2 are controlled by different mechanisms and are independent of each other. One of the methods of determining T_2 is the spin echo method, which circumvents the dephasing. It accomplishes this by first applying a $90°$ pulse followed by a sequence of $180°$ pulses. The effect of the $180°$ pulse is to reverse the direction of the spin precession so that the out-of-phase spins move back together to restore phase coherence as the FID signals decay with the time constant T_2.

Imaging by NMR

Damadian's early work provided no spatial information as to the location of a diseased tissue. The breakthrough was provided by Paul Lauterbur in 1973. He proposed the introduction of a weak magnetic field gradient superimposed on the uniform external field $\mathbf{B}_0$. The gradient field provides the slice or the localization necessary for imaging. This started the revolution that is still continuing. After the Lauterbur invention, the major initial advances in imaging were made in Britain by research groups in Nottingham, Aberdeen, and the Hammersmith Hospital in London.

The imaging is the process of generating a mosaic of pixels that are characterized by intensity and position. In MRI, the main magnetic field is uniform throughout the imaging volume and is on continuously. This is provided by a superconducting magnet cooled by liquid helium surrounded by a liquid nitrogen bath and insulation. In addition, three pairs of gradient coils are mounted inside the bore of the main solenoid to identify x, y, and z coordinates from which the signals originate. These gradient coils are the key elements in the imaging system. The fundamental design difference between NMR spectroscopy and MRI is that in the former, uniformity of the field is of utmost importance, whereas in the latter, field gradients are purposely introduced as the key elements for imaging.

In a uniform magnetic field all the spins precess at the same Larmor frequency given by Eq. (1). When rf pulse signal of this frequency is applied, all the spins are in resonance and can absorb photons and flip by, say, a 90° angle. Then the signals from their free induction decay do not carry any information as to where they come from. When the small gradient field is added to the otherwise uniform field, spins precess at different frequencies depending on where they are located. Let us say that the gradient field and the primary field are both in the z direction, which can be expressed as

$$\mathbf{B} = \mathbf{B}_0 + mz, \qquad (2)$$

where the origin $z = 0$ is located at the midpoint of the solenoid that supplies the uniform field $\mathbf{B}$, and m is the measure of the field gradient (or $m = d\mathbf{B}/dz$). Spins at each slice of z axis now have different Larmor frequency. When a pulse of frequency ω_1 is sent, the spins that are first excited, and subsequently send back the free induction decay (FID) signal, are located at location $z = z_1$, which is given by

$$z_1 = \frac{\omega_1 - \gamma\mathbf{B}}{m\gamma}. \qquad (3)$$

Since the field gradient m is designed into the instrument and can be measured accurately by the standard technique of magnetic field measurement, ω_1 gives the location of "slice" perpendicular to the z axis. Thus the magnetic field gradient provides slice selection. This is not sufficient for the localization of the pulse, that is, to identify the location where the particular signal originated. To localize a signal as arriving from a particular volume element (voxel) we must know its x and y coordinates as well.

The most common clinical method for localization and imaging is the 2DFT method, or two-dimensional Fourier transform method. This involves firing of y- and x-gradient pulses sequentially following the z-gradient pulse. The data thus acquired are analyzed by two-dimensional Fourier transformation using a dedicated computer that also controls pulse sequencing and data acquisition. The typical sequencing of the pulses is as follows: (1) The uniform field is on in the z direction, which is the transaxial direction (along the length of a patient). (2) z-gradient field pulse and the 90° rf signal pulse are simultaneously turned on to excite the spins and turned off; FID starts immediately thereafter. (3) As soon as the FID begins, the y gradient is pulsed to localize the pixels by column. Since, up to this step, all the spins in a single target slice are in phase, the y-gradient pulse has the effect of varying the phase of precessing spins along the vertical line within the slice, that is, phase encoding the spins. The size of the y gradient determines the degree of coarseness with which one can distinguish parallel lines in the x direction. A strong y-gradient allows one to distinguish two horizontal lines closer together than a weak gradient because it gives rise to larger phase difference between adjacent rows of spins. In other words, selection of a y-gradient value selects out a certain corresponding single spatial frequency, or, in the parlance of MRI, phase encoding selects out spatial frequency. (4) Immediately following the phase encoding pulse, x gradient is turned on. This is the frequency-encoding horizontal gradient that gives rise to phase shift between two adjacent spins on a horizontal line. In actual practice, it is not the FID but spin echo that is acquired. This is due to the fact that a single "life" of an FID is so short that it is impossible to perform all the necessary gradient switching during that time span.

The steps described so far yield one signal acquisition. In order to image a single slice, we must typically perform 128 or 256 acquisitions by changing the size of the phase-encoding gradient to obtain sufficient resolution. Then this process is repeated for as many slices as one needs. The time interval between each acquisition (TR) is limited by the electronics of the instrument, and this limitation dictates the length of time a patient needs to remain inside the MRI magnet. In order to shorten this time, one can take advantage of the long waiting time required for a single TR to perform multislice imaging by using rf pulses of different frequencies to work on different slices.

Many different methods have been devised to shorten the total scan time and, at the same time, improve the image contrast and resolution.

All these data acquisitions and analyses are performed digitally. The standard techniques for digital data handling apply here for MRI as well. MRI shares basic concepts with fields such as digital audio recording and reproduction. These include issues such as sampling and quantization. For these, ample references are available in the general area of electrical engineering, and specifically in digital data processing and audio technology, including pulse code modulation (PCM). The field of Fourier transform is an active subfield of applied or engineering mathematics.

See also: CAT SCAN; FOURIER SERIES AND FOURIER TRANSFORM; LIQUID HELIUM; NUCLEAR MAGNETIC RESONANCE; RABI, ISIDOR ISAAC

Bibliography

ARMSTRONG P., and KEEVIL, S. F. "Magnetic Resonance Imaging 1: Basic Principles of Image Production." *British Medical Journal* **303**, 35 (1991).

BRACEWELL, R. N. *The Fourier Transform and Its Applications*, 2nd rev. ed. (McGraw-Hill, New York, 1986).

BUSHONG, S. C. *Magnetic Resonance Imaging, Physical and Biological Principles.* (C. V. Mosby, St. Louis, MO, 1988).

WAUCH, J. S., ed. *Advances in Magnetic Resonance.* (Academic Press, San Diego, CA, 1982).

CARL T. TOMIZUKA

MAGNETIC SUSCEPTIBILITY

The magnetic susceptibility χ is a fundamental property of all materials whose measurement provides much information about the electronic structure and presence of any ordered magnetic states in the material. If M is defined as the magnetization and H as the applied magnetic field, then

$$\chi = M/H. \tag{1}$$

Positive and negative values of χ are referred to as paramagnetism and diamagnetism, respectively. Since, in general, M is not a linear function of H, a differential susceptibility is sometimes defined as DM/dH.

Paramagnetic Susceptibility

The susceptibility of paramagnetic substances is determined by the alignment of permanent magnetic dipole moments μ by the field. A classical theory gives

$$M = N\mu L(\mu H/k_B T), \tag{2}$$

where N is the number of moments per unit volume, k_B is Boltzmann's constant, T the absolute temperature, and $L(\mu H/k_B T)$ is the Langevin function, where $L(x)$ is defined as coth $x - (1/x)$. For large values of $\mu H/k_B T$, $L \to 1$ so that all the moments are aligned with the field. For small values of $\mu H/k_B T$, $L(\mu H/k_B T) \approx \mu H/3k_B T$. Thus in small fields

$$M = N\mu^2 H/3k_B T \tag{3}$$

and

$$\chi = \frac{N\mu^2}{3k_B} \frac{1}{T} \equiv \frac{C}{T}, \tag{4}$$

where C is the Curie constant and Eq. (4) is called Curie's law. A plot of measured susceptibility that is inversely proportional to T thus implies the presence of permanent moments whose magnitude can be estimated from the data.

In the case of metals, the susceptibility often is found to be independent of temperature. The reason for this is that, in Eq. (4), the effective number of electrons that can be aligned with the field is reduced by the ratio k_BT/E_F; this gives

$$N_{\text{eff}} \approx (k_BT/E_F)N, \qquad (5)$$

where E_F is the Fermi energy. Thus $\chi \approx N\mu^2/3E_F$. This qualitative effect arises because of the Pauli exclusion principle which limits the number of electrons in a given quantum state to, at most, one. This temperature-independent form of paramagnetism for metals is called Pauli paramagnetism.

Ferromagnetic Susceptibility

All ferromagnetic materials are paramagnets above their ordering or Curie temperatures T_c. Below T_c the moments align as if there were an internal field. In the simple Weiss molecular-field theory it is assumed that the total internal field is $H \rightarrow H + \lambda M$, where λ is called the molecular field or exchange constant. Then, just above T_c, Eq. (3) gives

$$M = \frac{N\mu^2}{3k_BT}(H + \lambda M) = \frac{C}{T}(H + \lambda M). \qquad (6)$$

It follows that

$$\chi = \frac{M}{H} = \frac{C}{T - \Theta_p}, \qquad (7)$$

where $\Theta_p = C\lambda$. This equation, called the Curie–Weiss law, implies, as $T \rightarrow \Theta_p$, that the susceptibility diverges to infinity, that is, a finite M exists in the absence of a field. Figure 1 shows the temperature dependence of the susceptibility for several important cases. In the case of an antiferromagnetic material, in which, below the ordering or Neél temperature T_N, there are two opposing networks of moments that in zero field cancel each other, the susceptibility above T_N is given by

$$\chi = \frac{C}{T + \Theta_p}. \qquad (8)$$

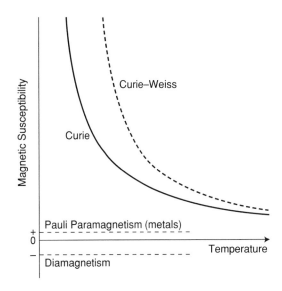

Figure 1 The temperature dependence of the susceptibility for several important cases.

In both the ferromagnetic and antiferromagnetic cases, the temperature parameter θ_p is proportional to a sum of the exchange interactions between neighboring magnetic moments. These exchange interactions arise from quantum mechanical effects which cause the moments to align in a parallel (antiparallel) manner for the ferromagnetic (antiferromagnetic) case.

Susceptibility Measurements

Several of the most common methods for measuring χ are outlined briefly.

The Faraday method relies on the force exerted on a magnetized body in an inhomogeneous field. A sample is hung from a wire connected to an electrobalance so that the sample is in the field of an electromagnet. The pole pieces of the magnet are shaped so that $\mathbf{H} \approx H_x\hat{\mathbf{i}}$ and the field changes mostly in the vertical (z) direction. Then the force $\mathbf{F}$ on the sample of moment $\boldsymbol{\mu}$ is

$$\mathbf{F} = \boldsymbol{\nabla}(\boldsymbol{\mu} \cdot \mathbf{H}) \approx \boldsymbol{\nabla}(\boldsymbol{\mu}_x H_x) \approx \mu_x\boldsymbol{\nabla}H_x. \qquad (9)$$

The z component of the force is then

$$F_z \approx \mu_x\frac{dH_x}{dz} = \chi V\left[H_x\frac{dH_x}{dz}\right], \qquad (10)$$

where V is the volume of the sample. After calibration of the term in square brackets with a known sample, the susceptibility may then be measured.

The alternating current susceptibility χ_{ac} is measured by placing the sample in one of the cores of a secondary coil which, in turn, is inside a primary coil with an alternating current in it. The voltage output of the secondary coil is

$$V_0 = A \, dB/dt = A \frac{d}{dt} (H + 4\pi M), \qquad (11)$$

where A is the area of the coil and B is the magnetic induction inside the coil. If the two secondary coils are wound in series opposition and are balanced in area turns, the first term vanishes and

$$V_0 \propto \frac{dM}{dt} = \frac{dM}{dH} \frac{dH}{dt} \propto \frac{dM}{dH} \cos \omega t, \qquad (12)$$

where ω is the frequency of the exciting current. Thus V_0 is a measure of the differential susceptibility dM/dH. In the case of a ferromagnetic material below T_c, the internal field is given by $H_i = H - NM$, where NM is the demagnetizing field that can be thought of as resulting from magnetic poles induced on the surface of the sample. Then

$$\chi_{ac} = \frac{dM}{dH}$$

$$= \frac{dM}{d(H_i + NM)}$$

$$= \frac{dM/dH_i}{1 + N \, dM/dH_i}$$

$$= \frac{\chi_i}{1 + N\chi_i}, \qquad (13)$$

where χ_i is the intrinsic susceptibility. Since $\chi_i \to \infty$ as $T \to T_c$, $\chi_{ac} \to N^{-1}$ as $T \to T_c$. This sharp flattening of the ac susceptibility at the Curie temperature is a useful way of measuring T_c in ferromagnets.

Additional popular methods for measuring magnetizations and susceptibilities are the vibrating-sample magnetometer, superconducting quantum interference device (SQUID) magnetometer, and alternating-gradient force magnetometer. All of these modern devices can be obtained commercially for approximately \$100,000.

See also: ANTIFERROMAGNETISM; DIAMAGNETISM; FERROMAGNETISM; FIELD, MAGNETIC; MAGNETIC MATERIAL; MAGNETIZATION; PARAMAGNETISM; PAULI'S EXCLUSION PRINCIPLE; SUPERCONDUCTING QUANTUM INTERFERENCE DEVICE

Bibliography

CULLITY, B. D. *Introduction to Magnetic Materials* (Addison-Wesley, Reading, MA, 1972).

JILES, D. *Introduction to Magnetism and Magnetic Materials* (Chapman and Hall, London, Eng., 1991).

KITTEL, C. *Introduction to Solid-State Physics*, 6th ed. (Wiley, New York, 1986).

DAVID J. SELLMYER

MAGNETIZATION

Different materials react to the presence of a magnetic field in different ways. Magnetization is the term that is used in describing these reactions. Technically, magnetization is the magnetic moment per unit volume for a given material. Practically, magnetization is the name used to describe the ability of a piece of material to act like a magnet. If the magnetization is high, the material acts like a magnet attracting paper clips. If the magnetization is low or zero, the material will not even affect a compass needle.

Any material, when placed in a magnetic field, will become magnetized. That is, it will acquire at least some magnetic moment. That magnetic moment will depend on the size and shape of the object placed in the magnetic field. In order to discuss the effect of the material, independent of the shape of the material, one divides the magnetic moment by the volume to get the magnetization in a way analogous to obtaining the density of an object by dividing its mass by its volume so that one can compare, for example, aluminum to lead without being forced to have exactly the same amount of both.

In most cases the magnetization, usually denoted by **M,** is proportional to the magnetic field, usually denoted by **B.** Thus

$$\mathbf{M} = k\mathbf{B},$$

where k is a proportionality factor. It should be noted that both **M** and **B** are vectors with direction as well as magnitude. One of the goals of physics is to express k in terms of the atomic structure of the material. This is an exceedingly difficult task.

The equation is deceptively simple. In general k is at least dependent on temperature. Sometimes it is even negative. For one class of material, the ferromagnetics, it is dependent on **B** so that the equation is nonlinear. In the case of some crystalline materials, k is a tensor so that **M** does not even point in the direction of **B**. For an important class of materials, the permanent magnets, **M** will be nonzero even if **B** is zero.

Materials can be classified into groups depending on k. If k is negative, the material is called diamagnetic. All materials have a diamagnetic component. Many materials also have a positive component to their k. If the component is large enough to make k greater than zero, these materials are called paramagnetic. If k is positive and large, the material is called ferromagnetic.

In the case of most materials, k is small—so small in fact that the magnetization is essentially zero and the material has no effect on the external magnetic field. The ferromagnetic case, where k is large and nonlinear, is the most interesting. It is an experimental challenge just to measure k. The factor not only depends on the value of **B,** but is dependent on the past magnetic history of the material.

Magnetization, as a precisely defined quantity, plays the same role in magnetic theory as does polarization in electrical theory.

See also: ANTIFERROMAGNETISM; DIAMAGNETISM; ELECTROMAGNETISM; FERRIMAGNETISM; FERROMAGNETISM; FIELD, MAGNETIC; MAGNET; MAGNETIC MOMENT; PARAMAGNETISM; POLARIZATION

Bibliography

GRIFFITH, D. J. *Introduction to Electrodynamics,* 2nd ed. (Prentice Hall, Englewood Cliffs, NJ, 1989).
HECHT, E. *Physics* (Brooks/Cole, Pacific Grove, CA, 1996).

A. F. BURR

MAGNETOHYDRODYNAMICS

Magnetohydrodynamics, or MHD (hydromagnetics, magnetofluiddynamics, etc.), refers to the dynamics of an electrically conducting fluid, for example, a liquid metal or an ionized gas (plasma), permeated by a magnetic field **B.** The essential feature is that over a sufficiently large scale l the magnetic field is constrained to move with the fluid. The magnetic field is said to be "frozen in" to the fluid and is carried along and deformed by any nonuniformity in the fluid motion. The evolution of the magnetic field in a nonuniform flow of conducting fluid is most effectively visualized in terms of the magnetic field lines (sometimes called the magnetic lines to force), which move with the fluid. In effect, the field slides freely along the field lines but carries the lines with any transverse motion. If the field is sufficiently weak, it has no sensible effect on the fluid motion, so the result is entirely hydrodynamical with the field transported as an incidental passive entity. Nonuniform fluid motions stretch and wrap the field lines into complicated patterns, the way a line of ink would be smeared and stretched in the swirling fluid.

Magnetic fields contain their own internal (Maxwell) stresses, of course, so that if the field is not weak, the field reacts against the fluid motion, modifying the flow. The force exerted by the field on the fluid is called the Lorentz force, with the fluid exerting an equal and opposite force on the field. The result is a fully magnetohydrodynamical flow, with the fluid motions strongly modified by the presence of the magnetic field and the magnetic field strongly modified by the fluid motion.

Magnetohydrodynamics manifests itself in the terrestrial laboratory in the confinement and manipulation of plasmas. However, the limited scale of the laboratory provides only for transient effects. It is over the huge scales of astronomical objects that the full effect of MHD emerges, in the long-lived magnetic fields of Earth and other planets, sunspots, solar flares, solar coronal plumes and coronal mass ejections, the interplanetary magnetic field, and the magnetic field of the Milky Way Galaxy.

The theoretical basis for MHD is the insignificance of the electric field in the frame of reference moving locally with the highly conducting fluid; any electric field within the fluid would immediately produce an electric current, canceling out the electric charges associated with the field, thereby reducing the electric field to negligible levels. But if the fluid were to move relative to the magnetic field, there would be an electric field induced in the fluid. Since that cannot happen, it follows that the fluid and field move together so that no electric field is induced.

To treat these concepts quantitatively, note that in a frame of reference moving with velocity $\mathbf{v}$ (assumed to be small compared to the speed of light c) relative to the laboratory frame, the electric field $\mathbf{E}'$ within the fluid is related to the electric field $\mathbf{E}$ and magnetic field $\mathbf{B}$ in the laboratory frame of reference by the Lorentz transformation $\mathbf{E}' = \mathbf{E} + \mathbf{v} \times \mathbf{B}/c$. Then if $\mathbf{E}' \cong 0$ in the fluid, it follows that $\mathbf{E} = -\mathbf{v} \times \mathbf{B}/c$ in the laboratory. Faraday's induction equation $\partial\mathbf{B}/\partial t = -c\boldsymbol{\nabla} \times \mathbf{E}$ becomes the MHD induction equation

$$\partial\mathbf{B}/\partial t = \boldsymbol{\nabla} \times (\mathbf{v} \times \mathbf{B}), \tag{1}$$

from which it can be shown that the magnetic field $\mathbf{B}$ is transported bodily by the fluid.

The Lorentz force $\mathbf{F}$ exerted on the fluid by the magnetic field can be written in terms of the electric current density $\mathbf{j}$ as $\mathbf{F} = \mathbf{j} \times \mathbf{B}/c$. Ampère's law $4\pi\mathbf{j} = c\boldsymbol{\nabla} \times \mathbf{B}$ asserts that there are always electric currents associated with magnetic fields and vice versa, and one may ask whether $\mathbf{j}$ or $\mathbf{B}$ is the fundamental physical quantity in MHD. In the laboratory, the magnetic field is commonly created by driving an electric current through a coil of wire so that the current is clearly the cause of the magnetic field. The energy is introduced via the current and the applied emf. In astronomical settings, and in many circumstances in the plasma laboratory, the system is driven by the fluid motions that do work against the Lorentz force. The magnetic field receives its energy from the fluid motion, and the energy needed to drive the current (against whatever slight electrical resistivity there may be in the fluid) comes from the magnetic field. Hence, the magnetic field is the cause of the current. There is no externally applied emf, and the electric current contains no significant energy or stress, so it is only a secondary physical quantity fed from the magnetic energy density $B^2/8\pi$. It is convenient, therefore, to use Ampère's law to eliminate $\mathbf{j}$ from the Lorentz force, yielding $\mathbf{F} = (\boldsymbol{\nabla} \times \mathbf{B}) \times \mathbf{B}/4\pi$. Using a well-known vector identity this can be rewritten as

$$\mathbf{F} = -\boldsymbol{\nabla}\frac{B^2}{8\pi} + \frac{(\mathbf{B} \cdot \boldsymbol{\nabla})\mathbf{B}}{4\pi}. \tag{2}$$

The first term on the right-hand side represents the gradient of the isotropic magnetic pressure $B^2/8\pi$. The second term represents the divergence of the tension $B^2/4\pi$ along the magnetic field, where B is measured in gauss. If B is given in tesla (1 T = 10^4 G), replace 4π by μ_0. It is possible, of course, to write all these expressions in terms of $\mathbf{j}$, using the Biot–Savart integral

$$\mathbf{B}(\mathbf{r}) = \frac{1}{c}\int \frac{d^3\mathbf{r}'(\mathbf{r}' - \mathbf{r}) \times \mathbf{j}(\mathbf{r}')}{|\mathbf{r}' - \mathbf{r}|^3}$$

but the resulting complexity of the equations is forbidding. The essential point is that MHD concerns the dynamical interaction between the fluid motions and the magnetic stresses, and it is best formulated directly in terms of those quantities.

There is always some slight electrical resistivity in a liquid metal or ionized gas as a consequence of collisions of the conduction electrons with the ions, so the ideal condition of a frozen-in field is achieved precisely only in the limit of large $l\sigma$, where l is the characteristic scale of variation of the magnetic field and σ is the electrical conductivity of the fluid. The conductivity is in the range 10^5–10^7 mho/m for liquid metals and $2 \times 10^{-3}T^{3/2}$ mho/m for ionized hydrogen at an absolute temperature T. The ideal MHD induction equation is modified by the fact that $\mathbf{E}' = \mathbf{j}/\sigma$ instead of $\mathbf{E}' = 0$, yielding

$$\frac{\partial\mathbf{B}}{\partial t} = \boldsymbol{\nabla} \times (\mathbf{v} \times \mathbf{B}) - \boldsymbol{\nabla} \times (\eta\boldsymbol{\nabla} \times \mathbf{B}) \tag{3}$$

in place of Eq. (1), where the resistive diffusion coefficient η is defined as $c^2/4\pi\sigma$. For uniform η and $\mathbf{v} = 0$, this reduces to the diffusion equation $\partial\mathbf{B}/\partial t = \eta\nabla^2\mathbf{B}$, where $\eta \sim 10^3$–10^5 cm/s in molten metal and $\eta \sim 4 \times 10^{12}/T^{3/2}$ cm²/s in ionized hydrogen. The diffusion $\eta\nabla^2\mathbf{B}$ allows the field to diffuse with a characteristic speed on the order of η/l. Thus, if the fluid velocity v is to dominate over diffusion, the ratio vl/η, called the magnetic Reynolds number, must be large compared to 1. Putting it another way, the characteristic dissipation time t is on the order of l^2/η. Thus, for instance, a plasma column of radius 10 cm and 10^6 K extending along a strong magnetic field disperses in a characteristic time of 20 ms, since $\ell \sim 10$ cm and $\eta \sim 4 \times 10^3$ cm²/s. The plasma column spreads out through the field, achieving a characteristic radius of the order of $(\eta t)^{1/2}$ after a time t. The same applies to the complementary configuration where a column of magnetic field is confined in a plasma of high pressure p ($>> B^2/8\pi$).

A plasma with pressure p can be confined by a magnetic field **B** only if the magnetic pressure $B^2/8\pi$ exceeds p, of course. In fact, unless $B^2/8\pi >> p$, it turns out that, on the small scales available in the laboratory, the plasma develops internal small-scale dynamical instabilities and manages to wiggle out of the field much more rapidly than it escapes by resistive diffusion.

In astronomical settings, the scale l is so vast that resistive effects and small-scale plasma instabilities have relatively little effect in most circumstances. The field and fluid move together on a quasi-permanent basis. Thus, for instance, the resistivity η of the solar photosphere is relatively large, on the order of 10^{10} cm^2/s, while a few hundred kilometers below the visible surface it is 10^8 cm^2/s or less. Thus, the smallest visible sunspots (pores), with a radius of 700 km, have a resistive life on the order of 10^8 s $\sim$ 3 yr, determined by the gas beneath the visible surface. The pore may appear and disappear in the course of a few hours, over which time the field and fluid move together to good approximation.

The basic MHD concepts and their wide application to plasmas in the laboratory and in astronomical settings were first pointed out explicitly by Hannes Alfven, who was awarded the Nobel Prize in 1970 in recognition of the fundamental importance of MHD. Curiously, by that time he had repudiated the notion of frozen-in field lines. But the important fact of a frozen-in field remains, and the phenomenon is exploited in the plasma laboratory to hold an extremely hot plasma (10^6–10^8 K) in a magnetic field away from metal walls. The expectation is to achieve a sufficiently dense hot plasma of ionized hydrogen to initiate nuclear fusion, in which the heavier isotopes of hydrogen (deuterium and tritium) combine to form helium, with the release of large amounts of nuclear energy. The ultimate goal is to generate electrical power from this nuclear energy while avoiding some of the radioactive waste disposal problems associated with the fission of heavy nuclei. The major obstacle to success has been the development of small-scale instabilities. Larger dimensions and proper configuration of the magnetic field and plasma are the means for overcoming this limitation.

Most astronomical bodies (planets, stars, and galaxies) produce their own magnetic fields. It is also a fact that their outer atmospheres (their coronas or halos) are exceedingly hot (10^6–10^8 K), leading to x-ray emission when the gas density is high enough. These extraordinary temperatures seem to be directly related to the magnetic field. For instance, the magnetosphere of Earth contains populations of electrons and ions with energies equivalent to 10^8 K. The halos of galaxies show 10^7 K. These facts raise the two fundamental questions of the origin of the magnetic fields and the dissipation of magnetic energy into heat.

The answer to the question of the origin was initiated by Walter M. Elsasser in 1946 in connection with the magnetic field of Earth (the only known astronomical magnetic field except for sunspots at that time). The magnetic field of Earth originates in the liquid metal core, with radius $l \sim 3 \times 10^8$ cm and $\eta \sim 10^4$ cm^2/s. The resistive decay time is, therefore, $l^2/\eta \sim 10^5$ yr. This is so short compared to the age of Earth (4×10^9 yr) that it must be concluded that the magnetic field is sustained, somehow, by the fluid motions ($\sim$1 mm/s) in the liquid core. Elsasser showed that the nonuniform rotation of the core shears the north-south field (the dipole field) of 0.5 G that we see at the surface to provide a strong east-west (azimuthal) magnetic field within the liquid core. It was subsequently pointed out that the cyclonic convection in the core twists the east-west field into meridional loops that diffuse together over periods of 10^3 yr to reinforce the north-south field, thereby completing the dynamo process that sustains the magnetic field of Earth. The system is driven by the liberation of heat deep in the liquid core. It appears that all the other planets except Venus (which rotates extremely slowly), Mars (which has only a very small liquid core), and coreless Pluto have magnetic fields produced by similar internal MHD effects. The process is called an $\alpha\omega$-dynamo, with α referring to the cyclonic form of the convection and ω referring to the nonuniform rotation. It appears that the periodic magnetic field of the Sun, responsible for the 11-yr magnetic sunspot cycle, is produced by the same basic MHD processes.

The conversion of magnetic energy into heat, suggested by the intimate association of 10^6–10^8 K gases with the magnetic fields of astronomical bodies such as the Sun, is an effect that is not fully understood at the present time, largely for lack of crucial observations of the very-small-scale fields and fluid motions. The problem is that hot gases have low resistivity η, so the dissipation times l^2/η are extremely long. For instance, with $T = 2 \times 10^6$ K, appropriate for the x-ray emitting regions of 10^2 G on the Sun ($l \sim 10^9$–10^{10} cm), the resistive dissipation time is 10^{15}–10^{17} s or 3×10^7–3×10^9 yr. That is to say, the resistive heating is negligible. If intense high frequency (periods $\lesssim 5$ s) Alfven waves with small transverse scales are generated in the convection beneath the surface, it might be possible to account for the heating (of about 10^7 erg/cm^2) by the viscous and resistive dissipation of the waves. But there is no reason to expect a sufficient supply of high-frequency waves. An alternative possibility arises from the fact that the magnetic fields extending above the visible surface of the Sun are continually mixed and deformed by the convection (~ 1 km/s) below the surface. It can be shown that the combined magnetic pressure and magnetic tension have the curious property of producing surfaces of magnetic discontinuity as the field relaxes toward equilibrium. The magnitude of the field is continuous across the surface, but the direction of the field rotates abruptly. The characteristic thickness l of the change in direction declines asymptotically toward zero and is prevented from reaching zero only by the rapid dissipation of magnetic energy by whatever small resistivity may be present. The phenomenon is driven by the Maxwell stresses and is referred to as rapid reconnection or neutral point reconnection, vigorously converting magnetic free energy of the deformed fields into heat. It appears that the dissipation of magnetic energy sometimes proceeds so rapidly as to produce the outbursts referred to as flares. But, of course, this all needs to be checked out by observations of the field and fluid motions on small scales before we can be sure.

The active magnetic field of the Sun serves as the MHD paradigm for the magnetic fields of the other stars, which are seen only as amorphous blobs of light in a telescope. Therefore, it is important for all astrophysics to resolve the questions concerning the coronal heating and x-ray emission of the Sun.

See also: ALFVEN WAVE; AMPÈRE'S LAW; FIELD, MAGNETIC; FIELD LINES; LORENTZ FORCE; LORENTZ TRANSFORMATION; MAGNETOSPHERE; PLANETARY MAGNETISM; PLASMA; SUN

Bibliography

PARKER, E. N. *Cosmical Magnetic Fields* (Clarendon Press, Oxford, Eng., 1979).

PARKER, E. N. *Spontaneous Current Sheets in Magnetic Fields* (Oxford University Press, New York, 1994).

ROBERTS, P. H. *An Introduction to Magnetohydrodynamics* (Elsevier, New York, 1967).

E. N. PARKER

MAGNETON

See BOHR MAGNETON

MAGNETO-OPTICAL EFFECTS

When light (or more generally, electromagnetic radiation) interacts with a magnetized medium there may be an interaction between the medium and the radiation that we describe in terms of magneto-optical effects (MOE).

Transmission

Consider the propagation of light through a homogeneous block of material with plane surfaces normal to the beam. In the first case it is assumed to have no birefringence and no dichroism. This means that light with any polarization can propagate through it without a change in polarization state, though the intensity may decrease with increasing thickness.

If into this block of material one introduces a uniform magnetization **M** parallel to the propagation direction, it is found by solution of Maxwell's equations that only the two circular polarizations (CP) can propagate. Generally these two "normal modes" travel at different velocities and experience different attenuations. The velocity difference means that the two refractive indices n_+ and n_- are different, and this constitutes a circular birefringence. Since it arises from the magnetization, it is termed magnetic circular birefringence (MCB). Similarly the difference in the attenuation constants of the two normal modes constitutes a magnetic circular dichroism (MCD).

Note that we describe the polarization state by the shape of the curve swept out by the tip of the light wave **E** vector in a plane perpendicular to the propagation direction. Suppose the incident polarization is linear, as indicated by the double-headed arrow at the left-hand side of Fig. 1a. On entering the sample this would be resolved onto the two normal modes indicated by the two circles on the right-hand side: these travel through the sample with different velocities and attenuations. Upon exiting, the field vectors of the two normal modes produce an elliptical polarization. It is elliptical rather than linear because one CP was attenuated more than the other (MCD); its major axis is rotated because one CP travels with greater velocity than the other, and so there is a shift in relative phase of the two waves (MCB) on emergence from the crystal.

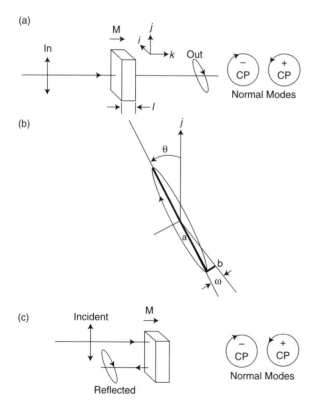

(a)

(b)

(c)

Figure 1 (a) Linearly polarized light transmitted by a sample with **M** along the axis may emerge with a rotation and an ellipticity. (b) Sketch showing the angles θ and ω used to describe specify rotation and ellipticity. (c) Rotation and ellipticity on reflection in the polar geometry, polar Kerr effect. The incident and reflected beams are displaced for clarity.

The polarization state represented by the output ellipse in Fig. 1b can be characterized by two angles and a plus or minus sign to indicate sense. First there is the azimuth of the major axis of the ellipse measured from where it would be if there were no magnetization. This is magneto-optical rotation (MOR), often called the Faraday rotation. The shape of the ellipse can be measured by an angle of ellipticity given by

$$\omega = \tan^{-1}\frac{b}{a}.$$

The value of ω is 0 for linear polarization and increases to 45° as the polarization approaches circular. Since alternatives are in use, a sign convention must be stated clearly in specifying the sense of the ellipse.

To a very good approximation it is found that in materials in which there is no other birefringence, the rotation θ of a uniformly magnetized sample is proportional to the fractional component of magnetization along the axis of propagation. In vector notation, we may write

$$\theta = l\,\frac{\mathbf{M}\cdot\mathbf{k}}{|\mathbf{M}|}\Theta(\text{material},\,T,\,\lambda).$$

Here, l is the thickness of the sample, **M** the magnetization vector, and **k** a unit vector along the axis. The specific rotation Θ varies with the material, the temperature T, and wavelength λ. The wavelength dependence is often very strong. The specific rotations of transparent magnetic crystals range up to hundreds of thousands of degrees/centimeter. Relative to the axis system, the rotation is the same whether the light travels toward the right or toward the left. For example if the light travels from left to right it undergoes a rotation θ. If it then encounters a mirror and is reflected back through the magnetic sample, the final polarization azimuth would be 2θ. *However it is particularly important to note that the sign of the rotation changes if the magnetization is reversed.* Such behavior is said to be antireciprocal and contrasts with the reciprocal rotation seen, for example, in sugar solutions and in crystalline quartz. In general the transmitted light is also elliptical because there is an MCD, but in many practical cases this ellipticity can be ignored.

These MOE on transmission have been widely used in research. The MOR makes it possible to see magnetic domains in transparent samples, and many details of domain behavior can be observed directly. Antireciprocity makes possible optical isolators, devices that are transparent for light traveling in one direction, but opaque for the opposite direction. These are crucial components in modern optical communications technology. If **M** is proportional to the magnetic field **H,** it is possible to construct MO sensors of **H** in which field modulation is transformed into polarization azimuth modulation. These are being increasingly used by the electric power industry.

The magnetization represented by **M** can be the spontaneous magnetization of a ferromagnet or of a sublattice in a ferrimagnet. It can be the net magnetization of an unsaturated magnet in a multidomain state. Alternatively, it can be the magnetic field induced magnetization of a paramagnet, though here the effects are usually much smaller.

Consider now the geometry of Fig. 2 in which the magnetization is perpendicular to the optical path. It turns out that the normal modes of propagation are the two linear polarizations parallel and perpendicular to **M.** If light with either of these polarizations enters the sample it is transmitted without change of polarization state, but the two velocities differ. If light of some other polarization state impinges on the sample (e.g., linear at 45° as illustrated), it is resolved onto the two normal modes traveling with different velocities. Upon exiting, the resultant polarization is, in general, elliptical. The velocity difference corresponds to a refractive index difference, $n_\perp \neq n_\parallel$, a linear birefringence. Since this is associated with the magnetization, it is called magnetic linear birefringence (MLB). As in the circular case, a difference in the attenuation of the two normal modes constitutes a magnetic linear dichroism (MLD). Analysis shows and experiment confirms that MLB and MLD depend on $\mathbf{M}^2$, thus reversing **M** does not change either quantity. When **M** is induced by a magnetic field the MLB is sometimes called the Cotton Mouton effect. Spontaneous MLB and MLD can be seen in ferromagnets, ferrimagnets, and antiferromagnets. They are used in the spectroscopic study of these materials, and in investigations of critical phenomena. In the study of magnetic domain structure by transmitted light, the MLB makes possible a contrast between domains in which the magnetization is along different directions in the plane normal to the axis of the microscope.

Figures 1 and 2 are concerned with uniformly magnetized samples. It will be seen that when a light ray is transmitted by a magnet broken up into domains, the normal modes and the refractive indices may change from domain to domain. Some of the light may be scattered or diffracted out of the beam, thus decreasing the transmitted intensity. We term this effect magneto-optical scattering. When a saturating field is applied the transmitted intensity increases to its saturated value.

Reflection

Figure 1a deals with the transmitted light, but part of the light is also reflected from the surface in which **M** is normal, as sketched in Fig. 1c. Experiment, as well as the solution of Maxwell's equations, for this problem show that the reflected light is elliptically polarized, and the major axis is rotated. Changes in polarization state on reflection from a magnetized surface are called magneto-optical Kerr effects. For the geometry described here, we speak of the polar MO Kerr effect, the polar Kerr rotation, and the polar Kerr ellipticity. There are two other traditional geometries whose description we must omit for brevity. Here the rotation and ellipticity change sign if **M** is reversed, so the effects are antireciprocal. The MO Kerr effects are of interest in ferro- and ferrimagnets. They are used in the study of domain structure in opaque magnetic materials (e.g., Fe, Ni, and Co). The rotations in the visible are quite small, usually less than 0.5°. Nevertheless they form the basis for a technologically important magneto-optic memory technol-

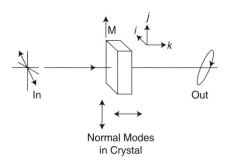

Figure 2 If **M** is perpendicular to the propagation direction, the normal modes are linear parallel and perpendicular to **M.** In general, any other linear polarization becomes elliptical on transmission (e.g., the oblique linear polarization marked "In" goes to the ellipse "Out").

ogy. It is remarkable that a significant industry (about one billion dollars in 1995, and growing rapidly) has been built on MO polar Kerr rotations of less than 0.4° from thin films with compositions such as amorphous TbFeCo.

In addition to the MOE just described, there are a number of other more subtle effects. These include phenomena in which some properties of a magnetic sample are changed by irradiation with light.

Applications

In addition to the technological applications of the MOE mentioned above, they have been widely used in research on the properties of magnetic materials and of critical phenomena. From an educational point of view, the fact that the MOE enables direct observation of magnetic domains makes possible straightforward experiments illustrating the fundamentals of magnetism. Such experiments are suitable for high school and college students.

See also: BIREFRINGENCE; DICHROISM; FERRIMAGNETISM; FERROMAGNETISM; MAXWELL'S EQUATIONS; POLARIZED LIGHT; SCATTERING, LIGHT

Bibliography

EISENSTEIN, B.; MILLMAN, S.; and PALLRAND, G. "Introducing the Physics of Technology into the High School Curriculum." *Phys. Today* **44** (Jan.), 46–50 (1991).

HECHT, E., and ZAJAC, A. *Optics,* 2nd ed. (Addison-Wesley, Reading, MA, 1987).

LANDAU, L. D., and LIFSHITZ, E. M. *Electrodynamics of Continuous Media,* 2nd ed. (Elsevier, New York, 1984).

MORRISH, A. H. *The Physical Principles of Magnetism* (Krieger, New York, 1980).

JOSEPH F. DILLON JR.

MAGNETOSPHERE

"Magnetosphere" is the term applied to the extended volume of space around a planet or star that is dominated by the magnetic field **B** of the planet or star. All of the planets except Venus, Mars, and Pluto have magnetic fields strong enough to provide magnetospheres in the presence of the solar wind. The outer boundary of a planetary magnetosphere is called the magnetopause, representing a relatively thin transition layer ($\sim 10^2$ km thick) in which the ions and electrons that constitute the solar wind are deflected by the magnetic field and directed back into space. The magnetopause is located where the pressure $B^2/8\pi$ of the planetary magnetic field falls to the level of the impact pressure of the solar wind, so that **B** does not have the strength to push farther out into space.

A planetary magnetic field declines outward approximately as $1/r^3$, where r is radial distance from the center of the planet, except insofar as the outer weaker portions of the magnetic field are compressed or stretched by the passing solar wind. Using Earth as the primary example, the magnetic field has an intensity of about 0.3 G at the equator on the surface of Earth (0.6 G at the poles). Therefore, if undisturbed by the solar wind, the field strength in the equatorial plane at a radial distance of 10 R_E Earth's radii ($R_E = 6.4 \times 10^8$ cm) would be 3×10^{-4} G with a pressure $B^2/8\pi$ of about 4×10^{-9} dyn/cm^2. In the presence of the solar wind, the field is pushed in and compressed on the sunward side, approximately doubling the field strength at the magnetopause so that the magnetic pressure $B^2/8\pi$ of the magnetopause at 10 R_E would be about 1.6×10^{-8} dyn/cm^2.

The solar wind, composed of N atoms/cm^3, each atom of mass M with wind velocity v, has an impact pressure on the order of NMv^2. A typical quiet-day wind of 5 hydrogen atoms/cm^3 ($M \cong 1.7 \times 10^{-24}$ g) and a speed of 400 km/s provides an impact pressure on the order of 1.4×10^{-8} dyn/cm^2. Thus, under these conditions, the sunward magnetopause, where the solar wind impacts squarely against the magnetopause, lies at a radial distance of about 10 R_E. Elsewhere around the magnetopause the impact is oblique so that the wind does not press as hard. Consequently, the distance to the magnetopause increases rapidly away from the subsolar point to about 20 R_E in the dawn and dusk directions. The magnetosphere extends indefinitely far in the midnight (anti-solar) direction because it is shielded from the impact of the solar wind by the sunward magnetosphere. The net result is a comet shaped magnetosphere, with a long tail extending in the anti-solar direction, with a radius on the general order of 40 R_E.

The impact pressure NMv^2 of the solar wind varies with time, causing the dimensions of the magnetosphere to vary. In particular, there are blast waves from the Sun occasionally exerting as much as 10 times the usual impact pressure for an hour or so, pushing the sunward magnetopause in to a distance on the order of 7 R_E.

The interior of a planetary magnetosphere is an active region. Not only is the magnetosphere subject to fluctuating compression, but there is also an overall convective circulation of the entire magnetosphere, including the electrically conducting ionosphere. Furthermore, the electrons, protons, and heavier ions within the magnetosphere are trapped by the magnetic field, being restricted to circular (cyclotron) motion about the magnetic field. Thus, for instance, an energetic proton of 10 keV in the equatorial plane of Earth at a radial distance of 4 R_E has a cyclotron period of about 0.1 s and a cyclotron radius of about 30 km. At the same time, these trapped particles drift in longitude as a consequence of the outward decrease of the magnetic field. The positively charged ions drift westward and the electrons eastward. The net effect is a westward electric current, called the ring current, which varies enormously with the large changes in trapped particle population.

With these facts in mind, consider the phenomenology of the major fluctuations, called magnetic substorms and storms, observed within the magnetosphere. We begin with a brief account of the principal driver of magnetospheric convection. The essential point is that there is a magnetic field in the solar wind with a mean value of about 5×10^{-5} G. The field direction fluctuates strongly. The average north-south component of the field is zero, but over periods of minutes or hours the field may be principally northward or southward. When the field has a strong southward component, it reconnects rapidly with the northward geomagnetic field at the sunward magnetopause. The geomagnetic field that becomes connected into the field in the solar wind is then carried by the wind back into the geotail. The rapid accumulation of magnetic field in the geotail compresses the ambient field and plasma already present and initiates what is called a magnetic substorm. The increasing magnetic pressure in the nightside magnetosphere pushes the magnetic field nearer Earth in the sunward direction. However, the magnetic field is tied to the massive (relatively speaking) ionosphere and is not free to flow sunward.

So the accumulation of magnetic field in the geotail evidently goes on until the electric current required by Ampère's law ($4\pi\mathbf{j} = c\mathbf{\nabla} \times \mathbf{B}$) around the periphery of the accumulating magnetic flux becomes so intense that the tenuous ambient plasma ($\lesssim$ 1 electron/cm³) cannot carry the current without being forced by a strong electrical field $\mathbf{E}_{\parallel}$ along the magnetic field lines. This leads to electrical potential differences of 10^4 V or more, producing intense aurora and decoupling the magnetic field from the ionosphere (i.e., the field is not "frozen in" to the tenuous plasma in the presence of a strong $\mathbf{E}_{\parallel}$). The accumulated field in the geotail then returns rapidly to the sunward magnetosphere, relieving the enhanced pressure in the geotail and providing the expansion phase of the substorm. Substorms usually recur at irregular intervals of an hour or more, depending upon the variations in the orientation of the magnetic field in the solar wind.

Then there is the occasional magnetic storm, arising when a blast wave with a southward magnetic field impacts the magnetosphere, increasing the field intensity at the surface of Earth by one part in 10^3 or more in several minutes. The initial (compressive) phase may last from one to several hours, accompanied by a rapid sequence of severe substorms. Solar wind plasma is injected into the magnetosphere along the reconnected field lines, inflating the nightside magnetosphere. The substorms also accelerate electrons, protons, and ambient magnetospheric ions, further inflating the magnetosphere. After a few hours, the inflation of the magnetic field reduces the horizontal component at the surface of Earth by as much as one part in 10^2 at the equator. This is called the main phase of the magnetic storm, which may last ten hours. The extended recovery phase (one day or more) follows as the inflating ions and electrons are lost by charge exchange, wave scattering, and magnetospheric convection.

It is important to note that the magnetic storm is sometimes so severe, producing rapid magnetic transients at the surface of Earth, as to disrupt telegraph and telephone communications and occasionally knock out the electrical power grid over extended geographical areas. The physical hardship and economic loss in an extensive power outage is enormous.

As already mentioned, the general activity of the magnetosphere accelerates electrons, protons, and so on, and the continuing cosmic ray bombardment of the upper atmosphere introduces additional high-speed electrons and protons. The net result is the accumulation of energetic charged particles in the magnetosphere at altitudes high enough to avoid rapid collision with the upper atmosphere. This semi-permanent population constitutes the Van Allen radiation belts. The intensity of the trapped particles varies with the general activity of the Sun and with the solar wind (which drives the activity of the magnetosphere) and is sometimes greatly augmented in the polar regions by energetic particles (1–100 MeV) from outbursts (flares) on the Sun. The energetic particle intensity can sometimes rise to such high levels as to damage the electronic chips in spacecraft as well as posing a physical threat to astronauts. The essential point is that the variable space weather within the magnetosphere affects the successful functioning of technology at the surface of Earth and the operation of communication satellites and other spacecrafts. The reliable anticipation by even a few hours of violent weather in space is essential for the effective operation of modern technology.

The magnetospheres of the other planets all have their own special features. Jupiter is perhaps the most interesting example because of the immense size and rapid rotation of the Jovian magnetosphere and because of neutral atoms continually introduced into the magnetosphere by the moon Io. Jupiter is itself about 10 times the diameter of Earth, with a magnetic field nearly 10 times stronger (4–5 G) at the surface of the planet. The pressure of the solar wind is only about 1/25 what it is at Earth (Jupiter being 5 times farther from the Sun). The net result is a sunward magnetopause at about $35R_J$ ($R_J \cong 6 \times 10^9$ cm) or about 350 times the dimensions of the terrestrial magnetosphere. The rapid rotation of Jupiter (10-hour day) means that the sunward magnetopause slides around at about 4×10^3 km/s, approximately 10 times the speed of the solar wind. The enormous variability of the Jovian magnetosphere is at least in part a consequence of the high rate of rotation. The net result is a lethal trapped energetic particle intensity that quickly degrades solid-state electronic components in passing spacecrafts. There are also intense plasma oscillations and associated radio emission.

Turning to the magnetospheres of stars, it must be understood that they are activated from the star within rather than by an external wind. The magnetospheres of intensely magnetized, rapidly spinning neutron stars (pulsars) are a subject in themselves. We treat only the magnetosphere of the Sun, called the heliosphere, with the outer boundary called the heliopause where the heliosphere meets the interstellar medium. The heliosphere is presumably typi-

cal of the wind region of the average star. The heliosphere is created by the solar wind, arising from the gradual outward hydrodynamic expansion of the solar corona to supersonic speeds at large distance from the Sun. The solar wind is highly supersonic, so no acoustic or magnetohydrodynamic signals propagate inward from interstellar space. The velocity of the wind is more or less constant once the gas is well away from the Sun, with a typical value of 400 km/s, and with a density of five hydrogen atoms at the orbit of Earth. The density of the wind, and therefore the impact pressure, fall off as $1/r^2$ with increasing distance r from the Sun. The wind sweeps away the interstellar gas and magnetic field out to the distance R, where the wind pressure becomes equal to the interstellar pressure of about 10^{-12} dyn/cm^2. This occurs somewhere in the vicinity of 100 AU. It is expected that the supersonic wind passes through a shock transition to subsonic velocity at that distance and is swept away with the interstellar wind ($\sim$ 20 km/s) beyond to form a comet-shaped heliosphere.

The general magnetic field of the Sun, with a mean value on the order of a few gauss, is stretched outward in the solar wind. The radial component of the extended field falls off as $1/r^2$, with a value of about 5×10^{-5} G at the orbit of Earth, where the magnetic field has an energy density of only about 10^{-2} of the kinetic energy density of the solar wind. The 25-day rotation period of the Sun shears the radial field, producing an azimuthal component comparable to the radial component at the orbit of Earth and declining only as $1/r$ beyond. Thus, the dominant magnetic field beyond Earth is the azimuthal component, whose pressure declines outward as $1/r^2$, remaining about 10^{-2} of the total solar wind impact pressure or kinetic energy density all the way out to the heliopause. At $r = 100$ AU, the magnetic field in the solar wind is about 0.5×10^{-6} G, to be compared with the interstellar magnetic field, estimated at 2×10^{-6} G. It should be noted, however, that the number density of the solar wind at 100 AU is only about 5×10^{-4} atoms/cm^3, compared to the local interstellar gas density of perhaps 0.1 atom/cm^3. Passing through the shock transition at 400 km/s boosts the temperature of the gas to something on the order of 5×10^6 K and the density to perhaps 1.5×10^{-3} atoms/cm^3, providing gas pressure on the order of 10^{-12} dyn/cm^2 and compressing the azimuthal magnetic field to $1 - 2 \times 10^{-6}$ G. The outer heliosphere and the heliotail extending away in the interstellar wind are expected to be very hot, very tenuous, relatively strongly magnetic, and hence, magnetically active.

See also: AURORA; FIELD, MAGNETIC; IONOSPHERE; MAGNETOHYDRODYNAMICS; NEUTRON STAR; PLANETARY MAGNETISM; SOLAR WIND; VAN ALLEN BELTS

Bibliography

JACOBS, J. A. ed. *Geomagnetism,* Vol 4. (Academic Press, London, 1987).

KIVELSON, M. G., and RUSSELL, C. T. *Introduction to Space Physics* (Cambridge University Press, Cambridge, MA, 1994).

RUSSELL, C. T. "Planetary Magnetospheres." *Rep. Prog. Phys.* **56,** 687–732 (1993).

E. N. PARKER

MAGNETOSTRICTION

When a piece of ferromagnetic material is magnetized, any change in the net magnetization is accompanied by a change in the material's physical dimensions. This phenomenon was first recognized by the British physicist James Prescott Joule in 1842. When the changes in magnetization are caused by changes in an externally applied magnetic field, the associated changes in dimension are known as magnetostriction. Measurements of magnetostriction are often made by observing the small changes in size of a ferromagnetic rod when an external magnetic field is applied along its length. The change in the length of the rod is called longitudinal magnetostriction, while the change in the rod's diameter is called transverse magnetostriction. When the magnetization in the rod is not saturated, the combined effects of longitudinal and transverse magnetostriction cause (to first order) no net change in the volume of the material. When the magnetization in the rod is saturated by an applied magnetic field, both the transverse and longitudinal magnetostriction have the same sign; the resulting change in the volume of the rod is called volume magnetostriction.

The total energy of a ferromagnetic substance includes four principal contributions. Exchange energy is reduced when neighboring magnetic moments in a ferromagnetic crystal align parallel to one another. Magnetostatic energy results from the interaction of the magnetic moments with their internal and any externally applied magnetic field. In zero applied field, the magnetostatic energy is minimized when the field leaking from the ferromagnet is minimized. Magnetocrystalline anisotropy energy is the result of spin-orbit coupling. The result of anisotropy energy is the introduction of easy axes or preferred directions of magnetization within the crystal. In a crystal such as body-centered cubic iron, there are six easy axes of magnetization along the main crystal axis directions. The elastic energy depends upon the state of strain within the ferromagnet. There is a magnetization-independent contribution that is minimized when the ferromagnet is unstrained and a magnetostrictive contribution to the energy that depends upon the direction of magnetization. In the presence of anisotropy, minimizing the total energy within the crystal results in an equilibrium state in which the crystal lattice is slightly distorted from the unstrained state. This is a direct consequence of the linear dependence of the magnetoelastic energy on strain and the quadratic dependence of the elastic energy on strain. In nickel, for example, below the Curie temperature where spontaneous magnetization persists, there will be a lattice distortion that compresses the lattice along the direction of magnetization while expanding the lattice perpendicular to the magnetization direction. The fractional contraction along the field direction for polycrystalline nickel due to longitudinal magnetostriction is about 30 ppm, while the fractional change in volume above a field of 1,000 Oe is about 0.1 ppm.

In an unmagnetized remnant state, a ferromagnet will consist of numerous uniformly magnetized regions pointing along the easy axes (of magnetization) of the crystal. When an external magnetic field is applied, the domains begin to align along the field direction. The saturation magnetostriction depends upon the orientation of unit vectors along the magnetization direction $\hat{\alpha} = (\alpha_1, \alpha_2, \alpha_3)$ and along the measurement direction $\hat{\beta} = (\beta_1, \beta_2, \beta_3)$ with respect to the crystal axes. The subscripts refer to the projection of these unit vectors (direction cosines) along the three principal axes. To lowest order, the fractional change in length due to magnetostriction for a cubic material such as iron is

$$\delta l/l = (3/2)\lambda_{100}\{\alpha_1^2\beta_1^2 + \alpha_2^2\beta_2^2 + \alpha_3^2\beta_3^2 - 1/3\}$$
$$+ 3\lambda_{111}\{\alpha_1\alpha_2\beta_1\beta_2 + \alpha_2\alpha_3\beta_2\beta_3 + \alpha_1\alpha_3\beta_1\beta_3\},$$

where λ_{100} and λ_{111} are the longitudinal saturation magnetostrictions along the [100] and [111] directions. The saturation magnetostrictions are related to fundamental magnetoelastic coupling constants. The isotropic form is $\delta l/l = (3/2)\lambda_s \{\cos^2 \theta - 1/3\}$, where θ is the angle between the magnetization and the direction of observation of $\delta l/l$, and λ_s is the saturation magnetostriction. Some room temperature values for λ_s are -3.4×10^{-5} for nickel, -7×10^{-6} for iron, and 2×10^{-3} for TbFe$_2$ alloy. These values decrease with increasing temperature and vanish at the Curie temperature.

See also: ANTIFERROMAGNETISM; CURIE TEMPERATURE; FERROMAGNETISM; MAGNET; MAGNETIC DOMAIN; MAGNETIC MATERIAL; MAGNETIZATION

Bibliography

BOZORTH, R. M. *Ferromagnetism* (Van Nostrand, New York, 1951).

KITTEL, C. "Physical Theory of Ferromagnetic Domains." *Rev. Mod. Phys.* **21** (4), 541–583 (1949).

LACHEISSERIE, E. T. *Magnetostriction: Theory and Applications of Magnetoelasticity* (CRC Press, Boca Raton, FL, 1993).

LEE, E. W. "Magnetostriction and Magnetomechanical Effects." *Rep. Prog. Phys.* **18**, 185–229 (1955).

MICHAEL R. SCHEINFEIN

MAGNIFICATION

Magnification with an optical instrument is the process whereby the final image of an object is made larger than it would be without the instrument. There are two basic optical instruments used to magnify: the microscope and the telescope. With the microscope the object is placed nearby, whereas with the telescope the object is generally located far away.

The simplest example of a microscope is a single converging lens, also called a simple magnifier or magnifying glass. Without a magnifying glass, one would move an object as close as possible to the eye to make it appear large but not blurry; the image on the retina of the eye is then relatively large and clear. How closely the object can be moved to the eye depends on a person's vision; a nearsighted person at least has the advantage that his or her near distance d is less than that of a farsighted person. For a person with normal vision, d might be about 25 cm or 10 in. What the simple magnifier does is allow the object to be brought still closer to the eye because when placed inside the focal length f of the magnifier the light from the object appears to be coming from an enlarged image (a virtual image) farther from the magnifier than the object itself, and it is this light which is focused on the retina as a larger image than without the aid of the magnifier. For the largest possible magnification, the position of the object should therefore be adjusted until the virtual image is at a distance d from the magnifier. Under this circumstance the magnification turns out to be $m = 1 + d/f$. For example, if $f = d/3$ (about 8 cm), then $m = 4$. There are in fact limits to the magnification of a simple magnifier since a very short focal length lens has too much spherical aberration to produce a clear image. (For more relaxed viewing the virtual image may be adjusted to be far away rather than at the distance d; the formula is then $m = d/f$, so the magnification is slightly less in the above example—3 rather than 4.)

The compound microscope works on the same principle, except that there are two lenses (or lens systems), the objective and the eyepiece. The objective o forms an enlarged real image of the object and the eyepiece e further magnifies this image to produce a large virtual image, the overall magnification being $m = f_o/f_e$ if the virtual image being viewed by the eye is adjusted to be at infinity.

A telescope works much the same way (except that the object is essentially at infinity); indeed the magnification formula is the same as for a compound microscope. This is true whether the telescope is astronomical (also called refracting) or reflecting; in the first case the objective is a lens, and in the second it is a mirror. An objective of larger diameter D is generally ground to have a larger focal length f_o than one of smaller diameter; the larger the ratio D/f_o the shorter the time exposure necessary to photograph a faint object. As for a magnifying glass, there is a limit on f_e even for an eyepiece consisting of more than one lens; a 25-mm eyepiece is often better in quality than a 7-mm one. In a professional or amateur telescope, eyepieces are generally interchangeable in order to vary the magnification. However, magnification is not the most important consideration in buying a portable telescope, since more magnification sometimes just means a bigger, blurrier image. The quality of the optics, the sturdiness of the mount, and the presence of a clock drive are more important considerations.

See also: LENS; LENS, COMPOUND; TELESCOPE

Bibliography

HALLIDAY, D., and RESNICK, R. *Fundamentals of Physics,* 3rd ed. (Wiley, New York, 1988).

MORGAN, J. *Introduction to University Physics* (Allyn & Bacon, Boston, 1964).

THOMAS WINTER

MAGNON

A magnon is an elementary quantum of excitation of a spin wave, which may be excited when interacting magnetic atoms are each located at equivalent lattice positions in a crystalline array. Individual atoms may have electronic configurations in their ground states that endow each of them with spin magnetic moments. These atoms may then interact in various ways if they are incorporated in a spatially periodic fashion into the unit cells making up a regular crystalline array. The electrons of each atom are of course identical to those of its neighbors, and the Pauli exclusion principle has an important effect on the wave functions of electrons belonging to neighboring atoms. This effect results in an energetic interaction between such magnetic atoms that is called an "exchange interaction," which is measured by the value of an integral that involves the overlap of the atoms' electronic wave functions. When the exchange interaction between nearest neighbor atoms is negative, the ground state of the complete collection of magnetic atoms that make up the crystalline array corresponds to all of them lining up in the same direction, and this gives a single domain of ferromagnetism. If the exchange interaction between nearest neighbors is positive, then the ground state of the system essentially corresponds to an alternation in the magnetic spin directions of adjacent magnetic atoms, and this results in an antiferromagnetic domain.

Reversing the direction of any one atomic spin magnetic moment in the crystal requires the application of a torque that twists that atomic moment against the exchange forces that determined its direction in the ground state of the crystal. Such a reversal of a single spin direction requires a relatively large amount of excitation energy. However, if instead of completely reversing the direction of a single magnetic spin, consider that the directions of adjacent spins are made to twist slightly away from their original equilibrium direction. Suppose these adjacent spins successively deviate from that direction according to a sinusoidal function with a certain wavelength along a spatial direction perpendicular to the original equilibrium spin direction. If the amplitude of the sinusoidal deviation from the equilibrium direction is indeed slight, then the total amount of excitation energy is correspondingly also slight.

The disturbed distribution of atomic magnetic spins described above may then propagate through the crystal with the time dependence of a wave, and with a total excitation energy that depends on the amplitude of deviation from the equilibrium direction. The wave motion of such a spin wave is formally equivalent to that of a harmonic oscillator, as far as its total energy is concerned. When the collective wave motion of the magnetic spins is quantized according to the rules of quantum mechanics, the energy of excitation is given by the number of elementary spin wave quanta, or magnons, which may be present for any given wavelength. The energy carried by the magnons is the quantized excitation energy of the collective motion of the interacting atomic magnetic spins, and represents the low-lying energy levels of the magnetic spin system.

See also: MAGNETIC MOMENT; PAULI'S EXCLUSION PRINCIPLE; SPIN; WAVE FUNCTION

Bibliography

CHAIKIN, P. M., and LUBENSKY, T. C. *Principles of Condensed Matter Physics* (Cambridge University Press, New York, 1955).

KITTEL, C. *Quantum Theory of Solids* (Wiley, New York, 1963).

MICHAEL J. HARRISON

MASER

A maser is a source of coherent, monochromatic, electromagnetic radiation. The name is an acronym for *m*icrowave *a*mplification by *s*timulated *e*mission of *r*adiation. Sources of coherent light in the visible and near-visible wavelength range are called lasers.

Stimulated emission is the heart of masing action. Atomic, molecular, and solid-state systems are made up of a series of energy levels (states). If energy is added, the system can make a transition from a low-energy to an excited (high-energy) state. It can then be stimulated to make a transition back to a lower energy state by a photon with an energy equal to the difference between the upper and lower states. The system emits a photon with properties identical to the photon that caused the transition.

In masing, the medium acts as a photon amplifier by stimulated emission. A single input photon causes a second, identical photon to be emitted. These two photons stimulate two more, and the number of photons grows exponentially. For the stimulated emission process to lead to masing action, there must be more population in the upper state than in the lower state (population inversion).

A maser consists of an appropriate amplifying medium in a microwave cavity tuned to the masing resonance. A population inversion in the medium must be created (pumped) in some manner. Masing action can begin either by spontaneous emission or by seeding from an extenal microwave source.

The maser was predicted by Charles H. Townes in 1951 and was first demonstrated in an ammonia

(NH_3) beam by him, James P. Gordon, and Herbert J. Zeiger in 1954. Nikolai G. Basov and Aleksandr M. Prokhorov proposed a maser around the same time. Ammonia is made up of three hydrogen atoms as the base of a pyramid with the nitrogen atom at the peak. This is shown schematically in Fig. 1. The frequency difference ($\Delta\nu$) between the two energy levels is 24 GHz. In a classical sense, one can think of this as the oscillation frequency of the N atom through the plane made by the H atoms. Quantum mechanically, the ammonia atom has two states at different energies $\Delta E = \hbar\Delta\nu$, where $\hbar$ is Planck's constant. If the population is larger in the higher energy state, masing action occurs.

A beam of ammonia atoms typically has more atoms in the lower energy state. The different polarizability of the two states means that an electric field can be used to separate them. Before the ammonia beam enters the microwave cavity, it passes through an electrostatic filter to separate the atoms in the two energy states. Only the atoms in the higher energy state enter the cavity, leading to population inversion and masing.

Ammonia masers have a number of drawbacks that are typical of gas masers and lasers, including low gain and no frequency tuning. Solid-state masers have higher power output and are tunable over a wider range. A solid-state material with unpaired electrons is paramagnetic. When a magnetic field is applied to the system, the electron spin can be aligned parallel or antiparallel to it (see Fig. 2) The energy of the state is higher when the electron spin is parallel to the magnetic field and the energy difference between the two states is proportional to the strength of the magnetic field. Under normal conditions, more electrons are in the antiparallel state. When there are more spins in the parallel state, masing can occur. Various techniques can be used to create a population inversion in paramagnetic materials.

Ammonia and the simplest example of the paramagnetic maser are two-level masing systems. The disadvantage of such systems is that the population inversion is difficult to maintain because the stimulated emission process fills up the lower energy level, quenching the masing action. Thus, two-level masing systems operate in a burst mode or through the use of an atomic beam. Nicolaas Bloembergen proposed the three-level paramagnetic maser. When there are more than two energy levels, a population inversion can be maintained between two of the states by

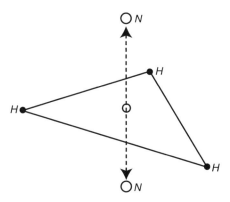

Figure 1 Schematic structure of ammonia, consisting of three hydrogen atoms as the base of the pyramid with the nitrogen atom at the peak.

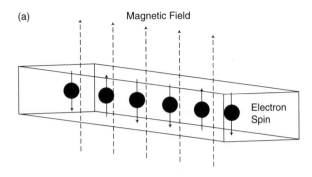

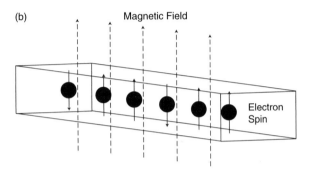

Figure 2 Electron spin in a magnetic field: (a) normal configuration and (b) inverted configuration.

pumping population through a third state. This technique can be extended to systems with four or more levels and is commonly used in laser systems.

The earliest masers produced output of approximately 10^{-10} W at a frequency of approximately 24 GHz (~1 cm wavelength) with a frequency stability of better than one part per billion. The frequency stability is related to the low-noise-amplification properties of the maser. The power output from masers was increased to the microwatt range by using solid-state materials. The primary applications of masers are as stable, low-noise, radio frequency sources, amplifiers, or extremely stable clocks. One application was in the amplification of the weak microwave signals measured in radio telescopes.

In 1965 a group of unexpected emission lines of the hydroxyl (OH) molecule were observed in radio emission from the Great Nebula of Orion. This emission was due to masing action in the interstellar gas. Stars send energy into the interstellar region through the emission of radiation and energetic particles. The energy is absorbed by the molecules, exciting them and creating a population inversion. Microwave emission from the star can then be amplified as it passes through the medium. Astrophysi-

cal masers have also been observed in water and silicon monoxide. In the mid-1990s the first astrophysical laser (visible emission) was observed.

In the mid-1980s a micromaser similar to the original ammonia masers was developed. The microwave cavity is made of superconducting material, giving it a very high quality factor (Q), or long decay time. A beam of atoms in the excited state passes through the cavity, allowing amplification of field. The difference from the original maser is that there is at most one atom in the cavity at a time. The high Q means that even with no atoms in the cavity, the maser field remains. Micromasers allow a detailed study of stimulated emission and of other quantum optics effects.

See also: LASER; LASER, DISCOVERY OF; PARAMAGNETISM; SPIN, ELECTRON

Bibliography

BERTOLOTTI, M. *Masers and Lasers: An Historical Approach* (Adam Hilger, Bristol, Eng., 1983).

WEBER, J. *Masers* (Gordon and Breach, New York, 1967).

DAVID D. MEYERHOFER

MASS

The word "mass" is derived from the Latin word *massa* meaning "lump of dough." However, through the years, as scientists began to understand better what mass conceptually meant, the working definition became "quantity of matter." You may know a quantity of matter when you see it, but it is much more difficult to come up with a physically precise definition of what a quantity of matter is. Isaac Newton had a large role in conceptually defining mass, and Albert Einstein had a large role in linking the different, seemingly unconnected definitions together.

Newton's laws of motion operationally define what is called inertial mass, the concept that the more massive an object, the larger a force needed to accelerate it. Specifically, if an object of inertial mass M_{I1} has a force F exerted on it, the inertial mass is found by taking the ratio of the exerted force and the acceleration a of the object:

$$M_{I1} = \frac{F}{a}. \qquad (1)$$

It follows that, if two objects of two different inertial masses M_{I1} and M_{I2} have identical forces exerted on them, the ratio of the accelerations a_1 and a_2 of M_{I1} and M_{I2}, respectively, is given by

$$\frac{a_1}{a_2} = \frac{M_{I2}}{M_{I1}}. \qquad (2)$$

That is, the larger a mass is, the smaller its acceleration will be for a given force. Both Eq. (1) and Eq. (2) are forms of Newton's second law. Inertial mass can also be linked directly to momentum. For instance, if a compressed spring is placed between the two inertial masses, the final velocities of each mass could be found after the spring decompresses. Since both of the masses are initially at rest, the final velocities are proportional to the accelerations, and the ratio of the v_1 (the final velocity of M_{I1}) and v_2 (the final velocity of M_{I2}) is given by

$$\frac{|v_1|}{|v_2|} = \frac{M_{I2}}{M_{I1}}. \qquad (3)$$

The universal gravitational law, also attributed to Newton, leads to the second definition of mass. This law states that between any two objects in the universe there is a force of attraction with a magnitude of

$$F = \frac{Gm_{G1}m_{G2}}{r^2}, \qquad (4)$$

where G is an experimentally determined constant, m_{G1} and m_{G2} are the gravitational masses of the two objects, and r is the distance between the two objects. If m_{G1} is the mass on which we want to find the gravitational force due to m_{G2}, a distance r away, Gm_{G2}/r^2 is called the gravitational field caused by m_{G2} at r. A gravitational field has units of acceleration (distance per time squared) and is often called the acceleration due to gravity.

If an object is near the earth's surface, where the force of attraction due to the earth's gravitational mass is towards the earth, and the gravitational field near the earth's surface is experimentally found to equal 9.81 m/s^2, the gravitational force on an object of gravitational mass m_G due to the earth may be written as

$$F_G = m_G g, \qquad (5)$$

where $g = 9.81$ m/s^2.

The gravitational mass of the object is a property of the object no matter where the object is located. The gravitational force on an object, however, depends not only on the gravitational mass of the object but also on the gravitational field at the object's position due to the gravitational mass causing the force.

For example, the gravitational field due to the Moon's mass at the surface of the Moon is roughly 1/6 to Earth's gravitational field at Earth's surface. If an object of gravitational mass m_G is placed on the Moon's surface, the gravitational force on it is $m_G \cdot 1.6$ m/s^2: 1/6 the magnitude of the force on the object at Earth's surface. Its gravitational mass, however, would be the same on the Moon and on Earth.

The gravitational force on an object is called its weight. As long as an object is on Earth, its weight and gravitational mass are directly proportional. If, however, the object is taken away from Earth and placed near some other gravitational mass, the object's weight changes, since the force of attraction changes, but its gravitational mass does not change. The gravitational mass is a property of the object, but the gravitational force is a property of the position of the object as well as of the object.

Although Galileo Galilei might not have realized it, he did one of the first experiments to verify that inertial mass and gravitational mass are experimentally indistinguishable. Objects of two masses dropped from rest from identical heights strike the ground at the same time, indicating that their accelerations are identical. The experimentally measured acceleration for a free-fall experiment such as this is found to be 9.81 m/s^2. From Newton's second law of motion, $F_{net} = M_I a$, where $F_{net} = m_G \cdot 9.81$ m/s^2, clearly the two types of masses must be the same value. In the early 1900s, Roland Baron Eötvös did a classic experiment experimentally testing to see if there is a distinguishable difference between the gravitational and inertial masses. Eötvös found that the two masses are the same to within 5 parts in 10^9.

Before the twentieth century, there was no physical argument that could be made to justify why the gravitational mass and the inertial mass should be the same property of the object. Classically, it would have been less surprising if they had not been. For example, some of the properties of a proton are its charge and its gravitational mass. Coulomb's law generates an equation that describes the force on a charged particle due to another charged particle. The universal law of gravitation generates an equation that describes the force on a gravitational mass due to a gravitational mass. Charge is a property that is not related at all to inertial mass, yet gravitational mass is experimentally identical to inertial mass. Until the twentieth century, there was no physical argument to justify why *either* of these two properties would or should be identical to the property of the object linked with its inertia.

This experimental evidence that inertial and gravitational masses are indistinguishable without any theoretical physical argument bothered Einstein and helped him develop his general theory of relativity. The general theory of relativity states, in part, that no physical experiment can distinguish between being in a gravitational field and being accelerated. For example, suppose you are in a closed elevator. If you are at rest in a gravitational field, you feel the floor of the elevator pushing on you, and if you drop a ball from rest with respect to you, it strikes the floor after some time t. If however, rather than being in a gravitational field, you are in the elevator and it is accelerating upward with an acceleration exactly the same magnitude as the gravitational field's magnitude, you would feel the same push due to the floor, and if you dropped the ball from rest with respect to you, it would hit the floor after the same amount of time. One consequence of the general relativity theory is the inability to distinguish between a gravitational field and an equal acceleration. If there is equivalence between a force in a gravitational field and a force due to an acceleration, there must be equivalence between inertial and gravitational mass. The gravitational field is multiplied by the gravitational mass to obtain a force, and the inertial mass is multiplied by the acceleration (which in magnitude equals the gravitational field) to get the same force.

In the early 1900s (before his general theory of relativity), Einstein revolutionized physics with the theory of special relativity. One of the consequences of his postulate of the constancy of the speed of light in special relativity was that, as energy is transferred to a rapidly moving object, the object's speed can increase neither indefinitely nor linearly. As the speed of an object increases, it takes a larger force to change the speed by the same increment compared to the force needed on the same object at a smaller speed. Yet from the work-energy theorem, the kinetic energy of the object must increase. If the speed does not change significantly, then the inertial mass of the object must increase, since it clearly resists a change in velocity more than it did at low speeds [see Eqs. (1) and (2)]. This leads to a third definition of mass:

$$m = \frac{m_0}{\sqrt{1 - v^2/c^2}}, \tag{6}$$

where m_0 is the rest mass, the inertial mass of the object when it is at rest. Thus, the inertial mass of an object is *not* a constant; it depends on the object's speed.

In addition, it can be shown by using Einstein's special theory of relativity that the kinetic energy for any object (even at low speeds) is given by

$$KE = mc^2 - m_0c^2, \tag{7}$$

where m is defined by Eq. (6), m_0 is the rest mass, and c is the speed of light. This leads to perhaps one of the most recognized equations in modern physics:

$$E = mc^2, \tag{8}$$

where E is the total energy of the object—its kinetic energy plus its rest mass energy, m_0c^2. This equation contains a lot of information. It states that the inertial mass and energy are not separate entities; they are different manifestations of the same entity. An object, merely because it possesses mass, has a certain amount of energy associated with it.

The inertial mass-energy equivalence has been verified or used in many different ways. In fission, when large atoms are split apart, the smaller constituents that are formed have less mass than the original, larger constituent. The difference in the mass before and after the fission is freed in the form of energy. This energy has been used to create electricity for homes in nuclear power plants. It has also been used in atomic bombs. In fusion, which occurs in stars and in laboratories on Earth, when small atoms are combined with other small atoms, the larger atom created has less mass than the constituents that formed it. Again the difference in mass is freed in the form of energy. At particle accelerators such as those found at Fermilab in Illinois and CERN in Switzerland, by colliding small particles that move very fast, scientists are able to create particles that are more massive than the particles that collide originally were, again creating mass from energy. Two photons (mass = 0) can collide to form particles where the lost energy (energy of the photons minus the kinetic energy of the new particles) equals the mass of the new particles times the speed of light squared. In chemistry, the mass of an atom with bound electrons is found to be less than the mass of the atom's nucleus and free electrons. The increased mass of the components equals the energy needed to free the electrons from the bound state. The equivalence of inertial and gravitational mass for matter consisting of composite objects such as atoms then clearly shows that all forms of energy are a source of gravitational field, not just rest mass.

Scientists are still testing the equivalence of inertial and gravitational mass. Improvement of Eötvös's experimental work has shown that the two types of masses must be identical to one part is 10^{11}. Some physicists have theorized the possibility of a fifth force in nature. This force is one of repulsion between gravitational masses. If this force exists, then gravitational and inertial masses would *not* be identical.

Once general relativity linked inertial and gravitational mass, the three apparently distinct definitions of mass were unified. However, in understanding the unification, our comprehension of mass changed. One can no longer think of mass as merely a quantity of matter, for the quantity depends on the speed of the matter. And, even more complicated, the lines that were previously drawn between mass and energy have been blurred. They appear to be different manifestations of the same thing.

See also: CENTER OF MASS; CENTER-OF-MASS SYSTEM; COULOMB'S LAW; FIELD, GRAVITATIONAL; INERTIAL MASS; MASS, CONSERVATION OF; MASS-ENERGY; NEWTON'S LAWS; RELATIVITY, GENERAL THEORY OF; RELATIVITY, SPECIAL THEORY OF; REST MASS; WEIGHT

Bibliography

FEYNMAN, R. P.; LEIGHTON, R. B.; and SANDS, M. *The Feynman Lectures on Physics*, Vol. 1 (Addison-Wesley, Reading, MA, 1963).

JAMMER, M. *Concepts of Mass in Classical and Modern Physics* (Harvard University Press, Cambridge, MA, 1961).

SWARTZ, C. "A Morality Tale for Physics Teachers." *Phys. Teach.* **25,** 154–155 (1987).

TSAI, L. "The Relation Between Gravitational Mass, Inertial Mass, and Velocity." *American Association of Physics Teachers* **54,** 340–342 (1986).

MARIE BAEHR

MASS, CENTER OF

See CENTER OF MASS

MASS, CONSERVATION OF

The persistence of matter is one of the basic properties of nature. Our confidence in this property allows us to make sense of our surroundings and begin to study the laws of physics. When objects move around or collide, or when liquid is poured from one container to another, the amount of material does not change. To be quantitative about the study of matter we need to define a quantity that measures the amount of matter, independent of its shape or form. The quantity that we first notice in this respect is weight. Once basic mechanics has been understood, we recognize that weight is a derived quantity; a property not of the matter alone but of the interaction of the matter with the gravitational field of the earth. Mass, rather than weight, provides the intrinsic measure of matter.

It was found that mass appears conserved in all physical processes. Even when matter changes state, from solid to liquid or from liquid to gas (or conversely, from gas to liquid or liquid to solid), the mass of the matter is unchanged. The recognition of the mass of a gas was an important step to understanding the nature of this often invisible form of matter. The law of conservation of mass was one of the early conservation laws used by scientists to organize their understanding of the world around them. However, as we shall see, it is only an approximate law. Our modern understanding of matter includes processes that completely change our view of the persistence of matter.

The law of conservation of mass is often taught as one of the laws of chemistry. This is inaccurate and creates a very confusing situation. Chemical processes either give off heat or require heat to proceed. Thus, energy must flow in or out of the system in such processes. Where does this energy come from or disappear to? There is a contradiction between exact conservation of mass in chemical reactions and the law of conservation of energy. Very careful experiments show that conservation of energy is an exact law, while conservation of mass is only approximate.

Mass is simply one of the forms of energy. It enters into the accounting of energy through Albert Einstein's relation $E = mc^2$. This expresses the amount of energy E stored as mass in an object of mass m, where c is the speed of light. Thus, the energy absorbed or released in chemical changes is due to very small changes of the mass of the material. These changes are a tiny fraction of the masses of the atoms involved. They are small compared to the accuracy with which mass is usually measured in the chemistry laboratory, so the law of conservation of mass is correct within the accuracy of such measurements.

Chemical and Physical Processes: Conservation of Atoms

Today we recognise that the approximate conservation of mass in chemical and physical processes is a consequence of an exact conservation law: conservation of the number of atoms of each element in the process. The bookkeeping of the atoms forms the basis of the equations for chemical processes. Each atom has a particular mass, chiefly due to the masses of the protons and neutrons in its nucleus. This mass is characteristic for each element (or more precisely, each isotope has a definite mass). Since the atoms are unchanged in physical and chemical processes, the bulk of the mass is conserved.

When atoms bind to form molecules, the mass of the molecule is slightly less than the sum of the masses of the atoms it contains, because there is a negative electrical potential energy between the atoms in the molecule (negative relative to the situation where the atoms are far apart). This energy is called the binding energy of the molecule. The mass difference m between the sum of the atomic masses and the mass of the molecule is related to the binding energy E by $E = mc^2$. This is why the molecule cannot simply fall apart into its constituent atoms; it does not have enough energy to do so.

In a chemical process, the atoms are rearranged into different molecules. The binding energies of the starting molecules are not the same as the binding energies of the resulting molecules. If the sum of the masses of the starting molecules is less than the sum of the resulting molecule masses, some energy must be provided for the reaction to occur. This is called an endothermic reaction, meaning one that requires heat to occur. If the sum of starting masses is greater than the sum of resulting masses, then energy is released in the process and appears as kinetic energy of the products (that is, as heat in a bulk process); this is called an exothermic reaction, meaning one that gives off heat.

Even in physical changes, such as melting, there are tiny changes in mass because the binding energy of atoms or molecules that form a solid is different from that of the liquid form, and the liquid is different from a gas. The latent heat of melting or evaporation is the energy required for the change of state, and it corresponds to a tiny change of mass. When a spring is stretched, it has additional energy stored in it and thus a greater mass than when in it is at its relaxed length.

Nuclear Processes: Conservation of Neutrons Plus Protons

At the level of nuclear processes, the law of conservation of atoms fails. Some atoms undergo fission, in which the nucleus breaks apart to form two or more other atomic nuclei. In other cases, two nuclei can be made to combine to form a third one; this is known as fusion. Just as for the molecules, in a nucleus there is negative potential energy, or binding energy. Here it is due to the interactions between the protons and neutrons. For any nucleus the sum of the masses of the protons and neutrons is a little more than the mass of the nucleus they comprise. The difference is related to the binding energy of the nucleus by $E = mc^2$. In a fission process the initial nucleus has more mass than the sum of the product nuclei, so energy is released in the process. In a fusion process the initial separate nuclei have more mass than the single final nucleus, so again there is energy released.

Most of the mass of atomic nuclei comes from the sum of the masses of the neutrons and protons within the nuclei. The total number of these do not change in either a fission or a fusion process. Only the combined number of neutrons plus protons is truly a conserved quantity in nuclear decays. Radioactive transitions in which a neutron becomes a proton, or vice versa, are observed. (Conservation of energy and electric charge are maintained by the other very low mass particles, such as electrons, produced in the transition.)

Nuclear binding energies are typically larger than molecular binding energies, so the amount of energy released in a fission or fusion reaction (and hence the mass change via $E = mc^2$) is considerably larger than that in chemical processes. However, the changes in mass are still only a small percentage of the total mass involved, so even in nuclear processes mass is approximately conserved due to the conservation of the sum of the number of protons plus the number of neutrons.

Antimatter: Conservation of Baryon Number

Conservation of the number of the neutrons plus protons is found to fail in particle processes seen at accelerator experiments. The possibility of producing antimatter particles adds a new twist to particle counting. It is found that an exact conservation law still applies, but it requires that antiprotons, antineutrons, and the like must be counted as negative contributions to the sum. At high-energy accelerators, protons and neutrons can be produced when none were there before, but they always appear accompanied by a matching number of antimatter particles (for example, antiprotons or antineutrons). The inverse process also can occur; if a particle meets its matching antiparticle the two can disappear (physicists say they annihilate each other). The energy that was present before they met appears in some new form, for example, as electromagnetic radiation.

Processes involving the production and annihilation of particles and antiparticles are frequently observed in high-energy physics experiments, and they are well understood. All experiments are consistent with an exact conservation of the quantity known as baryon number. Baryons are a class of particles that include protons and neutrons as well as a number of similar, more massive particles (all of them unstable). For each type of baryon there is a corresponding type of antibaryon. Baryon number is the total number of baryons minus the total number of antibaryons. It is unchanged in all particle processes observed to date.

Now we come to a startling realization. If there were any antibaryons around, then the baryons that ran into them would be likely to disappear. Matter only persists because there is almost no antimatter around for it to meet. One of the great puzzles of the universe is to understand why it contains baryons and no antibaryons. If the numbers were equal, we would not be here to think about it, because by now all the baryons and antibaryons in the universe would have annihilated each other.

So what remains of the original idea that matter persists, which led to the law of conservation of mass? Today it is explained by the conservation of baryon number. We recognize that mass is not a conserved quantity.

Physicists speculate that there may be very rare processes that violate even this conservation law. The proton is the least massive type of baryon, so conservation of energy tells us that, if baryon number is conserved, an isolated proton can never decay. What do we know about the stability of the proton? Its observed half-life is greater than about 10^{32} years, which is many times the age of the universe. (By looking at very many protons one can have a very high probability that one would have seen one decay if the half-life were less than this value, even though one has looked for only a few years.) No evidence for proton decay has yet been found. The conservation of baryon number, if not exact, is at least very nearly so. That, and the rarity of antibaryons in the universe, are the reason for the persistence of matter.

See also: ATOMIC MASS UNIT; BARYON NUMBER; CONSERVATION LAWS; FISSION; FUSION; INERTIAL MASS; MASS; MASS-ENERGY; PARTICLE MASS; REST MASS

Bibliography

FEYNMAN, R. P. *The Character of Physical Law* (MIT Press, Cambridge, MA, 1967).

HELEN R. QUINN

MASS DEFECT

When an electron is attracted to a proton to form a hydrogen atom, energy is released (ordinarily as light). This shows that the energy of the bound system—a hydrogen atom—is lower than the energy of a free electron and proton. This energy difference is called the binding energy, B. In accordance with the well-known mass-energy equivalence, $E = mc^2$, the mass of the hydrogen atom is therefore less than the sum of the masses of a free electron and proton, and this mass difference, Δm, is called the mass defect. Thus,

$$B = \Delta m c^2. \qquad (1)$$

In the case of the hydrogen atom, the attractive force is relatively weak so the mass defect is only 1.4×10^{-8} times the mass of the atom, which is immeasurably small. It is much easier to measure B directly than to measure Δm and apply Eq. (1).

In the early years of nuclear physics, that situation was reversed. Mass defects are nearly 1 percent of the mass of nuclei, and techniques of mass spectroscopy were well developed by the 1930s. At that time, mass defect measurements with application of Eq. (1) were the principal method of determining binding energies; the results were very interesting and historically important. This is exemplified by Fig. 1, a plot for nuclei found in nature (and thus available for mass spectroscopy) of binding energy per nucleon, B/A, versus the number of nucleons (i.e., neutrons and protons) in the nucleus, A. The fact that B/A is smaller by about 0.8 MeV for uranium ($A = 235$–238) than for nuclei of half its mass suggests that if a uranium nucleus is split in half, a process called fission, the binding energy of the system would be increased by 190 MeV, and that large amounts of energy would be released. Figure 1 also shows that if lighter nuclei are combined to form ^{4}He, a process called fusion, the binding energy is greatly increased, and that energy is released. The observation from Fig. 1 that B/A is roughly the same—about 8 MeV—for all but the lightest nuclei was very important in developing an early understanding of nuclei. Fitting the curve by a semi-empirical mathematical formula gave a much more detailed understanding of nuclear structure. The observation from Fig. 1 that B/A has a maximum in the iron-nickel region has very important consequences in astrophysics and in explaining the abundances of the elements.

While mass spectroscopy techniques continued to improve until about 1960, techniques for direct measurement of energies of particles involved in reaction and decay processes improved much more rapidly and have been the principal basis for determining binding energies since the early 1950s.

Perhaps the simplest and most direct manifestation of mass defect is that the masses of neutrons and protons are 1.008 u, which means that the mass of A free neutrons and protons is $1.008 \times A$, but the mass of nuclei is close to $1.000 \times A$; this difference of $0.008 \times A$ is the mass defect due to the binding of the nucleus by nuclear forces.

See also: FISSION; FUSION; MASS-ENERGY; NUCLEAR BINDING ENERGY; NUCLEON

Bibliography

COHEN, B. L. *Concepts of Nuclear Physics* (McGraw-Hill, New York, 1971).

BERNARD L. COHEN

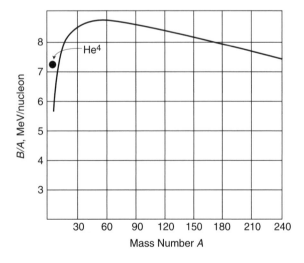

Figure 1 Binding energy per nucleon, B/A, for nuclei found in nature.

MASS-ENERGY

One of the most startling consequences of the theory of special relativity is that mass is a form of energy, the energy equivalent of an object of mass m

being given by Albert Einstein's famous equation $E = mc^2$, where c is the speed of light. Why is this so?

Conservation of Momentum and the Principle of Relativity

A fundamental principle of physics is that the total momentum of an isolated system of particles is conserved. In Newtonian mechanics, a particle with velocity **v** and mass m has momentum **p** = m**v**, and we simply add these particle momenta (as vectors) to find a system's total momentum. Centuries of experiments have shown that the law of conservation of Newtonian momentum is valid for systems of particles moving at nonrelativistic speeds. However, conservation of Newtonian momentum proves to be inconsistent with the theory of relativity when applied to systems of particles moving with relativistic speeds.

As a concrete example of the problem, imagine that we observe the following decay process in our laboratory. A subatomic particle of known mass M moving in the $+x$ direction with speed $V = \frac{3}{5}c$ decays into two particles of unknown masses $\underline{m}$ and m. After the decay, the first particle is observed to be at rest (its speed u is zero), while the second particle is observed moving in the $+x$ direction with a speed $v = \frac{4}{5}c$. This decay process is illustrated in Fig. 1a.

A Newtonian physicist would analyze this process as follows. The x component of the law of conservation of momentum in this situation implies that

$$MV_x = \underline{m}u_x + mv_x \Rightarrow M(+\tfrac{3}{5}c)$$

$$= m(+\tfrac{4}{5}c) + 0 \Rightarrow m = \tfrac{3}{4}M. \qquad (1)$$

Moreover, the mass of the two product particles should add up to the original particle's mass, so

$$M = \underline{m} + m \Rightarrow \underline{m} = M - m$$

$$= M - \tfrac{3}{4}M \Rightarrow \underline{m} = \tfrac{1}{4}M. \qquad (2)$$

Therefore, as long as $m = \frac{3}{4}M$ and $\underline{m} = \frac{1}{4}M$, both momentum and mass will be conserved in the laboratory frame.

Now consider the same process viewed in an inertial reference frame moving with the original particle. In this frame, the original particle is at rest ($V' = 0$) while the particle with mass $\underline{m}$ (which was at rest in the lab frame) is now seen to move in the $-x$ direction with speed $u' = \frac{3}{5}c$. Because of the relativistic speeds involved, we need to compute the speed of the third particle in this frame using the relativistic velocity transformation formula (which directly follows from the basic Lorentz transformation equation). This equation reads

$$v_x' = \frac{v_x - \beta}{1 - \beta v_x/c^2} = \frac{\frac{4}{5}c - \frac{3}{5}c}{1 - (\frac{4}{5}c)(\frac{3}{5}c)/c^2}$$

$$= \frac{\frac{1}{5}c}{1 - \frac{12}{25}} = \tfrac{1}{5}c(\tfrac{25}{13}) = \tfrac{5}{13}c, \qquad (3)$$

where v_x and v_x' are the x components of the particle's velocity in the lab and the original particle's frames, respectively, and β is the speed of the parti-

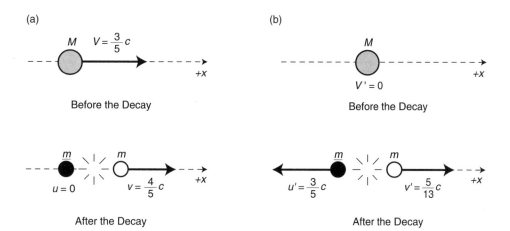

(a)

M $V = \dfrac{3}{5}c$

$+x$

Before the Decay

$\underline{m}$ m

$u = 0$ $v = \dfrac{4}{5}c$ $+x$

After the Decay

(b)

M

$+x$

$V' = 0$

Before the Decay

$\underline{m}$ m

$u' = \dfrac{3}{5}c$ $v' = \dfrac{5}{13}c$ $+x$

After the Decay

Figure 1   The decay process for a subatomic particle.

cle's frame relative to the lab frame. The decay process as seen in the original particle's frame is thus as shown in Fig. 1b.

The product particles must have the speeds shown in Fig. 1b to be consistent with relativity and our observations in the lab frame, but if they do, Newtonian momentum is not conserved in this frame. The original particle's momentum is zero in this frame, but

$$\underline{m}u_x' + mv_x' = \tfrac{1}{4}M(-\tfrac{3}{5}c) + \tfrac{3}{4}M(+\tfrac{5}{13}c)$$
$$= (-\tfrac{3}{20} + \tfrac{15}{52})Mc \neq 0. \quad (4)$$

This is a problem, because the founding assumption of relativity is the principle of relativity, which says that the laws of physics are the same in every inertial reference frame. Here we see that if we make the law of conservation of momentum work in the lab frame, it fails in the original particle's frame, contrary to the principle of relativity. Yet the law of conservation of momentum is one of the most basic laws of physics. How can we resolve this problem?

Relativistic Momentum

It turns out that we only need to redefine momentum slightly. The components of a particle's Newtonian momentum are

$$p_x \equiv mv_x = m\frac{dx}{dt}, \, p_y \equiv m\frac{dy}{dt}, \, p_z \equiv m\frac{dz}{dt}. \quad (5)$$

In this definition, we divide the particle's infinitesimal coordinate displacement dx, dy, dz in a given frame by the infinitesimal time interval dt required for that displacement, as measured in the same frame. However, instead of dividing by dt, we could have divided by the infinitesimal proper time $d\tau$ required for that displacement as measured by a clock traveling with the particle itself. In Newtonian mechanics, dt would be equal to $d\tau$, but in special relativity they are related by the time dilation formula

$$d\tau = dt\sqrt{1 - v^2/c^2}, \quad (6)$$

where v is the speed of the particle in question.

We define the components of a particle's relativistic momentum to be

$$P_t \equiv m\frac{cdt}{d\tau}, \, P_x \equiv m\frac{dx}{d\tau}, \, P_y \equiv m\frac{dy}{d\tau}, \, P_z \equiv m\frac{dz}{d\tau}, \quad (7)$$

where m here is a velocity-independent quantity describing an intrinsic property of the particle. (Most modern treatments of relativity use this definition of mass: the older approach that treated m as velocity-dependent is not wrong so much as awkward in most applications.) We include a time component P_t in this set of components because in relativity theory time and space are considered to be part of a unity called spacetime, and they are generally treated equally in relativistic equations (the extra factor of c in P_t makes its units consistent with those of the other components). Because this relativistic generalization of momentum has four components instead of three, we call it the particle's four-momentum.

Eq. (6) implies that

$$P_t \equiv m\frac{cdt}{dt\sqrt{1 - v^2/c^2}} = \frac{mc}{\sqrt{1 - v^2/c^2}},$$

$$\qquad (8)$$

$$P_x \equiv m\frac{dx}{dt\sqrt{1 - v^2/c^2}} = \frac{mv_x}{\sqrt{1 - v^2/c^2}},$$

and similarly for P_y and P_z. Note that if $v \ll c$, then $\sqrt{1 - v^2} \approx 1$, and $P_x \approx mv_x$ (similarly, $P_y \approx mv_y$, $P_z \approx mv_z$). Therefore, the spatial components of a particle's relativistic momentum become essentially the same as those for Newtonian momentum at nonrelativistic speeds. This means that experiments with nonrelativistic particles cannot distinguish between conservation of Newtonian and relativistic momentum. Might it really be relativistic momentum that is conserved?

Conservation of Relativistic Momentum

Consider again the decay process shown in Fig. 1a. Conservation of the x component of relativistic momentum implies that in the lab frame

$$\frac{MV_x}{\sqrt{1 - V^2/c^2}} = \frac{mu_x}{\sqrt{1 - u^2/c^2}} + \frac{mv_x}{\sqrt{1 - v^2/c^2}}$$

$$\Rightarrow \frac{M(\frac{3}{5}c)}{\sqrt{1 - (\frac{3}{5})^2}} = 0 + \frac{m(\frac{4}{5}c)}{\sqrt{1 - (\frac{4}{5})^2}}$$

$$\Rightarrow \tfrac{3}{4}M = \tfrac{4}{3}m \Rightarrow m = \tfrac{9}{16}M. \tag{9}$$

(The y and z components of relativistic momentum are trivially conserved in this situation.) Note that m has a significantly different value here than we found in our Newtonian analysis.

We now have a choice. We can either make the masses of the product particles add up to the original particle's mass (in which case $m = \frac{7}{16}M$) or we can assume that the t component of the relativistic momentum, like the x component, is conserved, which implies that

$$\frac{Mc}{\sqrt{1 - V^2/c^2}} = \frac{mc}{\sqrt{1 - u^2/c^2}} + \frac{mc}{\sqrt{1 - v^2/c^2}}$$

$$\Rightarrow \frac{Mc}{\sqrt{1 - (\frac{3}{5})^2}} = \frac{mc}{\sqrt{1 - 0^2}} + \frac{mc}{\sqrt{1 - (\frac{4}{5})^2}}$$

$$\Rightarrow \tfrac{5}{4}Mc = mc + \tfrac{5}{3}\tfrac{9}{16}Mc \Rightarrow m = \tfrac{5}{16}M. \tag{10}$$

If you take the first option, you will find that in the reference frame of the original particle, the x component of the system's total relativistic momentum is zero before the decay and $-\frac{3}{32}Mc$ afterward and thus is not conserved, violating the principle of relativity. On the other hand, if you choose m so that P_t as well as P_x is conserved in the lab frame, then you will find that P_t and P_x are also conserved (with total values of Mc and zero, respectively) in the original particle's frame, consistent with the principle of relativity. For example, the x component of the original particle's momentum is clearly zero in this frame; after the decay it is

$$\frac{mu_x'}{\sqrt{1 - u'^2/c^2}} + \frac{mv_x'}{\sqrt{1 - v'^2/c^2}} = \frac{\frac{5}{16}M(-\frac{4}{5}c)}{\sqrt{1 - (\frac{4}{5})^2}}$$

$$+ \frac{\frac{9}{16}M(\frac{5}{13}c)}{\sqrt{1 - (\frac{5}{13})^2}} = Mc(-\tfrac{5}{4}\tfrac{5}{16}\tfrac{3}{5} + \tfrac{13}{12}\tfrac{9}{16}\tfrac{5}{13}) = 0. \tag{11}$$

In short, we can make conservation of relativistic momentum consistent with the principle of relativity but only if its time and spatial components are *simultaneously* conserved.

Relativistic Energy

So what is this mysterious new conserved quantity P_t? According to the binomial approximation, $(1 + x)^a \approx 1 + ax$ if $x \ll 1$. If we multiply the definition of P_t by c and use the binomial approximation, we find that P_t becomes

$$P_t c \equiv \frac{mc^2}{\sqrt{1 - v^2/c^2}} = mc^2\left[1 + \left(\frac{v^2}{c^2}\right)\right]^{-1/2}$$

$$\approx mc^2\left[1 - \tfrac{1}{2}\left(-\frac{v^2}{c^2}\right)\right] = mc^2 + \tfrac{1}{2}mv^2 \tag{12}$$

for nonrelativistic particles ($v \ll c$). We see that in this low-speed limit, $P_t c$ is equal to mc^2 plus the particle's kinetic energy. This means that in low-speed Newtonian interactions that preserve a system's mass, we would interpret conservation of P_t as conservation of kinetic energy. Because of this, we identify $P_t c$ as being the particle's relativistic energy E and conservation of $P_t c$ as the relativistic generalization of Newtonian conservation of energy.

The interesting implication of this is that even a particle at rest has relativistic energy:

$$E_{\text{rest}} = mc^2. \tag{13}$$

We call this the particle's mass-energy. We can think of a particle's total relativistic energy as being a sum of two terms, its mass-energy mc^2 and a kinetic energy part K that is zero when the particle is at rest and grows as its velocity increases:

$$E = mc^2 + K, \; K = mc^2\left(\frac{1}{\sqrt{1 - v^2/c^2}} - 1\right), \tag{14}$$

where $K \approx \frac{1}{2}mv^2$ at low speeds. Relativity only requires that a system's total E be conserved; there is no requirement that mc^2 and K be conserved separately. This suggests that there exist processes that convert mass-energy into kinetic energy or vice versa.

Our example decay process does just that! The mass of the original particle was M; the total mass of the two product particles is $m + \underline{m} = (\frac{9}{16} + \frac{5}{16})M = \frac{7}{8}M$. Before the decay, the system has no kinetic energy, but afterwards it has nonzero kinetic energy. This process thus converts $\frac{1}{8}M$ of mass-energy into the final kinetic energy of the system. (Indeed, our analysis showed that assuming that the mass does *not* change got us into trouble with the principle of relativity.)

Equation (13) is the famous equation that has come to symbolize the revolutionary implications of the theory of relativity. It asserts the equivalence of mass and energy (more precisely, that mass is simply one form of energy) and implies that mass-energy may be freely converted to other forms of energy or vice versa.

Processes That Convert Mass-Energy

Our example decay process was idealized to make calculations simple, but real decay processes are qualitatively the same. For example, a subatomic particle known as the kaon (whose mass is 975 electron masses) at rest decays into two neutral pions (whose masses are 264 electron masses each) that fly symmetrically away at speeds of $0.84c$. This process converts 46 percent of the kaon's mass-energy into kinetic energy.

An extreme example of conversion of mass-energy is matter–antimatter annihilation. An electron and a positron (antielectron) at rest have mass-energy but no kinetic energy. If these particles approach each other closely enough, they annihilate, producing two photons of light that have energy but zero mass. This process thus converts the particles' mass-energy completely into light energy. Physicists have also observed the inverse process; two photons with the right energy can collide and produce an electron-positron pair at rest, converting pure energy completely into mass-energy.

In nuclear fission, a uranium or plutonium nucleus is induced by incoming neutrons to split into two fragments with slightly smaller total mass than the original nucleus. The process ends up converting roughly 0.1 percent of the original mass-energy of the nucleus into energy. In a nuclear fusion process, two heavy hydrogen nuclei ($^2_1$H and $^3_1$H) collide to form a helium nucleus and a neutron. This process converts about 0.4 percent of the mass-energy of the original nuclei into energy.

Unlike decay processes involving subatomic particles, these nuclear reactions do not involve creation or destruction of particles but rather the rearrangement of protons and neutrons into configurations having lower potential energy. According to Eq. (12), the total mass of a nucleus at rest is its total relativistic energy divided by c^2. Its total energy, in turn, is the sum of the energies of all its constituent particles, including the potential energies of their interactions. Decreasing the potential energy of a system, even without changing its constituent particles, will thus decrease its mass. Similarly, chemical reactions rearrange atoms into configurations with lower potential energy. However, the potential energies involved in molecules are so much smaller than those involved in nuclei that the mass change in chemical reactions is typically 10^6 to 10^7 times smaller.

It does not take conversion of very much mass-energy to produce enormous amounts of other forms of energy. Complete conversion of 1.0 kg of mass-energy would release an amount of energy $E = mc^2 = (1.0\text{ kg})c^2 = (1.0\text{ kg})(3.0 \times 10^8\text{ m/s})^2 = 9.0 \times 10^{16}$ J, comparable to the electrical energy that a major power plant produces in three years. A nuclear bomb typically converts only a few grams of mass-energy into other forms of energy. Burning a kilogram of gasoline, by contrast, converts only about 10^{-7} g of mass-energy into energy.

See also: ENERGY; ENERGY, CONSERVATION OF; ENERGY, KINETIC; FISSION; FISSION BOMB; FRAME OF REFERENCE, INERTIAL; FUSION; FUSION BOMB; FUSION POWER; KAON; MASS; MOMENTUM, CONSERVATION OF; RELATIVITY, GENERAL THEORY OF; SPACETIME

Bibliography

ADLER, C. G. "Does Mass Really Depend On Velocity, Dad?" *Am. J. Phys.* **55** (8), 739–743 (1987).

MOORE, T. A. *A Traveler's Guide to Spacetime* (McGraw-Hill, New York, 1995).

TAYLOR, E. F., and WHEELER, J. A. *Spacetime Physics,* 2nd ed. (W. H. Freeman, New York, 1992).

THOMAS A. MOORE

MASS NUMBER

Briefly, the mass number (A) of an atom is an integer equal to the number of protons (Z) plus the number of neutrons (N) contained in the atom's nucleus ($A = Z + N$).

During the latter half of the nineteenth century, chemical atomic weight measurements became available for a number of elements. Based on a scale where the atomic weight of oxygen was defined to be 16.00, some elements were observed to have non-integral atomic weights. This fact was at odds with the prevailing hypothesis first put forward in 1815 by William Prout that the atoms of elements were formed from combinations of integral numbers of hydrogen atoms. While well-known and abundant species like carbon (atomic weight = 12.00), fluorine (atomic weight = 19.00), and sodium (atomic weight = 23.00) provided support for this idea, the atomic weights of magnesium (atomic weight = 24.32) and chlorine (atomic weight = 35.46) were puzzling exceptions, calling for a new description of the structure of atoms. In a serendipitous proposal made a few years before the turn of the nineteenth century, the British physicist William Crookes anticipated a twentieth-century discovery by suggesting that an element with a nonintegral atomic weight really consisted of atoms having differing integer atomic weights (we now call them isotopes) in various proportions. He went on to suggest that calculating an atomic weight averaged over all the atoms of that element would yield the experimental atomic weight.

It was not until 1932, when the British researcher James Chadwick discovered the neutron, that the picture of isotopes and the structure of atoms became clear. Earlier work by Ernest Rutherford and his collaborators, using the scattering of alpha particles from thin metallic foils, had shown that the positive charge of an atom was concentrated in a very small body called the nucleus. This was extended a few years later by the Danish physicist Niels Bohr, who showed that an atom's mass is primarily located in its nucleus. When the neutron appeared on the scene and was discovered to have virtually the same mass as the proton, the concept of isotopes arose naturally as being atoms with the same number of protons and differing numbers of neutrons.

The mass number (A) of an isotope is an integer defined to be the number of protons (Z), plus the number of neutrons (N), and is the integer closest to the isotope's atomic weight. Since protons and neutrons are known as nucleons, the mass number is identical to the nucleon number. Using the hypothesis of Crookes, the fact that chlorine atoms turn out to be either isotopes of mass number 35 (about 75%) or 37 (about 25%) easily explains the atomic weight of chlorine atoms found in nature. Crookes would calculate the average as $35 \times 0.75 + 37 \times 0.25 = 35.5$, which agrees with experiment.

See also: ATOMIC WEIGHT; ELEMENTS; ISOTOPES; NEUTRON; NUCLEON; NUCLEUS; PROTON

Bibliography

KRANE, K. S. *Introductory Nuclear Physics* (Wiley, New York, 1988).

CARY N. DAVIDS

MASS SPECTROMETER

A mass spectrometer (MS) is any one of a large range of instruments which is used to determine the masses (or more specifically the mass to charge ratio, M/Z) of atomic or molecular ions. The ionic masses are determined by dispersing (separating) the ions in a magnetic, electrostatic, or combined electrostatic/magnetic field. The methods used to ionize the atoms and molecules are as varied as the type of mass spectrometer employed to determine their masses. The mass analysis region of the mass spectrometer is almost always at low pressure so that the flight of an ion is uninterrupted by collisions with a background gas. The sample in the ion source region may be a solid, liquid, gas, or a plasma (a mixture of positive and negative ions). The method of introducing the sample to the mass spectrometer is also important. For example, a gas chromatograph (GC) is often used to separate gas mixtures by their different mobilities through a column prior to introduction into the mass spectrometer. The ion-detection region of the mass spectrometer consists of either a simple current collecting device (Faraday cup/electrometer), an electron multiplier (device for greatly amplifying a single ion count), or (historically) a photographic plate. In the latter case, the mass spectrometer is called a mass spectrograph. Thus a mass spectrometer can be divided into four basic regions: (1) sample introduction, (2) ion source, (3) mass analyzer, and (4) ion detection. The two most important of these elements are the ion source and the mass analyzer.

Ionization Source

Great advances have been made in the development of new methods to ionize atoms and molecules. The analysis of solid and liquid samples is of particular importance in the area of environmental and biomolecular mass spectrometry.

Electron impact ionization (EI). Electrons with kinetic energies above the minimum energy required to remove an electron from an atom or molecule (ionization potential) can produce a parent positive ion (an ion whose mass is the same as that of the atom or molecule minus the mass of the electron). If the electron energy is above the energy required to both ionize and fragment (dissociate) a molecule, fragment ions can result. Also, when the electron energy is sufficiently high, multiply charged positive ions may be produced.

The cross section for electron impact ionization of an atom or molecule is maximized at an electron energy corresponding to roughly three times the ionization potential. (The cross section is the "effective area" of an atom or molecule for ionization expressed in units of square angstroms, $\overset{\circ}{A}^2$, or 10^{-16} cm^2.) Since the ionization potentials of atoms and molecules range from 3.89 eV (cesium) to 24.6 eV (helium), an optimum electron impact ionization energy is roughly 75 eV. Electron impact ionization cross sections at the maximum are in the neighborhood of one-tenth to one-hundredth of a square angstrom.

Electron attachment. Low energy electrons can be directly attached to molecules which possess an affinity for an electron; that is, if the electron can be bound to the atom or molecule (positive electron affinity) or the lifetime of the ion is long enough to survive mass analysis and ion detection. When free electrons are directly attached to an atom or molecule the resulting anion must lose this excess energy by emitting radiation, or by energy exchanging collisions in the source in order to become stable. In some cases, the anion is large enough to accommodate this excess energy within the molecule and the lifetime of the anion is long enough to survive mass analysis. Dissociative electron attachment can also occur in which a stable fragment negative ion is produced. Unlike the case of positive ions, not all molecules are capable of

forming a stable negative ion (e.g., H_2 N_2, benzene, etc.). However, when electron attachment is possible the cross sections can be very large, often more than 1,000 $\overset{\circ}{A}^2$.

Chemical ionization (CI). Chemical ionization represents a gentle way to ionize molecules in which a proton (positive hydrogen ion) is exchanged between a donor hydride ion and a neutral molecule. This method of ionization is much less violent than electron impact ionization and often results in the parent ion (also containing an extra proton) and minimal fragmentation. Positive ion chemical ionization reactions transfer a proton from a more abundant "reagent" ion (e.g., CH_5^+) to a sample molecule forming a protonated molecule (i.e., a positive parent ion plus a hydrogen atom). Negative ion chemical ionization, in which a parent negative ion (or modified parent ion) is produced, is also employed in mass spectrometry and can be highly sensitive and selective.

Photoionization. Ionization of atoms and molecules by photons takes advantage of the relatively large cross section for ionization at threshold (i.e., for energies just above the ionization potential). With the advent of high power, high photon energy lasers this method of ionization has become very popular. Some lasers have sufficiently high power such that many photons can interact with a single molecule in a sufficiently short time period to induce what is called multiphoton ionization (MPI). Furthermore, if the frequency of the laser can be "tuned" to a resonant electronic frequency of the molecule, the multiphoton ionization can be resonantly enhanced, thereby greatly increasing the ionization probability. This method, called resonantly enhanced multiphoton ionization (REMPI), has the potential for ionizing most of the molecules which happen to be in the laser volume at the time the laser pulse arrives. REMPI mass spectrometry has the advantage of providing further molecular identification since every molecule has a distinct optical absorption spectra. In principle, isomers (different molecules possessing identical mass) can be distinguished using REMPI mass spectrometry.

Desorption ionization. Many samples of interest in mass spectroscopy cannot be easily introduced into the gas phase or often exist in minute quantities on surfaces. A number of methods have been developed which involve the desorption of ions from surfaces. The simplest of these techniques involves laser desorption, in which a pulsed laser is incident on a surface ejecting ions into the gas phase.

In some cases, the sample is mixed with a photoabsorbing substance (matrix) and placed on a surface. The laser light and matrix are selected to interact in such a way as to enhance the ejection of positive or negative ions into the vacuum. This method is called matrix assisted laser desorption ionization or MALDI. MALDI mass spectrometry has become very popular in the biomolecular mass spectroscopy community. Peptides with $M/Z > 300,000$ have been detected.

If the laser power density incident on the surface is significantly increased over that required to desorb surface species, the surface material can be "ablated" into the gas phase. Some of this material will be ionic and can be mass analyzed resulting in a technique called laser ablation mass spectroscopy. Positive or negative ions can also be ejected from surfaces by fast atom bombardment (FAB) or by secondary ion mass spectroscopy (SIMS). One particular method involves the bombardment of a surface with the high energy fragments resulting from the nuclear fission of the radioactive californium atom.

Strong electric fields can also directly "jerk" ions from a contaminated surface. If molecules are placed on a conductive surface containing a large number of very sharp points (e.g., graphite dendrites) those molecules which happen to find themselves sitting on one of these points experience a very large electric field provided that a large electrostatic voltage is applied between the surface and the entrance grid to a mass spectrometer. This extremely large field can induce positive or negative ions to "fly" from the surface to be mass analyzed. This method is called field desorption mass spectroscopy (FDMS) and is found to be very efficient at producing the parent ion in some cases. The method works for both positive and negative ions.

Liquid phase ionization. An important new ionization technique called electrospray has been developed recently by John Fenn at Yale University in which ions preformed in solution are ejected into the vacuum through a small hole in the mass spectrometer vacuum chamber. Multiply charged positive and negative ions are detected for the parent molecules, which may have been protonated or deprotonated in solution. The fact that most of the ions are multiply charged has the advantage that the mass to charge ratio, M/Z, is conveniently reduced into a region where mass analysis is more feasible (e.g., a triply charged ion with a mass corresponding to 15,000 Da is reduced to an M/Z of 5,000).

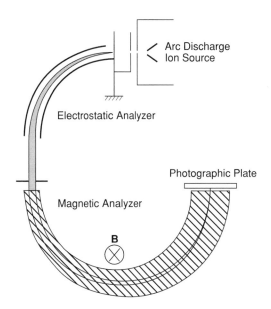

Figure 1 Diagram of the Dempster mass spectrograph. The combination of electrostatic and magnetic focusing is referred to as a double focusing mass spectrometer. Most mass spectrometers employ an exit slit followed by an electron multiplier detector in place of the photographic plate.

Mass Analyzer

The mass analyzer section of the mass spectrometer serves to distinguish the ions according to their mass. The mass separation can occur in space, time, or frequency.

Magnetic analyzers. The operation of the magnetic sector mass analyzer is best illustrated by the original double focusing mass spectrograph built by A. J. Dempster in 1922 and illustrated in Fig. 1. In this instrument, ions are extracted from a discharge ion source operating at high voltage (~20,000 V) with respect to ground and directed through the electric and magnetic field analyzers as shown in Fig. 1. The electric and magnetic fields are arranged to focus ions of the same mass but different kinetic energies along a single line on the photographic plate. The mass spectrum then consists of a series of lines.

The ionic masses are separated in the magnetic field as follows. The force F on an ion of charge Zq traveling with velocity v perpendicular to the magnetic field of strength B is given by

$$F = BZqv. \qquad (1)$$

As a result of this force the ions will follow a circular path of radius R in which the centripetal force $BZqv$ is just balanced by the centrifugal force Mv^2/R, that is,

$$BZqv = \frac{Mv^2}{R},$$ (2)

where v = ion velocity. The kinetic energy of the ions is simply the ionic charge times the ion source voltage with respect to ground ZqV,

$$\frac{Mv^2}{2} = ZqV.$$ (3)

Combining Eqs. (2) and (3) gives a relation between the mass to charge ratio, the magnetic field, and the acceleration voltage,

$$\frac{M}{Z} = q\,\frac{B^2R^2}{2V}.$$ (4)

Thus a mass spectrum is generated by sweeping the magnetic field or source voltage and recording the ion signal at an exit slit corresponding to a fixed radius. There are numerous mass spectrometers employing the double focusing geometry. For example, the Mattauch–Herzog design employs a 31.8° cylindrical electric sector and 90° magnetic sector, while the Nier–Johnson geometry has 90° electric sector and 60° or 90° magnetic sector.

Quadrupole mass filter. The mass analyzer of a quadrupole mass spectrometer (QMS) consists of a time-varying electric field (radiofrequency field) along with dc voltages applied to four long rods arranged as shown in Fig. 2. The work of Wolfgang Paul laid the foundation for the development of the QMS, a subject for which he shared the Nobel Prize in 1989.

Ion trap mass spectrometer (ITMS). The ITMS operates on principles similar to the QMS, however; the ions are trapped in a radio-frequency field provided by hyperbolic-shaped ring electrodes together with hyperbolic-shaped end caps as shown in Fig. 3.

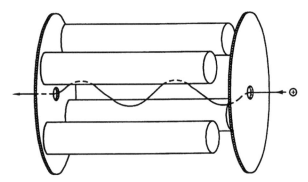

Figure 2 Schematic of the quadrupole mass filter showing the ion trajectories for a selected M/Z and a set of electrostatic voltages and radiofrequencies applied to the poles. The mathematical description of the ion trajectories in this mass spectrometer are quite complicated. A mass spectrum is obtained by sweeping the pole voltages and recording the ion signals exiting the end of the quadrupoles.

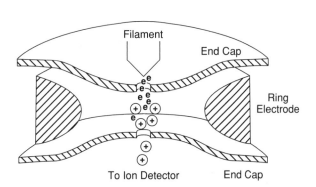

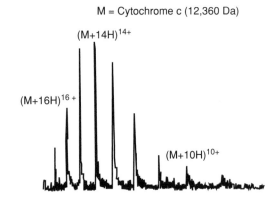

Figure 3 Diagram of the ITMS. Application of a voltage across the two end caps ejects an ion from the trap. A mass spectrum is generated by recording the ejected ion signal as a function of this voltage. The mass spectrum represented is that of electrospray ionization of the molecule cytochrome C showing multiple proton additions.

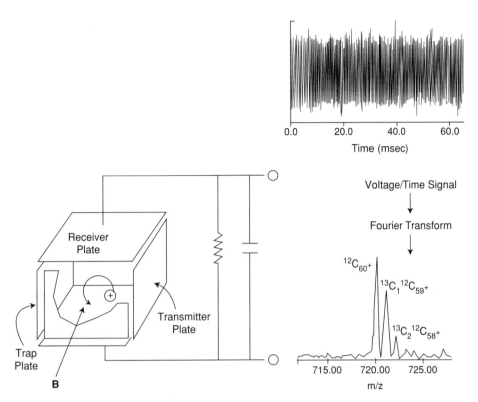

Figure 4 Schematic diagram of the FTMS. The data shown are that for the C_{60} fullerene or "Buckyball" molecule. Although ^{13}C is only 1.1 percent in natural abundance on the earth, the probability that a sixty-atom carbon cluster contains one ^{13}C is 66 percent.

Fourier transform ion cyclotron resonance mass spectrometry (FT/ICR MS or FTMS). The FTMS was originally developed in 1974 by A. G. Marshall and M. B. Comisarow. The method consists of trapping ions in a strong magnetic field (1 to 7 T, 1 T is 10,000 G, the earth's magnetic field is ~ 0.18 G). Ions of mass M having charge Q in this strong magnetic field undergo cyclical motion with an angular frequency ω given by the cyclotron frequency

$$\omega = \frac{QB}{M}. \tag{5}$$

A typical FTMS ion source geometry is shown in Fig. 4. A radiofrequency pulse is first applied to the transmitter plate in order to excite all of the ions of a given mass to move together coherently. The motion of these ions is detected on the receiver plates as a voltage wave exhibiting a complicated wave motion. If ions of only one mass are present, the wave is a sine wave with the frequency given by the cyclotron frequency [Eq. (5)]. When ions of many different masses are present, the wave form (voltage versus time) is complex. The wave form may be converted to the frequency domain using a Fourier transform. Frequency, and therefore mass, can be measured very accurately. Ions can be injected into the FTMS by essentially all of the ionization methods mentioned above.

Time-of-flight mass spectrometry (TOFMS). The time-of-flight mass spectrometer operates on the principle that ions of different mass traveling with the same kinetic energy $Mv^2/2$ can be separated by their time of flight over a given flight path. The method can be illustrated by reference to the Wiley–McLaren geometry ion source as shown in Fig. 5.

Ions of charge Zq created by a pulsed electron beam, for example, are accelerated to an energy Zq times the voltage drop V that the ion experiences in going from its point of creation to the entrance to the ion flight tube. All ions travel with this same ki-

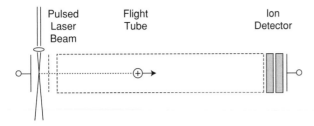

Figure 5 Schematic diagram of a typical linear TOFMS.

netic energy down the flight tube. The ion kinetic energy is equal to the energy gained falling through the acceleration voltage,

$$\frac{Mv^2}{2} = ZqV \qquad (6)$$

and the ion velocity, v, can then be written as

$$v = \sqrt{\frac{2ZqV}{M}}. \qquad (7)$$

The time-of-flight, T, of an ion of mass M along the flight length L is then

$$T = \frac{L}{v} = \frac{L}{\sqrt{2V}} \cdot \sqrt{\frac{M}{Zq}}. \qquad (8)$$

The time-of-flight mass spectrum is proportional to the square root of the mass (or $\sqrt{M/Q}$). The Wiley–McLaren geometry (dual acceleration grids) shown in Fig. 5 helps to compensate for the spread in the kinetic energy of ions created at different positions in the source region, thereby improving the mass resolution. Those ions created farther from the detector will gain more kinetic energy and will overtake and pass those which are created closer to the detector. The Wiley–McLaren geometry allows one to adjust the point at which these ions just cross to be at the detector. Further improvements to the mass resolution are possible using the reflectron TOF mass spectrometer. The TOF mass spectrometer is becoming a major part of the mass spectrometer community as a result of its ability to analyze at high mass. Also, TOF mass spectrometers are often used with pulsed lasers as shown in Fig. 5.

The mass spectrometer played a central role in the Manhattan Project of the United States during World War II in the form of the Calutron mass separators. These mass separators were very similar to the mass spectrometer shown in Fig. 1, but were designed to collect large ion currents of the minor uranium isotope ^{235}U needed for construction of the atom bomb.

See also: ELECTRON; ION; IONIZATION; IONIZATION POTENTIAL; LINE SPECTRUM; SPECTROPHOTOMETRY; SPECTROSCOPY, MASS

Bibliography

COMISAROW, M. B., and MARSHALL, A. G. "Fourier Transform Ion Cyclotron Resonance Spectroscopy." *Chem. Phys. Lett.* **25,** 282 (1974).

COTTER, R. J. *Time-of-Flight Spectrometry* (American Chemical Society, Washington, DC, 1994).

MCLAFFERTY, F. W., and TURCEK, F. *Interpretation of Mass Spectra* (University Science Books, Mill Valley, CA, 1993).

WATSON, J. T. *Introduction to Mass Spectroscopy,* 2nd ed. (Raven Press, Lancaster, CA, 1985).

WILEY, W. C., and MCLAREN, I. H. "Time-of-Flight Mass Spectrometer With Improved Resolution." *Rev. Sci. Inst.* **26,** 1150 (1955).

ROBERT N. COMPTON

MASS SPECTROSCOPY

See SPECTROSCOPY, MASS

MATERIALS SCIENCE

The emergence of materials science as a scientific discipline is a post–World War II phenomenon that grew out of long-term roots in metallurgy and ceramics. The study of metallurgy goes back to ancient times and is intertwined with ancient art forms, spiritual rites, and the fabrication of utensils for practical use. Ceramics, which developed independently from metallurgy, and also contributed to art, culture, and utensils of practical use by ancient civilizations.

During World War II, metallurgists and ceramists became involved with solid state physicists and other scientists in large interdisciplinary projects, such as the Manhattan project, and were much influenced by the role that science could play in controlling and enhancing the properties of materials. Emerging from this experience was a young generation of metallurgists and ceramists with a broader vision of materials, seeing common issues and principles across different classes of materials. These considerations, and the appearance of new classes of interesting materials such as polymers and semiconductors, led to the creation of the field of materials science in the 1960s.

While metals and ceramics remain as members of the major classes of materials, they are joined by polymers, composites, and semiconductors as important materials classes examined under the scope of materials science. For many practitioners, the science of materials is closely coupled with applications and the processing of materials to achieve desired properties, and for this reason, it is common for practitioners to describe this field more broadly as materials science and engineering.

Characteristics of the Field

The intellectual framework that binds materials science together is the structure—properties issues common to the five major classes of materials enumerated above: metals, ceramics, polymers, composites, and semiconductors. The field of materials science is highly interdisciplinary, and couples closely to other fields of materials research, including physics, chemistry, geology, electrical engineering, mechanical engineering, and chemical engineering, among others.

Some of the fundamental structural issues common to all materials classes at the atomic level include atomic bonding, such as ionic, covalent, and metallic bonding, and the packing and coordination of atoms in the condensed phase. Periodic structures such as crystalline phases are found in various materials classes, and are studied in terms of their ideal crystalline structures, including examples of common space groups, crystallographic determinations, densities of crystalline structures, packing and atomic coordination, and anisotropy. Since the properties of materials are highly sensitive to mate-

rial defects, materials science focuses on the study of various types of materials defects, such as point defects, interstitials, deformation, edge and screw dislocations, slip, plastic deformation, and grain boundaries. Defects are studied from both the viewpoint of thermodynamics and kinetics, and the role of solid state diffusion in the transport of defects is widely investigated. Since many important materials are highly disordered and noncrystalline, materials science is concerned with the structural properties of glasses, amorphous solids, polymers, elastomers, and rubbers.

The explicit properties of materials are often linked to their microstructure, which describes the local structure of materials on a nanometer scale. Since the microstructure of materials tends to be dominated by the coexistence of numerous stable and metastable phases, materials science strongly focuses on gaining an understanding of phase diagrams and phase equilibrium between various solid and liquid phases. Because of the numerous components present in specially designed materials for the optimization of specific properties, phase diagrams can become complex. Kinetic studies of microstructure tend to emphasize the characteristics of phase nucleation, and whether the nucleation is homogeneous or heterogeneous on the desired spatial scale.

Since materials science focuses on structure-property relations, the properties of materials form a cornerstone to materials research. Most important are the mechanical properties, although attention is also given to the study of the electrical, optical, dielectric, magnetic, and thermal properties of materials. Study of the mechanical properties includes attention to hardness, elasticity, deformation, ductile and brittle fracture, stress-strain relations, fatigue fracture, and time-dependent issues (e.g., environmental effects and creep in metals and ceramics). Ensuring superior mechanical reliability, failure prevention, decreased friction and wear, and nondestructive testing are some of the current topics of interest in materials processing. Electrical, optical, and dielectric properties studies are most popular in semiconductors, ionic conductors, and conducting polymers; such investigations often make close contact with condensed matter physics on scientific issues and with electrical engineering on device fabrication, performance, and modeling. These property measurements are, however, powerful characterization tools with much broader interest to materials science.

The basic magnetic properties are significant to materials science, particularly for materials that are ferromagnetic, antiferromagnetic, and ferrimagnetic. Such materials are of great importance to magnetic applications, including permanent magnets, transformer cores, and magnetic storage disks and heads. Superconducting materials have recently started to attract increased attention, with the growth of the superconductivity industry for superconducting magnets and with the appearance of potential new classes of devices exploiting high-T_c superconductors.

Materials science has become increasingly concerned with processing issues and the use of composites and artificially structured materials to achieve specific desired properties. Examples of composite materials include fiber-matrix composites, metal-matrix composites, polymer-matrix composites, and ceramic-matrix composites. Of particular interest are improvements in elastic modulus, strength, hardness, ductility, and the suppression of fracture, fatigue, creep, and wear. Artificially structured multilayers and superlattices are of great interest for electronic, semiconductor, optoelectronic, and magnetic applications. However, for optimum performance, issues of epitaxial growth, lattice matching, and the avoidance of misfit dislocations must be addressed, thus leading to strong coupling between materials scientists and electrical engineers in the semiconducting electronics, optoelectronics, and magnetics industries.

Techniques

The strength of the materials science field comes from interdisciplinary research. In this context, materials scientists contribute strongly to materials processing, which involves materials preparation to achieve desired microstructures and properties. In addition, materials scientists in interdisciplinary teams are the ones who carry out most of the materials characterization, including both structural and properties measurements.

One dominant technique is x-ray determinations of crystal structures, of the coexistence of various solid state phases and microstructure, small angle x-ray scattering for the study of porosity of materials, x-ray fluorescence studies for compositional analysis, and extended x-ray absorption fine structure (EXAFS) analysis for the determination of near-neighbor atomic distances. For analysis of light elements, neutron diffraction is often used. Material scientists are the members of a materials research team who focus on studies of microstructure. Scanning electron microscopy (SEM) and transmission electron microscopy (TEM) provide sensitive tools for these studies, from submicron length scales all the way to atomic resolution. SEM studies are especially useful for providing a big picture of the microstructure and allowing various mechanical tests to be performed in situ on well-characterized samples, while TEM provides access to bright field images, selected area diffraction patterns, and dark field images, allowing the identification of microstructure with specific locations on the sample. Scanning tunneling microscopy (STM) has made possible the study of surface topology, surface structures, and surface microstructure with unprecedented resolution and discrimination at the atomic level. Atomic force microscopy (AFM) permits related studies on nonconducting materials, while scanning tunneling spectroscopy (STS) allows measurements of the electronic properties through high-resolution current-voltage measurements. New materials characterization tools of increasing resolution and discrimination are now becoming available at a rapid rate.

Materials scientists specialize in the mechanical characterization of materials, studies of surfaces and interfaces, and a variety of electrical, optical, dielectric, elastic, and other properties characterization. Because of the wide variety of characterization tools needed to characterize materials under investigation for structure-properties relations, materials characterization is usually carried out in shared materials research facilities which acquire and maintain the wide variety of sophisticated and expensive equipment needed to characterize materials. Such facilities are commonly available in universities, industry, and national laboratories. In addition, national laboratories operate the most sophisticated of the user facilities for materials characterization, such as synchrotron radiation, neutron scattering, small angle scattering, and to some extent high-resolution TEM facilities. These facilities are all becoming important tools for state-of-the-art materials science research.

Conclusion

Materials provide an enabling technology for many fields of science and technology. As materials scientists become increasingly successful with the discovery of new types of materials, the manipulation of chemical constituents to vary materials properties, the control of microstructure and the preparation of composites, multilayers, and other deliberately structured materials, the goal of materials tailoring for specific applications becomes closer to reality. Such materials research in tailoring materials is one current goal of materials science. Increased understanding of basic materials science issues and better characterization tools are essential ingredients for reaching this goal.

See also: CREEP; CRYSTAL DEFECT; CRYSTAL STRUCTURE; CRYSTALLOGRAPHY; DIELECTRIC PROPERTIES; MAGNET; METAL; METALLURGY; PHASE TRANSITION; POLYMER; SCANNING TUNNELING MICROSCOPE; SCANNING PROBE MICROSCOPIES; SEMICONDUCTOR

Bibliography

CHRISTIAN, J. W. *The Theory of Transformations in Metals and Alloys: An Advanced Textbook in Physical Metallurgy*, 2nd ed. (Pergamon, Oxford, Eng., 1975).

KINGERY, W. D.; BOWEN, H. K.; and UHLMANN, D. R. *Introduction to Ceramics*, 2nd ed. (Wiley Interscience, New York, 1976).

SUTTON, A. P., and BALLUFFI, R. W. *Interfaces in Crystalline Materials* (Clarendon, Oxford, Eng., 1995).

VAN VLACK, L. H. *Elements of Materials Science and Engineering*, 6th ed. (Addison-Wesley, Reading, MA, 1989).

MILDRED S. DRESSELHAUS

MATHEMATICAL CONSTANTS

In principal, every number is a mathematical constant, but only the most useful ones are given names. Every integer has a name, and all, particularly the ones near zero, are useful mathematical constants. Likewise, every rational number p/q (where p and q are integers with $q \neq 0$) has a "name" (the same p/q) and is a mathematical constant.

Of course, not all numbers are rational. Of the irrationals, simple square roots like $\sqrt{2}$ and $\sqrt{3}$ occur often. For instance, $\sqrt{2}$ is the length of a diagonal of a unit square, and $\sqrt{3}$ is the length of a diagonal of a unit cube. Another interesting irrational number is the golden ratio: $\phi = (1 + \sqrt{5})/2$. If a line segment ℓ is divided into two parts a and b (with $\ell = a + b$, $a > b$) such that the ratio of the whole to the larger part is the same as the ratio of the larger part to the smaller, then the ratio $\phi = \ell/a = a/b$ is the golden ratio. The golden ratio, called the divine proportion by Johannes Kepler, appears in mathematics and nature with surprising regularity.

Square roots are examples of algebraic numbers, which by definition satisfy some algebraic equation

$$a_n x^n + a_{n-1} x^{n-1} + \ldots + a_1 x + a_0 = 0 \ (n \geq 0, a_n \neq 0),$$

where $a_n, \ldots a_0$ are integers. Numbers that are not algebraic are called transcendental. The two most important are π and e.

The number π, which was known to the ancients, is the ratio of the circumference to the diameter of a circle. Consequently, π is intimately related to the trigonometric functions and appears in formulas like $\sin(\pi/6) = 1/2$, $\tan(\pi/4) = 1$, and so on. The decimal expansion of π (3.14159 …) has been carried out to hundreds of millions of places, with no discernible sign of regularity. Some interesting expansions for π are

$$\frac{\pi}{4} = 1 - \frac{1}{3} + \frac{1}{5} - \frac{1}{7} + \ldots,$$

$$\frac{\pi^2}{6} = \frac{1}{1^2} + \frac{1}{2^2} + \frac{1}{3^2} + \ldots,$$

and

$$\frac{\pi}{2} = \frac{2}{1} \cdot \frac{2}{3} \cdot \frac{4}{3} \cdot \frac{4}{5} \cdot \frac{6}{5} \cdot \frac{6}{7} \cdot \frac{8}{7} \cdot \frac{8}{9} \ldots.$$

The number π occurs throughout analytic geometry and analysis. The circumference and area of a circle of radius r are $2\pi r$ and πr^2. The surface area and volume of a sphere of radius r are $4\pi r^2$ and $\frac{4}{3}\pi r^3$. Analogous formulas hold for higher dimensional spheres. The number π is seen frequently in tables of definite integrals whenever trigonometric functions occur and even when they are not evident, as in $\int_0^\infty dx(1 + x^2)^{-1} = \pi/4$, since the indefinite integral of $(1 + x^2)^{-1}$ is $\tan^{-1}(x)$. More surprising is its appearance in the Gaussian integral $\int_0^\infty dx e^{-x^2} = \sqrt{\pi}/2$. (A geometric basis for this occurrence of π is evident in Carl Friedrich Gauss's method for evaluating this integral.) A related occurrence is in the gamma function $\Gamma(n) = \int_0^\infty dx\, x^{n-1} e^{-x}$ as $\Gamma(1/2) = \sqrt{\pi}$. Also unexpected is the occurrence of π in integrals like $\int_0^1 dx\, \ln(1 - x)/x = -\pi^2/6$ (which is related to the series for $\pi^2/6$ given above).

Another use of π is in the measure of angles. The natural unit of angular measure, the radian (or rad), is defined so that the angle in radians θ subtended by a circular arc of length s is the ratio of the arc length to the radius r: $\theta = s/r$. Consequently, 2π radians is the angle describing a complete rotation, and π rad $= 180°$.

The number e, defined as the sum of the rapidly converging series

$$e = 1 + \frac{1}{1!} + \frac{1}{2!} + \frac{1}{3!} + \dots + \frac{1}{n!} + \dots,$$

[where n factorial is $n! = n \cdot (n-1) \dots 3 \cdot 2 \cdot 1$] or as the limit

$$e = \lim_{n \to \infty} \left(1 + \frac{1}{n}\right)^n$$

has the value $e = 2.71828\ldots$. The notation for e was established by Leonhard Euler.

If we consider the exponential function $f(x) = a^x$, where a is some constant, then the slope df/dx of f is proportional to f itself, and the proportionality constant depends on a. The number of e is the unique constant for which the slope of the exponential function equals the function itself:

$$\frac{d}{dx} e^x = e^x.$$

If we expand e^x in a power series, and require the derivative relation above (using $dx^n/dx = nx^{n-1}$), then this definition of e tells us that the series must be

$$e^x = 1 + \frac{x}{1!} + \frac{x^2}{2!} + \frac{x^3}{3!} + \dots + \frac{x^n}{n!} + \dots.$$

For $x = 1$, this is Euler's e.

The number e forms the base of the natural logarithms. That is, $y = \ln(x)$ when $e^y = x$. The natural logarithm has the agreeable property that $dy/dx = 1/x$ with no constant factors. (This follows from $de^y/dy = e^y$ and the theorem on derivatives of inverse functions.) Of course, the usual properties of logarithms, such as $\ln(ab) = \ln(a) + \ln(b)$, hold in this base as in any other.

Several other constants are worthy of note. The Euler constant $\gamma_E = 0.57722\ldots$ is defined through

$$\gamma_E = \lim_{n \to \infty} \left\{ 1 + \frac{1}{2} + \frac{1}{3} + \dots + \frac{1}{n} \ln n \right\}.$$

Catalan's constant is $G = \sum_{n=0}^\infty (-1)^n (2n + 1)^{-2} = 0.91597\ldots$. These two occur frequently in analysis, as in $\int_0^\infty dx\, e^{-x} \ln(x) = -\gamma_E$ and $\int_0^1 dx\, \tan^{-1}(x)/x = G$. Special values of the polylogarithm

$$Li_n(x) = \frac{x}{1^n} + \frac{x^2}{2^n} + \frac{x^3}{3^n} + \dots,$$

such as $\ln(2) = -\mathrm{Li}_1(-1)$ and $\zeta(3) = \mathrm{Li}_3(1) = 1.20206\ldots$ (where $\zeta(n) = \mathrm{Li}_n(1)$ is the Riemann zeta function), appear frequently in the analysis of physical problems.

Not all of the useful mathematical constants are real numbers. In particular, the unit imaginary number $i = \sqrt{-1}$ is of fundamental importance. Properties of the complex numbers $a + ib$ (where a and b are real), such as $(a + ib)(c + id) = (ac - bd) + i(ad + bc)$, are based on $i^2 = -1$. The five most important mathematical constants are combined in Euler's relation $e^{i\pi} + 1 = 0$.

See also: AVOGADRO NUMBER; BOLTZMANN CONSTANT; COSMOLOGICAL CONSTANT; DIELECTRIC CONSTANT; ELASTIC MODULI AND CONSTANTS; FINE-STRUCTURE CONSTANT; GAS CONSTANT; GRAVITATIONAL CONSTANT; HUBBLE CONSTANT; PLANCK CONSTANT; RYDBERG CONSTANT

Bibliography

BOYER, C. B. *A History of Mathematics* (Wiley, New York, 1968).

COURANT, R., and ROBBINS, H. *What Is Mathematics?* (Oxford University Press, London, 1941).

GRADSHTEYN, I. S., and RYZHIK, I. M. *Table of Integrals, Series, and Products* (Academic Press, New York, 1980).

HUNTLEY, H. E. *The Divine Proportion* (Dover, New York, 1970).

LEWIN, L. *Polylogarithms and Associated Functions* (Elsevier, New York, 1981).

GREGORY S. ADKINS

MATRIX MECHANICS

The quantum theory, developed by Max Planck, Albert Einstein (better known for his work on relativity theory), and Niels Bohr in the early years of the twentieth century, had run its course by 1925. The "old" quantum theory, as it is now known, could account for many observed properties of atoms and molecules by supplementing classical (Newtonian) mechanics with strange "quantization rules." However, these rules seemed arbitrary, and the theory did not adequately handle many complex problems in the microscopic domain of atoms. There was a clear need to devise a new kind of mechanics, radically generalizing classical mechanics, but meshing with it when dealing with macroscopic bodies, which obey Newton's laws of motion (Bohr's correspondence principle).

The impasse, which had frustrated scientists of the older generation for years, was finally broken with the remarkable discovery of matrix mechanics by the twenty-four year old Werner Heisenberg, who was a junior faculty member at the University of Göttingen, Germany. Within the following two years, wave mechanics (created by Erwin Schrödinger) joined matrix mechanics as an alternate way of dealing with the problems of atomic physics. Shortly thereafter quantum mechanics (through the work of P. A. M. Dirac) became the overarching and enduring theoretical framework for explaining physics at the microscopic level. Having initially served as a gateway to quantum mechanics, matrix mechanics now is best thought of as an indispensable part of that general theory. Its strongest point is its close similarity to the laws of classical mechanics.

Heisenberg came upon the concepts of matrix mechanics by looking for a theory that is more logically consistent than the patchwork quilt of the old quantum theory, resembles mathematically the equations of classical (Newtonian) mechanics, and focuses on quantities that are accessible to observation and measurement. Ironically, these objectives led to a theory that is mathematically far more abstract than any previous physical theory.

The laws of matrix mechanics can indeed be cast in a form that makes them look like the equations of classical physics. For example, in matrix mechanics the motion of a particle is described by a formula that emulates Newton's second law:

$$\frac{d\mathbf{p}}{dt} = \mathbf{F}, \tag{1}$$

but in the new mechanics, physical quantities like the momentum **p**, and the force **F**, denote *matrices* instead of just ordinary real numbers.

How to Work with Matrices

A matrix is a mathematical quantity that was invented in the nineteenth century to express linear transformations simply and concisely. As an example, consider a vector in a plane, and let x and y be its two components. A homogeneous linear transformation is defined by four constants, a_{11}, a_{12}, a_{21}, a_{22} which map (x, y) into a new vector (x', y') with components

$$\begin{aligned} x' &= a_{11}x + a_{12}y, \\ y' &= a_{21}x + a_{22}y. \end{aligned} \tag{2}$$

A second similar mapping takes (x', y') into (x'', y''):

$$\begin{aligned} x'' &= b_{11}x + b_{12}y, \\ y'' &= b_{21}x + b_{22}y. \end{aligned} \tag{3}$$

By substitution, we see that (x, y) can be directly related to (x'', y'') according to the formulas

$$\begin{aligned} x'' &= (b_{11}a_{11} + b_{12}a_{21})x + (b_{11}a_{12} + b_{12}a_{22})y \\ &= c_{11}x + c_{12}y, \\ y'' &= (b_{21}a_{11} + b_{22}a_{21})x + (b_{21}a_{12} + b_{22}a_{22})y \\ &= c_{21}x + c_{22}y. \end{aligned} \tag{4}$$

These equations suggest that we write the four coefficients defining these transformations in a square array, called a matrix. For the transformation in Eq. (2) we write the matrix

$$\mathbf{A} = \begin{pmatrix} a_{11} & a_{12} \\ a_{21} & a_{22} \end{pmatrix}. \tag{5}$$

The entries a_{ij} are called the matrix elements of **A**. Analogous expressions hold for the matrices **B** and **C** representing transformations in Eqs. (3) and (4).

The rule spelled out in Eq. (4), which gives the matrix elements of **C** in terms of those of **A** and **B**, is used to define what we mean by multiplying two matrices:

$$\mathbf{C} = \mathbf{BA} \tag{6}$$

or, written out explicitly

$$\begin{pmatrix} c_{11} & c_{12} \\ c_{21} & c_{22} \end{pmatrix} = \begin{pmatrix} b_{11}a_{11} + b_{12}a_{21} & b_{11}a_{12} + b_{12}a_{22} \\ b_{21}a_{11} + b_{22}a_{21} & b_{21}a_{12} + b_{22}a_{22} \end{pmatrix}$$

$$= \begin{pmatrix} b_{11} & b_{12} \\ b_{21} & b_{22} \end{pmatrix} \begin{pmatrix} a_{11} & a_{12} \\ a_{21} & a_{22} \end{pmatrix}. \tag{7}$$

The most striking feature of this strange multiplication rule is that it is noncommutative, which means that generally

$$\begin{pmatrix} a_{11} & a_{12} \\ a_{21} & a_{22} \end{pmatrix} \begin{pmatrix} b_{11} & b_{12} \\ b_{21} & b_{22} \end{pmatrix} \neq \begin{pmatrix} b_{11} & b_{12} \\ b_{21} & b_{22} \end{pmatrix} \begin{pmatrix} a_{11} & a_{12} \\ a_{21} & a_{22} \end{pmatrix}. \tag{8}$$

The order of the factors in the matrix product *does* make a difference. To appreciate what is involved here, the reader might work out a product of two matrices, for example

$$\mathbf{A} = \begin{pmatrix} 2 & 0 \\ -1 & 3 \end{pmatrix} \tag{9a}$$

and

$$\mathbf{B} = \begin{pmatrix} 1 & 1 \\ 4 & -2 \end{pmatrix} \tag{9b}$$

in either order. Finally, adding two matrices is done simply by the natural rule

$$\begin{pmatrix} a_{11} & a_{12} \\ a_{21} & a_{22} \end{pmatrix} + \begin{pmatrix} b_{11} & b_{12} \\ b_{21} & b_{22} \end{pmatrix} = \begin{pmatrix} a_{11} + b_{11} & a_{12} + b_{12} \\ a_{21} + b_{21} & a_{22} + b_{22} \end{pmatrix}. \tag{10}$$

A rotation in the xy plane about the origin by an angle θ is a special example of a two-dimensional linear transformation. It is represented by a matrix

$$\begin{pmatrix} a_{11} & a_{12} \\ a_{21} & a_{22} \end{pmatrix} = \begin{pmatrix} \cos\theta & -\sin\theta \\ \sin\theta & \cos\theta \end{pmatrix}. \tag{11}$$

With a little knowledge of trigonometry, it is easy to verify that the result of performing two successive rotations is properly given by the law (6) of matrix multiplication. (The rules given above for 2×2 matrices can also be generalized for larger square matrix arrays.)

Rules of Matrix Mechanics

Heisenberg and his collaborators Max Born and Pascual Jordan proposed a set of postulates for physical quantities, such as the coordinate $\mathbf{x}$, the momentum $\mathbf{p}$, and the energy $\mathbf{E}$, all of which are now represented by matrices (with infinitely many rows and columns instead of just two).

I. The energy is expressed in terms of $\mathbf{x}$ and $\mathbf{p}$ just as in classical mechanics. For the important example of the linear harmonic oscillator this means

$$\mathbf{E} = \frac{\mathbf{p}^2}{2m} + \frac{m\omega^2}{2}\mathbf{x}^2. \tag{12}$$

II. Coordinate and momentum matrices do not commute and must satisfy a quntization condition

$$\mathbf{xp} - \mathbf{px} = i\hbar\mathbf{1}, \tag{13}$$

where $\hbar$ denotes Planck's constant and i is the imaginary unit $\sqrt{-1}$. The identity matrix $\mathbf{1}$ is a special case of a diagonal matrix

$$\begin{pmatrix} a_{11} & 0 & 0 & \cdots \\ 0 & a_{22} & 0 & \cdots \\ 0 & 0 & a_{33} & \cdots \\ \cdots & \cdots & \cdots & \cdots \end{pmatrix}. \tag{14}$$

For the identity matrix $\mathbf{1}$, all diagonal elements are equal to unity.

III. Now, if we can find two matrices $\mathbf{x}$ and $\mathbf{p}$ that satisfy postulate II [Eq. (13)] and if the matrix $\mathbf{E}$ calculated from them by application of Eq. (12) is a diagonal matrix, then the diagonal elements of $\mathbf{E}$ are the set of possible outcomes of a measurement of the energy E for the harmonic oscillator with mass m and frequency ω.

The problem posed in postulate III is solved for the harmonic oscillator by the following coordinate and momentum matrices:

$$\mathbf{x} = \sqrt{\frac{\hbar}{2m\omega}} \begin{pmatrix} 0 & \sqrt{1} & 0 & 0 & \cdots \\ \sqrt{1} & 0 & \sqrt{2} & 0 & \cdots \\ 0 & \sqrt{2} & 0 & \sqrt{3} & \cdots \\ 0 & 0 & \sqrt{3} & 0 & \cdots \\ \cdots & \cdots & \cdots & \cdots & \cdots \end{pmatrix},$$

(15)

$$\mathbf{p} = \sqrt{\frac{\hbar m\omega}{2}} \begin{pmatrix} 0 & -i\sqrt{1} & 0 & 0 & \cdots \\ i\sqrt{1} & 0 & -i\sqrt{2} & 0 & \cdots \\ 0 & i\sqrt{2} & 0 & -i\sqrt{3} & \cdots \\ 0 & 0 & i\sqrt{3} & 0 & \cdots \\ \cdots & \cdots & \cdots & \cdots & \cdots \end{pmatrix}.$$

It is easy to check that multiplication of $\mathbf{x}$ and $\mathbf{p}$ satisfies condition in Eq. (13). When these matrices are substituted in Eq. (12) and the rules of matrix algebra invoked, the energy matrix is found to be the infinite dimensional diagonal matrix:

$$\mathbf{E} = \begin{pmatrix} \frac{1}{2}\hbar\omega & 0 & 0 & \cdots \\ 0 & \frac{3}{2}\hbar\omega & 0 & \cdots \\ 0 & 0 & \frac{5}{2}\hbar\omega & \cdots \\ \cdots & \cdots & \cdots & \cdots \end{pmatrix}.$$

(16)

The numbers in the diagonal are the discrete allowed (quantized) energy values for a harmonic oscillator, a result known to be correct from the old quantum theory, but obtained there only by *ad hoc* arguments. Note that the lowest possible energy of an oscillator in quantum physics is $\frac{1}{2}\hbar\omega$, not zero as in classical physics.

This was only the first of many triumphs of matrix mechanics, which provided a consistent and satisfactory basis for the new (quantum) mechanics. In particular, matrix mechanics contained a prescription for the calculation not only of atomic energy levels, but also of the intensities of spectral lines emitted in quantum transitions between two levels. Matrix elements of various physical quantities now belong to the everyday language of physics. In 1927 Heisenberg deduced his uncertainty principle from the commutation relation in Eq. (13).

See also: ATOMIC PHYSICS; BOHR, NIELS HENRIK DAVID; BORN, MAX; CORRESPONDENCE PRINCIPLE; ENERGY LEVELS; HEISENBERG, WERNER KARL; OSCILLATOR, HARMONIC; QUANTIZATION; QUANTUM; QUANTUM MECHANICS; QUANTUM THEORY, ORIGINS OF; SCHRÖDINGER, ERWIN; WAVE MECHANICS

Bibliography

FEYNMAN, R. P.; LEIGHTON, R. E.; and SANDS, M. *The Feynman Lectures,* Vol. 3 (Addison-Wesley, Reading, MA, 1965).

JAMMER, M. *The Conceptual Development of Quantum Mechanics* (McGraw-Hill, New York, 1966).

JORDAN, T. E. *Quantum Mechanics in Simple Matrix Form* (Wiley, New York, 1986).

MERZBACHER, E. *Quantum Mechanics,* 2nd ed. (Wiley, New York, 1970).

EUGEN MERZBACHER

MAXWELL, JAMES CLERK

b. Edinburgh, Scotland, June 13, 1831; *d.* Cambridge, England, November 5, 1879; *electromagnetic theory, kinetic theory, statistical mechanics, control theory, colorimetry, theory of Saturn's rings.*

Maxwell's investigations in statistical mechanics and electromagnetism make his the founding mind of modern theoretical physics, which originated not as is often thought around the year 1900, but somewhere much further back in the 1850s and 1860s.

Maxwell's father, born John Clerk, took the name Maxwell after inheriting as younger son a 2,000-acre estate in Galloway (southwest Scotland) that had come into his family, the Clerks of Penicuik, through intermarriage in the eighteenth century with the Maxwells of Middlebie, illegitimate descendants of the ninth Lord Maxwell. This property, Glenlair, descended to Maxwell, who did much of his scientific work there.

The Clerks had long been prominent. The second John Clerk was chief Scottish financial officer of the 1707 Union with England. Maxwell's uncle George Clerk, M.P., sixth baronet, was a cabinet minister under Robert Peel. The family also had notable artistic and intellectual gifts. Maxwell's father, though trained as an attorney, occupied himself mainly in inventions. He was a Fellow of the Royal Society of Edinburgh. His one published paper, written in 1831, the year Maxwell was born, was a proposal for an automatic feed printing press. He was also an architect. The kirk at Parton overlooking Loch Ken, where he, Maxwell, and their wives are buried is his design. These interests he passed to his son, as appears from Maxwell's ingenious scientific instruments and the design of the old Cavendish Laboratory at Cambridge.

Marrying late, Maxwell's parents left Edinburgh for Galloway in 1826. They had a daughter, who died young, making Maxwell in effect an only child. Mrs. Maxwell, born Frances Cay, was a woman of great force. She personally educated her son, imparting to him her own deep love of English literature and the Bible. Her death at forty-eight in 1839 when he was eight was devastating. Eventually he was sent to Edinburgh for schooling, and for several uneasy years Maxwell shuttled back and forth between Glenlair and two rival Edinburgh aunts.

Maxwell's first blossoming was as a poet, and his first publication at fourteen was not scientific but a poem printed in the *Edinburgh Courant* in 1846. Almost immediately thereafter, however, he wrote a clever paper on oval curves and through his father met many of Edinburgh's leading scientists. At sixteen, the normal age of entry to Scottish universities, he began attending Edinburgh University, where he remained three years before transferring to Cambridge. He read voraciously, being particularly influenced by two men, the physicist-geologist James D. Forbes in whose laboratory he worked, and the metaphysician Sir William Hamilton. He wrote two more scientific papers and with Forbes began studying color vision.

Entering Cambridge in 1850, Maxwell rapidly acquired a reputation for effortless universal genius, an opinion he unwisely shared. In the intensely competitive Mathematical Tripos examination he ended in second place as Second Wrangler, behind Edward J. Routh, a painful blow. Cambridge gave him much, however, in college friendships and intellectual guidance. Withal he maintained a personal independence, not inaptly symbolized by his lifelong refusal to soften his thick Gallovidian accent.

In 1856, Maxwell was elected professor at Marischal College, Aberdeen. While there he married Katherine Mary Dewar, who later joined in several of his experiments. She was seven years his senior; they had no children. In 1860, he moved to King's College, London, from which in 1865 he re-

tired to Glenlair in order to enlarge his house and compose his *Treatise on Electricity and Magnetism*. Reviving his connection with Cambridge in 1866, he was appointed in 1871 first Cavendish professor of experimental physics, founding the Cavendish Laboratory. Eight years later at forty-eight, full of vigor and new ideas, he unexpectedly died of cancer, leaving as intellectual heritage a fine laboratory, five books, and some 120 scientific papers.

Color Vision and Color Photography

Maxwell created the science of colorimetry, took the first color photograph, established Thomas Young's three-receptor theory of color vision, proved that color blindness originates in the ineffectiveness of one or more of Young's receptors, and with his wife made other notable visual discoveries. All this began in 1849 when he and Forbes found on spinning a disk with adjustable colored sectors, that no combination of the artist's primary pigments (red, blue, and yellow) gave white. The correct mix was red, blue, and green.

For accurate colorimetry, Maxwell invented his "colour box." Picture white light from point A passing through a prism to form a spectrum. Now reverse this. Put the eye at A and pass three beams of white light back through the prism from slits at different points in the spectrum. The eye will see those colors, adjustable in intensity by varying the width of the slits. Another was a "double monochromator," using not white light but colored light from a second prism to illuminate the slits. Not until the 1920s were data obtained surpassing Maxwell's.

Maxwell's color photograph (1861), projected through red, green, and blue filters, was amusing as well as impressive. Taken with plates insensitive to red, it should not have worked. The "red" image was obtained with ultraviolet light.

Saturn's Rings

The topic of the 1856 Adams' Prize at Cambridge, announced a year after Maxwell's failure to become Senior Wrangler, was the structure of Saturn's rings. Maxwell's winning essay (1859) established that among all imaginable forms, solid, liquid, or gaseous, only rings comprising large numbers of separate, mutually attracting and colliding bodies would be stable. Final spectacular confirmation of this came with the flyby of NASA's Voyager spacecraft in 1981.

Maxwell's four-year study of Saturn seems a digression, but it counted for much. Solving this "stiff but curious" problem was a triumph and a settling of his Tripos account with Cambridge.

Electromagnetic Theory

Writing in 1950, Robert Andrews Millikan set Maxwell's *Treatise on Electricity and Magnetism* alongside Newton's *Principia*, "the one creating our modern mechanical world, the other our modern electrical world." Rightly. Maxwell's equations have outlasted Newton's; the *Treatise* remains brilliantly alive to this day.

Preceding Maxwell's was another all-embracing electrical theory, Wilhelm Weber's (1845). It rested on two principles: (1) electric currents comprise two equal, opposed flows of oppositely charged particles; and (2) between the particles is an attractive-repulsive force, acting instantaneously but varying in magnitude with the particles' relative velocities and relative accelerations along the line joining them. Odd as it now seems, Weber's theory embodied one vital perception. The ratio of magnetic to electric forces constituted a velocity, seemingly arbitrary, to be fixed experimentally.

Maxwell found Weber's theory repellent. Guided by William Thomson (the future Lord Kelvin), he read Michael Faraday. Here was another world: many experiments, no equations, and rich intuitions about lines of magnetic and electric force in space from which Maxwell in 1855 began framing his own theory. After that for six years nothing. Saturn, marriage, gas theory, colorimetry, and the move to London were enough. Then swiftly, he wrote three long papers, each quite different, that utterly revolutionized electromagnetism—and physics.

Besides thinking geometrically, Faraday had made a daring physical conjecture. Suppose lines of force have an innate tendency to shorten themselves and repel each other sideways. That would explain magnetic forces. Developing this idea, Maxwell in 1861 pictured space as filled with an "ether" of tiny vortex molecules separated by electrical "idle wheels." Having thus explicated magnetism and electromagnetism he extended the model to electrostatics by making his ether elastic. Now elastic bodies transmit waves. Maxwell's wave velocity c was just twice Weber's v, now known from measurements by Friedrich Wilhelm Georg Kohlrausch. To his amazement, Maxwell found that c was exactly the velocity of light. Light and electromagnetism must be one and the same.

How strange that no one before Maxwell, not Weber, not Kohlrausch, not Thomson, not even Gustav Kirchhoff, who came nearest, seized on Weber's ratio. Still, as Charles J. Monro said (October 1861): "A few such results are wanted before you can get people to think that every time an electric current is produced, a little file of particles is squeezed along between rows of wheels."

Fundamental to each of Maxwell's next two papers (1863, 1864) was his work for the British Association Committee on Electrical Standards chaired by Thomson. To this committee we owe the names and definitions of many familiar electrical quantities. Another goal was experimental, to establish an absolute standard of resistance, the ohm. Together these two intensely practical concerns led Maxwell to a profound work of philosophical reconstruction, establishing the existence of electromagnetic waves from electromagnetic equations alone.

He began phenomenologically. From dimensional reasoning he proved that ratios of naturally defined electrical units yield c, the velocity of light, not Weber's v. That was one new foundation. The second was to recognize, following Thomson, the power of energy principles in electromagnetic calculations, specifically in the experiment to standardize the ohm. Maxwell's 1861 paper had entangled two novelties. First general, that electromagnetic energy is disseminated throughout space. Second detailed, the vortex machinery. Concentrating on the former Maxwell rebuilt his theory on a combination of electrical and very general dynamical arguments and "cleared the electromagnetic theory of light from all unwarrantable assumption."

When Maxwell died fifteen years later his ideas had only just begun to sink in. His *Treatise*, freighted with surprises, lay there massive and alone, almost a sacred text to be deciphered without benefit of the master's ironic advice. Compounding these logistical difficulties was a philosophical one. More than a new theory, Maxwell's was a new kind of theory, relational not explanatory. Instead of explaining light or electromagnetism it established relations between them. Instead of laws or axioms like Newton's, its central conceptual structure was an elaborate system of relations among physical quantities, Maxwell's equations. Many physicists remained wary.

Nevertheless this was the way forward. With philosophy came discovery. Radio, relativity, and the electron all originated here.

Gases, Molecules, and Statistics

Chance made Maxwell a statistician—reading a paper by Rudolf Clausius about colliding molecules as he pondered in 1859 the colliding bodies of Saturn's rings.

The idea that gases exert pressure kinetically by impacts of fast-moving molecules rather than by static repulsion had been revived earlier, only to meet in 1857 a telling objection. Kinetic theory requires very high molecular velocities—about 1 km/s. Why then, asked C. H. D. Buys-Ballot, do pungent odors spread so slowly? Because, answered Clausius, molecules in air constantly collide like billiard balls, going on average only a very short distance (the mean free path λ) before bouncing off in new directions. Clausius crudely assigned all molecules the same speed.

First with breathtaking casualness Maxwell introduced a velocity distribution function expressing the proportions of molecules having at any instant a given speed. Here were large perplexities. Is Maxwell's the only stable distribution? How should averages be defined for molecules repeatedly changing speed and direction? Ludwig Boltzmann would later (1868) address these questions by studying time histories of individual molecules, and Maxwell (1878) by ensemble averaging, that is, by averaging simultaneously across all possible states of motion. Meanwhile, rushing past conceptual difficulties to physics, Maxwell derived linked equations governing three phenomena never previously connected—viscosity, heat conduction, and molecular diffusion—obtaining two distinct estimates of η in air at room temperature. Each was around 0.00006 mm.

Six years later (1865), Johann J. Loschmidt exploited similar data to provide the first intelligent estimate of molecular diameters.

Astonishingly, kinetic theory made viscosity η independent of pressure. Working together, Maxwell and his wife confirmed this experimentally. Contrary to theory, however, η varied with temperature not as $T^{1/2}$ but almost linearly. Clausius's billiard ball picture was too simple. With wonderful insight Maxwell reconstructed the mathematics to incorporate long-range forces between molecules, developing equations that became, with extensions by Boltzmann, key to modern investigations of transport processes in solids and liquids as well as gases. General solutions came later; the one case Maxwell could solve (inverse fifth-power repulsion) made η exactly proportional to T. Finally, brilliantly, a paper on rarefied gases (1879) laid the foundation of aerodynamics.

Meanwhile, embedded in the theory was a deadly problem, equipartition, whose gravity Maxwell alone saw. In 1859 he had proved two surprising statistical equalities. First, in mixed gases lighter and heavier molecules adjust speed to make their mean energies of translation equal. Second, for nonspherical molecules the mean internal (rotational) and external (translational) energies also become equal. This is equipartition, afterwards greatly extended by Boltzmann. External equipartition neatly explained several things; internal equipartition was disaster. Gases have two specific heats, at constant pressure and constant volume, whose ratio involves equipartition. For polyatomic gases, theory predicted a ratio of 1.333. Experiment gave 1.408.

At the famous British Association meeting at Oxford in 1860, where Thomas Huxley thunderingly proclaimed that he would sooner owe descent to a monkey than a bishop, Maxwell in the stillness of a small room wrecked classical physics. This one tiny anomaly, he remarked, "overturns the whole hypothesis." Thereafter, every extension of equipartition deepened the disaster. Others sought glib explanations. Not Maxwell. His parting prescription (1877) was "that thoroughly conscious ignorance that is the prelude to every real advance in knowledge." The needed advance was quantum mechanics.

Control Theory

In fixing the ohm Maxwell and his British Association colleagues had noticed a strange thing. Their rotating coil, controlled by a "governor" resembling James Watt's steam engine governor, sometimes varied in speed between two limiting values, sometimes went unstable. Adding friction helped, but why? After studying many different working governors, Maxwell developed in 1868 general stability criteria for systems up to the third order. He had founded control theory.

Undergraduate rivalries linger. Stability of motion was the topic for the 1876 Adams' Prize at Cambridge, set by Maxwell. The magisterial winning treatise, the next great advance in control theory, was by Routh.

Relative and Absolute

"I must say," wrote Maxwell's physician, "he is one of the best men I have ever met, and a greater merit than his scientific attainments is his being, so far as human judgment can discern, a most perfect example of a Christian gentleman." Others were intimidated. Young Henry Rowland felt "somewhat discouraged, for here I met a mind whose superiority was almost oppressive."

Culture, universal knowledge, concentration, a terrifying memory—it was overwhelming. One wonders how his wife stood it. Beyond all was his weird intellectual clairvoyance. He it was thirty years before Albert Einstein who imported the term "relativity" from philosophy into physics and wrote this passage, comparable in poetic, scientific, and religious feeling to Plato's vision of the cave:

> There are no landmarks in space; one portion of space is exactly like every other portion, so that we cannot tell where we are. We are, as it were, on an unruffled sea, without stars, compass, soundings, wind, or tide, and we cannot tell in what direction we are going. We have no log which we can cast out to take a dead reckoning by [Maxwell, (1876) 1920, sect. 102, p. 81].

Position, motion, energy, force all are relative. Most striking is gravity, distinguishable from inertial forces only by gradient effects. That is general relativity.

But science also is relative. The cosmic background radiation, discovered in 1965, has supplied physicists against all expectation with an apparent absolute reference frame.

See also: BOLTZMANN, LUDWIG; CLAUSIUS, RUDOLF JULIUS EMMANUEL; COLOR; ELECTROMAGNETISM; ELECTROMAGNETISM, DISCOVERY OF; EQUIPARTITION THEOREM; FARADAY, MICHAEL; KELVIN, LORD; KINETIC THEORY; MAXWELL–BOLTZMANN STATISTICS; MAXWELL'S DEMON; MAXWELL'S EQUATIONS; MAXWELL SPEED DISTRIBUTION; VISION

Bibliography

BRUSH, S. G.; EVERITT, C. W. F.; and GARBER, E. *Maxwell on Saturn's Rings* (MIT Press, Cambridge, MA, 1983).

CAMPBELL, L., and GARNETT, W. *Life of James Clerk Maxwell* (Macmillan, London, 1882).

EVERITT, C. W. F. *James Clerk Maxwell: Physicist and Natural Philosopher* (Scribners, New York, 1975).

EVERITT, C. W. F. "Maxwell's Scientific Creativity" in *Springs of Scientific Creativity*, edited by R. Aris, H. T. Davis, and R. H. Stuewer (University of Minnesota Press, Minneapolis, 1983).

GARBER, E.; BRUSH, S. G.; and EVERITT, C. W. F. *Maxwell on Molecules and Gases* (MIT Press, Cambridge, MA, 1986).

GARBER, E.; BRUSH, S. G.; and EVERITT, C. W. F. *Maxwell on Heat and Statistical Mechanics* (Lehigh University Press, Bethlehem, PA, 1995).

GOLDMAN, M. *The Demon in the Aether: The Story of James Clerk Maxwell* (Paul Harris in association with Adam Hilger, Edinburgh, 1983).

HARMAN, P. M. *Scientific Letters and Papers of James Clerk Maxwell*, 2 vols. (Cambridge University Press, Cambridge, Eng., 1990, 1995).

HENDRY, J. *James Clerk Maxwell and the Theory of the Electromagnetic Field* (Adam Hilger, Bristol, Eng., 1986).

MAXWELL, J. C. *Matter and Motion* (Society for Promoting of Christian Knowledge, London, 1876, reprinted 1920 with notes by J. Larmor).

SIEGEL, D. M. *Innovation in Maxwell's Electromagnetic Theory: Molecular Vortices, Displacement Current and Light* (Cambridge University Press, Cambridge, Eng., 1991).

TOLSTOY, I. *James Clerk Maxwell: A Biography* (University of Chicago Press, Chicago, 1982).

C. W. F. EVERITT

MAXWELL–BOLTZMANN STATISTICS

The foundation of modern statistical mechanics was laid towards the end of the nineteenth century by James Clerk Maxwell, Ludwig Boltzmann, and Josiah Willard Gibbs, before the advent of quantum mechanics. However, statistical physics becomes simpler if one can appeal to some quantum concepts.

Our starting point is the idea that one can count the number of available states of a system. In principle, these are discrete quantum states. For a large system the states will be very closely spaced. In classical mechanics we describe a microscopic system by specifying the coordinates and momenta of the particles. The values that these can have form a continuum. A measure of how many states are available with momentum between p and $p + dp$, coordinates between q and $q + dq$ is $dp\, dq/h$, where h is Planck's constant (6.626×10^{-34} J·s). By the correspondence between quantum and classical mechanics, established by Bohr–Sommerfeld quantization, the counting of classical and quantum states will then agree in the limit where classical mechanics governs the dynamics. The number of possible states with energy between E and $E + dE$ is

$$W(E) = g(E)\,\delta E, \qquad (1)$$

where $g(E)$ is the density of states. Next, consider a closed system with fixed volume V, number of particles N, and energy E. In order to avoid problems associated with the discreteness of the quantum states we take the energy to be specified within a tolerance δE. This tolerance should be chosen so that for a large system the precise value of δE does not matter. We do not know in which of the W allowed states the system finds itself. In fact, our fundamental assumption is that at equilibrium our ignorance in this matter is complete and that all the $W(E, V, N)$ possible states are equally likely (i.e., all memory of how the system was initially prepared is lost, except for the values of the energy, volume, and number of particles).

We define the entropy as

$$S = k_B \ln W(E, N, V). \tag{2}$$

Consider next an infinitesimally small change from an equilibrium state E, V, N to another, slightly different, equilibrium state $E + dE$, $V + dV$, $N + dN$. The change in the entropy is then

$$dS = \frac{\partial S}{\partial E} dE + \frac{\partial S}{\partial V} dV + \frac{\partial S}{\partial N} dN. \tag{3}$$

The change in the energy in this process would is given by

$$dE = dQ + dU, \tag{4}$$

where since the process takes one from one equilibrium state to another, the heat given to the system is $dQ = TdS$, while the work done on the system is $dU = -PdV + \mu dN$, where P is the pressure and μ is the chemical potential. Hence, $dE = TdS - PdV + \mu dN$, or

$$dS = \frac{1}{T} dE - \frac{\mu}{T} dN + \frac{P}{T} dV. \tag{5}$$

For our statistical definition of entropy to agree with the thermodynamic definition, we must now define the temperature, pressure, and chemical potential as follows:

$$T = \left(\frac{\partial S}{\partial E} \right)_{N,V}^{-1}$$

$$\mu = -T \left(\frac{\partial S}{\partial N} \right)_{E,V} \tag{6}$$

$$P = T \left(\frac{\partial S}{\partial V} \right)_{N,E}.$$

To establish the equivalence of these definitions and the conventional thermodynamic ones we shall make contact with the zeroth law of thermodynamics. This law has an analogy with mechanics, where in equilibrium the forces are balanced. In particular,

$$T = \text{const.} \rightarrow \text{thermal equilibrium,}$$

$$P = \text{const.} \rightarrow \text{mechanical equilibrium,} \tag{7}$$

$$\mu = \text{const.} \rightarrow \text{chemical equilibrium.}$$

The zeroth law has a fairly straightforward statistical interpretation and this will allow us to establish the equivalence between the thermodynamic and statistical definitions.

Consider two systems that are free to exchange energy but are isolated from the rest of the universe by an ideal insulating surface. The particle numbers N_1, N_2 and volumes V_1, V_2 are fixed for each subsystem. The total energy will be constant under our assumptions and we assume further that the two subsystems are sufficiently weakly interacting that

$$E_T = E_1 + E_2, \tag{8}$$

where E_1 and E_2 are the energies of the subsystems. Assume that the densities of state $g(E)$, $g_1(E)$, $g_2(E)$ are coarse grained so that $W = g(E)\delta E$, $W_1 = g_1(E1)\delta E$, and $W_2 = g_2(E_2)\delta E$. We then have

$$g(E) = \int dE_1 g_2(E_T - E_1) g_1(E_1). \tag{9}$$

If the subsystems are sufficiently large, the product $g_2(E_T - E_1)g_1(E_1)$ will be a sharply peaked function of E_1. The reason for this is that g_1 and g_2 are rapidly increasing functions of E_1 and $E_T - E_1$, respectively. From the definition of the entropy we note that it is a monotonically increasing function of g and that the product $g_1 g_2$ will be at a maximum when the total entropy

$$S(E, E_1) = S_1(E_1) + S_2(E - E_1) \qquad (10)$$

is at a maximum. The most likely value $\langle E_1 \rangle$ of E_1 is the one for which

$$\frac{\partial S_1}{\partial E_1} + \frac{\partial S_2}{\partial E_2}\frac{\partial E_2}{\partial E_1} = 0. \qquad (11)$$

Since $\partial E_2 / \partial E_1 = -1$, we find using Eq. (6) that

$$\frac{1}{T_1} - \frac{1}{T_2} = 0, \qquad (12)$$

or $T_1 = T_2$. The most probable partition of energy between the two systems is the one for which the two temperatures are the same.

We shall see later that our definitions of S, T, and P agree with the conventional ones for the special case of an ideal gas. Consider next an arbitrary system in thermal contact with an ideal gas reservoir. If the system is sufficiently large, it is overwhelmingly probable that the partition of energy between the system and the reservoir will be such as to leave the temperatures effectively the same, and we conclude that our definition of temperature is in agreement with thermodynamics.

It is easy to show from our definition of the pressure that the most likely partition of volume between two subsystems separated by a movable wall is the one for which the pressure in the two systems is the same. Similar argument can be made if the wall is fixed, but porous, so that the two subsystems can exchange particles. The most likely partition will be the one for which the chemical potential is the same. Our definition of chemical potential and pressure will therefore agree with the conventional one if they agree for the ideal gas.

We next wish to establish the concept of a Boltzmann factor, which provides a convenient method of calculating the probability $P(E)\,dE$ that a system in contact with a heat reservoir at temperature T will have energy between E and $E + dE$. Again, consider two systems in thermal contact in such a way that the volume and particle number in each subsystem are held fixed. We now assume that subsystem 2 is very much larger than subsystem 1. The probability

$p(E_1)\,dE_1$ that subsystem 1 has energy between E_1 and $E_1 + dE_1$ is

$$p(E_1)\,dE_1 = \frac{g_1(E_1)\,g_2(E - E_1)\,dE_1}{\int dE_1 g_1(E_1)\,g_2(E - E_1)}. \qquad (13)$$

We have

$$g_2(E - E_1)\,\delta E = \exp\left(\frac{S_2(E - E_1)}{k_B}\right). \qquad (14)$$

Since $E_1 \ll E$, we may expand S_2 in a Taylor series:

$$S_2(E - E_1) = S_2(E) - E_1\frac{\partial S_2}{\partial E} + \frac{1}{2}E_1^2\frac{\partial^2 S_2}{\partial E^2} + \dots. \qquad (15)$$

The temperature of the large system is T, and

$$\frac{\partial S_2}{\partial E} = \frac{1}{T},$$
$$\frac{\partial^2 S_2}{\partial E^2} = \frac{\partial(1/T)}{\partial E} = -\frac{1}{T^2}\left(\frac{\partial T}{\partial E}\right)_{V_2, N_2} = -\frac{1}{T^2 C_2}, \qquad (16)$$

where C_2 is the heat capacity of system 2 at constant V and N. Since the second system is very much larger than the first, $E_1 \ll C_2 T$ and

$$g_2(E - E_1) = (\text{const.})\exp\left(\frac{-E_1}{k_B T}\right). \qquad (17)$$

With the notation $\beta = 1/(k_B T)$ we thus find that (2) can be rewritten

$$p(E_1) = \frac{1}{Z}\,g_1(E_1)\exp(-\beta E_1). \qquad (18)$$

The normalizing term

$$Z = \int dE_1 g_1(E_1)\exp(-\beta E_1) \qquad (19)$$

is called the partition function and is, as we shall see, closely related to the Helmholtz free energy, which has the thermodynamic definition $A = E - TS$. The term $\exp(-\beta E)$ is called the Boltzmann factor.

In the case of a system in contact with a heat bath, the energy and entropy fluctuate about their mean values. We can calculate the mean energy by taking an ensemble average over all possible values of the energy:

$$\langle E \rangle = \frac{\int dE E\, g(E) \exp(-\beta E)}{\int dE\, g(E) \exp(-\beta E)}$$

$$= -\frac{\partial \ln Z}{\partial \beta}. \qquad (20)$$

The mean-square fluctuation in the energy is given by

$$\langle (E - \langle E \rangle)^2 \rangle = \langle E^2 \rangle - \langle E \rangle^2$$

$$= -\frac{\partial \langle E \rangle}{\partial \beta} = k_B T^2 \frac{\partial \langle E \rangle}{\partial T} = k_B T^2 C_{V,N}, \qquad (21)$$

where $C_{V,N}$ is the heat capacity at constant N and V and is proportional to the size of the system, as is the energy. The root-mean-square (rms) fluctuation in the energy will thus be proportional to $\sqrt{N}$. The mean fluctuation is therefore large for a large system, but is only a small fraction of the total energy.

The relationship in Eq. (21) between the response function $C_{V,N}$ and the mean-square fluctuation of the energy is a special case of a very general result known as the fluctuation-dissipation theorem.

When the energy and entropy fluctuate and T is fixed we give the statistical definition of the free energy as

$$A = \langle E \rangle - T \langle S \rangle. \qquad (22)$$

We next show that

$$A = -k_B T \ln Z. \qquad (23)$$

To see this, let us rewrite the partition function

$$Z = \int dE\, g(E, V, N) \exp\left(\frac{-E}{k_B T}\right)$$

$$= \int \frac{dE}{\delta E} \exp\left[-\frac{1}{k_B}\left(\frac{E}{T} - S(E, V, N)\right)\right]. \qquad (24)$$

If the system is large, it is overwhelmingly probable that the energy will be closed to the most likely value $\langle E \rangle$ given by

$$-\frac{1}{k_B}\left(\frac{\langle E \rangle}{T} - S(\langle E \rangle, V, N)\right) = \text{maximum}. \qquad (25)$$

We expand the exponent around its maximum value at $\langle E \rangle$:

$$-\frac{1}{k_B}\left(\frac{E}{T} - S\right) = -\frac{1}{k_B}\left(\frac{\langle E \rangle}{T} - \langle S \rangle\right)$$

$$+ \frac{1}{2k_B}\left(E - \langle E \rangle\right)^2 \frac{\partial^2 S}{\partial E^2}\Bigg|_{E = \langle E \rangle} + \dots \qquad (26)$$

Using Eq. (19), we find that

$$Z \approx \exp\left[-\beta(\langle E \rangle - T\langle S \rangle)\right] \int \frac{dE}{\delta E} \exp\left(-\frac{(E - \langle E \rangle)^2}{2 C k_B T^2}\right)$$

$$\approx \frac{\sqrt{2\pi k_B T^2 C}}{\delta E} \exp\left[-\beta(\langle E \rangle - T\langle S \rangle)\right] \qquad (27)$$

or

$$-k_B T \ln Z = \langle E \rangle - T \langle S \rangle - k_B T \ln \frac{\sqrt{2\pi k_B T^2 C}}{\delta E}. \qquad (28)$$

We may now choose the tolerance δE so that the logarithmic term is small compared to the other terms and we have the desired result.

Consider an infinitesimal process in which the volume changes from one equilibrium configuration to another by dV, the temperature by dT, and the number of particles by dN. The change in the free energy is

$$dA = dE - d(TS) = -SdT - pdV + \mu dN. \qquad (29)$$

We can use this result to calculate

$$\langle S \rangle = \frac{\partial k_B T \ln Z}{\partial T}, \qquad (30)$$

$$\langle P \rangle = \frac{\partial k_B T \ln Z}{\partial V}, \qquad (31)$$

$$\langle \mu \rangle = -\frac{\partial k_B T \ln Z}{\partial N}. \qquad (32)$$

We finally show how we can calculate the partition function for an ideal gas and demonstrate that this partition function yields expressions for the energy and pressure in agreement with the ideal gas laws. In addition we find expressions for the entropy and chemical potential which cannot be obtained directly from the ideal gas laws. Consider a system consisting of a single particle in contact with a heat bath. From our state counting rules the number of states with momentum components between p_x and $p_x + dp_x$, p_x and $p_x + dp_x$ is $V dp_x dp_y dp_z / h^3$ and

$$Z_1 = \frac{1}{h^3} V \int_{-\infty}^{\infty} dp_x \int_{-\infty}^{\infty} dp_y \int_{-\infty}^{\infty} dp_z \exp\left(\frac{-p_x^2 + p_y^2 + p_z^2}{2 m k_B T}\right)$$

$$= \frac{V}{h^3}\left(\frac{2\pi m}{\beta}\right)^{3/2}. \qquad (33)$$

From Eq. (20) we find the familiar result for the mean energy per particle

$$\langle E \rangle = -\frac{\partial \ln Z}{\partial \beta} = \frac{3}{2\beta} = \frac{3}{2}k_B T. \qquad (34)$$

Using Eq. (31) we find mean contribution per particle to the pressure to be $k_B T/V$ or

$$P = N k_B T \qquad (35)$$

again a familiar result. Getting the entropy or the chemical potential is a bit more tricky. The problem is that the free energy of N particles is not just N times the free energy per particle. If the particles in the gas are indistinguishable, quantum mechanics tells us that we must not treat as distinct states those that only differ by which particles are in which state. Instead we find the number of states where there are particles in the system with momenta between $p_{x,i}$ and $p_{x,i} + dp_{x,i}$, $p_{y,i}$ and $p_{y,i} + dp_{y,i}$, $p_{z,i}$ and $p_{z,i} + dp_{z,i}$, for $i = 1,...N$:

$$\frac{V^N}{h^{3N} N!} \Pi_i^N dp_{x,i} dp_{y,i} dp_{z,i}. \qquad (36)$$

If the factor $N!$ had not been included we would have counted the state where particle 1 has momentum p_1 and particle 2 has momentum p_2 as distinct from a state in which it is particle 2 that has momentum p_1 and particle 1 has momentum p_2. The need for the inclusion of the $N!$ term goes under the name Gibbs paradox. Before the introduction of quantum mechanics this term was hard to justify.

We find for the partition function of an ideal gas of N particles,

$$Z_N = \frac{Z_1^N}{N!} = \frac{V}{N! h^3}\left(\frac{2\pi m}{\beta}\right)^{3/2}. \qquad (37)$$

Using Stirling's formula $\ln N! \approx N \ln N - N$, we find the free energy

$$\langle A \rangle = -N k_B T \left(\ln \frac{V(2\pi m/\beta)^{3/2}}{N h^3} + 1\right). \qquad (38)$$

Differentiation of the expression for A, using Eq. (30), we obtain the so-called Sackur–Tetrode formula

$$\langle S \rangle = N k_B \ln \frac{V(2\pi m/\beta)^{3/2}}{N h^3} + \frac{5}{2} N k_B, \qquad (39)$$

and

$$\langle \mu \rangle = k_B T \ln \left(\frac{N h^3}{V(2\pi m/\beta)^{3/2}}\right). \qquad (40)$$

In the case of molecules containing more than one atom there will be corrections to the entropy and chemical potential from vibrational and rotational degrees of freedom.

In summary the reader should take note of our concept of equilibrium. We do not define equilibrium as the state which obtains dynamically after a long time. Instead we define it from the statistics obtained by counting states. There are a number of steady-state systems in which there is energy flowing through the system (all living organisms have a metabolism, and our planet receives a steady supply of energy from the sun). These systems need not develop dynamically toward states that obey Maxwell–Boltzmann statistics, and we consider them to be nonequilibrium systems. One often quoted diagnostic used to establish lack of equilibrium is failure of detailed balance, which means that the relative probability of two states is not given by the ratio of their Boltzmann factors. Another diagnostic is failure of a fluctuation dissipation theorem. We saw an example in the fluctuation expression for the specific heat.

See also: BOLTZMANN, LUDWIG; BOLTZMANN DISTRIBUTION; ENERGY, FREE; ENTROPY; EQUILIBRIUM; GIBBS, JOSIAH WILLARD; MAXWELL, JAMES CLERK; MAXWELL SPEED DISTRIBUTION; QUANTUM MECHANICS; STATISTICAL MECHANICS; THERMODYNAMICS

Bibliography

KITTEL, C., and KRÖMER, H. *Thermal Physics,* 2nd ed. (W. H. Freeman, New York, 1980).

PLISCHKE, M., and BERGERSEN, B. *Equilibrium Statistical Physics,* 2nd ed. (World Scientific, Singapore, 1994).

REIFF, F. *Fundamentals of Statistical and Thermal Physics* (McGraw-Hill, New York, 1965).

BIRGER BERGERSEN

MAXWELL'S DEMON

Maxwell's demon has intrigued generations of physicists and others since the idea first appeared in James Clerk Maxwell's *Theory of Heat* in 1871. Although Maxwell introduced the idea with a rather specific example, we now think rather broadly of a Maxwell's demon as any agent, intelligent or mechanical, whose aim is to sort a random arrangement of particles into a more ordered one. If the demon can accomplish this while doing zero or at most some negligible amount of work on the system, then it will violate the second law of thermodynamics. Thus, an alternate and even broader definition of Maxwell's demon is any device designed to operate in violation of the second law.

Maxwell's Original Conception

In Maxwell's original conception, two chambers labeled *A* and *B* are identical in all respects, with each containing an ideal gas in equilibrium at some temperature *T*. Between the two chambers is a small door which can be opened and closed by the demon without net expenditure of work. In equilibrium the individual molecular speeds *v* are governed by the Maxwell distribution

$$f(v) = C v^2 \exp(-mv^2/2kT),$$

where *C* is a constant, *m* is the molecular mass, and *k* is Boltzmann's constant. The demon's strategy is to observe individual molecules in both chambers, and open the door briefly when an unusually fast molecule (so that its energy is greater than the mean molecular energy) in *A* is approaching the door, allowing it to pass into *B*. The same procedure is followed when a slow molecule in *B* approaches the door, and it is allowed to pass into *A*. The demon could then repeat the process as often as feasible, and in the end the temperature of *A* will be less than the temperature of *B*. In effect, a refrigerator would be running with no energy cost, in violation of the second law.

This result troubled Maxwell to the extent that the creature that could in theory violate the second law was later dubbed Maxwell's demon by William Thomson (Lord Kelvin). In the end Maxwell concluded that his thought experiment proved that while the second law holds with a high degree of probability for large numbers of particles, it can in fact be violated if one can carefully follow and direct individual molecules. This conclusion is made more palatable by the fact that it is certainly possible for a distribution of particles to move *spontaneously* to a less probable configuration, and that the likelihood of such a spontaneous transformation increases as the number of particles in the sample decreases.

The Demon's History After Maxwell

Since Maxwell's time the demon has been studied and written about widely. Those addressing the subject include such well-known physicists as Thomson, Leo Szilard, Leon Brillouin, and Richard Feynman.

Numerous conceptions of the demon have been imagined. For example, the sorting of molecules may be accomplished by a mechanical device such as a spring-loaded trap door. In another conception the motions of individual molecules could be used to do mechanical work, also in violation of the second law. In an ingenious device described by Szilard, a single molecule could be used in a cyclic process to lower the entropy of a heat reservoir.

The Modern View of Maxwell's Demon

All the hypothetical devices mentioned above, and all others found in the literature, have been found to fail, and the second law of thermodynamics remains inviolate. The reasons for those failures are often subtle, however, and cannot be discussed fully in the context of this article. It is a fair summary of current thinking to say that, in contrast to Maxwell's view, it is impossible to violate the second law even on a molecular or atomic scale.

The most comprehensive way of explaining the general failure of Maxwell's demons relies on ideas from information theory and was expounded in the 1960s and 1970s principally by Rolf Landauer and Charles Bennett. Maxwell's original conception relied on the demon's ability to gather information about the motions of individual molecules, and then use that information. Szilard, Brillouin, and others have focused on the demon's physical intervention in the system and attempted to show that such interaction cannot be accomplished with arbitrarily little expenditure of energy. If the energy expenditure is sufficiently high, such reasoning goes, there will be a corresponding entropy increase that more than offsets the entropy reductions generated by the demon's rearrangement of particles.

Ultimately if the information about individual particles is to be put to use, it must be stored in a memory, which must be a real physical device with its own thermodynamic state. Landauer showed that reading and writing information—essentially the information gathering steps—can be done reversibly, that is, with arbitrarily little dissipation of energy and therefore no entropy increase. He showed, however, that memory erasure, essentially the discarding of information, results in an unavoidable entropy increase. Unless this last step of discarding information is accomplished, the demon has not performed a complete thermodynamic cycle and returned the system to its original state, and therefore the accounting of energy and entropy cannot be complete. Bennett later used Landauer's ideas to develop a theory of reversible computation and to further elucidate why the various conceptions of the demon must fail to violate the second law.

Summary

In spite of the demon's ultimate failure to violate the second law, its various incarnations remain as useful illustrations of the range, beauty, and utility of that law. Students of statistical mechanics or thermodynamics are well advised to study the demon in its numerous contexts, for it continues to provide important foundations for study in and bridges between a number of disciplines.

See also: BOLTZMANN CONSTANT; EQUILIBRIUM; EQUIPARTITION THEOREM; GAS; HEAT; KELVIN, LORD; MAXWELL, JAMES CLERK; MAXWELL SPEED DISTRIBUTION; MOLECULAR SPEED; STATISTICAL MECHANICS; TEMPERATURE; THERMODYNAMICS; THERMODYNAMICS, HISTORY OF

Bibliography

LANDAUER, R. "Irreversibility and Heat Generation in the Computing Process." *IBM J. Res. Dev.* **5,** 183–191 (1961).

LEFF, H. S. and REX, A. F. "Resource Letter MD–1: Maxwell's Demon." *Am. J. Phys.* **58,** 201–209 (1990).

LEFF, H. S. and REX, A. F. *Maxwell's Demon: Entropy, Information, Computing* (Princeton University Press, Princeton, NJ, 1990).

MAXWELL, J. C. *Theory of Heat* (Longmans, Green, & Co., London, 1871).

ANDREW F. REX

MAXWELL'S EQUATIONS

Maxwell's equations consist of a set of four differential equations that form the basis of the theory of electromagnetic phenomena. However, the significance of these equations goes far beyond the electromagnetic effects they were originally supposed to represent; many branches of physics, directly or indirectly, are dependent on Maxwell's equations.

History

The systematic quantitative study of electric and magnetic phenomena began late in the eighteenth century when Charles Coulomb, on the basis of careful measurements, postulated in 1785 his famous force law for electric charges

$$F = k \frac{q_1 q_2}{r^2},$$

where F is the force, k is a constant of proportionality, q_1 and q_2 are the charges, and r is the separation between them. The striking similarity of Coulomb's law with Newton's law of gravitation gave rise to theories of electricity and magnetism that adopted the mathematical apparatus previously developed for gravitational systems. In these action-at-a-distance theories, Coulomb's law for electric charges and a similar law for magnetic poles were regarded as the principal laws, and all electric and magnetic phenomena were thought to be deducible from them. However, the action-at-a-distance theories were not fruitful and provided little help toward a better understanding or use of electricity and magnetism. Drastic changes in the understanding of the nature of electric and magnetic phenomena were brought about by Michael Faraday. Between 1821 and 1848, he performed numerous electric and magnetic experiments that were unprecedented in scope, in the care of execution, and in the depth of interpretation. On the basis of his experiments, Faraday concluded that the carriers of electric and magnetic actions were regions of space around electric charges and magnets. Therefore, according to Faraday, these regions of space, or fields, as they were later named by James Clerk Maxwell, should be the primary subject of electromagnetic investigations, and the basic electric and magnetic phenomena should be explainable in terms of the properties of the electric and magnetic fields.

In 1855, Maxwell translated Faraday's ideas about electric and magnetic fields into a mathematical form. Later he succeeded in generalizing the basic facts of electromagnetism into a set of fundamental equations for electromagnetic fields. We call these equations "Maxwell's equations." A direct mathematical consequence of these equations were the equations indicating the existence of electromagnetic waves propagating with the velocity of light. In 1886, such waves were discovered by Heinrich Hertz, and this discovery was the first triumph of the Faraday–Maxwell field theory of electric and magnetic phenomena and of Maxwell's equations as the mathematical foundation of that theory.

Maxwell's Equations

In the now generally accepted vector notation (due to Heinrich Hertz and Oliver Heaviside), Maxwell's equations are written as the following four differential equations:

$$\nabla \cdot \mathbf{D} = \rho, \tag{1}$$

$$\nabla \cdot \mathbf{B} = 0, \tag{2}$$

$$\nabla \times \mathbf{E} = -\frac{\partial \mathbf{B}}{\partial t}, \tag{3}$$

and

$$\nabla \times \mathbf{H} = \mathbf{J} + \frac{\partial \mathbf{D}}{\partial t}, \tag{4}$$

where ∇ is the vector-analytical operator "del," $\partial/\partial t$ is the partial derivative with respect to time, $\mathbf{E}$ is the electric field vector, $\mathbf{B}$ is the magnetic flux density (magnetic induction) vector, $\mathbf{D}$ is the electric displacement vector, $\mathbf{H}$ is the magnetic field vector, $\mathbf{J}$ is the electric current density vector, and ρ is the density of the electric charge.

In Maxwell's equations, electric and magnetic fields are linked together in an intricate manner, and neither field is explicitly represented in terms of its sources ρ and $\mathbf{J}$. In order to solve Maxwell's equations, they must be supplemented by the so-called auxiliary, or constitutive, equations describing the properties of the media under consideration. For fields in a vacuum, the constitutive equations are

$$\mathbf{D} = \varepsilon_0 \mathbf{E} \tag{5}$$

and

$$\mathbf{B} = \mu_0 \mathbf{H}, \tag{6}$$

where ϵ_0 is a constant called the permittivity of space, and μ_0 is a constant called the permeability of space.

For linear isotropic conducting media, the constitutive equation is

$$\mathbf{J} = \sigma \mathbf{E}, \tag{7}$$

where σ is the conductivity of the medium (this equation represents Ohm's law in vector notation).

Maxwell's equations can also be expressed in an integral form, which is equivalent to the differential form but gives a clearer view of the physical significance of the equations:

$$\oint \mathbf{D} \cdot d\mathbf{S} - \int \rho dv = q_{\text{enclosed}}, \tag{8}$$

$$\oint \mathbf{B} \cdot d\mathbf{S} = 0, \tag{9}$$

$$\oint \mathbf{E} \cdot d\mathbf{l} - \frac{\partial}{\partial t} \int \mathbf{B} \cdot d\mathbf{S}, \tag{10}$$

and

$$\oint \mathbf{H} \cdot d\mathbf{l} = \int \mathbf{J} \cdot d\mathbf{S} + \frac{\partial}{\partial t} \int \mathbf{D} \cdot d\mathbf{S}. \tag{11}$$

The first of these equations represents Gauss's law of electricity, according to which the net electric flux passing through a closed surface is always equal to the electric charge enclosed by that surface. The second equation is Gauss's law of magnetism, according to which the net magnetic flux passing through a closed surface is always zero (because no "magnetic charges," or magnetic monopoles, are known to exist, although some scientists believe that such charges will eventually be discovered, in which case this law will be modified). The third equation represents Faraday's law of electromagnetic induction, according to which the voltage measured along a closed loop is always associated with a change of the magnetic flux passing through the loop and is equal to the rate of change of that flux (the minus sign in this equation reflects Lenz's law, according to which the current induced in a conducting loop produces a magnetic field that tends to counteract the initial magnetic flux variation). The fourth equation is Maxwell's generalization of Ampère's law to time-dependent systems. According to Ampère's law, in time-dependent systems, the line integral of the magnetic field $\mathbf{H}$ taken along a closed loop is always equal to the electric current (conduction current plus convection current) passing through this loop. However, Maxwell determined that for time-dependent fields the value of the line integral of $\mathbf{H}$ depends not only on the conduction and convection current, but also on the rate of change of the electric flux passing through the loop [the last term in Eq. (11)]. He named this term the displacement current [and, similarly, he named the last term in Eq. (4) the displacement current density].

Maxwell's theoretical discovery of the displacement current (in 1862) was crucial for asserting the field theory of electric and magnetic phenomena as the definitive electromagnetic theory. With the introduction of the displacement current, the theory not only explained most of the known electromagnetic phenomena and yielded the laws governing these phenomena, but also predicted phenomena not yet discovered. For example, the very important empirical law of the conservation of electric charge

$$\mathbf{\nabla} \cdot \mathbf{J} = -\frac{\partial \rho}{\partial t} \tag{12}$$

became a consequence of Eq. (4), and Kirchhoff's laws, which are indispensable for designing and analyzing electric circuits, became a consequence of Eqs. (3) and (4). But, most important, with the introduction of the displacement current, Maxwell's equations clearly indicated the existence of the yet to be experimentally discovered electromagnetic waves, indicated that light was an electromagnetic phenomenon, and indicated that the laws of optics were deducible from the laws of electromagnetism.

Wave Equations

By means of standard vector operations, Maxwell's equations can be transformed into the "inhomogeneous wave equations" (shown here for a vacuum)

$$\mathbf{\nabla} \times \mathbf{\nabla} \times \mathbf{E} + \frac{1}{c^2} \cdot \frac{\partial^2 \mathbf{E}}{\partial t^2} = \mu_0 \frac{\partial \mathbf{J}}{\partial t}, \tag{13}$$

and

$$\mathbf{\nabla} \times \mathbf{\nabla} \times \mathbf{H} + \frac{1}{c^2} \cdot \frac{\partial^2 \mathbf{H}}{\partial t^2} = \mathbf{\nabla} \times \mathbf{J}, \tag{14}$$

where

$$c = \sqrt{\varepsilon_0 \mu_0}. \tag{15}$$

In free space, where there are no charges or currents, Eqs. (13) and (14) become, after simplification,

$$\mathbf{\nabla}^2 \mathbf{E} - \frac{1}{c^2} \cdot \frac{\partial^2 \mathbf{E}}{\partial t^2} = 0, \tag{16}$$

and

$$\mathbf{\nabla}^2\mathbf{H} - \frac{1}{c^2} \cdot \frac{\partial^2\mathbf{H}}{\partial t^2} = 0. \tag{17}$$

These equations constitute a mathematical representation of an electromagnetic wave propagating in space with the velocity c. In the electromagnetic wave, the electric field vector $\mathbf{E}$ and the magnetic field vector $\mathbf{H}$ are perpendicular to each other, and also perpendicular to the direction of the propagation of the wave (transverse wave). Maxwell determined that the velocity of the electromagnetic waves was equal to the velocity of light [this follows from Eq. (15)] and concluded that light consisted of electromagnetic waves. However, his equations and conclusions did not have many supporters until the electromagnetic waves were actually produced and studied in a laboratory by Heinrich Hertz (in 1886).

Solutions of Maxwell's Equations

The general solutions of Eqs. (13) and (14) and therefore the general solutions of Maxwell's equations themselves (for fields in a vacuum) can be written as

$$\mathbf{E} = -\frac{1}{4\pi\varepsilon_0}\int \frac{\left[\mathbf{\nabla}'\rho + \frac{1}{c^2} \cdot \frac{\partial\mathbf{J}}{\partial t}\right]}{r} dv' \tag{18}$$

and

$$\mathbf{H} = \frac{1}{4\pi}\int \frac{[\mathbf{\nabla}' \times \mathbf{J}]}{r} dv'. \tag{19}$$

The square brackets in these equations are the retardation symbol indicating that the quantities between the brackets are to be evaluated for the "retarded" time $t' = t - (r/c)$, where t is the time for which $\mathbf{E}$ and $\mathbf{H}$ are evaluated, ρ is the electric charge density, $\mathbf{J}$ is the current density, r is the distance between the field point x, y, z (point for which $\mathbf{E}$ and $\mathbf{H}$ are evaluated) and the source point x', y', z' (volume element dv'), and c is the velocity of light. The integrals in both equations are extended over all space.

Two alternative solutions of Maxwell's equations that are more lucid than Eqs. (18) and (19) are ob-

tained by eliminating the spatial derivatives from Eqs. (18) and (19). They are

$$\mathbf{E} = \frac{1}{4\pi\varepsilon_0}\int\left\{\frac{[\rho]}{r^2} + \frac{1}{rc} \cdot \frac{\partial[\rho]}{\partial t}\right\}\mathbf{r}_u dv'$$
$$- \frac{1}{4\pi\varepsilon_0 c^2}\int\frac{1}{r} \cdot \left[\frac{\partial\mathbf{J}}{\partial t}\right]dv' \tag{20}$$

and

$$\mathbf{H} = \frac{1}{4\pi}\int\left\{\frac{[\mathbf{J}]}{r^2} + \frac{1}{rc} \cdot \frac{\partial[\mathbf{J}]}{\partial t}\right\} \times \mathbf{r}_u dv', \tag{21}$$

where $\mathbf{r}_u$ is a unit vector directed from dv' to the field point. If desired, Eq. (21) can be written in terms of $\mathbf{B}$ by using Eq. (6).

Equation (20) shows that the time-dependent electric fields have three causative sources: the charge density ρ, the time derivative of ρ, and the time derivative of the current density $\mathbf{J}$. Equation (21) shows that the time-dependent magnetic fields have two causative sources: the electric current density $\mathbf{J}$ and the time derivative of $\mathbf{J}$. Since both equations contain the time derivative of $\mathbf{J}$, in time-dependent systems, an electric and a magnetic field are always simultaneously created by the same time-variable electric current. Therefore the actual cause of electromagnetic induction is not a changing magnetic field (as was previously thought by some scientists), but the changing electric current which always simultaneously creates both the electric field $\mathbf{E}$ and the magnetic field $\mathbf{H}$ (or $\mathbf{B}$).

For time-independent systems, the solutions of Maxwell's equations, Eqs. (20) and (21), reduce to the familiar electrostatic Coulomb field law

$$\mathbf{E} = \frac{1}{4\pi\varepsilon_0}\int\frac{\rho}{r^2}\mathbf{r}_u dv', \tag{22}$$

and to the magnetostatic Biot-Savart law

$$\mathbf{H} = \frac{1}{4\pi}\int\frac{\mathbf{J} \times \mathbf{r}_u}{r^2}dv'. \tag{23}$$

An important consequence of the solutions of Maxwell's equations, Eqs. (20) and (21), is that electromagnetic waves can be generated by antennas, in which electric charges and currents are made to oscillate at a rapid rate. This is the basis for the wireless transmission of radio, television, and radar signals.

Related Equations and Concepts

Maxwell's equations are closely associated with three other electromagnetic equations that, together with Maxwell's equations, constitute the core of Maxwell's theory of electromagnetic phenomena. They are the electromagnetic energy equation

$$U = \tfrac{1}{2}\int (\mathbf{E} \cdot \mathbf{D} + \mathbf{H} \cdot \mathbf{B})\, dv', \qquad (24)$$

the electromagnetic energy transport equation (the Poynting vector)

$$\mathbf{P} = \mathbf{E} \times \mathbf{H}, \qquad (25)$$

and the electromagnetic momentum equation

$$\mathbf{G} = \frac{1}{c^2}\int (\mathbf{E} \times \mathbf{H})\, dv'. \qquad (26)$$

Equation (24) expresses that electric and magnetic fields are carriers of energy; Eq. (25) shows that electromagnetic energy can be propagated through free space; and Eq. (26) expresses that the electromagnetic field is a carrier of momentum, so that the conservation of momentum law (but not necessarily the law of action and reaction) can be satisfied in all electromagnetic interactions.

Maxwell's equations form a remarkable set of physical equations. Their validity and applicability are truly universal. Their simplicity is contrasted by the tremendous range and complexity of phenomena which they represent. To many physicists they are not only the embodiment of the perfect mathematical representation of the laws of physics but also an object of aesthetic admiration.

See also: ACTION-AT-A-DISTANCE; AMPÈRE'S LAW; CHARGE; COULOMB, CHARLES AUGUSTIN; COULOMB'S LAW; CURRENT, DISPLACEMENT; ELECTRIC FLUX; ELECTROMAGNETIC INDUCTION; ELECTROMAGNETIC INDUCTION, FARADAY'S LAW OF; ELECTROMAGNETIC WAVE; ELECTROMAGNETISM; FARADAY, MICHAEL; FIELD, ELECTRIC; FIELD, MAGNETIC; GAUSS'S LAW; LENZ'S LAW; MAGNETIC FLUX; MAGNETIC MONOPOLE; MAXWELL, JAMES CLERK; MOMENTUM, CONSERVATION OF; NEWTON'S LAWS; OHM'S LAW

Bibliography

HEALD, M. A., and MARION, J. B. *Classical Electromagnetic Radiation*, 3rd ed. (Saunders, Fort Worth, TX, 1995).

JEFIMENKO, O. D. *Electricity and Magnetism*, 2nd ed. (Electret Scientific, Star City, WV, 1989).

MAXWELL, J. C. *A Treatise on Electricity and Magnetism*, 3rd ed. (Clarendon, Oxford, 1892).

NIVEN, W. D., ed. *The Scientific Papers of James Clerk Maxwell* (Cambridge University Press, Cambridge, Eng., 1890).

WHITTAKER, E. T. *A History of the Theories of Aether and Electricity*, 2nd ed. (Nelson and Sons, London, 1951).

OLEG D. JEFIMENKO

MAXWELL SPEED DISTRIBUTION

Consider a gas at a given temperature T. The molecules of the gas will move about with different speeds in random directions. The likelihood that a molecule has a given speed can be represented by a distribution function $P(v)$ according to which the probability that a given molecule has speed between v and $v + dv$ is $P(v)\, dv$. The Maxwell speed distribution is

$$P(v) = v^2 \sqrt{\frac{2}{\pi}} \left(\frac{m}{k_B T}\right)^{3/2} \exp\left(-\frac{mv^2}{2k_B T}\right), \qquad (1)$$

where m is the mass of a molecule and $k_B = 1.38041 \times 10^{-23}$ J/K is the Boltzmann constant. The distribution is normalized, that is,

$$\int_0^\infty P(v)\, dv = 1. \qquad (2)$$

The mean speed is

$$\langle v \rangle = \int_0^\infty v P(v)\, dv = \sqrt{\frac{8k_B T}{\pi m}}. \qquad (3)$$

Here $\langle\ \rangle$ denotes average over the distribution. The most probable speed v_p of the distribution is the po-

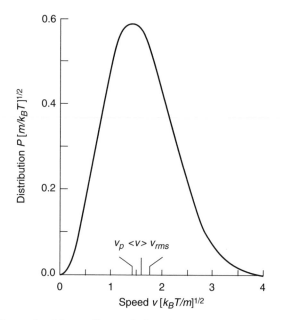

Figure 1 Maxwell speed distribution: v_p is the most probable speed, $\langle v \rangle$ is the mean speed and v_{rms} is the root mean square speed.

sition of the peak (see Fig. 1) and can be obtained by differentiating $P(v)$ and putting the result equal to zero. One finds

$$v_p = \sqrt{\frac{2k_B T}{m}}. \tag{4}$$

The mean square speed is

$$\langle v^2 \rangle = \int_0^\infty v^2 P(v)\, dv = \frac{3k_B T}{m}. \tag{5}$$

The mean speed and the root-mean-square (rms) speed are often used in kinetic theory to describe a typical speed of a molecule.

To derive the result in Eq. (1) we must assume that the gas molecules are in thermal equilibrium and that they obey classical (Newtonian) mechanics. In particular, the only velocity dependence of the energy of a molecule is the kinetic energy $mv^2/2$.

Maxwell's original derivation of the distribution was published in 1860 and was based on the notions of isotropy of space and the assumption that the probability distributions for velocity components in orthogonal directions should be uncorrelated (i.e., independent of each other). Let $f(v_x)\, dv_x$ be the

probability that the x component of the velocity is between v_x and $v_x + dv_x$ and let $f(v_y)$ and $f(v_z)$ be the corresponding distributions for the y and z components. The probability that a particle has velocity components between v_x and $v_x + dv_x$ and also between v_y and $v_y + dv_y$ and v_z and $v_z + dv_z$ will then be

$$f(v_x)f(v_y)f(v_z)\, dv_x dv_y dv_z. \tag{6}$$

The orientation of the coordinate axes is arbitrary, and therefore the product $f(v_x)f(v_y)f(v_z)$ must depend only on the speed v which can be obtained from $v^2 = v_x^2 + v_y^2 + v_z^2$. We therefore can define a function ϕ by

$$f(v_x)f(v_y)f(v_z) = \phi(v_x^2 + v_y^2 + v_z^2). \tag{7}$$

Since f and ϕ are probability distributions they must be positive, and can be written as the exponential of a function. Since directions do not matter, $f(v_x) = f(-v_x)$. We must thus be able to express f in terms of the square of the velocity component. Therefore,

$$f(v_x) = e^{h(v_x^2)}$$

$$\phi(v_x^2 + v_y^2 + v_z^2) = e^{\psi(v_x^2 + v_y^2 + v_z^2)} \tag{8}$$

or

$$h(v_x^2) + h(v_y^2) + h(v_z^2) = \psi(v_x^2 + v_y^2 + v_z^2), \tag{9}$$

implying that ψ must be a linear function. We also require that the probability becomes small for large speeds and find that the distribution must be on the form

$$f(v_x) = Ce^{-x^2/a}$$

$$\phi(v^2) = C^3 e^{-v^2/a}. \tag{10}$$

From the normalization condition

$$1 = \int_{-\infty}^\infty dv_x f(v_x) = C\sqrt{\pi a}, \tag{11}$$

$C = 1/\sqrt{\pi a}$. We can also calculate the mean kinetic energy by evaluating

$$\frac{m\langle v^2 \rangle}{2} = \frac{3m}{2} \int_{-\infty}^{\infty} dv_x v_x^2 f(v_x) = \frac{3a^2}{2}. \qquad (12)$$

The constant a is therefore closely related to the mean kinetic energy of the gas. If $m\langle v^2 \rangle/2 = 3k_BT/2$, $a = 2k_BT/m$. To calculate to the speed distribution from the distribution for the velocity components we go to polar coordinates $v_z = v \cos \theta$, $v_y = v \sin \theta \cos \phi$, $v_x = v \sin \theta \sin \phi$:

$$dv_x dv_y dv_z = v^2 dv \sin \theta\, d\theta\, d\phi. \qquad (13)$$

The speed distribution can be obtained by integrating over the angles:

$$P(v) = v^2 \int_0^{\pi} \sin \theta\, d\theta \int_0^{2\pi} d\phi C^3 e^{-v^2/a}$$

$$= 4\pi v^2 C^3 e^{-v^2/a}. \qquad (14)$$

Substituting the values for C and a we finally obtain Eq. (1).

A more "modern" way of deriving Eq. (1) is to appeal to Maxwell–Boltzmann statistics according to which the probability distribution for the speed of a molecule is

$$P(v) = \frac{g(v) \exp\left(\frac{-E(v)}{k_BT}\right)}{\int dv\, g(v) \exp\left(\frac{-E(v)}{k_BT}\right)}, \qquad (15)$$

where $g(v)\, dv$ is the number of states available to the system with speed between v and $v + dv$, and the denominator guarantees that the probabilities add up to unity. We assume that the number of states with velocity x component between v_x and $v_x + dv_x$ is proportional to dv_x and find by going to polar coordinates (as above) that the number of states $g(v)$ with speed between v and $v + dv$ is proportional to $4\pi v^2 dv$. If we substitute $E(v) = mv^2/2$ and normalize the distribution we recover Eq. (1).

It is worth noting that the derivation does not require that the gas be ideal, as long as the potential energy associated with interaction between the gas molecules is independent of particle speed. The energy can then be written as a sum of terms that de-

pend only on speed and only on the spatial position of the molecules in the gas. The Boltzmann factor for the total energy will then be a product of terms involving speed only and spatial coordinates only. The position and speed of a particle will then be statistically independent.

It is also instructive to note that the derivation using the Boltzmann factor allows one to extend the result to the case of a relativistic gas. It is now easier to work with the momentum rather than the velocity. The number of particle states with momentum between p and $p + dp$ will be proportional to $p^2 dp$, where

$$p = \frac{m_0 v}{\sqrt{1 - v^2/c^2}} \qquad (16)$$

and m_0 is the rest mass and c the speed of light. The probability distribution for momentum will then be proportional to $4\pi p^2 \exp(-\sqrt{m_0 c^4 + c^2 p^2}/k_BT)$ with a proportionality constant that can be determined by normalizing the distribution. The original derivation of the Maxwell velocity distribution will not work in this case, since it is assumed in Eq. (7) that the probability distribution for the three components of the momentum or of the velocity are independent of each other. This will no longer be the case with the more complicated relationship between energy and momentum or speed in the special theory of relativity.

There will be significant corrections to the Maxwell speed distribution at low temperatures when quantum effects are important, and one must correct for the Fermi–Dirac or Bose–Einstein characters of the particles in a gas.

The Maxwell velocity distribution has been verified by a number of scientists for monatomic gases of metal atoms produced in a furnace. A system of slits and shutters allows selection of atoms moving in a particular direction with speed in a certain range and their number can be measured by allowing the atoms to condense on a cold surface. An alternative method is through elastic scattering of light. The scattered light will be Doppler shifted, if the particles that scatter light are moving. By studying the line shape one can deduce the velocity distribution.

Finally, it must be stressed that thermal equilibrium has been assumed. The ionized gas in the outer atmosphere of the Sun, and in other situations in astrophysics or plasma physics, is typically produced in a nonequilibrium setting. Significant deviations from the Maxwell velocity distribution are known to occur in such situations.

See also: BOLTZMANN DISTRIBUTION; DISTRIBUTION FUNCTION; DOPPLER EFFECT; KINETIC THEORY; MAXWELL, JAMES CLERK; MAXWELL–BOLTZMANN STATISTICS; MOLECULAR SPEED

Bibliography

BAUMAN, R. P. *Modern Thermodynamics with Statistical Mechanics* (Macmillan, New York, 1992).

FLOWERS, B. H., and MENDOZA, E. *Properties of Matter* (Wiley, New York, 1978).

HARMAN, P. M. *The Scientific Letters and Papers of James Clerk Maxwell*, 2 vols. (Cambridge University Press, Cambridge, Eng., 1990, 1995).

TABOR, D. *Gases, Liquids, and Solids*, 3rd ed. (Cambridge University Press, Cambridge, Eng., 1991).

BIRGER BERGERSEN

MAYER, MARIA GOEPPERT

b. Kattowicz, Upper Silesia, June 28, 1906; *d.* La Jolla, California, February 20, 1972; *atomic physics, chemical physics, nuclear physics, statistical mechanics.*

Mayer was born the only child of Friedrich and Maria (Wolff) Goeppert. In 1910 the family moved to Göttingen, Germany, where her father was a professor of pediatrics. After attending a private girls' school, Mayer entered Göttingen University in 1924, completing her Ph.D. in theoretical physics under Max Born in 1930.

Shortly before completing her dissertation, she married the American chemist Joseph E. Mayer. They subsequently moved to the United States, where his career took them to Johns Hopkins University (1930–1939), Columbia University (1939–1945), and the University of Chicago (1946–1959). While they were at Johns Hopkins, their two children were born, Maria Ann in 1933 and Peter in 1938.

Mayer's work during the early years covered a broad range of physics. In her doctoral thesis she calculated the probability of double-photon emission or absorption using P. A. M. Dirac's theory of matter and radiation. This work has been cited as a theoretical basis for the development of lasers. Several years later she applied the same techniques to a calculation of the probability of double-beta decay in the nucleus.

During her years at Johns Hopkins, Mayer worked extensively in the field of chemical physics in collaboration with her husband and Karl F. Herzfeld. The textbook *Statistical Mechanics* (1940) resulted from the first of these collaborations. At the same time, she made one of the first calculations of the electronic spectrum of an organic molecule. During World War II, Mayer was a member of the Manhattan Project, working on problems of isotope separation.

After 1945 Mayer turned her attention to nuclear physics. She discovered that nuclei with certain "magic numbers" of protons or neutrons—20, 28, 50, 82, or 126—are unusually stable. Her recognition that a strong spin-orbit coupling effect in nucleons can account for these numbers led her to formulate a shell model of the nucleus. This model was based on the idea that nucleons occupy discrete energy levels in the nucleus in the same way that electrons occupy energy levels in the atom. Her theory was first published in 1949, coinciding with the publication of a report by J. H. D. Jensen, who had formulated the same model simultaneously and independently. The two subsequently collaborated on a book, *Elementary Theory of Nuclear Shell Structure* (1955), and shared half of the 1963 Nobel Prize in physics for their work; the other half of the prize was awarded to Eugene P. Wigner.

It was not until 1959 that Mayer herself was offered a full-time paid position. This offer was made by the University of California at San Diego, where she and her husband would spend the rest of their careers.

Throughout her career, Mayer was active in several scientific organizations, including the Federation of Atomic Scientists, the National Academy of Sciences, and the American Academy of Arts and Sciences. In addition to her accomplishments within the field of physics, however, Mayer was known for her gardening, having created lush and extensive gardens at all of their homes. The Mayers, who had opened their home to scientists fleeing Nazi Germany in the 1930s, were also known in the scientific community for their generous hospitality.

See also: BORN, MAX; CHEMICAL PHYSICS; DIRAC, PAUL ADRIEN MAURICE; NUCLEAR SHELL MODEL; NUCLEAR PHYSICS; STATISTICAL MECHANICS

Bibliography

DASH, J. *A Life of One's Own: Three Gifted Women and the Men They Married* (Harper & Row, New York, 1973).

JOHNSON, K. E. "Maria Goeppert Mayer: Atoms, Molecules, and Nuclear Shells." *Physics Today* **39,** 44–49 (1986).

MAYER, M. G., and JENSEN, J. H. D. *Elementary Theory of Nuclear Shell Structure* (Wiley, New York, 1955).

MAYER, M. G., and MAYER, J. E. *Statistical Mechanics* (Wiley, New York, 1940).

SACHS, R. "Maria Goeppert Mayer" in *Remembering the University of Chicago: Teachers, Scientists, and Scholars,* edited by E. Shils (University of Chicago Press, Chicago, 1991).

KAREN E. JOHNSON

MEAN FREE PATH

The kinetic theory of gases treats molecules as point-like particles, so intermolecular collisions are not considered. The fundamental relationship between the average microscopic kinetic energy of an individual molecule and a gas's macroscopic temperature T. The equation $(1/2)m\overline{v^2} = (3/2)kT$ is derived from kinetic theory assuming that collisions occur only with the walls of the gas's container. Actual gases obviously consist of molecules of finite size. Collisions between these molecules take place at a rate of several billion per second, and each collision will change the speed and direction of both molecules participating in the collision. An individual molecule executes a random walk as it moves through the gas, where between collisions the molecules can be treated as moving in straight line paths. The average distance between the collisions of these molecules is known as the mean free path.

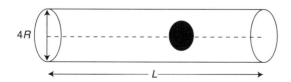

Figure 1 Molecular movement along the symmetry axis of a cylinder.

In estimating the mean free path the molecules are slightly more realistically modeled as identical, hard spheres. If the radius of each sphere is R, then there can be no interaction between molecules unless the center of an approaching molecule is within a distance $2R$ of the center of the target molecule. The approaching molecule can be visualized as moving along the symmetry axis of a cylinder of radius $2R$, as shown in Fig. 1. A collision occurs whenever the center of a target molecule passes within this cylinder. If in a given time the approaching molecule has moved a distance L, the volume of the corresponding cylinder is $\pi(2R)^2L$. The number of collisions occurring within this volume is given by $\pi(2R)^2Ln$, where n is the number of target molecules per unit volume whose centers fall within the volume of the cylinder and therefore undergo collisions. The number of collisions occurring per unit length of the cylinder is then $\pi(2R)^2n$, and the distance between these collisions is $\lambda = 1/[\pi(2R)^2n]$, where λ is known as the mean free path. A typical value for the mean free path of a gas at 1 atm pressure and room temperature is on the order of 1×10^{-7} m. A related quantity, the collision frequency f (the number of collisions per unit time) is obtained by expressing the length L of the cylinder as $\overline{v}t$, where $\overline{v}$ is the mean molecular speed and t is the time of travel, and then dividing the expression for the number of collisions by the time:

$$\frac{\pi(2R)^2\overline{v}tn}{t} = \pi(2R)^2\overline{v}n.$$

The preceding estimate of λ did not consider the motion of the target molecules. The relative speed between the two colliding molecules would in most cases be greater than $\overline{v}$, which would have the effect of increasing the collision frequency and decreasing the mean free path. A calculation which takes into account the actual speed distribution of the molecules reduces the mean free path to

$$\lambda = \frac{1}{\sqrt{2}\,\pi(2R)^2n}.$$

Intermolecular collisions are important in transport phenomena, such as heat transport and interdiffusion of gases. As an example, a determination of the rate of diffusion of one gas through another is dependent on the mean free path of the diffusing gas.

See also: COLLISION; DIFFUSION; HEAT TRANSFER; KINETIC THEORY; MOLECULE

Bibliography

OHANIAN, H. C. *Physics* (W. W. Norton, New York, 1989).

SEARS, F., and SALINGER, G. L. *Thermodynamics, Kinetic Theory, and Statistical Mechanics* (Addison-Wesley, Reading, MA, 1975).

CYNTHIA GALOVICH

MEDICAL PHYSICS

The subject matter that constitutes the field of medical physics significantly overlaps both medicine and physics. It is primarily an applied branch of physics and has several distinct subdivisions. For example, most medical physicists in the United States work in the field of radiation oncology physics, which involves the use of radiation in the treatment of cancer patients. Medical physicists working in diagnostic imaging and nuclear medicine form the next largest group, involving the use of radiation in the diagnosis of disease as well as the use of radionuclides in medicine. Another significant subdivision of medical physics involves radiation protection of patients, workers, and the general public and is referred to as medical health physics. Medical physicists are also involved with the use of nonionizing radiation such as ultrasound, ultraviolet, radiofrequency, and laser radiation. They also play a significant role in the application of computer science and electronics to medical problems.

While medical physicists collaborate mostly with physicians in radiation oncology and diagnostic radiology, they also participate, to some degree, in almost all fields of medicine: for example, bioelectrical investigations of the brain and heart (electroencephalography and electrocardiography), the application of heat for cancer treatment (hyperthermia), and lasers for laser surgery. In addition, they work closely with other hospital support groups such as clinical engineering.

Although it is difficult to pinpoint an exact date for the beginning of medical physics as we know it today, one could reasonably point to Wilhelm Röntgen's discovery of x rays in 1895 and Antoine Becquerel's subsequent discovery of radioactivity in 1896. These events were immediately followed by the application of ionizing radiations to the diagnosis and treatment of disease. Thus in the early part of the twentieth century, a small number of physicists began to apply their skills exclusively to clinical problems. These problems lay principally in radiation oncology (the treatment of cancer patients with radiation) and radiation protection but also included diagnostic imaging and radiation biology. By the middle of the twentieth century, the use of radionuclides was introduced for both diagnostic and therapeutic purposes. Physicists were employed increasingly by hospitals and clinics as radiological physicists whose duties encompassed all three areas: therapy, and diagnostic and nuclear medicine. These physicists contributed to the improvement of radiological equipment and techniques, radiation dosimetry, and radiation safety. With the introduction of the cobalt-60 therapy machine, medical linear accelerators, and more advanced imaging techniques, the number of medical physicists increased rapidly during the next several decades, and the field became even more specialized. Today, nearing the end of the twentieth century, the number of medical physicists in the United States exceeds 4,000 and continues to expand.

The day to day work of medical physicists can generally be divided into three broad areas: clinical support, research and development, and teaching, with clinical support being by far the major component of the work. In fact, the word "medical" in medical physics is sometimes replaced with the word "clinical," particularly if the job is closely connected with patient problems in hospitals (i.e., clinical physics). Clinical physicists are heavily involved in the treatment of cancer patients with radiation. Their physician colleagues, the radiation oncologists, are often in need of their consultation. One important example is in the planning of radiation treatment for a cancer patient using external radiation beams or internal radioactive sources. In this case, the radiation oncologist is faced with the difficult problem of prescribing a treatment regimen with a radiation dose large enough to potentially cure or control the disease but which will not cause serious normal tissue complications. This task is a difficult one as tumor control and normal tissue effect responses are very steep functions of radiation dose; that is, a small change in the dose delivered ($\pm5\%$) can result in a dramatic change in the local response of the tissue ($\pm20\%$). Moreover, the prescribed curative doses are often, by necessity, very close to the maximum doses tolerated by the normal tissues. Thus, for optimum treatment, the radiation dose must be planned and delivered with a high degree of accuracy and the physicist plays a key role in this effort. Specifically, the medical physicist is responsible for the accurate measurement of the radiation output from radiation sources employed in medical use, acceptance testing of equipment and periodic testing of equipment performance, design and implementation of quality assurance programs, design of radiation facilities, and control of radiation hazards. Invariably, the medical physicist is called on to contribute scientific advice and mount the necessary resources to solve complex physical problems related to clinical care that arise continually in many specialized medical areas.

Academic medical physicists are also actively involved in medical research. For example, exciting new developments for improving dose delivery in radiation oncology are occurring as a result of the efforts of medical physicists to implement three-dimensional conformal radiation therapy; that is, radiation therapy in which the dose distribution is conformed to the target volume (cancerous cells) in three dimensions, thus sparing sensitive normal structures. These developments are based on advanced computer hardware and software technology and include developing 3-D radiation therapy treatment planning systems that essentially create a virtual model of the patient and the dose distribution prior to actual treatment, computer-controlled treatment machines, multileaf collimators for beam shaping, and online electronic portal imaging devices for treatment verification. These new technologies provide for radiation therapy techniques that are likely to improve therapeutic ratios that cannot be achieved using traditional planning, delivery, and verification methods.

Medical physicists are also involved in the research and development of new instrumentation and technology for use in diagnostic imaging. These include the exciting areas of computed tomography and magnetic resonance images for displaying detailed cross-sectional images of the anatomy and also imaging using ultrasound and infrared nonionizing radiation.

While radiation oncology and diagnostic imaging research occupys the majority of medical physics research efforts, a small number of medical physicists are also concerned with research in many other areas of medicine. Examples of such research include the application of computers and information theory in medicine, the measurement of blood flow and oxygenation in heart disease research, the measurement and interpretation of bioelectric potentials in mental illness research, measurement of radioactivity in the human body and foodstuffs, and study of the spatial and temporal distribution of radioactive substances in the body.

Typical examples of the various research areas presently under active investigation by medical physicists may be found in scientific journals dedicated to this field, such as *Medical Physics* and *Physics in Medicine and Biology.*

The third area in which the medical physicist may devote time and effort is in providing education and training in medical physics for physicians, beginning medical physicists, and medical technicians, such as radiation therapists and nurses.

Medical physicists are required to have a broad background of science education and experience. Close working relationships with physicians and medical scientists require the medical physicist to be familiar with basic medical science, such as anatomy and physiology, in addition to the need for a sound education in physical science. The minimum degree requirement for a practicing medical physicist is a master's degree. However, university academic appointments and senior positions at many hospitals most often require a doctorate degree.

A bachelor's degree in physics or the equivalent is required for an entrant to the field of medical physics training. The medical physicist must be primarily educated as a physicist to be competent to apply physics to the continually evolving problems and situations that occur in medicine. Hence the bachelor's degree program must be equivalent to one of a physics major at a major college or university. Such programs usually require that over half of the course work be composed of traditional physics and mathematics courses, including advanced courses in electromagnetic theory, electronics, optics, classical mechanics, quantum mechanics, nuclear and modern physics, and calculus (through partial differential equations).

In the second stage of training of a medical physicist, a graduate degree (master's or doctorate) must be obtained. Most often this is in a traditional field of physics (e.g., solid-state or nuclear physics). However, several U.S. universities now offer academic programs in medical physics that lead to a master's or doctorate degree. Typically, two to three calendar years are required for a master's program, whereas a doctoral program—with its emphasis on developing the research capabilities of the individual—will typically involve four or six years of post-baccalaureate education.

In the last phase of training of a medical physicist, the physicist with a master's or doctorate degree typically receives on-the-job training through an informal apprenticeship under established medical physicists. This on-the-job apprenticeship approach is gradually being replaced with organized clinical physics training programs similar to physician residency programs. This practical training is needed to ensure adequate competence of medical physicists when they are hired to provide clinical services. During this phase of training, an individual whose graduate degree is in a traditional field of physics will also need to obtain additional postgraduate training in specialized clinical areas such as anatomy, physiology, and radiation biology.

The majority of medical physicists work in hospitals or other medical care facilities. Frequently, the hospital is associated with a university medical school, and the physicists are members of the academic staff. In many private hospitals, physicists hold professional appointments in one of the clinical departments, frequently radiation oncology or radiology, and are considered members of the professional staff of the hospital. Medical physicists work with a wide variety of health care professionals. In addition to physicians, they frequently work with biologists and engineers. The physicians with whom they work may be specialists in radiation oncology, diagnostic radiology, nuclear medicine, neurosurgery, and so on.

Several scientific and professional societies have been organized to represent medical physics. These include the American Association of Physicists in Medicine (AAPM) and the American College of Medical Physics (ACMP). Membership is also available in several medical societies, such as the American College of Radiology (ACR), the American Society of Therapeutic Radiology and Oncology (ASTRO), the Radiological Society of North America (RSNA), the Radium Society, the American Roentgen Ray Society, and the Society of Nuclear Medicine. Certification boards, such as the American Board of Radiology and the American Board of Medical Physics, award a professional diploma to physicists who pass a set of examinations (written and oral) specific to a medical physics specialty such as radiation oncology physics.

In summary, medical physics is a creative, rewarding, and well-paid profession. While radiation oncology continues to be the major field of employment for medical physicists, new developments and greater sophistication in equipment and procedures for diagnostic radiology and nuclear medicine have brought about a substantial increase in positions available in these two fields. There is no doubt that medical physicists will continue to contribute significantly to technological advances in many diverse areas of medicine, and that these new technologies will require the special skills of medical physicists in their application to patient care.

See also: CAT Scan; Magnetic Resonance Imaging; Nuclear Medicine

Bibliography

Hobbie, R. K., ed. *Medical Physics* (American Association of Physics Teachers, College Park, MD, 1986).

Khan, F. *The Physics of Radiation Therapy,* 2nd ed. (Williams and Wilkins, Baltimore, MD, 1994).

Mould, R. F. *A Century of X-Rays and Radioactivity in Medicine* (Institute of Physics Publishing, Bristol, Eng., 1993).

Purdy, J. A., ed. *Advances in Radiation Oncology Physics: Dosimetry, Treatment Planning, and Brachytherapy* (American Institute of Physics, Woodbury, NY, 1992).

James A. Purdy

MEITNER, LISE

b. Vienna, Austria, November 7, 1878; *d.* Cambridge, England, October 27, 1968; *radioactivity, nuclear physics.*

Meitner's work spanned the development of twentieth century atomic physics from radioactivity to nuclear fission. Albert Einstein called her "our Marie Curie." It is true that Meitner, like Curie, was a brilliant experimental physicist of exceptional prominence.

She was born to intellectual, politically liberal parents, the third of eight children of Philipp (a lawyer) and Hedwig Meitner. Although the family background was Jewish, Judaism played no role in the children's upbringing, and all were baptized as adults, Lise as a Protestant in 1908.

At the time, schooling for Austrian girls ended at age fourteen, but in 1897 Austria opened its universities to women. Lise attended university from 1901 to 1906, the second woman to receive a physics doctorate in Vienna (the first was Olga Steindler in 1903). There she learned physics from Ludwig Boltzmann, a brilliant teacher who gave her the "vision of physics as a battle for ultimate truth" (Frisch, 1970, p. 406). Her doctoral research was experimental; she was introduced to radioactivity by Stefan Meyer in 1906.

There were no jobs for women physicists, however, and in 1907 Meitner went to Berlin to study under another great theoretical physicist, Max Planck, who became her mentor and friend. For research she found a partner in Otto Hahn, a radiochemist just her age. Berlin became her professional home, and she stayed thirty-one years.

With Hahn she found new radioactive species, studied beta decay and beta spectra, and in 1918 discovered protactinium, element 91.

Between 1920 and 1934 Meitner, independent of Hahn, pioneered in nuclear physics. From studies of beta-gamma spectra, she clarified the radioactive decay process by proving that gamma radiation follows the emission of alpha (or beta) particles; her studies of the absorption of gamma radiation verified the formula of Oskar Klein and Yoshio Nishina and indirectly the relativistic electron theory of P. A. M. Dirac. Meitner was among the first to determine the mass of neutrons and to observe the formation of electron-positron pairs.

In 1934 Meitner began studying the products formed when uranium is bombarded with neutrons; the investigation, led by Meitner and including Hahn and chemist Fritz Strassmann, culminated in the discovery of nuclear fission in December 1938. Five months earlier, however, Nazi racial policies forced her to flee Germany, and although she conducted an intense scientific correspondence with Hahn from exile in Stockholm, she was not credited with her share of the discovery. In early 1939 she and her physicist nephew Otto Robert Frisch published the first theoretical explanation for the process and named it fission. Hahn was awarded the 1944 Nobel chemistry prize alone, an injustice that clouded Meitner's reputation in later years.

Meitner had a talent for friendship and a deep love for music and the outdoors. She served in the Austrian army as an x-ray nurse in World War I and retained her Austrian citizenship all her life. In 1943 she was asked to join the atomic bomb project at Los Alamos but refused on principle; years later she said that her "unconditional love for physics" had been damaged by the knowledge that her work had led to nuclear weapons.

See also: FISSION; FISSION BOMB; NUCLEAR PHYSICS; PLANCK, MAX KARL ERNST LUDWIG; RADIOACTIVITY

Bibliography

FRISCH, O. R. "Lise Meitner, 1878–1968." *Biographical Memoirs of the Fellows of the Royal Society* **16,** 405–420 (1970).

MEITNER, L. "Looking Back." *Bulletin of the Atomic Scientists* **20** (11), 2–7 (1964).

MEITNER, L., and FRISCH, O. R. "Disintegration of Uranium by Neutrons: A New Type of Nuclear Reaction." *Nature* **143,** 239–240 (1939).

SIME, R. L. "The Discovery of Protactinium." *Journal of Chemical Education* **63,** 653–657 (1986).

SIME, R. L. "Lise Meitner's Escape from Germany." *Am. J. Phys.* **58,** 262–267 (1990).

SIME, R. L. "Lise Meitner and Fission: Fallout from the Discovery." *Angewandte Chemie, International English Edition* **30,** 943–953 (1991).

SIME, R. L. *Lise Meitner: A Life in Physics* (University of California Press, Berkeley, 1996).

WATKINS, S. A. "Lise Meitner and the Beta-Ray Energy Controversy." *Am. J. Phys.* **51,** 551–553 (1983).

RUTH LEWIN SIME

METAL

Metals are chemical elements with characteristic physical and chemical properties. The physical properties most closely identified with metals are high electrical and thermal conductivity (low resistivity), metallic luster and opacity, and ductility. The property often used to define metals is that the electrical resistivity increases with increasing temperature. The chemical properties characteristic of metals are that they readily lose electrons and thus form positive ions in solution. In the periodic table of the elements, groups IA and IIA (except hydrogen), groups IB-VIIIB, group IIIA (Al and below), Sn and Pb in group IVA, Sb and Bi in group VA, the lanthanides and the actinides, are all considered metals. Mixtures of one or more metals displaying metallic properties are called alloys; some organic compounds with metallic electrical properties are called "organic metals."

Many of the properties of metals are a direct result of their electronic structure. In the terminology used by chemists, atoms of metals have relatively few valence (outer) electrons, which they give up more or less easily, thus attaining a stable rare gas electronic configuration. In the solid state, the metal atoms are arranged on the points of a lattice, often face- or body-centered cubic (atoms on the corners of a cube with added atoms at the centers of all the cube faces or at the center of the cube, respectively). In the simplest view (the Drude–Lorentz free electron theory), the outer electrons of these atoms are not localized on the atoms, but are free to move anywhere in the solid. The solid thus consists of a lattice of positive ions and an "electron cloud"; the ability of the electrons to move freely gives rise to the high electrical and thermal conductivity. The energy of such a free electron is continuously variable; it thus absorbs or reflects light of all wavelengths (above a threshold related to the plasma frequency), accounting for the opacity. This type of bonding is called "metallic," in contrast to the ionic bonding in ionic salts (e.g., sodium chloride, iron oxide), covalent bonding in silicon, diamond, or organic solids, or molecular (van der Waals) bonding in the rare gas solids.

A more rigorous quantum mechanical analysis of the electronic structure shows that the periodic potential of the lattice ions restricts the energies of the electrons to certain bands, separated by energy gaps (forbidden energies). The bands corresponding to the inner electrons are full and the electrons in them do not conduct electricity because they cannot absorb energy from the electric field. When the band is partly empty, like the outer bands of metals, the electrons can absorb energy from an electric field and conduct electricity or heat, or absorb light. The electron waves are not scattered by a perfect lattice, but the thermal motion of the ions gives rise to scattering. The amplitude of this motion increases with increasing temperature, so the scattering and hence the electrical resistivity increase as well.

Ductile metals deform plastically under stress, instead of fracturing in a brittle manner, like, say, minerals or semiconductors. In all of these materials, plastic deformation takes place by the motion of line defects, called dislocations, in the crystal lattice. Dislocations move with less difficulty in metals because the interatomic bonds are neither directional, as in silicon or diamond, nor is there a strong repulsion between ions of the same charge, as is encountered during the motion of a dislocation in an ionic crystal.

See also: ALLOY; CRYSTAL; ELECTRICAL CONDUCTIVITY; ELECTRICAL RESISTIVITY; ELECTRON; ELECTRON, CONDUCTION; ELEMENTS; ION; METALLURGY; STRESS; TEMPERATURE

Bibliography

ABRIKOSOV, A. A. *Fundamentals of the Theory of Metals* (North-Holland, Amsterdam, 1988).

ASHCROFT, N. M., and MERMIN, N. D. *Solid-State Physics* (Holt, Rinehart and Winston, New York, 1976).

KITTEL, C. *Introduction to Solid-State Physics* (Wiley, New York, 1996).

STEVEN J. ROTHMAN

METALLURGY

Metallurgy is the art and science of metals and alloys. The different branches of metallurgy cover the subject from ore processing to the quantum mechanics of electrons in alloys. Extractive, chemical, or process metallurgy is the science and technology of ore beneficiation and of the smelting of metals from ores. Mechanical metallurgy deals with the shaping and the strength of metals and alloys. Physical metallurgy covers the structure and properties of metals and alloys, and is based on the fundamentals of metal physics. The basic questions of metallurgy are: What structure gives rise to the desired properties? How should the metal or alloy be processed to obtain that structure.

History of Metallurgy

The profession of metallurgy goes back to neolithic times; the caveman who first discovered a smelted alloy in a campfire was the first metallurgist. The accidental smelting of bronze, harder than native copper and therefore more useful, developed into the purposeful alloying of tin with copper to make bronze, after which the Bronze Age (ca. 3000–1100 B.C.E.) was named. The manufacture of wrought iron and steel, which was more difficult because of its higher melting temperature, developed more slowly but was well enough known by 1100 B.C.E. to usher in the Iron Age. The metallurgists of these times were craftsmen who made decorative and utilitarian objects from metal, and whose science was based on empirical observation. Experimental studies basic to metallurgy were carried out by the alchemists in the course of their attempts to transmute base metals into gold. Books on actual metallurgical practice appeared in the sixteenth century; the book *De Re Metallica* by Georgius Agricola is a fine early example. The mass production of steel began with Henry Bessemer's furnace in 1856. More reactive metals such as titanium or uranium were commercially produced in the twentieth century. Scientific metallurgy was greatly aided by the application of the microscope to the study of metals (metallography) in the late-nineteenth century. This allowed the study of metallic structures down to the 1 μm (1 μm $= 10^{-6}$ m) level. Another important discovery was that of x-ray diffraction early in the twentieth century; this allowed the crystal structure of metals to be elucidated. On the theoretical side, atomic theory and quantum mechanics laid a sound scientific foundation for the study of metals and alloys, allowing quantitative explanations of some aspects of metallic behavior. The resolution of the optical microscope was greatly exceeded by that of the electron microscope, and the application of this instrument to thin metal foils in transmission allowed the study of defects in metals on an atomic scale.

Because metals have been the most important structural materials since prehistoric times, the scientific study of metals and alloys was well established earlier than the study of other classes of materials, such as ceramics, semiconductors, or polymers. In the last three decades, the same scientific methods have been applied to these materials as were applied earlier to metals, leading to a unified description of materials in terms of the relations among processing, properties, and structure. This field is now called materials science.

The Education of a Metallurgist

Unlike chemists or physicists, whose university courses are concentrated in their major field, metallurgy students must take many courses in chemistry, physics, and engineering, in addition to specialized courses in metallurgy. Beyond the basic curriculum, students planning careers in process metallurgy are also likely to take courses in geology, mining, and civil engineering. A physical metallurgist is more likely to take advanced courses in physics, chemistry, and mechanical engineering. Students with bachelors degrees usually go into industrial jobs; those more interested in research get graduate degrees. Metallurgists can be registered as professional engineers. There are many technical societies for metallurgists worldwide; these hold meetings where the latest technological and scientific discoveries are presented.

The Work of a Metallurgist

Metallurgists work at mines, in the crushing and grinding of the ore and the concentration of metal-bearing minerals. They work in the primary metal producing industries, such as steel mills, where metals are smelted from the ore concentrates, and in the rolling mills where the rods, beams, or sheet are formed. Metallurgists are now very involved in the development of environmentally benign production processes, and the remediation of old metallurgical sites. They work in secondary metallurgical industries, such as the auto, aircraft, and machinery industries, where parts made of metal, such as engine blocks, gears, and tools are produced. The nuclear industry employs metallurgists in the development, fabrication, and reprocessing of nuclear fuel elements. In all these industries, metallurgists do technical work, such as design, process development, testing, quality control, or research, but with many chances to get their hands dirty. Metallurgists also teach in technical colleges and universities, and do research in industrial, government, or university laboratories, but even academic metallurgists keep in contact with metallurgical industry, the basis of the profession.

See also: ALLOY; DIFFRACTION; MATERIALS SCIENCE; METAL; POLYMER

Bibliography

AGRICOLA, G. *De Re Metallica*, translated by H. C. and L. H. Hoover (Dover, New York; [1556] 1950).

DENNIS, W. H. *Metallurgy in the Service of Man* (Pitman, New York, 1961).

SULLIVAN, J. W. W. *The Story of Metals* (American Society for Metals, Cleveland, OH, 1951).

STEVEN J. ROTHMAN

METEOR SHOWER

The term "meteor" applies to the streak of light (lasting typically one-half second) produced when an interplanetary stone (a "meteoroid") enters the upper atmosphere of the earth. Meteoroids enter the atmosphere of earth with relative speeds between 11 and 74 km/s, and are decelerated by the atmosphere of earth (usually at heights between 110 and 70 km). The meteoroid rapidly heats to the boiling point ($\approx$ 2,000 K) and evaporates. The light produced by the meteor is a result of atomic collisions between the evaporated meteoroid material and atmospheric atoms and molecules. Due to their high speeds, even small meteoroids can produce impressive meteors; for example, an object the size of a pea can produce a meteor comparable in apparent brightness to the brighter stars. Those few meteors that are large (and strong) enough to survive atmospheric flight reach the ground as "meteorites."

The meteoroid complex consists of two parts: a "stream" component made up of particles in highly correlated orbits, and a random "sporadic" component. It is believed that most meteoroid streams are formed by the release of dust by a parent comet. When the orbit of the earth intersects that of the meteoroid stream, a "meteor shower" is observed. Meteor showers recur on approximately the same date because the earth is in the same position in its orbit. Some meteor showers vary extensively in activity from year to year because the dust is not spread equally around the orbit. A particularly intense meteor shower (with visual rates of thousands of meteors per hour) is termed a "meteor storm." The last truly impressive meteor storm occurred on the night of November 16–17, 1966, when rates of the order of 60,000 visual meteors per hour were recorded. Collisions with other interplanetary dust particles and the effects of the radiation of the sun gradually disperse stream meteoroid orbits, and shower meteors become sporadic meteors. Therefore the division between shower and sporadic meteors is not precise, and there continues to be debate regarding the validity of a large number of minor meteor showers.

A perspective effect causes the parallel orbits of shower meteors to appear to diverge from a single point (similar to the apparent convergence of straight railway tracks). The constellation in which this point of convergence, or "radiant," seems to lie is used in naming the meteor shower. For example, the impressive August Perseid shower has a radiant within the constellation Perseus. It is important to realize that this is only a perspective effect, and the meteors have no real association with the stars of that constellation. Sometimes the name of a particular star in the constellation (e.g., Eta Aquarid) is added to differentiate between two showers with radiants in the same constellation. The activity of a meteor shower is expressed by the zenithal hourly rate (ZHR), which is the number of meteors a single observer would see in one hour from a dark-sky location, assuming the shower has a radiant directly overhead. Light from the moon or from artificial sources dramatically reduces the meteor rate. The observed meteor rate is highest when the radiant is near the zenith (i.e., directly over-

Table 1   Major Annual Nighttime Meteor Showers

Shower	Date	ZHR (no./h)	Duration (days)	Velocity (km/s)	Parent (? = uncertain)
Quadrantid	Jan. 3	85	0.5	41	Comet Machholz?
Lyrid	Apr. 21	15	2	48	Comet Thatcher
Eta Aquarid	May 5	30	6	65	Comet Halley
S. Delta Aquarid	July 29	20	8	41	?
Perseid	Aug. 12	100	5	60	Comet Swift–Tuttle
Orionid	Oct. 21	20	5	66	Comet Halley
Taurid	Nov. 3	15	30	29	Comet Encke
Leonid	Nov. 17	10	4	71	Comet Tempel–Tuttle
Geminid	Dec. 13	95	3	35	Asteroid Phaethon
Ursid	Dec. 23	20	0.8	34	Comet Tuttle?

head), since this corresponds to perpendicular projection of the stream cross section on the surface of the earth.

Table 1 lists the major nighttime showers. With the exception of the Quadrantid shower, these showers are named after the constellation containing the radiant (the Quadrantid shower is named after the obsolete constellation Quadrans Muralis, in the area of the constellation Bootes). The second column gives the mean date of maximum activity, while the third column provides the zenithal hourly rate. Keep in mind that for some showers (such as the Lyrid and Leonid) this rate varies extensively from year to year. The duration (in days) of appreciable activity is given in column 4. Some showers (e.g., Quandrantid) are highly concentrated with strong displays lasting only a few hours, while others (e.g., Taurid) are spread over weeks. Short-duration showers are more likely of recent formation. Each shower will be visible only when the radiant is above the horizon. The peak of the Eta Aquarid shower occurs during daytime although some of the meteors are observable during the predawn period. The geocentric (relative to Earth) speed of the meteors in each shower is given in kilometers per second.

By comparing orbits, we can frequently determine the parent object for a meteor shower. As shown in Table 1, most meteor showers are derived from comets, although there is a clear link between the Geminid shower and the asteroid Phaethon. One might conclude that most meteors are cometary in origin, and this is probably true for the majority of visual meteors. However, asteroids contribute many of the largest meteoroids, and it is believed that almost all meteorites come from asteroids (cometary material does not have the strength to survive atmospheric flight).

Bibliography

Bone, N. *Meteors* (Sky Publishing, Cambridge, MA, 1993).

Hawkes, R. L. "Meteors" in *Observer's Handbook 1996*, edited by R. L. Bishop (University of Toronto Press, Toronto, Can., 1996).

Kronk, G. W. *Meteor Showers: A Descriptive Catalog* (Enslow Publishers, Hillside, NJ, 1988).

McKinley, D. W. R. *Meteor Science and Engineering* (McGraw-Hill, New York, 1961).

Roggemans, P., ed. *Handbook for Visual Meteor Observations* (Sky Publishing, Cambridge, MA, 1989).

Robert L. Hawkes

METROLOGY

Metrology is the science of measurement. Its origins go back 200 years to the first serious attempts to create a complete system of physical units that would be self-consistent, precisely defined, accessible, and widely used. The pursuit of this goal has led to the creation and continual improvement of the International System of Units (abbreviated SI, for *Système International d'Unités*).

Our understanding of the physical world is based on measurement: "[W]hen you can measure what you are speaking about, and express it in numbers, you know something about it, but when you cannot express it in numbers, your knowledge is of a meager and unsatisfactory kind" (usually attributed to Lord Kelvin). Metrology attempts to meet the measurement needs of science and technology so that knowledge will be neither "meager" nor "unsatisfactory." It is often the case that metrologists are also physicists, chemists, or engineers.

The International System of Units

The SI at present has seven base units, the meter, kilogram, second, ampere, kelvin, mole, and candela. All other physical quantities may be derived from these. Each unit must be unambiguously *defined*. Equally important, the definition must be *realized* experimentally; metrologists must be able to translate the words of the definition into useful procedures and measuring devices. Finally, in order that results obtained by different laboratories be accurate and therefore directly comparable, experimenters must know that their measuring instruments are *traceable* to the appropriate unit definitions. To improve accuracy, the metrology community may put into place new definitions and their practical realizations. The general public is usually unaware of such changes to the SI because the effect of such changes is smaller than the precision of the measurements we make in daily life.

All physical quantities may be measured in terms of the base units, but this is not often practical. For example, 1 V is equal to $1 \text{ kg} \cdot \text{m}^2 \cdot \text{s}^{-2} \cdot \text{A}^{-1}$ yet precise measurement of voltage should not require an instrument capable of simultaneously measuring mass, length, time, and current! Instead, metrologists develop secondary standards, not just for voltage but also for the myriad other physical quantities of scientific and technological importance. Metrologists refer to the best of these secondary standards as *representations* of a particular unit.

Metrology Institutes

In order to provide traceability and hence measurement accuracy to physicists, engineers, and other professionals in the exact sciences, industrialized countries have established national laboratories devoted to metrology. The United States created the National Bureau of Standards (now known as the National Institute of Standards and Technology, or NIST) in 1901. In 1875, the Treaty of the Meter laid the basis for international metrology and continues to provide a mechanism for amending the SI upon the advice of committees of international experts. The treaty also created the International Bureau of Weights and Measures (Bureau International des Poids et Mesures, or BIPM), located in a suburb of Paris.

Metrologists in national laboratories have the task of ensuring that measuring instruments traceable to the definition are available to people who need them. Subdisciplines such as engineering metrology and legal metrology help to meet specialized industrial and commercial needs. In the United States, the customary units still used in commerce and some areas of technology are defined by exact conversion factors to their SI analogs.

Fundamental Constants

At first, artifact standards defined the base units: the meter was defined as the distance between two parallel scratches engraved on a particular bar of platinum–iridium, and the mass of a particular cylinder of the same platinum–iridium alloy defined the kilogram (and, indeed, still does). Metrologists at the BIPM provided national laboratories with copies of the artifact standards and organized periodic checks on the stability of the copies with respect to the international standard. Among its services, the BIPM still provides checks for the unit of mass, since the kilogram is the last base unit defined by an artifact standard.

An artifact, no matter how well-made, may suffer changes. Some physical quantities, however, we believe to be constants of nature. Historically, Newton's universal constant of gravitation was among the first of these to be measured. The concepts of physics that emerged in the first part of the twentieth century postulated or predicted many other fundamental constants, among them the speed of light in a vacuum c, the charge of an electron e, the mass of an atom of ^{12}C, and Planck's constant h. Physicists continually test such postulates and predictions to ever greater precision, and metrological institutes collaborate to provide access to the most accurate standards. Standards based on fundamental constants have replaced artifact standards whenever metrologists have overcome the problems of practical realization and traceability. The meter, for example, is now defined as the distance light travels in vacuum during a specified fraction of a second and is traceable through the wavelength of stabilized lasers. The second no longer depends on the motion of the earth for its definition, but is now derived from a "clock" based on the vibrations of cesium atoms.

An Example

The evolution of modern voltage standards provides a good example of the relationship between the physical sciences and metrology. Thirty years ago, electrochemical cells were the most precise representations of the volt. Complicated chemical equilibria determine the voltage, which is not exactly the same for each cell. In addition, the voltage difference between seemingly identical cells drifts with time.

In 1962 the physicist Brian D. Josephson theorized that microwaves acting on a particular class of superconducting devices should induce voltages quantized in exact steps of $hf/(2e)$, where f is the frequency of the microwaves and h and e are fundamental constants. To check this remarkable prediction, researchers first observed the quantized voltage and then worked closely with metrology laboratories to determine its precise value. The quantized voltage proved to be more stable than the electrochemical cells used to measure it, and Josephson shared the 1973 Nobel Prize in physics. Metrology laboratories continue to improve the precision of representations of the volt based on the Josephson effect.

Improvements in metrology are not always as dramatic as this, but progress is closely linked to the exploitation of the latest developments in science and technology.

See also: ACCURACY AND PRECISION; ATOMIC CLOCKS; JOSEPHSON EFFECT; SI UNITS

Bibliography

CROVINI, L., and QUINN, T. J., eds. *Metrology at the Frontiers of Physics and Technology* (North-Holland, Amsterdam, 1992).

INTERNATIONAL ORGANIZATION FOR STANDARDIZATION (ISO). *International Vocabulary of Basic and General Terms in Metrology,* 2nd ed. (International Organization for Standardization, Geneva, Switzerland, 1993).

PETLEY, B. W. *The Fundamental Constants and the Frontier of Measurement* (Adam Hilger, Bristol, Eng., 1988).

RICHARD S. DAVIS

MICHELSON, ALBERT ABRAHAM

b. Strzelno, Poland, December 19, 1852; *d.* Pasadena, California, May 9, 1931; *optics.*

Michelson, the eldest child of Samuel and Rosalie, was born in Prussian-occupied Poland in 1852. Samuel was a merchant, and in 1856 he emigrated with his wife and several children to the United States, where he set up business in the mining town of Murphy's Camp, California. While attending high school in San Francisco, Michelson developed a deep interest in the science of optics. Upon graduation, he applied to his congressman for nomination to the U.S. Naval Academy at Annapolis, where he hoped to pursue these studies. Although he was not initially successful in his request, persistence and a personal interview with President Ulysses S. Grant eventually gained him admission.

After Michelson completed his studies, he became an instructor at the Academy. The facilities at the Academy allowed him to begin his lifetime of experimentation in optics, and he made his first measurement of the speed of light in 1878. He later repeated this experiment many times in different circumstances, and on his death in 1931 he was still involved in a determination of the speed of light in an evacuated tube a mile long.

The most remarkable experiment of Michelson's career was performed in 1887, after he had left Annapolis to become the first professor of physics at the Case School of Applied Science (now Case Western Reserve University) in Cleveland, Ohio. There he collaborated with chemistry professor Edward W. Morley in an extraordinarily precise measurement designed to determine whether the speed of light depended on the direction in which it was traveling.

The motivation for the Michelson–Morley experiment concerned the wave nature of light. In 1887 it was thought that a light wave could only travel if there were some material medium to carry the wave. This substance was called the luminiferous ether. Michelson argued that Earth, in its passage around the Sun, must encounter a "headwind" as it travels through the ether. This wind would have the effect of slowing down any ray of light traveling upwind and speeding up a ray going downwind. The ingenious apparatus first split a beam of light into two parts by means of a partially silvered mirror, which reflected some light and transmitted the rest. After traveling off in different directions, these two beams were then brought back together by a system of mirrors. By looking for the patterns formed when two light waves interfere with each other, the experimenters were able to see if there was any difference in the time taken by the two beams. To their astonishment, they found no difference at all!

This result challenged all the existing theories of the nature of light. It was not explained until 1905, when Albert Einstein published his special theory of relativity and showed that the whole concept of an ether was unnecessary. It is ironic that when in 1907 Michelson became the first American scientist to win a Nobel Prize, he was honored not for the seminally important Michelson–Morley experiment but for the more mundane achievement of comparing the wavelength of light with the length of the standard meter.

Michelson's dedication to his science left him little time for other activities, although he was an enthusiastic amateur violinist and, after some turbulent early years, a devoted family man. Michelson fought his way back from mental illness to make his finest achievements, and he stands as a giant in the history of optics. The instrument he invented to perform the Michelson–Morley experiment remains a valuable analytical tool in laboratories all over the world.

See also: ETHER HYPOTHESIS; LIGHT, SPEED OF; LIGHT, WAVE THEORY OF; MICHELSON–MORLEY EXPERIMENT

Bibliography

JAFFE, B. *Michelson and the Speed of Light* (Doubleday, New York, 1960).

LIVINGSTON, D. M. *The Master of Light: A Biography of Albert A. Michelson* (University of Chicago Press, Chicago, 1973).

MICHELSON, A. A. *Light Waves and Their Uses* (University of Chicago Press, Chicago, 1903).

MICHELSON, A. A., and MORLEY, E. W. "Influence of Motion of the Medium on the Velocity of Light." *American Journal of Science* **31,** 377–386 (1886).

MICHELSON, A. A., and MORLEY, E. W. "On the Relative Motion of the Earth and the Luminiferous Ether." *American Journal of Science* **34,** 335–339 (1887).

MICHELSON, A. A., and MORLEY, E. W. "On a Method of Making the Wavelength of Sodium Light the Actual and Practical Standard of Length." *Am. J. Sci.* **34,** 429 (1887).

SHANKLAND, R. S. "Michelson–Morley Experiment." *Am. J. Phys.* **32,** 16–35 (1964).

LAWRENCE M. KRAUSS

PHILIP TAYLOR

MICHELSON–MORLEY EXPERIMENT

The Michelson–Morley experiment, performed in 1887 by Albert A. Michelson, then of the Case School of Applied Science, and Edward Morley, of the neighboring Western Reserve University, both in Cleveland, Ohio, established that the speed of light in various different directions was unaffected by the motion of Earth, and provided definitive empirical evidence against the existence of an "ether" that was assumed to be necessary to propagate electromagnetic disturbances. The ether was later shown to be theoretically untenable by Albert Einstein in his special theory of relativity (1905), in which he demonstrated that the speed of light is a universal constant, unaffected by the motion of observer or source. This fact is incompatible with electromagnetic waves being considered as disturbances propagating in a background medium such as the ether.

The Michelson–Morley experiment followed on earlier work by Michelson during his graduate studies in Berlin, under Hermann von Helmoltz. There, in 1880, he developed the Michelson interferometer, and at the same time performed a preliminary experiment in Potsdam designed to detect the effects of Earth's motion through the ether that laid the basis of the experiment performed seven years later. The basis of the experiment was simple. If two light rays were to travel identical path lengths, one parallel to Earth's motion through the ether, and one perpendicular to it, the rays should take different times to complete the round trip. If the speed of light is c, the round-trip path length is $2L$, and the assumed speed of Earth through the ether is $v \ll c$, then the approximate time difference between the two round-trip travel times is Lv^2/c^3. By starting out with a single light beam, which was then split to perform two separate traverses, and then recombined, the effect of the time difference would be that one beam would arrive slightly out of phase with the other. As a result, any interference fringes present when the beams were in phase should be observed to shift slightly. If the apparatus were turned around so that the beam that originally traveled parallel to Earth's motion became perpendicular, the fringe shift should be maximal. The expected shift in this case was calculated to be .04 times as large as the distance between adjacent fringes—an effect that should have been measurable. In his Potsdam experiment, Michelson observed no such shift.

After moving to Case, Michelson met Morley, and in 1884 the two began a collaboration that successfully repeated an experiment by Armand Fizeau designed to probe the effects on the velocity of light when it passes through a moving stream of water. At the urging of Lord Rayleigh to redo Michelson's earlier Potsdam experiment, Michelson and Morley

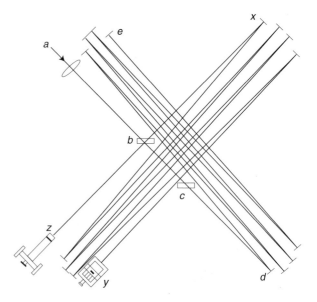

Figure 1 Arrangement of the sixteen mirrors for the Michelson–Morley experiment.

began their famed experiment in April 1887. Two important improvements were made in the experiment. First, the heavy sandstone slab bearing the optical components was floated in a mercury bath. This not only allowed the apparatus to rotate smoothly but also isolated it from outside vibrations, something that had been a major problem in the Potsdam experiment. Next, the light path was extended to 36 ft by using sixteen mirrors instead of two (see Fig. 1). With a path that was almost 10 times longer, the expected shift due to the motion of the earth through the ether was now predicted to be about 0.4 times the distance between fringes.

The experiment was performed between July 8 and 12, and Michelson reluctantly reported the result to Rayleigh in a letter written shortly thereafter:

The Experiments on relative motion of the earth and ether have been completed and the result is decidedly negative. The expected deviation of the interference fringes from zero should have been 0.40 of a fringe—the maximum displacement was 0.02 and the average much less than 0.01—and then not in the right place. As displacement is proportional to the squares of the relative velocities it follows that if the ether does not slip past [earth] the relative velocity is less than one sixth of the earth's velocity.

The results of the Michelson–Morley experiment filtered through the physics community and caused great confusion. Hendrik Lorentz wrote to Rayleigh in 1892: "I am totally at a loss to solve the contradiction. . . . Can there be some point in the theory of Mr. Michelson's experiment which has yet to be overseen?" The Irish physicist George FitzGerald was perhaps the first to accept the results of the experiment at face value. In 1889 he wrote: "I would suggest that almost the only hypothesis that can reconcile this . . . is that the length of material bodies changes, according as they are moving through the aether . . . by an amount depending on the square of the ratio of their velocities to that of light."

FitzGerald had in fact hit upon what would be the ultimate solution of the Michelson–Morley paradox. However, the origin of the length contraction would not be clear for sixteen more years, when Albert Einstein published his special theory of relativity. According to this theory, the speed of light measured by any two observers in constant relative motion should be the same. This not only banished the ether as a background-fixed-medium in which light might travel, but it also required both the length

contraction postulated by FitzGerald and time dilation. While Einstein is said to have had no conscious recollection of the Michelson–Morley experiment when developing his theory, he could have asked for no better experimental verification.

See also: ETHER HYPOTHESIS; INTERFEROMETRY; LIGHT; MICHELSON, ALBERT ABRAHAM; RELATIVITY, SPECIAL THEORY OF; TIME DILATION

Bibliography

MICHELSON, A. A., and MORLEY, E. W. "On the Relative Motion of the Earth and the Luminiferous Ether." *American Journal of Science* **34**, 335–339 (1887).

LIVINGSTON, D. M. *The Master of Light: A Biography of Albert A. Michelson* (University of Chicago Press, Chicago, 1973).

SHANKLAND, R. S. "Michelson–Morley Experiment." *Am. J. Phys.* **32**, 16–35 (1964).

LAWRENCE M. KRAUSS

MICROPHONE

A microphone is a transducer that produces an electrical signal when actuated by sound waves. Microphones may be designed to respond to either variations in air pressure due to the sound wave or to variations in particle velocity (or pressure gradient) as the sound wave propagates. Most microphones in common use are pressure microphones.

Important parameters for describing the characteristics of a microphone include sensitivity, linearity, frequency response, and directivity. The sensitivity of a microphone may be expressed in terms of either the voltage output or the power output for a given sound pressure (usually taken to be either 1 μbar = 0.1 Pa or else 10 μbar = 1 Pa). Linearity refers to the extent to which the output is directly proportional to the input; any nonlinearity in the microphone leads to distortion. A high-quality microphone will operate over a wide range of dynamic levels (as much as 80 to 100 dB).

The on-axis frequency response of a microphone is an important property of microphones. The microphone in a sound level meter or other measuring instrument should have a flat (uniform) response over a wide frequency range. On the other hand, a microphone used for speech reproduction might very well have a reduced response at low frequency to reduce low-frequency background noise.

The directivity or directional response also is an important parameter. For many uses, the microphone should be omnidirectional; it should receive sound equally well from all directions. For other applications, it is desirable to have a microphone with maximum sensitivity in one direction only. One such application is in sound systems in which acoustic feedback may lead to oscillation if the microphone picks up sound from a loudspeaker. A unidirectional microphone with a heart-shaped pickup pattern, popularly known as a cardioid microphone, is widely used in entertainment applications. It is relatively difficult to obtain a cardioid response pattern over a wide range of frequencies, however, and many cardioid microphones have an uneven frequency response, especially a roll-off at bass frequencies. For even greater directivity, line microphones or reflector microphones can be used. A reflector microphone consists of a microphone element at the focus of a parabolic reflector, giving it a very high directivity.

A pressure microphone generally has a thin diaphragm connected to some type of electrical generator, which may be a piezoelectric crystal (crystal microphone), a moving coil (dynamic microphone), or a variable capacitor (condenser microphone). Crystal microphones use piezoelectric crystals that generate a voltage when deformed by a mechanical force. Crystal microphones are inexpensive and give a relatively large electrical output voltage, making them convenient for use in portable sound equipment.

In a dynamic or magnetic microphone, an electrical signal is generated by the motion of a conductor in a magnetic field. The electrical voltage generated by a dynamic microphone is generally quite small, but the impedance of the voice coil also is small, so the power generated is not necessarily small. The low source impedance is a distinct advantage when the microphone is located a substantial distance from the amplifier. Dynamic microphones are rugged, have a broad frequency response, and are able to withstand the high sound levels that occur in popular music.

In a condenser or capacitor microphone, one plate of the capacitor is usually the thin movable diaphragm and the other plate is the fixed backing plate. A condenser microphone has a very high source impedance, and for this reason a preamplifier is often incorporated into the microphone itself. The diaphragm can be very light, and thus a condenser microphone is capable of excellent response at high frequencies. In an electret-condenser microphone, the diaphragm consists of a thick plastic foil with a very thin layer of metal. The foil has been given a permanent electrical charge by a combination of heat and high voltage or by electron bombardment during manufacture. Electret-condenser microphones avoid the need for a high-voltage bias that is required in condenser microphones with an air dielectric.

See also: PRESSURE, SOUND; SOUND; TRANSDUCER

Bibliography

BERANEK, L. L. *Accoustics* (Acoustical Society of America, New York, 1986).

OLSON, H. F. *Modern Sound Reproduction* (Van Nostrand-Reinhold, New York, 1972).

ROSSING, T. D. *Science of Sound,* 2nd ed. (Addison-Wesley, Reading, MA, 1990).

THOMAS D. ROSSING

MICROSCOPE

See ELECTRON MICROSCOPE; SCANNING TUNNELING MICROSCOPE

MICROWAVE SPECTROSCOPY

See SPECTROSCOPY, MICROWAVE

MILKY WAY

The Milky Way is one of nature's most beautiful sights. Seen from a dark place, away from the lights of a city, it looks like a brilliant, luminous, wandering path of light that splits the heavens in two. Unfortunately it has become very difficult to see the Milky Way well, owing to the intense lights of civilization, which light up the sky and obscure our view of the cosmos. In much of the world, even farm country, there is bright illumination that prevents people from having the experience of beholding it. In the mountains, the forest, or at sea, one can often see the Milky Way well and can note its broad extent, its bright condensations, and its peculiar dark lanes, bays, and islands.

The Milky Way is most conspicuous in the direction of the constellation Sagittarius, visible to the South in summer from the Northern Hemisphere. People fortunate enough to live in the Southern Hemisphere can have a glorious view of the bright Milky Way in Sagittarius, which crosses overhead in the winter. Also bright is the complex structure in Carina (in the South) and the fainter, but interesting Milky Way pathway through Cygnus and in Cassiopeia (in the North). Another well-known Milky Way segment passes through the bright equatorial constellation of Orion.

Nature of the Milky Way

The Milky Way was given its name by classical scientists, especially the Greeks and Romans, who called it the Via Galactica ("road of milk" in Latin). To these scientists and philosophers, its nature was a mystery but its appearance was like that of a stream of milk. Galileo was the first human to see the true nature of the Milky Way. He was the first to turn a telescope toward the sky and when he looked in the direction of the Milky Way, what had before appeared to be a diffuse, hazy luminosity was now resolved into thousands of individual stars. He was astounded and delighted by this discovery. In his book *The Starry Messenger* published in 1610, he said that the discovery resolved "all the disputes that have vexed philosophers through so many ages." With a telescope he was able to prove that the "galaxy" was made up of stars and clusters of stars, with the number of faint ones "quite beyond calculation."

Another milestone in the history of our understanding of the Milky Way occurred in the late eight-

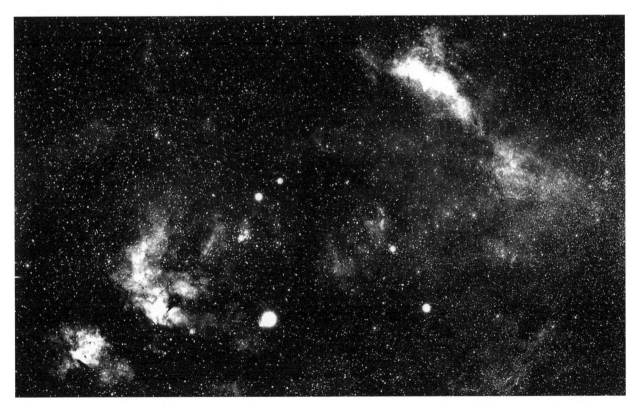

Figure 1 The Milky Way in Cygnus, showing myriad faint stars, as well as star clusters, dust lanes, and gas clouds (courtesy of Rajiv Gupta).

eenth century when the astronomers William and Caroline Herschel used their new, large telescopes to explore the sky. Among their many accomplishments was the mapping of star densities across the width of the Milky Way. In this way they endeavored to establish the true shape of the Milky Way Galaxy, by assuming that the brightnesses of the counted stars were a measure of their individual distances. The first complete map of the galaxy was constructed, which showed it to be a rather flat disk of stars with the sun near to the center. In one direction the map indicated a split into two layers, corresponding to the rift in the Milky Way visible in the direction of Cygnus through Scorpius.

This kind of exploration of the Milky Way continued into the nineteenth and early twentieth century. For example, the astronomer Jacobus Kapteyn used the apparent motions of stars to gauge their distances and built what became known as the "Kapteyn universe," a model of the cosmos based on a stellar census that was refined from the Herschel approach by the addition of information on the stellar distances. This model of the universe was rather similar in appearance to that of the previous century and was quite wrong for two remarkable reasons.

The first reason for the inadequacy of the Kapteyn universe was the lack of recognition (before about 1920) that the Milky Way is heavily shrouded in obscuring dust. Rather than seeing the totality of stars, we can see only the nearest ones in the Milky Way region. There is so much dust that our view comprises only about 10 percent of the stellar system to which the sun belongs. This fact was first demonstrated by Harlow Shapley, who measured the distances to giant star clusters, called "globular clusters," which surround the Milky Way disk in a nearly spherical halo. Shapley's work showed that the true center of the galaxy is off toward the constellation Sagittarius, some 30,000 light-years from the sun. The previous misapprehension about the apparent central location of the sun was caused by the surrounding dust, which permitted astronomers to look out only a few hundred light years in all directions along the Milky Way (away from the plane of the Milky Way in the sky, there is little or no dust, so we can see much farther). The dust is not uniformly distributed, but rather occurs in lanes, clouds, and complexes. The deep dust lanes explain Herschel's bifurcation; rather than a lack of stars in these areas, there are thick obscuring dust clouds.

The second reason for the failure of the Kapteyn universe is that before 1925 most astronomers thought that the Milky Way comprised the entire universe. With the discovery of external galaxies by Edwin Hubble and his contemporaries, however, it was found that our galaxy is just one of millions of similar stellar systems and that the universe extends out from the sun for billions of light-years.

Further understanding of the Milky Way resulted from a particularly clever analysis of stellar motions performed by Jan Oort, who found that the pattern of stellar velocities near the sun could be best explained in terms of a rotation of the Galaxy. He found that the sun and its stellar neighbors are orbiting around the center of the Milky Way at velocities of about 250,000 m/s. This kind of motion was also found to be present in distant external galaxies, such as the Andromeda Galaxy, which have a spiral structure. Oort suggested that our Milky Way might be a spiral galaxy, but optical recognition of such a structure was not forthcoming because of the curtain of obscuring dust.

Structure of the Milky Way

The true structure of the Milky Way Galaxy was not demonstrated until the mid-1950s, when spectroscopic (involving the use of wavelength-separated images of stars) results on the distances of highly luminous stars in physical groups ("stellar associations") showed the presence of three spiral arms near the sun's position. Only the nearest fragments of these arms could be seen optically, but their positions and shapes strongly suggested that a large-scale spiral shape probably exists for the Milky Way, as it does for so many nearby galaxies.

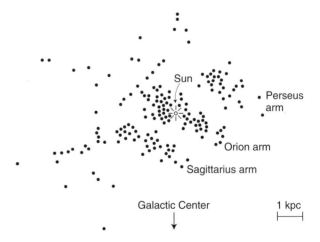

Figure 2 Spiral arms seen optically near the Sun.

An even more convincing demonstration of structure occurred when it became possible in the late 1950s to map the neutral hydrogen gas in the Milky Way. Because of its long wavelength (0.21 m), the radiation emitted by cool neutral hydrogen can penetrate through the thick dust, presenting us with an unobscured picture of our galaxy. Unfortunately, it is not possible to measure the distance to the individual gas clouds that make up the disk of the Milky Way, but measuring the velocities using the Doppler effect allows astronomers to calculate kinematic distances that are statistically useful. The picture presented by the hydrogen data is that of a beautiful spiral with long sweeping arms.

In terms of the Hubble classes of galaxy types, our galaxy is an Sbc spiral with the presence of a mild central bar. It is approximately 100,000 light-years in diameter, making it one of the larger galaxies in nearby space. It has a thin disk of young stars, gas, and dust that is surrounded by a thicker disk of older stars. All of this is enclosed in a halo of very old (15 billion-year-old) stars, which is rotating very slowly compared to the rapid rotation of the disk.

The detailed structure of the Milky Way Galaxy is still only partly known, largely because of our position in the dusty disk. We cannot see the forest very well because we are among the trees. To understand the structure of galaxies, it is better to study external examples, such as M33 or M51, where the clearly spread out spiral arms facilitate detailed study. There are two areas, however, where our galaxy is very important for a physical understanding of galaxies: the content and the dynamics of galaxies, which we can explore fairly thoroughly in spite of the optical obscuration in the disk.

Contents and Dynamics of the Milky Way

The four important components of our galaxy (which are also found in most other galaxies) are stars, gas, dust, and dark matter. The wide variety of kinds of stars of the Milky Way, their physical structure, chemical composition, evolution, and destiny are now well known in most cases. The location of different kinds of stars in the Milky Way is correlated

Figure 3   NGC 7822, a large gaseous nebula in the northern Milky Way in Cepheus (courtesy of Rajiv Gupta).

with their properties. The oldest stars, for example, are located primarily in the halo and the youngest stars are in the spiral arms. The halo stars are almost pure hydrogen and helium, while the disk stars contain significant amounts of heavier elements. These kinds of correlations are explained in terms of the evolutionary history of the Milky Way; the first stars to form in the young Milky Way formed when it was still a nearly spherical cloud of hydrogen and helium gas. As it collapsed because of its rotation, more stars formed and with time the prestellar material became contaminated with heavier elements (mainly carbon, nitrogen, and oxygen, but also small amounts of the other elements) formed by the evolution and death of massive stars. At various times in the history of the Milky Way there may have been other events; some evidence suggests that "we" have captured nearby smaller galaxies, which have thus contributed to the complexities of our galaxy's structure. Examples of such galactic cannibalism are found out among the distant galaxies.

In addition to the single (and multiple) stars, there are also dense groups of stars known as "star clusters." The largest of these are the globular clusters, which can contain as many as 100,000 stars and can be as large as 100 light-years in diameter. These clusters are made up of old stars and they inhabit the galactic halo. Smaller and usually younger clusters (called "open" or "galactic" clusters) are found in the plane of the Galaxy. The youngest are concentrated in the spiral arms, where they were formed, while older examples often lie in-between the arms, where they have drifted over their lifetimes. The globular clusters have highly elliptical orbits around the galactic center, while the open clusters orbit it nearly circularly, like the sun.

Most of the gaseous component of our galaxy is hydrogen. Neutral atomic hydrogen is easiest to detect and is common throughout the disk of the Milky Way. The total mass of hydrogen is about 5 percent of the mass of stars. Molecular hydrogen also exists, especially in areas thick with dust where new stars are being formed. These areas also contain many other molecules, dozens of which have been detected, including carbon dioxide, carbon monoxide, water vapor, ammonia, methyl alcohol, and many others. The gas generally rotates around the center of the Milky Way Galaxy with the disk stars, and it extends outward from the center at least as far as the most distant disk stars, probably farther.

The dust content of the Galaxy is much smaller, about 1 percent of the gas content by mass. It lies in the thin plane of the Milky Way and is concentrated in positions close to the spiral arms. It is heavily involved in the physics of star formation.

The "dark matter" content of the Milky Way remains a mystery. First discovered in the 1960s, the nature of dark matter is not known. Several suggestions have been proposed; for example, one possibility is that it is some new kind of elementary particle. It may make up most of the Galaxy's mass, which is measured (from its rotation rate) to be almost one million million times the mass of the sun. The mysterious dark matter is also found in external galaxies. It probably makes up most of the universe and its nature is one of the great unsolved puzzles of science.

See also: DARK MATTER; GALAXIES AND GALACTIC STRUCTURE; GALILEI, GALILEO; HUBBLE, EDWIN POWELL; STARS AND STELLAR STRUCTURE

Bibliography

BOK, B. J., and BOK, P. F. *The Milky Way,* 5th ed. (Harvard University Press, Cambridge, MA, 1981).

MIHALIS, D., and BINNEY, J. *Galactic Astronomy,* 2nd ed. (W. H. Freeman, San Francisco, 1981).

WHITNEY, C. F. *The Discovery of Our Galaxy* (Knopf, New York, 1971).

PAUL HODGE

MILLIKAN, ROBERT ANDREWS

b. Morrison, Illinois, March 22, 1868; *d.* Pasadena, California, December 19, 1953; *electron physics, atomic physics, cosmic rays.*

Millikan liked to portray his family as deriving from "typical American pioneer" stock. His grandfather, Daniel Franklin Millikan, had once been a shoemaker and tanner in Stockbridge, Massachusetts, before moving his family west to seek the opportunities of the frontier. Millikan's father, Silas Franklin Millikan, was a Congregational minister; his mother, the former Mary Jane Andrews, at one time served as a dean at Olivet College. These three personas of Millikan's heritage, pioneer, divine, and administrator, can serve to define the main characteristics of

Millikan's life in science: a brilliant and innovative researcher; a fervent proponent of science as a means to moral, spiritual, and social progress; and a visionary yet effective institutional manager. In playing these varied roles with both aplomb and energy, Millikan made himself one of the premier American scientists of the first half of the twentieth century.

From an early age, Millikan imbibed the self-reliant attitude and deep sense of personal responsibility central to his Protestant heritage. These ideals motivated him throughout his life. After a first-class education at his local high school in Maquoketa, Iowa, Millikan entered his parents' alma mater, Oberlin College. In 1891 he took his B.A. degree in physics, remaining at Oberlin for two additional years to earn his M.A. in the college's new "science curriculum" before beginning his doctoral studies in physics at Columbia University in New York City in the fall of 1893. Owing to the acknowledged leadership of Europe in the study of physics, Millikan traveled to the Continent, after completing his Ph.D. in 1895, to acquaint himself with the current research frontiers being pursued there. During his year there, Millikan met leading figures of modern physics including Max Planck, J. J. Thomson, and Wilhelm Röntgen.

When Millikan returned to the United States in 1896, a position as an instructor at the University of Chicago awaited him. Millikan proved to be an excellent teacher, spending most of his first decade establishing a physics curriculum at the new university and writing what became the standard textbook for an entire generation of American physicists. Having made his reputation as a teacher, Millikan began around 1908 to pursue a research agenda that would culminate more than a decade later in the Nobel Prize for physics.

The head of Chicago's physics department during this time was Albert A. Michelson, who viewed precise experimental measurement of phenomena as the appropriate means to new discoveries. Following in this tradition, Millikan chose to attack a fundamental problem that seemed to call for this approach: determining the unit of electric charge. Although it was widely accepted at the time that the electron was the particle of electric charge, the value of its charge was somewhat uncertain.

Millikan set about finding a way to determine this value precisely, undertaking a series of experiments in which he progressively refined the measurement techniques applied to the problem by previous researchers. The means that Millikan settled on was a falling oil-drop method, in which he observed the rate of descent of a single electrically charged oil droplet in an electric field that could be switched on and off. By measuring the difference in the rate of the droplet's descent under gravity and when slowed by the retarding force of the field, Millikan was able to demonstrate that the charge on the oil drop was a multiple of a quantity e. He determined this constant, the charge of a single electron, with increasing accuracy throughout the 1910s.

Millikan then moved on to a research project with a similar motivation on a related topic: a "direct" determination of h, Planck's constant relating the energy and wavelength of electromagnetic radiation. Millikan, who was committed to the classical view of electromagnetic radiation and the wave theory of light, hoped these experiments would challenge Einstein's explanation of the photoelectric effect and its underlying assumption of the wave–particle duality of light. His results, however, confirmed Einstein's formulation and yielded a value of h that remained the standard for more than two decades. For these two projects, Millikan was awarded the Nobel Prize for physics in 1923.

Prior to winning the prize, Millikan had been lured in 1921 to the California Institute of Technology in Pasadena by George Ellery Hale and Arthur A. Noyes to assume dual roles as the director of Caltech's newly established Norman Bridge Laboratory of Physics and as the chairman of the Institute's executive council. At Caltech, Millikan set about the task of transforming the Institute into a major scientific research center. By securing funds from wealthy Southern California philanthropists and recruiting top scientists to positions in the physical, chemical, biological, and earth sciences, Millikan rapidly won for Caltech both the financial resources and the scientific talent to match other American scientific research universities.

At Caltech, Millikan continued his experimental work while serving in effect as the Institute's president. Millikan's first research project concentrated on the spectra of ultraviolet radiation produced by the lighter elements, especially those stripped of valence electrons. These investigations revealed features of spectra that contributed to the elaboration of the concept of electron spin. A second program focused on what he termed "cosmic rays." Through a series of experiments on ionization at high altitudes and below the surface of two lakes, Millikan proposed that cosmic rays must be photons of enormously high energy originating in space. Millikan later gave an account of the origin of cosmic rays as part of the "atom-building" process, or what he liked to call the "birth cries" of atoms. In this theory of the creation of matter, Millikan quite explicitly felt he had come to a closer understanding of the Creator. Although his explanation of cosmic rays and his atom-building theory were quickly discredited, largely by Arthur H. Compton, Millikan maintained his deep faith in the agreement of the findings of science with religious belief.

In the course of his career, Millikan became the most visible and one of the most influential scientists in the United States. He was instrumental in the formation of the National Research Council, and he served as President of the American Association for the Advancement of Science in 1929. Upon his death in 1953, Millikan left behind a legacy of public service, institutional leadership, and scientific excellence that forms the basis for his lasting reputation as a one of America's greatest scientists.

See also: CHARGE, ELECTRONIC; COSMIC RAY; PHOTO-ELECTRIC EFFECT; PLANCK CONSTANT; WAVE–PARTICLE DUALITY, HISTORY OF

Bibliography

DuBRIDGE, L. A., and EPSTEIN, P. "Robert Andrews Millikan." *National Academy of Sciences Biographical Memoirs* **33,** 241–282 (1959).

GOODSTEIN, J. *Millikan's School: A History of the California Institute of Technology* (W. W. Norton, New York, 1991).

KARGON, R. H. *The Rise of Robert Millikan* (Cornell University Press, Ithaca, NY, 1982).

MILLIKAN, R. A. *Science and Life* (Pilgrim Press, Boston, 1924).

MILLIKAN, R. A. *Evolution in Science and Religion* (Yale University Press, New Haven, CT, 1927).

MILLIKAN, R. A. *Science and the New Civilization* (Scribners, New York, 1930).

MILLIKAN, R. A. *The Autobiography of Robert A. Millikan* (Prentice Hall, New York, 1950).

DAVID A. VALONE

MIRROR, PLANE

A mirror is a highly reflecting smooth surface that possesses the ability to form optical images. Early mirrors were made of polished copper and bronze.

A mirror in perfect condition was unearthed near the pyramid of Sesostris II (ca. 1900 B.C.E.) in the Nile Valley of Egypt. Plane mirrors are now commonly fabricated by coating a flat sheet of glass or plastic with a thin layer of silver or aluminum. The surface should be smooth compared to the wavelength of visible light, that is, 400–700 nm. If a beam of light strikes a mirror, the angle of reflection (angle between reflected beam and perpendicular to the surface) equals the angle of incidence (angle between incident beam and perpendicular) according to the law of reflection.

Mirrors are very useful because of their ability to form images of both luminous and nonluminous objects. The ray diagram in Fig. 1a demonstrates this for a plane mirror M. Light rays diverge from the point object O in all directions. After reflection by the mirror the light rays continue to diverge; however, if the reflected rays are extended behind the mirror, simple geometry shows that they appear to eminate from a corresponding image point I. This image is positioned directly across the mirror from the object and as far behind the mirror as the object is in front of the mirror. This is referred to as a vir-

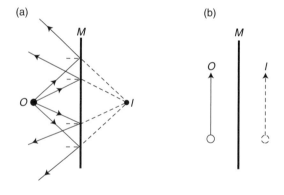

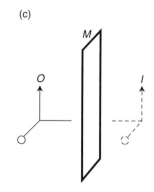

Figure 1   Image formation in a plane mirror.

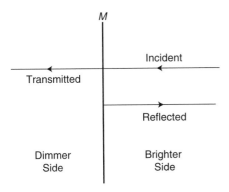

Figure 2 A half-silvered mirror.

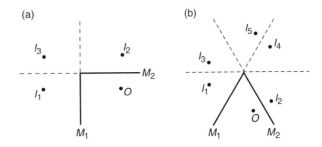

Figure 3 Images formed when two plane mirrors are positioned (a) perpendicular to each other and (b) at a 60° angle to each other.

tual image since light rays do not actually pass through I. An observer can view this type of image, however, it cannot be projected on a screen. For an extended two-dimensional object, applying the above rules to each point on the object leads to an upright virtual image of the same size as the object (Fig. 1b). Applying the same rules to an extended three-dimensional object produces the virtual image shown in Fig. 1c. The only additional feature of this image is a right-to-left inversion, that is, the image of a right hand in the mirror is a left hand.

A half-silvered mirror is a plane mirror containing an extremely thin metal coating such that only about 50 percent of the incident light is reflected, the remaining 50 percent being transmitted. If such a mirror M separates two spaces of unequal illumination, it behaves much like a normal mirror to a viewer on the brighter side, but to a viewer on the dimmer side it appears more like a window (Fig. 2). So-called reflection glass is commonly used in modern office buildings because of its high energy efficiency properties.

If two plane mirrors are placed together, multiple reflections can lead to the formation of multiple images, the number depending on the angle of inclination. If two mirrors M_1 and M_2 are placed together perpendicularly, three virtual images are produced for a point object O: I_1 by single reflection from M_1, I_2 by single reflection from M_2, and I_3 by double reflection from M_2 and M_2 (Fig. 3a). If the two mirrors are tilted by 60° as in a simple kaleidoscope Fig. 3b, five virtual images are formed: I_1 and I_2 by single reflection, I_3 and I_4 by double reflection, and I_5 by triple reflection. If two plane mirrors are parallel, for example, on opposite walls of an elevator, an infinite number of images will be formed in both directions along a line perpendicular to the mirrors. If three plane mirrors are positioned so that they are mutually perpendicular, seven virtual images are formed by various multiple reflections. Such an arrangement of mirrors is called a "corner reflector" and possesses the interesting property of reflecting any incident beam of light back in the direction it came from. Astronauts placed corner reflectors on the moon so that a laser beam from Earth could be reflected back to Earth.

See also: IMAGE, OPTICAL; IMAGE, VIRTUAL; LIGHT; REFLECTION; WAVELENGTH

Bibliography

FALK, D. S.; BRILL, D. R.; and STORK, D. G. *Seeing the Light* (Wiley, New York, 1986).

HECHT, E. *Optics*, 2nd ed. (Wiley, New York, 1987).

HEWITT, P. G. *Conceptual Physics*, 7th ed. (HarperCollins, New York, 1993).

ROBERT MORRISS

MKS UNITS

See SI UNITS

MODELS AND THEORIES

The words "theory" and "model" are used in physics in a variety of ways. For example, the Bohr atom is a model that is discussed both in the foundations of atomic theory and in the origins of quantum theory. The electromagnetic theory of light is part of classical field theory. There is also the wave theory of light, and wave and particle models for light. The contemporary theory of gravitational attraction is provided by the general theory of relativity, one of whose models is designated the Schwarzschild black hole. The general theory of relativity is often regarded as a generalization of the special theory of relativity. All these usages of "model" and "theory" share a role in the explanatory structure of physics. One account of this explanatory structure that organizes most of these usages is as follows.

At the broadest and highest level, activity in physics in most any era is carried out under the informal rubric of what has been called a paradigm, a general research tradition. Such a tradition is really a whole world view, including a commitment to certain forms of order and fundamental kinds of explanations.

At the next level of structure within a paradigm are one or more theoretical frameworks. A theoretical framework is based on a specific mathematical approach to a certain set of physical properties. For instance, the two theoretical frameworks that dominated physics for most of the twentieth century are the spacetime framework, associated with the general theory of relativity, and the framework associated with quantum theory. The spacetime framework characterizes all physical objects by means of geometrical fields—scalar, vector, and tensor—defined on a four-dimensional manifold whose points are the locations of possible idealized events and whose metric structure is described explicitly. The quantum theoretical framework treats the states of physical systems as vectors in an infinite-dimensional, complex-valued Hilbert space. Observables are operators on this space, and system evolution is characterized by the Schrödinger equation. The thus-far irreconcilable difference between these two encompassing theoretical frameworks is one of the great enigmas of contemporary physics.

Specific theories reside within theoretical frameworks. Thus contemporary Newtonian theory is the spacetime theory that posits the unique separation of spacetime into three-dimensional Euclidean space and one-dimensional time, with physical object fields defined independently. (Historical Newtonian theories of mechanics and gravity, still described in outline in many physics textbooks, fall under the Newtonian synthesis, the paradigm that guided physics from the seventeenth through the nineteenth centuries.) The special theory of relativity posits non-unique separability (the relativity of global simultaneity) of flat spatial sections. The general theory of relativity links physical object fields to the metric structure of spacetime itself through Einstein's field equations, so commits itself to no specific global geometry.

Models of theories result from specification of theory components. Thus, a model of contemporary Newtonian particle theory might specify a set of particle trajectories on Newtonian spacetime. Models of general relativity theory are often characterized as solutions of Einstein's field equations and are associated with specific physical systems. The Schwarzschild model is identified as the specific relativistic spacetime corresponding to a spherically symmetric, isolated mass. Actually, this is a model-type. A completely specified Schwarzschild model with properties (such as light paths) that can be compared with measurements on a real physical system (such as our solar system or an astronomical candidate for a black hole) requires a specification of mass. Similarly, in quantum mechanics, models are often identified as solutions of the Schrödinger equation. Thus, one might have the quantum mechanical model of a particle of a particular mass and charge scattered in an electromagnetic field of a particular configuration. Or one might have the quantum mechanical model of a one-dimensional harmonic oscillator of particular mass and frequency.

In this account of the structure of physics, the cutting edge of physical explanation—and of much theory testing—lies at the junction of models and the properties of identifiable real systems. Consider the claim, for instance, that the Sun and Mercury form, to a high degree of approximation, a Newtonian two-particle system. At the very least, what is asserted here is that the positions and velocities of Mercury relative to the Sun are very similar to those of a specific two-particle Newtonian model. If the claim is to be taken as genuinely explanatory, however, other properties of the model must be taken to represent properties of the Mercury-Sun system, for example, the instantaneously transmitted central force between the particles that respond to that force according to Newton's second law. Still other aspects of the model are clearly dissimilar to the Sun-Mercury system: the point-mass character of the particles and the absence of other interacting particles. Use of models with more similarities to the real system—extended bodies and additional interacting bodies—would be expected to improve the closeness of fit between the orbital similarities. The process of thus improving the fit between data and theory is a crucial part of theory testing. It was partially the failure of such Newtonian models to fit the data of Mercury's perihelion shift that gave impetus to the general theory of relativity and its claim that the geometry of spacetime external to the Sun is that of a particular Schwarzschild model.

Of course Newtonian models are still used in many solar system calculations, despite an acknowledgment that forces that act instantaneously at a distance do not exist in nature. Models are often used in this way simply as calculational models. Models may also be used to probe aspects of real systems insufficiently understood for the positing of deep explanatory similarities. Such models are called heuristic models. Even when no great explanatory insight is claimed on their basis, the formal relationships within the model provide guides to the search for corresponding relationships in the physical system.

In general, then, a theoretical hypothesis in physics asserts that some real system is similar to some theoretical model in particular respects, to indicated degrees. And, for the practical purposes of explanation and testing, a theory is just a whole collection of models, or sets of models, together with hypotheses linking the models with real physical systems.

See also: EXPERIMENTAL PHYSICS; NEWTONIAN SYNTHESIS; RELATIVITY, GENERAL THEORY OF, ORIGINS OF; SPACETIME; THEORETICAL PHYSICS

Bibliography

GIERE, R. *Explaining Science* (University of Chicago Press, Chicago, 1988).

HESSE, M. *Models and Analogies in Science* (University of Notre Dame Press, Notre Dame, IN, 1966).

KUHN, T. *The Structure of Scientific Revolutions,* 2nd ed. (University of Chicago Press, Chicago, 1970).

E. ROGER JONES

MOE

See MAGNETO-OPTICAL EFFECTS

MOLECULAR PHYSICS

Although it is not entirely straightforward to define the meaning of molecular physics, it is probably fair to say that it comprises any of the numerous physical phenomena for which the role of the molecule is essential to their understanding. Thus, the most fundamental function of molecular physics is to try to explain and to describe the nature and the features of those special aggregates of atoms that are called molecules. Excluded in this discussion is the greatest part of the near-continuum sequence of "soft" aggregates called clusters, which often exhibit some of the features of the more usual molecules. Rather, the historically accepted definition of molecules refers to aggregates of atoms that are bound to each other by energy values markedly larger than those which such aggregates can exchange (usually very rapidly) with the environment in which they are immersed. In other words, one may say that the classical definition of a molecule is that of a "stable" aggregate of atoms (with respect to their local "bath"), which then shapes and influences most of the physical behavior of the macroscopic objects of which it is an "elementary" component.

This is an important distinction in the sense that molecular physics often aims at the understanding of macroscopic phenomena where the role of such elementary components is central, and where their structural features as individual objects ultimately play a major role. This means that the behavior of condensed matter, for example, molecular liquids and molecular solids, and any aggregate state in between, is often explained sufficiently well by that set of characteristics which come from them being an ensemble of large "particles" and, therefore, less crucially related to the specific structural details of such "particles." The central object of molecular physics is the study of the very great variety of properties of matter that chiefly and ultimately are seen to depend on the characteristics of its molecular components.

The idea of molecules is, of course, a very old one and it has been part of the development of the foundations of science for several centuries. However, it is only during the twentieth century, and possibly not much earlier than the 1920s, that experimental science has been able to "see" molecules and to relate specific results of measurements to the existence of stable atomic aggregates that behave differently from their individual components. The main tool of those earlier observations has been the absorption of light by different molecules, and finding that, in various regions of the electromagnetic spectrum, there exist specific "fingerprints" of each molecule that correspond to light absorption at characteristic frequency values and to transparency within other ranges of frequency. Such observations have had tremendous success over the years and have led to the very rich field of fundamental science that goes under the name of molecular spectroscopy and of optical interactions with molecules. The spectral features of each molecular system (better if observed under conditions of near-isolation) tell us a great deal about the structure of the molecule, that is, the relative positions of its component atoms in three-dimensional space; and about the quantum units of energy storage within the molecule itself, distributed among the various degrees of freedom originating from the relative motion of both the bound component atoms of the molecule as well as the atomic electrons.

As a result of the increased variety of possible phenomena and possible processes that are ultimately related to these assemblies of atoms that we call molecules, molecular science has become a very diverse field that spans a broad range of research areas and applications, including most of chemistry and significant portions of biology. Narrowing the focus, one might define molecular physics as the study of molecules; of small clusters and of molecular ions, including their structure and properties; optical interactions; collisions and interactions with electrons; external fields; and solids and surfaces. In particular, this important area of physics attempts to study the details of molecular shapes and of molecular interactions at the resolved quantum state level, that is, to isolate the elementary steps of the often very complex chain of phenomena that link the specific properties of a given molecule to the macroscopically observed behavior of the aggregate, either dilute or condensed. This means today that modern molecular physics often strives to make observations of ultrafast phenomena in order to use the experimental apparatus for taking individual snapshots of the molecular process under study. In this sense, one may say that "molecular physics," as a specific name for the more general area of molecular science, can be seen simultaneously as a basic and an "enabling" science that answers questions about the behavior of matter and energy in molecular systems that we can precisely probe, control, and manipulate. It focuses on the common building blocks of

the world around us and on phenomena that mostly occur in the ranges of temperature and energy, which are characteristics of daily human activities.

As a basic science, molecular physics provides answers to fundamental questions about the physical world and can accurately test basic physical theories such as quantum electrodynamics, quantum measurements, relativity, time reversal, and the *CPT* invariance.

As an "enabling" science, molecular physics has, throughout its extensive history, contributed to the technological strength and knowledge base of many industries. The rapid pace of new discoveries and of unexpected developments in molecular science can be attributed to the continued invention and implementation of new techniques for the control and manipulation of atoms, molecules, and light and to generate light with well-defined characteristics. This aspect of molecular physics strongly underscores the impact that it has had, and keeps having, in important applications in all branches of science, engineering, and technology. As an example, the world's most accurate measurements occur chiefly in molecular science because time and frequency, which are the most accurately measurable physical quantities, fall primarily in the domain of molecular systems. The quest in this field for improved measurement techniques and accuracy has resulted in inventive new instrumentation, including new sources of light, and technologies that find application in areas ranging from industrial manufacturing of new materials and processing to medicine and environmental monitoring.

The marriage between modern optical science and molecular physics has been one of the key developments for our understanding of the microscopic role of molecules in the broad variety of processes and phenomena that have been mentioned. In the earlier days of classical optics, the advances in the design of lenses, gratings, and light sources had allowed pioneers, such as Gerhard Herzberg and Robert S. Mulliken during the 1940s and 1950s, to first provide a unified description of the behavior of molecules under light stimulation and absorption in terms of the molecular orbital quantum description of their bonding features, and of the excitation mechanisms for both electronic and nuclear motions. The invention of the laser, one of the most remarkable products of twentieth-century science and technology, has directly evolved from basic atomic and molecular physics concerning light and its interaction with diverse matter. It was the independent research of Nikolai G. Basov and Alexei M. Prokhorov in the Soviet Union and of Charles Townes in the United States on stimulated emission of radiation in the microwave and optical regions of the spectrum that led to the development of masers and lasers and to their being awarded the Nobel Prize in 1964. Interestingly enough, Mulliken and Herzberg were in turn awarded the Nobel Prize for their theoretical and experimental work on the properties of molecules in 1966 and in 1971, respectively. Since the invention of the laser, further research and development activities have resulted in the discovery of many different classes of lasers that, taken together, provide an enormous range of output wavelengths, pulse lengths, and power levels. All of these also provide the scope for an enormous variety of observations and of experimental instruments that can be used extensively to study the physics of molecules and their chemical interactions. Thus, it is fair to say that today's studies of molecular systems, of their individual properties, of their interactions, of their exchanges of energy, and of their mutations during a chemical process (i.e., a chemical reaction) are nearly always carried out by combining some high-vacuum technology of sample preparation with a laser source as "analyzer" of the molecular behavior.

With the advent of laser radiation new methods for molecular photoexcitation have become avail-able that had not been possible with the "classical" light sources. Figure 1 compares the old and new techniques for selective molecular photoexcitation in schematic form. The classical, or pre-laser, photochemical method (Fig. 1a) was based on a single-step excitation of an electronic state of the molecule. It had the rather serious disadvantage that comes from the fact that most molecules, particularly polyatomic ones, present broad, structureless electronic absorption bands at room temperature, and therefore one cannot use the above scheme to select one particular final state of the molecule for further study or simply for separating it from all the other states. Only for a few simple molecules, in fact, is the typical absorption line narrow enough for selecting states that, for instance, differ in their isotopic atomic components. On the other hand, excitations of electronic states are known to provide high quantum yield for the photochemical processes in which they may be involved.

Using classical radiation one could also consider the single-step excitation of a specific vibrational molecular state and then perform photochemical processes in the ground electronic state. This process, in fact, features rather high excitation selectivity for both simple and complex molecules. The main disadvantage of this method, however, is that the fast relaxation of vibrational excitation to heat leads to a low quantum yield of the subsequent pho-

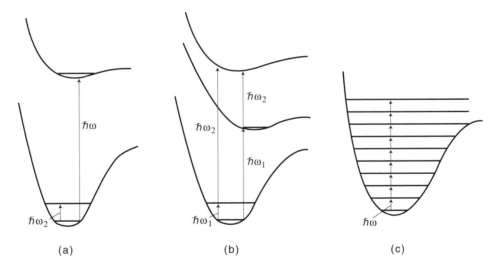

Figure 1 Selective molecular photoexcitation: (a) single-step excitation of electronic or vibrational states, (b) two-step excitation of an electronic state through intermediate vibrational or electronic states, and (c) multiple-photon excitation by infrared radiation.

tochemical processes. In any case, such studies could only be used for the further analysis of processes with low activation energy as very little extra energy could be provided to the excited molecular state that was being selected.

On the other hand, the new laser sources have allowed for multistep excitations of molecular electronic states through one or several intermediate states (usually electronic and/or vibrational) by the combined action of sources of different wavelengths. In the simpler cases of two-step excitations (Fig. 1b) it is possible to separate the functions of selective excitation by having the molecule acquire rather little energy from an infrared (if) photon and then absorb the much greater amount of energy via an ultraviolet (uv) photon by the selectively excited molecule. This kind of two-step photoexcitations have added a tremendous variety of experimental architectures by combining sufficiently high selectivity of the final species with the traditionally high quantum yield that is necessary in the field of molecular photochemical processes. The more general case is shown schematically in Fig. 1c, where the multiple absorption of infrared photons of the same frequency can transfer to a molecule an amount of energy comparable to the typical energy of an electronic excitation. With such a process one can therefore realize at the same time the excitation selectivity for isolating a specific molecular state with the necessary high quantum yield necessary to subsequently carry out further photochemical processes. Although such a method is still rather limited by the fact that it can be applied only to polyatomic molecules with high density of excited vibrational levels in the specific electronic state that is being selected, it has opened up in recent years a great deal of possibilities for selective photochemical processes like photodissociation, photoexcitation into optically forbidden states, and photoselective chemical reactions or isotopic separation. Such methods, in fact, can only be carried out with laser radiation because the intermediate quantum levels need to be highly populated. Conventional, incoherent light sources have too low a radiation temperature for all processes except for the single step of Fig. 1a and, furthermore, have usually higher efficiency for exciting electronic states than for doing the same to vibrational ones.

Another area in which the modern molecular physics experiments have struck a very successful collaboration with other fields of science is that of the study of fundamental mechanisms of chemical reactions. We indeed know that molecules that modify each other by strongly interacting and by perturbing each other's equilibrium states are the basic actors during chemical processes, a ubiquitous class of processes forever occurring around and inside us. However, to study in detail how these modifications and changes can come about at the molecular level can indeed be likened to the task of trying to make a slow-motion picture of the various steps of a given reaction. The trouble with achieving this goal, however, comes from the fact that too many would-be actors tend to be around and, to follow this analogy, wish to come upon the stage of the main action without proper cue and insist on mumbling their lines too rapidly for them to be understood in order to follow the plot. Interactions and reactions between molecules occur with the ease of striking a match and at a speed so fast (on the subpicosecond time scale for the making of new bonds and the breaking of old ones) as to be a severe challenge to the movie maker who would like to record as many as possible of the individual frames describing the play.

The advent of the laser, through its monochromaticity, tunability, and intensity, has once more changed the experimental architectures that can study such important processes at the molecular level, and has brought students of kinetics close to realizing the above goal. Laser interaction, in fact, makes possible the preparation of reagents in specific excited internal states, serving then as effectors of molecular changes, and further allows the identification of the final molecules and the analysis of their internal states of energy distribution, thus performing the function of detectors of molecular changes. A third use of the laser is further being tested and experimented with success: it can provide radiation fields intense enough to influence intermolecular interactions during the subpicosecond intervals in which the molecules that are reacting are intimately interlocked. Nonresonant, laser-assisted collisional processes are then found to occur. Since the mid-1970s, therefore, sophisticated microscopic experimental techniques and large-scale theoretical computations, which model the elementary processes observed with the above techniques, have significantly enhanced our understanding of the detailed dynamics and of the detailed structural properties of molecular systems considered in their isolation. The development of huge computer codes that are able to perform lengthy and detailed computations has made it possible to clarify the connection between the first principles of the quantum world and the behavior of substances in the real world. Modern spectroscopy, laser technology, and the use of molecular-beam methods allow us to observe molecules either as isolated species or interacting during single collisions with other atoms, ions, photons, and other molecules. These feats from modern experimental equipment have enabled us to observe these new entities, the molecules, with ever-increasing detail.

One further consequence of such intimate knowledge of the physics of molecules has been the discovery and the observation of the properties of novel "molecular" species, which were either too unstable to be observed or not different enough at the coarser level of observation of the early 1980s. Since that time, in fact, it has become possible to manipulate molecules with astonishing facility, again using light, and to produce very cold molecular species by trapping them at the focus of a laser beam. Thus, after much experimentation and efforts, in the mid-1980s neutral atoms and molecules were successfully cooled to less than 0.001 K with a configuration of laser beams that was termed "optical molasses." Because of the Doppler shift resulting from the molecular thermal motion, the light in this configuration acts as a viscous medium, and atoms imbedded in the "molasses" are slowed down to very low velocities. The possibilities that originate from such conditions for the study of molecular properties can be easily imagined. Another example comes from the discovery of the C_{60} molecule, today well-known as the "fullerene" molecule. In the early 1980s Richard Smalley and associates at Rice University in Texas developed a powerful mass spectrometry technique for studying the refractory clusters found in the plasmas produced by focusing a pulsed laser on a solid target. The technique seemed to offer a way of simulating the chemistry in a carbon star, and therefore a joined Rice–Sussex experiment with Smalley and Howard Kroto revealed, by using a graphite target, a novel molecular state of carbon corresponding to sixty atoms in a closed cage structure made out of pentagons and hexagons of carbon atoms. It was named Buckminster-fullerene to commemorate the look-alike geodesic dome pioneered by the architect Buckminster Fuller.

The recent decades have indeed represented a period of great development in our knowledge of molecules, in the sense that we have learned a tremendous amount about the properties and behavior of such building blocks of the world, which

surround us and which, ultimately, we are also made of. Because of its intrinsic diversity, molecular physics is often at the cutting edge of various scientific advances and easily interfaces with other fields of science, engineering, and practical applications.

See also: ATOM; ATOMIC PHYSICS; ELECTROMAGNETIC SPECTRUM; LASER, DISCOVERY OF; MOLECULE; QUANTUM ELECTRODYNAMICS; SPECTROSCOPY

Bibliography

ATKINS, P. W. *Molecular Quantum Mechanics* (Oxford University Press, Oxford, Eng., 1987).

BRANDSDEN, B. H., and JOACHAIN, J. C. *The Physics of Atoms and Molecules* (Longman, London, 1983).

COULSON, C. A. *The Shape and Structure of Molecules* (Clarendon Press, Oxford, Eng., 1982).

LEVINE, R. D., and BERNSTEIN, R. B. *Molecular Reaction Dynamics* (Clarendon Press, Oxford, Eng., 1987).

F. A. GIANTURCO

MOLECULAR SPEED

To understand what is meant by the term "molecular speed," it is helpful to discuss the internal motions of a sample of gas that is in thermal equilibrium. At the atomic level the gas is a chaos of motion and collisions among an extraordinarily large number of molecules. Under these conditions, what are the expected speeds of the molecules trapped within a container?

To start, consider only one component of a molecule's velocity vector, say the x component, v_x. Each molecule interacts with all the others. The intuitive expectation, to be confirmed by experiment, is that collisions randomize v_x; that is, each molecule will eventually move with every allowed speed. Most students first encounter a random distribution while studying error analysis. Random errors of measurement result in a distribution that takes on the familiar bell-shaped, or Gaussian, curve.

Maxwell Distribution

The average value of v_x for the molecules must be zero since there is no net motion of the gas out of its container. Similarly, the distribution of v_x must be symmetric about $v_x = 0$ since there can be no dependence on the arbitrarily chosen left-right directions. A Gaussian curve of the form $\exp(-v_x^2)$ fits these requirements. Of course, the y and z components of the motion will have similar distributions. The speed v of a molecule is $v^2 = v_x^2 + v_y^2 + v_z^2$. In 1859 James Clerk Maxwell showed that a gas, composed of molecules of mass m and at temperature T, has a speed distribution given by

$$f(v)\,dv = 4\pi(m/2\pi kT)^{3/2}v^2 e^{-(m/2kT)v^2}dv,$$

where k is Boltzmann's constant. The Maxwell speed distribution $f(v)\,dv$ gives the fraction of all the molecules that have speeds in the range v to $v + dv$. This function was derived again by Ludwig Boltzmann using statistical methods, and thus it is often called the Maxwell–Boltzmann speed distribution.

In Fig. 1, $f(v)$ is plotted for three different temperatures. The distributions peak at speed v_{mp}, the most probable speed among the molecules; v_{mp} is indicated in Fig. 1 for the distribution at temperature T. The most probable speed is easily calculated from $df/dv = 0$:

$$v_{\text{mp}} = \sqrt{2kT/m}.$$

Two other commonly used speeds are the average, v_{ave}, and the "root-mean-square" speed, v_{rms}. These speeds, also indicated in Fig. 1, are defined by

$$v_{\text{ave}} = \int_0^\infty vf(v)\,dv$$

$$= \sqrt{8kT/\pi m}$$

$$= 1.128\, v_{\text{mp}}$$

and

$$v_{\text{rms}} = \left[\int_0^\infty v^2 f(v)\,dv\right]^{1/2}$$

$$= \sqrt{3kT/m}$$

$$= 1.225\, v_{\text{mp}}.$$

Experimental Proof

The apparatus shown in Fig. 2 is typical of that used to verify the Maxwell–Boltzmann speed distribution. A beam of molecules is allowed to escape through a small opening. The process is called effusion: the opening must be so small that equilibrium in the container is not significantly disturbed. Essentially the opening is smaller than the mean free path, the average distance between molecular collisions. For air under room conditions the mean free path is $\sim 10^{-7}$ m. Rotating shutters allow only molecules of a predetermined speed to reach a detector, where the number of molecules is counted. Careful measurements over many speeds confirm the validity of the Maxwell–Boltzmann speed distribution for the gas within the container.

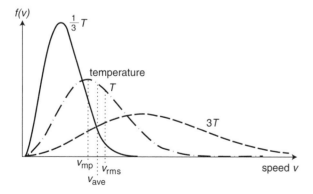

Figure 1 Speed distributions at three temperatures. The most probable, the average, and the root-mean-square speeds are shown for temperature T.

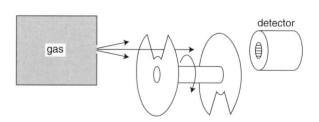

Figure 2 The Maxwell–Boltzmann speed distribution for a gas can be experimentally verified by sampling a molecular beam. Molecules reaching the detector are restricted to a small speed range by the action of rotating shutters.

Applications

As a numerical example, consider molecular hydrogen at room temperature. The relevant parameters are $k \cong 10^{-23}$ J/K, $T \cong 300$ K, and $m \cong 3 \times 10^{-27}$ kg, giving $v_{mp} \cong 1,600$ m/s $\cong 3,500$ mph.

In a mixture of gases, lower mass molecules have higher speeds. In a process similar to effusion, described above, molecules with higher speeds will escape at a greater rate; this variation with mass is especially important in isotope separation since conventional chemical methods fail. The separation of uranium isotopes ^{235}U and ^{238}U during World War II was first accomplished this way.

See also: BOLTZMANN DISTRIBUTION; DISTRIBUTION FUNCTION; EQUILIBRIUM; MAXWELL–BOLTZMANN STATISTICS; MAXWELL SPEED DISTRIBUTION

Bibliography

FEYNMAN, R. P.; LEIGHTON, R. B.; and SANDS, M. *The Feynman Lectures on Physics*, Vol. 1 (Addison-Wesley, Reading, MA, 1966).

REIF, F. *Fundamentals of Statistical and Thermal Physics* (McGraw-Hill, New York, 1965).

SEARS, F. W., and SALINGER, G. L. *Thermodynamics, Kinetic Theory, and Statistical Thermodynamics*, 3rd ed. (Addison-Wesley, Reading, MA, 1975).

STOWE, K. *Introduction to Statistical Mechanics and Thermodynamics* (Wiley, New York, 1984).

THORNTON, S. T., and REX, A. *Modern Physics for Scientists and Engineers* (Saunders, Fort Worth, TX, 1993).

JOHN MOTTMANN

MOLECULE

Everyday materials are made up of atoms or molecules. A molecule is the smallest piece of a substance that retains its chemical identity. (The Latin word *molecula* means "small mass.") A molecule is formed whenever two or more atoms are linked by a chemical bond. Since atoms are neither created nor destroyed in chemical reactions, molecules can be labeled by their constituent atoms in a chemical formula. For example, two hydrogen (H) atoms and an oxygen (O) atom form the water molecule H_2O. If the substance is an element, then the atoms in the molecule are identical (a homogeneous molecule), such as H_2. If the substance is a compound, the molecule contains different atoms (a heterogeneous molecule), as in NaCl.

Molecules vary enormously in size and shape. Even molecules with the same chemical formula can have different shapes; these are called structural isomers. Small molecules often have a particular symmetry—for example, the methane molecule CH_4 is tetrahedral (i.e., like a pyramid with a triangular base, with the C atom at its center), and the benzene molecule (C_6H_6) consists of six CH units in a ring. Molecules that repeat a certain pattern, called a unit, in a long chain are called polymers. These can be either man-made, such as polyethylene $(CH_2)_n$, or biological, such as DNA. They can consist of very large numbers of atoms. Furthermore, crystalline solids might be considered as very large molecules, such as metallic sodium (Na). Molecules that have several repeating units in more than one direction are called clusters. Examples include Na_{128} and buckyballs (C_{60}). A buckyball is a large molecule consisting of sixty carbon atoms forming a surface in the shape of a soccer ball. Buckyballs were named after the late Buckminster Fuller, who popularized this shape in architecture through his use of the geodesic dome.

A chemical bond that holds two (or more) atomic nuclei together in a molecule is formed by the rearrangement of electrons as the nuclei come together. The electrons that rearrange themselves, called valence electrons, are responsible for many of the chemical and physical properties of the substance. (The electrons in each atom that change little when the bond is formed are called core electrons.) If a molecule is to exist, this bond must lower the energy of the molecule relative to the separated atoms. The difference between the energy of the molecule and its separated constituents is called the atomization energy of the molecule (or the cohesive energy of a solid).

The particular way in which the electrons rearrange determines the type of chemical bond. In an ionic bond, a single valence electron is transferred from one nucleus to the other. For example, NaCl consists of an Na^+ ion and a Cl^- ion, because an electron (with its single negative charge) has been donated by the Na to the Cl. Since unlike charges attract, the resulting molecule is held together by electrostatic forces, the strongest kind of chemical bond. In a covalent bond, the valence electrons are shared between the two constituent atoms. This sharing reduces the energy somewhat, but not as much as in the ionic case. Thus covalent bonds are much weaker than ionic bonds, that is, covalently bonded molecules have much smaller atomization energies. Finally, in metal aggregates, such as solids or large clusters, the atoms are held together by a metallic bond, in which the electrons are highly mobile.

The forces between atoms in molecules are called intramolecular forces. There are also forces between molecules, called intermolecular forces. An important example is the hydrogen bond, a weak dipolar force between two molecules containing protons, such as that between two water molecules. Another example is the long-range fluctuating van der Waals force between molecules in a molecular crystal. The energies due to these forces are much smaller than those of ionic or covalent bonds. In some cases, molecules are held together by these weak forces, such as the two strands of the DNA double helix.

Modern technology allows scientists to determine quickly the chemical structure of a molecule. Mass spectroscopy can help to deduce the chemical formula, and nuclear magnetic resonance can show how specific atoms are arranged in the molecule.

Because the electron has a much smaller mass and typically a higher velocity than the nuclei in a molecule, when the nuclei move the cloud of electrons usually remains in its ground state, that is, it distorts adiabatically. Thus the nuclear degrees of freedom can be separated from the electronic degrees of freedom. For a molecule with N nuclei, there are $3N - 3$ degrees of freedom, which are often classified as either rotations or translations. For example, a diatomic molecule looks like a dumbbell. It can undergo two independent rotations, each about the midpoint, or a single vibration in which the bond length is stretched. A molecule with a large number of atoms has many such degrees of freedom, each of which has its own characteristic frequency. Optical and electron spectroscopy may be used to probe these rotations and vibrations.

All these properties can be modeled by finding the ground state of the electrons in a molecule for a given arrangement of the nuclei. Traditionally, quantum chemists have done this by solving the Schrödinger equation for the wave function of the electrons. Today, density functional theory, the modern version of the Thomas–Fermi theory, is used because it can be applied to much larger molecules at a fraction of the computational cost.

See also: ATOM; COVALENT BOND; DEGREE OF FREEDOM; ELECTRON; ELECTROSTATIC ATTRACTION AND REPULSION; HYDROGEN BOND; ION; IONIC BOND; SPECTROSCOPY; VAN DER WAALS FORCE

Bibliography

LEICESTER, H. M. *The Historical Background of Chemistry* (Dover, New York, 1956).

McQUARRIE, D. A., and ROCK, P. A. *General Chemistry* (W. H. Freeman, New York, 1984).

PAULING, L. *The Nature of the Chemical Bond,* 3rd ed. (Cornell University Press, Ithaca, NY, 1960).

KIERON BURKE

MOMENTUM

The idea of "quantity of motion" of an object is defined precisely by the object's linear momentum, angular momentum, and kinetic energy. Linear and angular momentum are vector quantities, having both magnitude and direction.

Linear Momentum

Linear momentum, or simply momentum, is associated with translational motions in space and may include one-dimensional (in a line), two-dimensional (on a surface), or three-dimensional motions. The momentum **p** is defined as

$$\mathbf{p} = \gamma m \mathbf{v},$$

where m is the mass of the object, **v** is the vector velocity, and γ is the Lorentz factor ($\gamma = 1/\sqrt{1 - v^2/c^2}$). In the Lorentz factor, v is the speed of the object and c is the speed of light. The direction of the linear momentum vector is the direction of the velocity. We learn from relativity that momentum is related to energy by $E^2 = p^2 c^2 + m^2 c^4$.

The Lorentz factor is approximately equal to 1.0 (within one-half percent) for speeds less than $0.1 c$, which is true of nearly all objects we observe in everyday life. So for anything traveling at ordinary speeds, to a very good approximation, $\mathbf{p} = m\mathbf{v}$ ($v \ll c$).

On the other hand, light and other forms of electromagnetic radiation carry momentum and always travel at the speed of light. The photons that carry the energy and momentum of these radiations have zero mass, but γ is infinite, so we have to calculate their momentum by another means: The direction of momentum is still the direction of travel of the photon or particle, and the magnitude of momentum is $p = hf/c$, where h is Planck's constant and f is the frequency of the radiation. This relation is entirely consistent with the first definition given for **p** above, since the two can be connected directly by using the principles of special relativity and the Einstein relation for the energy of a photon, $E = hf$. These same relations allow one to calculate the momentum of other subatomic particles of zero mass that also travel at the speed of light.

The above definition of momentum states that the magnitude of the momentum, or quantity of motion, depends on the product of mass and speed. So the momentum of a very massive object moving slowly will be large due to its mass, while an object of tiny mass must move very fast in order to have large momentum. This is intuitively reasonable if you think about how much effort is required to add to or subtract from the motion of an automobile or a truck by pushing it and consider how adding or subtracting that same quantity of motion to a bicycle, for example, by expending the same effort, would change its speed. Clearly, both mass and speed participate in the quantity of motion.

A two-person jet airplane of mass 24,000 kg taxiing very slowly at about 10 km/h (2.8 m/s) has momentum $24,000 \times 2.8 = 67,000$ kg·m/s. An automobile of mass 1,500 kg would have to be moving at a speed of 44 m/s (i.e., 160 km/h or 99 mph) to have the same momentum as this small jet plane. If this jet plane flies at 400 m/s (1,440 km/h or 899 mph), its momentum is 9.6×10^6 kg·m/s.

We experience many different magnitudes of momentum in our everyday lives. For example, a 125-lb. woman ($m = 57$ kg) walking briskly at 5 mph (2.2 m/s) has momentum $p = 57 \times 2.2 = 130$ kg·m/s. A 1,500 kg car moving at 100 km/hr (28 m/s) has momentum $p = 1,500 \times 28 = 42,000$ kg·m/s. But a photon of yellow sunlight (wavelength 590 nm) carries a momentum of only 1.1×10^{-27} kg·m/s. On the other hand, a medical diagnostic x-ray photon would have 15,000 to 45,000 times greater momentum than the photon of sunlight. And an electron accelerated from rest through 30,000 V in a television tube will have momentum 9.5×10^{-23} kg·m/s,

still very tiny, but about 84,000 times greater than the momentum of the photon of sunlight. The speed of this electron is nearly one-third the speed of light, and relativistic effects are apparent. If we accelerate the electron to a speed about three times faster, say $v = 0.999c$, then its momentum will be 6.1×10^{-21} kg·m/s, 64 times larger than before for just triple the speed. This illustrates that relativistic effects can be dramatic.

The role of mass in the quantity of motion is apparent in head-on collisions of hard, elastic objects. If the objects are of very different masses, like a baseball (about 0.145 kg) and bat (about 1.1 kg), the slower bat will cause the faster moving, but much less massive, baseball to reverse its direction. This reversal of direction gives a very large change of momentum because momentum is a vector quantity, having direction as well as magnitude. Therefore, we must take into account the algebraic sign (+ or −) of the momentum when comparing momentum before and after the collision. If a ball going 40 m/s (144 km/hr or 90 mph) in one direction ends up going the same speed but in the opposite direction after a collision, its momentum will have changed from mv to $-mv$, a difference of $2mv$ or $2 \times 0.145 \times 40 = 11.6$ kg·m/s. If momentum were simply equivalent to velocity, the bat (or the system of bat plus batter) would have to be going much faster than the ball when they collided, in order to impart that much motion to the ball.

In quantum mechanics, momentum is an operator whose values are quantized. How accurately we can measure the momentum of atoms, nuclei, or elementary particles is limited by the Heisenberg uncertainty principle.

Momentum of a System

We often refer to the set of objects we are studying (e.g., bodies or particles or even radiation) as a system. For example, a human body, made up of a very large number of molecules, might constitute a system. Or two colliding automobiles might constitute a system. Other systems might be a set of microscopic particles (like a proton and an electron forming a hydrogen atom), a beam of photons (light quanta), the Earth–Sun system, or a set of colliding billiard balls.

The momentum of a system is the total momentum found by adding up all the individual momenta vectorially, since momentum is a vector quantity.

To find the vector sum of individual momenta, we could take the sum of all x components of individual momenta to obtain the x component of the system's total momentum, and similarly for y and z components. If N individual particles (or subsystems) have respective momenta $\mathbf{p}_k$, then the total system momentum $\mathbf{P}$ can be written $\mathbf{P} = \mathbf{p}_1 + \mathbf{p}_2 + \mathbf{p}_3 + ... + \mathbf{p}_N = \sum \mathbf{p}_k$, where the addition is vector addition.

Newton's Laws of Motion and Linear Momentum

For systems that are much more massive than atoms and that do not travel at speeds faster than about $0.1c$, Newton's laws of motion can be used to describe motion accurately. Very small systems must be treated by the principles of quantum mechanics, and very fast ones must be treated by the principles of relativity. But most of our direct experience is with systems that function within the domain of Newton's laws. These laws are best expressed in terms of momentum, as Isaac Newton did originally.

Newton's first law states that in the absence of external interactions the momentum of a system will not change (i.e., $d\mathbf{p}/dt = 0$ in the absence of external interactions). Newton's second law, usually written $\mathbf{F} = m\mathbf{a}$, is expressed with greater generality as $\mathbf{F} = d\mathbf{p}/dt$. This formulation allows for the study of systems with varying mass, such as a rocket ship that expels gases from the ignition of its fuel, and even for the inclusion of special relativistic effects in which γ varies with speed. Finally, Newton's third law can be expressed simply as the conservation of momentum for a system of interacting particles that have no interactions with anything outside the system under study.

Angular Momentum

Just as translational motions involve linear momentum, rotational motions, including motions with some rotational component such as those along curved paths, entail angular momentum about some axis. In addition, if even a single particle moves in such a way that the line joining the particle to the chosen axis rotates, even if the particle itself is moving along a straight line, then there is angular momentum in the motion of the particle. A system of several particles will likewise have angular momentum about an axis. Most physical systems move with *both* translational and rotational components to their motion, but we can analyze these components separately.

We encounter objects with angular momentum very commonly in our lives: the rotating wheels of automobiles, a bicycle turning a corner, the regular rotation of Earth and its revolutions about the Sun, the revolution of the Moon about Earth, and hurricanes or tornadoes. We also observe angular momentum in the spiral shapes of rotating galaxies or in the properties of atoms and nuclei. Subatomic particles, like electrons, protons, and neutrons, have intrinsic angular momentum, sometimes called spin, in addition to the angular momentum of their orbital motion. Ice skaters, divers, gymnasts, and other athletes make use of angular momentum in their sports. A system composed of two ice skaters moving on separated parallel straight paths has angular momentum. A star collapsing to a neutron star will produce a high-speed pulsar by conserving its angular momentum. Accurate predictions of day lengths, the seasons, the tides, and eclipses of the Moon and the Sun all depend on our understanding of angular momentum.

Angular momentum is a measure of the rotational motion of an object or a system about some origin. It is a vector; it has both magnitude and direction. For example, the magnitude of Earth's angular momentum of rotation about its axis determines the length of the day, 23 h 56 min 4.09 s (the sidereal day), and the direction of Earth's angular momentum of rotation about its axis determines that the Sun rises in the east and sets in the west.

Definition of Angular Momentum

Angular momentum ℓ for a single particle is defined in terms of the linear momentum $\mathbf{p}$ of the particle and the vector $\mathbf{r}$ from the origin to the particle, as $\ell = \mathbf{r} \times \mathbf{p}$ (a cross, or vector, product). For motions much slower than c, the speed of light, the magnitude of the angular momentum is just the mass times twice the area swept out by the position vector $\mathbf{r}$ as it moves with the particle. This relationship links angular momentum with Kepler's second law of planetary motion.

The direction of the angular momentum vector is given by a right-hand rule: ℓ is perpendicular to both $\mathbf{r}$ and $\mathbf{p}$ and is in the direction pointed by the thumb of your right hand when you point the fingers of your right hand along **r** and curl them into the direction of **p**.

Newton's Laws of Motion and Angular Momentum

To write Newton's laws of motion in angular form, we take the vector cross product of the position vector **r** with the momentum **p** to relate angular momentum to the new quantities obtained from the **r** × operation. For Newton's first and second laws, $F = d\mathbf{p}/dt$ becomes $\mathbf{r} \times \mathbf{F} = \mathbf{r} \times d\mathbf{p}/dt = d\ell/dt$. (There is another term in this development, $d\mathbf{r}/dt \times \mathbf{p}$, which is identically zero.) Not surprisingly, the angular analog for force that appears here has a name of its own, torque: $\boldsymbol{\tau} \equiv \mathbf{r} \times \mathbf{F}$. Now, Newton's first and second laws for angular motion are written $\boldsymbol{\tau} = d\ell/dt$.

Torque is an angular analog for force. Torques are forces acting through a moment arm to cause a body to rotate about some axis. There are also angular analogs for mass and velocity: moment of inertia and angular velocity. And, of course, angular momentum is an angular analog for momentum. In the classical domain, where Newton's laws apply, our definition of angular momentum can be written $\ell = \mathbf{r} \times m\mathbf{v}$. The angular velocity $\boldsymbol{\omega}$ is defined by $\mathbf{v} = \boldsymbol{\omega} \times \mathbf{r}$, so $\ell = m\mathbf{r} \times (\boldsymbol{\omega} \times \mathbf{r})$. This expression can be written in terms of a moment of inertia tensor and angular velocity, so angular momentum is expressed as a product of inertia and velocity, $\ell = \mathbf{I} \cdot \boldsymbol{\omega}$, analogously to linear momentum $\mathbf{p} = m\mathbf{v}$. This complicated expression in its full generality requires a tensor moment of inertia with dimensions of mass × length^2. The moment of inertia involves the distribution of mass in the system; for example, how close it is to the axis of rotation. Angular velocity also involves the distance from the axis of rotation, since points in a rotating body at different radii from the axis of rotation move at different speeds. In the general case, the direction of angular momentum ℓ need not be the same as the direction of angular velocity $\boldsymbol{\omega}$.

For a simple example that avoids complicated vector algebra, consider a concentrated mass m traveling in a circular path on a horizontal frictionless surface and attached to a center point by a very light string. In this case, **v** and **r** (vector from the axis of rotation) are perpendicular, and $\boldsymbol{\omega}$ is perpendicular to both **r** and **v**. The angular momentum has magnitude $\ell = mr^2\omega = I\omega$, and the directions of ℓ and ω are both perpendicular to the plane of motion and directed according to the right-hand rule. Here, $I = mr^2$, and $\omega = v/r$.

Angular Momentum of a System

We are often interested in the angular momentum not of a single particle but of a system like a rotating body, say a wheel, a spinning ice skater, a gymnast, a rotating binary star system, or a rotating molecule. The angular momentum of the system is the total angular momentum found by adding up all the individual angular momenta vectorially, since angular momentum is a vector quantity. If N individual particles (or subsystems) have respective angular momenta ℓ_k, then the total system angular momentum **L** can be written $\mathbf{L} = \ell_1 + \ell_2 + \ell_3 + \ldots + \ell_N = \sum \ell_k$, where the addition is vector addition.

Quantization of Angular Momentum

Like many other physical quantities, angular momentum is allowed to have only certain values; it is quantized. Quantization of angular momentum is particularly important because the allowed values are given by universal rules that apply to all physical systems. The set of possible allowed values is not determined by the detailed makeup of the system.

Angular momentum quantization has a profound impact on the nature of our physical universe. It provides the order in the chemical periodic table and is fundamental to atomic, molecular, and nuclear structure. The analysis of angular momentum has deep consequences for much of our understanding of nature.

See also: ACCELERATION, ANGULAR; CENTER OF MASS; CONSERVATION LAWS; CONSERVATION LAWS AND SYMMETRY; ENERGY, KINETIC; INERTIA, MOMENT OF; KEPLER'S LAWS; MOMENTUM, CONSERVATION OF; MOTION, ROTATIONAL; NEWTONIAN MECHANICS; NEWTON'S LAWS; QUANTIZATION; RELATIVITY, SPECIAL THEORY OF; VELOCITY, ANGULAR

Bibliography

ASIMOV, I. *Understanding Physics* (George Allen & Unwin, London, 1966).

FEYNMAN, R. P. *The Character of Physical Law* (MIT Press, Cambridge, MA, 1965).

FEYNMAN, R. P., LEIGHTON, R. B., and SANDS, M. *The Feynman Lectures on Physics*, Vol. 1 (Addison-Wesley, Reading, MA, 1963).

HALLIDAY, D., and RESNICK, R. *Fundamentals of Physics*, 3rd ed. (Wiley, New York, 1988).

WILLIAM E. EVENSON

MOMENTUM, CONSERVATION OF

The principle of conservation of momentum can be stated as follows: The total momentum of an isolated system remains constant. In this usage, conservation means the quantity does not change in time. Both linear momentum ($\mathbf{p} = \gamma m\mathbf{v}$, where m is mass, $\mathbf{v}$ is vector velocity, and γ is the Lorentz factor $1/\sqrt{1 - v^2/c^2}$ and angular momentum ($\boldsymbol{\ell} = \mathbf{r} \times \mathbf{p}$, where $\mathbf{r}$ is the position vector from the origin to the moving body) are conserved for isolated systems.

We often refer to the set of objects we are studying (e.g., bodies or particles or even radiation) as a system. For example, a human body, made up of a very large number of molecules, might constitute a system. Colliding automobiles, billiard balls, or bowling ball and pins might also constitute systems. Other systems might be colliding microscopic particles (like a proton and an electron), a beam of photons (light quanta), or a cloud of stars.

A system is isolated if there are no interactions with objects outside the system. In the case of colliding automobiles, only the automobiles that participate in the collision make up the system, not uninvolved automobiles, the road, or any other external objects. To eliminate interactions outside the system, so we can work with an isolated system, we define the system only during a restricted time period: after the automobiles are established in motion with all passengers aboard, while they collide with each other, and before they hit anything else in the vicinity. The upward contact force on the tires of the automobiles from the road just balances the weight of the automobiles, so there is no net force acting on the system from outside. Of course, there are interactions between the automobiles during the collision, but these are internal forces, entirely within the system, so this carefully specified system of colliding automobiles is isolated. This idealization of the colliding automobiles as isolated is a good approximation only for a very short time from immediately before until immediately after the collision. Nevertheless, it allows an accurate description of the motions during the short instant of the collision itself. Friction certainly comes into play before the instant if the drivers are braking and after that instant as the vehicles skid to a stop.

Conservation of momentum implies that even though the individual automobiles collide, change speed and direction, and therefore change momentum, the *total* momentum of the system remains constant. That is, the sum of the individual momenta of the automobiles remains constant. It is important to emphasize that the individual momenta of the bodies making up the isolated system are not conserved—they change as the bodies interact through collisions, for example. But the total *vector* momentum of all objects in the system is conserved. This total momentum is found by adding up all the individual momenta vectorially, since momentum is a vector quantity. The independence of the x, y, and z components of momentum implies that any component of total momentum for an isolated system is conserved separately if there is no external interaction in that direction.

Conservation of Linear Momentum

Examining conservation of momentum more precisely for our isolated system of automobiles, we calculate the total momentum by adding up the individual momenta before the collision: $\sum_k \mathbf{p}_{before,\,k} = \mathbf{P}_{before}$. This total momentum vector $\mathbf{P}_{before}$ has the same magnitude and direction as the total momentum vector obtained by adding up the individual momenta after the collision (or at any other time while the system remains isolated): $\sum_k \mathbf{p}_{after,\,k} = \mathbf{P}_{after}$, and $\mathbf{P}_{before} = \mathbf{P}_{after}$. If we have sufficient information about the speeds and directions of the objects in the isolated system, this rule allows us to predict the results of collisions or to fill in unknown data about one participant in a collision, whether of automobiles, subatomic particles, a comet and a planet, two stars, or any other particles making up (to a reasonable approximation) an isolated system.

What about the apparent creation of momentum as in a bullet, initially at rest, fired from a gun? Or a rocket ship, initially at rest, taking off into space? In the case of the bullet, the system must consist of both gun and bullet, with internal forces between the two setting the bullet in motion. If the gun is suspended freely, its recoil momentum, in the opposite direction to and just balancing the bullet's momentum, will be apparent. We find that after the bullet leaves the gun, $m_{bullet}\mathbf{v}_{bullet} + m_{gun}\mathbf{v}_{gun} = \mathbf{0}$, where m is the respective mass and $\mathbf{v}$ is the respective vector velocity. The mass of the gun is much greater than the mass of the bullet, so their respective velocities will differ greatly, but their momenta add to zero. The initial momentum in this case was zero, with both gun and bullet at rest. The final momentum is also zero, with the bullet's positive momentum adding to the gun's negative momentum. Hence, momentum is conserved. A similar analysis works for the rocket ship if we remember that the expanding gases ejected out the back of the rocket must be included in the system we analyze.

We could also ask about the apparent loss of momentum in a head-on collision in which two particles collide and come to rest. Here the net initial momentum is zero for two particles of equal masses approaching each other at equal speeds. This is consistent with the final momentum of zero when the particles are at rest. Again, total momentum is conserved.

Because of its generality, the principle of conservation of momentum is widely used in physics. It was first stated by Christiaan Huygens, a Dutch physicist, John Wallis, an English mathematician, and Christopher Wren, an English architect and scientist, in papers they presented at the request of the Royal Society of London in 1668, a few years before Newton published his laws of motion. Huygens wrote privately about this principle as early as 1652.

Relationship to Newton's Laws of Motion

Newton's laws are closely related to the conservation of momentum. Even when Newtonian mechanics is not valid, as in the relativistic or quantum mechanical domains, the conservation of momentum is found experimentally to hold, and this principle is believed to be true much more generally and fundamentally than are Newton's laws of motion. So we use it as a starting point to derive Newton's laws in the domain in which they hold.

Consider first an isolated system consisting of only one particle. Conservation of momentum immediately implies Newton's first law: The momentum of the particle will not change.

Now, if the particle interacts with something external to it, the system of one particle is no longer isolated, and the momentum will change. Thus, there is some nonzero rate of change of momentum with time: $d\mathbf{p}/dt$. This is the content of Newton's second law, $\mathbf{F} = d\mathbf{p}/dt$, that is, that external interactions change the momentum, with the force $\mathbf{F}$ defined as the rate of change.

Finally, consider an isolated system of two mutually interacting bodies, labeled A and B. Conservation of momentum for this system is expressed by $\mathbf{p}_A + \mathbf{p}_B$ = constant. This means that the rate of change of $(\mathbf{p}_A + \mathbf{p}_B)$ is zero, so $(d/dt)(\mathbf{p}_A + \mathbf{p}_B) = 0$. Equivalently, we can write $d\mathbf{p}_A/dt = -d\mathbf{p}_B/dt$, which is just $\mathbf{F}_A = -\mathbf{F}_B$, using Newton's second law. We have found the forces to be equal and opposite (action and reaction) and recognize Newton's third law, now a consequence of conservation of momentum.

Conservation of Momentum and Position Symmetry

The reason that momentum conservation is such a general principle becomes apparent when we see the relationship of this principle to the symmetry of physical laws with respect to location in space, sometimes called position symmetry. All experimental evidence indicates that space itself is homogeneous, (i.e., it has the same properties everywhere) and physics does not depend on where an experiment is carried out. Therefore, if we perform some experiment on an isolated system at one point in space and then again at another point, the same physical laws must apply to both experiments. So the laws of physics for an isolated system of particles cannot depend on the absolute position of these particles with respect to some fixed reference point. This independence is referred to as the translational frame invariance of the laws of physics, a direct consequence of position symmetry. Using a principle of least action, which follows naturally from the principles of quantum mechanics or which must be assumed within classical mechanics, we find that position symmetry of the laws of physics due to the homogeneity of space leads directly to the conservation of momentum.

We can obtain an elementary glimpse into this relationship between position symmetry and conservation of momentum by considering how the colliding automobiles could be affected if position symmetry did not hold. Suppose that Newton's second law had a different coefficient in different regions of space: $\mathbf{F} = \zeta m\mathbf{a}$, with ζ a function of position. Then let two automobiles of equal masses and equal but opposite velocities collide head-on at a point where ζ is larger to the right of the collision than to the left (so greater force would be required to provide the same acceleration to a particle on the right than on the left). The total momentum of this system before the collision is zero: $\mathbf{P}_{before} = -m\mathbf{v}$

+ $m\mathbf{v}$ = **0.** But if Newton's third law holds at the collision point, then the forces imparted to the automobiles leaving the collision are equal. Transmitting the force through the finite size of the automobiles would then impart different total velocities to them. The automobile on the left would finally rebound with greater speed than the automobile on the right because of the peculiar properties of space that we have assumed. So the final total momentum would not be zero, and momentum would not be observed. This somewhat artificial thought experiment thus illustrates on an elementary level the relationship between position symmetry and the conservation law.

Other fundamental symmetries give rise to other conservation laws. For example, the isotropy of space (i.e., its sameness in every direction) leads to rotational invariance of the laws of physics and, hence, to conservation of angular momentum. Likewise, homogeneity of time implies conservation of energy. These are deep and wonderful relationships, linking physical symmetries with conservation of seemingly unrelated physical quantities.

Conservation of Angular Momentum

To this point we have considered only linear momentum. Linear momentum is associated with translational motions in space and may include one-dimensional, two-dimensional (as for the automobiles), or three-dimensional motions. Rotational motions lead to the concept of angular momentum. These include motions with some rotational component, such as those along curved paths. In addition, if even a single particle moves in such a way that the line joining the particle to a chosen, fixed origin rotates, even if the particle itself is moving along a straight line, then there is angular momentum in the motion of the particle. Most physical systems move with both translational and rotational components to their motion, but we can analyze these components separately.

The principle of conservation of angular momentum is stated very much like the principle of conservation of linear momentum: The total angular momentum of a system isolated from external torques remains constant.

Torques are forces acting through a moment arm to cause a body to rotate about some axis. Torque $\boldsymbol{\tau}$ is related to force by $\boldsymbol{\tau} = \mathbf{r} \times \mathbf{F}.$ Our system might be a rotating body, like a wheel, which spins about its axis. Other systems might be an upright spinning top; a spinning ice skater, gymnast, or diver; a pair of ice skaters moving on separated parallel straight paths; a rotating binary star system; a rotating molecule; or subatomic particles with intrinsic angular momentum or spin.

In the case of the rotating wheel, there are internal forces between the particles of matter that make up the wheel. But if the wheel's axis is fixed in a specific horizontal position and the wheel itself is closely symmetrical, then all forces and torques on the wheel are balanced. (The isolation of this system is only approximate because of friction between the wheel and its axis. If the bearings are good, we can neglect this external torque for reasonably short time intervals.) The system consisting of the rotating wheel alone is isolated, with net zero total force and net zero total torque exerted upon it. For this system, the total angular momentum remains constant in both magnitude and direction. This is just Newton's first law in angular form, sometimes called the first law of rotational motion.

Just as for conservation of linear momentum, conservation of angular momentum applies to *total* angular momentum, obtained by adding up vectorially the angular momenta of all the individual particles in the system.

As a consequence of the constant total angular momentum of an isolated system, we can understand and predict interesting motions. Angular momentum depends on the angular analog of mass, namely moment of inertia I. I is proportional to mr^2, where r is some average distance of the mass from the axis of rotation. So moment of inertia depends on the distribution of mass in the system (i.e., how close it is to the axis of rotation). Therefore, a spinning ice skater can change rotational speed by extending or drawing in arms and legs, because changing the distribution of mass about the axis of rotation changes the moment of inertia and, hence, requires a change in angular speed to keep the angular momentum constant.

Conservation of angular momentum is related to Newton's laws of motion similarly to conservation of linear momentum. Since we take conservation of linear momentum to be fundamental and derive Newton's laws from this principle, we also derive conservation of angular momentum from the principle of conservation of linear momentum. For an isolated system, conservation of momentum means $(d/dt) \sum \mathbf{p}_k = \mathbf{0}.$ Therefore,

$$dL/dt = \frac{d}{dt} \sum r_k \times p_k$$

$$= \sum v_k \times p_k + \sum r_k \times dp_k/dt$$

$$= 0 + \sum r_k \times F_k$$

$$= \tau$$

$$= 0.$$

Since the system is isolated, the total torque is zero, and **L** does not change with time; that is, angular momentum is conserved.

The deeper connection of conservation of angular momentum to the rotational symmetry of the physical laws mentioned earlier in this entry supports the view that this conservation law, too, is even more general than Newtonian mechanics. Indeed, experiments on relativistic and quantum mechanical systems have shown this to be the case. Angular momentum is quantized, so its conservation in quantum mechanical systems is an important specifier of the quantum mechanical state of the system.

See also: CONSERVATION LAWS; CONSERVATION LAWS and SYMMETRY; KEPLER'S LAWS; MOMENTUM; NEWTONIAN MECHANICS; NEWTON'S LAWS

Bibliography

ASIMOV, I. *Understanding Physics* (George Allen & Unwin, London, 1966).

DUGAS, R. *Mechanics in the Seventeenth Century* (Éditions du Griffon, Neuchatel, Switzerland, 1958).

FEYNMAN, R. P. *The Character of Physical Law* (MIT Press, Cambridge, MA, 1965).

HALLIDAY, D., and RESNICK, R. *Fundamentals of Physics,* 3rd ed. (Wiley, New York, 1988).

LIBOFF, R. L. *Introductory Quantum Mechanics,* 2nd ed. (Addison-Wesley, Reading, MA, 1992).

WILLIAM E. EVENSON

MOON, PHASES OF

Over the course of a lunar month (29.5 days), the Moon changes its visible appearance in a systematic fashion. As the Moon revolves about Earth, its geo-metric position with respect to the Sun-Earth direction also changes. In consequence, observers on Earth see a constantly changing proportion of the lunar surface illuminated by the Sun.

Changing Phases

When the Moon appears in the sky in the general direction of the Sun, it is invisible; this time is called either new moon or the dark of the moon. As time passes, the orbital motion of the Moon takes it eastward across the sky and within one or two days it may be first detected as a thin crescent in the western sky immediately after sunset. Day by day the Moon continues to move further around the sky, increasing its eastward distance from the solar position. The visible portion of the illuminated lunar surface also grows fatter, or waxes, until two weeks after new moon it appears as a fully illuminated disk (full moon), now rising over the eastern horizon as

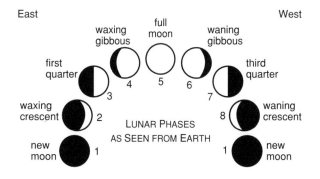

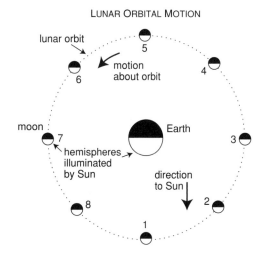

Figure 1   Lunar phases.

the Sun sets in the west, directly opposite the lunar direction. After full moon the process is reversed, the Moon becoming thinner day by day (waning moon) until only a narrow crescent is again apparent. The Moon, having moved eastwardly around the sky much faster than the Sun, has now caught up to the position of the Sun. It once again disappears into the brightness of the morning sky to become another new moon.

Halfway between new moon and full moon occur the quarter moons. The first quarter moon occurs when it has moved exactly 90° away from the Sun's position to the east; thus, it is seen to rise at noon when the Sun is on the meridian and it will set at midnight. The third quarter occurs when the Moon has moved halfway again around the sky, and is now viewed 90° west of the Sun in the sky. Rise and set times are now reversed, with the Moon coming up above the eastern horizon at midnight and setting below the western horizon at noon.

The cause of this changing pattern of the appearance of the Moon was recognized as early as the time of ancient Greece. Anaxagoras of Clazomenae made the first clear statement that the Moon shines by reflection of light it receives from the Sun. As only one-half of the lunar sphere is illuminated at any time, differing proportions of the illuminated and the dark hemispheres are seen from earth depending on the geometric orientation of Earth relative to the Sun and the Moon. Anaxagoras seems to have understood the origin of the phases, although their explanation was more clearly described in the later works of Aristotle, who was fully aware that the observed pattern of phases could occur only if the Moon were truly a spherical object.

Lunar Calendars

In human history, observation of the phases of the Moon has served an important calendrical function in dividing the year into shorter periods of time. The month comes directly from awareness of the lunar phase cycle, and the idea of a week consisting of seven days possibly originated as the best approximation to the time between lunar quarter phases. Direct use of a lunar-solar calendar, however, does involve some difficulties, as an integral number of lunar months (29.5 days) do not fit within a year (365.25 days). Twelve alternating months of 29 and 30 days, to keep lunar months and calendar months in phase with each other, produces a "year" of 354 days, some 11 days short of the real

length of the year. At the end of a second year, the calendar is off by 22 days. If a thirteenth month is added to (or intercalated into) a third year, the calendar is brought back to close, but not perfect, agreement with the true year. The ancient Greeks thus adopted more complicated schemes such as that of Solon. His oktaeteris calendar of 5 years with 12 months interspersed with 3 years of 13 months produced an error of only three-quarters of a day at the end of an 8-year period. The 19-year cycle of Meton, with 7 years of 13 months and 12 years of 12 months, was off by less than one-half a day at the end of that time.

Western culture inherited a modified calendar from the ancient Romans, in which months were divorced from the lunar cycle by simply defining the year to contain exactly 12 months. In our calendar, the setting of Easter remains as a holdover from an ancient lunar-solar calendar: Easter is always the first Sunday after the fourteenth day of the Moon which happens on, or next after, the occurrence of the vernal equinox.

Other cultures of today, such as that of Islam, continue to rely on a lunar calendar in which the start of the months are determined by the first appearance of the crescent moon.

Planetary Phases

Like the Moon, the planets Mercury and Venus also show a full pattern of phases from new to full. Mars in its orbit about the Sun is close enough to Earth that at times we can view a small portion of its nighttime hemisphere; hence, it may appear gibbous. The more distant planets are sufficiently far away that they present only their sunlit hemispheres for observation.

Misconception

A modern misconception held by many people is that lunar phases are caused by the Moon passing through Earth's shadow. The view of Earth and the Moon obtained by the Galileo unmanned spacecraft as it left Earth on its journey to Jupiter showed that this is not the case. The Sun illuminated one-half of the spherical surfaces of both objects. The spacecraft, looking back at an angle of 90° to the solar direction, saw both Earth and the Moon in quarter phase. The (invisible) shadow of Earth, when extended, entirely missed the position of the Moon.

See also: ECLIPSE; KEPLER'S LAWS; NEWTON'S LAWS; PLANETARY SYSTEMS

Bibliography

KRUPP, E. C. *Echoes of the Ancient Skies* (Harper & Row, New York, 1983).

O'NEIL, W. M. *Time and the Calendars* (Manchester University Press, Manchester, Eng., 1976).

CHARLES J. PETERSON

MÖSSBAUER EFFECT

The Mössbauer effect can best be understood by the history of discovery itself. A nucleus in an excited state can decay to a lower energy state by a number of processes, one of which is an emission of gamma ray photon. An exited state of a nucleus has a finite lifetime (τ) and thus has a nonzero natural linewidth (Γ_0) according to the uncertainty principle $\tau\Gamma_0 = \hbar$. Thus if the excited state has energy E_0 above the ground state, experimental determination of energy by absorption of gamma rays should reveal that the absorption curve has a sharp peak at E_0 with the peak falling off to one half of its maximum height at $E_0 \pm \Gamma_0$, that is, both sides of the peak. If the levels concerned are atomic energy levels, one can observe absorption of a photon emitted by another identical atom making a transition between these two states. This is called resonance fluorescence. Resonance fluorescence, however, is not observed for nuclear decay and excitation.

When a photon is emitted there is a recoil of the emitting parent body just as a cannon recoils when it fires a shell. The atomic recoil energy is very small compared to the natural linewidth of atomic spectra. The nuclear recoil energy, on the other hand, is large compared to Γ_0 and the emitted gamma ray photon no longer has sufficient energy to excite an identical nucleus from its ground state to the excited state. In other words, there is no nuclear resonance fluorescence.

Energy (E) of the photon can be approximated by the expression

$$E \approx E_0 - E_0/2Mc^2,$$

where M is the mass of the recoiling nucleus and c is the speed of light. To make the matter worse, the absorbing nucleus also needs the recoil energy when it absorbs a photon. The energy a photon must carry to be absorbed is then approximately

$$E_0 + E_0/2Mc^2.$$

This means that the nuclear resonance fluorescence does not happen unless the linewidth of the nuclear exited state is much broader than $2 \times E_0/2Mc^2$ or E_0/Mc^2. A lifetime of 10^{-8}–10^{-12} s, which corresponds to $\Gamma_0 = h/\tau$ of approximately 10^{-7}–10^{-3} eV (1 eV = 1.6×10^{-19} J). Combined nuclear recoil energy (of decay and absorption) might be typically 10^{-1} eV. The recoil energy for nuclear decay overshadows the natural linewidth of nuclear levels. (See schematic representation in Fig. 1.) One possible recourse was to use Doppler effect to induce line broadening to create an overlap between the photon energy and the excited level. The thermal motion of the nucleus provides Doppler broadening and gamma ray resonance fluorescence appeared to be a possibility.

Based on earlier results of some significance, in 1958 Rudolf Mössbauer began an attempt to detect resonance fluorescence by thermal Doppler broadening and used 129-kV gamma ray of radioisotope source Ir^{191} embedded in a crystal. Mössbauer cooled the source and expected to see decrease in overlap. What he observed was an increase in absorption and actually what one would expect in the absence of recoil and broadening. He realized that Ir^{191} was not recoiling as an isolated nucleus but

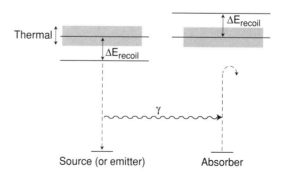

Figure 1   Recoil energy for nuclear decay.

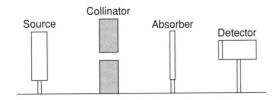

Figure 2 Typical set-up for study of the Mössbauer effect.

Bibliography

FRAUENFELDER, H. *The Mössbauer Effect* (W. A. Benjamin, New York, 1962).

CARL T. TOMIZUKA

the entire crystal was participating in the recoil. Since the recoil energy is E_0/Mc^2, where M stands for the recoiling mass, replacing the nuclear mass by the mass of the crystal, which is larger by a factor of 10^{23} or so, makes the recoil energy entirely insignificant. It is as though one replaced a light barrel of a cannon with a massive one; the energy of the shell lost in recoil is reduced substantially and the muzzle velocity of the shell would increase. (Actually only a fraction of the transitions are recoil free.)

The original experiment of Mössbauer required a cryostat to maintain low temperature. Since then 14.4-kV gamma ray photon from Fe^{57} decay has been most frequently used for various experiments on Mössbauer effect as the effect can be readily observed at room temperature. To date, more than 100 nuclear transitions of about 90 isotopes have been known to exhibit the Mössbauer effect.

Application of the Mössbauer effect ranges from the study of solids, materials science, and chemistry to general relativity. The typical experimental layout for Mössbauer effect study is shown schematically in Fig. 2. The radioactive source mount is coupled to a electromechanical drive to execute a triangular or harmonic motion. In a transmission mode a stationary absorber is placed between the source and a gamma-ray detector.

The natural linewidth of 14.4-keV Fe^{57} isotope is so narrow that a small relative velocity between the absorber (containing Fe^{57}) and the source, as small as $10\mu/s$, can be detected in the laboratory configuration described above. Because this "spectroscopy" is extremely sensitive, one can detect the change in a nuclear transition caused by subtle changes in the crystal lattice, in chemical bonding, or even a gravitational redshift.

See also: DECAY, GAMMA; DECAY, NUCLEAR; DOPPLER EFFECT; EXCITED STATE; FLOURESCENCE; RESONANCE; SPECTROSCOPY

MOTION

Among the most fundamental concepts in physics are those of distance and time, and how they relate to define motion. Most simply put, motion is the continuous change of position of a body relative to some frame of reference. Motion is characterized by the instantaneous position of the body, its velocity (rate of change of position with time), and its acceleration (rate of change of velocity with time). How the motion of a body evolves in time is determined by the pertinent laws of motion. The study of motion is basic to any understanding of the physical world; thus, it has occupied the greatest minds of science for millennia and continues to do so to the present day.

In the written record of western culture, this lineage can by traced back perhaps most notably to ancient Greece and the philosopher Aristotle. The primary elements of nature (air, fire, earth, and water), which were believed to account for the observed motion of bodies, were laid out in his great work *Physics*. Aristotle believed that the world was kept in equilibrium by the opposing tendencies of air and fire to move upward, and earth and water downward. The fifth element, ether, accounted for the motion of the stars and planets.

Also of great and longstanding influence was the view of the motions of the heavenly bodies devised by the astronomer Claudius Ptolemaeus of Alexandria and codified in his *Almagest*. Ptolemy believed that the stars, planets, Moon, and Sun revolved about Earth in orbits consisting of a succession of circles upon circles, known as epicycles. The complexity of the geocentric system of epicycles was necessary to predict the rather complicated paths traced by these bodies through the sky. The form of this mathematical invention was prompted by his conviction that only circular orbits conformed to the presumed perfection of the universe.

The Aristotelian and Ptolemaic worldviews dominated scientific thought for nearly two thousand years, standing as great monuments of observation and intellectual accomplishment. The overturning of these ideas by Nicolaus Copernicus, Johannes Kepler, Galileo Galilei, Isaac Newton, and others represents the heart of the renaissance of science that took place during the fifteenth to seventeenth centuries in Europe.

For example, at the risk of excommunication and imprisonment, Copernicus challenged the authority of the Church by placing the Sun rather than Earth at the center of the known universe. This Copernican Revolution was furthered by Kepler through the formulation of his laws of planetary motion. In particular, Kepler's laws state that

1. The planets move in elliptical orbits with the sun at one of the foci;
2. The radius of the orbits sweep out equal areas in equal times; and
3. The square of the orbital period of each planet is proportional to the cube of its mean radius from the Sun.

Galileo supported the Copernican–Keplerian model of the solar system and made important observations of the motion of bodies. He contributed greatly to the field of kinematics, for example, by studying the motion of falling bodies. He clarified the concepts of velocity and acceleration, and proposed a principle of relative motion ("the laws of motion are invariant under translations, rotations, and changes of velocity"). After Galileo was subjected to the Inquisition and forced to recant his views, he died in 1642, the same year Newton was born.

Even in his own time, Newton was seen as bringing an all encompassing order to the universe through his mathematical description of motion and his conceptions of absolute space and absolute time. He invented, contemporaneously with Gottfried Wilhelm Leibniz, calculus and used this mathematical tool to define the relationship of distance (or position), velocity, and acceleration. For example, if $f(x,t)$ is a function representing the time history of the position x of a body, then the velocity v is given by

$$v = (d/dt)f(x,t), \qquad (1)$$

and the acceleration a is given by

$$a = (d^2/dt^2)f(x,t) = (d/dt)v, \qquad (2)$$

where d/dt and d^2/dt^2 are the first and second time derivatives, respectively.

With this mathematical foundation, Newton rigorously explained the mechanics of motion and showed how the same laws governing the motion of terrestrial objects, such as apples and cannon balls, also governed the motion of heavenly bodies, in one of the greatest works of science—*Philosophiae Naturalis Principia Mathematica*. The fundamental laws expounded there form the basis of what has come to be known as classical or Newtonian mechanics and are referred to as Newton's laws. They state that

1. A body will remain at rest or travel with a constant velocity in a straight line unless acted on by an external force [also known as the law of inertia, enunciated by Galileo as a negation of Aristotlian mechanics].
2. A body's acceleration (related to the change in momentum, momentum being equal to mass × velocity) is proportional to the sum of the external forces F on it and inversely proportional to the mass m of the body [this law expresses Newton's famous equation $F = ma$, which can also be written $F = (d/dt)mv$—an example would be Newton's law of universal gravitation $F_{gravity} = GMm/r$, where G is the Newtonian gravitational constant, M is the mass of one object (say, the earth), m is the mass of the other object (say, an apple), and r is the separation of their centers].
3. The force that is exerted on one body by a second body is equal in magnitude and opposite in direction to the force exerted by the second body on the first [the law of action and reaction—every action has an equal and opposite reaction].

Later, Joseph Louis Lagrange and William Rowan Hamilton provided mathematical elaborations of Newton's principles, which have been significant in the development of modern formulations of mechanics and other physical theories.

By the late nineteenth century, developments of the theory of electromagnetic radiation began to challenge the Newtonian view of absolute space and time. Albert Einstein stands as the focal point for this clash, having developed the special (or restricted) theory of relativity in 1905 and the general theory of relativity in 1915. The special theory main-

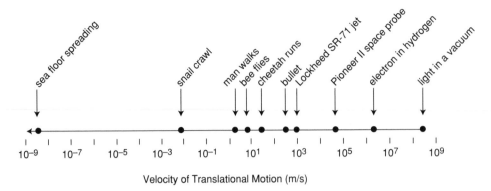

Figure 1 Translational velocities of bodies illustrating the range of this form of motion.

tains that the speed of light traveling in a vacuum is the limiting velocity that any physical body can attain, and that physical laws are invariant for inertial reference frames (a frame in which Newton's first law describes the motion of a body upon which no external force acts). Two bodies can therefore only interact with one another in accordance with this speed limit. In contrast, essential to the mechanics of Galileo and Newton was the instantaneous interaction of bodies (so-called action-at-a-distance). As a consequence of these new laws, the addition of velocities cannot be accomplished by the Galilean–Newtonian prescription, but rather requires the transformations derived by Einstein and by Hendrik Antoon Lorentz.

Furthermore, Einstein showed that space and time are not in fact disparate, but rather manifestations of a single concept—spacetime. Einstein's general theory of relativity is a geometric theory of gravitation. In this view, gravity is a result of the curvature of spacetime caused by mass and energy (which in turn are simply two aspects of a single phenomena, mass-energy, expressed by his famous equation $E = mc^2$). This theory is a relativistic generalization of Newton's theory of gravitation and reduces to it in the limit of weak gravitational fields and velocities that are small compared to that of light. General relativity has resulted in the modern view of the origin of the universe in the so-called big bang, and the theory of very massive objects such as neutron stars and black holes. Along with quantum mechanics, which is the theory of the interactions of physical bodies on the microscopic scale of the atom and subatomic particles, general relativity is the most modern generally accepted theory of motion.

To provide some orientation to the scope of motions of physical bodies, the most common forms may be classified as translational, rotational, and vibrational motions. Examples of vibrational motion are the oscillation of a molecule or the atoms within a solid. Sound is a manifestation of vibration of a medium such as air. In bulk objects, such motion gives rise to heat. Rotational velocities range from the very slow rotation of our galaxy ($\approx 10^{-15}$ rad/sec) to more moderate rotations such as that of Earth upon its axis ($\approx 10^{-4}$ rad/sec) to very fast rotations such as that of an atomic nucleus ($\approx 10^{20}$ rad/sec). The velocity of light limits the translational velocity of any object as required by the special theory of relativity, and the range of translational velocities of some well-known objects is displayed in Fig. 1.

See also: COPERNICAN REVOLUTION; COPERNICUS, NICOLAUS; EINSTEIN, ALBERT; KEPLER'S LAWS; LORENTZ, HENDRICK ANTOON; MOTION, ROTATIONAL; NEWTON'S LAWS; RELATIVITY, GENERAL THEORY OF; RELATIVITY, SPECIAL THEORY OF; SPACETIME; TIME; VELOCITY

Bibliography

ABELL, G. *Exploration of the Universe* (Holt, Rinehart and Winston, New York, 1964).

HALLIDAY, D., and RESNICK, R. *Fundamentals of Physics* (Wiley, New York, 1988).

SEARS, F. W.; ZEMANSKY, M. W.; and YOUNG, H. D. *University Physics* (Addison-Wesley, Reading, MA, 1983).

WEAVER, J. H. *The World of Physics*, Vol. 1: *The Aristotelian Cosmos and the Newtonian System* (Simon & Schuster, New York, 1987).

DAVID R. SCHULTZ

MOTION, BROWNIAN

In 1828 botanist Robert Brown described the ceaseless "swarming" motion of pollen dispersed in water; the small particles he saw in his microscope were constantly in motion. In May 1905 Albert Einstein, twenty-six years old and working as a patent examiner for the Swiss Government at Berne, submitted a brief paper to the *Annalen der Physik* that explained this Brownian motion quantitatively as following from fluctuating forces on the particles from collisions with molecules. However, even the existence of molecules had not then been demonstrated to the satisfaction of all scientists. But a few years later Jean Perrin used Einstein's results to determine the mass of molecules from his measurements of the Brownian motion, and, hence, demonstrate unequivocally their existence.

While Brown—and Perrin—studied small particles in water, we discuss the simpler Brownian motion in air. The left-hand diagram of Fig. 1 shows a cube bombarded by oxygen and nitrogen molecules in the air where, for simplicity, we consider only motion in one direction. The pressure on the cube comes from the collisions of the molecules. Let us consider the collisions of molecules with the cube over a very short time such that there are only 100 collisions with each face on the average. But there will be statistical fluctuations in the number of colli-

sions; sometimes there will be more, sometimes less. From simple statistical theory, the average fluctuation in the number of collisions will be about $\sqrt{100} = \pm 10$, giving an average fluctuation in the pressure on each face of about 10 percent. Under such fluctuating unbalanced forces, the cube will move; sometimes to the left, sometimes to the right, sometimes more, sometimes less. But if the same pressure was caused by 10,000 collisions of much smaller molecules over the same time, the proportional fluctuations in the pressure would be 10 times smaller; the mean fluctuation in pressure would be only 1 percent and the cube would move about 10 times less. Hence, the magnitude of the fluctuations in the pressure and of the Brownian motion that ensues depends upon the number of particles that make up the mass of a given volume of gas.

When a particle is given a kick by a fluctuation in the pressure, the particle will move initially and then slow down through the effects of viscous friction; the larger that viscosity, the smaller the displacement distance. Then the next fluctuation in pressure will move the particle again. But that next random displacement will be as likely to subtract from the first as add to it. The particle then performs a "random walk." Since the movement is equally probable in each direction, the many displacements over time will tend to cancel and the average displacement of a set of particles will be zero. But the squares of the deflections, $(\Delta x)^2$, are necessarily positive and accu-

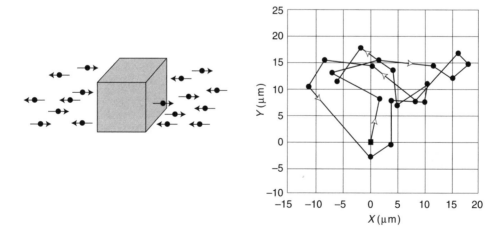

Figure 1 To the left, a particle in air bombarded by gas molecules that generate the air pressure on the particle. To the right, simulated positions of a mastic particle in water at 30-s intervals.

mulate with time; indeed the total square deviation must be proportional to time. Einstein found that for a sphere of radius a,

$$\overline{(\Delta x)^2} = \frac{RT/N_A}{3\pi\eta a}\,\Delta\tau, \qquad (1)$$

where η is the coefficient of viscosity, $\Delta\tau$ the elapsed time, R is the gas constant, and T the absolute temperature. Here $RT = PV$, where P is the pressure and V the volume of a mole of gas that contains N_A molecules. We notice that $\overline{(\Delta x)^2}$, the mean-square-deviation, is inversely proportional to the viscosity η, and inversely proportional to N_A, the number of molecules in a mole of gas.

In his primary experiment on the Brownian motion, Perrin observed the motion in water of individual mastic spheres, marking their positions every 30 seconds. The spheres were small, with a diameter about equal to the wavelength of blue light, but their mass was about 1 billion times that of an oxygen or nitrogen molecule. The right-hand diagram of Fig. 1 shows a computer generated simulation of the kind of results that Perrin found from his microscopic measurements. The points represent the observations; the lines connecting the points only show the order—the particles drift randomly and erratically from point to point.

If Perrin had been able to make measurements 100 times faster—each 0.3 s—the pattern would have had the same random-walk character plotted on a one-tenth scale. If he had made measurements every 3,000 s, 100 times slower, the pattern would again be similar plotted on a 10-times-larger scale. The Brownian movement is fractal-like—the patterns are the same for any scale. As the scale gets smaller, the average point-to-point drift velocity is greater by the inverse of the scale factor. But that velocity is still much smaller than the average instantaneous velocity v of the spheres. For any particle in temperature equilibrium in a gas or liquid, the average kinetic energy, $\frac{1}{2}m\overline{v^2} = \frac{3}{2}RT/N_A = \frac{3}{2}kT$, where $k = R/N_A$, is Boltzmann's constant. Hence, the average kinetic energy of the mastic sphere was equal to the mean energy of a molecule in the air. Calculated from this relation, the mean instantaneous velocity of the mastic spheres was about 1 cm/s; but such a sphere drifted, typically, less than 1/1,000 of a centimeter in 30 s.

From a 10-min set of 20 measurements such as shown in Fig. 1, Perrin could determine 20 values of

$(\Delta x)^2$ — and 20 of $(\Delta y)^2$ that are equally useful. From a determination of the mean value of $(\Delta x)^2$ from measurements of many particles over many sets of 30-s intervals, he found, using Eq. (1), that there were $N_A \approx 6 \times 10^{23}$ molecules in a mole of gas, where N_A is called Avogadro's number (or Loschmidt's number). Perrin, by following Einstein's theoretical development, had proved that molecules existed and determined their mass. He received the Nobel Prize in 1926 for this work.

See also: EINSTEIN, ALBERT; QUANTUM MECHANICAL BEHAVIOR OF MATTER; QUANTUM MECHANICS; QUANTUM MECHANICS, CREATION OF

Bibliography

BORN, M. *The Restless Universe* (Dover, New York, 1951).

EINSTEIN, A. *Investigations on the Theory of the Brownian Movement* (Dover, New York, 1956).

FEYNMAN, R. P.; LEIGHTON, R. B.; and SANDS, M. *The Feynman Lectures on Physics,* Vol. 3 (Addison Wesley, Reading, MA, 1965).

ROBERT K. ADAIR

MOTION, CIRCULAR

Humans have been intrigued with circular motion since before the time of Aristotle. The Greek philosophers thought that circular motion was the most perfect kind of motion. Necessarily, they argued, the motions of the heavenly bodies must be circular. While Aristotle believed that the Sun and known planets revolved about Earth, Copernicus convinced us that the planets go around the Sun in orbits that are almost circular and that the daily rotation of Earth makes the stars appear to travel in circular paths through the sky. Copernicus, however, did not answer the questions "What keeps Earth in orbit about the Sun?" "What keeps the Moon in orbit about Earth?" That was left for Isaac Newton to study more than 100 years later.

Each of us, whether it be our first ride on the merry-go-round at the local community park or that exhilarating experience on the latest version of the roller coaster, has had first-hand experience with and is fascinated by circular motion. These experiences have raised the following questions: "What keeps us in our seat on the roller coaster when it does the loop-de-loop and we suddenly find ourselves upside down?" "Why do we have a sudden sinking feeling in our stomach when riding in a car that goes over a sudden rise in the road or when the Ferris wheel goes past its highest point?" These questions are all answered using the concepts of circular motion.

One of the first to give a rational explanation for why the Moon stays in a nearly circular orbit about Earth was Isaac Newton when he proposed that the Moon was actually falling about Earth. According to Newton, an object moving without external forces acting on it will move with a constant speed in a straight line. However, if the object is subjected to a force that is always perpendicular to the direction of travel, then the object will travel in a circle of some radius r with a constant speed v. This special type of force, required to keep the object going in a circle, is called a centripetal force. The word "centripetal" simply means that the force is acting toward the center of the circle. If an object of mass m is to travel in a circle of radius r with constant speed v, then the centripetal force needed is mvv/r or mv^2/r. For the case of the Moon in orbit about Earth, this force is provided by the gravitational "pull" between the Moon and Earth. If instead we had considered the circular motion of a ball on the end of a string, then the tension that the string exerts on the ball provides the centripetal force. Finally, for the case of a car moving on a circular track, the friction force between the tires and the road provides the required centripetal force.

The time that it takes for a body to make one complete trip around the circle is called the period T. The period is equal to the circumference of the circle (two times the radius r) divided by the speed v. That is, $T = 2 r/v$. For the case of the Moon revolving about Earth, the Moon's period is about 28 days. Earth's period (as it revolves about the Sun) is about 365 days.

Since the force required to keep an object moving in a circle of a specified radius depends upon the mass of the object, we can use a device called a centrifuge to separate a mixture of masses. When a mixture of different masses is rotating at some speed, the heavier particles move to an outer radius while the lighter particles stay closer to the center. The more massive red and white blood cells, for example, may be separated from the serum using centrifuges. Large centrifuges have been designed to simulate the effects of gravity and have proven to be very useful in the training of astronauts in the space program. Very large centrifuges have even been used to simulate the stresses and strains experienced by dams, buildings, and other large structures.

Physicist Gerard K. O'Neill used circular motion concepts when suggesting that space could be colonized by placing in orbit huge spheres (approximately 500 m in diameter) that would serve as space habitats. The sphere would be set into rotation at just the right speed so that the gravitational force at the spaceship's equator would be the same as that on Earth. O'Neill envisioned space communities with all the comforts of Earth (ballet, swimming pools, gardening, hiking) but with a variable gravitational force depending on the radius at which one resides in the rotating sphere.

See also: CENTRIFUGE; CENTRIPETAL FORCE; COPERNICUS, NICOLAUS; MOTION; NEWTON, ISAAC; NEWTON'S LAWS

Bibliography

GLASHOW, S. L. *From Alchemy to Quarks* (Brooks/Cole, Pacific Grove, CA, 1993).

HALLIDAY, D.; RESNICK, R.; and WALKER, J. *Fundamentals of Physics*, 4th ed. (Wiley, New York, 1993).

O'NEILL, G. K. *The High Frontier: Human Colonies in Space* (William Morrow, New York, 1977).

JAMES H. STITH

MOTION, HARMONIC

Harmonic motion, or simple harmonic motion (SHM), is the most fundamental periodic motion. It is common to many physical systems, including a simple pendulum, a tuning fork, the strings of a violin, the air column in an organ pipe, a torsional pendulum, an alternating current in an electrical circuit, and the atom vibrating about its equilibrium position in a crystal lattice. Mathematically, it is characterized by a sinusoidal function of time t with a fixed frequency v. For a physical system subject to a change $\mathbf{q}$ (e.g., a displacement from an equilibrium position, an increase or a decrease in length, or twisting of a rod around its axis), an SHM may re-

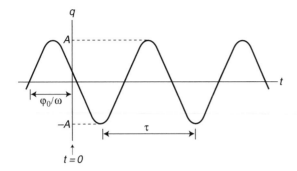

Figure 1 $q(t) = A \sin(\omega t + \varphi_0)$.

sult if the system is subject to a restoring force $\mathbf{Q}_q$ (e.g., mechanical force, pressure, or torque) that is linearly proportional to the amount of change, or

$$\mathbf{Q}_q = -k\mathbf{q}, \qquad (1)$$

where k is a proportional constant called the force constant. Equation (1) is also known as Hooke's law, after Robert Hooke.

By combining Eq. (1) with Newton's second law, an object of mass m subject to an SHM can be described alternatively by

$$\frac{d^2 q}{dt^2} + \omega^2 \mathbf{q} = 0, \qquad (2)$$

where $\omega = (k/m)^{1/2}$ is known as the angular frequency of the SHM. Following Eq. (2), an SHM may also result if the acceleration of an object is proportional to its displacement $\mathbf{q}$ and is pointing in the direction opposite to $\mathbf{q}$. For a one-dimensional system, Eq. (2) is satisfied by the solution

$$q(t) = A \sin(\omega t + \varphi 0), \qquad (3)$$

shown graphically in Fig. 1. The frequency, ν—the number of complete oscillations per second, in hertz (Hz)—is related to the angular frequency ω by $\omega = 2\pi\nu$. The period, τ—the time required to complete one oscillation—is given by $\tau = 1/\nu$. The object undergoing an SHM is confined within its limits $q = \pm A$, where A is the amplitude. The initial phase angle, φ_0, is determined by the initial displacement q_0 and the initial velocity v_0. The frequency ν (or the angular frequency ω) is independent of the amplitude A and the initial phase angle φ_0. In general, any periodic motion can be described by a linear super-

position of a set (infinite in size, if necessary) of SHMs of various frequencies with appropriate amplitudes, following a mathematical procedure known as Fourier analysis.

The conservation of mechanical energy provides a convenient starting point leading to some of the general features of the SHM. For example, the restoring force of a massless spring, $-kq$, is represented by a potential energy $V = \frac{1}{2}kq^2$, where q is the displacement from its equilibrium position. The total mechanical energy E_t for an object of mass m, attached to the end of the spring, sliding horizontally on a frictionless surface with a speed v (or a momentum $p = mv$), is given by the sum of the kinetic energy $E_k = \frac{1}{2}(p^2/m)$ and the potential energy $V = \frac{1}{2}m\omega^2 q^2$. That is,

$$E_t = \frac{1}{2}\frac{p^2}{m} + \frac{1}{2}m\omega^2 q^2. \qquad (4)$$

The potential energy V as a function of the displacement q is given by the solid parabola in Fig. 2. The total energy is represented by a solid horizontal line. The amplitude A of the SHM is defined by the intersections between the horizontal line and the parabola. The motion of the object is limited to values of q between A and $-A$. At the turning points (i.e., when $v = 0$ at $q = \pm A$), E_t equals V. At the equilibrium position (i.e., at $q = 0$), the speed v reaches its maximum value, v_{max}. As the mass m oscillates back and forth between the turning points, the energy changes from potential energy to kinetic energy and back again.

In general, for an object subject to a force $\mathbf{Q}$, its motion is characterized by a potential energy V defined by $\mathbf{Q} = -\nabla V$. Whenever the object is displaced slightly from a local minimum in V—that is, from its stable equilibrium position $r = r_e$, the restor-

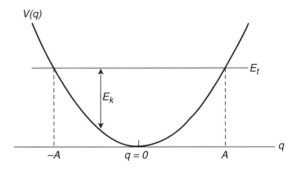

Figure 2 $E_t = E_k + V.$

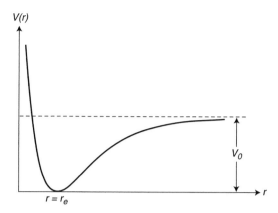

Figure 3 Morse potential.

ing force is almost always proportional to the displacement, $q = r - r_e$, and the motion of the object is well represented by an SHM if q is sufficiently small. For example, the potential energy for a diatomic molecule is often approximated by the Morse potential. That is,

$$V(r) = V_0(1 - e^{-a(r-r_e)})^2, \qquad (5)$$

which can be expressed by a Taylor series in terms of small q. That is,

$$V(q) = V_{q=0} + \sum_{n=1}^{\infty} \frac{q^n}{n!} \left(\frac{d^n V}{dr^n} \right)^{q=0} \simeq \tfrac{1}{2} k q^2, \qquad (6)$$

where the force constant $k = 2a^2 V_0$. As a result, a diatomic molecule vibrating around its stable equilibrium position r_e (shown in Fig. 3) can be described approximately by an SHM with an angular frequency $\omega = (2a^2 V_0/m)^{1/2}$, where m is the reduced mass of the system.

A system undergoing an SHM can also be represented, in both classical and quantum mechanics, by a Hamiltonian:

$$H = \tfrac{1}{2} \frac{p^2}{m} + \tfrac{1}{2} m\omega^2 q^2. \qquad (7)$$

The corresponding Schrödinger equation in quantum mechanics is

$$-\frac{\hbar}{2m} \frac{d^2\psi}{dq^2} + \frac{m\omega^2}{2} q^2\psi = E\psi, \qquad (8)$$

where $\hbar$ is Planck's constant divided by 2π. The energy eigenvalue E is given by

$$E_n = (n + \tfrac{1}{2})\hbar\omega. \qquad (9)$$

For the ground state, $n = 0$, and for the excited state, n equals a positive integer. The corresponding eigenfunction ψ_n is given by

$$\psi_n(q) = \left[\frac{\alpha}{\pi^{1/2} 2^n n!} \right]^{1/2} \exp\left[-\tfrac{1}{2} \alpha^2 q^2 \right] H_n(\alpha q), \quad (10)$$

where H_n is the Hermite polynomial of degree n and $\alpha^2 = \omega m/\hbar = [mk/\hbar^2]^{1/2}$.

See also: HOOKE'S LAW; OSCILLATOR, HARMONIC; PENDULUM; WAVE MOTION

Bibliography

FEYNMAN, R. P.; LEIGHTON, R. B.; and SANDS, M. *The Feynman Lectures on Physics*, Vol. 1 (Addison-Wesley, Reading, MA, 1963).

TIPLER, P. A. *Physics for Scientists and Engineers* (Worth, New York, 1990).

TU-NAN CHANG

MOTION, PERIODIC

Periodic motion is the regular repetition of the trajectory of a moving body. The time interval to complete one full cycle is the period.

Periodic motion attracted the attention of ancient civilizations all over the world since it affected day-to-day life to a large extent. Without detailed understanding of the matter, observation of dawn and dusk, the lunar cycle, and the seasons revealed that characteristic repeat times for the motion of the Sun, the Moon, and Earth are approximately one day, one month, and one year, respectively. Astronomers in Babylonia (about 1000 B.C.E.) recorded minute observations of the motion of planets and stars in the night skies. In ancient Greece, Ptolemaeus (150 C.E.) applied his extensive knowledge of geometry to a description in which celestial objects revolve in, sometimes very complicated, orbits around Earth.

It took centuries before Niklas Koppernigk (Copernicus) from Toruń, Poland, realized that accepting a world view in which the planets, including Earth, revolve around the Sun greatly simplifies the description of planetary motion. By modern standards this merely means a revision of the coordinate system. However, in those days Copernicus met fierce opposition against his theory that abandoned Earth as the center of the universe, especially from the clergy. In his heliocentric view of the solar system, the planets move in periodic orbits around the Sun, with times for the completion of one full cycle ranging from 88 days for Mercury to 248 years for Pluto. In the sixteenth and seventeenth centuries Tycho Brahe and Johannes Kepler further refined knowledge about the heliocentric world view on a purely empirical basis. Isaac Newton put the crown on this work when he accurately accounted for the periodic motion of the planets within his theory of gravitation.

Periodic motion, such as that of the planets orbiting around the Sun, is characterized by a time interval or period T after which the trajectory of the moving body repeats. The reciprocal of the period is the frequency, $f = 1/T$, related to the angular frequency ω through $\omega = 2\pi f$. Periodic motion is a common occurrence. Perhaps the simplest example of periodic motion is provided by a pendulum, where a bob is suspended at the end of a string. Another example of periodic motion is an artificial communication satellite orbiting our planet along periodic trajectories relative to Earth's surface. Newton's laws predict these orbits. Newtonian mechanics also explains the geostationary satellite, orbiting the Earth at an altitude such that the period of its orbit matches Earth's own rotational motion. Hence, the geostationary satellite is located above a fixed point on Earth's surface. At the microscopic level, the wave-like behavior of electrons in atoms and molecules, as well as the propagation of electrons through three-dimensional, regular crystal lattices are manifestations of periodic motion.

The elementary form of periodic motion is realized in a harmonic oscillator. Harmonic motion can be mathematically related to circular motion through the projection of the trajectory on a straight line. Position x along the line as a function of time t is given by an expression involving a goniometric function: $x = A \cos(2\pi f t + \delta)$, where A is the amplitude, corresponding to the radius of the circle, and δ

is the so-called phase shift, which equals the angular displacement with respect to the line at $t = 0$. The Fourier transform of any complex periodic trajectory yields a decomposition in terms of simple harmonic components. For example, the periodic motion of a point moving along a non-circular orbit, when projected onto a straight line and subsequently Fourier transformed, will decompose into the sum of several contributions, characterized by different amplitudes, frequencies, and phase shifts.

In the real world around us, periodic motion usually is damped. Therefore the total mechanical energy of, for instance, the oscillating pendulum is no longer a constant of motion. Energy will be dissipated as a result of frictional forces acting on the pendulum. This means that the amplitude of the oscillations decreases as a function of time, and the pendulum will eventually come to rest. Tidal forces of the Moon (and to a lesser extent of the Sun) are acting on Earth, and it is conceivable that the plastic deformation of the inner layers of the earth provide a mechanism for the dissipation of rotational kinetic energy as heat. This damping effect may account for the empirical observation that Earth's rotational frequency decreases by small, but significant, amounts. At the microscopic level, electrons in a non-ideal crystal are scattered by imperfections in the crystal lattice that destroy the perfect periodicity. The perturbation of the periodic motion of conduction electrons in a crystal leads to electric resistivity in metals, whereby energy is dissipated via Joule heating.

In contemporary physics, chaotic motion receives much attention. The text book example of a system where chaotic behavior occurs is the driven, damped harmonic oscillator. Under most circumstances the system will still exhibit periodic motion. However, for certain values of the parameters describing the driving and frictional forces, the motion is no longer periodic; it never repeats. The driven, damped harmonic oscillator is in a chaotic state of motion. Such behavior is directly related to the fact that the restoring force is no longer linear in the displacement from the equilibrium position. Nonlinear terms in the equation of motion for the system may change the dynamics of the oscillator from periodic to chaotic.

See also: CHAOS; COPERNICAN REVOLUTION; DAMPING; JOULE HEATING; KEPLER'S LAWS; MOON, PHASES OF; NEWTONIAN MECHANICS; OSCILLATION; OSCILLATOR, HARMONIC; PENDULUM; SEASONS; TIDES

Bibliography

DURHAM, F., and PURRINGTON, R. D. *Frame of the Universe* (Columbia University Press, New York, 1983).

TIPLER, P. A. *Physics for Scientists and Engineers,* 3rd ed. (Worth, New York, 1991).

JOHANNES VAN EK

MOTION, ROTATIONAL

Rotational motion occurs widely and consistently in the physical world, so it is not surprising that the theoretical study of rotations has been a central element in both classical and quantum physics. The study of motion begins with the abstract notion of a point particle that follows a trajectory in three-dimensional space. But for a real object, one with a certain size and shape, motion consists of rotation superimposed on a trajectory followed by its center of mass. A punted football goes through the same kind of parabolic arc as a batted baseball, but because of the football's elongated shape we can see that it is simultaneously spinning end over end.

For a rigid body (an object that does not change its size or shape), rotation can be described by giving the axis about which rotation occurs and the rate of rotation. For a bicycle wheel the axis is a line through the wheel's center and perpendicular to the plane of the wheel. For the spinning Earth, the axis is the line connecting the North Pole and South Pole. The rate of rotation is called angular velocity, the angle through which the object turns per unit time. Earth's angular velocity is 360° in 24 hours, or 0.0042° per second.

The angular velocity of a rotating object is not necessarily constant in time. Think of a wheel, free to rotate around a fixed axis, with a cord wrapped around its rim. The cord has a stone attached to its end, and as the stone falls downward the wheel rotates faster and faster. This demonstrates angular acceleration, defined as the rate of change of angular velocity. The cause of angular acceleration in this example is the weight of the stone, which creates a torque on the wheel. The equation that determines angular acceleration is analogous to Newton's second law of motion: Angular acceleration is equal to the torque divided by the moment of inertia. The moment of inertia is a property of the mass, size, and shape of the object and is a measure of its reluctance to change its angular velocity.

Rolling is a familiar combination of linear and rotational motion. Angular velocity in degrees per second is, in this case, equal to $(180/\pi)(v/R)$. This is known as the rolling condition, where v is the linear velocity and R is the radius. Rolling occurs when there is enough friction to prevent slipping between the rotating object (say, a rubber ball) and the surface. On a bowling alley friction is low, so a bowling ball often starts out not rolling; it may be rotating, but its angular velocity is less than the rolling condition. As the ball moves down the alley its angular velocity will increase (while its linear velocity decreases), and before the ball gets to the end of the alley it may reach the rolling condition.

An important parameter of rotation is the following product: moment of inertia times angular velocity; this is called angular momentum. If an object has no torque on it, its angular momentum remains constant. More generally, if a system of several interacting objects has no external torque acting on it, the system's total angular momentum is constant, or conserved. Angular momentum conservation is one of the most fundamental principles of physics and plays a crucial role in systems from the astronomical level to the atomic and subatomic levels.

Rotation is a salient characteristic of the solar system. The revolution of the planets around the Sun, the spin of each planet on its own axis, the spin of the Sun, the revolution of the satellites of the planets, and the spin of the satellites are all rotations. With very few exceptions, they are all in approximately the same direction: counter-clockwise as viewed from a point above the plane of the ecliptic. This observation is a key element pointing to the origin of the solar system as a single large rotating cloud of gas.

At a larger scale, the Sun and some 10^{11} stars make up the Milky Way galaxy, which also rotates, as do other galaxies. Here the rotation is not that of a rigid body; it consists of individual orbits of stars around the galaxy's center. At still greater distances, astronomers observe clusters of galaxies, again exhibiting rotational motion about each other.

At the opposite distance scale, rotation and angular momentum conservation play a central role in the behavior of molecules and atoms. In the earliest successful explanation of atomic spectra, Niels Bohr built upon Ernest Rutherford's "planetary" model of the atom, in which electrons orbit a nucleus. While modern quantum mechanics supplants the notion of the electron's orbit (because the uncertainty principle does not allow the electron's position to be precisely known), it remains true that the quantization of angular momentum determines the possible energy states of the electron in the atom. Planck's constant, h, equal to 6.63×10^{-34} J·s, is a measure of angular momentum, and the Bohr theory postulated that the electron's angular momentum had to be an integer times $h/2\pi$. This condition leads to the quantization of the possible energy values of the electron. More modern quantum theory takes account of additional angular momentum intrinsic to the electron (not to its "orbit" around the nucleus), which must have the value one-half of the fundamental unit, $h/2\pi$. This is called electron spin, although we cannot properly envisage the electron as a spinning ball (there is no evidence for a finite size of the electron). By developing rules for combining orbital and spin angular momentum, atomic scientists successfully account for the properties of atoms in many experimental situations, including emission and absorption of light and collisions between atoms.

When several atoms combine to form a molecule, a new kind of rotational motion becomes possible, rotation of the molecule about some fixed axis through it. These rotations are also quantized, and, together with vibrational motions, serve to define the molecule's possible energy states.

At the subatomic level, the nucleus itself consists of nucleons (protons or neutrons) held together by strong nuclear forces. Nucleons also have spin equal to $(1/2)(h/2\pi)$, and within the nucleus these particles move in complex patterns, creating orbital angular momentum. In some cases a nucleus looks approximately like an inert core with one or two nucleons in orbit around it. In this case, an atom-like model accounts for the properties of the nucleus by combining the orbital and spin angular momentum of the few orbiting nucleons. In other cases, the nucleus looks like a deformed, nonspherical object, rotating as a unit. A molecule-like model accounts for the properties of such a nucleus.

See also: ACCELERATION, ANGULAR; INERTIA, MOMENT OF; MILKY WAY; MOTION; NUCLEON; QUANTUM THEORY, ORIGINS OF; RIGID BODY; SOLAR SYSTEM; SPIN; SPIN, ELECTRON; VELOCITY, ANGULAR

Bibliography

EDMONDS, A. R. *Angular Momentum in Quantum Mechanics* (Princeton University Press, Princeton, NJ, 1960).

EISENBERG, J. M., and GREINER, W. *Nuclear Models* (North-Holland, Amsterdam, 1975).

SYMON, K. R. *Mechanics*, 2nd ed. (Addison-Wesley, Reading, MA, 1960).

ZEILIK, M., and GAUSTAD, J. *Astronomy: The Cosmic Perspective* (Harper & Row, New York, 1983).

MICHAEL I. SOBEL

MOTOR

A motor is a device that is designed to transform chemical or electrical energy into mechanical energy of motion. The traditional chemical energy source is called the fuel for the motor. The oxidation (burning) of the fuel results in a conversion of the chemical energy contained in the fuel into the motion of the molecules resulting from the high temperatures of the burning process. These molecules in turn are constrained by the motor design to transfer their energy to moving cylinders or turbines on drive shafts that provide the output source for the mechanical energy available from the motor. The internal combustion engine is a motor in which the combustion of the hydrocarbon fuel produces high temperature exhaust gases that transfer their kinetic energy to the piston and the drive train system of a vehicle.

An electric motor is designed to transform electromagnetic energy to mechanical motion. This transformation is a result of the force on a current-carrying wire in a magnetic field. The simplest electric motor is a current-carrying rod in a constant magnetic field (directed perpendicular to the rod) as illustrated in Fig. 1.

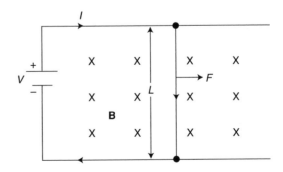

Figure 1 Simple dc motor. The force on the current-carrying rod produces motion in the direction of the force.

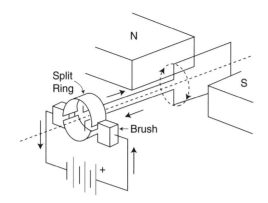

Figure 2 Schematic diagram of a dc electric motor.

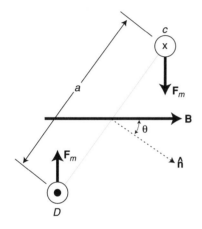

Figure 3 A side view of the loop in the magnetic field *B*.

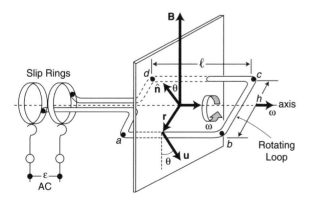

Figure 4 A simple ac motor with slip rings rotating in the direction of ω.

The typical dc motor consists of a coil of wire carrying current (I), called the armature, in a magnetic field (B) as shown in Figs. 2 and 3. The torque on each loop of the coil (with an area = A) making an angle θ with the direction of B is given by $BIA \sin \theta$. The total torque for N loops producing clockwise rotation of the coil in the plane of the page is $NBIA \sin \theta$. The coil is connected to a dc voltage supply by a split commutator ring that completes the circuit to the dc power source by the contact with the brush set. This design maintains the torque in the same direction by reversing the direction of the current through the coil when the coil enters and leaves the magnetic field region.

An ac motor could use a set of slip rings instead of a commutator to provide a synchronous armature speed based on the frequency of the ac power supplied to the motor. This simple ac motor would not start by itself because of the alternating current direction that produces canceling torques in the constant magnetic field, but once it is rotated at the synchronous speed so that the torques are in the same direction throughout the rotation, it would continue to run at that speed. Figure 4 illustrates this simple ac motor.

A common ac motor is based on the interaction between induced currents in a rotating coil with a rotating magnetic field. This rotating field is created by ac currents flowing in at least two coils with a phase difference (the time when the ac current is a maximum is not the same in the different coils). This type of motor is called an induction motor.

See also: CURRENT, ALTERNATING; CURRENT, DIRECT; ENERGY, MECHANICAL; FORCE

Bibliography

POLLACK, H. W. *Applied Physics* (Prentice Hall, Englewood Cliffs, NJ, 1971).

ROGERS, E. *Physics for the Inquiring Mind* (Princeton University Press, Princeton, NJ, 1960).

RICHARD M. FULLER

MUON

The muon is an unstable particle with a mass 206.8 times that of the electron and a half-life of 1.52 μs. It is a member of the lepton family of elementary particles and is thus a relative of the electron. Like the electron, it has a spin of 1/2 and carries one unit of electric charge. The negative muon is considered a particle and the positive one an antiparticle.

The muon was the first of the unstable subatomic particles to be discovered. They were found in cosmic rays in 1937 by Carl Anderson. At that time, a theory of nuclear forces proposed by Hideki Yukawa called for a particle called a mesotron (later shortened to meson) of roughly the muon's mass and half-life, so the name meson and symbol μ (mu) was assigned. In 1947 M. Conversi, E. Pancini, and O. Piccioni found that this particle did not interact strongly with nuclei, and thus could not be Yukawa's meson. Shortly thereafter, C. Lattes, G. P. S. Occhialini, and C. F. Powell discovered that the μ was the product of the decay of a somewhat heavier meson they named the pi (π, for "primary") meson, a particle that did interact strongly with nuclei. Today the term "meson" is reserved for strongly interacting particles, and the name "mu meson" has fallen out of use, replaced by the name muon.

Once the character of the muon had been established, its decay posed a puzzle. Since there appeared to be no difference between a muon and an electron other than mass, it should be able to decay into a gamma ray and an electron. Instead, it always decays into three particles, an electron of the same charge as the muon plus one neutrino and one antineutrino. It was suggested that this was because the muon did have some internal property (now called *flavor*) that distinguished it from the electron, which must be conserved during the decay by transferring it to one member of the neutrino-antineutrino pair. If we assign a flavor number $+1$ for particles and -1 for antiparticles, then this theory says there are neutrinos that carry electron flavor $+1$ and others that carry muon flavor $+1$, and that the total flavor number for both electrons and muons is unchanged in the decay.

This idea, known as the "two-neutrino hypothesis," was confirmed in 1962 by Leon Lederman, Melvin Schwartz, and Jack Steinberger, and now physicists recognize three flavors of neutrino, the e neutrino (ν_e), the mu neutrino (ν_μ), and the tau neutrino (ν_τ). A negative muon decays into a mu neutrino, an electron, and an e antineutrino, while the positive muon decays into a mu antineutrino, a positron, and an e neutrino. In a collision with a nucleus, a mu neutrino can be transformed into a muon, while an e neutrino can only be transformed into an electron. In similar fashion, a negative pi meson decays into a muon and a mu antineutrino, while a positive pi decays into a positive muon and a mu neutrino. Thus all of these reactions conserve electron and muon flavor, as do all processes observed to date.

Muons are the principal constituents of cosmic rays at sea level. They are produced by the decay of pi mesons high in the atmosphere. The pi mesons are produced by collisions of very energetic nuclei that impinge on Earth from space. The pi mesons and other strongly interacting particles are absorbed in nuclear collisions before reaching sea level.

See also: ELEMENTARY PARTICLES; FLAVOR; INTERACTION, STRONG; LEPTON; LEPTON, TAU; NEUTRINO; NEUTRINO, HISTORY OF; YUKAWA, HIDEKI

Bibliography

HALZEN, F., and MARTIN, A. D. *Quarks and Leptons: An Introductory Course in Modern Particle Physics* (Wiley, New York, 1984).

ROBERT H. MARCH

MUSIC

See SOUND, MUSICAL

MUTUAL INDUCTANCE

See INDUCTANCE, MUTUAL

N

NAMBU–GOLDSTONE BOSON

See Boson, Nambu–Goldstone

NATURAL FREQUENCY

See Frequency, Natural

NAVIER–STOKES EQUATION

The Navier–Stokes equation describes the motion of a viscous fluid. It was independently developed by Cloude Louis Marie Henri Navier in 1822 and George Stokes in 1845. Navier's approach came from an incorrect molecular viewpoint, whereas Stoke's method assumed a fluid could be treated as a continuous material. Although the Navier–Stokes equation is merely a fluid version of Newton's Second Law, the equation is a nonlinear, partial differential equation that defies solution except in the most simplified or approximated cases. The following analysis describes a method of arriving at (but not solving) the Navier–Stokes equation.

Newton's second law, $F = ma$, describes the acceleration of a single particle of mass m experiencing an applied force, F. However, consider the motion of a small element of fluid with mass dm which is actually comprised of many individual particles. It is necessary to employ Leonhard Euler's model of a velocity field to describe the acceleration of the element, since it is not an individual particle. The concept of a velocity field results from velocity vectors that have time-dependent components in each direction (x, y, and z). Thus the velocity in the x direction is $v_x = v_x(x, y, z, t)$, with similar terms for v_y and v_z. The acceleration **a** is

$$\mathbf{a} = \frac{D\mathbf{v}}{Dt} = v_x\frac{\partial \mathbf{v}}{\partial x} + v_y\frac{\partial \mathbf{v}}{\partial y} + v_z\frac{\partial \mathbf{v}}{\partial z} + \frac{\partial \mathbf{v}}{\partial t}$$

$$= (\mathbf{v} \cdot \boldsymbol{\nabla})\mathbf{v} + \frac{\partial \mathbf{v}}{\partial t},$$

where the capital D's indicate a special derivative known as a substantial or material derivative. The right-most expression is a condensed version of the middle term using the gradient ($\boldsymbol{\nabla}$), a mathematical function common in vector calculus which describes how the velocity components change in regions surrounding the point of study. There are two "parts" to the acceleration. The term $(\mathbf{v} \cdot \boldsymbol{\nabla})\mathbf{v}$ or "convective acceleration" occurs because the element may travel into regions of higher or lower velocity. The term $\partial \mathbf{v}/\partial t$ appears if the flow changes in time and is called "local acceleration."

Forces on the element include those due to pressure p from surrounding fluid, viscosity (F_{vis}), and other external forces (F_{ext}) such as gravity. Detailed theoretical physics gives the viscous force

$$f_{vis} = \eta \nabla^2 v + (\tfrac{1}{3}\eta + \eta') \nabla (\nabla \cdot \mathbf{v}),$$

where f_{vis} is the force per unit volume, η is the viscosity, and η' is the volume coefficient of viscosity. This form of the viscous force generally produces an unsolvable equation. However, if the fluid is assumed to be incompressible (i.e., constant density) then the second term in the above equation is zero, leaving $f_{vis} = \eta \nabla^2 v$. Here again the gradient term indicates a movement of particles to a region of different velocity but this now occurs irreversibly because of the nonconservative viscous force.

Applying Newton's second law $dF = dm \times a$ gives the Navier–Stokes equation for an incompressible fluid:

$$\rho \frac{D\mathbf{v}}{Dt} = \rho\left(\frac{\partial \mathbf{v}}{\partial t} + (\mathbf{v}\cdot\nabla)\mathbf{v}\right) = -\nabla p + f_{ext} + \eta\nabla^2\mathbf{v},$$

where p is the pressure, ρ is the density, and f_{ext} is the external force per unit volume. This is the fundamental equation in fluid dynamics. For just the x component, the Navier–Stokes equation reduces to

$$\rho \frac{Dv_x}{Dt} = \rho\left(v_x\frac{\partial v_x}{\partial x} + v_y\frac{\partial v_x}{\partial y} + v_z\frac{\partial v_x}{\partial z} + \frac{\partial v_x}{\partial t}\right)$$
$$= -\frac{\partial p}{\partial x} + f_x + \eta\left(\frac{\partial^2 v_x}{\partial x^2} + \frac{\partial^2 v_x}{\partial y^2} + \frac{\partial^2 v_x}{\partial z^2}\right).$$

Assuming one can solve (integrate) the Navier–Stokes equation, then the motion of an incompressible fluid would be completely predicted. Currently, numerical (computer generated) solutions provide some reasonable solutions in simplified cases. For certain values of parameters in the equation, the solution becomes chaotic. The Navier–Stokes equation reduces to the Euler equation for ideal (nonviscous) fluids if $\eta = 0$.

See also: FLUID DYNAMICS; VISCOSITY

Bibliography

BATCHELOR, G. K. *An Introduction to Fluid Dynamics* (Cambridge University Press, Cambridge, Eng., 1977).

FEYNMAN, R. P.; LEIGHTON, R. B.; and SANDS, M. *The Feynman Lectures on Physics,* Vol. 2 (Addison-Wesley, Reading, MA, 1964).

FOX, R. W., and MCDONALD, A. T. *Introduction to Fluid Mechanics,* 3rd ed. (Wiley, New York, 1985).

LIGHTHILL, J. *An Informal Introduction to Theoretical Fluid Mechanics* (St. Edmundsbury Press, Suffolk, Eng., 1989).

KENNETH D. HAHN

NEGATIVE ABSOLUTE TEMPERATURE

Begin with a hypothetical system made up of atoms or other entities, such as nuclear spins, that have only two quantum states: the ground state and an excited state. The entities, which we shall call "elements" of the system, do *not* interact with one another. In the limit of the absolute zero of temperature (approached through positive temperatures), all the elements of the system occupy the ground state. As the temperature is raised, some of the elements acquire sufficient energy to jump to the excited state. Thus, the occupation ratio of the elements is a function of temperature.

If the population of the ground state is N_1 and that of the excited state is N_2, the temperature dependence of the ratio is given by $N_2/N_1 = \exp(-\Delta E/k_B T)$, where k_B is the Boltzmann constant and ΔE is the positive energy difference between the excited state and the ground state. This equation may be taken as the definition of temperature for the two-level system. When the temperature is low, such that $0 < T \ll \Delta E/k_B$, the ratio $\Delta E/k_B T$ in the exponent is very large, and the exponential of a large negative number is very small. Thus, the ratio N_2/N_1 is very small. In other words, very few atoms are excited from the ground state to the excited state. When the temperature is high, meaning $T \gg \Delta E/k_B$, the ratio N_2/N_1 becomes larger and finally, at very high temperatures, approaches the value 1. Then the population of the excited state approaches that of the ground state.

When atoms or nuclear spins are "pumped" continuously by an external supply of energy and are forced into the excited state, N_2 sometimes becomes larger than N_1. This situation is called population inversion. When this happens, the ratio N_2/N_1 becomes larger than 1. If we take the equation displayed above to provide a definition of temperature for a two-level system, then population inversion corresponds to a situation with negative temperature because the expression $\exp(-\Delta E/k_BT)$ cannot equal or exceed 1 as long as T is positive.

The foregoing is a way to look at temperature from a mathematical point of view, with the physical interpretation given in the later section. Note that, in the system described below, the elements *do* interact with one another. That interaction is experimentally crucial.

First Experiment

Experiments can produce temperatures below absolute zero: negative absolute temperatures. The first such experiment was performed by Edward M. Purcell and Robert V. Pound in 1951; their physical system was the assembly of nuclei in a crystal of lithium fluoride. In more detail, the system was the nuclei acting as though they were tiny spinning bar magnets: the nuclear spin system. That orientational aspect of the nuclei was adequately decoupled from the nuclei acting as though they were vibrating masses, having ordinary kinetic energy; the latter aspect cannot exhibit a negative temperature. By swiftly reversing an external magnetic field and achieving internal magnetic thermal equilibrium before allowing a second, slow reversal, Purcell and Pound produced a final spin temperature of -350 K.

One would like a microscopic picture of how the negative temperature was achieved. The interactions of the nuclear magnetic moments among themselves are crucial. In a crystal, a magnetic moment interacts not only with an external magnetic field but also with the neighboring magnetic moments. The energy associated with the latter interaction is called the spin-spin energy. Initially, the spin system is at a positive temperature with the magnetic moments pointed predominantly parallel to the external field, an orientation of negative potential energy. The first field reversal is so fast that the spins cannot change their orientation during the reversal. Consequently, the spins find themselves pointed predominantly antiparallel to the reversed field, an orientation of positive potential energy. In a short time, some of this energy is transferred, by mutual interactions, to spin-spin energy. The energy associated with the mutual interactions becomes relatively large. In fact, that energy becomes larger than what one would predict at any positive temperature.

Because the interaction between the spin system and the lattice vibrations is weak, the spin system holds onto that relatively large amount of energy, even as the crystal is transferred, for purposes of observation, from the small external field to a large external magnetic field. There is, in fact, so much spin-spin energy present that the magnetic moments "prefer" to line up antiparallel to the new external field, a position of positive potential energy, rather than line up in the usual parallel position, the position of negative potential energy. There is enough energy present so that a spin can easily line up in what is ordinarily an energetically unfavored orientation. This occurs no matter how the crystal, as a macroscopic object, is rotated during the physical transfer from one magnetic field in the laboratory to the other.

Indeed, the examination of the crystal in the large field was quite directly a test for the amount of energy possessed by the nuclear spin system. Purcell and Pound applied electromagnetic radiation at a frequency chosen so that lithium-7 nuclei could flip from an orientation of low to high potential energy by the absorption of a photon. Instead of absorption, they found—as they had hoped—stimulated emission of radiation at that frequency. The spin system had so much energy that spins predominantly flipped to low potential energy orientations and simultaneously emitted a photon. From the direct observation of stimulated emission of radiation—rather than absorption—one infers both that the total magnetic moment is antiparallel to the external field and that the nuclear spin system is at a negative temperature.

At $T = -350$ K, Purcell and Pound's system of nuclei was hotter than anything at a positive temperature. That is, it would transfer energy to anything at a positive temperature. The experimental test for a negative temperature relied on that property: The nuclei now emitted energy when probed with radio-frequency waves. At ordinary room temperature, they had absorbed such waves. Indeed, all negative temperatures are hotter than all positive temperatures.

As their system cooled down slowly, its temperature went from -350 K to $-1,000$ K to $-\infty$ K, which is physically equivalent to $+\infty$ K, and continued through $+1,000$ K to $+300$ K, ordinary room temperature. The system did not pass through absolute zero. So far as one knows, one can not achieve absolute zero by approaching it from either side. Figure 1 provides a picture.

The coldest temperatures are just above 0 K on the positive side. The hottest temperatures are just below 0 K on the negative side.

Theory

For a theoretical understanding of negative absolute temperature, one can turn to the connection among entropy S, energy E, and absolute temperature T. In one of its many equivalent formulations, the second law of thermodynamics states that

$$\Delta S \geq \frac{\left(\begin{array}{c}\text{energy input} \\ \text{by heating}\end{array}\right)}{T}, \quad (1)$$

where the equal sign applies provided the process occurs slowly (or, more technically, occurs reversibly), which we specify henceforth. If we think of the entropy as a function of the energy E, then Eq. (1) implies

$$\frac{\Delta S}{\Delta E} = \frac{1}{T}, \quad (2)$$

provided external parameters, such as an external magnetic field, are kept constant. Thus, on a graph of S versus E, the slope gives $1/T$. Eq. (2) provides an extremely general definition of absolute temperature.

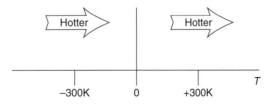

Figure 1 A depiction of hotness as a function of the absolute temperature. The vertical line at $T = 0$ represents both a barrier for the arrows and an unattainable value. One cannot cool a system to absolute zero from above nor heat it to zero from below.

For simplicity, we specialize now to a system consisting of many magnetic moments, each moment associated with a spin of $\frac{1}{2}\hbar$. The moments, we specify further, interact with an external magnetic field but not among themselves. As a function of the energy E, the entropy may be calculated as

$$S(E) = k_B \ln \left(\begin{array}{c}\text{number of distinct spin} \\ \text{arrangements for given}\end{array}\right). \quad (3)$$

Figure 2a displays this function.

Maximum energy occurs when all moments are antiparallel to the magnetic field and thus can be achieved by only one spin arrangement. Minimum

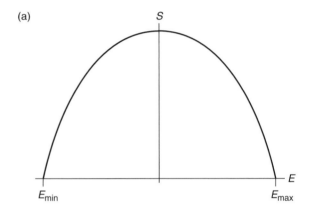

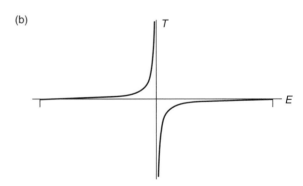

Figure 2 (a) A graph of entropy S versus energy E for the simplified magnetic system. At any energy, the slope gives $1/T$. As E approaches the extremes, the slopes become infinite, and T approaches zero, but the difference in the sign of the slopes implies approaches to absolute zero from opposite sides. (b) The absolute temperature as deduced from the slope of the preceding graph: $T = 1/\text{slope}$.

energy requires all spins to be parallel to the field and again can occur in only one way. Any allowed energy near zero energy can be achieved in many ways by having various spins parallel to the field and others antiparallel. Hence $S(E)$ has a central maximum.

For any energy in the range $0 < E \leq E_{max}$, the slope of S versus E is negative and hence, by Eq. (2), $1/T$ is negative; thus, the absolute temperature is negative.

Thermodynamically, the larger that the ratio $\Delta S/\Delta E$ is, the colder the system (because the system so readily absorbs energy, for thereby the system enjoys a large increase in entropy). In Fig. 2a, the end near E_{min} has the maximum slope and will be the coldest (as T tends to zero from above). The maximum negative slope occurs as E tends to E_{max}. That approach gives the hottest system and corresponds to T approaching zero from below.

See also: HEAT; TEMPERATURE; TEMPERATURE SCALE, KELVIN; THERMODYNAMICS

Bibliography

PROCTOR, W. G. "Negative Absolute Temperatures." *Sci. Am.* **239** (2) 90–99 (1978).

PURCELL, E. M., and POUND, R. V. "A Nuclear Spin System at Negative Temperature." *Phys. Rev.* **81**, 279–280 (1951).

RAMSEY, N. F. "Thermodynamics and Statistical Mechanics at Negative Absolute Temperatures." *Phys. Rev.* **103**, 20–28 (1956).

ZEMANSKY, M. W. *Temperatures Very Low and Very High* (Van Nostrand, Princeton, NJ, 1964).

RALPH BAIERLEIN

NEUTRINO

Neutrinos are elementary particles that, to the best of our present knowledge, are structureless. That class is further subdivided into particles that obey a fundamental constraint limiting their total number and particles that are freely created and destroyed in the absence of that constraint. The former are said to be described mathematically by Fermi–Dirac statistics and are referred to as fermions, while the latter are described by Bose–Einstein statistics and are referred to as bosons. These historically significant names delineate the two basic classes inclusive of all elementary particles. Within the class of fermions are two further sub-classes: particles, known as quarks, capable of interacting with great strength (strongly) with one another and those, known as leptons, that interact relatively feebly (weakly) with quarks and one another. All are, in principle, capable of interacting with electrically charged matter and electric and magnetic fields, as well as gravity.

Neutrinos are weakly interacting fermions, which, as their name implies, are electrically neutral and have very small, perhaps zero, mass. They are to be distinguished from the massive, also electrically neutral, neutrons, which with protons are the constituents of nuclei. Neutrinos are produced in the spontaneous disintegrations of radioactive nuclei and other elementary particles and may disappear through the inverse processes, that is, through being captured by a nucleus or an elementary particle. To satisfy the empirically observed conservation of leptons in all physical processes, neutrinos are produced or disappear in conjunction with another (anti) lepton, for example, an antineutrino. In counting leptons, for example, the neutrino and antineutrino in effect cancel each other and leave the lepton number count unchanged.

Within the family of neutrinos, we now know of only three types, each identified by the name of the electrically charged lepton with which it is associated in a lepton sub-family. Thus, there is an electron-type neutrino (ν_e); a muon-type neutrino (ν_μ); and a tauon-type neutrino (ν_τ). The ν_e, electron (e^-), and their antiparticles ($\bar{\nu}_e$ and e^+) comprise one sub-family; the others are ($\nu_\mu, \mu^-, \bar{\nu}_\mu, \mu^+$) and $\nu_\tau, \tau^-, \bar{\nu}_\tau, \tau^+$). Within each sub-family the so-called separate lepton numbers (i.e., e-type, μ-type, and τ-type numbers) appear also to be conserved. For example, in the spontaneous disintegration of the muon, $\mu^+ \rightarrow e^+ + \nu_e + \bar{\nu}_\mu$, the μ-type number is conserved as well as the e-type number, since μ^+ and $\bar{\nu}_\mu$ are both defined to be antiparticles and one disappears when the other appears; e^+ and ν_e are antiparticle and particle, respectively.

The apparent failure to conserve energy and momentum in the beta decays of a number of radioactive elements induced Wolfgang Pauli to propose the existence of the neutrino in 1930 as the means of retaining the conservation of those quantities. The first direct detection of neutrinos (actually $\bar{\nu}_e$), carried out in 1953 by Frederick Reines and Clyde

Cowan, was followed by the experimental recognition of the ν_μ and ν_τ in 1962 and 1973, respectively.

The unique property of the neutrino is that it is the only known elementary particle with only weak and gravitational interactions with matter, the latter observable only on cosmic scales. This overall weakness in its interactions accounts for the fact that all fundamental properties of the neutrino, other than those specified above, are known only by their limiting values (e.g., mass, charge, magnetic dipole moment, etc.). Furthermore, we are not aware of any reason that lepton number should be rigorously conserved, or that neutrino masses should be identically zero. Accordingly, intensive searches for a violation of lepton number conservation and for precise values of neutrino masses are an important part of present particle physics and astrophysics experimentation.

As a result of their ability to penetrate great thicknesses of dense matter, neutrinos probe the interior of stars at crucial periods in their lifetimes. They constitute the principal direct evidence for nuclear fusion in the Sun, and they are a potential thermometer to provide a precision measurement of the temperature of the solar core. They are the coolant of neutron stars resulting from the collapsed core of a massive star on its way to becoming a supernova; they have been directly observed as products of the supernova SN 1987A. With the detection of solar and supernova neutrinos since the 1970s, neutrino astrophysics has become a recognized discipline.

In the cosmological big bang model, neutrinos decoupled relatively early in the evolution of the universe. They should, therefore, have a substantial relic abundance today, if they are stable, with a number density approximately the same (≈ 300 cm^{-3}) as the photons that constitute the background electromagnetic radiation. Their intrinsic properties (e.g., number of types and total mass of all types) have upper limits placed on them by our knowledge of the evolution of the universe. In particular, they are limited by the observed abundance of the light elements when compared with the semi-empirical theory of nucleosynthesis. In turn, improved knowledge of neutrino properties would constrain cosmological models of the early universe. For example, if neutrinos are massive, with masses in the electron-volt region, they may contribute to the experimentally indicated, but not yet identified, dark matter and perhaps to the total mass of the universe.

See also: ELEMENTARY PARTICLES; FERMIONS and BOSONS; INTERACTION, WEAK; LEPTON; NEUTRINO, HIS-TORY OF; NEUTRINO, SOLAR; NUCLEOSYNTHESIS; PAULI, WOLFGANG

Bibliography

ALLEN, J. S. *The Neutrino* (Princeton University Press, Princeton, NJ, 1958).

BAHCALL, J. N. *Neutrino Astrophysics* (Cambridge University Press, Cambridge, Eng., 1989).

ALFRED K. MANN

NEUTRINO, HISTORY OF

The idea of the neutrino was conceived in 1930 by the Austrian-Swiss theoretical physicist Wolfgang Pauli as a possible solution to two vexing problems confronting a widely accepted model of the structure of the atomic nucleus, which made use of the two elementary constituents of matter then known: the electron and the proton. The neutral atom of mass number A and atomic number Z was supposed to contain A protons and an equal number of electrons, A-Z of these in the nucleus, with the remaining Z making up the electron shells of the atom. One said that matter was made of electricity, positive protons and negative electrons; this was an economical picture. Supporting it was the observation that protons were knocked out of light nuclei in experiments using alpha particles from radioactive decay of heavier nuclei, while in the beta-decay form of radioactivity, electrons emerged from the nucleus.

However, there were puzzles concerning the nuclear electrons. In beta decay, the nuclear charge Z increased by one unit and changed its internal energy by a definite amount ΔE; but the electron generally came out with varying (lesser) amounts of energy, so that a part of ΔE appeared to be lost. Another puzzle was due to the fact that electrons and protons have a property known as spin angular momentum (by analogy with a spinning top) of quantity $\frac{1}{2}$ (in units $h/2\pi$, where h is Planck's quantum of action). In those nuclei where the total number of nuclear particles (protons plus electrons) is odd, such as the common element nitrogen ($_7$N^{14}), the total nuclear angular momentum should be half-integral, but in the case of $_7$N^{14} it was shown to be 1. Nevertheless, most physicists were willing to accept the electron-proton model, even though it contradicted the well-known laws of conservation of energy and angular momentum, believing that different physical laws might hold within the tiny space of the nucleus. Indeed, they had recently learned to accept the fact that the puzzling laws of quantum mechanics were valid within the atom. The influential physicist Niels Bohr stated his belief that the law of conservation of energy was valid only in a statistical sense, like the law of increase of entropy in statistical mechanics.

Pauli, however, was unwilling to give up the conservation laws and conjectured the existence of a new particle, which he called (at first) the neutron, to solve the two difficulties mentioned. This was a neutral particle of spin $\frac{1}{2}$ with a mass "not larger than 0.01 proton mass," as Pauli suggested in a famous letter, sent on December 4, 1930, to nuclear physicists at a meeting in Tübingen, Germany. For each electron in the nucleus, there was to be a neutron as well, thus solving the problem of $_7N^{14}$ and analogous cases. In beta decay, a neutron would emerge with each electron, carrying the energy that appeared to be "lost." Pauli's neutron would have been almost undetectable.

Aside from private letters and lectures, Pauli let the matter rest there, permitting publication of his idea first in the proceedings of an international conference held in Brussels in October 1933. There he referred to his particle as a neutrino and said it might be massless. The change of name was suggested by the Italian Enrico Fermi, since the nuclear particle that we now call the neutron was discovered earlier in 1932 by the Englishman James Chadwick. A Russian, Dmitri Iwanenko, had suggested that Chadwick's neutron was a kind of neutral proton, a massive elementary particle of spin $\frac{1}{2}$, and that the nucleus contained only neutrons and protons, no electrons. In beta decay, an electron and a neutrino were created, as a photon is created in an ordinary atomic transition.

Shortly after the Brussels conference, Fermi made a successful theory of beta decay using the neutrino. With some important later modifications, Fermi's theory forms a part of the modern electroweak theory that unifies electromagnetism with the weak nuclear interaction, of which beta decay is an example. As a result of Fermi's success and Chadwick's discovery of the neutron, nuclear models without electrons (and their puzzles) became acceptable; eventually, even Bohr agreed that the conservation laws were valid in each individual event.

The neutrino was theoretically indispensable, but it was still necessary to try to detect it directly. This formidable task took two decades to accomplish, since the neutrino can pass through light-years of matter without interacting. It was first observed in 1956 by a group led by Clyde L. Cowan and Frederick Reines of Los Alamos National Laboratory, who used the enormous flux of neutrinos from a nuclear reactor at the Savannah River Plant in South Carolina and a very large detector filled with a liquid scintillator. In 1962, at Brookhaven National Laboratory in New York, Leon M. Lederman, Melvin Schwartz, and Jack Steinberger used a large detector of spark chambers and scintillator to show the existence of a second type of neutrino. This second type accompanies an elementary particle called the muon in the decay of particle called the pion, a process first observed in cosmic rays in 1947. The pion belongs to the class of particles with strong nuclear interaction called hadrons. The muon is a lepton, the general name for particles whose nuclear interaction is weak (like the neutrino itself); it decays in about a microsecond into an electron and two neutrinos, one of each kind, called electron neutrino and muon neutrino. A third kind, the tau neutrino, is assumed to exist, corresponding to a third charged lepton, the tau, discovered in 1975 by a group at the Stanford Linear Accelerator Center led by Martin Perl.

Beta decay plays an essential role in the nuclear reaction cycles that produce energy in stars, and neutrinos, with their high degree of penetration, can carry information to scientists on Earth about processes occurring deep in stellar interiors. Neutrinos from the Sun have been monitored by Raymond Davis Jr. of Brookhaven National Laboratory and his collaborators for a quarter century using a large tank of cleaning fluid (C_2Cl_4) located 1,500 m underground to reduce cosmic ray background. Neutrinos absorbed by chorine nuclei convert the nuclei to argon at a measurable rate. The number of solar neutrinos detected this way has been smaller than theoretically expected; the reason for this is an open question. Other large underground installations are used for detecting very-high-energy neutrinos from outer space (neutrino astronomy). In 1987 two of these detectors observed neutrinos produced by a supernova in the southern skies.

Since it was shown in 1962 that high-energy neutrinos from large particle accelerators can be detected, neutrino beams have been used as effective probes of the proton and neutron, supplementing the use of high-energy electron beams. Both types of beams produce simpler, more easily interpretable interactions than beams of hadrons. Neutrino experiments carried out at CERN in Geneva, Switzerland, and at Fermilab in Batavia, Illinois, beginning in 1972, formed the basis for the electroweak theory. Other neutrino experiments have helped to establish the color quark model, the other sector of the standard model of elementary particle interactions that has dominated the theory.

See also: COLOR CHARGE; DECAY, BETA; HADRON; LEPTON; LEPTON, TAU; MUON; NEUTRINO; NEUTRINO, SOLAR; PAULI, WOLFGANG

Bibliography

BROWN, L. M. "The Idea of the Neutrino." *Phys. Today* **31** (9), 23–28 (1978).

LEDERMAN, L. M. "Resource Letter Neu-1 History of the Neutrino." *Am J. Phys.* **38**, 129–136 (1970).

PAIS, A. *Inward Bound* (Oxford University Press, New York, 1986).

REINES, F., and COWAN, C. L. "Neutron Physics." *Physics Today* **10** (8), 12–18 (1957).

WU, C. S. "The Neutrino" in *Theoretical Physics in the Twentieth Century,* edited by M. Fierz and V. F. Weisskopf, (Wiley Interscience, New York, 1960).

LAURIE M. BROWN

NEUTRINO, SOLAR

The energy radiated by the Sun and the mechanical stability of the Sun are both products of nuclear fusion in the Sun's core. As far as we know, this is true of all stable, luminous stars. Nuclear fusion is the process, involving several nuclear reactions, in which four hydrogen nuclei are fused and thereby transmuted into a helium nucleus (and two positively charged electrons) with the simultaneous release of a large quantity of energy. It is this energy that gives rise to the light emitted from the surface of the Sun, provides the expansive force that balances the contractive force of gravity, and results in the mechanical stability of the Sun.

A part of the energy generated in the fusion reactions in the Sun's core (approximately $1/1,000$ of its volume) is in the form of neutrinos. These are elementary particles that, as their name implies, are electrically neutral and of very small, perhaps zero, mass; they are to be distinguished from the massive, also electrically neutral, neutrons, which with protons are the constituents of nuclei. Neutrinos are not a part of the spectrum of electromagnetic radiation; they are not created or absorbed in electric or magnetic phenomena. Neutrinos are produced in the spontaneous disintegrations of radioactive nuclei and elementary particles; they may disappear through the inverse processes, that is, through being captured by a nucleus or an elementary particle. Neutrinos belong to a general class of elementary particles that interact with all other elementary particles and all other matter through what is known as the Fermi force, named for the Italian physicists, Enrico Fermi, who early described its properties. The force so intrinsically weak that it is also referred to as the weak force.

Because of their electrical neutrality and the weakness of their interactions with matter, neutrinos generated in the core of the Sun emerge directly from that source without significant loss of energy or intensity. As a consequence, they provide a direct probe of the central region of the Sun and serve to test the detailed description of the solar interior embodied in a mathematical model of the Sun known as the standard solar model.

Another reason for interest in solar neutrinos is the desire to learn more about the intrinsic properties of the neutrinos themselves. We are able to study their basic wave and particle properties through indirect observation of their behavior in traversing the dense matter of the Sun and the long distance to Earth.

The intensities and energies of the neutrinos generated in the $p - p$ chains and the carbon-nitrogen-oxygen (CNO) cycle in the Sun are shown in Fig. 1. It can be seen that the most abundant, but lowest energy, source is the primary fusion reaction $p + p \rightarrow d + e^+ + \nu_e$, in which p stands for proton, d stands for deuteron (the hydrogen isotope of mass number 2), e^+ stands for positively charged electron (positron), and ν_e stands for electron-type neutrino. As Fig. 1 shows, approximately 6×10^{10} ν_e from that reaction strike every square centimeter of Earth's surface every second. This statement is correct even when the Sun is on the opposite side of the earth, since neutrinos penetrate the earth with negligible attenuation.

Parts of the solar neutrino spectrum in Fig. 1 have been observed by four different neutrino telescopes

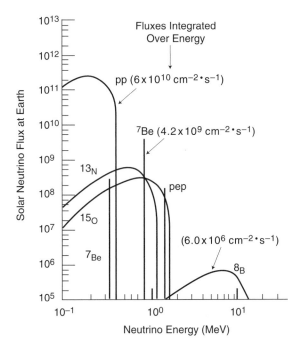

Figure 1 Solar neutrino fluxes predicted by the standard solar model. The energy thresholds for the gallium, chlorine, and Kamioka neutrino dectectors are 0.233, 0.814, and 7.5 MeV, respectively.

in Earth. Such telescopes, unlike the usual astronomical telescopes, are very massive and located deep underground for shielding against cosmic rays from Earth's atmosphere. The pioneer solar neutrino telescope, first operated in 1970, consists of 100,000 gallons of perchlorethylene (a chlorine based cleaning fluid) in the Homestake Gold Mine in Lead, South Dakota. It detects neutrinos principally from almost the highest energy, lowest intensity, solar source, the beta decay of an isotope of boron, ^{8}B in Fig. 1. A second solar neutrino telescope, located in the Kamioka tin-zinc mine in Gifu prefecture in Japan, observed neutrinos from the same ^{8}B solar source beginning in 1987. This telescope functions in real time and with directional capability, so the detected neutrinos are tracked back to the Sun as their origin despite the continuous motion of the Kamioka telescope with respect to the Sun. Most recently, two telescopes, one in the Gran Sasso auto tunnel under the Apennine Mountains in Italy, and one in Baksan in the Caucasus Mountains in Russia, have observed the ν_e from the primary $p + p$ reaction by means of multiton detectors employing the element gallium. All of these neutrino telescopes, except the one in Japan,

are based on a radiochemical method in which the solar neutrinos from ^{8}B transmute chlorine nuclei into radioactive argon nuclei (as in the Homestake telescope); or solar neutrinos from the $p + p$ reaction transmute gallium into radioactive germanium (as in Gran Sasso and Baksan).

The subject of solar neutrinos is one of continuing intense observation and experimentation because solar neutrinos still have much to teach us about the interior of the Sun and about the intrinsic nature of the neutrinos themselves.

See also: FUSION; NEUTRINO; NEUTRINO, HISTORY OF

Bibliography

BAHCALL, J. N. *Neutrino Astrophysics* (Cambridge University Press, Cambridge, Eng., 1989).

DAVIS, R., JR.; MANN, A. K.; and WOLFENSTEIN, L. *Annual Review of Nuclear and Particle Science*, Vol. 39: *Solar Neutrinos* (Annual Reviews, Inc., Palo Alto, CA, 1989).

ALFRED K. MANN

NEUTRON

The neutron, a neutral subatomic particle discovered by James Chadwick in 1932, is one of the two major components of the atomic nucleus. The neutron has a mass that is comparable to, though slightly larger than, the positively charged proton, the other major nuclear component. The neutron and proton are collectively known as nucleons. The neutron is sometimes viewed as being an elementary particle, but while this view has some justification, it is more appropriate to consider the neutron as a composite structure made up of quarks. In the quark model, the neutron is made of one up quark, with a $+\frac{2}{3}$ electron charge, and two down quarks, each having a $-\frac{1}{3}$ electron charge.

When liberated from an atomic nucleus, the neutron is unstable and radioactively decays with a half-life of about 11 min. Neutrons can be liberated from nuclei through nuclear reactions, most notably fission and fusion. The production and absorption of neutrons in the fission process are essential features of self-sustaining nuclear chain reactions. Nuclear reactors and nuclear weapons are both examples of such nuclear chain reactions and are copious sources of free neutrons. As a health physics hazard, the neutron is considered to be a penetrating radiation whose biological effects are unique and are very sensitive to the neutron energy. Neutron beams can serve as extremely useful probes for the study of solids and liquids and many modern high-intensity neutron sources have been built specifically for this purpose.

Properties of the Neutron

The neutron is thought to have zero electric charge, and very stringent experimental limits have been set on a possible neutron charge. Although the neutron has no *net* electric charge, it has, on average, an unequal distribution of positive and negative charge over its finite size (about 10^{-15} m). Its core is relatively more positive, and its periphery is more negative. The neutron has spin angular momentum with a spin of $S = \frac{1}{2}$, and it has a finite magnetic moment opposite in sign to that of the proton. The mass of the neutron, which is very accurately known, is greater than the sum of the proton and electron masses. This implies that it is energetically possible for the neutron to decay, through the process called beta decay, into a proton, electron, and antineutrino. The origin of the neutron-proton mass difference is not well understood. When the existence of the neutron was first hypothesized, it was proposed as a very tightly bound state of a proton and an electron. Such a view is inconsistent with the observed neutron, proton, and electron masses. Table 1 gives accepted values for several important neutron properties.

The Neutron and Nuclei

The atomic nucleus is a system of protons and neutrons bound together by the so-called strong nuclear force. This force is much stronger than the weak nuclear force, which is responsible for beta decay. The strong force totally dominates the weak force in determining nuclear structure. To some extent, nuclear structure resembles the electronic structure of atoms in having well-defined energy levels and a shell structure for nucleons that influences both nuclear stability and the spectra of nuclear excitations. A nucleus is classified by its proton number Z, its neutron number N, and its mass number $A = Z + N$.

Nuclei having the same proton number but different neutron number are said to be different isotopes of the same element. A common method for the production of new isotopes is neutron capture. Because the neutron has no charge, it is not repelled by the electric field of the nucleus. Thus, relatively-low-energy neutrons can enter nuclei and, if captured, form a new isotope ($A \rightarrow A + 1$). Neutron capture is an important method of production of special isotopes required for medical diagnostics, radiation treatment, and a variety of technological applications. Isotope production by neutron capture is typically carried out in the core of a nuclear reactor having a high neutron density. In many cases, the isotope produced by neutron capture is radioactive. As a result, it will eventually undergo beta decay, changing its proton number Z and forming a new chemical element. This is the principle method by which significant quantities of transuranic elements are produced. The production of radioactive isotopes by neutron capture is also employed in the chemical analysis method known as neutron activation analysis. Neutron activation analysis provides one of the most sensitive methods for the detection of trace elements and is of importance in a wide variety of fields, including chemistry, biology, environmental studies, archeology, art history, and forensic studies.

The neutron, though unstable as an isolated particle, may be stable (or exist for times that are much longer than the free neutron lifetime) when in a nucleus. As noted above, the instability of the neutron results from its being more massive than the proton and the electron together. The beta decay of a neutron within a nucleus will manifest itself as the overall beta decay of that nucleus. For a nucleus to be beta unstable, it is necessary for its mass to exceed the mass of the daughter nucleus *plus* an electron mass. If this condition is not met, it is energetically impossible for the original nucleus to undergo beta decay; thus, it will be stable. A significant contribu-

Table 1 Neutron Properties

Mass	1.674928×10^{-27} kg
	1.0086649 a.u.
	939.5656 MeV
Magnetic Moment	0.966237×10^{-26} J·T^{-1}
Half-Life	652 s
Charge	experimental limit $< 10^{-21}$ e
Electric Dipole Moment	experimental limit $< 10^{-25}$ e cm

tion to the masses of nuclei arises, via the Einstein mass-energy relation ($E = mc^2$), from the binding energy that maintains the nucleons in the nucleus. The most significant contribution to nuclear binding energies results from the strong nuclear interactions between constituent nucleons. Thus, the stability, or instability, of neutrons within a nucleus depends on the nuclear environment in which the neutron resides.

In a reaction analogous to beta decay, a nucleus can capture an atomic orbital electron and emit a neutrino to become a different nuclei species. This process, known as electron capture, will only proceed if the total masses of the initial and final particles are consistent with the conservation of energy.

Additional Discussion

The neutron and its properties are of considerable importance to a variety of scientific problems in addition to those mentioned above. In cosmology, the lifetime of the neutron sets the time scale for the production of light elements during the first few minutes of the big bang expansion. In stellar astrophysics, the major energy producing reactions in the Sun and other stars is closely related to neutron decay. Neutron stars, thought to be the remnants of supernova explosions, are gravitationally bound collections of neutrons having stellar masses and nuclear densities. The neutrons in neutron stars are prevented from decaying by intense gravitational forces.

As a scientific and technological research tool, neutrons are unique as probes for the study of condensed matter. In neutron radiography, an object is placed in a collimated beam of neutrons and an image of the intensity pattern of the transmitted beam is made on photographic film (or a position sensitive detector). Neutron radiography is complementary to more familiar x-ray radiography in its differing response to different elements within materials. Neutrons suffer relatively low loss on transmission through common metals and allow the imaging of light elements, especially hydrogenous materials within a matrix of heavier elements. In neutron scattering, a sample is placed in a low-energy neutron beam and the spatial pattern and energy change of the neutrons scattered out of the beam is measured. Neutron scattering provides information about the microscopic structure and dynamics of the sample at the molecular level.

The neutron, like any quantum particle, has both wave-like and particle-like characteristics. In a remarkable series of experiments, the wave-like characteristics of the neutron have be exploited in creating neutron interferometers that are the matter-wave equivalent of optical interferometers. Such devices have been used to demonstrate explicitly some of the novel quantum mechanical properties of matter.

See also: CHADWICK, JAMES; DECAY, BETA; ELEMENTARY PARTICLES; ISOTOPES; MASS-ENERGY; NEUTRON, DISCOVERY OF; NEUTRON STAR; NUCLEAR SHELL MODEL; QUARKS, DISCOVERY OF

Bibliography

BROWN, L. M., and RECHENBERG, H. "Nuclear Structure and Beta Decay (1932–1933)." *Am. J. Phys.* **56,** 982–988 (1988).

CHADWICK, J. "The Existence of a Neutron." *Proc. R. Soc. London, Ser. A* **136,** 692–708. (1932).

GEOFFREY GREENE

NEUTRON, DISCOVERY OF

The discovery of the neutron by James Chadwick at the Cavendish Laboratory, Cambridge, England, occupies a central place in the history of nuclear physics. The 1932 discovery is often portrayed as a straightforward and exemplary story of unambiguous theoretical prediction followed by experimental investigation, leading ultimately to a decisive, confirmatory experiment. According to this canonical account, the existence of the neutron was predicted by Ernest Rutherford in 1920. A number of unsuccessful attempts were made to detect the new particle, until in 1932 Chadwick was finally able to provide conclusive proof of its existence. For this work, he was awarded the Nobel Prize in physics in 1935.

Historians have come to see such a portrayal as simplistic and misleading, however, and now view the discovery of the neutron as part of a much more complex historical process. The discovery is linked to a wider series of changes in laboratory instrumentation and theoretical practice in the emergent discipline of nuclear physics in the 1920s and early

1930s. To understand the discovery of the neutron, therefore, it is important to understand the experimental and theoretical framework within which nuclear physics developed in this period.

Ideas about nuclear structure first began to take shape soon after the postulation of the nuclear atom by Rutherford in 1911. The Rutherford–Bohr atomic model (1913) assumed that the positively charged nucleus contained the bulk of an atom's mass, orbited at a relatively large distance by the atom's electrons. By 1914 it was generally assumed that the nucleus consisted of positive electrons (hydrogen nuclei), negative electrons, and doubly charged alpha particles, the latter understood as being composed of positive and negative electrons in a particularly stable combination. Applied to the heavy elements, these assumptions allowed satisfactory explanation of the phenomena of radioactive emission and isotopy. By analogy, they were taken also to hold for nuclei of the light elements.

In 1917 Rutherford obtained evidence indicating that the nuclei of light elements such as nitrogen could be disintegrated under impact by high-energy Radium C [^{214}Bi] alpha particles to release their constituent hydrogen nuclei—protons, as he would christen them in 1920. Later he understood that the reaction followed the scheme:

$$^{4}He + {}^{14}N \rightarrow {}^{17}O + {}^{1}H,$$

with other elements, such as oxygen, and disintegration products of mass 2 and 3—which were less easy to identify—also being observed. These results were taken to vindicate earlier speculations about nuclear structure, and laid the basis for all subsequent work in the field.

In his 1920 Bakerian Lecture to the Royal Society of London, Rutherford elaborated some of his ideas about nuclear structure in the light of these new results. Drawing support from the evidence of isotopy in the light nuclei then beginning to emerge from Francis Aston's mass spectrograph, Rutherford took the disintegration product of mass 3—which he now designated X_3^{++}—to be an isotope of helium. Rutherford supposed it to consist of three protons bound together by one electron. From this hypothesis, he reasoned by analogy to the possibility that one electron could bind two protons (giving an isotope of hydrogen), and perhaps even one proton (giving an electron-proton compound with zero charge, "a kind of neutral doublet").

Rutherford's tentative speculations on nuclear structure shaped the research agenda of the Cavendish Laboratory in the 1920s. During 1921 and 1922, as the Cavendish turned under Rutherford's direction to the systematic investigation of nuclear structure, for example, a series of researchers, among them J. L. Glasson, Keith Roberts, and Chadwick, looked for experimental evidence of the hypothetical neutral particles—called neutrons by analogy with protons—in gas discharges and elsewhere, but without success. By 1923, in any case, it had become clear that Rutherford's 1920 speculations were based on a false premise, for new experimental evidence showed that the X_3^{++} particles were not, as he had originally thought, disintegration products, but long-range alpha particles from the radioactive source. The experimental search for Rutherford's neutrons came to an end, though neutral particles remained part of the everyday grammar of qualitative nuclear theory throughout the 1920s.

During the mid-1920s, the nucleus became contested territory. Researchers at the Cavendish Laboratory became involved in controversies with other laboratories entering the field of nuclear research for the first time. In particular, an extended dispute with workers in the Vienna Radium Institute over the results of disintegration experiments and their theoretical interpretation cast doubt on the principal technique used in the disintegration experiments—the optical counting by human observers of scintillations caused by the impact of disintegration fragments on a zinc sulphide screen. During the late 1920s, several groups of workers in Paris, Berlin, Washington, D.C., and elsewhere began to repeat the disintegration experiments in an attempt to clarify the discrepancies between Cambridge and Vienna. This rapid expansion of the disciplinary field was accompanied by significant changes in experimental technique. From about 1928, partly in response to the Cambridge-Vienna controversy, which had brought to light a number of difficulties associated with the scintillation technique, and partly because of the availability of new kinds of instrumentation as a spin-off from the booming radio industry of the 1920s, electrical counting methods (such as the Geiger counter and valve amplifier) began to replace the obsolescent scintillation method. Likewise, as deployed by Patrick Blackett and others, the cloud chamber began to acquire a new prominence in nuclear research.

These changes in instrumentation and technique elicited new phenomena for laboratory study. In August 1930, in a series of experiments using the new electrical counting apparatus, for example, the German physicist Walther Bothe and his student Herbert Becker, working in Berlin, found that under the impact of polonium α particles, beryllium nuclei apparently emitted not the expected protons, but penetrating gamma rays. This was not unexpected, for work by the Russian theoretician George Gamow and the German experimentalist Heinz Pose in Halle had suggested that disintegration particles could leave the nucleus with a number of discrete energies, often accompanied by gamma rays. In the case of boron, for example, disintegration protons occurred in three discrete groups, understood to correspond to the reactions:

$$^4\text{He} + {}^{11}\text{B} \rightarrow {}^{14}\text{C} + {}^1\text{H},\qquad \text{(Group I)}$$

$$^4\text{He} + {}^{10}\text{B} \rightarrow {}^{13}\text{C*} + {}^1\text{H} \rightarrow {}^{13}\text{C} + \gamma + {}^1\text{H},\quad \text{(Group II)}$$

$$^4\text{He} + {}^{10}\text{B} \rightarrow {}^{13}\text{C} + {}^1\text{H}.\qquad \text{(Group III)}$$

In Group II, alpha-particle bombardment produces an excited nucleus of a carbon isotope (^{13}C*), which reverts to a stable state by emission of nuclear gamma radiation. In the case of beryllium, no disintegration protons were observed, but there was excitation of similar gamma radiation, which could then be explained by the analogous reaction scheme:

$$^4\text{He} + {}^9\text{Be} \rightarrow {}^{13}\text{C} + \gamma.$$

These experiments were quickly taken up by other researchers, among them Frédéric and Irène Joliot-Curie in Paris. Using a much stronger polonium alpha-particle source than that available to the Germans, the Joliot-Curies confirmed the results reported by Bothe and Becker, and added the observation that the penetrating gamma radiation was able to eject protons from paraffin and other hydrogeneous substances. They drew on familiar phenomena to interpret this result. Likening the new radiation to high-energy cosmic rays, they suggested that the new phenomenon be understood as a modified version of the Compton effect. These findings were published in the Académie des Science's *Comptes Rendus* in December 1931.

The Joliot-Curies admitted that their tentative interpretation would imply a violation of the principle of conservation of energy. Nevertheless, they had good reasons for considering such a prospect. Between 1926 and 1931, theoretical physicists had increasingly come to feel that nuclear theory contained a number of fundamental contradictions, especially in regard to the spin statistics of certain light nuclei and the explanation of radioactive beta decay. They also found it difficult to apply the new wave mechanics to the nucleus, not least because of the presumably necessary presence of nuclear electrons. In the face of these difficulties, some theoreticians—among them Niels Bohr—were beginning to question the applicability of the conservation of energy to nuclear processes.

The beryllium experiments of Bothe and Becker and of the Joliot-Curies were taken up by Chadwick at the Cavendish Laboratory early in 1932. Chadwick refused to accept the violation of the principle of conservation of energy implied in the Joliot-Curies' interpretation of their experiments. In a few weeks in January and February 1932, using a small air-filled ionization chamber and valve amplifier as a detector, Chadwick and his assistant Horace Nutt repeated the Joliot-Curies' work. After a series of measurements, Chadwick concluded that the penetrating radiation produced by alpha-particle bombardment of beryllium could be understood as streams of a new, uncharged nuclear particle of approximately protonic mass:

$$^4\text{He} + {}^9\text{Be} \rightarrow {}^{12}\text{C} + {}^1n.$$

Chadwick quickly identified the new particle as the close proton-electron compound predicted by Rutherford in 1920. On February 17, 1932, he sent a short letter entitled "Possible Discovery of a Neutron" to the journal *Nature,* where it appeared ten days later. Chadwick's work was also widely reported in the press, where it brought nuclear physics to the attention of a large public. On May 10 he submitted a full account of his work to the *Proceedings of the Royal Society.*

Chadwick's corpuscular interpretation of the penetrating radiation from beryllium was accepted almost immediately by all those involved in nuclear research. While this was partly because the hypothesis allowed the conservation laws to be saved in the beryllium experiments, there were other, equally pragmatic reasons for its rapid adoption. For experimentalists, the changes in technique that had taken place over the previous few years meant that nuclear physics laboratories were already equipped with the materials and instruments—polonium, electronic counting equipment, and so on—needed to produce and detect neutrons, so that Chadwick's work could be replicated relatively quickly. The new particles soon found use as projectiles in nuclear disintegration experiments, leading to Fermi's exploitation of the Joliot-Curies' discovery of the phenomenon of artificial radioactivity in 1934.

Chadwick's interpretation also provided a useful new resource for theoretical physicists, who could use the new particle to explore the outstanding difficulties in nuclear theory. In the summer of 1932, for instance, Werner Heisenberg and others articulated and elaborated proton-neutron models of the nucleus, demonstrating the new particle's usefulness to nuclear theorists.

Thus, within months of Chadwick's announcement, the new particle had been taken up as a working tool by experimentalists and theoreticians alike, and was transforming material practice and theoretical understanding in the emergent field of nuclear physics. At the same time, it is important to remember that the discovery of the neutron was only one of several innovations in nuclear physics in the period from 1931 to 1934: The disclosure of a heavy isotope of hydrogen, the artificial splitting of the lithium nucleus by John Cockcroft and Ernest Walton, and the discovery of the positron all played a part in reshaping the understanding of the nucleus in the early 1930s.

Despite the neutron's rapid adoption into experimental and theoretical practice, it would also be wrong to suppose that the new particle immediately resolved the difficulties facing nuclear physicists, for a number of contradictions persisted in nuclear theory. Nor was there an immediate consensus among nuclear physicists as to the nature and role of the new particle, for different workers attributed different characteristics to it. So while the neutron quickly became naturalized as a nuclear constituent, its character and properties were hotly contested. Disagreement quickly arose among theoreticians, for example, as to whether the neutron was the complex proton-electron compound described by Rutherford and Chadwick, or whether it was in fact a new elementary particle.

Among experimentalists, too, there was disagreement about the nature of the neutron. Chadwick assumed that the neutron consists of a proton and an electron in close combination, giving a net charge of 0 and a mass which should be slightly less than the mass of the hydrogen atom. Chadwick gave a value of 1.0067 mass units (or, taking into account experimental errors, a value within the range 1.005 to 1.008 mass units) for the mass of the neutron. This was less than the sum of the masses of the proton and the electron (1.0078 mass units), implying a binding energy equivalent to 1–2 MeV, apparently providing convincing evidence for Chadwick's interpretation. In 1933, however, Ernest O. Lawrence suggested the low value of 1.0006 mass units for the mass of the neutron, while the Joliot-Curies proposed the much higher value of 1.011 mass units.

The ensuing debate was not resolved until 1934, when Chadwick and Maurice Goldhaber undertook a new measurement of the neutron's mass by the photodisintegration of the deuteron. Their value of 1.0080 mass units was greater than the sum of the masses of the proton and the electron, providing convincing evidence that Chadwick's neutron was not the particle "predicted" by Rutherford in 1920, but a new elementary particle. This conclusion was reinforced by Fermi's 1934 theory of beta decay, which brought together Chadwick's neutron and Wolfgang Pauli's neutrino to offer a theoretical justification for the exclusion of electrons from the nucleus, allowing many of the earlier contradictions of nuclear theory to be resolved. Physicists continued to argue about the character of the neutron after 1934, of course, but by that time the particle itself, whatever its nature, had become firmly embedded in experimental and theoretical practice.

From this account, it is clear that much of the neutron's actual significance for physics in the 1930s lay in its adoption as a useful working tool by experimentalists and theoreticians alike, and in its effect on material practice and theoretical understanding in the emergent field of nuclear physics. Later, too, the use of the neutron in techniques such as neutron diffraction led to the development of important new modes of physical analysis. From the perspective of the second half of the twentieth century, however, the discovery of the neutron has assumed an additional historical significance colored by the particle's role in the discovery of nuclear fission in 1938, the development of the atomic bomb during World War II, and the inauguration of the nuclear age.

See also: ATOM, RUTHERFORD–BOHR; BOHR, NIELS HENRIK DAVID; CHADWICK, JAMES; NEUTRON; NUCLEAR BOMB, BUILDING OF; NUCLEUS; PAULI'S EXCLUSION PRINCIPLE; PROTON; RUTHERFORD, ERNEST

Bibliography

BROMBERG, J. L. "The Impact of the Neutron: Bohr and Heisenberg." *Historical Studies in the Physical Sciences* **3,** 307–341 (1971).

BROWN, L. M., and RECHENBERG, H. "Nuclear Structure and Beta Decay (1932–1933)." *Am. J. Phys.* **56,** 982–988 (1988).

CHADWICK, J. "Possible Existence of a Neutron." *Nature (London)* **129,** 312 (1932a).

CHADWICK, J. "The Existence of a Neutron." *Proc. R. Soc. London, Ser. A* **136,** 692–708 (1932b).

CHADWICK, J. "Personal Notes on the Discovery of the Neutron" in *Cambridge Physics in the Thirties,* edited by J. Hendry (Adam Hilger, Bristol, Eng., [1962] 1984).

FEATHER, N. "Chadwick's Neutron." *Contemporary Physics* **15,** 565–572 (1974).

FEATHER, N. "The Experimental Discovery of the Neutron" in *Cambridge Physics in the Thirties,* edited by J. Hendry (Adam Hilger, Bristol, Eng., [1962] 1984).

STUEWER, R. H. "The Nuclear Electron Hypothesis" in *Otto Hahn and the Rise of Nuclear Physics,* edited by W. Shea (Reidel, Dordrecht, 1983).

STUEWER, R. H. "Rutherford's Satellite Model of the Nucleus." *Historical Studies in the Physical Sciences* **16,** 321–352 (1986).

STUEWER, R. H. "Mass-Energy and the Neutron in the Early Thirties." *Science in Context* **6,** 195–238 (1993).

JEFF HUGHES

NEUTRON STAR

In 1932 British physicist James Chadwick demonstrated that the nucleus of an atom is composed not only of protons but also of neutrons. Two years later, astronomers Walter Baade and Fritz Zwicky proposed the existence of neutron stars that would have very high density, very small radius, and much higher gravitational binding energy than ordinary stars; they even suggested that neutron stars would form in supernova explosions. After decades of investigation, astronomical observations have confirmed their prescient suggestion. As ordinary massive stars evolve, they build up a core of iron that eventually collapses when its mass reaches the Chandrasekhar limit, about 1.4 times the mass of the Sun. The collapse of the iron core forms a neutron star and triggers the explosion of the rest of the original star. This catastrophic event can be observed as a bright new star and is called a supernova.

The first studies of neutron star models were performed by J. Robert Oppenheimer and George M. Volkoff in 1939. After this early theoretical work, the interest in neutron stars faded, as calculations showed that they would be too faint to observe at astronomical distances with optical telescopes, given their small size and low surface brightness.

The situation changed dramatically in 1967 when Jocelyn Bell Burnell and Antony Hewish discovered regularly pulsing radio sources, which they initially called "little green men" and ultimately named pulsars. Upon further investigation, pulsars were shown to be magnetized, spinning neutron stars that produce a rotating beam of radiation. As a neutron star rotates, the beam of radiation can be seen as a series of regularly spaced pulses similar to that of a lighthouse.

The study of pulsars transformed neutron stars from a theoretical curiosity to a familiar astronomical phenomenon. The discovery of pulsars in supernova remnants confirmed Baade and Zwicky's suggestion for their origin; a prime example is the Crab pulsar, discovered in the center of the Crab nebula. The discovery of binary pulsars by Russell A. Hulse and Joseph H. Taylor in 1974 also led to important advances (for which Hulse and Taylor were awarded the Nobel Prize in physics in 1993). The Hulse–Taylor pulsar is a system of two neutron stars rotating around each other with a period of 0.059 s. This binary system is a perfect celestial clock and has been used to determine the masses of both stars in the system (about 1.4 times the mass of the Sun) and to validate Einstein's general theory of relativity.

The detailed radiation mechanism responsible for pulsars is still a subject of debate. Most models make use of the very strong magnetic fields around a neutron star to accelerated particles to very high energies that ultimately radiate a beam of optical, radio, and x-ray photons. The magnetic field around some neutron stars has been estimated to reach 10^{12} G or higher, which is 10^{12} times stronger than that of the surface of Earth. The strength of the magnetic fields are so extreme that relativistic particles are trapped by the field in what is called the magnetosphere of the star.

Neutron stars are extremely dense objects, composed mostly of self-gravitating neutrons, with relatively smaller amounts of protons and electrons. With a mass of 1.4 times the mass of the Sun, a neutron star has a radius of about 10 km, and a mean density of about 7×10^{14} g/cm^3. A sugar cube of neutron star matter weights more than all human beings on Earth. At these extreme densities, the space occupied by a single neutron in ordinary atomic nuclei has to be shared by approximately two neutrons. The gravitational field on the surface of a neutron star is also quite extreme. To escape from the surface of a neutron star a rocket must reach a speed of about 1.9×10^{10} cm/s, more than half the speed of light.

At such extreme densities and gravitational fields, astrophysicists need to take both nuclear interactions and general relativity into account to understand the detail structure of neutron stars. The theory of nuclear interactions at these densities is incomplete and hard to test in terrestrial laboratories. Therefore, the study of the structure of neutron stars via astronomical observations may ultimately lead to a better understanding of nuclear interactions.

Like ordinary stars, neutron stars can be described by an onion-like structure of different density regimes. The outer layers have densities below those of the atomic nucleus and are more precisely studied. As the internal layers get denser, nuclear theory becomes less complete. The detail description of the core region of neutron stars where densities reach several times nuclear density is at present an active area of research.

The outermost layer of a neutron star is called the outer crust and spans a range in density from the surface to 4×10^{11} g/cm^3. This density range is similar to that of a white dwarf compressed from 10^4 km to about 1 km. The outer crust is a solid lattice of heavy nuclei coexisting with a relativistic degenerate electron gas.

Inside the outer crust, there is the inner crust that spans the density range between 4×10^{11} g/cm^3 and 2×10^{14} g/cm^3. The inner crust consists of a lattice of neutron rich nuclei together with a superfluid of neutrons and an electron gas.

At densities higher than 2×10^{14} g/cm^3, nuclei overlap and dissociate into mostly neutrons and some protons. This inner region is called the neutron liquid interior and consists mostly of superfluid neutrons with some superfluid protons and normal electrons. The superfluid and superconducting properties of this region suggest a complex dynamical coupling between the inner layers and the crust, which are presently being studied through models of the observed pulsar glitches.

Finally, the core of a neutron star can reach densities that are too high for present nuclear theory to address adequately. Among the proposed states of matter at these densities, there is the possible disintegration of neutrons into their quark components, such that matter at the innermost part of a neutron star would be better described as quark matter. Other proposals include the condensation of pions and kaons, particles composed of a quark and an antiquark, that may form a condensate at these high densities. A better handle on the composition of the core awaits progress in observations of neutron stars as well as in experiments at terrestrial laboratories.

See also: PULSAR; PULSAR, BINARY; STARS and STELLAR STRUCTURE; STELLAR EVOLUTION; SUPERNOVA

Bibliography

SHAPIRO, S. L., and TEUKOLSKY, S. A. *Black Holes, White Dwarfs, and Neutron Stars* (Wiley, New York, 1983).
SHU, F. *The Physical Universe* (University Science Books, Mill Valley, CA, 1982).

ANGELA V. OLINTO

NEWTON, ISAAC

b. Woolsthorpe, England, December 25, 1642; *d.* London, England, March 20, 1727; *mechanics, orbital dynamics, optics, cosmology.*

Newton was the offspring of a yeoman family of southwestern Lincolnshire, none of whom before him could read; their steadily increasing prosperity indicates that his forebears were not short of ability, however. Newton's father, also named Isaac, died in October 1642, two months before his only child was born.

When his mother remarried, Newton was left in the care of his maternal grandparents at Woolsthorpe. His mother's family, the Ayscoughs, had a tradition of education. Consequently, following primary instruction in local day schools, Newton began attending grammar school in Grantham in 1655, after his mother, now with three more children, Newton's half-brother and half-sisters, had been widowed a second time. In June 1661, Newton enrolled in Trinity College, Cambridge, the college of his mother's brother.

From the notebooks that he kept, we know a good deal about Newton's undergraduate education and how, around 1664, he discovered the new natural philosophy of René Descartes, Pierre Gassendi, and others. The influence of a friend and patron from Lincolnshire, highly placed in Trinity, led the college to elect him to a scholarship in April 1664 and a fellowship in 1668 upon the completion of his M.A. degree. The influence of Isaac Barrow added the Lucasian Professorship of Mathematics in 1669. Trinity remained Newton's home for the following twenty-seven years; all of his achievements in science stem from the years in Cambridge.

The period 1665–1666 has been called Newton's *annus mirabilis,* his wonderful year. It was, first of all, the period of his greatest achievement in mathematics, leading up to an essay composed in October 1666 that set down the first full exposition of what he called the fluxional method, what is now called calculus. Calculus became and remains the basic tool of modern physics.

During the *annus mirabilis,* as he pursued the new natural philosophy, Newton also began the work that would transform physics. He took up the science of mechanics and explored the law of impact and the dynamics of circular motion. Substituting the relations of Johannes Kepler's third law into his formula for the radial force in circular motion, Newton concluded that the radial force in the solar system varies inversely as the square of the planets' distances from the Sun. He also compared the force acting on the Moon with the force of gravity at Earth's surface and found a very rough approximation to the inverse square relation. No one any longer accepts the myth that with this calculation Newton effectively discovered the law of universal gravitation in 1666. The radial force in orbital motion, as he then conceived of it, was centrifugal rather than centripetal. In comparison with what would follow twenty years later, Newton's early work in mechanics was relatively crude. Nevertheless, it marked an important step in his development.

His early work in optics was more conclusive. Around 1665 Newton became interested in the phenomena of colors. He began to entertain the then novel ideas that light, as it comes from the Sun, is not simple and pure but heterogeneous and that phenomena of colors arise from the separation of the mixture into its components. He had seen the colored fringes around bodies observed through a prism. If rays differ both in their degree of refrangibility (refraction) and in the sensation of color they provoke, his idea would explain this appearance.

He first tested the notion by a simple experiment with a prism and then proceeded to a more elaborate one. Making a small hole in the shutter to his window, he admitted a narrow beam of light into the otherwise darkened room and refracted the beam through a prism onto a wall about twenty-two feet away. He saw, as he expected to see, an elongated spectrum about five times as long as it was broad. Newton went on to devise a variety of further experiments with prisms, of which two were especially important. When he intercepted the diverging spectrum with a board and isolated the rays of one color, which passed through a hole in the board to be refracted through a second prism, he observed no further dispersion, but different parts of the spectrum did refract different amounts in exact proportion with the refractions at the first prism. In the second experiment he played the diverging spectrum onto a lens, which brought the rays back to a focus. When a screen was placed at the focus, the spot appeared white, the sensation caused by the mixture of rays that comes from the Sun. When the screen was moved beyond the focus, the spectrum reappeared in reverse order.

By 1669 Newton had worked out the full implications of his theory for presentation in his first series of Lucasian lectures. To avoid the problems of chromatic aberration present in refracting telescopes, he constructed the first reflecting one about this time. He published a brief statement of the theory in the *Philosophical Transactions* of the Royal Society in 1672, and some thirty years later he rewrote the material of the lectures as Book I of his *Opticks* (1703).

He had also to explain the colors of solid bodies if his theory were to be complete. Newton was convinced that differential reflections can also separate the heterogeneous mixture of sunlight into its components. In the phenomena of thin transparent films he found a means to investigate this notion. Pressing a lens of known curvature against a flat sheet of glass and playing a beam of sunlight onto the apparatus, he observed a pattern of colored rings, which continue to be called Newton's rings, reflected from the film of air between the lens and the sheet. From the measured diameters of the rings, the geometry of circles allowed him to calculate the thickness of the film that corresponded to individual rings and individual colors, the first successful venture of mathematical physics into quantities this small. In 1703 this investigation became Book II of the *Opticks*.

Newton's work in optics was not as determinative for the future as his work in mechanics and cosmology would be. His corpuscular conception of light had to be replaced by the wave theory in the early nineteenth century before optics could move on, and the details of his explanation of the colors of solid bodies have not survived. Nevertheless, he effected an immense step forward. He established the heterogeneity of light—extended in the nineteenth century to the heterogeneity of electromagnetic radiation—apparently for all time, and his measurements of Newton's rings first demonstrated the periodicity of an optical phenomenon, though he did not consider that periodicity was a property of light itself.

About 1670, Newton's interest shifted away from the topics we associate with his name to alchemy and theology, which dominated his life for roughly fifteen years. He pursued alchemy with great intensity, though his work did not affect the development of chemisty; his study of theology, which he kept very secret, led him to doubt the divinity of Christ and the doctrine of the trinity. In August 1684, a visit from Edmond Halley with a question about orbital dynamics brought Newton back to physics. The steadily expanding enquiry that Halley's question opened led to his 1687 masterpiece, *The Mathematical Principles of Natural Philosophy* (or *Principia*, from the key word in its Latin title).

In the *Principia*, Newton returned to the science of mechanics. The work opened with a system of dynamics, the three laws of motion still taught as the foundation of modern physics. In Book I he then proceeded to apply the laws to problems of orbital motion, developing for the first time a successful orbital dynamics. He enunciated the concept of centripetal force, reversing the centrifugal force that he and other scientists had associated with bodies in circular motion, and he showed that orbital motion reduces to two essential elements, an inertial motion that is constantly deflected by a centripetal force sufficiently strong to hold a body in a closed circuit. Whenever a body in inertial motion is diverted from its rectilinear path by any centripetal force, Johannes Kepler's law of areas is obeyed. Orbital motion in an ellipse with a centripetal force directed to one focus requires the force to vary inversely as the square of the distance, and a system of satellites orbiting a center that attracts them with an inverse square force must obey Kepler's third law, which relates periods to orbital radii.

After Book II, which examined motions in and of material media, Book III applied the conclusions of Book I to the observed phenomena of the heavens. Newton demonstrated that planets orbiting the Sun and satellites orbiting Jupiter and Saturn, all three systems in accordance with Kepler's third law, require the presence of inverse square attractions toward the central bodies. There must likewise be some attraction toward Earth that holds the Moon in an orbit about it. He compared the centripetal acceleration of the Moon, calculated from its observed orbit, with the acceleration of gravity on Earth's surface, measured by experiments with pendulums. This was the comparison he had made in 1666, but now he had an accurate measurement of Earth's size, and the correlation agreed with a high degree of accuracy. He concluded then that the inverse square force present in the cosmos is identical to the force that causes bodies to fall to the earth, and in Proposition 7 of Book III Newton pronounced the law of universal gravitation.

The rest of Book III applied the concept of universal gravitation with quantitative precision to a series of known phenomena that had not been used in its derivation—the perturbations of the Moon's orbit, the tides, a conical motion of Earth's axis that gives rise to an appearance called the precession of the equinoxes, and finally comets, which he showed to be planet-like bodies orbiting the Sun. He succeeded in defining the orbit of the great comet of 1681–1682, treating it as a parabola in the first edition and as a greatly elongated ellipse in subsequent ones.

Newton's *Principia* was received at once as an epochal achievement in England. His newly won prominence led Newton to abandon Cambridge in 1696 for London, where he was first Warden and then Master of the Mint and, after 1703, President of the Royal Society, positions that he filled until his death. On the continent, the work met opposition from the entrenched Cartesian natural philosophy. Nevertheless, there too Newtonian science was triumphant well before the middle of the eighteenth century. It shaped and dominated the development of the physical sciences until the beginning of the twentieth century, and even then such developments as relativity and quantum mechanics were possible only because Newtonian science had prepared the way for them.

See also: ABERRATION, CHROMATIC; CENTRIFUGAL FORCE; CENTRIPETAL FORCE; DISPERSION; GRAVITA-

TIONAL FORCE LAW; INVERSE SQUARE LAW; KEPLER, JO-
HANNES; KEPLER'S LAWS; NEWTONIAN MECHANICS; NEW-
TONIAN SYNTHESIS; NEWTON'S LAWS

Bibliography

COHEN, I. B. *The Newtonian Revolution with Illustrations of the Transformation of Scientific Ideas* (Cambridge University Press, Cambridge, 1980).

DOBBS, B. J. T. *The Janus Faces of Genius* (Cambridge University Press, New York, 1992).

WESTFALL, R. S. *The Life of Isaac Newton* (Cambridge University Press, Cambridge, 1993).

RICHARD S. WESTFALL

NEWTONIAN MECHANICS

Mechanics is the branch of physics that deals with the forces of interaction between objects and with how these forces affect the motions of objects. Newtonian mechanics refers specifically to the system of mechanics that was developed from the application of Newton's laws of motion and from the assumptions inherent within these laws.

Newtonian mechanics provided the foundation for the dominant physical world view from its establishment late in the seventeenth century until the introduction of relativistic and quantum mechanics near the beginning of the twentieth century. The great achievement of Newtonian mechanics was the assertion that, given the initial positions and momenta of any system of objects, if the forces of interactions between these objects are known, then one can predict the positions and momenta of these objects at any time in the future. This resulted in a mechanistic, cause and effect, view of the universe and encouraged a philosophy of determinism for both inanimate and animate objects.

At the heart of Newtonian mechanics are Newton's three laws of force and motion, originally formulated by Isaac Newton in his 1687 book *Philosophiae Naturalis Principia Mathematica (Mathematical Principles of Natural Philosophy)*. Newton's first law, known also as the law of inertia, is based on inertial concepts developed previously by Galileo Galilei and René Descartes. The first law states that if an object is free of externally applied forces, then

it will remain motionless if it is already at rest. If the object is currently in motion, then it will continue to move in a straight line at constant speed. Such motion is commonly referred to as natural motion; the motion exhibited by an object free of any imposed forces from the outside. Newton's first law also states that the uniform motion of an object will be disrupted if forces are exerted on the object from the outside, but the law does not describe the nature of the resulting motion in such situations.

A consequence of Newton's first law concerns the selection of a reference frame for use in the observation of physical events. It follows from the first law that all reference frames moving at constant velocity with respect to each other offer equally valid perspectives for making observations. No experiment can be performed that would indicate there is a preferred reference frame to be used, although the actual motions observed will vary from one reference frame to another. This result is in contradiction to Aristotelian mechanics, the preeminent mechanical system prior to Newtonian mechanics. In Aristotelian mechanics it was thought that the natural tendency for objects was to be at rest. In such a case, the rest frame of the object would be the preferred reference frame for making observations.

Newton's second law states that the change in momentum of an object is proportional to the net force exerted upon the object. Momentum is defined to be the product of the mass times the velocity of an object. The net force on an object is defined to be the vectorial sum of all the forces exerted upon the object. The predicted momentum change is in the direction of the exerted force. In principle, Newton's second law completely prescribes the evolution of the motion of any object provided that the forces exerted on the object are known.

If the mass of a particular object does not change over time, then the equation of motion for the object that corresponds to Newton's second law can be written as $\mathbf{F} = m\mathbf{a}$, where m is the mass of the object and $\mathbf{a}$ is the acceleration of the object. The form of this equation indicates that the mass of an object determines how resistant an object is to a change in its motion; the greater the mass, the greater the inertia or resistance to changes in motion. Acceleration is defined to be the time rate of change of the velocity $\mathbf{v}$, and velocity is defined to be the time rate of change of the position $\mathbf{r}$: $\mathbf{a} = d\mathbf{v}/dt$, and $\mathbf{v} = d\mathbf{r}/dt$.

Newton's second law is a quantitative and more general statement about motion than is Newton's first law. The first law can be derived from the second law by considering the special case where there is no net force being exerted upon an object. In this situation, the mathematical expression for Newton's second law reduces to **v** = constant. This mathematical expression is equivalent to Newton's law of inertia.

Newton's third law deals with the forces of interaction that may exist between any two objects. It is stated that for every force exerted on object 1 by object 2 there will always be another force exerted on object 2 by object 1. These associated forces are equal in magnitude but oppositely directed, and they are commonly referred to as action-reaction pairs.

Kinematics and Dynamics

Newton's first and second laws completely describe the motion or the changes in the motion of an object in any situation if the forces of interaction are known. In situations where the forces exerted on an object remain constant in time, then Newton's second law becomes $\mathbf{a} = d\mathbf{v}/dt = \text{constant}/m$. The motion of any object can be determined by solving this same mathematical equation whenever the net force remains constant. The forms of the solutions to this equation are called the kinematic equations: $\mathbf{v} = \mathbf{v}_0 + \mathbf{a}t$, and $\mathbf{r} = \mathbf{r}_0 + \mathbf{v}_0 t + \frac{1}{2}\mathbf{a}t^2$. These equations predict the position and velocity of the object as a function of the time or the acceleration, but they do not explicitly contain reference to the forces involved. The study of the motion of objects without reference to the causes of the motion is a branch of mechanics called kinematics.

When the forces exerted on an object do vary over time, then the mathematical form previously stated for Newton's second law must be made more general. For these situations, the differential form of the second law must be used to describe the motion and is written as $\mathbf{F} = d\mathbf{p}/dt$, where **p** stands for the momentum of the object.

The equation above is the most general mathematical formulation of Newton's second law. It allows for the possibility that the forces on an object may vary over time or that the mass of the object might be changing. Impulse is defined from this equation to be the integration of the force over time, and it equals the change in momentum of an object. Mechanical work is defined to be the integral of the scalar product of the net force and the displacements along the trajectory of the object.

The motion of the object is found by solving the differential equation, which is usually accomplished by finding the impulse or mechanical work done. Solving for the motion of an object as a function of the forces exerted on the object is a branch of mechanics known as dynamics.

Range of Validity of Newton's Laws

Newtonian mechanics is found to be consistent with experimental measurements only when measurements are taken within the physical limits of normal human experience. Outside of those limits Newtonian mechanics becomes measurably invalid. Consequently, within certain domains Newton's laws must be modified to fit experimental results; at other times, they must be completely discarded. The discrepancies between Newtonian mechanics and experimental measurements exist primarily because of the implicit assumptions underlying Newton's laws, which are now known to be valid only within certain ranges of measurement.

The validity of Newton's third law is contingent upon the assumption of action-at-a-distance forces. The assumption behind action-at-a-distance forces is that the force from one object can be instantaneously transmitted to another object some distance away. Without action-at-a-distance, action-reaction pairs cannot exist, because in general an object will have moved between the time it imposes a force on another object and the time that the second object can exert a force back. The new location of the objects will result in action-reaction pairs that in general are not of the same magnitude or in opposite directions. Since forces propagate between objects at the speed of light, Newton's third law is only an approximation, and as the distances between objects becomes greater, or as the relative motion between two objects becomes greater, the approximation becomes progressively worse.

Newton's second law is partly based on three assumptions: (1) The amount of energy (and therefore the amount of momentum) that an object may possess is continuous. There are no values of momentum that an object is forbidden to have. (2) In principle, both the position and the momentum of an object can be known to within any precision desired (assuming the appropriate measuring device is available). (3) Momentum is equal to the product of the mass times the velocity of the object.

For velocities approaching the speed of light, Newton's second law remains valid only if a more general, relativistic form of momentum is used: $\mathbf{p} = \gamma m\mathbf{v}$, where **v** is the velocity of the object, c is the velocity of light in a vacuum, and γ is given by $\gamma = 1/\sqrt{1 - v^2/c^2}$. The relativistic definition for momentum reduces to the classical Newtonian definition when the velocity of the object is much less than the speed of light.

For microscopic particles roughly the size of molecules or smaller, Newton's second law becomes inappropriate for almost all applications. In this regime the problem is basically twofold. The uncertainty principle of quantum mechanics states that the position and momentum of a particle cannot both be known simultaneously to within arbitrary accuracy. Since it is impossible to know the exact initial conditions of any particle, it is impossible to use Newton's second law to predict with certainty how the position and momentum of a particle will evolve over time. Quantum mechanics also shows that the possible energy states of particles experiencing forces are discreet and not continuous. Thus, particles cannot exist in a continuum of states of position and momentum as Newton's second law implies and predicts. In this domain, quantum mechanical approaches must be used to find the location and energies of particles.

See also: ACTION-AT-A-DISTANCE; KINEMATICS; NEWTON, ISAAC; NEWTONIAN SYNTHESIS; NEWTON'S LAWS

Bibliography

HUMMEL, C. *The Galileo Connection* (Inter Varsity Press, Downers Grove, IL, 1986).

NEWTON, I. *Mathematical Principles of Natural Philosophy,* trans. by F. Cajori (University of California Press, Berkeley, [1687] 1934).

OLENICK, R.; APOSTOL, T.; and GOODSTEIN, D. *The Mechanical Universe* (Cambridge University Press, Cambridge, Eng., 1985).

SYMON, K. *Mechanics,* 3rd ed. (Addison-Wesley, Reading, MA, 1971).

JAMES JADRICH

NEWTONIAN SYNTHESIS

The concept of Newtonian synthesis has at least three different meanings. On the most general level,

it refers to the new synthesis of learning at the end of the seventeenth century that replaced the defunct medieval synthesis. The so-called medieval synthesis, which had found its fullest expression in the writings of St. Thomas Aquinas in the thirteenth century, had drawn together learning and society under the aegis of Christianity and the church and subsumed every human activity under the ultimate end of eternal salvation. By the close of the seventeenth century, nearly everything had changed. First and foremost, the new synthesis was secular. Politically and economically, European life had thrown off Christian tutelage; equally, the new synthesis of learning did not pursue Christian purposes. Although every scientist who contributed significantly to the Scientific Revolution was a Christian and frequently a devout one, the close filiation between the theories of science and the doctrines of Christianity had ceased to exist. Only the Catholic Church still attempted—and it half-heartedly and in regard to astronomy alone—to make science conform to the words of Scripture.

Where the earlier universe had been geocentric, with Earth as the stage for the cosmic drama of salvation, the new one was heliocentric. Instead of treating nature on the model of living beings, investigators approached scientific problems as issues in mechanics. The new science was convinced of the uniformity of nature, whereas medieval natural philosophy had applied different rules to the mundane earthly world and the perfect heavens. The law of universal gravitation, which united heaven and earth, was the ideal embodiment of the new conception. Physics had become heavily mathematical and pursued questions by experimental enquiry instead of Aristotelian logic. In a word, Newtonian synthesis was the first fully fledged incarnation of the world in which we continue to live.

On a somewhat lower plane of generality, Newtonian synthesis refers to Newton's position as the culmination of the Scientific Revolution, weaving together into one coherent fabric the central strands of the transformation of science during the previous century and a half. Nicolaus Copernicus and his reordering of the solar system stood at the dawn of the Scientific Revolution. At the end of the sixteenth century, Johannes Kepler at once embraced and transformed heliocentric astronomy, replacing its collection of perfect circular motions with elliptical orbits on which the planets moved with nonuniform velocities. Newtonian synthesis derived Kepler's three laws of planetary motion from the laws of motion. In more general terms, Isaac Newton showed that only a Keplerian heliocentric system is compatible with the universal principles of dynamics.

He embraced more than Keplerian heliocentrism. Once Earth was allowed to spin daily on its axis, the principal argument for the finitude of the universe collapsed. Moreover, the telescope, invented in 1609, revealed not only hitherto unknown bodies in the solar system but an apparently unlimited number of stars at varying and immense distances. From the time when the heliocentric system was first proposed, the lack of observed parallax of the fixed stars, that is, the fact that their apparent positions did not shift when they were viewed from opposite sides of the large terrestrial orbit, had been a major argument against the system. To overcome it, astronomers had had to accept the placement of the fixed stars at distances that at first seemed incredible. During the seventeenth century, astronomers and scientists in general ceased to find the distances incredible, and the continuing failure to observe parallax, even through the most advanced telescopes of the late seventeenth century, now seemed to offer means to calculate distance, that is, the least possible distance of the nearest stars. In turn, that distance, many times larger than distances that a century earlier had seemed incredible to most, fed a growing conviction that the universe is infinite. Newtonian science, though it did not demonstrate the infinity of the universe, unambiguously embraced it.

Even more than the absence of parallax, the problem of motion, posed primarily by the asserted diurnal rotation of Earth, offered the greatest obstacle to heliocentric astronomy. How was it possible that no one observed the immense velocity with which everyone is moving if the heliocentric system were correct? How was it possible that other phenomena of motion on Earth, such as the vertical drop of heavy objects, happen as they were observed to happen? The questions raised by the heliocentric system helped to inaugurate a re-examination of the nature of motion, in which Galileo Galilei was the leading figure. Not only did he effectively refute this objection to the Copernican system by redefining motion (which had been held to require the constant presence of a mover) and proposing a concept almost identical to the principle of inertia, he also carried the investigation of motion further and founded a new science of mechanics. If bodies on ideal horizontal planes move inertially, those falling vertically move with uniformly accelerated motion. Galileo asserted that all bodies on Earth fall with the same acceleration, and he worked out the mathematical consequences (i.e., the relations of time, distance, and velocity) entailed by the concept of uniformly accelerated motion. Newtonian science embraced Galilean mechanics as fully as it embraced Keplerian astronomy and made it one of the pillars of the new synthesis.

In the first half of the seventeenth century, as the Aristotelian system of natural philosophy collapsed, René Descartes (and others such as Pierre Gassendi) proposed a natural philosophy built on completely different foundations, what became known as the mechanical philosophy. It held that the physical universe is composed solely of inert matter that lacks any source of spontaneous action and is divided into particles that move where the motions of other similar particles that impinge on them force them to move. The mechanical philosophers admitted the existence of spirit, but they confined spirit to God on the one hand and human souls on the other. The overwhelming majority of natural phenomena were the mechanical result of inert particles of matter interacting with each other in multiple impacts. Where the earlier natural philosophy had treated physical nature on the model of organic beings, the mechanical philosophy insisted that animals and plants are merely complicated machines.

Newton's relation to the mechanical philosophy was complex. On the one hand, it introduced him to the world of the new science; on the other hand, he revised it profoundly, even while remaining true to its basic tenets. What he revised was the notion that bodies can interact by impact alone. His concept of action-at-a-distance—best known in universal gravitation, but for Newton present also in a number of short range attractions and repulsions—appeared to other mechanical philosophers as a repudiation of the basic principle of their program. Nevertheless, Newton presented a picture of nature as a system of interacting particles of matter. The essence of his major 1687 work, *The Mathematical Principles of Natural Philosophy* (known as the *Principia*), was the treatment of planetary motion solely as a problem in mechanics. Like mechanical philosophers, he accepted the uniformity of nature. The same matter present on Earth is present in the heavens; the same laws that apply to the heavens govern terrestrial motions. The principle of universal gravitation, an attraction present in all matter as matter, is perhaps the ultimate statement of the most basic outlook of the mechanical philosophy. Most important, his concept of action-at-a-distance, quantitatively defined, made the mechanical philosophy fruitful by projecting mathematics into it.

Mathematics was another dimension of Newtonian synthesis. From the beginning, especially in the work of Copernicus, Kepler, and Galileo, but not solely in them, the Scientific Revolution had elevated the status of mathematics in natural philosophy. Newton elevated its status still further with the *Principia*, in which he presented science in terms of the most complex mathematics. Newton confirmed once and for all that mathematics would be the language of modern science.

From the dawn of the seventeenth century, a new methodology, experimental procedure, had begun to appear in science. Newton also wove this into his synthesis. Although the *Principia* devotes itself primarily to working out the consequences of Newton's laws of motion mathematically, it also contains a considerable number of experiments that connect the principles with empirical reality. Newton's other work, *Opticks* (1704), is not only a book of experimental science, but it contains, in Newton's investigation of phenomena of colors, one of the great experimental investigations of the seventeenth century, which helped to make experimentation the characteristic procedure of science.

The third meaning of Newton synthesis, which relates it to the work of Galileo and Kepler, is the most restricted. Near the beginning of the seventeenth century, Kepler and Galileo had realized two of the fundamental advances of the Scientific Revolution, the kinematics of celestial motions and the kinematics of terrestrial motions. With the three laws of motion that introduce the *Principia*, Newton knit the two together by providing their common dynamic foundation. That is, at the heart of Newtonian synthesis was a mathematical system of dynamics from which both the celestial kinematics of Kepler and the terrestrial kinematics of Galileo emerged as necessary consequences.

Newtonian synthesis had little to say about the life sciences, which, though much pursued, were not at the heart of the Scientific Revolution. It did draw together all of the basic features of the seventeenth century's new approach to the physical sciences, combining them in the pattern that has characterized them ever since. It is then small wonder that almost from the day of publication of the *Principia*, Newtonian science has exercised immense influence. The Enlightenment of the eighteenth century

almost deified Newton, and it took his law of universal gravitation as the model of the natural laws it pursued in every facet of social life. The French scientist Joseph Lagrange said that Newton was the most fortunate man who ever lived; there is only one universe, and he was the one who discovered its basic law. Newton's influence continued undiminished through the nineteenth century, and even the efflorescence of the biological sciences at that time, though not based on Newtonian concepts, took place within the boundaries of his synthesis. If the revolution in physics in the twentieth century has shown that Newtonian science is not after all the final word, nevertheless the entire structure of modern science, perhaps the central element of modern civilization, would be impossible had not Newtonian synthesis constructed its foundation.

See also: COPERNICAN REVOLUTION; COPERNICUS, NICOLAUS; GALILEI, GALILEO; KEPLER, JOHANNES; KEPLER'S LAWS; NEWTON, ISAAC; NEWTONIAN MECHANICS; NEWTON'S LAWS

Bibliography

COHEN, I. B. *The Birth of a New Physics*, rev. ed. (W. W. Norton, New York, 1985).

KOYRÉ, A. *Newtonian Studies* (Harvard University Press, Cambridge, MA, 1965).

WESTFALL, R. S. *The Construction of Modern Science* (Cambridge University Press, Cambridge, Eng., 1977).

RICHARD S. WESTFALL

NEWTON'S LAWS

"[T]he whole burden of philosophy," Isaac Newton wrote in the preface to his *Principia Mathematica* (1686), "seems to consist in this: *from the phenomena of motions* to *investigate the forces of nature,* and then *from these forces* to *demonstrate the other phenomena.*" He specified the tools required to effect this program in his celebrated "Axioms or Laws of Motion." (What Newton called "philosophy" is now called "physics.")

For an understanding of Newton's laws a few comments about the concepts of motion and of force on an object may be helpful. It is important to recognize that the concept of the motion of an object is meaningless unless the frame of reference used for the description of the motion is specified. In different frames the object has different motions. In particular, characteristics of a motion at some instant, such as the velocity or acceleration, are reference frame dependent; they are not properties intrinsic to the object.

About force on an object: Any physical influence which changes the characteristics of a motion of an object from what they would be in the absence of the influence is said to be produced by a force on that object; moreover the concept of force as used by Newton assumes that force can only be produced by observable physical objects in the general environment of the object (either in contact or at a distance).

To recognize the effects of forces on a motion one must first know what characteristics the motion would have in the absence of forces (when the environment contains no sources which might produce forces on the object). These characteristics are specified by Newton's first axiom or law of motion.

First Law of Motion

There exist special frames of reference (called Galilean) relative to which an object will move with constant velocity (zero acceleration) if there is no force on it.

Note that if a reference frame is Galilean, any frame with constant velocity relative to it is also Galilean; constant velocity in one such frame implies constant velocity in all of them. More significantly, the acceleration of an object has the same value in all Galilean frames. In the following considerations of the laws of motion, "acceleration" always refers to its unique measure in a Galilean frame.

It follows from the first law and the meaning of force that if an object has an acceleration a force must be acting on it and that some effective physical source for that force exists in its environment. As yet the force concept is not well defined. Newton completes the concept in his second law. It is convenient to state the law in two parts.

Second Law of Motion

(1) The environment of an object *uniquely* determines its acceleration. (2) If the acceleration of an object of mass m at some instant is **a,** the force **F** acting on the object is defined as

$$\mathbf{F} = m\mathbf{a}. \qquad (1)$$

It is the first part of this law which contains its physical content. If it were to be observed that on different motions the same environment produced different accelerations, the second law would be false.

The two sides of Eq. (1) refer to very different physical features: The acceleration is a kinematic property of the motion while the force is a property of the environment of the object. It is the second law that permits the investigation of the forces of nature by means of the phenomena of motion (accelerations) and also the demonstration of the other phenomena. The measurement of the acceleration of an object at some instant in the course of its motion specifies, by Eq. (1), the force exerted on it by its environment at that instant. The correlation of observed accelerations with the environments that produce them leads (for some environments at least) to the induction of a general law of force from which the force for any of the environments the object might meet in its motions may be computed. Once such a general force function has been found it may be used, by application of the second law, to *predict* new motions. From the analysis of planetary and other motions (the Moon around Earth, the motions of the satellites about other planets, and other gravitational motions) Newton guessed that the law of force between any two mass elements m_1 and m_2 separated by a distance d is

$$F = \frac{Gm_1 m_2}{d^2},$$

where G is a constant. With this force law Newton went on to demonstrate the other phenomena. He showed that the motions of comets, the tidal motions of the seas, and, in the most remarkable of the applications of his force law, the quantitative features of the precession of the earth's axis, all followed mathematically from the force law he had induced. His law implied that the force on a particular planet arose not only from the sun but also from all the other planets in the solar system. In consequence, the planets do not move exactly on Keplerian orbits. The small deviations of the planetary motions from the Keplerian orbits predicted by Newton's law of gravitation were discovered when telescopic observations became available.

In the foregoing it was assumed that the object under consideration has, at each instant of time, a definite position, hence also a definite velocity and acceleration. But this is true only for a dimensionless particle. An extended object may be considered as a collection of point particles held together by strong forces. The different particles in the object will generally have different velocities and accelerations. On any particular particle of the body, we may distinguish two sources of force: "external forces" which are produced by sources outside of the body and "internal forces" produced by sources inside the body. In a rapidly spinning ball in a force free external environment, for example, the external force on a particle in the object is zero while the internal force on it is towrd the axis of the rotation and varies with the distance from the axis and the magnitude of the spin.

How do these internal forces influence the motion of the body as a whole? The fact that Newton's laws work as well as they do without consideration of the possible influence of internal forces (which may, in some motions, exceed the external forces) indicates that the internal forces must have very little influence on the general motion of the object. Newton recognized that, for the internal forces to have so little effect on the motion as a whole, there must be special relations between them. He tried to discover, by experiment, what these relations might be. To this end he studied properties of the motions of two small objects interacting through collision and a system of three mutually attracting objects. (Incidentally, these are the only experiments Newton cites in support of his laws of motion.) He summarized his findings, which must have been more than half expected by him before he began, in his third law.

Third Law of Motion

The sum of *all the internal forces* on the different particles in an extended object vanishes.

Consequently, the *total* force on an object (the sum of all internal *and* external forces on all the particles in the object) is equal to the *total external* force on the object.

Note that nothing has been said about the form of the object under consideration. The object merely provides a separation of what is taken as internal from what is considered external. Any collection of particles or bodies may be considered as an object. The sum of all the internal forces on the particles of such a collection must vanish according to the third law. Suppose the collection is taken as composed of two partial collections, A and B. Since the total force on the collection, by the third law, is zero, the action (the net force) of part A on part B is equal and exactly opposite to the reaction of B on A. (This is, effectively, Newton's formulation of his third law.)

From the third law it is a simple matter to prove that the total external force, $\mathbf{F}_{ex}$, on an extended object is equal to the product of the total mass M, of the object with the acceleration $\mathbf{A}$, of its center of mass: $\mathbf{F}_{ex} = M\mathbf{A}$. If there are no external forces, $\mathbf{F} = 0$, and the center of mass of the object moves with constant velocity. These results show why the first and second laws could be obtained without consideration of the influence of internal forces.

The procedure for finding the forces of nature from the phenomena of motions is straightforward *if* a Galilean frame of reference is given. The accelerations of motions relative to the Galilean frame are measured and the associated forces are then obtained by the use of the second law. But *real* observations are made relative to a frame fixed to the earth. To describe the observed motions in a Galilean frame and to compute the accelerations of these motions relative to this frame, the motion of a Galilean frame relative to the earth frame must be known. Newton's laws do not provide a procedure for finding this motion. How, then, are forces to be found from the analysis of motions? In reference to this question Newton wrote, "It is a matter of great difficulty to distinguish the 'true' motions [motions relative to a Galilean system] from the 'apparent' [motions relative to a non-Galilean system].... Yet the thing is not altogether desperate: for we have some arguments to guide us."

It can be shown that if, relative to a chosen frame, a force function for an environment is obtained (by whatever means) that, with the use of Eq. (1), successfully predicts all motions, the chosen frame is Galilean. However, only a limited selection of motions can ever be tested, and these only within certain limits of error. Hence, observation can only establish that a frame is *approximately* Galilean for the *selected range* of tested motions. In a frame fixed relative to the earth, small motions, suitably limited in space and time, are predicted with fair accuracy by the use of Newton's second law with the force function, $\mathbf{F} = m\mathbf{g}$ (where $\mathbf{g}$ is the acceleration due to gravity). This "earth frame," therefore, is approximately Galilean for the range of motions specified. It is because of the (approximate) Galilean property of a frame fixed to the earth that Newton's laws may be used to discover unknown forces by measurements of the accelerations of motions as observed in a laboratory fixed to the earth.

For motions over longer intervals of time and space, predicted motions, relative to the earth frame, depart radically from observed motions even with the use of the force law obtained from Newton's law of gravitation. Observations on the Foucault pendulum suggest the use of a frame relative to which the earth rotates once a day. With this frame and Newton's law of force, motions are predicted successfully over ranges in space and time of the order of earth dimensions and days or months. (The statement that the earth rotates means only that it rotates relative to Galilean frames.)

Kepler had discovered, long before Newton, a frame of reference relative to which the motions of the planets had remarkably simple properties. As was noted earlier, it was partly through an analysis of the motions implied by Kepler's laws that Newton arrived at his law of gravitation. Relative to Kepler's frame Newton's force law successfully predicts, with great accuracy, the motions of the planets, of the satellites now wandering through the solar system, and many other motions. Kepler's frame, therefore, is Galilean to a high degree of accuracy.

One further quote from Newton may serve as a concluding remark, "I wish we could derive the rest of the phenomena by the same kind of reasoning from mechanical principles, for I am induced by many reasons to suspect that they may all depend on certain forces.... These forces being unknown, philosophers have hitherto attempted the search of nature in vain; but I hope that the principles here laid down will afford some light either to this or some truer method of philosophy." His hope has been abundantly fulfilled.

See also: ACCELERATION; FORCE; GALILEAN TRANSFORMATION; GRAVITATIONAL ATTRACTION; GRAVITATIONAL CONSTANT; GRAVITATIONAL FORCE LAW; KEPLER, JOHANNES; KEPLER'S LAWS; KINEMATICS; MOTION; NEWTON, ISAAC; PARTICLE; PENDULUM, FOUCAULT; SOLAR SYSTEM; VELOCITY

Bibliography

EISENBUD, L. "On the Classical Laws of Motion." *Am. J. Phys.* **26**, 144–159 (1958).

NEWTON, I. *Principia Mathematica,* 3rd ed., trans. A. Motte, rev. F. Cajori (University of California Press, Berkeley, [1727] 1962).

LEONARD EISENBUD

NICOL PRISM

The Nicol prism provides wide separation between the two linear polarizations present in unpolarized light. William Nicol, a Scottish physicist, first published his description of the prism's construction in 1829. It is the earliest example of a number of prism polarizers that employ the birefringence of calcite (calcium carbonate) crystals. Birefringence in calcite and other crystals had long been known to cause double images of objects. Thomas Young and Augustin Fresnel determined early in the nineteenth century that the two images consisted of two perpendicular polarizations of light. The total angular divergence between the rays making up the images was quite small, however. Separation of the two polarizations using birefringence alone required a very large crystal. Nicol introduced the use of total internal reflection to greatly increase the angular separation and efficiently obtain very pure linearly polarized light from unpolarized sources using modestly sized crystals.

Construction

Calcite is a negative uniaxial crystal with n_o (the index of refraction for the ordinary ray) approximately 1.65 and n_e (the index of refraction for the extraordinary ray) approximately 1.49 across the visible spectrum. A Nicol prism is actually two complementary triangular prisms of calcite cemented together along their common hypotenuse (see Fig. 1). This cement is normally Canada balsam with an index of refraction of 1.53. The entrance and exit faces make an angle of approximately 20° with the mechanical axis. Refraction at the entrance face, causes the ordinary and extraordinary rays to deviate approximately 8.0° and 6.7°, respectively, from their original direction. The common hypotenuse

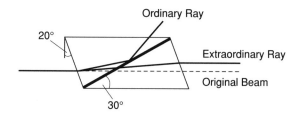

Figure 1 Nicol prism.

makes an angle greater than approximately 60° with the mechanical axis so that the ordinary ray is totally internally reflected from the Canada balsam layer. Since n_e is less than the index of refraction of the Canada balsam, the cement layer transmits most of the extraordinary ray. It passes through the second complementary calcite triangle where it is again refracted at the exit face. The exit beam consists entirely of the linearly polarized extraordinary ray, parallel to the original beam direction but displaced a few millimeters. This design cleanly separates the two linear polarizations present in the original, unpolarized beam.

Similar Devices

Two similar devices use birefringence and total internal reflection to efficiently separate perpendicular linear polarizations. In the Glan–Thompson and Glan–Taylor prism polarizers, the entrance and exit faces are perpendicular to the mechanical axis, resulting in no refractive deviation of either polarization. The exit beam is still parallel to the entrance beam, but it is not displaced. The Glan–Thompson prism uses Canada balsam cement to join the common hypotenuse of the prism pair, now cut at an angle greater than 68° to the mechanical axis. This angle achieves total internal reflection of the unrefracted ordinary ray but results in higher reflection losses of the extraordinary ray. The Glan–Taylor prism replaces the cement with a narrow gap of air or vacuum by rigidly mounting the prism pair in a fixture. The polarizer can then withstand much higher powers for laser applications. The hypotenuse must be cut at an angle greater than 37° and less than 42° to ensure total internal reflection of the ordinary ray but not of the extraordinary ray.

Uses

Commercially available optical quality calcite polarizing prisms routinely achieve a linear polarization purity of 99.999 percent. This is much higher than the 50 to 80 percent typical of polaroid-type polymer film polarizers, which polarize by absorption rather than reflection. Prism polarizers, especially of the Glan–Taylor type, withstand high incident light intensity without damage. They find many uses in the control of light polarization for basic research and in precision measurements of the concentration of compounds in solution (sugar in water, for example) through their effect on the polarization of transmitted light.

See also: BIREFRINGENCE; FRESNEL, AUGUSTIN-JEAN; POLARIZATION; POLARIZED LIGHT; REFRACTION; REFRACTION, INDEX OF; YOUNG, THOMAS

DAVID W. DUQUETTE

NOBEL PRIZE WINNERS

The Nobel Prizes were instituted by Alfred Nobel, who developed dynamite for industrial use. Nobel referred to five prizes in his 1895 will. Among these, the first was to be awarded to "the person who shall have made the most important discovery or invention within the field of physics," and the second to "the person who shall have made the most important chemical discovery or improvement." The remaining three prizes were to be awarded in the fields of physiology or medicine, literature, and peace. The first Nobel Prizes were awarded in 1901. From the start the Nobel Prizes were recognized as the highest award in science, mainly because of the large sums of money involved; a prize was equivalent to the annual salaries of twenty professors. In the 1990s the prize amount was more than one million dollars.

Selection Procedures

The prizes in physics and chemistry are awarded by the Royal Swedish Academy of Sciences. The academy elects the Nobel Committee for Physics and the Nobel Committee for Chemistry, each having five members. As a first step in the process of deciding the year's prize winners, the Nobel committees solicit nominations for the prizes from scientists around the world. Although a person cannot be considered for the prize without having been nominated, the committees are not obliged to base their decisions on the number of nominations the candidates receive. Nominators for the prizes either have a permanent right to submit proposals or are invited to do so for a given year. Nobel laureates in physics and chemistry represent the most important category of permanent nominators. When the nominating period is closed on February 1, the Nobel committees carry out the evaluations of the candidates. In the early autumn they recommend one or more candidates to the academy. The final decision is made by the academy meeting in a plenary session. The decisions concerning the prize winners in physics and chemistry are announced in October each year and the prize is awarded at a ceremony that takes place in Stockholm on December 10 (the day of Alfred Nobel's death). In connection with the ceremony, the prize winners fulfill the statutory obligation to give a public lecture (Nobel lecture) on the work for which the prize was awarded.

Discovery, Invention, and Improvement

Most of the works recognized by the prizes in physics and chemistry fall in the category of "discovery," that is, they are the results of basic research. During the more than ninety years that the prizes have been awarded, the definitions of discovery have undergone important change. In the early years, the discoveries rewarded were overwhelmingly in experimental physics. They concerned such new phenomena as x rays, radioactivity, and the electron (e.g., prizes for Wilhelm Conrad Röntgen, Antoine Henri Becquerel, Pierre and Marie Curie, and Joseph John Thomson). In chemistry, discovery often came to mean the isolation of new elements, for example, the noble gases (William Ramsay), fluorine (Henri Moissan), radium (Marie Curie), and radioactive isotopes (Frederick Soddy and Francis William Aston).

In the early 1920s, the Nobel Committee for Physics overcame its initial hesitancy about theoretical work and awarded prizes to Albert Einstein and Niels Bohr. That its view of discovery was still experimentalist is shown by the prize to Einstein that was awarded to him not for the special theory of relativity but for the law of the photoelectric effect. Since the 1920s the prizes in physics have alternated between experimental and theoretical work.

Relatively few prizes have been awarded for "inventions," that is, the results of industrial research or technology. Although proposed for the physics prize, prominent inventors such as Thomas A. Edison (the telephone) and the Wright brothers (the airplane) had to live out their days without awards. An important category of prizes in the area of "invention," broadly conceived, has concerned breakthroughs in the area of instrumentation and research technology. The first prize in this category was awarded to Albert Michelson in 1907 for his optical precision instruments used (although this was not mentioned in the award citation) in the Michelson–Morley experiment. Since then there has been a series of prizes in the areas of spectroscopy (Karl Manne Georg and Kai M. Siegbahn, Arthur L. Schawlow and Bertram Brockhouse), cyclotrons (Ernest Orlando Lawrence), and particle detectors (Patrick Maynard Stuart Blackett, Donald A. Glaser, and Georges Charpak). Choices in the area of technology with proven industrial use belong mainly to the period before World War I (Guglielmo Marconi and Carl Ferdinand Braun for the development of wireless telegraphy and Nils Gustaf Dalén for the automatically regulated lighthouse). In the post–World War II era, there have been a few prizes in the area of technology although the inventions honored were not developed for industrial use at the time of the award. Such prizes have concerned semiconductors and transistors (William Shockley, John Bardeen, and Walter Houser Brattain) and superconductivity (J. Georg Bednorz and K. Alexander Müller).

Physics and Chemistry as Defined in Successive Prize Decisions

Although Nobel's will made use of the broad term "physics," prizes have tended to cluster in the fields of atomic, nuclear, and particle physics as well as relativity and quantum mechanics. By contrast, there have been relatively few prizes in astrophysics and hardly any in geophysics. When works in these fields have been rewarded, they have also been of significance for atomic or particle physics. Thus when Victor Franz Hess received the prize for the discovery of cosmic radiation in 1936, he shared it with Carl David Anderson who had established the existence in cosmic radiation of a previously unknown positive particle that became known as the positron.

Throughout the history of the prizes, the prize awarders have had to grapple with the difficulty of deciding whether work on the borderlines of physics and chemistry belonged to one or the other of these prize domains. In order to settle such matters, the Nobel committees have generally relied on the precedence created by prize decisions in the early period. One such precedence concerned physical chemistry (in particular, electrochemistry and thermochemistry) that was the subject of a string of awards early in the century (Jacobus Hernicus van't Hoff, Svante August Arrhenius, Wilhelm Ostwald, and Walther Hermann Nernst). That these and subsequent awards were made in chemistry rather than in physics was due to the strong influence exerted by Arrhenius, the Swedish physical chemist, over the Nobel Committee for Chemistry and the opposition that Arrhenius's own prize (1903) had encountered in the Nobel Committee for Physics.

Precedence has also determined whether work in radioactivity should be awarded with the physics or chemistry prize. The first precedence for considering such work as falling within the domain of chemistry was set by the prize to Ernest Rutherford in 1908 for the disintegration of the elements and the chemistry of radioactive substances. This was confirmed when Marie Curie received her second prize (1911) for the isolation of radium. (Her first prize had been awarded in physics in 1903—together with her husband and Becquerel—for pioneering work in radioactivity.) Henceforth research on radioactive substances was considered as falling within the domain of radiochemistry and referred to the Nobel Committee for Chemistry for evaluation and eventual prize recommendation. Examples include the awards to Soddy and Aston (1921), Frédéric Joliot and Irène Joliot-Curie (1935) and George de Hevesy (1943). However, research on radioactive substances and radioactivity was often undertaken because of its significance for atomic theory. This problem first arose when Rutherford was awarded the prize in chemistry in 1908. It recurred with the discovery of nuclear fission in 1939 for which Otto Hahn was awarded the 1944 prize. Although both Hahn and Lise Meitner were repeatedly proposed for the prize in physics by leading atomic physicists, the evaluation of the work was carried out by the Nobel Committee for Chemistry on the grounds that the discovery was based on the use of radiochemical methods.

Number, Nationality, and Gender

By 1995, 271 persons had received prizes in physics and chemistry. The much larger number of prize winners compared to the number of years the prizes have been awarded is due to the practice of dividing the prizes. According to the statutes of the Nobel Foundation, at most three individuals can share a prize. Given this restriction it is not possible to reward every member of the large research teams behind the discoveries in high energy physics; only the leaders of such teams will receive the prize. In two cases, a person has received a second Nobel Prize in the same domain but for new work. Bardeen was awarded his first prize in physics in 1956 for the

discovery of transistors and the second in 1972 for the theory of superconductivity. Frederick Sanger alone received the prize in chemistry in 1958 "for his work on the structure of proteins, especially that of insulin." In 1980 he shared a second prize with Paul Berg and Walter Gilbert "for their contributions concerning the determination of base sequences in nucleic acids."

The nationality of the prize winners has reflected the leadership positions of major science-producing nations during the twentieth century. Counts of the number of Nobel Prizes are frequently used as an indicator of the scientific performance of a country despite the fact that the small numbers of prizes would seem to limit the validity of this theory. Up to World War II, slightly more than 30 percent of the prizes were awarded to German physicists and chemists. Great Britain occupied second position with 18 percent and France third with 15 percent of the prize winners. By contrast, the United States only accounted for 10 percent of the prize winners. After World War II, slightly more than 50 percent of the awards have gone to Americans. In the early postwar period a fair number of prizes were won by naturalized Americans who had fled Nazism (e.g., Otto Stern, Wolfgang Pauli, Max Born, and Eugene P. Wigner). Compared to the larger nations, the number of prizes awarded physicists and chemists working in smaller countries (e.g., the Netherlands, Switzerland, and Sweden) are out of proportion with their populations. However, some very large nations in terms of population have very few prize winners. During the first seventy-five years of its existence, only seven physicists and one chemist in the USSR were awarded prizes, while no Chinese scientist *living in China* was honored with a prize.

Women are notably underrepresented among prize winners in physics and chemistry. Only four prizes have been won by women, two of these by the same person (Marie Curie). The other two have been awarded to Dorothy Crowfoot Hodgkin and Maria Goeppert Mayer.

Nobel Prize Winners as an Ultra Elite

The position of the Nobel Prizes at the top of the hierarchy of prizes and other rewards (membership in scientific societies, honorary doctorates, etc.) conferred on scientists has given Nobel Prize winners a high degree of prestige and public visibility. Their status is thus one of an ultra elite in science. Especially in the United States they are at the very top of the hierarchy of scientists in terms of influence, authority, power, and prestige.

There are important consequences for the individual prize winners who, of course, benefit from the considerable amounts of money involved. It has been found, however, that the prize winners' scientific productivity often declines in the years immediately following the award due to the many extraneous demands made on their time. There is no doubt that the prize almost always has beneficial effects for the prize winner's laboratory, department, or even university. This is so because the prize attracts new funding and makes the laboratory or department more attractive to graduate and post-doctoral students. Boasting about the number of Nobel Prize winners at major American universities has also been used in fund-raising campaigns at these universities.

There is no reason to believe that the status of Nobel Prize winners as an ultra elite will diminish in the twenty-first century.

See also: BARDEEN, JOHN; BOHR, NIELS HENRIK DAVID; BORN, MAX; BROGLIE, LOUIS-VICTOR-PIERRE-RAYMOND DE; CHADWICK, JAMES; COMPTON, ARTHUR HOLLY; CURIE, MARIE SKLODOWSKA; DIRAC, PAUL ADRIEN MAURICE; EINSTEIN, ALBERT; FERMI, ENRICO; FEYNMAN, RICHARD PHILLIPS; HEISENBERG, WERNER KARL; LANDAU, LEV DAVIDOVICH; LAWRENCE, ERNEST ORLANDO; LORENTZ, HENDRIK ANTOON; MAYER, MARIA GOEPPERT; MICHELSON, ALBERT ABRAHAM; MILLIKAN, ROBERT ANDREWS; PAULI, WOLFGANG; PLANCK, MAX KARL ERNST LUDWIG; RABI, ISIDOR ISAAC; RÖNTGEN, WILHELM CONRAD; RUTHERFORD, ERNEST; SAKHAROV, ANDREI DMITRIEVICH; SCHRÖDINGER, ERWIN; THOMSON, JOSEPH JOHN; YUKAWA, HIDEKI

Noble Lectures

Nobel Lectures in Chemistry, 1901–1970, 4 vols., edited by the Nobel Foundation staff (Elsevier, Amsterdam, 1964–1972).

Nobel Lectures in Chemistry, 1971–1980, edited by S. Forsen (World Scientific, Singapore, 1993).

Nobel Lectures in Chemistry, 1981–1990, edited by B. G. Malmström (World Scientific, Singapore, 1992).

Nobel Lectures in Physics, 1901–1970, 4 vols., edited by the Nobel Foundation staff (Elsevier, Amsterdam, 1964–1972).

Nobel Lectures in Physics, 1971–1980, edited by S. Lundqvist (World Scientific, Singapore, 1992).

Nobel Lectures in Physics, 1981–1990, edited by G. Ekspong (World Scientific, Singapore, 1993).

Bibliography

CRAWFORD, E. *The Beginnings of the Nobel Institution: The Science Prizes 1901–1915* (Cambridge University Press, Cambridge, Eng., 1984).

MAGILL, F. N., ed. *The Nobel Prize Winners: Physics* (Salem Press, Pasadena, CA, 1989).

MAGILL, F. N., ed. *The Nobel Prize Winners: Chemistry* (Salem Press, Pasadena, CA, 1990).

STAHLE, N. K. *Alfred Nobel and the Nobel Prizes* (The Nobel Foundation and the Swedish Institute, Stockholm, 1993).

TAUBES, G. *Nobel Dreams: Power, Deceit, and the Ultimate Experiment* (Random House, New York, 1986).

ZUCKERMAN, H. *Scientific Elite: Nobel Laureates in the United States* (Free Press, New York, 1977).

ELISABETH CRAWFORD

NOBEL PRIZE WINNERS IN CHEMISTRY

The year listed is that *for* which the prizes were awarded, not necessarily that *in* which the prizes were awarded. An asterick indicates that there is a separate biographical entry for a given prize winner. The nationality listed for each winner is that determined by the recipient at the time of the award.

1901　JACOBUS HENRICUS VAN'T HOFF (1852–1911), Netherlands, for his discovery of the laws of chemical dynamics and osmotic pressure in solutions.

1902　HERMANN EMIL FISCHER (1852–1919), Germany, for his synthetic experiments in the sugar and purine groups of substances.

1903　SVANTE AUGUST ARRHENIUS (1859–1927), Sweden, for his electrolytic theory of dissociation.

1904　WILLIAM RAMSAY (1852–1916), Great Britain, for the discovery of the inert gaseous elements in air and his determination of their place in the periodic system.

1905　JOHANN FRIEDRICH WILHELM ADOLF VON BAEYER (1835–1917), Germany, for his work concerning organic dyes and hydrocarbon hydroaromatic compounds.

1906　HENRI MOISSAN (1852–1907), France, for his investigation and isolation of fluorine and for his work on the Moissan electric furnace.

1907   EDUARD BUCHNER (1860–1917), Germany, for his biochemical researches and his discovery of cell-free fermentation.

1908   ERNEST RUTHERFORD* (1871–1937), Great Britain, for his investigations into the disintegration of the elements and the chemistry of radioactive substances.

1909   WILHELM OSTWALD (1853–1932), Germany, for his work on catalysis and for his investigations into the fundamental principles governing chemical equilibria and rates of reaction.

1910   OTTO WALLACH (1847–1931), Germany, for his pioneer work in the field of alicyclic compounds.

1911   MARIE SKLODOWSKA CURIE* (1867–1934), France, for the discovery of radium and polonium and for the isolation and study of radium.

1912   VICTOR GRIGNARD (1871–1935), France, for the discovery of the Grignard reagent, which greatly advanced the progress of organic chemistry.
        PAUL SABATIER (1854–1941), France, for his method of hydrogenating organic compounds in the presence of finely disintegrated metals, greatly advancing the study of organic chemistry.

1913   ALFRED WERNER (1866–1919), Switzerland, for his work on the linkage of atoms in molecules.

1914   THEODORE WILLIAM RICHARDS (1868–1928), United States, for his exact determinations of atomic weights of a large number of chemical elements.

1915   RICHARD MARTIN WILLSTÄTTER (1872–1942), Germany, for his researches on plant pigments, especially chlorophyll.

1916   No award.

1917   No award.

1918   FRITZ HABER (1868–1934), Germany, for the synthesis of ammonia from its elements.

1919   No award.

1920   WALTHER HERMANN NERNST (1864–1941), Germany, for his work in themochemistry.

1921   FREDERICK SODDY (1877–1956), Great Britain, for his study of radioactive substances and his investigations into the origins and nature of isotopes.

1922   FRANCIS WILLIAM ASTON (1877–1945), Great Britain, for his discovery, by means of his mass spectrograph, of isotopes in a large number of nonradioactive elements and for his enunciation of the whole-number rule.

1923   FRITZ PREGL (1869–1930), Austria, for his invention of the method of microanalysis of organic substances.

1924   No award.

1925   RICHARD ADOLF ZSIGMONDY (1865–1929), Germany, for his demonstration of the heterogeneous nature of colloid solutions.

1926   THE [THEODOR] SVEDBERG (1884–1971), Sweden, for his work on disperse systems.

1927   HEINRICH OTTO WIELAND (1877–1957), Germany, for his investigations of the constitution of the bile acids and related substances.

1928   ADOLF OTTO REINHOLD WINDAUS (1876–1959), Germany, for his research into the constitution of the sterols and their connection with the vitamins.

1929   ARTHUR HARDEN (1865–1940), Great Britain, for investigations of the fermentation of sugar and fermentative enzymes.
        HANS KARL AUGUST SIMON VON EULER-CHELPIN (1873–1964), Sweden, for investigations of the fermentation of sugar and fermentative enzymes.

1930   HANS FISCHER (1881–1945), Germany, for his researches into the constitution of hemin and chlorophyll and especially for his synthesis of hemin.

1931   CARL BOSCH (1874–1940), Germany, for the invention and development of chemical high-pressure methods.
        FRIEDRICH BERGIUS (1884–1949), Germany, for the invention and development of chemical high-pressure methods.

1932   IRVING LANGMUIR (1881–1957), United States, for his discoveries and investigations within the field of surface chemistry.

1933   No award.

1934   HAROLD CLAYTON UREY (1893–1981), United States, for his discovery of heavy hydrogen.

1935   FRÉDÉRIC JOLIOT (1900–1958), France, for the synthesis of new radioactive elements.
        IRÈNE JOLIOT-CURIE (1897–1956), France, for the synthesis of new radioactive elements.

1936   PETRUS [PETER] JOSEPHUS WILHELMUS DEBYE (1884–1966), Netherlands, for his investigations on dipole moments and on the diffraction of x rays and electrons in gases.

1937   WALTER NORMAN HAWORTH (1883–1950), Great Britain, for his investigations on carbohydrates and vitamin C.

PAUL KARRER (1889–1971), Switzerland, for his investigations on carotenoids, flavins, and vitamins A and B$_2$.

1938 RICHARD KUHN (1900–1967), Germany, for his work on carotenoids and vitamins.

1939 ADOLF FRIEDRICH JOHANN BUTENANDT (1903–), Germany, for his work on the mammalian sex hormones.

LEOPOLD RUZICKA (1887–1976), Switzerland, for his work on polymethylenes and higher terpines.

1940 No award.

1941 No award.

1942 No award.

1943 GEORGE DE HEVESY (1885–1966), Hungary, for his work on the use of isotopes as tracers in the study of chemical processes.

1944 OTTO HAHN (1879–1968), Germany, for his discovery of the fission of heavy nuclei.

1945 ARTTURI ILMARI VIRTANEN (1895–1973), Finland, for his research and discoveries in the field of agricultural and nutrition chemistry, particularly his method of preserving animal fodder.

1946 JAMES BATCHELLER SUMNER (1887–1955), United States, for his discovery that enzymes can be crystallized.

JOHN HOWARD NORTHROP (1891–1987), United States, for the preparation of enzymes and virus proteins in pure form.

WENDELL MEREDITH STANLEY (1904–1971), United States, for the preparation of enzymes and virus proteins in pure form.

1947 ROBERT ROBINSON (1886–1975), Great Britain, for his investigations on plant products of biological importance, especially the alkaloids.

1948 ARNE WILHELM KAURIN TISELIUS (1902–1971), Sweden, for his research work on electrophoresis and on analysis by adsorption and, in particular, for his discoveries concerning the heterogeneous nature of the proteins of the serum.

1949 WILLIAM FRANCIS GIAUQUE (1895–1982), United States, for his contributions in the field of chemical thermodynamics, particularly concerning the behavior of substances at extremely low temperatures.

1950 OTTO PAUL HERMANN DIELS (1876–1954), Germany, for the discovery and development of the synthesis of dienes.

KURT ALDER (1902–1958), Germany, for the discovery and development of the synthesis of dienes.

1951 EDWIN MATTISON MCMILLAN (1907–1991), United States, for discoveries in the chemistry of the transuranium elements.

GLENN THEODORE SEABORG (1912–), United States, for discoveries in the chemistry of the transuranium elements.

1952 ARCHER JOHN PORTER MARTIN (1910–), Great Britain, for the invention of partition chromatography.

RICHARD LAURENCE MILLINGTON SYNGE (1914–), Great Britain, for the invention of partition chromatography.

1953 HERMANN STAUDINGER (1881–1965), Germany, for his discoveries in the field of macromolecular chemistry.

1954 LINUS CARL PAULING (1901–1994), United States, for his research into the nature of the chemical bond and its application to the structure of complex substances.

1955 VINCENT DU VIGNEAUD (1901–1978), United States, for his work on biochemically important sulfur compounds, and especially for the first synthesis of a polypeptide hormone.

1956 CYRIL NORMAN HINSHELWOOD (1897–1967), Great Britain, for research into the mechanics of chemical reactions.

NIKOLAY NIKOLAEVICH SEMENOV (1896–1986), USSR, for research into the mechanics of chemical reactions.

1957 ALEXANDER R. TODD (1907–), Great Britain, for his work on nucleotides and nucleotide coenzymes.

1958 FREDERICK SANGER (1918–), Great Britain, for his work on the structure of proteins, especially that of insulin.

1959 JAROSLAV HEYROVSKY (1890–1967), Czechoslovakia, for his discovery and development of the polarographic methods of analysis.

1960 WILLARD FRANK LIBBY (1908–1980), United States, for his method to use carbon-14 for age determination in archaeology, geology, geophysics, and other branches of science.

1961 MELVIN CALVIN (1911–), United States, for his research on the carbon dioxide assimilation in plants.

1962 MAX FERDINAND PERUTZ (1914–), Great Britain, for studies of the structures of globular proteins.

JOHN COWDERY KENDREW (1917–), Great Britain, for studies of the structures of globular proteins.

1963 KARL ZIEGLER (1898–1973), Germany, for discoveries in the field of chemistry and technology of high polymers.

GIULIO NATTA (1903–1979), Italy, for discoveries in the field of chemistry and technology of high polymers.

1964 DOROTHY CROWFOOT HODGKIN (1910–1994), Great Britain, for her determination by x-ray techniques of the structures of important biochemical substances.

1965 ROBERT BURNS WOODWARD (1917–1979), United States, for his achievements in the art of organic synthesis.

1966 ROBERT S. MULLIKEN (1896–1986), United States, for his fundamental work concerning chemical bonds and the electronic structure of molecules by the molecular orbital method.

1967 MANFRED EIGEN (1927–), West Germany, for studies of extremely fast chemical reactions.

RONALD GEORGE WREYFORD NORRISH (1897–1978), Great Britain, for studies of extremely fast chemical reactions.

GEORGE PORTER (1920–), Great Britain, for studies of extremely fast chemical reactions.

1968 LARS ONSAGER (1903–1976), United States, for the discovery of the reciprocal relations fundamental for the thermodynamics of irreversible processes.

1969 DEREK H. R. BARTON (1918–), Great Britain, for contributions to the development of the concept of conformation and its application in chemistry.

ODD HASSEL (1897–1981), Norway, for contributions to the development of the concept of conformation and its application in chemistry.

1970 LUIS F. LELOIR (1906–1987), Argentina, for his discovery of sugar nucleotides and their role in the biosynthesis of carbohydrates.

1971 GERHARD HERZBERG (1904–), Canada, for his contributions to the knowledge of electronic structure and geometry of molecules, particularly free radicals.

1972 CHRISTIAN B. ANFINSEN (1916–1995), United States, for his work on ribonuclease, especially concerning the connection between the amino acid sequence and the biologically active conformation.

STANFORD MOORE (1913–1982), United States, for contributing to the understanding of the connection between chemical structure and catalytic activity of the active center of the ribonuclease molecule.

WILLIAM H. STEIN (1911–1980), United States, for contributing to the understanding of the connection between chemical structure and catalytic activity of the active center of the ribonuclease molecule.

1973 ERNST OTTO FISCHER (1918–), West Germany, for pioneering work on the chemistry of the organometallic, so-called sandwich compounds.

GEOFFREY WILKINSON (1921–), Great Britain, for pioneering work on the chemistry of the organometallic, so-called sandwich compounds.

1974 PAUL J. FLORY (1910-1985), United States, for his fundamental achievements, both theoretical and experimental, in the physical chemistry of the macromolecules.

1975 JOHN WARCUP CORNFORTH (1917–), Australia, for his work on the stereochemistry of enzyme-catalyzed reactions.

VLADIMIR PRELOG (1906–), Switzerland, for his research into the stereochemistry of organic molecules and reactions.

1976 WILLIAM N. LIPSCOMB JR. (1919–), United States, for his studies on the structure of boranes illuminating problems of chemical bonding.

1977 ILYA PRIGOGINE (1917–), Belgium, for his contributions to nonequilibrium thermodynamics, particularly the theory of dissipative structures.

1978 PETER D. MITCHELL (1920–1992), Great Britain, for his contribution to the understanding of biological energy transfer through the formulation of the chemiosmotic.

1979 HERBERT C. BROWN (1912–), United States, for development of temporary chemical links for complex molecules.

GEORG WITTIG (1897–1987), Germany, for development of temporary chemical links for complex molecules.

1980 PAUL BERG (1926–), United States, for his fundamental studies of the biochemistry of nucleic acids, with particular regard to recombinant DNA.

WALTER GILBERT (1932–), United States, for contributions concerning the determination of base sequences in nucleic acids.

FREDERICK SANGER (1918–), Great Britain, for contributions concerning the determination of base sequences in nucleic acids.

1981   KENICHI FUKUI (1918–), Japan, for theories concerning the course of chemical reactions.

ROALD HOFFMANN (1937–), United States, for theories concerning the course of chemical reactions.

1982   AARON KLUG (1926–), Great Britain, for his development of crystallographic electron microscopy and his structural elucidation of biologically important nucleic acid–protein complexes.

1983   HENRY TAUBE (1915–), United States, for his studies of the mechanisms of electron transfer reactions, particularly of metal complexes.

1984   ROBERT BRUCE MERRIFIELD (1921–), United States, for his methodology for chemical synthesis on a solid matrix.

1985   HERBERT A. HAUPTMAN (1917–), United States, for development of direct methods for the determination of crystal structures.

JEROME KARLE (1918–), United States, for development of direct methods for the determination of crystal structures.

1986   DUDLEY R. HERSCHBACH (1932–), United States, for contributions concerning the dynamics of elementary chemical processes.

YUAN T. LEE (1936–), United States, for contributions concerning the dynamics of elementary chemical processes.

JOHN C. POLANYI (1929–), Canada, for contributions concerning the dynamics of elementary chemical processes.

1987   DONALD J. CRAM (1919–), United States, for work on synthetic molecules that can mimic vital chemical reactions of life processes.

JEAN-MARIE LEHN (1939–), France, for work on synthetic molecules that can mimic vital chemical reactions of life processes.

CHARLES J. PEDERSEN (1904–1989), United States, for work on synthetic molecules that can mimic vital chemical reactions of life processes.

1988   JOHANN DEISENHOFER (1943–), West Germany, for unraveling the three-dimensional structure of a complex of proteins essential to photosynthesis.

ROBERT HUBER (1937–), West Germany, for unraveling the three-dimensional structure of a complex of proteins essential to photosynthesis.

HARTMUT MICHEL (1948–), West Germany, for unraveling the three-dimensional structure of a complex of proteins essential to photosynthesis.

1989   SIDNEY ALTMAN (1939–), United States and Canada, for the discovery that RNA can act as an enzyme.

THOMAS R. CECH (1947–), United States, for the discovery that RNA can act as an enzyme.

1990   ELIAS J. COREY (1928–), United States, for the development of a technique of retrosynthetic analysis for synthesizing complex molecules on the basis of their natural organizing pattern.

1991   RICHARD R. ERNST (1933–), Switzerland, for his major role in applying nuclear magnetic resonance spectroscopy to chemical analysis.

1992   RUDOLPH A. MARCUS (1923–), United States, for his mathematical analysis of electron transfer between molecules in solutions.

1993   KARY B. MULLIS (1944–), United States, for inventing a technique for making many copies of a single gene fragment.

MICHAEL SMITH (1932–), Canada, for inventing a technique for splicing foreign components into gene molecules.

1994   GEORGE A. OLAH (1927–), United States, for the development of methods to form, stabilize, study, and recombine positively charged fragments of hydrocarbon molecules.

1995   PAUL CRUTZEN (1933–), Germany (Dutch citizen), for work in atmospheric chemistry, particularly concerning the formation and decomposition of ozone.

MARIO MOLINA (1943–), United States, for work in atmospheric chemistry, particularly concerning the formation and decomposition of ozone.

F. SHERWOOD ROWLAND (1927–), United States, for work in atmospheric chemistry, particularly concerning the formation and decomposition of ozone.

NOBEL PRIZE WINNERS IN PHYSICS

The year listed is that *for* which the prizes were awarded, not necessarily that *in* which the prizes were awarded. An asterick indicates that there is a separate biographical entry for a given prize winner. The nationality listed for each winner is that determined by the recipient at the time of the award.

1901 WILHELM CONRAD RÖNTGEN* (1845–1923), Germany, for the discovery of Röntgen rays (x rays).

1902 HENDRIK ANTOON LORENTZ* (1853–1928), Netherlands, for work on the influence of magnetism on radiation.
PIETER ZEEMAN (1865–1943), Netherlands, for work on the influence of magnetism on radiation.

1903 ANTOINE HENRI BECQUEREL (1852–1908), France, for his discovery of spontaneous radioactivity.
PIERRE CURIE (1859–1906), France, for study of radiation phenomena discovered by Becquerel.
MARIE SKLODOWSKA CURIE* (1867–1934), France, for study of radiation phenomena discovered by Becquerel.

1904 LORD RAYLEIGH [JOHN WILLIAM STRUTT] (1842–1919), Great Britain, for investigation of densities of gases and the discovery of argon.

1905 PHILIPP EDUARD ANTON VON LENARD (1862–1947), Germany, for his work with cathode rays.

1906 JOSEPH JOHN THOMSON* (1856–1940), Great Britain, for research on the conduction of electricity by gases.

1907 ALBERT ABRAHAM MICHELSON* (1852–1931), United States, for his optical precision instruments and the spectroscopic and metrological investigations carried out with their aid.

1908 GABRIEL LIPPMANN (1845–1921), France, for his method of reproducing colors photographically based on the phenomenon of interference.

1909 GUGLIELMO MARCONI (1874–1937), Italy, for contributions to the development of wireless telegraphy.
CARL FERDINAND BRAUN (1850–1918), Germany, for contributions to the development of wireless telegraphy.

1910 JOHANNES DIDERIK VAN DER WAALS (1837–1923), Netherlands, for work with the equation of state for gases and liquids.

1911 WILHELM WIEN (1864–1928), Germany, for his discoveries regarding the laws governing the radiation of heat.

1912 NILS GUSTAF DALÉN (1869–1937), Sweden, for his invention of automatic regulators for use in conjunction with gas accumulators for illuminating lighthouses and buoys.

1913 HEIKE KAMERLINGH ONNES (1853–1926), Netherlands, for his investigations on the properties of matter at low temperatures, which led to the production of liquid helium.

1914 MAX VON LAUE (1879–1960), Germany, for his discovery of diffraction of Röntgen rays by crystals.

1915 WILLIAM HENRY BRAGG (1862–1942), Great Britain, for analysis of crystal structure by means of x rays.
WILLIAM LAWRENCE BRAGG (1890–1971), Great Britain, for analysis of crystal structure by means of x rays.

1916 No award.

1917 CHARLES GLOVER BARKLA (1877–1944), Great Britain, for his discovery of the characteristic Röntgen radiation of the elements.

1918 MAX KARL ERNST LUDWIG PLANCK* (1858–1947), Germany, for his discovery of energy quanta.

1919 JOHANNES STARK (1874–1957), Germany, for his discovery of the Dopler effect in canal rays and the decomposition of spectral lines in electric fields.

1920 CHARLES ÉDOUARD GUILLAUME (1861–1938), Switzerland, for the discovery of anomalies in nickel steel alloys.

1921 ALBERT EINSTEIN* (1879–1955), Germany, for his services to theoretical physics, especially for his discovery of the law of the photoelectric effect.

1922 NIELS HENRIK DAVID BOHR* (1885–1962), Denmark, for his investigations of the structure of atoms and the radiation emanating from them.

1923 ROBERT ANDREWS MILLIKAN* (1868–1953), United States, for his work on the elementary charge of electricity and on the photoelectric effect.

1924 KARL MANNE GEORG SIEGBAHN (1886–1978), Sweden, for his discoveries and research in the field of x-ray spectroscopy.

1925 JAMES FRANCK (1882–1964), Germany, for the discovery of the laws governing the impact of an electron on an atom.
GUSTAV HERTZ (1887–1975), Germany, for the discovery of the laws governing the impact of an electron on an atom.

1926 JEAN BAPTISTE PERRIN (1870–1942), France, for his work on the discontinuous structure of matter, and especially for his discovery of sedimentation equilibrium.

1927 ARTHUR HOLLY COMPTON* (1892–1962), United States, for his discovery of the Compton effect.
CHARLES THOMSON REES WILSON (1869–1959), Great Britain, for his method of making the paths of electrically charged particles visible by condensation of vapor.

1928 OWEN WILLANS RICHARDSON (1879–1959), Great Britain, for work on the thermionic phenomenon and the Richardson law.

1929 LOUIS-VICTOR-PIERRE-RAYMOND DE BROGLIE* (1892–1987), France, for his discovery of the wave nature of electrons.

1930 CHANDRASEKHARA VENKATA RAMAN (1888–1970), India, for his work on the scattering of light and for the discovery of the Raman effect.

1931 No award.

1932 WERNER KARL HEISENBERG* (1901–1976), Germany, for the creation of quantum mechanics.

1933 ERWIN SCHRÖDINGER* (1887–1961), Austria, for the discovery of new productive forms of atomic theory.
PAUL ADRIEN MAURICE DIRAC* (1902–1984), Great Britain, for the discovery of new productive forms of atomic theory.

1934 No award.

1935 JAMES CHADWICK* (1891–1974), Great Britain, for the discovery of the neutron.

1936 VICTOR FRANZ HESS (1883–1964), Austria, for the discovery of cosmic radiation.
CARL DAVID ANDERSON (1905–1991), United States, for the discovery of the positron.

1937 CLINTON JOSEPH DAVISSON (1881–1958), United States, for the discovery of the diffraction of electrons by crystals.
GEORGE PAGET THOMSON (1892–1975), Great Britain, for the discovery of the diffraction of electrons by crystals.

1938 ENRICO FERMI* (1901–1954), Italy, for his demonstrations of the existence of new ra-dioactive elements produced by neutron irradiation and for his related discovery of nuclear reactions brought about by slow neutrons.

1939 ERNEST ORLANDO LAWRENCE* (1901–1958), United States, for the invention and development of the cyclotron, and for results obtained with it, especially with regard to artificial radioactive elements.

1940 No award.

1941 No award.

1942 No award.

1943 OTTO STERN (1888–1969), United States, for his contribution to the development of the molecular-ray method and his discovery of the magnetic moment of the proton.

1944 ISIDOR ISAAC RABI* (1898–1988), United States, for his resonance method for recording the magnetic properties of atomic nuclei.

1945 WOLFGANG PAULI* (1900–1958), Austria, for the discovery of the Pauli exclusion principle.

1946 PERCY WILLIAMS BRIDGMAN (1882–1961), United States, for the invention of an apparatus to provide extremely high pressures and for the discoveries he made therewith in the field of high-pressure physics.

1947 EDWARD VICTOR APPLETON (1892–1965), Great Britain, for his investigations of the physics of the upper atmosphere, especially for the discovery of the Appleton layer.

1948 PATRICK MAYNARD STUART BLACKETT (1897–1974), Great Britain, for his development of the Wilson cloud-chamber method and his discoveries therewith in the fields of nuclear physics and cosmic radiation.

1949 HIDEKI YUKAWA* (1907–1981), Japan, for his prediction of the existence of mesons, based on theoretical work on nuclear forces.

1950 CECIL FRANK POWELL (1903–1969), Great Britain, for his development of the photographic method of studying nuclear processes and his discoveries regarding mesons made with this method.

1951 JOHN DOUGLAS COCKCROFT (1897–1967), Great Britain, for pioneer work on the transmutation of atomic nuclei by artificially accelerated atomic particles.
ERNEST THOMAS SINTON WALTON (1903–1995), Ireland, for pioneer work on the transmutation of atomic nuclei by artificially accelerated atomic particles.

1952   FELIX BLOCH (1905–1983), United States, for the development of new methods for nuclear magnetic precision measurements and related discoveries.

EDWARD MILLS PURCELL (1912–   ), United States, for the development of new methods for nuclear magnetic precision measurements and related discoveries.

1953   FRITS [FREDERIK] ZERNIKE (1888–1966), Netherlands, for his demonstration of the phase-contrast method, especially for his invention of the phase-contrast microscope.

1954   MAX BORN* (1882–1970), Great Britain, for his fundamental research in quantum mechanics, especially for his statistical interpretation of the wave function.

WALTHER BOTHE (1891–1957), Germany, for the coincidence method and his related discoveries.

1955   WILLIS EUGENE LAMB (1913–   ), United States, for his discoveries concerning the fine structure of the hydrogen spectrum.

POLYKARP KUSCH (1911–1993), United States, for his precision determination of the magnetic moment of the electron.

1956   WILLIAM SHOCKLEY (1910–1989), United States, for research on semiconductors and the discovery of the transistor effect.

JOHN BARDEEN* (1908–1991), United States, for research on semiconductors and the discovery of the transistor effect.

WALTER HOUSER BRATTAIN (1902–1987), United States, for research on semiconductors and the discovery of the transistor effect.

1957   CHEN NING YANG (1922–   ), China, for the investigation of the parity law, which led to important discoveries regarding the elementary particles.

TSUNG-DAO LEE (1926–   ), China, for the investigation of the parity law, which led to important discoveries regarding the elementary particles.

1958   PAVEL A. CHERENKOV (1904–   ), USSR, for the discovery and interpretation of the Cherenkov effect.

ILYA M. FRANK (1908–1990), USSR, for the discovery and interpretation of the Cherenkov effect.

IGOR TAMM (1895–1971), USSR, for the discovery and interpretation of the Cherenkov effect.

1959   EMILIO GINO SEGRÈ (1905–1989), United States, for the discovery of the antiproton.

OWEN CHAMBERLAIN (1920–   ), United States, for the discovery of the antiproton.

1960   DONALD A. GLASER (1926–   ), United States, for the invention of the bubble chamber to study elementary particles.

1961   ROBERT HOFSTADTER (1915–1990), United States, for his pioneering studies of electron scattering in atomic nuclei and for the resulting discoveries concerning the structure of the nucleons.

RUDOLF L. MÖSSBAUER (1929–   ), Germany, for his researches concerning the resonance absorption of gamma radiation and his discovery in this connection of the Mössbauer effect.

1962   LEV DAVIDOVICH LANDAU* (1908–1968), USSR, for his pioneering theories for condensed matter, especially liquid helium.

1963   EUGENE P. WIGNER (1902–   ), United States, for his contributions to the theory of the atomic nucleus and the elementary particles, particularly through the discovery and application of fundamental symmetry principles.

MARIA GOEPPERT MAYER* (1906–1972), United States, for discoveries concerning nuclear shell structure.

J. HANS D. JENSEN (1907–1973), Germany, for discoveries concerning nuclear shell structure.

1964   CHARLES HARD TOWNES (1915–   ), United States, for fundamental work in the field of quantum electronics, which led to the construction of oscillators and amplifers based on the maser-laser principle.

NIKOLAI G. BASOV (1922–   ), USSR, for fundamental work in the field of quantum electronics, which led to the construction of oscillators and amplifers based on the maser-laser principle.

ALEKSANDR M. PROCHOROV (1916–   ), USSR, for fundamental work in the field of quantum electronics, which led to the construction of oscillators and amplifers based on the maser-laser principle.

1965   SIN-ITIRO TOMONAGA (1906–1979), Japan, for fundamental work in quantum electrodynamics.

JULIAN SCHWINGER (1918–   ), United States, for fundamental work in quantum electrodynamics.

RICHARD PHILLIPS FEYNMAN* (1918–1988), United States, for fundamental work in quantum electrodynamics.

1966 ALFRED KASTLER (1902–1984), France, for the discovery and development of optical methods for studying Hertzian resonances in atoms.

1967 HANS ALBRECHT BETHE (1906–   ), United States, for his contributions to the theory of nuclear reactions, especially his discoveries concerning the energy production of stars.

1968 LUIS W. ALVAREZ (1911–1988), United States, for his decisive contributions to elementary particle physics, in particular the discovery of a large number of resonance states.

1969 MURRAY GELL-MANN (1929–   ), United States, for his contributions and discoveries concerning the classification of elementary particles and their interactions.

1970 HANNES ALFVÉN (1908–1995), Sweden, for fundamental work and discoveries in magnetohydrodynamics with applications in different parts of plasma physics.
LOUIS NÉEL (1904–   ), France, for fundamental work and discoveries concerning antiferromagnetism and ferrimagnetism, which have important applications in solid-state physics.

1971 DENNIS GABOR (1900–1979), Great Britain, for his invention and development of the holographic method.

1972 JOHN BARDEEN* (1908–1991), United States, for the development of the Bardeen–Cooper–Schrieffer theory of superconductivity.
LEON N. COOPER (1930–   ), United States, for the development of the Bardeen–Cooper–Schrieffer theory of superconductivity.
JOHN ROBERT SCHRIEFFER (1931–   ), United States, for the development of the Bardeen–Cooper–Schrieffer theory of superconductivity.

1973 LEO ESAKI (1925–   ), Japan, for experimental discoveries regarding tunneling phenomena in semiconductors and superconductors.
IVAR GIAEVER (1929–   ), United States, for experimental discoveries regarding tunneling phenomena in semiconductors and superconductors.
BRIAN D. JOSEPHSON (1940–   ), Great Britain, for his theoretical predictions of the properties of a supercurrent through a tunnel barrier, in particular those phenomena that are generally known as the Josephson effect.

1974 MARTIN RYLE (1918–1984), Great Britain, for pioneering research in radio astrophysics and for his observations and inventions, in particular in the aperture synthesis technique.
ANTONY HEWISH (1924–   ), Great Britain, for pioneering research in radio astrophysics, particularly his decisive role in the discovery of pulsars.

1975 AAGE BOHR (1922–   ), Denmark, for the discovery of the connection between collective motion and particle motion in atomic nuclei and the development of the theory of the structure of the atomic nucleus based on this connection.
BEN R. MOTTELSON (1926–   ), Denmark, for the discovery of the connection between collective motion and particle motion in atomic nuclei and the development of the theory of the structure of the atomic nucleus based on this connection.
JAMES RAINWATER (1917–1986), United States, for the discovery of the connection between collective motion and particle motion in atomic nuclei and the development of the theory of the structure of the atomic nucleus based on this connection.

1976 BURTON RICHTER (1931–   ), United States, for pioneering work in the discovery of a heavy elementary particle of a new kind (J/psi).
SAMUEL C. C. TING (1936–   ), United States, for pioneering work in the discovery of a heavy elementary particle of a new kind (J/psi).

1977 PHILIP W. ANDERSON (1923–   ), United States, for fundamental theoretical investigations of the electronic structure of magnetic and disordered systems.
NEVILL F. MOTT (1905–   ), Great Britain, for fundamental theoretical investigations of the electronic structure of magnetic and disordered systems.
JOHN H. VAN VLECK (1899–1980), United States, for fundamental theoretical investigations of the electronic structure of magnetic and disordered systems.

1978 PYOTR LEONIDOVICH KAPITSA (1894–1984), USSR, for his basic inventions and discoveries in the area of low-temperature physics.

ARNO A. PENZIAS (1933–), United States, for the discovery of cosmic microwave background radiation.

ROBERT W. WILSON (1936–), United States, for the discovery of cosmic microwave background radiation.

1979 SHELDON L. GLASHOW (1932–), United States, for contributions to the theory of the unified weak and electromagnetic interaction between elementary particles.

ABDUS SALAM (1926–), Pakistan, for contributions to the theory of the unified weak and electromagnetic interaction between elementary particles.

STEVEN WEINBERG (1933–), United States, for contributions to the theory of the unified weak and electromagnetic interaction between elementary particles.

1980 JAMES W. CRONIN (1931–), United States, for the discovery of violations of fundamental symmetry principles in the decay of neutral K-mesons.

VAL L. FITCH (1923–), United States, for the discovery of violations of fundamental symmetry principles in the decay of neutral K-mesons.

1981 NICOLAAS BLOEMBERGEN (1920–), United States, for contributions to the development of laser spectroscopy.

ARTHUR L. SCHAWLOW (1921–), United States, for contributions to the development of laser spectroscopy.

KAI M. SIEGBAHN (1918–), Sweden, for his contribution to the development of high-resolution electron spectroscopy.

1982 KENNETH G. WILSON (1936–), United States, for his theory for critical phenomena in connection with phase transitions.

1983 SUBRAHMANYAN CHANDRASEKHAR (1910–1995), United States, for his theoretical studies of the physical processes of importance to the structure and evolution of the stars.

WILLIAM A. FOWLER (1911–), United States, for the theoretical and experimental studies of the nuclear reactions of importance in the formation of the chemical elements of the universe.

1984 CARLO RUBBIA (1934–), Italy, for decisive contributions that led to the discovery of the field particles W and Z, communicators of the weak interaction.

SIMON VAN DER MEER (1925–), Netherlands, for decisive contributions that led to the discovery of the field particles W and Z, communicators of the weak interaction.

1985 KLAUS VON KLITZING (1943–), West Germany, for the creation of economic models and their application to the analysis of economic fluctuations and economic policies.

1986 ERNST RUSKA (1906–1988), West Germany, for his fundamental work in electron optics and for the design of the first electron microscope.

GERD BINNIG (1947–), West Germany, for the design of the scanning tunneling microscope.

HEINRICH ROHRER (1933–), Switzerland, for the design of the scanning tunneling microscope.

1987 J. GEORG BEDNORZ (1950–), West Germany, for the discovery of high-temperature superconductors.

KARL ALEXANDER MÜLLER (1927–), Switzerland, for the discovery of high-temperature superconductors.

1988 LEON M. LEDERMAN (1922–), United States, for pioneering work in creating beams of neutrinos for use as research tools.

MELVIN SCHWARTZ (1932–), United States, for pioneering work in creating beams of neutrinos for use as research tools.

JACK STEINBERGER (1921–), United States, for pioneering work in creating beams of neutrinos for use as research tools.

1989 NORMAN F. RAMSEY (1915–), United States, for methods of studying the structure of atoms.

HANS G. DEHMELT (1922–), United States, for methods of studying the structure of atoms.

WOLFGANG PAUL (1913–1993), West Gemany, for methods of studying the structure of atoms.

1990 JEROME I. FRIEDMAN (1930–), United States, for experimental verification of the existence of quarks.

HENRY W. KENDALL (1926–), United States, for experimental verification of the existence of quarks.

RICHARD E. TAYLOR (1929–), Canada, for experimental verification of the existence of quarks.

1991 PIERRE-GILLES DE GENNES (1932–), France, for his mathematical discoveries about the ordering of molecules, which provide a general physical description of many different phenomena.

1992 GEORGES CHARPAK (1924–), France, for the development of the multiwire proportional-chamber particle detector.

1993 RUSSELL A. HULSE (1950–), United States, for investigation of gravitational forces exerted by pulsars.

JOSEPH H. TAYLOR (1941–), United States, for investigation of gravitational forces exerted by pulsars.

1994 BERTRAM N. BROCKHOUSE (1918–), Canada, for the development of neutron-scattering techniques for exploring the atomic structure of matter.

CLIFFORD G. SHULL (1915–), United States, for the development of neutron-scattering techniques for exploring the atomic structure of matter.

1995 MARTIN L. PERL (1927–), United States, for the discovery of the tau lepton.

FREDERICK REINES (1918–), United States, for the detection of the neutrino.

NOISE

Noise N is an unwanted component of a signal S that does not convey information about the phenomena being studied. It is present in all signals and can have many sources. Because of noise the measured signal M is actually a sum of signal S and noise N. Noise prevents perfection of the measuring process.

A recording of a bird song might contain static—electrical noise created by the recording process and hiss—acoustical noise caused by background wind. A photographic image of a bug might show fog, excessive film grain, and other defects. A CAT scan displayed on a television screen might be plagued with "snow." A needle deflection on a meter used to indicate a voltage, brightness, temperature, weight or speed, might quiver slightly about an average non-

zero value. Some of the clicks from a Geiger counter might be caused by cosmic rays or electrical spikes and not radioactive events. The counting procedure used to measure the number of fish in a pond, trees in a forest, lightning flashes during a thunderstorm might count events not related to the signal.

The static and hiss, fog, film grain, extra clicks and counts, nonzero values and oscillations are called noise N. Whenever noise is present the measured value M is not the true signal S, it is a mixture of the signal S plus noise N. $M = (S + N)$. If we know N we can find $S = (M - N)$

Noise prevents an exact determination of the physical quantity being measured—it can introduce confusion (uncertainty) and cause errors. However, noise itself is neither uncertainty nor error. Sometimes the nature and amount of noise N can be determined exactly by measuring the "signal" when no signal S is present. Then $M = (S - N) = N$ when $S = 0$.

A tape recording of absolute silence will never be perfectly quiet. It will produce hiss, snap, crackle, pop, hum, and rumble. A photograph or a picture on a television screen, taken in complete darkness will never be absolutely black. It will show background fog, snow, and other specks of white. A needle on a completely disconnected meter will never read exactly zero or be perfectly stationary. It will have a small residual nonzero value and perhaps oscillate slightly due to residual electrical and thermal effects. A Geiger counter completely shielded from the outside world will not give zero counts. Residual radioactivity in the instrument itself and electrical effects will create extraneous counts. All measurement hardware is constructed of matter. Because of its discrete nature and its thermal energy at all temperatures above absolute zero, matter is noisy.

Thus these "zero signals," signals obtained from measurements made when the signal S is absent, are called noise N. The amount of noise N present in the measured signal M depends on the sensitivity and quality of the measuring equipment and on the measurement process. Noise can be time dependent, varying slowly or rapidly with time. It can be constant, continuous, periodic, erratic, larger than the signal to be measured, too small to matter or even too small to be measured with the instrument at hand. Noise signals, if detectable, should be measured carefully to determine their source, magnitude and time dependence. Measured noise N can be subtracted from the measured signal M to get the true signal $S = (M - N)$.

Bibliography

DERENIAK, E. L., and CROWE, D. G. *Optical Radiation Detectors* (Wiley, New York, 1984).

GRUM, F., and BECHERER, R. S. *Optical Radiation Measurements,* Vol. 1 (Academic Press, New York, 1979).

WILLIAM BICKEL

NONLINEAR PHYSICS

Forecasts for next week's weather are generally unreliable; hexagonal patterns form on the surface of a gently heated pot of tomato soup; some forms of fibrillating heart beats can be controlled by gentle electrical pulses; waves at an ocean beach crest and then crash. All of these phenomena can be understood by the recently developed branch of science known as nonlinear dynamics, or more generally, nonlinear science. In some ways nonlinear science is as old as physics. Nonlinear effects have been recognized since the time of Isaac Newton, and the French mathematician Henri Poincaré developed many mathematical tools for dealing with nonlinear systems in the late 1800s. However, the field progressed very slowly until computers and, in particular, computer graphics became widely available. After a brief introduction to the notion of nonlinearity, we will return to the question of why computers are so important in this field.

First we need to explain what nonlinear means in science. There are two ways to think about this. First, and perhaps more simply, we can think of the response of a system to a stimulus. Suppose we double the strength of the stimulus. If the response doubles in size, we say that the behavior of the system is linear. If the response to the doubled stimulus is either larger or smaller than twice the original response, we say that the behavior is nonlinear. Nonlinearity manifests itself in many ways. For example, nonlinearity in a stereo music system leads to distortion and "noise" in the sound when the input (stimulus) becomes too large, an effect often heard at loud rock concerts. Another example: your own vision becomes "bleached" and washed out if too much light enters your eye.

The second way of describing nonlinearity is more formal and looks at the nature of the mathematical equations used to describe the behavior of the system. In physics we use a mathematical model to describe how the variables that characterize the system change with time. For example, in describing the motion of a planet around the Sun, we might use the planet's position and velocity relative to the Sun as the appropriate variables. If the equations describing how those variables change in time involve the variables (and their time derivatives, using the language of calculus) just to the first power then we say that the mathematical model is linear. However, if the model involves the variables (other than time t itself) in some other mathematical form such as x^2 or $\sin x$, then we say that the model is nonlinear.

Scientists now recognize that most systems in nature are nonlinear. In fact, engineers and biologists have had to deal with nonlinearity all the time. Why this belated recognition, particularly in the basic physical sciences? The reason is primarily mathematical. We know that the equations used to describe nonlinear systems cannot generally be solved in terms of simple functions or formulas. On the other hand, the equations describing linear systems can usually be solved in what mathematicians call closed form. For example, the standard textbook example of a mass oscillating on a spring whose force varies linearly with distance gives the position $y(t)$ of the mass as $y(t) = A \sin 2\pi ft$, where A is the amplitude of the motion and f is the frequency (number of cycles per second). For most nonlinear models, no such formula exists. The lack of closed-form solutions means that each nonlinear problem needs to be treated by special methods often involving numerical computations, not readily done before the widespread availability of computers. Moreover, even the string of numbers from a numerical calculation does not often reveal in a simple way the important features of nonlinear behavior. With the advent of easy-to-use computer graphics, we are now able to see, both literally and figuratively, the similarities in the nonlinear behavior of many different kinds of systems. In fact, these common features are the hallmark of the contemporary study of nonlinear science. The kinds of nonlinear behavior seen in physical systems such as lasers and vibrating metal ribbons are the same as those seen in oscillating chemical reactions and the irregular beats of a malfunctioning heart. The contemporary study of nonlinear phenomena has focused on understanding and characterizing these common features and determining the extent and limitations of their universality and predictability.

Why fuss over the distinction between linear and nonlinear? In some sense a linear system is boring: no matter what kind of stimulus it receives, its response is either a larger or smaller version of the same kind of behavior. On the other hand, nonlinear systems often have dramatic changes in their behavior as the size of the stimulus changes. In some cases, their behavior can change from being very simple and regular to being very complex (what we now call chaotic behavior) with just a small change in the size of the stimulus. In other cases, nonlinear systems can "spontaneously" generate intricate ordered patterns even when none are present in the stimulus.

The most surprising feature of nonlinear systems is the type of behavior called chaos. Although it is only one of many types of nonlinear behavior, chaos has caught the attention of both scientists and the general public. When a system is behaving chaotically, its variables change in time in an apparently random, irregular fashion. Traditionally, scientists would attribute such irregular behavior to either noise or complexity. Noise means that the system is subject to many irregular outside influences (changes in temperature, air pressure, magnetic fields, building vibrations, and so on) that are essentially uncontrolled and perhaps uncontrollable. The net result of this external jostling is irregular behavior, like a tree limb buffeted by random gusts of wind. The other possible cause of irregular behavior might be the complexity of the system itself, made of many parts, each doing its own thing with consequently complex, apparently random behavior. (The irregular rise and fall of prices on the stock market is often described as the result of the independent decisions by millions of investors. We then attribute the random behavior of the stock prices to the complexity of the stock market.)

However, the study of nonlinearity has revealed another important cause of irregular behavior: the nonlinear behavior of only a few independent variables. The traditional prejudice was that simple systems (those with only a few independent variables) behave simply. In fact, we now know that it is possible for simple nonlinear systems to have behavior that is as complex as that of systems with thousands of variables. What is even more surprising is that systems whose behavior is chaotic seem to undermine the most basic goal of science: predicting the future behavior of a system based on what we know about its current state and the fundamental principles that

describe its behavior. Chaotic systems display what is called sensitive dependence on initial conditions: Even the slightest change in the current state of the system—perhaps due just to the flapping of a butterfly's wings—may quickly lead to dramatically different behavior. Under these conditions long-term prediction becomes almost impossible.

An Example

The best way to appreciate the novelty and power of nonlinear science is to explore a simple example: Suppose that we wish to understand how the population of some biological species such as mayflies, which are born and die within a day, varies from year to year in a specific location. Observations show that the number of mayflies born in a particular year increases and decreases in an apparently irregular fashion. We want to see if we can understand those variations. We will develop a simple mathematical model, which incorporates some basic ideas of how the population of mayflies changes with time.

We will build the model in steps. First, let us assume that the number of mayflies born in a particular year, say 1995, denoted by N_{1995}, depends only on the previous year's number of mayflies, denoted by N_{1994}. To make a formal statement, we say that the numbers are related as $N_{1995} = AN_{1994}$, $N_{1996} = AN_{1995}$, and so on. A is a number telling us something about the number of mayflies produced by the previous year's population. If $A < 1$, the number of mayflies will steadily decrease as the years go by. If $A > 1$, the population will increase. The parameter A is a simple way of incorporating information on the food supply, the weather, the number of predators, and the other factors that affect the population growth (or decline) of the mayflies. We generalize this model for the population in any year n by writing

$$N_{n+1} = AN_n. \qquad (1)$$

This is obviously a very simple model, but illustrates the basic idea of modeling the year-to-year changes in the population. Using the terminology introduced earlier, we say that Eq. (1) is a linear model, since the variable N appears to the first power in our model. For this linear model, the possible kinds of change for the population are quite limited: Assuming that the environment does not change from year-to-year (that is, the number A stays fixed), then the population either always increases or always decreases or stays the same (if $A = 1$). Not terribly interesting. Thus, if we want to describe more complicated behavior, we need a somewhat more complicated model.

As a first step in developing a more complicated model, we might try to include the notion that the population of mayflies cannot increase forever. Eventually the mayflies deplete the food resources, or predators find the large number of flies an attractive food resource, just to cite two possible effects that might limit the population. How do we include this idea mathematically? One possible model (and perhaps the simplest) would be to include a mathematical term that tends to decrease the population if the population gets too large. We could do that by writing

$$N_{n+1} = AN_n - BN_n^2. \qquad (2)$$

Note that the N^2 term in Eq. (2) will tend to decrease the population. If the parameter B is much smaller than A, then this term will have an effect only when N gets relatively large. Before analyzing the predictions of our extended model, let us put it into a somewhat simplified mathematical form. First, we note that the model has a maximum allowed population

$$N^{max} = A/B. \qquad (3)$$

If N exceeds this value, our model predicts a negative value for the number of may flies, clearly an absurdity. Furthermore, if N_n is less than this value, then it can readily be shown that N_{n+1} must also be less than this value. (In the special case $N_n = N^{max}$, the population in the next generation goes to zero and remains at zero thereafter, a dramatic reminder of the dangers of overpopulation.) It is then useful to divide both sides of Eq. (2) by N^{max} and to introduce a new variable

$$x_n = N_n/N^{max}, \qquad (4)$$

which gives the population relative to the maximum population allowed by the model. The variable x is then limited to lie between 0 and 1. With these changes, our model equation can be rewritten as

$$x_{n+1} = Ax_n(1 - x_n). \qquad (5)$$

Equation (5) is known as the logistic equation and has served as one of the best-known paradigms in the contemporary study of nonlinear science.

Now the prediction of the population from year to year is relatively simple. We start off with some initial value of the population x_0 in the year we call 0. (Of course, x_0 must be between 0 and 1.) We then insert that value into Eq. (5), multiply by the value for A to get the population x_1 for year 1. We then use that value of x_1 in Eq. (5) to find x_2 and so on. This repeated process is called iteration. The sequence of relative population values x_0, x_1, x_2, and so on, is sometimes called the trajectory or orbit in analogy to the sequence of position values that make up a planet's orbit.

What does our new model predict for the population changes from one year to the next? We answer that question by choosing a value for the parameter A, selecting some initial population, and then repeatedly using Eq. (5). (You can easily do this yourself with a pocket calculator or with a simple computer program.) For $A < 1$, we find that the sequence of x values tends toward 0 for any starting value of x between 0 and 1. The population dies out just as it did for the linear model with $A < 1$. For A between 1 and 3 the sequence of values always seems to approach the numerical value $x = 1 - 1/A$. For example, for $A = 2$ and starting with $x_0 = 0.1$, we get the sequence 0.100, 0.180, 0.295, 0.486, 0.499, ..., which rapidly approaches the value 1/2. In fact, starting with any x_0 value between 0 and 1 gives a sequence that approaches the value 1/2. We see that our nonlinear model gives a somewhat more satisfactory result than the linear model. At least for some values of A (corresponding to appropriate environmental conditions) the population settles down to a steady value, which is then maintained year after year.

But there are some surprises in store. If we choose a value of A just slightly larger than 3, we get some strange behavior. Figure 1 shows the trajectory for $A = 3.2$ and $x_0 = 0.1$. The population, after an

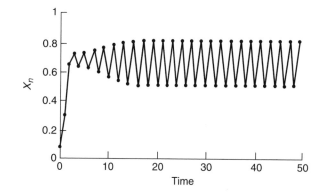

Figure 1 The sequence of x values with $A = 3.2$.

initial rise, seems to settle into oscillations between a higher value and a lower value. The population is high one year and low the next, and then high again. This sudden change in behavior, when A changes from just less than 3 to just greater than 3, is called a bifurcation. It turns out that any other starting value of x_0, as long as A stays the same, leads to the same kind of oscillation. This oscillation should be quite surprising. The environment, as modeled by the parameter A, is absolutely stable. Yet the population does not settle down to a stable value; it oscillates. This kind of spontaneous oscillation is one signature of nonlinearity.

Another surprise occurs when A is just larger than 3.4. We find that the oscillations become more complicated with the population cycling among a sequence of four different values. Figure 2 shows this behavior with $A = 3.5$. The change from two values to four values is called a period-doubling bifurcation because the pattern now takes four years (in our model's time) to go through the full sequence while

for A between 3.0 and 3.4 it takes just two years. The repetition period has doubled.

As a final example of the surprises contained in our simple model, let's try $A = 3.8$. Figure 3 shows that the population never seems to settle down, at least over a fifty-year time span. In fact, if you calculate the sequence for as long as you want, the population never settles down and never repeats itself exactly. This kind of behavior is called chaos, which is without a doubt the most important surprise that nonlinear science has brought to scientists and mathematicians. Understanding how a very simple model can lead to such complex behavior has been one of the goals of the contemporary study of nonlinear science. What is even more surprising is that many real physical, chemical, and biological systems exhibit bifurcations, period doublings, and chaotic behavior, just like the behavior seen in our simple model.

All of this behavior can be conveniently summarized in a so-called bifurcation diagram as shown in Fig. 4. There we plot, for each value of the parameter A, the sequence of x values that occurs after initial transients die away. In the region with A between 2.5 and 3.0, for example, for each value of A there is only one x value plotted, indicating that the population has reached a single steady-state value. Between 3.0 and 3.44, we get two x values, indicating an oscillating, period-two behavior. Near $A = 3.5$, there are four x values, corresponding to the trajectory shown in Fig. 2. The regions with a smear of dots have chaotic behavior. The bifurcation diagram is one example of the complex and intriguing patterns that can be produced by nonlinear systems.

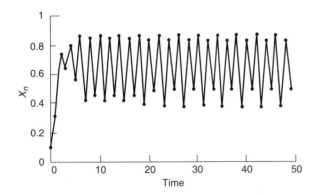

Figure 2 The sequence of x values with $A = 3.5$.

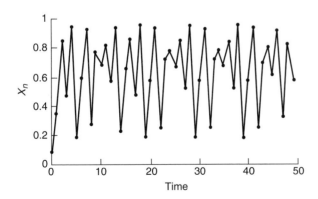

Figure 3 The sequence of x values with $A = 3.8$.

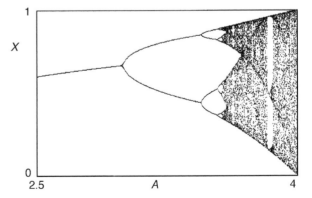

Figure 4 The bifurcation diagram for the logistic model. For each value of A, fifty values of the trajectory of x are plotted (initial transients are ignored).

Career Issues

As the introduction to this entry indicates, nonlinear physics is important to all branches of science and engineering and to all the subdisciplines of physics. Career opportunities are equally broad. Most physicists working in nonlinear science pursue a broad training in physics and mathematics with specialized work in the subdiscipline in which they are applying the ideas of nonlinear science.

See also: CHAOS; OPTICS, NONLINEAR; OSCILLATION

Bibliography

CRUTCHFIELD, J. P.; FARMER, J. D.; PACKARD, N. H.; and SHAW, R. S. "Chaos." *Sci. Am.* **245,** 45–57 (1986).

DEVANEY, R. L. *Chaos, Fractals, and Dynamics* (Addison-Wesley, Reading, MA, 1990).

GLEICK, J. *Chaos: Making a New Science* (Viking, New York, 1987).

KAUFFMAN, S. *At Home in the Universe: The Search for the Laws of Self-Organization and Complexity* (Oxford University Press, New York, 1995).

PEAK, D., and FRAME, M. *Chaos Under Control: The Art and Science of Complexity* (W. H. Freeman, New York, 1994).

STEWART, I. *Does God Play Dice? The Mathematics of Chaos* (Blackwell, New York, 1989).

ROBERT C. HILBORN

NORMAL FORCE

Why do we not fall through the floor? Conceptually, a question like this can be addressed by considering the atoms in each object (our feet and the floor) and their resistance to atomic compression in light of the uncertainty principle. However, to answer a question such as this, we must carefully analyze the quantum mechanical interactions between our feet and the floor—a difficult task even for most sophisticated computers. Thus, for simplicity, physicists have defined a macroscopic force, called the normal force, that encompasses all of these complex atomic interactions and describes their collective effect.

The normal force is so named because it acts perpendicular (or normal) to an object's surface. It is a contact force, in that the normal force exists only if two objects are in contact with each other. In our example, the floor exerts a normal force on our feet in a direction perpendicular to the surface of the floor. Similarly, because of Newton's third law, our feet exert a normal force on the floor, perpendicular to the surface of our feet.

The normal force depends upon other forces. When a 5 kg book rests on a flat, horizontal table (located at sea level on Earth), the table exerts a normal force of 49 N on the book. However, if we also push downward on the book with a force of 20 N, the normal force exerted by the table on the book compensates by increasing to 69 N.

One of the most common encounters with the normal force occurs, for example, when we step on a bathroom scale to determine our weight; we are actually using the scale to measure the normal force exerted on us by the scale. This normal force, which just happens to be equal to our weight, prevents us from "falling" through the scale. In fact, an object is colloquially said to be weightless when the normal force between it and another object goes to zero. For example, if a table with a book resting on it is dropped from a 10-m tower, the normal force between the book and table drops to 0 N—we say the book and table are in free fall. Similarly, astronauts orbiting Earth in the space shuttle are in free fall. Thus, if we measured the normal force experienced by astronauts "sitting" in their seats while in orbit, we would measure a normal force of 0 N.

The normal force is frequently used in calculating the *g*'s experienced by a pilot performing an acrobatic maneuver in an aircraft. For example, today's typical high-performance jet aircraft can sustain 8 to 9 *g*'s. However, the typical pilot can only withstand 3 to 4 sustained *g*'s without special equipment or physical preparation. Thus, it is extremely important for a pilot to know how many *g*'s he or she is pulling. The number of *g*'s experienced by a pilot is equal to the normal force acting on the pilot divided by the pilot's weight. In a typical inverted loop [radius = 2,700 ft (825 m)], an F-16 pilot will attempt to complete the maneuver with an air speed of 350 kn (180 m/s) at the bottom of the loop. At that point, the pilot is experiencing 5 *g*'s. In other words, the normal force on the pilot has increased to five times the pilot's weight.

See also: FORCE; FREE FALL; LIGHT; NEWTON'S LAWS; WEIGHT; WEIGHTLESSNESS

Bibliography

Feynman, R. P.; Leighton, R. B.; and Sands, M. *The Feynman Lectures on Physics,* Vol. 3 (Addison-Wesley, Reading, MA, 1965).

Rolf C. Enger

NOVA

A classical nova is a variable star characterized by a very rapid rise in brightness followed by a much slower decline to quiescence. It is a member of the class of cataclysmic variables. They are given this name because the outburst leaves the underlying stellar system undamaged and it can repeat after an appropriate interval of time. Other types of cataclysmic variables are dwarf novas, recurrent novas, and symbiotic novas. Some types of dwarf novas have extremely strong magnetic fields and are also called polars. The violence of the outburst increases from dwarf novas to classical novas.

A cataclysmic variable is a close binary stellar system in which one star orbits the other under the mutual influence of their gravity. The orbital periods of these systems range from 90 min to a few hours. Applying Kepler's Laws tells us, therefore, that the two stars are very close together. In some systems, such as GQ Mus, the entire binary could almost fit inside the Sun. One star is very much like the Sun both in size and internal structure. The other star is very different. It is called a white dwarf and has the mass of the Sun but its radius is about 5,000 km. Compressing so much gas into such a small volume causes the gas in the star to reach extremely high densities. A matchbox full of gas from the center of a white dwarf, for example, would weigh more than a ton. White dwarfs are the endpoints of the evolution of stars like the Sun and finding a white dwarf in a cataclysmic variable implies that the systems are very old. We also know that since a white dwarf has reached the end of its thermonuclear evolution, it has consumed all of its hydrogen fuel and only the "ashes" of this energy production process are left: either carbon and oxygen or oxygen, neon, and magnesium. A single white dwarf moving through space cannot obtain energy from thermonuclear fusion. It can only radi-

ate the heat produced before it became a white dwarf. This is not the case for the white dwarf in a cataclysmic variable system in which the solarlike star provides new fuel for nuclear fusion.

Because the two stars are close together and the solarlike star is much larger than its companion, tides formed on the solar-type star by the white dwarf cause hydrogen-rich gas to flow from off its surface into the region of space surrounding the white dwarf. Since the two stars are orbiting each other at high speeds and the stream of gas has the angular momentum of the solar-type star, the gas cannot flow directly onto the surface of the white dwarf. It is forced to flow into an accretion disk before falling onto the white dwarf. The gas gradually spirals through the accretion disk and, because of magnetic fields within the gas, falls onto the white dwarf. Over time, the gas from the companion slowly flows into the accretion disk and then onto the white dwarf where the intense gravity of the white dwarf compresses and heats it. The accreting gas forms a thicker and thicker layer and the bottom of this layer is gradually heated until the gas reaches thermonuclear fusion temperatures.

After more than 10^4 years of evolution, the bottom of the accreted layer has been heated to temperatures of 10^7 K and densities exceeding 10^4 g/cm^3. This gas has now become electron degenerate, which means that the nuclei and electrons have been squeezed to such a point that they act like a metal. When such a gas is heated, it is unable to expand and cool as occurs in an ordinary gas, such as our terrestrial atmosphere. Therefore, when this gas is heated by thermonuclear reactions it cannot expand and cool; it only becomes hotter.

The first types of fusion reactions that occur in the gas are those of the proton-proton chain. This is the same set of nuclear fusion reactions that occur in our sun. As the nova evolves and the temperatures in the nuclear fusion regime increase, the most important process becomes the carbon-nitrogen-oxygen bi-cycle that normally occurs in stars that are more massive than our own sun. These reactions are very temperature sensitive. If we double the temperature, for example, the energy production increases by a factor of 65,536 (i.e., 2^{16}). The white dwarf now has all of the ingredients necessary to produce a large explosion. The hydrogen-rich gas from the other star is the fuel, degeneracy acts as a container, and accretion onto the white dwarf, which causes the temperatures to increase, acts as the igniter. It takes more than 10^4 years for the temperatures in the accreted gas to reach 3×10^7 K. It takes only a few days more for the temperatures to climb to 10^8 K and only a few minutes longer for them to reach 3×10^8 K. Finally, the layers are no longer degenerate and the gas can expand in a violent explosion and begin to cool.

The evolution described above is sufficient to produce a slow nova outburst. A more violent class of explosion, or fast nova outburst, requires that the proportion of carbon and oxygen nuclei in the layers undergoing nuclear fusion be much larger than in the Sun. Carbon acts as a catalyst in the carbon-nitrogen-oxygen reaction cycle so that the larger the amount of carbon in the gas, the more energy produced at a given temperature and density. The fact that enriched carbon was necessary to produce a fast nova outburst was predicted in the early 1970s and later confirmed by observational studies that determined the elemental abundances in the gas ejected by nova explosions.

A nova explosion is so violent that an amount of gas equivalent to 10 times the mass of Earth (or more) is blown into space at speeds exceeding 1,000 km/sec. In a few hours the brightness of the nova increases by a million times its quiescent value. To the unaided eye, a star appears out of nowhere, therefore, the name nova. Astronomers have obtained the most important information about a nova by taking spectra throughout the outburst at wavelengths from the ultraviolet to the infrared. Earth satellite observatories have also provided a large amount of data on the explosion. If a nova is discovered early enough, it is found to be hot, but cooling rapidly. This is called the fireball phase analogous to the earliest stages of a hydrogen bomb explosion. Within a short time (sometimes less than a few hours) the nova will reach its maximum brightness at optical wavelengths, and its surface temperature will have dropped to 10^4 K. Spectra obtained at this time show the signature of singly ionized iron and this phase is called the iron curtain. The expanding gas now starts to become transparent and it becomes possible to see into deeper and hotter layers. Over the next few weeks, the spectrum evolves to that of a hotter gas but with emission lines. This is typical of a low-density gas heated by an underlying hot source. Analyses of the spectra shows that about 10^{-5} of the mass of the white dwarf has been blown into space and the abundances of the chemical elements in this gas are very peculiar. Helium, carbon, nitrogen, and oxygen are very enriched compared to an equivalent amount of material from the Sun. This is the case for most novas. In the past few years, however, astronomers have identified a new class of novas in which the gas is enriched, in addition, in neon, magnesium, aluminum, and even sulfur (in a few cases). This new class is called the oxygen-neon-magnesium class. It is believed that this type of nova occurs on a very massive white dwarf, about 40 percent more massive than the Sun. Studies of many novas suggest that they may be the sources of some light isotopes found in the solar system such as carbon-13 and nitrogen-15.

See also: KEPLER'S LAWS; STARS AND STELLAR STRUCTURE; SUN; SUPERNOVA

Bibliography

SHORE, S. N., and STARRFIELD, S. "Nova Cas 1993: A Nova for the Holidays." *Sky and Telescope* **87** (4), 42–44 (1994).

STARRFIELD, S. "The Classical Nova Outburst" in *Multiwavelength Astrophysics,* edited by F. A. Cordova (Cambridge University Press, Cambridge, Eng., 1988).

STARRFIELD, S. "Thermonuclear Processes and the Classical Nova Outburst" in *Classical Novae,* edited by M. Bode and A. N. Evans (Wiley, New York, 1989).

STARRFIELD, S., and SHORE, S. N. "Nova Cygni 1992: Nova of the Century." *Sky and Telescope* **87** (2), 20–25 (1994).

STARRFIELD, S., and SHORE, S. N. "The Birth and Death of Nova V1974 Cygni." *Sci. Am.* **272**, 76–81 (1995).

SUMNER STARRFIELD

N RAY

In 1903, at a time when the properties of x rays and radioactivity were under intense investigation, the French physicist René Blondlot announced the discovery of another new radiation. Blondlot, a distinguished professor at the University of Nancy in Lorraine in eastern France, named the radiations "N" to commemorate his city, his university, and Nicolas François, the seventeenth-century regent of Lorraine.

At least forty people corroborated the existence of N rays between 1903 and 1906, although some sixty or so scientists and medical doctors could not duplicate Blondlot's results. Scientific reports on N rays disappeared after 1906 following sensational reporting that they were a hoax. N rays became remembered as an example of bad science, pseudoscience, or fraud. Most recently their discovery has been revived by comparison with the discovery of "cold fusion" in 1989.

Before 1903 Blondlot's best-known work focused on measurement of the velocity of electric waves using the method of photographing electric sparks reflected by a rotating mirror. Establishing experimental values for the speed of electric waves very close to the speed of light, Blondlot provided further confirmation, after Heinrich Hertz's work, of the relationship between electromagnetic radiation and light.

After Wilhelm Röntgen's discovery of x rays in 1895, Blondlot searched for evidence that x rays are emitted from an electrical discharge tube in the form of a plane-polarized wave. As a detector, he used changes in the brightness of an electric spark when the detector's orientation was changed in relation to the discharge tube. However, when he found that the radiations were refracted by a quartz prism, he concluded that they could not be x rays because it was well established that a quartz prism will not bend x rays.

Initial reaction to Blondlot's announcement was mixed. Many investigators used photographs of spark lengths as detectors, but Blondlot's method was one of visual detection of changes in the brightness of an unstable spark less than 0.1 mm long. Blondlot's report that N rays were refracted by a quartz prism placed their wavelength in the infrared region, but Heinrich Rubens and other experts denied that infrared radiations could traverse aluminum or quartz, as had been reported by Blondlot.

Undeterred, Blondlot multiplied the sources of N rays and the means of detection. He reported that the rays were emitted by gas lamps, incandescent lamps, the Sun, and tempered steel. He further claimed that the humoral fluid of the eye stores N rays and re-emits them. More spectacularly, some of Blondlot's colleagues in the Sciences and Medical Faculties at Nancy reported that N rays were emitted by animals' nervous systems, including the brain.

Two new methods of detection were developed. In one method, a colloidal calcium-sulfide powder was spread out with ether and painted into spots on black cardboard. According to Blondlot, the luminous spots became distinct and brighter when an ironclad gas lamp with an aluminum N-ray window and a quartz focusing lens was brought near. In a variation of this method, a vertical thread was painted with luminous powder and moved through an N-ray spectrum created by an aluminum prism

that showed N rays to be in the far ultraviolet rather than the infrared.

In a second method, an electric spark device was moved back and forth manually from one side of a photographic plate, where it was exposed to N rays, to another side of the plate, where it was protected by a lead shield. The photographic plate registered the spark more brightly on the side exposed to N rays than on the other side.

However, objections continued to be made to reports confirming the discovery. Many observers perceived no difference in intensity of sparks or spots with or without N-ray illumination. Otto Lummer reported that he and Rubens duplicated the reported "discovery" simply by standing in a dark room, gazing at a screen coated with sulfide spots, and then looking slightly away from the screen so that the peripheral vision of the rods of the retina came into action. Critics demanded that the photographic method should be automated so that no unconscious bias entered the manual procedure of moving the spark device at equal time intervals.

On the whole, French scientists were the most positive investigators of N rays and German scientists the most negative. Some younger French scientists, especially Jean Perrin and Paul Langevin, were among the few French physicists who strongly questioned Blondlot's claims. The American physicist Robert Wood became embroiled in the controversy after attending the annual meeting of the British Association for the Advancement of Science in England where Rubens persuaded Wood to visit Blondlot's laboratory. Within a few days Wood penned a devastating letter to the widely read journal *Nature* claiming, among other things, that he had clandestinely removed the aluminum prism from Blondlot's apparatus with no effect on Blondlot's results.

Contrary to many reports, Blondlot did not resign in infamy or commit suicide in despair. But neither did he accept demands for controlled experiments in collaboration with his critics. In 1909, when he was sixty years old, Blondlot retired from teaching and continued to work occasionally on N rays.

The best explanation of this mistaken series of investigations lies not in hoax or fraud, but in a self-deception shared by Blondlot and a good number of other scientists, mostly at Nancy and a few other French institutions. The will to believe among many members of the French scientific community and the will not to believe among Germans is partly explained by the existence of strong antagonisms and rivalries among the two communities following the Franco-Prussian War of 1870–1871. French hopes pinned on the success of science at the University of Nancy played an important role in this dispute at a time when Germans occupied most of Alsace-Lorraine.

In addition, among the tiny elite that made up the French scientific community, many knew Blondlot personally and respected him for his work on electric waves. These friends were loathe to call Blondlot's judgment into question. Indeed members of the French Academy of Sciences awarded him a prize in the summer of 1904 for the ensemble of his work, only to be gravely embarrassed a few months later by Wood's exposé. Blondlot's suggestion that N rays might be a phenomenon that could be "seen" by some observers and not by others provided a means to assimilate N rays into spiritualism but not into science.

See also: X Ray; X Ray, Discovery of

Bibliography

Huizenga, J. R. *Cold Fusion* (Oxford University Press, Oxford, Eng., 1993).

Klotz, I. M. "The N-Ray Affair." *Sci. Am.* **242,** 168–175 (1980).

Langmuir, I. "Pathological Science" (1953; trans. and edited by R. N. Hall). Reprinted in *Phys. Today* **42,** 36–48 (1989).

Nye, M. J. "N-Rays: An Episode in the History and Psychology of Science." *Historical Studies in the Physical Sciences* **9,** 125–156 (1980).

Wood, R. W. "The N Rays." *Nature* **70,** 530–531 (1904).

MARY JO NYE

NUCLEAR BINDING ENERGY

Nuclear binding energy is defined as the energy required to break up a given nucleus into its constituent parts of N neutrons and Z protons. In terms of the masses $M(A,Z)$ of the neutral atom with atomic mass number $A = N + Z$, the binding energy $B(A,Z)$ is defined by

$$B(A,Z) = N \cdot M(H) c^2 + Z \cdot M(n) c^2 - M(A,Z) c^2,$$

where $M(H)$ is the mass of the hydrogen atom and $M(n)$ is the mass of the neutron. Atomic masses are usually tabulated in terms of the mass excess $\Delta(A,Z) \equiv M(A,Z) - uA$, where u is the atomic mass unit defined by $u = M(^{12}C)/12 = 931.494$ MeV/c^2. In terms of the mass excess the binding energy is given by

$$B(A,Z) = N\Delta(H) c^2 + Z\Delta(n) c^2 - \Delta(A,Z) c^2,$$

where $\Delta(H) = 7.2890$ MeV and $\Delta(n) = 8.0713$ MeV. Reactions or decays of the type

$$\sum_i [A_i, Z_i] \rightarrow \sum_f [A_f, Z_f]$$

are characterized by their Q value:

$$Q = \sum_i M(A_i, Z_i) c^2 - \sum_f M(A_f, Z_f) c^2$$

$$= \sum_f B(A_f, Z_f) - \sum_i B(A_i, Z_i).$$

Spontaneous decay involves a single initial nuclear state and is allowed if $Q > 0$. In the decay, energy is released in the form of the kinetic energy of the final products. Reactions involve two initial nuclei and are endothermic (a net loss of energy) if $Q < 0$; the reactions are exothermic (a net release of energy) if $Q > 0$.

Nuclei are bound due to the overall attractive strong interactions between nucleons. The strong interaction arises from the exchange of mesons. The interactions are short-ranged and occur mainly between neighboring nucleons. In addition, the nuclear interaction saturates, resulting in a nearly constant interior nucleon density and a surface radius approximately equal to $1.2\, A^{1/3}$. The analogy of this situation with a droplet of liquid, results in the liquid-drop model for the nuclear binding energies in which the binding energy is expressed in the form

$$B(A,Z) = \alpha_1 A - \alpha_2 A^{2/3} - \alpha_3 Z^2 A^{-1/3}$$
$$- \alpha_4 (N - Z)^2 A^{-1}.$$

The first term represents the nearest neighbor attractive interaction between nucleons, and the second term represents the correction due to the fact

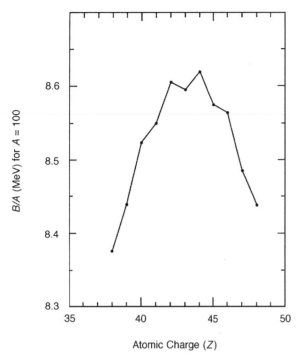

Figure 1 Experimental values nuclear binding energy versus atomic charge.

that the nucleons on the surface only interact with those in the interior. The third term is due to the Coulomb repulsion between protons. The fourth term arises because the proton-neutron strong interaction is on the average more attractive than the proton-proton or neutron-neutron strong interactions and because the total kinetic energy is minimized when $N = Z$. A reasonably good fit to all data is given by $\alpha_1 = 15.56$ MeV, $\alpha_2 = 17.23$ MeV, $\alpha_3 = 0.697$ MeV, and $\alpha_4 = 23.28$ MeV.

Experimental values $B(A,Z)/A$ versus Z are shown in Fig. 1 for $A = 100$. The form of this curve for $A = 100$ is similar to those for other A values. The binding energy has a maximum at $Z_{max} = 44$. Without the Coulomb interaction between protons, the maximum would occur at $N = Z$ due to the symmetry energy (the α_4 term). When the Coulomb interaction between protons (the α_3 term) is added, the peak is shifted to more neutron-rich nuclei. The oscillation in the binding energy curve in Fig. 1 is due to nuclear pairing interaction, which gives rise to the fact that nuclei with even numbers of protons or neutrons are more bound than their neighboring (odd) nuclei.

The binding energy per nucleon, $B(A,Z_{max})/A$, is shown in Fig. 2 for $A = 4$ to 250. Note the x axis

scale change at $A = 50$. The points $B(\text{deuteron})/2 = 1.112$ MeV and $B(^3\text{H})/3 = 2.827$ MeV for $A = 2$ and 3, respectively, are below scale. This particular set of nuclei are referred to as the those at the top of the ridge-of-stability of the binding energy or those at the bottom of the valley-of-stability of mass. The binding energy per nucleon is roughly constant (the α_1 term in the liquid-drop formula) due to the nearest neighbor interaction between nucleons. The binding energy per nucleon curve has irregular features (beyond liquid-drop model) due to the nuclear pairing interaction, the tendency of nucleons in the nucleus to clump together into alpha particles, and the gaps between orbitals in the shell-model structure. The maximum in the binding energy per nucleon occurs for ^{58}Fe ($Z_{max} = 26$). ^{58}Fe represents the most bound (lowest energy) state for nucleons. Thus, fusion of two light nuclei with a combined mass of $A < 58$ usually results in energy release. The fusion of deuterium and tritium is the main reaction being investigated for controlled fusion reactors. Other fusion processes are important for solar energy and for the creation of elements up to $A = 58$ in stellar environments. The falloff in binding energy per nucleon above $A = 58$ implies that most of these nuclei can spontaneously decay into lighter products. The most common of these decay processes are alpha decay, where a ^{4}He is emitted, and fission, where the nucleus breaks up into two roughly equal mass fragments. The fission products are usually accompanied by neutrons. Intermediate decay modes, where light fragments such as ^{14}C are emitted, are also possible and have re-

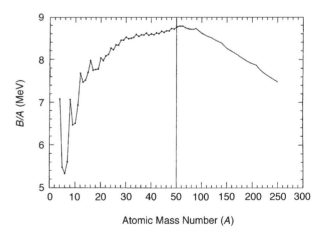

Figure 2 Binding energy per nucleon for $A = 4$ to 250.

cently been observed. All of these decay modes are hindered due to the exponential decay through the Coulomb barrier. Although most heavy nuclei have a positive Q value for spontaneous decay, many of them have lifetimes on the order of the age of the universe and thus exist in nature. Recently, the element $Z = 110$ was discovered at GSI (*Gesellschaft für Schwerionenforschung* in Darmstadt, Germany) with a half-life of about 0.0004 s due to alpha decay. Experiments are being planned to search for even heavier elements whose lifetimes will depend upon their alpha and fission decay probabilities.

The various decay modes described above account for all observations out to the proton and neutron drip lines. The neutron drip line occurs in the most neutron-rich nuclei and is defined as the path where $Q_n = B(A - 1, Z) - B(A,Z)$ changes sign from negative (where the nuclei are stable to neutron emission) to positive (where the time for neutron decay becomes very short, typically 10^{-21} s). Likewise, the proton drip line occurs for neutron-deficient nuclei and is defined as the path where $Q_p = B(A - 1, Z - 1) - B(A,Z)$ changes sign. Proton decay just outside the proton drip line is often hindered due to the Coulomb barrier. For medium and heavy mass nuclei, many nuclei that lie inside the drip lines have not yet been observed experimentally. For example, for $A = 100$, the liquid-drop model formula gives $Z = 33$ for the neutron drip line and $Z = 49$ for the proton drip line, which should compared with the range of $Z = 38$ to $Z = 48$ for the experimental binding energies shown in Fig. 1.

See also: ATOMIC NUMBER; DECAY, ALPHA; DECAY, NUCLEAR; ELECTRON; NEUTRON; NUCLEAR REACTION; NUCLEAR SHELL MODEL; NUCLEON; NUCLEUS; PROTON

Bibliography

WILLIAMS, W. S. C. *Nuclear and Particle Physics* (Clarendon Press, Oxford, Eng., 1991).

B. ALEX BROWN

NUCLEAR BOMB, BUILDING OF

Nuclear fission was discovered at the end of 1938 by Otto Hahn and Fritz Strassmann in Hahn's Berlin laboratory. In Sweden, on the basis of the liquid-drop model of the nucleus, Lise Meitner and her nephew, Otto Robert Frisch, immediately interpreted, correctly, the reaction to be a process in which a uranium atom is cleaved roughly in half after absorbing an incident neutron, releasing an enormous amount of energy—the signature of the event. Atom for atom, the energy release is more than 10 million times the amount of energy released when an atom of carbon in wood, coal, or oil is burned, or when an atom of nitrogen in TNT is oxidized.

The implications of this discovery for the controlled production of energy or for the manufacture of an explosive which, in an instant, releases unprecedented amounts of energy, were well understood worldwide. It was not obvious, however, that practical devices for such applications could be invented.

There were two questions that had to be answered before anyone could seriously entertain exploiting nuclear fission: First, are secondary neutrons released in the fissioning process that can themselves initiate fission in other uranium atoms? Second, which of the two uranium isotopes found in nature actually fissions? Is it the common $^{238}_{92}$U (92 protons and 146 neutrons) or the much rarer (1 in 142 $^{235}_{92}$U (92 protons and 143 neutrons)?

Programs to explore these questions were started in Germany, France, Great Britain, the Soviet Union, and Japan, but it was only in the United States that a sustained, well-financed effort emerged and was sustained throughout World War II.

Getting Started

Initially, the American program was driven by the fear that German scientists would provide Adolf Hitler with fission bombs that would make the Nazi war machine invincible. These fears were felt most keenly by scientists who had fled to America to escape Nazi persecution or control. The chief initiator among these scientists was the Hungarian-born physicist Leo Szilard, who organized his colleagues and who was instrumental in making certain that President Franklin Delano Roosevelt was aware of the potential threat.

Between the fall of 1939 and the fall of 1941, beginning with a research expenditures on uranium fission expanded from an initial $6,000 (a grant from special funds available to Roosevelt) to more than $300,000 per year. Almost from the start the work was done under a veil of secrecy. The funds were funneled through the National Bureau of Standards. The bulk of the work was conducted at University laboratories at Columbia, Princeton, Cornell, Iowa State, California (Berkeley), Chicago, and Illinois, among others. In June 1940, at the suggestion of Vannenar Bush (the president of the Carnegie Institute of Washington), Roosevelt created the National Defense Research Committee (NDRC) to initiate and coordinate development and production of new tools for waging war. Bush was appointed chairman of NDRC and, in effect, became Roosevelt's adviser on science and technology. At Bush's own suggestion, he inherited oversight of the research on fission. A year later, Bush created a new more comprehensive organization, the Office of Scientific Research and Development (OSRD) of which NDRC became a branch.

By 1939 researchers had already determined that, when a uranium nucleus fissions, it liberates, on average, more than 1 neutron. Thus a chain reaction would be theoretically possible. By spring 1940 it was proven that the rare isotope ^{235}U fissions, not ^{238}U. This meant that if one wanted to develop an explosive based on fission, it would be necessary to separate the two naturally occurring isotopes and concentrate ^{235}U. This knowledge defined the main lines of research in the United States during the years 1940 and 1941.

Isotope separation. The chemical characteristics of an element are defined by the number of protons in the nucleus. All isotopes of uranium are chemically identical—their nuclei all contain 92 protons. Therefore, it is not possible to separate isotopes of an element by chemical means. The defining difference between an element's isotopes is their atomic weights. The difference in weight between ^{235}U and ^{238}U is the weight of three neutrons, a difference of 1.3 percent. Isotope separation can be accomplished if a technique can be developed for exploiting this slight difference in their response to mechanical forces. During the years 1940 and 1941, a variety of separation techniques were explored, including distillation, electrolysis, electromagnetic separation, centrifuge, gaseous diffusion, and thermal diffusion. By the fall of 1941, none had produced definitive results but the most promising were electromagnetic separation, gaseous centrifuge, and gaseous diffusion. Electromagnetic separation relied on the difference in the two uranium isotopes' (in the form of uranium tetrachloride, UCl_4) response to a magnetic field. In both gaseous centrifuge and

gaseous diffusion, the active material was uranium hexafluoride (UF_6), an extremely reactive, corrosive, and dangerous gas.

Chain reaction. During 1940 and 1941 research on the possibilities of a chain reaction was carried out primarily at Columbia University by a team headed by Enrico Fermi, which also included Leo Szilard. There is only a certain probability that when a neutron strikes a ^{235}U nucleus, the nucleus will undergo fission. In order not to lose neutrons, Fermi and Szilard devised a uranium pile (later called a reactor), a three-dimensional matrix composed of lumps of uranium metal or uranium oxide embedded at regular intervals in a moderator—a substance that does not absorb neutrons.

The most likely candidate for a moderator was heavy water, that is, water in which ordinary hydrogen ($^1_1$H) is replaced by deuterium ($^2_1$H), the isotope of hydrogen whose nucleus contains one proton and one neutron. At the time, there was little heavy water available and its cost of manufacture was high. Both Fermi and Szilard were confident that ordinary carbon, in the form of graphite, would also prove to be an effective moderator if neutron-absorbing impurities could be removed. Indeed, successive pilot experiments with purer and purer graphite during 1940 and 1941 suggested this was the case.

Enter element 94. In early 1940, Berkeley physicist Edwin McMillan discovered that when ^{238}U absorbs neutrons, rather than being fissioned, the nucleus emits a beta particle, thereby converting the nucleus from element 92 to element 93, the first transuranic element. McMillan, who won a Nobel Prize for his discovery, named this element neptunium ($^{239}_{93}$Np). Theory predicted that the next transuranic element, element 94, should behave much like ^{235}U and fission after absorbing a neutron.

In 1941, working for the Uranium Committee, Glen Seaborg documented that neptunium, which had a half-life of 2.3 days, itself emits a beta particle, becoming element 94. Seaborg named this new element plutonium ($^{239}_{94}$Pu). Working with submicroscopic quantities of plutonium, Seaborg then confirmed the prediction that the new transuranic element fissioned, and in the process produces, on average, more than one neutron.

This meant that in the uranium pile, while some neutrons are fissioning ^{235}U, other neutrons are being absorbed by ^{238}U, which is thus ultimately transformed into ^{239}Pu. Not only could a uranium pile produce energy, it could also produce another nuclear fuel, plutonium.

To Build or Not to Build

By the spring of 1941, as the prospect of war loomed with greater and greater certainty, Vannevar Bush decided that the time had come to make a decision on whether or not to continue to pursue fission research. It was still not clear that practical, usable results could be obtained in the short term. Some of Bush's colleagues, especially his deputy, Harvard University President James B. Conant, urged Bush to stop work on fission technologies and concentrate available funds on those projects, such as radar, which were likely to play a role in the coming war.

Instead, perhaps because persistent and plausible rumors filtered out of Germany that scientists there were on the verge of manufacturing a fission bomb, Bush appointed a secret, blue-ribbon National Academy of Sciences (NAS) committee to recommend a course of action. The NAS committee was chaired by Nobel Laureate Arthur H. Compton, who was himself already heavily involved in NDRC-sponsored fission research. The committee's first recommendation, on May 18, 1941, was that research should be intensified for six months at which time another review should be undertaken. Neither Bush nor Conant were satisfied with this report and, after bolstering the NAS committee's makeup with several engineers, Bush asked it to meet again. To Bush's dismay, the second report, which was given to him on July 18, 1941, made much the same recommendations as the first report.

By this time, Bush had in hand the draft of a report by British scientists who had been investigating potential uses of uranium fission. The British committee, code-named MAUD, had concentrated on the possibilities of a fission bomb and had concluded that a pure ^{235}U bomb was feasible and could be produced within the next few years.

On October 9, 1941, just after commissioning a third NAS report, Bush met with Roosevelt and, after informing the president of the British report and of his own estimate of the situation, obtained Roosevelt's unconditional backing to do whatever Bush decided with regard to pursuing fission technologies. Roosevelt also instructed Bush to restrict deliberation of policy on these matters to only four other people: Conant, Vice President Henry Wallace, Secretary of War Henry Stimson, and U.S. Army Chief of Staff George Marshall.

On November 6, 1941, the NAS committee reported that a fission bomb could be built and estimated that the total cost would be $133 million (more than $1 billion in 1990s dollars). Even though isotope separation had not yet been proven feasible and a sustained chain reaction had not yet been achieved, Bush informed the president of his decision to go forward with a program to build a fission bomb.

Up to this time, the work on fission technologies had been supported with special funds the Congress had appropriated to the president to be used at his discretion. Given all the other technical projects these monies were funding, Bush's financial reservoirs were near exhaustion. For this reason, and as a way of not having to reveal details of the project to the Congress, Bush decided at Roosevelt's suggestion to turn over the construction associated with the venture to the Army Corps of Engineers, whose massive budgets could easily soak up the fission project costs without drawing undo attention from the Congress. In fact, during the war, Congress as a whole never knew the nature of the project or its cost. Ironically, at about the same time, the German military leadership had decided not to pursue a fission bomb on the grounds that they would win the war quickly—so quickly that the bomb would never be completed in time to be of any use.

The Manhattan Project

The fission program, which now focused on making an atomic bomb, was formally turned over to the U.S. Army Corps of Engineers in June 1942, under the direction of Colonel James Marshall. Marshall intended to locate his headquarters in New York City, which gave rise to the code name for the project—Manhattan Engineer District (MED)—often referred to as the Manhattan Project.

The technical work continued as before under the immediate jurisdiction of the scientific personnel. In June, the real estate office of the U.S. Army Corps of Engineers had recommended that the project locate in a 57-square-mile reservation, bordered by the Clinch River, near Knoxville, Tennessee, but as September approached Colonel Marshall, a good manager but one who moved with deliberate speed, had not yet acted on that recommendation.

Having committed himself to an expensive, secret project in the midst of a brutal war, Bush was concerned. He urged Colonel Marshall's superiors to replace Marshall with a more aggressive manager.

Colonel Leslie R. Groves was appointed to head MED on September 17, 1942. In many ways he was a natural choice. He was an extremely able civil engineer. For the two years prior, he had been in charge of all army construction in the United States and had shown amazing skill in expediting and in keeping track of a construction enterprise, which at the peak of mobilization in mid-1941 was growing at the rate of $600 million a month. Nothing escaped his notice. He was ruthless in ferreting out slothfulness and dishonesty. Groves's formula for successful administration was to delegate freely and then hold those responsible strictly accountable for performing to his expectations.

Groves had not wanted the MED assignment. He had hoped for a combat engineering regimental command. As a quid pro quo he was promptly promoted to brigadier general and given carte blanche. He reported directly to Chief of Staff General Marshall and Secretary of War Stimson.

Within three days of being appointed, Groves had moved on acquiring the Tennessee land (later known as Oak Ridge) and had secured the highest priority ranking for the acquisition of hard-to-get materiel and personnel, something that had eluded his predecessor. Groves then toured the major MED facilities—laboratories at Columbia University, the University of Chicago, and the University of California at Berkeley. Groves was shocked at how little was known by the scientists and engineers who had responsibility for solving the puzzles that obstructed completion of their assignment.

Tensions between the scientists and Groves were quite high. Groves prided himself on his common sense and no-nonsense practicality. He was flabbergasted at some of the habits of scientists, for example, their willingness to rely on order-of-magnitude estimates as opposed to the kinds of precise calculations necessary for any kind of construction project. For several years, the scientists had chaffed under the secrecy rules imposed by the project leadership. The majority of the scientists believed that progress in laboratory science depended on an open and sharing environment. Now, Groves exasperated many of the scientists by making it known that the secrecy rules would be made even more stringent.

Groves was somewhat reassured when he met Berkeley physicist and Nobel Laureate Ernest O. Lawrence, and theoretical physicist J. Robert Oppenheimer. Lawrence, the inventor of the cyclotron, headed up the electromagnetic isotope separation project. He displayed a confidence and enthusiasm for his work that Groves understood and appreciated. But Groves was also mindful that Lawrence's group had not yet achieved any reliable results.

Oppenheimer was the leader of a group studying what had become known as "bomb physics." He was a polyglot and a polymath—as equally at home with seventeenth-century French poetry or ancient sanskrit texts as with quantum mechanics. Groves was impressed with Oppenheimer's concentration on what Groves considered to be realistic approaches to designing an atomic bomb.

Groves's orders had been to take charge of construction, to assure security, and to stand in readiness to assist the scientists in any way possible. Groves interpreted his orders as being to oversee the building of the atomic bomb in the shortest time possible. He realized that the contributions of the scientists were absolutely crucial but he was now sure that the scientists, as a group, had little understanding of the scale of operation that would be required if their benchtop experiments were to be transformed into industrial processes. Between the end of October and the middle of December 1942, Groves took control of all phases of the project, and set a course to develop the weapon as quickly as possible.

Building the Bomb

Groves was not entirely free to act on his own. He had to answer to an oversight committee, the Military Policy Committee, chaired by Vannevar Bush, flanked by Admiral William R. Purnell and General Wilhelm Styer. Groves served as committee secretary. After the first few meetings, Groves's initial decisive actions had so gained him the committee's implicit trust that the committee became little more than a rubber stamp for his administrative decisions. To supplement his lack of scientific expertise, Groves appointed and relied on two scientific advisers, Harvard President James B. Conant, a distinguished chemist, and Professor Richard C. Tolman, a theoretical physicist, dean of the California Institute of Technology Graduate School, and a member of OSRD.

On December 2, 1942, Enrico Fermi and his group at the University of Chicago demonstrated a self-sustaining, controllable chain reaction in a carbon-moderated uranium pile; and by January 1, 1943, Groves had taken a number of decisive decisions:

- To immediately build industrial plants for two kinds of isotope separation—gaseous diffusion and electromagnetic separation—at Oak Ridge.

- To create a new factory complex including three nuclear reactors for the manufacture of plutonium, and to construct factories for the chemical separation of plutonium from the uranium out of which the plutonium was to be created. The site, yet to be chosen, would be a 750-square-mile reservation in south-central Washington State—Hanford.
- To create a separate, isolated complex at Los Alamos, New Mexico, directed by Oppenheimer, for fission bomb design and construction using the Hanford-produced plutonium and the Oak Ridge–produced ^{235}U.
- To enlist the services of some of the country's largest corporations, including Dupont, Eastman Kodak, and Union Carbide—over two hundred prime contractors in all.

None of these steps were taken without risk or controversy.

As 1943 began, Groves had already obligated expenditures totaling $500 million. Shortly thereafter, construction began on the Oak Ridge facilities even though the basic processes for isotope separation had not been perfected. By mid-1943, construction began at Hanford on the reactors and separation factories.

This was a project of gargantuan scale. Factories of colossal proportions were built in record time. In the summer of 1944, the time of most intense activity, the MED employed about 160,000 people at sites, large and small, all across the United States and Canada. By the summer of 1945 over half-a-million people had cycled through as employees. The cost had far exceeded the $133 million estimated in 1941; the project had cost more than $2 billion. Security was so good that almost no one working at the project's industrial sites had any idea concerning the project's true nature.

General Groves's driving style and masterly skill at running a large, far-flung industrial operation was a major reason that fissionable material from Hanford and Oak Ridge was being shipped to Los Alamos, in quantity, by the spring of 1945. Another key factor was the ingenuity of the scientists and engineers, who time after time were able to overcome seemingly insurmountable technical hurdles.

The first atomic bomb was tested in the New Mexico desert on July 16, 1945. On August 6, 1945, an atomic bomb containing about 200 lbs of ^{235}U destroyed the city of Hiroshima, Japan. On August 8, the Soviet Union declared war on Japan and invaded Manchuria. A day later, on August 9, an atomic bomb containing just 13 lbs of plutonium laid waste to the city of Nagasaki, Japan. Japan was besieged on all sides. It surrendered on August 14, 1945. Whether Japan would have surrendered had the atomic bomb not been used remains a hotly debated subject.

See also: ATOMIC WEIGHT; CHAIN REACTION; FERMI, ENRICO; FISSION; FISSION BOMB; FUSION; FUSION BOMB; LAWRENCE, ERNEST ORLANDO; MEITNER, LISE; NEUTRON; OPPENHEIMER, J. ROBERT

Bibliography

BADASH, L. *Scientists and the Development of Nuclear Weapons: From Fission to the Limited Test Ban Treaty, 1939–1963* (Humanities Press, Atlantic Highlands, NJ, 1995).

GOSLIN, F. G. *The Manhattan Project: Science in the Second World War* (U.S. Department of Energy, Washington, DC, 1990).

GROVES, L. M. *Now It Can Be Told: The Story of the Manhattan Project* (Harper, New York, 1962).

HEWLETT, R. G., and ANDERSON, O. E., JR. *The New World: A History of the United States Atomic Energy Commission, Volume I 1939–1946* (Pennsylvania State University Press, University Park, PA, 1962).

RHODES, R. *The Building of the Atomic Bomb* (Simon & Schuster, New York, 1986).

SERBER, R. *The Los Alamos Primer: The First Lectures on How to Build an Atomic Bomb* (The University of California Press, Berkeley, CA, 1992).

WALKER, M. *German National Socialism and the Quest for Nuclear Power, 1939–1949* (Cambridge University Press, Cambridge, Eng., 1989).

STANLEY GOLDBERG

NUCLEAR DECAY

See DECAY, NUCLEAR

NUCLEAR ENERGY

See ENERGY, NUCLEAR

NUCLEAR FORCE

The nuclear force is the force that holds protons and neutrons (nucleons) together inside an atomic nucleus. In the absence of this force, atomic nuclei would not exist, because the repulsive Coulomb force between charged protons would prevent nuclei from staying together. Despite substantial efforts, a complete understanding of the force between nucleons is still the fundamental unanswered question in nuclear physics.

In nuclei, the nuclear force acts primarily between pairs of nucleons. Much of our knowledge of the nucleon-nucleon interaction has been obtained by studying the reactions and scattering of nucleons over a wide range of energies. This includes proton-proton, neutron-neutron, and proton-neutron scattering, as well as the study of the simplest nucleus, the deuteron, which consists of a neutron and a proton bound together. Calculations are performed using a phenomenological nuclear potential, from which the force can be derived. The results of such calculations can then be compared with experimental data.

Properties of the Nuclear Force

Range. The nuclear force acts over an extremely short range; it is effective only over distances of a few femtometers (1 fm = 10^{-13} cm). We know this because, despite the repulsion of the Coulomb force, protons in atomic nuclei are held together with very small separation distances. More specifically, the nucleon-nucleon force is attractive over separations roughly between 0.5 and 2 fm, and repulsive at distances of less than 0.5 fm. The strength of the main attractive component of the nuclear force is central in nature, that is, it depends on the relative separation of the two nucleons. Outside the nucleus, at atomic distances, the nuclear force is negligibly weak, leaving the Coulomb force to determine the interaction between nuclei. A consequence of the short range of nuclear forces is the relative constancy of the binding energy of medium and heavy nuclei, even though the number of nucleon pairs is proportional to the square of the number of nucleons. This saturation of the nuclear force means that a given nucleon interacts with only a of its few nearest neighbors, rather than with every nucleon in the nucleus.

Spin dependence. Nucleons possess an intrinsic angular momentum, called spin. The fact that there is only one bound state of the neutron-proton system, in which the spins of the two nucleons are parallel to one another, means that the nucleon-nucleon force depends on the spatial orientation of the nucleon spins.

Charge symmetry and independence. It is convenient to think of the neutron and proton as two different states of a single particle—the nucleon—distinguished by a quantity called isospin (in analogy with spin). The nucleon-nucleon force depends on the total isospin of the pair, but it also exhibits charge symmetry, which means that for the same spin conditions, the neutron-neutron force is almost exactly equal to the proton-proton force, after subtracting the electromagnetic (charge) contribution to the total proton-proton interaction. In addition, the nuclear force is charge-independent to a very high degree. This means that for the same spin and isospin configuration, the neutron-proton force is identical to the neutron-neutron force and the proton-proton force (corrected for the electromagnetic interaction).

Tensor component. Careful studies of the magnetic and electric properties of the deuteron have revealed the presence of a noncentral, or tensor, component of the nuclear force. The manifestation of this small perturbation is the presence of more than one angular momentum component in the quantum mechanical description of the deuteron. The strength of the tensor force depends on the internucleon separation and the relative orientations of the nucleon spins.

Other properties. The nucleon-nucleon force is sensitive to the relative velocity or momentum of the two nucleons. The form of this component appears as an interaction between the relative angular momentum and the total spin of the two-nucleon system. This component is known as the spin-orbit force. In addition, the experimentally verified requirement that the two-body nuclear force exhibit both parity conservation and time-reversal invariance has been used to restrict the number and type of components that may appear in a phenomenological nuclear potential.

Exchange Nature of the Nuclear Force

A number of experimental observations, including the saturation of nuclear forces, indicate that the nucleon-nucleon force arises from the exchange of virtual particles between the nucleons. Hideki Yukawa, a Japanese physicist, was the first to work out the details of this idea quantitatively in 1935.

According to this model, the range of the nucleon-nucleon force is given by $R \sim \hbar/mc$, where m is the mass of the exchanged particle, c is the speed of light, and $\hbar$ is Planck's constant divided by 2π. With a range of 1.5 fm, a mass of 130 MeV is obtained. It is now believed that the exchanged particles responsible for the long-range part of the nucleon-nucleon force (distances between 1 and 2 fm) are the pi-meson family (pions), with masses of 140 MeV ($\pi^{\pm}$) and 135 MeV (π^0). Heavier mesons with a two-pion structure (ω, ρ) are involved at distances less than 1 fm. The tensor component of the nucleon-nucleon force arises naturally from the pion-exchange model.

Relation to the Quark Structure of the Nucleon

Although much has been learned about the nucleon-nucleon interaction since the mid-1930s, physicists are still far from a complete quantitative understanding of this force and how it is modified when the nucleons are located inside an atomic nucleus. One basic problem arises from the fact that nucleons themselves are not elementary particles, but are made up of three closely confined, pointlike objects known as quarks. Quarks interact with one another through the strong force, one of the four fundamental forces in nature. The study of quark-quark interactions, known as quantum chromodynamics, is a relatively new field, and very sophisticated computational techniques have been developed for calculations of quark interactions at high energies. Although the nucleon-nucleon force is almost certainly a manifestation of low-energy quark-quark interactions, it has not yet been possible to quantitatively approach a solution to the nucleon-nucleon problem by considering the interactions of six quarks.

See also: DEUTERON; INTERACTION, FUNDAMENTAL; NEUTRON; NUCLEAR STRUCTURE; NUCLEON; NUCLEUS; PROTON; QUARK; SPIN

Bibliography

KRANE, K. S. *Introductory Nuclear Physics* (Wiley, New York, 1988).

WONG, S. S. M. *Introductory Nuclear Physics* (Prentice Hall, Englewood Cliffs, NJ, 1990).

CARY N. DAVIDS

NUCLEAR MAGNETIC RESONANCE

Nuclear magnetic resonance (NMR) refers to the behavior of the magnetic moment of a nucleus placed in a suitably chosen applied external magnetic field. The magnetic field consists of two fields: (1) a large static (i.e., constant in time) magnetic field H_0 ($\sim 1-10$ T) and (2) a smaller perpendicular field H_1 ($\sim 10^{-2}-10^{-4}$ T) that oscillates in time at an appropriately chosen frequency f, normally in the radio-frequency (rf) region (i.e., 1–100 MHz). When a classical magnetic moment (e.g., a bar magnet) is placed in a static magnetic field H_0, it will precess about H_0 at a frequency f_0 that is proportional to H_0. Similarly, a nucleus having a magnetic moment due to its quantum mechanical spin angular momentum will also precess at a frequency proportional to H_0. The proportionality constant $\gamma/2\pi$ (where γ is called the gyromagnetic ratio) depends on the particular nucleus and has different values for different nuclear species (e.g., ^{1}H, ^{7}Li, ^{13}C). The effect of the rf magnetic field H_1 is to cause energy to be exchanged between the nuclear spins and the rf field provided the frequency f of the rf field is adjusted to equal the Larmor frequency f_0. This phenomenon is called NMR absorption. Since the Larmor frequencies of different nuclei are considerably different even in the same magnetic field, NMR can easily separate their contributions to the total magnetization and can be used to detect a particular nuclear species, even in the presence of other much stronger sources of magnetism.

Similar to NMR is electron paramagnetic resonance (EPR), which occurs when an electron is placed in a magnetic field. Since the electron's magnetic moment is of the order of 1000 times larger than that of a proton, EPR typically occurs at microwave rather than at rf frequencies. Another related phenomenon is nuclear quadrupole resonance (NQR), which occurs when a nucleus having an electric quadrupole moment is placed in an electric field gradient (i.e., a spatially varying electric field). NQR occurs for nuclei whose spin quantum number $I > 1/2$, and requires an rf field H_1 but not a large static field H_0.

Brief Early History

The first successful NMR experiments in condensed matter were performed simultaneously by the groups of Edward Purcell at Harvard and Felix

Bloch at Stanford and published in 1946. They also performed a number of important fundamental experiments, such as the demonstration of a negative spin temperature. A major advance occurred a few years later when Erwin Hahn observed that significant signal-to-noise and other advantages would result from the application of a pulsed, rather than steady, rf magnetic field $\mathbf{H}_1$. He also discovered that the magnetization decay following the turnoff of the $\mathbf{H}_1$ pulse could be reversed even after the magnetization had gone to zero, resulting in the spin-echo. Today, almost all NMR experiments, including magnetic resonance imaging (MRI) use pulsed NMR techniques based on those originally developed by Hahn.

Applications

Magnetometer. Since the Larmor frequency is proportional to the static magnetic field $\mathbf{H}_0$, a measurement of the resonance frequency f provides an exceedingly precise way to measure the external magnetic field and its uniformity. This application of NMR is useful primarily at larger magnetic fields where the NMR signal strength is greater.

Chemical applications. NMR is widely used in determining the chemical structure of molecules, particularly in liquids. The actual static magnetic field that a nucleus sees may be slightly different from $\mathbf{H}_0$ because of several other small internal magnetic fields. In particular there is a small internal magnetic field (the chemical shift field) due to the magnetic moment associated with an electron's orbital angular momentum. This field depends strongly on the immediate environment of the nucleus and may be different for different nuclei of the same species even in the same molecule. Accordingly, the NMR spectrum of a typical molecule consists of several chemically shifted lines. The intensity of each line is proportional to the number of equivalent nuclei and the frequency of the line provides information about the electron's orbital and wave function.

Chemical shift measurements have in recent years been extended to solids, where the shift would normally be obscured by the line broadening due to the dipolar local fields of neighboring nuclei. This problem can be circumvented by such techniques as magic angle spinning, spin-flip narrowing, two-dimensional Fourier transform NMR (not to be confused with MRI), and multiquantum coherence.

Physical applications. Physical applications of NMR have been largely, though not exclusively, limited to solids. Measurements of the Knight shift, which is the additional shift seen at the nucleus in a metal due to the magnetic moment of an unpaired electron, have been widely used to distinguish metals from insulators and to study properties of the conduction electrons in metals. These measurements have also been applied to superconductivity (including high T_c superconductors) and catalytic surfaces. Measurements of the dipolar and quadrupolar linewidths have been used to study the properties of disordered solids (glasses, amorphous systems, and incommensurate insulators). Double resonance techniques have allowed the detection of nuclei of low abundance or very small magnetic moments. Measurements of the relaxation times, T_1 and T_2, have allowed the investigation of dynamic properties like the diffusion of atoms or the rotational motions of molecules.

Biological applications. Chemical shift, Fourier transform, and double resonance techniques have been successfully used to determine the three-dimensional structures of some macromolecules (proteins and nucleic acids). Relaxation time studies have been used to study solvent-solute effects and dynamical effects in biological systems. The addition of a magnetic field gradient to $\mathbf{H}_0$ allows the regional study of separate portions of a macrospic specimen, like a living animal.

Industrial applications. NMR has been used in the oil industry to study the separation of oil and water in shale. It has also been applied to the study of porous media like cement.

See also: FIELD, MAGNETIC; MAGNETIC MOMENT; MAGNETIC RESONANCE IMAGING; PRECESSION, LARMOR

Bibliography

ABRAGAM, A. *The Principles of Nuclear Magnetism* (Clarendon, Oxford, 1961).

ERNST, R. R.; BODENHAUSEN, G.; and WOKAUN, A. *Principles of Nuclear Magnetic Resonance in One and Two Dimensions* (Clarendon, Oxford, 1987).

FUKUSHIMA, E., and ROEDER, S. B. W. *Experimental Pulse NMR-A Nuts and Bolts Approach* (Addison-Wesley, Reading, PA, 1981).

SLICHTER, C. P. *Principles of Magnetic Resonance,* 3d ed. (Springer-Verlag, Berlin, 1990).

DAVID C. AILION

NUCLEAR MEDICINE

Nuclear medicine is a medical specialty in which physicians use unsealed radioactive sources (radioactive tracers or radiopharmaceuticals) for the diagnosis and treatment of disease. An unsealed radioactive source is in a form that can readily disperse throughout the body or throughout an organ or space in the body. For example, a radioactive gas (xenon-133) can be inhaled to determine where air goes in the lungs. Thallous (thalium-201) chloride can be injected into a vein during exercise to evaluate the blood flow to the heart. Sources where the radioactive material is embedded in a solid (sealed sources), and therefore cannot readily disperse, are rarely used for diagnosis or treatment in nuclear medicine.

The distinctions between diagnostic radiology, radiation oncology, and nuclear medicine are sometimes poorly understood. In diagnostic radiology, an external source of radiation (an x-ray tube) is used to make an image (x-ray or radiograph) of the body. The underlying principle in obtaining diagnostic x rays is that dense structures (such as bones) absorb more of the x rays than do less-dense structures (soft tissues). X rays, therefore, provide information about the differential density of the body. From these studies, the size and shape (anatomy) of organs and abnormalities can be inferred. In nuclear medicine, the underlying principle used for diagnosis is very different. The accumulation from the radioactive tracers in various organs is related to the function of the organ (physiology) as opposed to its size and shape.

Nuclear medicine studies have two major potential advantages over diagnostic x rays. First, because changes in physiology usually occur before changes in anatomy, changes may be seen on a nuclear medicine study before they are seen on a conventional x ray. Some information provided by nuclear medicine studies (blood flow to the heart) cannot be obtained with simple diagnostic x rays. The second potential advantage of nuclear medicine over diagnostic x rays is that detecting abnormalities that may occur anywhere in the body can be easier with nuclear medicine. When a radioactive tracer is injected into a vein, it usually disperses throughout the body. For this reason, the entire body is often more easily studied with nuclear medicine. An example of a whole-body nuclear medicine study is bone imaging, which is performed with a radioactive tracer (technetium-99m–labeled phosphate compound) that accumulates in normal and abnormal bone.

Exciting new applications in nuclear medicine include the study of brain function and the use of antibodies to detect tumors. Special nuclear medicine studies using positron-emitting radionuclides (PET) can detect increases in blood flow and metabolism of parts of the brain which occur when subjects perform certain selected tasks. These studies have contributed to our understanding of how the brain works. Antibodies to tumor antigens are being developed and evaluated as an entirely new way to detect and treat tumors in the body.

In the past, only a small number of therapeutic procedures have been performed in nuclear medicine. The most common procedure is the use of radioactive iodine to treat an overactive thyroid gland or thyroid cancer. Because the thyroid gland greatly concentrates radioactive iodine, the thyroid gland receives a much larger radiation dose than any other organ in the body. The use of physiologic principles to deliver a large proportion of the radiation dose to the abnormal tissue or organ is quite unique. Conventional radiation therapy usually uses an external source of radiation. In order for this external source of radiation to reach the organ or tissue of interest, it first must pass through normal tissues. Recently, new therapeutic procedures have been developed which involve treatment of bone metastasis with an intravenous injection of a radiopharmaceutical, such as strontium-89.

Some people express concern about having radioactive tracers injected into their body. The radiation dose from diagnostic nuclear medicine is similar to the dose from many diagnostic radiology studies. In general, the benefits from the information obtained from the diagnostic studies greatly outweigh the risks from the radiation. The risks are the same whether or not the radiation comes from inside or outside the body.

See also: CAT SCAN; ISOTOPES; MAGNETIC RESONANCE IMAGING; MEDICAL PHYSICS

Bibliography

MCAFEE, J. G.; KOPECKY, R. T.; and FRYMOYER, P. A. "Nuclear Medicine Comes of Age: Its Present and Future Roles in Diagnosis." *Radiology* **174,** 609–620 (1990).

POSNER, M. I. "Seeing the Mind." *Science* **262,** 673–674 (1993).

HENRY D. ROYAL

NUCLEAR MOMENT

There are two classes of nuclear moments: electric moments, which describe the distribution of electric charge within a nucleus, and magnetic moments, which describe the distribution of electric current.

Electric moments can be classified as monopole, quadrupole, hexadecapole, and so on. In a spherically symmetric nucleus, all the electric moments are zero, except for the electric monopole moment, which is numerically equal to the total electric charge. The electric quadrupole moment of a nonspherical nucleus is positive if the charge distribution is prolate (football-shaped) and negative if the charge distribution is oblate (doorknob-shaped). Hexadecapole and higher moments describe more subtle deviations from spherical symmetry.

Magnetic moments can be classified as dipole, octupole, and so on. The nuclear magnetic dipole moment has two sources: the electric current associated with the motion of the protons in the nucleus and the intrinsic magnetic moments of the neutrons and proton. If two protons move in the same orbit but in opposite directions, their contributions to the nuclear magnetic dipole moment will cancel. Similarly, if their intrinsic spins are oppositely directed, their intrinsic contributions to the nuclear magnetic moment will cancel. Thus, the nuclear magnetic moment is determined by the unpaired neutrons and protons.

If a nucleus has only an electric monopole moment, the electrons moving around it experience a spherically symmetric nuclear electric field. The presence of a nonzero electric quadrupole implies that the nuclear charge distribution is nonspherical, which leads to a nonspherical component in the electric field experienced by the atomic electrons. Similarly, the existence of a nonzero nuclear magnetic dipole moment implies a magnetic coupling between the nucleus and the atomic electrons. These couplings can influence the atomic energy levels and, thus, the wavelengths of the spectral lines emitted during transitions between atomic energy levels. The effect is to introduce small wavelength differences between spectral lines whose wavelengths would be equal if the nucleus were spherically symmetric and non-magnetic. These differences are referred to as hyperfine structure. Their observation provided the first information about nuclear shapes and nuclear magnetism. This information has proved to be very useful in testing theories of nuclear structure.

The most accurate measurements of nuclear magnetic dipole moments are made using radio-frequency techniques. If the nucleus is placed in an external magnetic field, it will precess around this field with a frequency that depends on its magnetic dipole moment. This precession frequency can be determined by perturbing the nucleus with a radio-frequency field. When the perturbing frequency is close to the nuclear precession frequency, energy will be strongly absorbed from the perturbing field. Thus, the precession frequency can be determined, and from it the nuclear magnetic dipole moment. A similar technique is used in magnetic resonance imaging (MRI) devices. Here, one determines the concentration of hydrogen nuclei (protons) by looking for magnetic dipole moments equal to the known proton intrinsic magnetic moment. This technique has important medical applications, especially as a diagnostic tool, since hydrogen concentrations vary greatly between different types of tissues.

See also: DIPOLE MOMENT; ELECTRIC MOMENT; HYPERFINE STRUCTURE; MAGNETIC MOMENT; MAGNETIC RESONANCE IMAGING; SPIN

Bibliography

EISBERG, R., and RESNICK, R. *Quantum Physics of Atoms, Molecules, Solids, Nuclei, and Particles,* 2nd ed. (Wiley, New York, 1985).

KRANE, K. S. *Introductory Nuclear Physics* (Wiley, New York, 1988).

BENJAMIN F. BAYMAN

NUCLEAR PHYSICS

Nuclear physics is the subfield of physics concerned with the structure, interactions, and decay of atomic nuclei. Some studies of nuclei, which involve the properties of the constituents of nuclei, are on the borderline between nuclear and particle physics, other studies which involve the atomic electrons surrounding the nucleus are on the borderline between the subfields of atomic and nuclear physics.

Nuclear physics is often subdivided, as are other subfields of physics, into experimental nuclear physics, which consists of measurements on nuclei, and theoretical nuclear physics, which consists of calculations for the purpose of explaining or generalizing experimental results. As in other subfields of physics, the aim of the theory is to predict the results of measurements that have not yet been performed.

Another way of subdividing nuclear physics is to distinguish radioactivity (the spontaneous decay of nuclei), nuclear structure (the description of the nucleus in terms of its constituents), and nuclear interactions (scattering and reactions of various projectiles).

A third way of subdividing nuclear physics is based on the energy of the projectiles used to explore nuclei. Low-energy nuclear physics refers to projectiles of energy up to around 10 MeV per nucleon, medium-energy nuclear physics refers to energies between 10 and 100 MeV per nucleon, and high-energy or relativistic nuclear physics refers to energies above 100 MeV per nucleon, but the dividing points are not well defined.

Finally one can distinguish pure and applied nuclear physics. These two branches differ primarily in the aim of the study, that is, whether the purpose is to learn about nuclei or whether it is to obtain information for the application of nuclear physics to other subfields of physics or to practical applications in such areas as medicine, power generation, astrophysics, nondestructive testing, geophysical exploration, etc.

Nuclear physics began in 1896 with the discovery by Henri Becquerel of the radioactivity of uranium and later of other heavy elements, but it was not realized until 1911 that atoms consisted of nuclei and electrons and that radioactivity was a nuclear phenomenon. The discovery of the atomic nucleus by Ernest Rutherford, Hans Geiger, and Ernest Marsden was based on observations of the scattering (deflection) of alpha particles emitted by radioactive elements. The occurrence of very large deflections could be explained only by the assumption of a heavy charged nucleus at the center of an atom. In 1919 Rutherford found that alpha particles emitted by radioactive substances could induce nuclear reactions, that is, nuclei of one chemical element could be transformed into nuclei of another element.

Nuclear physics became a recognized subfield of physics in 1932 with the development of particle accelerators. Until 1932 nuclear interactions could be studied only with alpha particles emitted by radioactive substances. In 1932 John Cockcroft and Ernest Walton induced nuclear transmutations with protons accelerated through high voltages. At about the same time Ernest Lawrence built a cyclotron that permitted acceleration of particles to much higher energies, hence extended the range of energies at which nuclear transmutations and interactions could be studied. Also in 1932 James Chadwick found a new particle, the neutron, an uncharged constituent of the atomic nucleus.

In 1938 Otto Hahn and Fritz Strassmann found that neutrons could break the heaviest nucleus then known, uranium, into two, a phenomenon called fission. In the fission process a very large amount of energy is released, about 10 million times more than in a chemical reaction. Furthermore neutrons are released in the fission process so that a chain reaction was possible. This discovery occurred just before the beginning of World War II. The potential for the military application of this discovery was immediately realized, both for the purpose of power generation and for the development of explosives. During World War II most nuclear physicists in the United States (and many who came to the United States from Europe) participated in the so-called Manhattan project, which accomplished both these objectives. Nuclear physics was suddenly transformed from an abstract study in pure science to a discipline with many applications ranging from electric power generation to the diagnosis and treatment of disease.

In the postwar years nuclear physics became the most active subfield of physics. Every major physics department carried out nuclear research. Usually an accelerator was the center of the nuclear physics experimental effort. The best theoretical physicists attempted to interpret the measurements.

In the years immediately following World War II accelerators in most nuclear physics laboratories produced particles of a few mega-electron-volts of energy. Gradually more and more energetic particles became available. In addition there was an increasing interest in accelerating heavier ions including those of the heaviest elements, such as uranium. These larger accelerators are costly to build and maintain, hence their number is much smaller than that of the smaller accelerators previously used. Some of the larger accelerators are at or near universities, others are at national laboratories. Many experimental nuclear physicists now have to travel to these facilities for their research, a practice familiar to particle physicists.

Although some of the smaller accelerators are still used in nuclear physics research, some have been shut down, others are now used in applied nuclear physics for purposes such as the production of radionuclides for nuclear medicine or as analytical tools. Accelerator mass spectroscopy has become a powerful tool in fields ranging from archaeology to biology.

In the early days of nuclear physics it was discovered that chemical elements contain nuclei of different mass called isotopes. After the discovery of the neutron, nuclei were understood to consist of neutrons and protons; isotopes correspond to nuclides that have the same number of protons but different numbers of neutrons. The elements occurring on Earth contain about 300 different nuclides. With accelerators and nuclear reactors around 2,500 new nuclides have been produced. They do not occur in nature because they are radioactive, some live only for a fraction of a second. Generally their lifetime is the shorter the more their neutron number differs from that of a stable isotope. Additional nuclides are found at the rate of around thirty per year, especially with the use of heavy-ion accelerators. There has been a great interest in producing isotopes of elements that do not occur in nature, in particular transuranic elements. By 1984 all elements through number 109 (uranium is 92) had been found. There are theoretical arguments that a nuclide with 114 protons (element 114) and 184 neutrons should be relatively stable, but all efforts to find such a nuclide have been unsuccessful.

Much of what we know about the structure of atoms was learned from atomic spectroscopy, that is, from a study of the frequencies of light emitted by atoms. Such studies led to an understanding of atomic energy levels. Nuclear physicists likewise have spent much effort in studying nuclear energy levels and in determining their characteristics. Much of what we know about the structure of nuclei was derived from such studies.

Another method for studying the structure of nuclei is, in a way, a refinement of the alpha-particle scattering experiments that led to the discovery of the atomic nucleus. The most successful probes of the nucleus have been high-energy electrons. An electron accelerator serves the same function as a microscope used for visualizing objects much larger than nuclei. Electron scattering has given detailed information of the distribution of charge and magnetization of nuclei and remains an active area of nuclear research.

The building blocks of nuclei are protons and neutrons. From theoretical advances in particle physics we have learned that protons and neutrons consist of quarks and gluons. Under normal conditions it is impossible to find an unattached quark or gluon. There are, however, theoretical predictions that at very high temperatures and pressures quarks and gluons may be free and may move independently over substantial distances so as to form a quark-gluon plasma. In the Relativistic Heavy Ion Collider (RHIC) at Brookhaven National Laboratory, nuclei circulating in opposite directions will collide head-on and produce a region of enormous energy density. The hope is that in this small region a quark-gluon plasma will be formed. A quark-gluon plasma may also exist in the core of neutron stars, but on earth it represents a new state of matter. The purpose of RHIC is to study this state of matter in the laboratory. Creating a quark-gluon plasma is viewed as one of the most challenging problems of current nuclear research.

The nucleus is a much more complicated system than the atom. In the atom there is a heavy center of force, the nucleus, about which the electrons move; the force between the nucleus and the electrons and the force between electrons is the well-understood electrical force. The properties of atoms are determined by the electrons. By contrast the nucleus does not have a stable center of force, and the force between the constituents of the nucleus is complicated and cannot be expressed in terms of a simple force law. Even the basic properties of the lightest nuclides cannot be calculated reliably from basic principles. Efforts to predict nuclear properties usually are based on nuclear models. A nuclear model is generally based on a well-understood system in another field of physics. There is, for example, a liquid-drop model in which the nucleus is described in analogy to a water drop. This model is helpful in calculating nuclear masses. There is an optical model in which the nucleus is pictured as an optical medium that refracts and absorbs particle waves. This model is helpful for calculations of the interaction of particles with nuclei. Models emphasize one aspect of the nucleus, neglecting others, and contain parameters that are deduced from measurements rather from basic principles. Models are useful for interpolating between measurements and for extrapolating to conditions for which there are no measurements. In spite of the great effort that has gone into the development of nuclear models predictions based on model calculations are not reliable, and refining the models is a continuing endeavor of nuclear theorists.

See also: ACCELERATOR; ACCELERATOR, HISTORY OF; BASIC, APPLIED, and INDUSTRIAL PHYSICS; NUCLEAR STRUCTURE; QUARK; RADIATION PHYSICS

Bibliography

PATEL, S. B. *Nuclear Physics: An Introduction* (Wiley, New York, 1991).

PEARSON, J. M. *Nuclear Physics: Energy and Matter* (Adam Hilger, Bristol, Eng., 1986).

TIPLER, P. A. *Elementary Modern Physics* (Worth, New York, 1992).

H. H. BARSCHALL

NUCLEAR REACTION

The term "nuclear reaction" is applied to a variety of processes involving the collisions of atomic nuclei. In a typical reaction, an incident particle or nuclide I interacts with a target nucleus T to produce outgoing particles O and a product nucleus P. This process,

$$T + I \rightarrow O + P,$$

is usually written in more compact notation as

$$T(I, O)P.$$

Transfer Reactions

An example of a typical nuclear reaction is that of a proton with the stable isotope of fluorine (^{19}F) to produce a deuteron and an unstable fluorine-18 isotope. This is written $^{19}F(p, d)^{18}F$. Such a reaction is called single-nucleon transfer because a neutron transfers between the target and the incident projectile. It is also called a pickup reaction, since the proton "picks up" a neutron from the ^{19}F nucleus as it passes by. Pickup reactions can also involve multiple particle transfer such as $^{19}F(p, {}^3H)^{17}F$, $^{19}F(p, {}^3He)^{17}O$, and $^{19}F(p, \alpha)^{16}O$.

Another type of transfer reaction is called a stripping reaction, where nucleons are "stripped" from the projectile as it interacts with the target. An example of such a reaction would be $^{72}Ge(d, n)^{73}As$.

Stripping and pickup reactions have had useful application in nuclear physics as a means to determine the degree to which various excited states in a nucleus appear to be excitations of a single particle (or a few particles). In that case, the probability for one (or multiple) transfer reactions becomes large.

Elastic and Inelastic Scattering

In elastic scattering reactions, the initial and final particles are the same, for example, $^{20}Ne(\alpha, \alpha)^{20}Ne$. Primes and an asterisk are used to denote inelastic scattering in which the incident particle loses energy and the product nucleus is left in an excited state, for example, the $^{20}Ne(\alpha, \alpha')^{20}Ne^*$ reaction.

Multiple Particle Emission

More than two final particles may emerge from a reaction such as in the $^{18}O(n, 2n)^{17}O$ neutron knockout reaction, or in a spallation reaction such as $^{16}O(p, X)^7Li$. Here the nine outgoing particles are simply denoted as X. Such reactions generally occur at very high energies (≥ 100 MeV) where the process can be viewed as the result of an "intranuclear cascade" of collisions among individual nucleons.

Electromagnetic Interactions

Some nuclear reactions involve electromagnetic interactions. One type is called radiative capture. These reactions usually lead to a single product nucleus as a nucleon or nuclide is "captured" by the target nucleus followed by electromagnetic decay of the product via the emission of a γ ray. An example of such a process is the radiative capture of a proton on carbon-12 to produce ^{13}N and a γ ray. This is written $^{12}C(p, \gamma)^{13}N$. Radiative capture reactions play an important role in elemental nucleosynthesis. One type of radiative capture (called direct radiative capture) involves capture via an electromagnetic (rather than nuclear) interaction. Such reactions have a small probability but are useful as a spectroscopic tool.

Other electromagnetic interactions include photonuclear reactions in which incident γ rays are absorbed by a nucleus. If the target nucleus is excited this way above the particle emission threshold, then it may de-excite by the emission of one or more particles, such as the reaction $^{208}Pb(\gamma, p)^{207}Tl$. Energetic electron scattering from nuclei has also been extensively performed. Because the electron is not affected by the strong nuclear force it can be used as a means to determine the charge distribution in nuclei.

Heavy Ion Reactions

Reactions involving projectiles heavier than an α particle are generally referred to as heavy ion reactions. An example is $^{136}Xe(^{12}C, ^{7}Li)^{141}La$. Such reactions have been of considerable interest as a means to produce new elements and to explore the properties of bulk nuclear matter. Heavy ion reactions at relativistic energies are of interest as a means to search for the existence of a state of matter called quark–gluon plasma which is expected to exist at very high temperatures and densities.

Nuclear Reaction Mechanisms

In general, a given incident projectile upon a target nucleus can lead to a number of final product nuclei and outgoing particle combinations. Each possible combination is referred to as a channel. Different nuclear reaction mechanisms are characterized by representing the interaction between an incident nucleon (or nuclide) and a nucleus by a residual interaction plus a potential, which varies smoothly with target nucleus mass number and projectile energy. Elastic scattering occurs when the incident projectile interacts with the potential by itself. This is also called shape elastic or potential scattering. When there are interactions between the incident nucleon or nucleons and the nucleons in the target, the target nucleus becomes excited. If the incident nucleon (or nucleons) has relatively high energy, most of this excitation energy is quickly carried away by one or two emitted particles. This is called a direct reaction. However, if the incident particle or particles undergo multiple interactions, the incident energy becomes randomly distributed among many particles and the system loses memory of the initial collision. This is referred to as a compound nucleus reaction. This type of reaction tends to predominantly occur at low energies.

Nuclear Cross Sections

The probability for a nuclear reaction to occur is characterized by a nuclear cross section σ, which represents the effective projected area of a nucleus to an incoming projectile. That is, the number of reactions per second R of an incident beam of I particles per second impinging on a thin target foil comprised of N particles per square centimeter is given by

$$R = NI\sigma.$$

For a target nucleus of radius r, the geometric projected area is πr^2. In actual nuclear reactions, however, the measured cross section will be higher or lower than this depending upon the details of the nuclear states and interactions involved. However, for a typical nucleus, $r \approx 5 - 6 \times 10^{-13}$ cm, which implies that $\pi r^2 \approx 10^{-24}$ cm^2. This unit of area is referred to as a barn, 1 barn = 10^{-24} cm^2. The origin of the barn unit is said to be with the American colloquial expression "big as a barn," which was first applied to the cross sections for the interaction of slow neutrons with certain atomic nuclei. Subsequently, this term has been adopted as the official unit of measure for nuclear cross sections.

See also: DECAY, GAMMA; DEUTERON; ELECTROMAGNETIC RADIATION; FISSION; FUSION; ISOTOPES; NUCLEAR BINDING ENERGY; NUCLEAR PHYSICS; NUCLEOSYNTHESIS; PROTON; RADIOACTIVITY

Bibliography

FESHBACH, H. *Theoretical Nuclear Physics: Nuclear Reactions* (Wiley, New York, 1992).

HARVEY, B. G. *Introduction to Nuclear Physics and Chemistry*, 2nd ed. (Prentice Hall, Englewood Cliffs, NJ, 1969).

HEYDE, K. L. G. *Basic Ideas and Concepts in Nuclear Physics: An Introductory Approach* (Institute of Physics, Bristol, U.K., 1994).

JACKSON, D. F. *Nuclear Reactions* (Methuen, London, 1970).

PRESTON, M. A., and BHADURI, R. K. *Structure of the Nucleus* (Addison-Wesley, Reading, MA, 1975).

GRANT J. MATHEWS

NUCLEAR REACTOR

See REACTOR, NUCLEAR

NUCLEAR SHELL MODEL

Nuclei are made up of protons and neutrons (nucleons) held together by the strong interaction inside of a volume with a radius of a few fermis (fm), (1 fm = 10^{-15} m). One might expect that the motions of these nucleons in this closely packed system should be very complex because of the large number of frequent collisions. However, the early experimental studies of the properties of nuclear ground states revealed systematic properties, which in 1949 were interpreted by Maria Goeppert Mayer, and Otto Haxel, J. Hans D. Jensen, and Hans Suess in terms of the motion of one nucleon in a simple one-body potential. This one-body potential model is the starting point for the nuclear shell model. In the shell model, the quantum mechanical problem for the motion of one nucleon in a nucleus is similar to that for the motion of an electron in the hydrogen atom, except that overall scale is determined by the size of the nucleus (10^{-15} m) rather than the size of the atom (10^{-10} m). Another important difference between the atomic and nuclear potentials is that the dependence of the potential on the relative orientation of the intrinsic nucleon (electron) spin and its orbital angular momentum is much stronger and opposite in sign for the nucleon compared to that for the electron.

The single-particle potential has eigenstates that are characterized by their single-particle energies and their quantum numbers. The properties of a nucleus with a given number of protons and neutrons are determined by the filling of the lowest energy single-particle levels allowed by Pauli's exclusion principle, which must be obeyed in a system of identical fermions (the nucleons in this case). Pauli's exclusion principle allows only one proton or neutron to occupy a state with a given set of quantum numbers. The average nuclear potential arises from the short-ranged attractive nucleon-nucleon interaction and is determined by the shape of the nuclear density distribution.

Evidence for the validity of the nuclear shell model comes from the observation of shell effects in experimental observables such as binding energy, size, spin, and level density. In particular, the nuclear binding energy is not a smooth function of proton and neutron number, but exhibits small fluctuations. The deviation of the experimental binding energies from the liquid-drop model is shown in Fig. 1. The liquid-drop model binding energy is a smooth function of proton and neutron number. When the liquid-drop values for the binding energies are subtracted from the experimental values, the differences show peaks at the "magic numbers": $N_m = 28, 50, 82,$ and 126. The peak indicates that the nuclei with these magic number are more tightly bound than average. Those nuclei that are magic with respect to both neutron and proton numbers are referred to as doubly-magic; an example is the nucleus ^{208}Pb with $N = 126$ and $Z = 82$. Although not obvious from the data in Fig. 1, $N_m = 2, 8,$ and 20 are also magic numbers. The occurrence of magic numbers in nuclei is analogous to that ob-

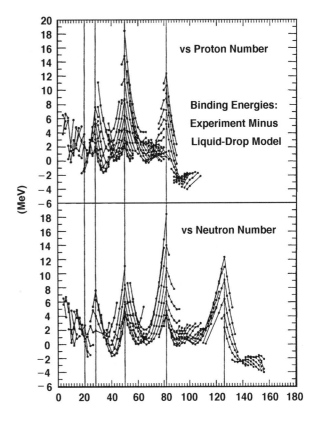

Figure 1 The differences between the experimental binding energies and those given by the liquid-drop model. The top part of the figure shows the difference plotted versus proton number Z with lines connecting the points for a given neutron number N. The bottom part of the figure shows the difference plotted versus neutron number with lines connecting the points for a given proton number. For clarity, only data for nuclei with even numbers of protons and neutrons are shown. The vertical lines show the positions of the magic numbers.

served for the properties of electrons in atoms. However, for electrons the magic numbers are 2, 10, 18, 54, and 86.

The calculated single-particle energy levels appropriate for the neutrons in ^{208}Pb are shown in Fig. 2. The potential arises from the average interaction of one neutron with the 207 other nucleons. Since

the nuclear force is short-ranged, the shape of potential is approximately proportional to the nucleon density in ^{208}Pb, which is experimentally known to be close to the Woods–Saxon shape of $\rho(r) \sim [1 + \exp(r - R)/a]^{-1}$, where $R \approx 1.2A^{1/3}$fm ($A = N + Z$) and $a \approx 0.60$ fm. The single-particle energy levels for a potential of approximately this shape and with

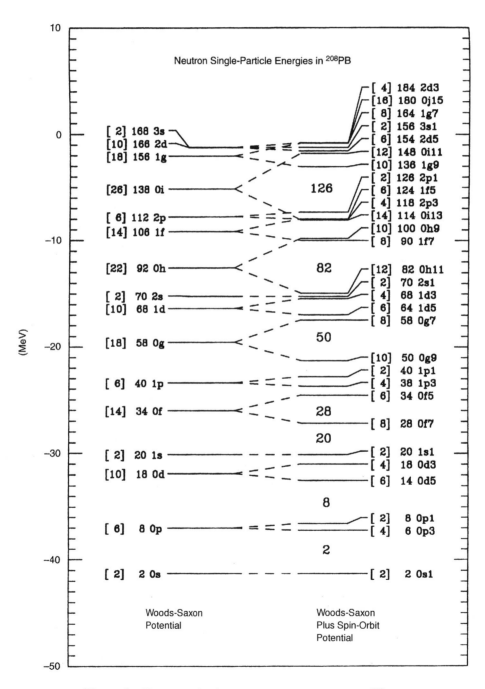

Figure 2 Neutron single-particle energy levels for ^{208}Pb.

a central depth of about -50 MeV are shown on the left-hand side of Fig. 2. The number of neutrons that are allowed by the Pauli principle to occupy one of these levels, the occupation number, is given by the number in square brackets. In addition, each level is labeled by its cumulative occupation number (the total number of neutrons needed to fill up to the given level) and its $n\ell$ value. n is the radial quantum number (the number of times the radial wave function changes sign) and ℓ is the angular momentum quantum number represented in the spectroscopic notations s, p, d, f, g, h, i, and j for $\ell = 0, 1, 2, 3, 4, 5, 6,$ and 7, respectively. Each ℓ value can have $2\ell + 1$ m states and each m state can contain a nucleon with spin up and spin down ($s_z = \pm 1/2$). The occupation number given by the Pauli principle is thus $N_o = 2(2\ell + 1)$.

The relative spacing of the neutron and proton levels for all nuclei are qualitatively similar to those shown in Fig. 2. (The overall spacing between levels goes approximately as $A^{-1/3}$.) According to the Pauli principle, as neutrons are added to nuclei they go into the lowest energy level not already occupied. When a nucleon is added to a nucleus in which the neutron number is equal to one of the cumulative occupation numbers, the neutron must be placed into a relatively high energy state. Thus the nuclei with the highest relative binding energy are those for which the proton or neutron number is equal to one of the cumulative occupation numbers. A magic number occurs when there is a relatively large energy gap above one of the cumulative numbers. The magic numbers are thus related to the bunching of energy levels. The Woods–Saxon potential gives the correct magic numbers for $N_m = 2, 8,$ and 20 but is incorrect for the higher values.

In 1949 Goeppert Mayer, Haxel, Jensen, and Suess postulated the existence of an additional strong spin-orbit potential that could account for the observed magic numbers. The spin-orbit potential has the form, $V_{so}(r)\vec{\ell} \cdot \vec{s}$, where $\vec{\ell}$ is the orbital angular momentum and $\vec{s}$ is the intrinsic spin angular momentum. With the spin-orbit potential, m and s_z are no longer good quantum numbers. The orbital and spin angular momentum must be coupled to a definite total angular momentum, $\vec{j} = \vec{\ell} + \vec{s}$. Eigenstates of the spin-orbit potential are determined by the total angular momentum quantum number $j = \ell \pm 1/2$ (except $j = 1/2$ for $\ell = 0$) and the quantum number m_j associated with its z component. Each m_j can contain one neutron (or proton) and the occupation number is thus $N_o = 2j + 1$. The

energy levels obtained when the spin-orbit potential is added to the Woods–Saxon potential are shown on the right-hand side of Fig. 2. The dashed lines that connect to the left-hand side of Fig. 2 indicate the effect of the spin-orbit potential in splitting the states of a given ℓ value. The strength of the spin-orbit potential has been determined empirically. Each level is labeled by the occupation number (in square brackets), the cumulative occupation number, and the values for n, ℓ, and $2j$ ($2j$ is twice the angular momentum quantum number j). The values of the neutron number for which there are large gaps in the cumulative occupation number now reproduce all of the observed magic numbers (as emphasized by the numbers shown in the energy gaps on the right-hand side).

The average nuclear potential can be calculated microscopically from the nucleon-nucleon interaction by using the Hartree–Fock theory together with the Breuckner theory for taking into account the repulsion at very short distances between the nucleons. The strength of the spin-orbit potential for nucleons is much larger and opposite in sign to spin-orbit potential for electrons atoms. The radial part of the spin-orbit potential, $V_{so}(r)$, is largest at the nuclear surface and is often taken to be proportional to the derivative of the Woods–Saxon form.

The essential physics behind the shell model is that the many nucleon collisions that might be expected are greatly suppressed in the nuclear ground states and low-lying nuclear levels because the nucleons would be scattered into states that are forbidden by the Pauli principle. At higher excitation energy the number of allowed states becomes much greater and the nuclear properties indeed become complex and chaotic.

The shell model in its simplest form is able to successfully predict the properties of nuclei that are one nucleon removed or added to the one of the magic number. The shell model can also be extended to include the more complex configurations that arise for the nuclei with nucleon numbers that are in between the magic numbers. For many applications these complex configurations can be taken into account exactly by the diagonalization of a Hamiltonian matrix. In other cases approximations must be used; these include the use of a deformed intrinsic single-particle potential, and the use of group theory to classify the configurations. Current theoretical investigations using the shell model focus on these complex configurations.

See also: LIQUID-DROP MODEL; NUCLEAR BINDING ENERGY; NUCLEON; PAULI'S EXCLUSION PRINCIPLE

Bibliography

BOHR, A., and MOTTELSON, B. R. *Nuclear Structure* (W. A. Benjamin, New York, 1969).

JOHNSON, K. E. "Maria Goeppert Mayer: Atoms, Molecules, and Nuclear Shells." *Phys. Today* **39** (Sept.), 44 (1986).

B. ALEX BROWN

NUCLEAR SIZE

The first measurements of the sizes of atomic nuclei were made by Ernest Rutherford soon after he inferred the existence of nuclei from the way alpha particles (helium nuclei) were scattered by gold foils. Rutherford compared the scattering of alpha particles by gold nuclei with the scattering that would have been produced by the electric field of a point charge. In this way, he concluded that an alpha particle could approach within about 3×10^{-14} m of the center of the gold nucleus and still experience no force other than an electrical force. This implied that an alpha particle 3×10^{-14} m from the center of gold nucleus was still outside the nucleus. In a later experiment in which alpha particles were scattered by hydrogen, Rutherford saw that non-electrical forces first came into play when a proton and alpha particle were separated by about 3×10^{-15} m, which thus provided an estimate of the sum of the radii of the proton and alpha particle.

Rutherford's idea of using deviations from the point-charge electric force law to determine nuclear radii is the basis for most measurement of the radial distribution of nuclear charge. The finite size of the nuclear charge distribution has a small but measurable effect on the energies of electronic states of atoms and, thus, on the energies of the observed photons emitted in atomic transitions. It has a larger effect on the energies of states in muonic atoms, since the fact that the muon mass is larger than the electron mass implies that it spends more time in the vicinity of the nucleus than does the electron and, thus, muonic energy levels are more sensitive to the details of the nuclear charge distribution.

The most precise information about nuclear sizes and charge distributions comes from electron scattering experiments. For example, an electron with a kinetic energy of 1,000 MeV has a de Broglie wavelength of about 10^{-15} m, which is small enough compared to nuclear dimensions to be able to resolve interesting structural details. The results of these experiments can be summarized as follows: The density of electric charge is nearly constant from the center of the nucleus out to a radial distance of about $1.2 \times A^{1/3} \times 10^{-15}$ m, where A is the number of nucleons (protons plus neutrons) in the nucleus. The charge density decreases from this constant value to nearly zero in a distance of about 10^{-15} m. Thus, for all but the lightest nuclei, the surface diffuseness occurs over a distance that is small compared to the nuclear radius, and for many purposes the nucleus can be approximated by a uniformly charged sphere or spheroid. The volume of a sphere of this radius would be $5.0 \times 10^{-45} \times A$ m^3, so that the volume per nucleon is 5.0×10^{-45} m^3, independent of A. These characteristics (relatively sharp surface and constant volume per particle) suggest that the nucleus can be modeled as a liquid droplet, and many of the properties of nuclei can in fact be described in this way.

Unfortunately, electron scattering gives very little information about the distribution of neutrons in nuclei, since neutrons are electrically neutral and thus have little effect on the motion of electrons. It is generally believed that the strong neutron-proton attraction causes the neutron and proton density distributions to be nearly proportional. Thus, even though a nucleus such as $^{208}_{82}\text{Pb}_{126}$ has 44 more neutrons than protons, the neutrons and protons seem to be distributed throughout the same volume, and there is no evidence of a neutron "halo" around a proton-rich core. Some light nuclei have been discovered, however, that have such large relative neutron excess that they do show signs of a neutron halo. An example is $^{11}_{3}\text{Li}_8$.

See also: BROGLIE WAVELENGTH, DE; ELECTRON; LIQUID-DROP MODEL; NEUTRON; NEUTRON, DISCOVERY OF; NUCLEAR STRUCTURE; NUCLEUS; PROTON; RUTHERFORD, ERNEST; SCATTERING, RUTHERFORD

Bibliography

EISBERG, R., and RESNICK, R. *Quantum Physics of Atoms, Molecules, Solids, Nuclei, and Particles*, 2nd ed. (Wiley, New York, 1985).

KRANE, K. S. *Introductory Nuclear Physics* (Wiley, New York, 1988).

BENJAMIN F. BAYMAN

NUCLEAR STRUCTURE

Atomic nuclei consist of protons and neutrons (collectively called nucleons). Each nucleon consists of three quarks. In every situation encountered so far, the quarks composing each nucleon remain confined within that nucleon, so it is permissible to describe the nucleus in terms of neutrons and protons without referring to this quark substructure.

The positive electric charge on each proton causes it to be repelled by every other proton in the nucleus. However, nucleons also attract each other by means of a strong, short-range force, which provides the cohesion needed to overcome the electrical repulsion. Because the range of this attractive force is so short, each nucleon only feels the attraction of its immediate neighbors. Thus, if a succession of nuclei of increasing size is considered, the amount of attractive interaction energy will increase in proportion to the number of nearest-neighbor pairs, which is proportional to the number of nucleons. However, because each proton can interact with every other proton via the longer-range electrical force, the amount of repulsive electrical energy increases in proportion to the total number of proton pairs, which is proportional to the square of the number of protons. Thus, as we get to bigger nuclei with more neutrons and protons, the electrical repulsion becomes relatively more important than the nuclear attraction. For sufficiently large nuclei (with more than about 240 nucleons), the disruptive effect of the electrical repulsion becomes so great that the nucleus is no longer stable. It undergoes fission into two smaller nuclei, which are driven apart by their electrical repulsion.

The short-range character of the nuclear attraction versus the long-range character of the electrical repulsion also plays an important role in the process of nuclear fusion. Consider, for example, the process in which two deutrons fuse to become an alpha particle. In so doing, a large amount of energy is released, as the attractive nuclear force pulls the two deutrons together. But in order to get the two deutrons close enough to feel this attractive force, it is necessary to overcome the electrical repulsion between the deutrons. Since the electrical repulsion extends to distances at which the attraction is very weak, it is necessary to supply some other form of energy to bring the deutrons together. If the deutrons are in a sufficiently hot plasma (of the order of millions of degrees) then some of them will have enough kinetic energy to overcome the barrier produced by the electrical repulsion, and they can then come close enough together to fuse. The main technical problem of fusion reactor design is that of creating and containing this hot plasma.

Nuclei have fairly constant density and binding energy per particle. In these respects, nuclei are similar to liquid droplets. Also, some nuclear excited states have droplet-like characteristics. For example, nuclear states are known in which the surface appears to vibrate. Other states have properties similar to a deformed liquid droplet, which rotates around a fixed axis. The process of nuclear fission can be modeled by an electrically charged liquid droplet, which becomes unstable when the disruptive effect of the electrical repulsion overcomes the cohesive effect of surface tension.

The motion of nucleons in a nucleus has many features in common with the motion of electrons in an atom. Both kinds of particles are fermions, and thus obey the Pauli exclusion principle. In both cases, the effect on any one particle of the forces exerted by all the other particles can be simulated by a self-consistent field. These principles imply that atoms with certain electron numbers (2, 10, 18, 36, 54, 86) will have very stable ground states. These are the so-called noble gases. The same principles imply that nuclei with certain numbers of protons or neutrons (2, 8, 20, 50, 82, 126) will have very stable ground states. These numbers are said to be magic. The existence of magic numbers has a strong effect on the relative abundance of different nuclei.

Knowledge of nuclear structure comes from observations of the electromagnetic radiation and particles they emit, as well as from the study of reactions that occur when nuclei collide. As in atoms, electromagnetic radiation is emitted when nuclei undergo a transition between two quantum states of the same system. The energy differences between atomic states, which equal the energies of the emitted photons (the quanta of electromagnetic radiation), are on the order of magnitude of 1 eV. These photons are in or near the visible part of the electromagnetic spectrum. However, the energy differences between nuclear states are about a million times greater, so that the photons emitted by nuclei have wavelengths about a million times shorter than visible photons. These nuclear photons are called gamma rays. Neutrons, protons, alpha particles, and fission fragments can also be emitted by nuclei in unstable states. In addition, it is possible for protons to transform into neutrons (and vice versa) with the emission of positrons and neutrinos (or electrons and antineutrinos), a process known as beta decay. All these emission processes are collectively called radioactivity.

Nuclear accelerators are devices that can be used to create beams of energetic protons, neutrons, or alpha particles, as well as heavier nuclei. These accelerated particles can collide with target nuclei, and if they have enough energy to overcome the electrostatic repulsion, various nuclear reactions will be initiated. In some cases, the projectile amalgamates with the target nucleus forming a compound nucleus, which may exist for such a long time that when it eventually decays it will have lost all "memory" or the process by which it was formed. In other cases, the reaction might occur without amalgamation of the projectile and target nuclei, as the projectile moves past the target on a peripheral trajectory. This is referred to as a direct reaction, since it does not proceed via the formation of an intermediate compound state.

It is generally believed that the universe immediately after the big bang consisted of a hot quark-gluon plasma. As this plasma cooled, a phase transition occurred, during which triplets of quarks condensed out of the plasma to form nucleons. This occurred about 10^{-6} s after the big bang. Attempts are now being made to reverse this phase transition by producing collisions of heavy nuclei at very high energy. It is hoped that we can thereby create conditions of such high density and temperature that the quarks in individual nucleons will be de-confined, and we will be able to study the matter as it existed in the primordial quark-gluon plasma.

See also: ACCELERATOR; BIG BANG THEORY; ELECTRON; ELECTROSTATIC ATTRACTION and REPULSION; LIQUID-DROP MODEL; NEUTRON; NUCLEAR SIZE; NUCLEON; NUCLEUS; PROTON; QUARK

Bibliography

EISBERG, R., and RESNICK, R. *Quantum Physics of Atoms, Molecules, Solids, Nuclei, and Particles*, 2nd ed. (Wiley, New York, 1985).

KRANE, K. S. *Introductory Nuclear Physics* (Wiley, New York, 1988).

BENJAMIN F. BAYMAN

NUCLEON

A nucleon is a constituent of the atomic nucleus made up of protons and neutrons. Protons carry a charge $+e$, while neutrons have zero charge. The discovery of the nuclear constituents was guided by Ernest Rutherford during the two decades that followed the pioneering experiments of Hans Geiger and Ernest Marsden in 1909 and 1910. Following publication of Bohr's model for hydrogen in 1913 (carried out in Rutherford's lab), the work of Henry G. J. Moseley in elucidating the sequence of atoms in the periodic table in terms of atomic number rather than atomic "weight" led Rutherford to recognition of the existence of the proton as a positively charged particle shortly before 1920. James Chadwick discovered the neutron in 1932 in Rutherford's lab. In fact, the nucleon number is approximately the atomic mass measured in atomic mass units (u). This occurs because the masses of the proton and the neutron are very close to 1 and to one another, 1.007276470 u and 1.008664904 u, respectively.

The exchange of the neutron for the proton would make no difference if the proton had no electric charge. In nuclei, electricity is responsible for repulsion of protons, so the existence of nuclei, electricity is responsible for repulsion of protons, so the existence of nuclei implies that nucleons are subject to a force stronger than this electric repulsion. The stronger interaction responsible for nuclear binding is known as the strong nuclear interaction; in comparison to the strong interaction, electricity is essentially negligible.

The designation of the nucleon as a generic nuclear constituent is consistent with the the ideas of nuclear symmetry, developed particularly by Eugene P. Wigner and called isospin symmetry. While the name isospin implies a similarity to particle spin, it merely recognizes that the formalism of spin, including analogs to spin multiplets, is reproduced. An exchange of a proton and a neutron constitutes a symmetry operation with respect to the strong interaction. This means that nuclei with equal nucleon numbers are degenerate (neglecting electricity) and form multiplets. Each member of the nuclear multiplets is distinguished from the others by the number and arrangement of protons in the nucleus. The real nuclei should be almost degenerate in ground state energy. For example, for $A = 14$ there are four quasi-stable nuclei: ^{14}O (mass = 14.0086 u), ^{14}N (mass = 14.00307 u), excited ^{14}N (mass = 14.00554 u), and ^{14}C (mass = 14.00324 u). The two ^{14}N states are predicted by isospin symmetry and correspond to states of different total isospin. Isospin symmetry also makes other predictions. For example, it relates (once one corrects for the effect of proton charge) the scattering probability of protons on protons to those of protons on neutrons and neutrons on neutrons at the same energy.

See also: ATOMIC MASS UNIT; CHADWICK, JAMES; ELECTRON; ELECTRON, DISCOVERY OF; NEUTRON; NEUTRON, DISCOVERY OF; NUCLEUS; NUCLEUS, ISOMERIC; PROTON; RUTHERFORD, ERNEST

GORDON J. AUBRECHT II

NUCLEOSYNTHESIS

Nucleosynthesis involves the production, on a cosmic scale, of the chemical elements that comprise the normal matter in stars and galaxies, and in the universe in general. While it is usual to speak of the chemical elements, it is rather *nuclear* reaction processes, operating both in the early universe and in the stellar components of galaxies, that are primarily responsible for the synthesis of the heavy elements. Nucleosynthesis studies seek to determine the physical mechanisms responsible for the production of the elemental abundance patterns observed in nature and to identify the astrophysical sites in which these mechanisms can operate.

The recognition of the fact that it is the nuclear properties of the elements that are crucial to understanding their abundances in nature also makes clear why it is that nucleosynthesis is such a young science. The occurrence in nature of natural radioactive elements, recognized early in the twentieth century, made clear the fact that the age of the elements is finite. However, significant advances in our knowledge of nuclear physics were still demanded for progress to be realized. It is significant to note that the identification, in the late 1930s, of the basic hydrogen burning sequences that power stars over the greater part of their lifetimes followed shortly on the discovery of the neutron in 1932.

An accurate and detailed knowledge of the abundances that characterize the stars and gas in our galaxy and in other galaxies also constitutes essential input to nucleosynthesis studies. The earliest spectral studies of the Sun and stars were able to establish only that the elements of which the stars and Earth are composed are *qualitatively* the same. It was therefore reasonable to assume the chemical composition of the universe to be uniform. Such uniformity strongly suggested a single-event theory of the origin of the elements; this led scientists to search within the framework of cosmology for a set of physical conditions that could produce the observed distribution of abundances.

In 1948, Ralph Alpher, Hans Bethe, and George Gamow proposed such a theory of nucleosynthesis tied to a cosmological big bang. This model envisioned the early state of matter as a compressed neutron gas. The expansion of this neutron fluid, following an initial explosion, would result in the decay of some of the neutrons into protons and electrons. Subsequent neutron captures, interspersed with beta decays, could then build up all of the heavier elements. In a continuation of this work, Alpher noted that the abundance peaks in the heavy-element region (mass number $A \geq 60$) corresponded to regions in which the neutron-capture probabilities were small. This correlation made evident the need for a neutron-capture process in the synthesis of the heavy elements, and such a process has indeed been incorporated into modern theories. It soon became clear, however, that cosmological synthesis of all the elements is not possible. Current calculations predict, rather, that big bang nucleosynthesis contributes substantially only to present day abundances of the light nuclei: hydrogen, deuterium, helium, and lithium.

Even as these studies proceeded, advances in our understanding of nuclear physics were preparing the way for revised and perhaps definitive theories of the origin of the elements. The defining role played by thermonuclear processes in providing an energy source sufficient to account for stellar lifetimes of billions of years was established in the late 1930s. The fusion of four protons to one helium nucleus (hydrogen burning), taking place in approximately 10 percent of the Sun's mass, provides several orders of magnitude more energy than has yet been released in gravitational contraction.

The recognition that nucleosynthesis is a continuing process in stellar interiors followed Paul Merrill's 1952 discovery of the presence of the element technetium in the atmospheres of red giant stars. Because technetium has no stable isotopes and because its longest-lived isotope has a half-life of less than two million years, its presence in stellar atmospheres indicates that element synthesis has taken place quite recently in those stars. The view that nucleosynthesis has proceeded primarily in stars over the lifetime of our galaxy received further support with the discovery, in the mid-1950s, that there exist abundance differences between certain broad classes of stars, in the sense that the ratio of the abundances of the heavy elements to hydrogen is variable and that this ratio is correlated with the age of the star. The heavy element (this usage in astronomy means elements more massive than helium) content of the Sun is approximately 2 percent by mass, while the oldest stars in the Milky Way Galaxy are now known to have less than one-thousandth of this solar value. Furthermore, the presence in stars of anomalous abundances of elements that are produced by specific nuclear-burning mechanisms provides somewhat more direct evidence for nucleosynthesis in stars. The presence of technetium in red giant stars, for example, implies that element synthesis by neutron capture has occurred, while the high abundances of carbon in carbon stars are attributed to helium burning in the interior.

These observations are consistent with a rather straightforward model of galactic evolution and nucleosynthesis. Assuming the primordial gas to be composed predominantly of hydrogen and helium, products of the cosmological big bang, the first generation of stars in our galaxy will be formed with this composition. The evolution of this first stellar generation will be characterized both by element synthesis during various stages of nuclear burning and, ultimately, by the return of processed matter to the in-terstellar gas by wind-driven mass loss or explosive (supernova) ejection. In this manner, the heavy-element content of the interstellar medium will be increased. Subsequent generations of stars will be formed from gas enriched in these heavy elements. Since the late 1950s, significant progress in our understanding of nucleosynthesis mechanisms has proceeded within this theoretical framework.

Cosmological Nucleosynthesis

The 1965 discovery by Arno Penzias and Robert Wilson of the 3 K background microwave radiation served to revive interest in the subject of the cosmological synthesis of the elements. The identification of this radiation as the relic of a cosmological big bang remains far the most satisfactory theoretical interpretation.

In the context of the standard model, the earliest moments of the universe are found to be characterized by extreme temperatures and densities. At initial temperatures above 10^{11} K, it follows that even weak interactions will achieve thermodynamic equilibrium. Equilibrium concentrations of neutrons and protons are maintained via the reactions

$$p + e^- \rightarrow n + \nu_e$$
$$n + e^+ \rightarrow p + \nu_e.$$

Following expansion and cooling, the critical temperature for nucleosynthesis of $\sim 10^9$ K is realized; the synthesis of heavy nuclei is restricted, at higher temperatures, by nuclear photodisintegrations and, at lower temperatures, by the inability of charged particles to penetrate their mutual Coulomb barriers on the required timescale. The neutron-to-proton ratio emerging at the onset of this era is approximately 1/7. Deuterium formation first proceeds by means of the reaction

$$n + p \rightarrow d + \gamma.$$

Nuclear reaction sequences leading to the production of ^{3}He and ^{4}He then take place rapidly. Subsequently, most of the initial neutrons are used in the production of ^{4}He, yielding a helium mass fraction ~ 0.25. However, extremely important contributions are made as well to the cosmic abundances of deuterium, ^{3}He, and ^{7}Li. In fact, the fragility of deuterium to destruction in stellar environments is such that it is currently believed that significant deuterium production can *only* take place in a cosmological context. It is also important to note that observations of deuterium, ^{3}He, ^{4}He, and ^{7}Li abundances in our galaxy and in quasi-stellar object (QSO) absorption line systems are now serving to impose extremely important constraints on the conditions in the early universe and on particle physics.

One rather firm conclusion follows from theoretical studies: Given the observational limit on the present universal baryon density of the universe, and assuming that the microwave radiation is correctly interpreted as a residual 2.7 K radiation temperature, no substantial production of nuclei heavier than ^{4}He is possible in the cosmological big bang. The production of heavier elements would require either a significantly higher present-day density or a lower present-day temperature. The triple-alpha reaction (3^4He $\rightarrow$ ^{12}C), which served successfully to bridge the mass gaps at $A = 5$ and $A = 8$ (no stable nuclei exist with these nucleon numbers) in stellar interiors, is ineffective at the prevailing densities in the expanding fireball at the appropriate burning temperature. It is currently believed that primordial nucleosynthesis contributes only to the present day abundances of hydrogen, deuterium, ^{3}He, ^{4}He, and ^{7}Li.

Stellar Energy Generation and Nucleosynthesis

Stellar evolution is characterized by a sequence of alternate stages of gravitational contraction to higher temperatures (and densities) and thermonuclear burning of the available fuels at these temperatures. The various burning stages in the lives of stars were first defined in 1957, in the now classic papers by Margaret Burbidge, Geoffrey Burbidge, William Fowler, and Fred Hoyle and by Alastair Cameron. By far the greatest fraction of the active burning lifetime of a star is spent converting hydrogen into helium (hydrogen burning), releasing approximately 7 MeV per nucleon. Nuclear transformation can provide at most only another ~1 MeV per nucleon, assuming burning proceeds all the way to ^{56}Fe. This is found to occur only in stars more massive than about 10 M$_\odot$ (solar masses). Less massive stars develop dense cores in which electron degeneracy pressure becomes dominant; following the loss of often a substantial fraction of their envelope mass,

they evolve to white dwarf stars of helium, carbon-oxygen, or oxygen-neon-magnesium composition.

Nuclear transformations in massive stars build toward ^{56}Fe in a succession of burning stages: (1) helium burning to ^{12}C and ^{16}O, proceeding via the reactions

$$^3\text{He} \rightarrow {}^{12}\text{C}$$

and

$$^{12}\text{C}(\alpha,\gamma)^{16}\text{O};$$

(2) carbon and oxygen burning, proceeding via the heavy-ion reactions

$$^{12}\text{C} + {}^{12}\text{C} \rightarrow {}^{20}\text{Ne} + {}^4\text{He}$$

$$\rightarrow {}^{23}\text{Na} + p$$

and

$$^{16}\text{O} + {}^{16}\text{O} \rightarrow {}^{28}\text{Si} + {}^4\text{He}$$

$$\rightarrow {}^{31}\text{P} + p$$

to intermediate mass nuclei; and (3) silicon burning, proceeding by a complex sequence of photonuclear and charged-particle-induced reactions to nuclei in the immediate vicinity of iron. Since ^{56}Fe is the most tightly bound nucleus (per nucleon), further processing of this matter cannot provide a nuclear energy source.

The formation of nuclei heavier than iron is due primarily to neutron-capture reactions. Two distinct environments are required, defined by the conditions that the rates of neutron capture are either slow (*s*-process) or rapid (*r*-process), compared to electron decay rates in the vicinity of the valley of beta stability. Production of the most neutron-rich isotopes of heavy nuclei and of nuclei heavier than lead, both require more substantial neutron fluxes (*r*-process), as may be realized in supernova environments. We now have observational confirmation of the occurrence of this *r*-process in massive star environments, as the oldest and most metal-deficient stars in our galaxy have been determined to have a pure *r*-process character. Less extreme neutron fluxes, realized during helium shell burning in red giant stars, can provide an appropriate s-process environment. The presence of technetium in the atmospheres of red giants testifies to the operation of this nucleosynthesis mechanism. In these red giant environments, neutron sources sufficient to drive heavy element synthesis are provided by the reactions $^{13}C(\alpha, n)^{16}O$ and $^{22}Ne(\alpha, n)^{25}Mg$.

From the point of view of nucleosynthesis, stable phases of stellar evolution serve as the likely source of the ^{12}C and ^{16}O existing in our galaxy and of the designated s-process heavy elements. Concentrations of nuclei from neon to iron, formed during stable burning phases, can be substantially altered as a result of their shock ejection in supernova events. It is therefore necessary to consider the mechanisms of nuclear transformations and associated nucleosynthesis that occur during the final, dynamic phases of stellar evolution.

Supernova Nucleosynthesis

Although the supernova phase constitutes only an extremely small fraction of the lifetimes of stars in restricted ranges of stellar mass, theoretical studies nevertheless suggest that supernova environments represent the likely site of formation of most of the heavy elements (mass number $A \gtrsim 20$) observed in nature. From the point of view of nucleosynthesis, one has the advantage that constructive nuclear transformations proceed in the shock-induced ejection of both core and envelope matter, ensuring that the resulting abundance distributions will not be distorted by subsequent evolution. Hydrodynamic studies of supernova ejection mechanisms predict very promising conditions for the operation of two distinct thermonuclear processes: (1) the synthesis of elements through the vicinity of iron by charged-particle reactions and (2) the neutron-capture synthesis of heavier nuclei.

The characteristics of these environments are apparent from a consideration of the final phases of evolution of massive stars ($\gtrsim 10$ M$_\odot$). Massive stars have been shown to proceed in stages of nuclear burning to a configuration consisting of an iron core, with overlying shells dominated by silicon, magnesium, oxygen, carbon, and helium, respectively, and an extended hydrogen envelope. Lacking further nuclear fuel, the iron core is compelled to shrink under gravity. The fate of massive stars is then dependent on the existence of some mechanism capable of using the gravitational energy re-

leased in core contraction to cause the ejection of the overlying matter. If this cannot occur, gravitational collapse of the entire star to a black hole is inevitable.

Until recently, definitive theoretical statements concerning nucleosynthesis in supernovas were simply not possible. Observationally, the identification of supernovas with massive star progenitors was suggested by their occurrence in regions of active star formation, and their association with neutron-star remnants was, in some instances, confirmed (e.g., Crab nebula pulsar). However, theoretical studies were unable to establish the precise nature of the mechanism of mass ejection and, therefore, specific predictions concerning the nucleosynthesis conditions accompanying supernova events were highly uncertain.

The recent outburst of Supernova 1987A in the Large Magellanic Cloud, the closest visual supernova event since that recorded by Johannes Kepler in 1604, has changed the situation greatly by providing an extraordinary opportunity to test theoretical models. One significant outgrowth of observational studies of this supernova has been the unambiguous identification of the stellar progenitor, the star Sanduleak -69 202. It is estimated that this supergiant star had an initial mass of approximately 20 M$_\odot$, of which $\sim$ 1.4 M$_\odot$ was incorporated into a neutron star remnant and $\sim$ 5 M$_\odot$ was ejected in the form of heavy nuclei, ranging in mass from oxygen to iron and nickel. Neutrino detections, indicating the release of $\sim 3 \times 10^{53}$ erg in neutrinos, confirmed the collapse to a neutron star. The exponentially falling light curve of SN 1987A, with a lifetime consistent with the decay lifetime of ^{56}Co ($\tau_{1/2}$ = 77.8 days), together with a knowledge of the total observed luminosity of the supernova and the distance to the Large Magellanic Cloud, allow the determination of the total mass ejected in the form of nuclei of mass $A = 56$. Approximately 0.075 M$_\odot$ was ejected in the form of ^{56}Ni ($\tau_{1/2}$ = 6.1 days), which decays through ^{56}Co to ^{56}Fe. In general, supernovas involving massive stars are expected to eject substantial amounts of matter in the form of such nuclei as oxygen, neon, magnesium, silicon, calcium, chromium, iron, and nickel, in essentially solar proportions, as a consequence of the nuclear burning of matter at temperatures in the range $\sim 2.6 \times 10^9$ K. For some range of stellar masses, expansion and cooling of matter from the immediate vicinity of the forming neutron star core may provide appropriate conditions of the (r-process) synthesis of the heavy elements. A single supernova event can therefore be responsible for the formation of a broad range of nuclear species from oxygen to uranium—in roughly solar proportions.

Cosmic Abundance Evolution

Recent observational studies, both with the Hubble Space Telescope and with large aperture ground-based telescopes, are providing increasing amounts of information concerning the spectroscopic and photometric properties of gas clouds and galaxies at high redshifts, which are serving to impose increasingly stringent constraints on models of galaxy formation, galactic evolution, and nucleosynthesis. Gas enriched by nucleosynthetic processes is now detected out to redshifts $z \simeq 4$, by the metal absorption lines it produces in the spectra of background QSOs. Redshifts measured in the absorption systems presumably sample the gas at different points in their roughly common evolutionary paths, and the abundance patterns in these systems offer further clues to nucleosynthesis at early epochs in our universe. The heavy element abundances found in the QSO absorbers span the metallicity range that overlaps the range found in galactic disk, thick disk, and halo dwarf stars. We are now poised on the brink of a new era of study of chemical evolution, this time on the scale of the universe itself.

See also: BARYON NUMBER; BIG BANG THEORY; COSMIC MICROWAVE BACKGROUND RADIATION; COSMOLOGY; ELEMENTS; NEUTRON STAR; PULSAR; QUASAR; STELLAR EVOLUTION; SUPERNOVA

Bibliography

ARNETT, W. D. "Explosive Nucleosynthesis Revisited: Yields." *Ann. Rev. Astr. Ap.* **33**, 115 (1995).

ARNETT, W. D.; BAHCALL, J.; KIRSHNER, R.; and WOOSLEY, S. E. "Supernova 1987." *Ann. Rev. Astr. Ap.* **27**, 269 (1989).

BETHE H. A., and BROWN, G. "How a Supernova Explodes." *Sci. A.* **252** (5), 60 (1985).

COWAN, J. J.; THIELEMANN, F.-K.; and TRURAN, J. W. "Radioactive Dating of the Elements." *Ann. Rev. Astr. Ap.* **29**, 447 (1991).

SCHRAMM, D. N. "The Age of the Elements." *Sci. Am.* **230** (1), 69 (1974).

TAYLOR, R. J. *The Origin of the Chemical Elements* (Wykeham, London, 1972).

TRIMBLE, V. "The Origin and Abundances of the Chemical Elements." *Rev. Mod. Phys.* **47**, 877 (1975).

TRURAN, J. W. "Nucleosynthesis." *Ann. Rev. Nucl. Part. Sci.* **34**, 53, (1984).

WHEELER, J. C.; SNEDEN, C.; and TRURAN, J. W. "Abundance Ratios as a Function of Metallicity." *Ann. Rev. Astr. Ap.* **27**, 279 (1989).

WOOSLEY, S. E., and WEAVER, T. "The Great Supernova of 1987." *Sci. Am.* **261** (2), 32 (1989).

JAMES W. TRURAN

NUCLEUS

Briefly, the nucleus of an atom is the small core in which is found almost all of the atom's mass and all of its positive charge.

Historical Perspective

The story of the atomic nucleus begins in 1896 with the discovery of natural radioactivity by the French physicist Henri Becquerel. He found by accident that well-wrapped photographic film on which he had placed some crystals of uranium salts showed a blackened image of the crystals when developed. He subsequently determined that the effect was independent of the chemical form of the crystal, and that the image strength depended on the quantity of uranium present. It was thus due, not to some chemical or molecular process, but to a specific atomic property of the uranium.

Soon other natural substances were found to produce the same effect, and the phenomenon was named radioactivity by the Polish physical chemist Marie Curie. The radioactivity manifested itself in the form of two so-called emanations, identified as alpha and beta rays. The alpha rays were characterized as soft because they were effectively blocked by a few layers of thin paper or several centimeters of air. The beta rays were substantially harder, since they could travel up to a few meters in air and penetrate thin metal foils.

Subsequent work by many others during the early 1900s resulted in the extraction from natural ores and identification of many new radioactive substances. Some had different radioactive properties, but when sufficient quantities were accumulated, they were found to have identical chemical properties. Further work showed that the differing radioactive properties were associated with different atomic weights. Thus in 1910 Frederick Soddy, working at Cambridge University, was led to point out that since some radioactive elements have atoms possessing different masses, which he called isotopes, the same might be true of stable elements.

The idea that each element contains an integral number of hydrogen atoms had been put forward by the English chemist William Prout early in the nineteenth century, but had been discredited when it was discovered that some common elements such as chlorine and magnesium had nonintegral atomic weights. Soddy's concept of isotopes showed that elements having nonintegral atomic weights could be accounted for by the appropriate mixture of isotopes. Resurrecting Prout's hypothesis was tempting, but the full story on whether atoms were composed of common constituents had to wait until several other crucial discoveries were made.

In 1897 the British physicist Joseph John Thomson discovered the electron, whose mass and negative charge were subsequently measured. The stage was set for the work by Becquerel that led to the identification of beta rays as electrons. This meant that atoms must contain electrons, since they are products of radioactivity. The electron mass turned out to be about 0.05 percent of that of the hydrogen atom, implying that most of the mass of an atom is associated with a positive charge, since atoms were known to be electrically neutral. Thomson put forth his idea of the structure of an atom—a small spherical cloud of positive electricity, inside of which is embedded the number of negative electrons required to balance the positive charge. The electrons would vibrate when disturbed, emitting light in the visible spectrum. There were problems with this model, and it was soon to be replaced by a new concept originated by the New Zealander Ernest Rutherford, working at Cambridge University in England.

Rutherford carried out experiments with the alpha particles that were emitted by radioactive materials. One of his first findings was that alpha rays were heavy charged particles, with a mass four times that of hydrogen ions and a charge twice that of hydrogen ions. In this way he conclusively identified alpha particles as helium ions. After measuring the velocity of the alpha particles from various radioactive substances and calculating their kinetic energies, he concluded that the amount of energy carried away by an alpha particle was several million times greater than the energy release of a typical chemical reaction. An entirely new form of energy was thus discovered, which was destined to have its impact later on in the twentieth century.

There followed a period of intense investigation of the interactions of alpha particles with matter. Rutherford initiated experiments in which a pencil-like beam of alpha particles in a vacuum was allowed to impinge upon a thin mica foil. He observed that the beam spread out slightly upon leaving the foil, as though the individual alpha particles had been scattered or deflected by material of the foil. More quantitative measurements were carried out by his assistants Hans Geiger and Ernest Marsden. They found that the most probable deflection angle for an alpha particle in passing through a gold foil of thickness 0.0005 mm was about 1°. However, they were also able to determine that one alpha particle out of 8,000 actually scatters by an angle greater than 90°. Such large deflection angles could not be explained by reflection from the foil's surface, since doubling the foil thickness also doubled the fraction of particles that scattered through large angles. Furthermore, the fraction of scattered alpha particles undergoing large deflection angles was so completely at odds with the predictions of the Thomson model of the atom that a new explanation was called for.

In a classic paper published in 1911, Rutherford first proposed the nuclear model of the atom. Here he suggested that the positive charge of an atom resides in a small dense core, which we now call the nucleus. Around the nucleus are arranged the electrons, needed to assure the atom's electrical neutrality. He demonstrated that the large alpha particle scattering angles observed could be explained if the diameter of the nucleus were less than 10^{-12} cm. In addition, the model provided quantitative predictions for the variation of alpha particle scattering intensity with angle, foil thickness, alpha particle kinetic energy, and nuclear charge of the foil atoms. All of these predictions were subsequently verified in experiments carried out by Geiger and Marsden. The success of Rutherford's nuclear model, combined with other observations using x rays, led to the conclusion that the number of electrons in the atom, equal to the number of positive charges in the nucleus, could be identified with the number assigned to the element in the periodic table of the elements devised by the Russian chemist Dmitri Ivanovich Mendeleev. Thus the nucleus has a charge of Ze, where $-e$ is the electron charge and Z is the symbol denoting the atomic number. There are also Z electrons surrounding the nucleus.

It was not until the neutron was discovered by James Chadwick in 1932 that a more complete understanding of the structure of the nucleus was reached. The neutron was found to have nearly the same mass as the hydrogen nucleus, or proton. With the discovery of the neutron, many puzzling phenomena were easily explained. The picture that emerged was a nucleus with mass number A consisting of Z protons and ($N = A - Z$) neutrons, surrounded by Z electrons. Isotopes were simply nuclei with the same number of protons and differing numbers of neutrons. Because they serve as the constituents of nuclei and possess similar masses, protons and neutrons are referred to as nucleons.

Properties of the Nucleus

In addition to possessing charge and mass, nuclei have other properties which have been extensively investigated. Nuclei are many-body quantal systems with well-defined properties, and since there are hundreds of different nuclei that can be found in nature or prepared in the laboratory, the nuclear physicist is presented with the unique opportunity to investigate the same phenomena under many different and controllable conditions.

Size. Most nuclei can be visualized as tightly packed spheres with approximately constant density. This implies a volume that is proportional to the number A of nucleons in the nucleus, and a radius that is given by the expression $R = R_0 A^{1/3}$, where $R_0 = 1.1$ to 1.2×10^{-13} cm. The unit of length 10^{-13} cm invariably turns up when referring to nuclear distances, and has been named the fermi (fm) in honor of the Italian physicist Enrico Fermi. The radii of nuclei are all less than 10 fm, extremely small when compared with the radii associated with atoms, namely, a few times 10^{-8} cm.

Spin. The spin, or total angular momentum, of the lowest energy state plays a very important role in the structure of nuclei. It is normally quoted, in units of $h/2\pi$, by the number I, where h is Planck's constant. In nuclei with an even number of nucleons (even A), I is 0 or an integer, while nuclei with an odd number of nucleons have half-integer values of I. For even-even nuclei, those having even values of Z and N, the ground state spin is always 0. Individual protons, neutrons, and electrons possess spin $I = 1/2$.

Magnetic moment. Associated with nuclear spin is a magnetic dipole moment, having its origin in the circulation of electric charge inside the nucleus. An electron with spin $I = 1/2$ has a magnetic dipole moment about 1,000 times greater than the typical magnetic moments found for nuclei. The nuclear magnetic moment manifests itself by producing very small modifications to the optical spectra of atoms.

Electric quadrupole moment. The electric charges contained within the nucleus can produce electric field components other than the pure radial one expected from a spherical charge. These arise from the fact that the time average of the charge distribution in many nuclei may differ slightly from sphericity. This means that the nucleus will exhibit higher multipole moments, in particular a quadrupole component. From symmetry considerations an electric dipole moment is forbidden. As in the case of magnetic moments, the electric quadrupole and higher moments manifest themselves by slightly modifying the optical emissions of the atom.

Parity. The nuclear ground state, being a quantum system, possesses a classification known as parity. It is an attribute relating to the symmetry of the mathematical function describing the state under reflection in three-dimensional space. Parity is a conserved quantity, like energy and angular momentum, and is useful for the description of nuclear reactions and decays.

Statistics. Another quantum attribute of the nuclear ground state is the type of statistics that governs its behavior. Odd-A nuclei conform to the rules for Fermi–Dirac statistics, while those having an even number of nucleons obey Bose–Einstein statistics. The most important difference between the two types of statistics are that the Pauli exclusion principle applies only to quantum systems governed by Fermi–Dirac statistics.

Excited states. All nuclei possess excited states in addition to their ground or lowest energy level. The study of the excitation modes of nuclei has yielded a wealth of information on the internal dynamics of nuclear systems. Nuclear excitations demonstrate collective phenomena such as rotation and vibration, as well as structure that can be attributed to the action of a single valence nucleon. Both behaviors can coexist in the same nucleus.

Conclusion

Nuclear physics came into its own as a science when the first artificially induced nuclear reactions were studied more than half a century ago. Although much has been learned about nuclei and their interactions in the intervening years, we still lack a complete understanding of the forces that hold nucleons together in the nucleus. This gap in our knowledge means that we are unable to make accurate predictions regarding basic properties of nuclei such as their masses and the energies of their excited states.

A complete theory of nuclear forces remains an elusive goal. In order to facilitate calculations, various nuclear models have been developed to help explain the experimental data in a phenomenological way. Each model in turn deals with rather limited parts of our experimental knowledge. In addition, even though neutrons and protons are the elementary constituents of nuclei, we now know that they themselves are not elementary particles. This means that a full understanding of nuclear structure must take into account the structure of the nucleons themselves, a topic which has received a great deal of attention in recent years. It is as if the study of nuclei resembles peeling an onion. Underneath each layer another layer is found. The search for understanding these microscopic objects remains one of the most challenging intellectual problems in science.

See also: EXCITED STATE; NEUTRON; NEUTRON, DISCOVERY OF; NUCLEAR MOMENT; NUCLEAR PHYSICS; NUCLEAR SIZE; NUCLEAR STRUCTURE; NUCLEON; NUCLEUS, ISOMERIC; PARITY; PAULI'S EXCLUSION PRINCIPLE; PROTON; RADIOACTIVITY, DISCOVERY OF; SPIN

Bibliography

KRANE, K. S. *Introductory Nuclear Physics* (Wiley, New York, 1988).

WONG, S. S. M. *Introductory Nuclear Physics* (Prentice Hall, Englewood Cliffs, NJ, 1990).

CARY N. DAVIDS

NUCLEUS, ISOMERIC

Briefly, isomeric nuclei are atomic nuclei possessing an excited state whose lifetime is longer than a few nanoseconds (1 ns = 10^{-9} s).

Atomic nuclei have many discrete quantum states, all having definite properties such as energy, lifetime, angular momentum, and parity. In a given nucleus the state having the lowest energy is called the ground state, and all others are denoted as excited states. Excited states will decay, that is, seek a lower total energy, by emitting their excess energy. They can do this by various means, depending on the individual situation. If sufficient energy is available, particle emission (mediated by the strong interaction) will take place. Otherwise, emission of gamma rays (mediated by the electromagnetic interaction) or beta decay (mediated by the weak interaction) will occur. All of these decay processes are subject to energy and angular momentum conservation laws, as well as selection rules governing changes in parity and angular momentum. The decay rate depends on the energy available for the decay, on the quantum properties of the initial and final states, and on the decay mode.

The lifetime or mean life of a state (denoted by the symbol τ) is defined as the average time that a state is likely to survive before it decays. The reciprocal of the lifetime is the decay rate, so a long-lived state implies a slow decay rate. The vast majority of excited nuclear states decay by gamma-ray emission with an angular momentum difference of one or two units, and with lifetimes typically less than 1 ns (i.e., $\tau < 10^{-9}$ s). Excited states with lifetimes longer than a few nanoseconds are known as isomeric states, or isomers, and nuclei possessing one or more isomers are called isomeric nuclei. Isomers arise because decays involving large changes in angular momentum or small changes in energy have reduced decay rates. Figure 1 shows an example of the decay of such a state, the isomeric state in ^{137}Ba, denoted as ^{137m}Ba. Here the m stands for "metastable," which is a synonym for isomeric. In the case of ^{137m}Ba, energy considerations permit only decay to the ground state by gamma emission. Even though the energy available for the decay is a rela-

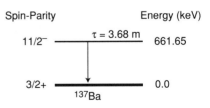

Figure 1 Decay of ^{137m}Ba.

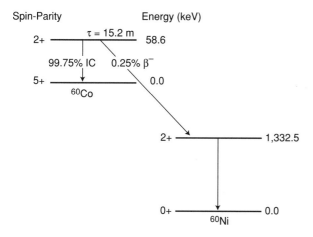

Figure 2   Decay of ^{60m}Co.

tively large 661.65 keV, the state has a lifetime of 3.68 min, extremely long when compared with similar energy changes in other nuclei. The long life is attributed to the fact that the angular momentum difference between initial and final state is four units.

In some cases the decay rate by gamma emission will be so retarded that other weaker decay modes become competitive. This situation is illustrated by the decay of ^{60m}Co, shown in Fig. 2, whose decay rate is retarded due to both a small energy difference and a large angular momentum change. With a lifetime of 15.2 min, the state decays to the ground state mostly via the electromagnetic process known as internal conversion. However, about once in every 400 times it decays by beta-particle emission to an excited state in ^{60}Ni.

See also: DECAY, BETA; DECAY, GAMMA; DECAY, NUCLEAR; EXCITED STATE; INTERACTION, WEAK; ISOTOPES; NUCLEUS; RADIOACTIVITY

Bibliography

KRANE, K. S. *Introductory Nuclear Physics* (Wiley, New York, 1988).

LEDERER, C. M., and SHIRLEY, V. S., eds. *Table of Isotopes,* 7th ed. (Wiley, New York, 1978).

CARY N. DAVIDS

OHMMETER

An ohmmeter is an instrument that is used to measure the electrical resistance of a resistor, or the resistance between two points of a circuit. The most common ohmmeter is that found as part of a multimeter or VOM (volt-ohmmeter), which is a multipurpose instrument that can measure voltage, resistance, and current.

The analog (or moving needle) VOM uses a galvanometer to indicate flowing electrical current. A VOM is accurate to a few percent and is useful to obtain an approximate value of resistance. It also is convenient for checking electrical continuity in a circuit. For example, an extension cord should be measured to have a low resistance between the ends of the same wire in the cord; however, a very large resistance (a reading of infinite ohms) should be measured between the two wires. An ohmmeter should never be used to test resistance in a circuit if there are batteries or other voltage sources present in the circuit. At best, the ohmmeter readings will be false, and at worst the meter may be damaged.

A VOM measures resistance through use of a battery in the meter that is connected in series with the galvanometer and an unknown resistance R_x. Then the current I that flows through R_x is related to its resistance, and the ohmmeter scale on the dial is marked to indicate the value of R_x. The basic ohmmeter circuit is shown in Fig. 1. Before measuring a resistance, the test leads are connected together and

the zero-adjust resistance r is set so that the meter reads zero ohms. This corresponds to maximum current flow, and the zero ohms marking is at full-scale needle deflection. Conversely, if the test leads are unconnected (open circuit), no current flows and the needle should point to infinite ohms at the left of the scale.

A digital multimeter (DMM) also includes a resistance measuring function. One DMM ohmmeter technique utilizes an electronic constant current source I_c located in the meter. When the meter is connected to an unknown resistance, the constant current passes through the resistance and a voltage V_{R_x} develops across the resistance: $V_{R_x} = I_c R_x$. The voltmeter function of the DMM is used by the meter

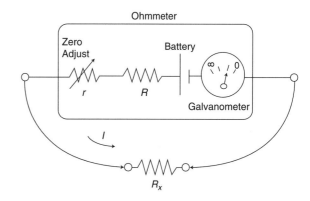

Figure 1 Ohmmeter circuit. The ohmmeter is shown connected to measure an unknown resistance R_x.

to read V_{R_x}, from which the unknown R_x can be displayed. For example, if I_c is 1 mA, the voltage reading is equal to the R_x resistance value in kΩ.

See also: AMMETER; ELECTRICAL CONDUCTIVITY; ELECTRICAL RESISTANCE; GALVANOMETER; VOLTMETER

Bibliography

BARTHOLOMEW, D. *Electrical Measurements and Instrumentation* (Allyn & Bacon, Boston, 1963).
JONES, L. *Electrical and Electronic Measurements* (Wiley, New York, 1983).

DENNIS BARNAAL

OHM'S LAW

The equation $V = IR$ is Ohm's law, where V is the potential difference across a circuit element, I is the electric current in the element, R is the resistance of the element, and R is independent of I. Ohm's law states that the relation between V and I is linear, and R is the proportionality constant in the equation. This law and the SI unit of resistance, the ohm (Ω), are named for George Simon Ohm.

If a circuit element obeys Ohm's law, it is called ohmic, whereas if the element does not obey this law, it is called nonohmic. For a particular element, suppose $I = 2$ A when $V = 20$ V in one set of measurements. Then, in another set, $I = 4$ A when $V = 40$ V. The element's resistance is

$$R = \frac{V}{I} = \frac{20 \text{ V}}{2 \text{ A}} = 10 \text{ Ω},$$

or

$$R = \frac{40 \text{ V}}{4 \text{ A}} = 10 \text{ Ω}.$$

Each set of data yields the same value of R for the element, which indicates that this element is ohmic. The relation between V and I is linear for this element, and the element's resistance is a property of the element.

For another element, suppose $I = 2$ A when $V = 20$ V, and then $I = 5$ A when $V = 40$ V. For the first data set,

$$R = \frac{20 \text{ V}}{2 \text{ A}} = 10 \text{ Ω}.$$

but for the second set

$$R = \frac{40 \text{ V}}{5 \text{ A}} = 8 \text{ Ω}.$$

The relation between V and I for this second element is nonlinear, which shows that the element is nonohmic. The resistance of a nonohmic element is not a characteristic property of the element because it changes when the current (or the potential difference) changes.

A resistor is an example of a circuit element that is ohmic. Resistors are used to limit the current in a circuit and the resistance of a resistor is labeled on the resistor with a color code. Also, wires that are used to connect circuit elements together are ohmic. A diode is an example of a circuit element that is nonohmic.

Strictly speaking, no material is ohmic because the linear relation between V and I can be made to break down by extending to larger and larger values. However, homogeneous materials are essentially ohmic over a wide range of values of V (or I). Consequently, Ohm's law is a very useful relation in dealing with electric circuits.

See also: CIRCUIT, AC; CIRCUIT, DC; ELECTRICAL RESISTANCE; ELECTRICITY; OHMMETER; RESISTOR

Bibliography

HALLIDAY, D.; RESNICK, R.; and WALKER, J. *Fundamentals of Physics,* 4th ed. (Wiley, New York, 1993).
KELLER, F. J.; GETTYS, W. E.; and SKOVE, M. J. *Physics: Classical and Modern,* 2nd ed. (McGraw-Hill, New York, 1993).
TIPLER, P. A. *Physics,* 2nd ed. (Worth, New York, 1982).

FREDERICK J. KELLER

OLBERS'S PARADOX

Why is the sky dark at night? This is the celebrated riddle that originated in the sixteenth century and is now known as Olbers's paradox. In 1823 Wilhelm

Olbers, a German astronomer, presented the riddle in its simplest terms: In an infinite universe, populated everywhere with stars of finite size, a line of sight in any direction, when extended out into space, must ultimately intercept the surface of a star. Hence stars should cover the entire sky. And if all stars are Sun-like, the sky at every point should blaze as bright as the disk of the Sun. Olbers has been incorrectly credited with the discovery of the riddle; he did, however, express it in this lucid form and showed that the riddle still holds even when stars are irregularly distributed in clusters.

A boundless universe of either flat or curved space, in which luminous stars stretch away endlessly, seems a not unreasonable cosmological picture. Yet it leads to a conclusion in total contradiction with experience. The night sky is dark and not a continuous blaze of bright stars. A calculation, first made in 1744 by the Swiss astronomer Jean-Philippe Loys de Chéseaux, proved that starlight should be 90,000 times more intense than sunlight at Earth's surface. Chéseaux showed that in an infinite universe, uniformly populated with stars, we see only a finite number of stars. Visible stars cover the sky out to a distance (the "background distance") where they fuse together and form a continuous background; the stars at greater distances are covered over by the foreground stars and cannot be seen.

A forest analogy helps us to understand the riddle. Every (horizontal) line of sight in an endless forest terminates at a tree trunk. Wherever we stand we find ourselves always surrounded by trees stretching away into a continuous background. The distance of the background is the average area occupied by a tree divided by the typical thickness of a tree trunk at eye level. (For example, the background distance is 50 m if trees are separated from one another by distances of 5 m and have trunk diameters of 0.5 m).

More than a dozen solutions have been proposed since the riddle was first discovered by Thomas Digges in 1576. The riddle has two interpretations, and the proposed solutions are therefore of two kinds.

First Interpretation

The first interpretation assumes the argument is correct and the sky is indeed covered with stars, most of which cannot be seen, and the riddle asks: What happened to the missing starlight? Both Chéseaux and Olbers adopted this interpretation and proposed that starlight is slowly absorbed as it travels large distances in space. The analogy in this case is the foggy forest; we see only foreground trees and fog obscures background trees. This solution fails because the interstellar absorbing medium quickly heats up and then emits the absorbed radiation. Hermann Bondi in the 1950s also adopted this first interpretation and argued that the light from distant stars is redshifted into invisibility by the expansion of the universe. The redshift solution contributed greatly to the popularity of cosmology and for several years it was believed that darkness at night proved that the universe is expanding. But Bondi's redshift solution is incorrect in a big bang universe and applies only to the steady-state expanding universe that was disproved in 1965 by the discovery of the radiation from the big bang.

Second Interpretation

The second interpretation assumes that the sky is not covered with stars, in agreement with observation, and therefore the assertion that every line of sight intercepts the surface of a star is misleading. The riddle, in effect, now asks: What do lines of sight intercept when we look at the dark gaps between stars? In the early seventeenth century Johannes Kepler adopted this second interpretation and argued that we look out between a finite number of stars and see a dark surface enclosing the universe. The analogy is a finite forest enclosed by a high wall. Another solution, consistent with this interpretation and popular with astronomers in the late nineteenth and early twentieth centuries, was the argument that our galaxy—the Milky Way—is the only galaxy in the universe, and we look out between the stars to an empty, dark, and infinite space beyond. The corresponding analogy is a finite forest, and we look out between the trees to a treeless plain beyond. In 1848 Edgar Allen Poe advanced the novel idea that the universe is not old enough for the light from very distant stars to have reached us; hence lines of sight that fail to intercept stars reach back to the darkness that prevailed before the birth of stars. This idea was investigated mathematically by Lord Kelvin in 1901. He assumed that stars are no older than 100 million years, and showed that out to a distance traveled by light in 100 million years, visible stars are insufficient in number to cover the whole sky. The missing stars are actually out there, but we look out in space and back in time to a prenatal era before the stars were born.

Modern calculations confirm and extend Kelvin's conclusions, and the second interpretation of Olbers's paradox is hence correct: most lines of sight do not intercept stars but extend back to the beginning of the universe. Light travels 300,000 km/s, and in a static universe of age between 10 and 20 billion years, the universe is not old enough for starlight to reach us from regions sufficiently distant for visible stars to cover the sky. If the night sky is dark in a static universe, then obviously in an expanding universe of the same age the night sky is even darker because of the red shift of starlight. Calculations show that the redshift effect is in fact relatively small.

The old riddle of why the sky is dark at night has been solved by modern cosmology. In our big bang universe of age 10 to 20 billion years, we cannot see sufficient stars to cover the sky. Instead, on looking out in space, we look far back in time to the beginning of the universe and see in all directions the big bang covering the whole sky. The expansion of the universe has reduced the incandescent glare of the big bang and redshifted its radiation into the invisible infrared.

See also: BIG BANG THEORY; COSMOLOGY; KELVIN, LORD; REDSHIFT; SPACE; UNIVERSE, EXPANSION OF

Bibliography

BONDI, H. *Cosmology,* 2nd ed. (Cambridge University Press, Cambridge, Eng., 1960).

HARRISON, E. R. *Cosmology: The Science of the Universe* (Cambridge University Press, Cambridge, Eng., 1981).

HARRISON, E. R. *Darkness at Night: A Riddle of the Universe* (Harvard University Press, Cambridge, MA, 1987).

EDWARD R. HARRISON

OPPENHEIMER, J. ROBERT

b. New York, New York, April 22, 1904; *d.* Princeton, New Jersey, February 18, 1967; *nuclear physics, particle physics.*

Oppenheimer grew up in splendid surroundings. From his boyhood home, an apartment high above New York City's Riverside Drive, he could gaze out over the Hudson River and watch the sun setting in the west. Hanging on the walls of his home were magnificent paintings by Van Gogh, Cézanne, and Gauguin. During the hot Manhattan summer months, Oppenheimer, his younger brother (Frank), and his parents (Julius and Ella) could escape to their shoreline home on Long Island, where two sailboats awaited.

Oppenheimer's interest in science came early. He collected minerals and, at age twelve, presented his first paper to the New York Mineralogy Club. He attended a private school, the Ethical Culture School, and was a superior student. In September 1922, upon completion of high school, Oppenheimer entered Harvard University, where he began a major in chemistry. Quickly, however, he was drawn to physics and, based on his extensive reading, was granted advanced standing in the physics department, where he was invited to work in Percy Bridgman's laboratory.

The laboratory was not fitting for the talents of Oppenheimer, but that was all he knew. In the 1920s American physics was recognized for its experimental prowess; theoretical physics was largely the domain of European physicists. Thus, when he approached his graduation, Oppenheimer acted on what he knew and applied for a position at Cambridge University with the eminent experimental physicist Ernest Rutherford, who, in 1911, discovered the atomic nucleus. When Rutherford rejected Oppenheimer's application, he wrote to another renowned experimentalist, J. J. Thomson, who had discovered the electron in 1897. Thomson accepted Oppenheimer, and in the summer of 1925 he traveled to Cambridge, England, to begin work in Thomson's laboratory.

Fortunately, for Oppenheimer and for physics, Cambridge was an active center and Oppenheimer quickly met other Cambridge physicists such as Paul Dirac and Ralph Fowler, as well as visiting physicists Niels Bohr and Paul Ehrenfest. All these men were theoretical physicists. At Fowler's suggestion, Oppenheimer began to learn Dirac's new quantum mechanics and apply the theory to a problem in spectroscopy. Oppenheimer was soon absorbed in his theoretical work and he was demonstrating a level of brilliance that had been masked by the laboratory. Max Born, a theoretical physicist from the University of Göttingen, was another visitor to Cambridge and he was so impressed by Oppenheimer that he invited Oppenheimer to come to Göttingen to do his doctoral dissertation research. Four months after his arrival at Göttingen, Oppenheimer submitted a paper that was his dissertation, and three months after that, he received his doctor of philosophy degree.

Oppenheimer returned to the United States in the summer of 1929, where he accepted a dual appointment at the University of California, Berkeley, and the California Institute of Technology. During the period from 1929 to 1942, Oppenheimer and his students worked at the cutting edge of physics. Owing to his command of theoretical physics, his wide-ranging erudition, and his personal charm, bright students flocked to Berkeley and Pasadena to work with Oppenheimer. A school of theoretical physics formed around him which, by the mid-to-late 1930s, had helped to launch American physics into the forefront of world physics.

Before he left Europe, however, Oppenheimer had written two very important papers. First, a short paper that Born, aghast at its brevity, expanded: This paper became known as the Born–Oppenheimer approximation and is the contemporary basis for the quantum mechanical study of molecules. In a second paper, Oppenheimer showed that electrons could be extracted from the surface of a metal by a weak electric field. The mechanism for this is quantum mechanical tunneling that, fifty-four years later, became the basis for the scanning tunneling microscope.

In 1930 Oppenheimer conducted research that fell just short of predicting the existence of the positron—the antiparticle of the electron. Two years later, the positron was discovered. Later in the 1930s Oppenheimer and his students applied the general theory of relativity to examine stellar collapse. In this work they showed that stars can collapse into both neutron stars and black holes.

Oppenheimer's physics is very impressive; however, it is for his work during World War II that he is more widely known. In 1942 Oppenheimer became the director of the Manhattan Project's laboratory in Los Alamos, New Mexico—the site where the atomic bomb was developed. Many of the world's greatest physicists worked under intense pressure from 1943 to mid-1945. With few exceptions, the participants in the project acknowledged Oppenheimer as a superb leader. His brilliance put him in command of every technical issue; his attention extended to the everyday problems of family members.

The first atomic explosion occurred in the predawn darkness on July 16, 1945, over the sand of Alomogordo desert. Each witness to this event, in his own way, was moved by the awesome sight. To Oppenheimer's mind came the words from the Hindu scripture in the *Bhagavadgita:* "I am become death, the shatterer of worlds."

After the end of World War II, Oppenheimer became a public figure. He advised presidents and was the chairman of high-level government committees. He had an international stage and, for a few years, he was universally admired—until 1954.

During the postdepression days of the 1930s, Oppenheimer had been sympathetic with the idealist ideas of socialist and even communist groups. During the time of the Manhattan Project, there were a few highly placed individuals who were suspicious of Oppenheimer. Their suspicions continued to fester after the war, and when the Science Advisory Committee, which Oppenheimer chaired, recommended against a crash program to develop the hydrogen bomb, these suspicions were activated. In 1954 a three-man board was formed to conduct a hearing to determine whether Oppenheimer's security clearance should be revoked. The hearing was brutal. Oppenheimer's character was called into question, and he was demeaned. In this hostile environment, he was unable to defend himself effectively. By a vote of 2 to 1, the board recommended revocation of Oppenheimer's clearance. With this decision, his public service came to an end.

In 1947 Oppenheimer had been named the director of the Institute for Advanced Study in Princeton, New Jersey. After the results of the 1954 hearing were announced, Oppenheimer's colleagues at the Institute, including Albert Einstein, proclaimed their confidence in Oppenheimer. The board of directors of the Institute reappointed Oppenheimer as director. In 1963 he received the Atomic Energy Commission's Enrico Fermi Award from President Lyndon B. Johnson. This award, in a small way, acknowledged the wrong that had been done to Oppenheimer in 1954.

In the spring of 1966 Oppenheimer announced his early retirement from the Institute effective June 30 of that year. Oppenheimer died less than eight months after his retirement.

See also: BORN, MAX; NUCLEAR BOMB, BUILDING OF; NUCLEAR PHYSICS; QUANTUM MECHANICS, CREATION OF; THEORETICAL PHYSICS

Bibliography

RHODES, R. *The Making of the Atomic Bomb* (Simon & Schuster, New York, 1986).

RIGDEN, J. S. "J. Robert Oppenheimer: Before the War." *Sci. Am.* **273** (July), 76–81 (1995).

STERN, P. M. *The Oppenheimer Case: Security on Trial* (Harper & Row, New York, 1969).

JOHN S. RIGDEN

OPTICAL FIBER

In a sense fiber optics, lasers, and microprocessors are the wires, gears, and levers of modern technology. In any given telecommunication, such as a telephone call, computer data transfer, or cable television broadcast, chances are that all three of these devices will be involved in the encoding, transmission, and decoding of digital voice, data, or video signals.

Rapid development and widespread use of fiber-optic technology has progressed, but some of the concepts underlying this means of communicating by guiding beams of light are surprisingly old. In 1854 John Tyndall demonstrated that light could be conducted through a jet of water presaging fiber-optic cables. Then in 1880 Alexander Graham Bell patented a design for a "photophone" in which sunlight was reflected off a diaphragm mounted on an acoustic horn and detected at a distance by a photocell connected to a speaker, anticipating optoelectronic modulators and receivers. Theoretical description of the propagation of light through dielectric cylinders was made by Hondros and Debye in 1910, and by 1920 the first experiments testing these ideas had been made. The invention of the laser in the early 1960s facilitated the advance of these studies, providing an intense, highly collimated light source overcoming the difficulties previously experienced due to power losses in the existing glasses. In 1970 the first low-loss silica fibers were produced by the Corning Glass Works. By the 1980s, routine production of very low-loss fibers was possible and applications of this means of sending signals over long distances became a commercially attractive alternative to conventional means.

The greatest advantage of transmitting a signal using photons through a fiber-optic waveguide is the increased bandwidth. Data can be transmitted at a higher rate the greater the carrier wave frequency. Light in the visible range has a frequency in the range of 10^{14} Hz compared to gigahertz for the highest band used in television broadcast. Unlike metallic wires, fiber-optic cables do not leak radiation that can interfere with other devices in their environment or which could be surreptitiously monitored, nor are they susceptible to electromagnetic interference from other cables or devices nearby. They are much smaller and lighter weight than copper wires, and fibers are very low loss, so that fewer repeaters and amplifiers are required than in metallic wire systems. Their principle disadvantages are the greater difficulty in handling and installing fiber-optic cables, which must have clean, efficient optical joints.

The term "fiber optics" refers to both the technology relating to transmitting signals over optical networks, and to the thin, usually flexible, glass or plastic fibers through which light is transmitted. An optical fiber contains a thin core of dielectric material of a relatively high index of refraction, typically 10–50 μm in diameter (compared to a human hair which is roughly 50 μm = 50×10^{-6} m thick). This core is surrounded by a concentric cladding of material of a lower index of refraction. The core and cladding together are often enshrouded by a protective layer.

Light is transmitted through an optical fiber by internal reflection. This is easiest to illustrate for a simple optical fiber consisting of a core of uniform index of refraction, n_1, as depicted in Fig. 1. We assume that the wavelength of the light transmitted is long compared to the diameter of the fiber to assure the validity of geometrical optics (i.e., the use of the "light ray" description of the propagation of light). In this picture, a ray propagating down the fiber will be totally internally reflected if the incident angle, θ_i, at which it strikes the boundary joining the core and the cladding (which has index of refraction n_2) is greater than some critical value given by $\theta_{\text{critical}} = \sin^{-1} n_2/n_1$. There will be some angle of acceptance outside of which light rays will not be transmitted down the optical fiber. The angle of acceptance may be shown to be determined by the relation

$$n_0 \sin \theta_a = \sqrt{n_1^2 - n_2^2}, \qquad (1)$$

where $n_0 (\approx 1)$ is the index of refraction of air. The quantity $n_0 \sin \theta_a$ is called the numerical aperture

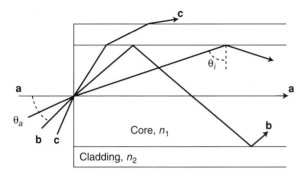

Figure 1 Schematic illustration of the propagation of several light rays in a simple optical fiber with index of refraction n_1 in the core and n_2 in the cladding. Rays **a** and **b** indicate the extremes of the angle of acceptance (θ_a). The ray labeled **c** is an example of a light ray that would be lost from the fiber since the angle of incidence (θ_i) is less than the critical angle for internal reflection.

and its square is a measure of the light-gathering power of the fiber.

The type of fiber described in Fig. 1 is called a step index, multimode fiber and is much like the rods and cones of the human eye or the elements of a fly's eye that channel light by internal reflection. The term "step index" indicates that the index of refraction is constant in the core and steps to another value in the cladding. "Multimode" refers to the fact that a number of different modes, corresponding to various rays with angles within the cone defined by θ_a, can be transmitted by the fiber. Even in a more complete treatment in which the wave nature of light is accounted for, there exists a finite, discrete number of modes in such a fiber. For a fixed difference between the core and cladding indices the number of allowed modes decreases as the diameter of the core is decreased. If the diameter becomes the same order as the wavelength of the light, only a single mode is transmitted, and the fiber is therefore termed "single mode."

The primary drawback of the step index, multimode optical fiber is the effect known as intermodal distortion. This effect arises because different light waves emerge from the distant end of the fiber out of synchronization, having been reflected a different number of times through their transmission. Rounded off to the nearest integer, the number of times that a ray will be reflected in a fiber of length l and core diameter d is given by

$$N = \frac{l \sin \theta_i}{d\sqrt{n_1^2 - \sin^2 \theta_i}} . \qquad (2)$$

Therefore, different rays cover different path lengths and take different travel times, spreading the initial pulse of light.

For fibers used simply to conduct light, say, for illumination, this kind of dispersion is not a serious problem. However, the primary use of fiber optics is the transmission of a signal. This is accomplished by modulating the light injected into the fiber. Typically a light emitting diode or solid-state laser is used to provide the bridge between the electronic devices used to create the signal, such as a video camera or computer, and the "photonic" elements of the system. These include not only the fiber-optic cables, but other necessary optoelectronic devices such as repeaters, amplifiers, and the photodetectors used to receive the optical signal and convert it back into an electronic signal.

To ameliorate the intermodal distortion of step index fibers, graded index fibers were developed. In these fibers the index of refraction changes as a function of the radial distance through the core. Therefore, the light rays travel in curved paths rather than straight lines. The rays with a large angle of incidence travel near the axis of the fiber, while the rays with small angle of incidence curve from near the cladding on one side of the axis to the other. The rays with the large oscillations spend part of their time in the lower index material and therefore travel more quickly, compensating their longer path length so that all modes remain synchronized.

Another difficulty has been to couple the largest possible fraction of the output power of the photoinjection device into the fiber. Typically as much as one-half the laser power can be lost at this stage. This has been countered by the invention of techniques that allow an aspheric microlens to be formed at the tip of the fiber so significantly more of the light can be focused into the fiber. Also deleterious is the attenuation of the transmitted signal due to scattering and absorption in the fiber. These effects come about from several sources: absorption of light that passes into the cladding, absorption of light by the core and by impurities (electronic excitation in the chemical bonds and vibrational excitation of molecules), and light scattering from microscopic fluctuations of the density or composition of the fiber (e.g., Rayleigh scattering). Understanding these attenuation processes allows one to predict which properties of the material composing a fiber will allow the maximum transmission efficiency to be determined. Recent experiments with fibers doped with rare Earth elements, for example, hold the promise of having very low loss for infrared wavelengths.

Fiber-optic technology has progressed rapidly, resulting in its widespread use especially in telecommunications and a range of other applications such as laser-assisted surgery, fiber-optic-based sensors, and image conducting and magnifying pipes and mosaics. Optoelectronics will certainly continue to be a mainstay of technology.

See also: FIBER OPTICS; FLY'S EYE EXPERIMENT; LASER; LASER, DISCOVERY OF; OPTICS, GEOMETRICAL; REFLECTION; REFRACTION; REFRACTION, INDEX OF

Bibliography

DESURVIRE, E. "Lightwave Communications: The Fifth Generation." *Sci. Am.* **266** (Jan.), 96–103 (1992).

DREXHAGE, M. G., and MOYNIHAN, C. T. "Infrared Optical Fibers." *Sci. Am.* **259** (Nov.), 110 (1988).

LINES, M. E. "The Search for Very Low Loss Fiber-Optic Materials." *Science* **226** (4675), 663–668 (1984).

SHUFORD, R. S. "An Introduction to Fiber Optics." *Byte* **9** (Dec.), 121 (1984).

UDD, E., ed. *Fiber Optic Sensors* (Wiley, New York, 1991).

DAVID R. SCHULTZ

OPTICAL IMAGE

See IMAGE, OPTICAL.

OPTICS

Optics is the physics of light. The term originally referred to vision and the improvement of the eye's ability to see with devices such as lenses. The field of optics also covers a broader range of detectors than just the eye, such as photographic film, which contains light-sensitive chemicals that react to the color and intensity of light, and charge-coupled devices (CCD), which transfer light into electrical signals. The field of optics also refers to a much broader range of the electromagnetic spectrum than just visible light, extending from short wavelength x rays to longer wavelength infrared and radio waves.

Although light is a well-studied subject, there is much innovation in the field of optics today. Light is used to study the extremes in our universe from the very large (astronomical telescopes that correct for turbulence in Earth's atmosphere during the observation) to the very small (optical microscopes with subwavelength resolution, able to distinguish individual molecules). Optics has also been extended to include information technology, such as storage and transmission of voice, video, and computer data in digital form (for example, lasers are used in supermarkets to scan bar codes on packages). The principles and applications of optics touch most disciplines of physics. They are used to study shock waves in the earth, to trap atoms in order to probe their fundamental characteristics with great precision, and for surgery in medicine.

There are three branches of optics: geometrical, physical, and nonlinear. Geometrical optics refers to principles guiding image formation using mirrors, lenses, prisms, and other optical devices. In geometrical optics, light is assumed to travel in straight lines, called rays, which undergo reflection upon encountering a mirror (responsible for our image in a mirror), refraction upon encountering a lens (responsible for focusing sunlight passing through a magnifying glass), and dispersion upon encountering a prism (responsible for rainbows, where raindrops in the sky act like prisms). In physical optics, light is treated as an electromagnetic wave, consisting of electric and magnetic fields that oscillate up and down in much the same way that water in ocean waves oscillates up and down. Physical optics is primarily concerned with the properties of light itself, such as polarization (the direction of oscillation of the electric field, discernible with polarizing sunglasses) and behaviors that require consideration of the wave-like nature of light, such as diffraction (the flaring out of light upon passing through an aperture, an effect not seen in ordinary circumstances since it becomes substantial only when the size of the aperture is comparable to the wavelength of light, about a millionth of a meter) and interference (responsible for the dazzling colors of soap bubbles). Nonlinear optics refers to the nonlinear response of matter to interaction with intense light, usually laser light. Such effects become apparent when the strength of the optical electric field approaches that of the Coulomb electric field that holds atoms together in material. An example of a nonlinear effect is second harmonic generation, in which, for example, red light passes through a special type of crystal, called a doubling crystal, and emerges as ultraviolet light.

Background

The ancient Greeks and Arabs knew that curved pieces of glass can be used to concentrate sunlight. They also knew how to calculate the direction of light reflected off a curved surface. Legend holds that Archimedes applied this knowledge in the third century B.C.E. for the defense of Syracuse by having soldiers use their shields as concave mirrors to reflect and concentrate sunlight onto enemy ships to set them afire. Yet an accurate explanation of vision does not appear until the early 1000s C.E. in the writings of Al Hazen ibn Al-Haytham, known as Alhazen. He proposed the idea that we see an object because light from the object enters our eye. The prevailing view on vision, which is now known to be incorrect, had been that we see an object because light from our eyes illuminates the object. Alhazen introduced many new scientific ideas, including the first known treatment of refraction at curved surfaces, which provided an explanation of how lenses focus light, and supported his conclusions with experimental methods somewhat similar to those of modern science.

By the late-thirteenth century, spectacles were being used in Europe to correct for both nearsighted and farsighted vision. However, it appears their invention was based on chance, for it was not until the 1600s that the foundations of the science of optics were established in Western culture. In 1604 Johannes Kepler published a book on optics in which he proposed that an object consists of points, each of which emits light in all directions. He then used this assumption to describe how light rays emitted from a single point of an object are bent as they pass through a lens, converging at a point called the image, and applied this concept to the eye, showing that light passing through the lens of the eye forms an inverted image on the retina. It is interesting to note that Kepler accepted this model of the physical mechanism of image formation in the eye, which predicted an inverted image, even though he could not explain how that inverted image was translated into right-side-up visual perception.

Although the origin of the telescope is obscure, Galileo Galilei first used one in 1610 for astronomical observation. His discoveries included that the Moon's surface has craters, and that the planet Venus undergoes phases like that of the Moon. The essential element of the telescope at that time was the lens, which refracts, or bends, light. In 1621 Willebrord Snell experimentally discovered a general relation quantifying the amount by which light is bent when traversing an interface between two media such as air and glass. This relation is called the law of refraction, or Snell's law.

The next major step in the development of optics came in 1704, when Isaac Newton published a comprehensive study of light, including refraction, dispersion, diffraction, and polarization. The observation that prisms give colored light dates at least as early as the fourth century B.C.E., however, the prevailing explanation, which we now know to be incorrect, was that the prism darkens white light by different amounts, depending on thickness of the glass, to give different colors. Newton realized, and experimentally verified, that a prism does not modify white light but physically separates it into its component colors. This phenomenon, known as dispersion, occurs because each color is refracted, or bent, by a different amount.

It took a hundred years before further outstanding contributions were made in the field of optics. In 1841 Carl Friedrich Gauss published a book on geometrical optics in which he characterized a lens by its focal length and provided formulas for calculating the position and size of an image for a lens with a known focal length. These ideas were extended over the next two decades to include characterization of five principal aberrations (lens distortions which limit the sharpness of an image), providing a mathematical foundation for lens design that was used until the invention of the computer in the mid-twentieth century. The computer revolutionized lens design, making possible creation of more complex devices, such as the zoom lens.

An understanding of the physical nature of light has been elusive. In the late 1650s, Christiaan Huygens postulated a theory of light as a wave, which provided an explanation for the phenomena of reflection and refraction, and gave physical meaning to the index of refraction, which characterizes the amount by which light is slowed down in a medium. The wave-like nature of light was not readily accepted since light does not display effects common to other waves such as water and sound waves. For instance, we can hear someone speaking around the corner from us, because sound waves bend around an obstacle; but since we cannot see around a corner, it appears that light does not bend around an obstacle. Thus, in 1704, Newton postulated that light consists of particles, in contrast with Huygens's wave theory. In an effort to resolve the controversy over the nature of light, Thomas Young performed an elegant experiment in the early 1800s, demonstrating that light displays interference effects, characteristic of waves, and thus supporting the wave-like nature of light, as postulated by Huygens. The culmination of classical optics occurred in 1864 with James Clerk Maxwell's theory of electromagnetism, which assumes light propagates as an electromagnetic wave. Maxwell's equations, in conjunction with the Lorentz equation, provide a foundation from which most of the laws of classical optics (reflection, refraction, lens formulas, and so on), can be derived.

In the twentieth century, new observations, such as the photoelectric effect and the spectrum of hydrogen, could not be explained in terms of the wave theory of light. Rather, interpretation of these observations requires a model of light as consisting of particles, called photons, somewhat as in Newton's theory. The controversy over whether to treat light as a wave or as a particle lead to a revolutionary new theory, quantum mechanics. Our understanding of light is that it has both a particle-like nature (the individual particles are called photons) and a wave-like nature. Most phenomena involving low-intensity radiation can now be explained using either Maxwell's theory or quantum mechanics. However, a general theory, valid for all conditions, is still lacking and thus is an area of continued research.

Recent Advances

The invention of the laser in 1960 had a huge impact on the field of optics, providing scientists with a powerful source of coherent light, which consists of waves with the same wavelength, or distance between successive crests, and with the same phase, meaning the crests of all the waves line up with each other. The laser initiated the field of nonlinear optics, opening the way for such achievements as

holography and interferometry. The imprint on many credit cards is a hologram, which is a record of the interference pattern between laser light reflected from an object and a reference beam. This interference pattern can be used to create a three-dimensional image. One scientific application of holograms, especially useful in biology, is holographic interferometry, in which comparison of successive holograms of a moving object reveals changes with minute detail.

The discovery of the laser, in conjunction with fundamental research in glass science, optics, and quantum mechanics, also brought on a technological revolution in telecommunications. Optical fibers, hair-thin strands of glass, transport information in the form of light pulses, replacing metal wires, which transport information in the form of electrical signals. The advantages of light are that the signal is less attenuated (light does not interact with matter as much as electrons do, and thus is altered less) and that optical circuits can carry a larger amount of information (the frequency of light exceeds the highest frequency of an electronic device by a factor of 10,000, and higher frequencies can carry more information; moreover, light of various frequencies can travel concurrently along a single fiber, and each frequency can carry a different signal). A fiber the thickness of a human hair can carry 35,000 telephone conversations simultaneously. Several million miles of optical fiber have been installed around the world, both on land and under the sea, and most long-distance telecommunication is carried by optical fibers.

Lasers are also used in recording audio and video signals and computer data on optical disks, such as the familiar compact disc (CD), which provides extremely high fidelity reproduction of sound signals. It is interesting to note that despite the huge impact lasers have had on technology, the primary motive for the invention of the laser was fundamental knowledge, and that for many years after its invention it was called "the solution looking for a problem." Lasers have become so important in communications and in information processing that a new technology, "photonics," has developed. Research is aiming to replace integrated electronic circuits with integrated optical circuits, which are thin films forming miniature lasers, lenses, switches, modulators, and detectors. One goal for many researchers is the development of an optical computer, expected to be many orders of magnitude faster than conventional electronic computers.

Since many optical beams can cross each other with no interference between them, unlike currents of electrons, optical computers may be designed in a completely different manner.

The science of imaging has made significant advances since the 1950s. Sophisticated microscopes are being developed which creatively exploit the wave properties of light to enhance resolution beyond what was originally thought possible. For instance, the near-field optical microscope not only provides images of single molecules, it does so nondestructively, thus enabling viewing of the molecule's behavior in real time. The confocal scanning optical microscope provides images in which out-of-focus portions disappear. The advantage of the confocal scanning optical microscope, when looking at transparent material common in biological applications, is that layers in front of or behind the layer of interest are not visible.

The "twinkling" of stars is frustrating for the astronomer taking quantitative measurements. One solution has been to place a telescope above the atmosphere (e.g., the Hubble Space Telescope). Another strategy is to improve ground-based telescopes by designing electro-optical systems that can compensate for turbulence in Earth's atmosphere in real time, as the observation is being made. Thus astronomical telescopes have evolved from passive collectors of light to interactive, computer-controlled instruments that change their optical characteristics as an observation is being made. These corrective capabilities are called adaptive optics.

The development of optical fibers is affecting medical technology. As early as 1966, optical fibers, or fiberscopes, were used to deliver light to a remote location in the body, such as the heart chambers, and thus provided doctors with a window into areas previously inaccessible except by major surgery. Surgeons use optical fibers to convey laser light into the body, for instance to cauterize blood vessels to stop intestinal bleeding. Optical fibers, inserted directly into the blood stream, are also used to perform chemical analysis of blood without the need to draw blood from a patient. Thus optical fibers provide a safer and less expensive approach to many medical procedures.

See also: ABERRATION; COHERENCE; COLOR; DIFFRACTION; ELECTROMAGNETIC WAVE; HOLOGRAPHY; HUYGENS, CHRISTIAAN; LASER, DISCOVERY OF; LENS; LIGHT; LIGHT, ELECTROMAGNETIC THEORY OF; LIGHT, WAVE THEORY OF; MAXWELL, JAMES CLERK; NEWTON, ISAAC; OPTICAL FIBER; OPTICS, ADAPTIVE; OPTICS, GEOMETRICAL; OPTICS, NONLINEAR; OPTICS, PHYSICAL; REFLECTION; REFRACTION; VISION; YOUNG, THOMAS

Bibliography

ABRAHAM, E.; SEATON, C. T.; and SMITH, S. D. "The Optical Computer." *Sci. Am.* **248** (Feb.), 85–93 (1983).

BLOEMBERGEN, N. "Optical Materials, Then and Now." *Annu. Rev. Mat. Sci.* **22**, 1–10 (1993).

FALK, D.; BRILL, D.; and STORK, D. *Seeing the Light: Optics in Nature, Photography, Vision, Holography, and Color* (Harper & Row, New York, 1986).

HARDY, J. W. "Adaptive Optics." *Sci. Am.* **270** (June), 60–65 (1994).

KATZIR, A. "Optical Fibers in Medicine." *Sci. Am.* **260** (May), 120–125 (1989).

RONCHI, V. *The Nature of Light: An Historical Survey* (Harvard University Press, Cambridge, MA, 1970).

TIEN, P. K. "Integrated Optics." *Sci. Am.* **230** (April), 28–35 (1974).

LAURIE KOVALENKO

OPTICS, ADAPTIVE

Adaptive optics is a technique for removing temporally and spatially varying optical wave front aberrations in real time. It has many applications in optics, lasers, and even medicine, but one of the most exciting is correcting blurred images in large ground-based astronomical telescopes.

An optical wave front propagating through the atmosphere becomes distorted by the randomly inhomogeneous variation in the refractive index of turbulent air. Even at the best mountain top astronomical sites, the atmosphere limits the angular resolution of large aperture (> 20 cm) telescopes to ~0.5 arcsec (or considerably worse) under good seeing conditions. In contrast, the diffraction limited resolution of a 10-m telescope at 0.5 μm is 0.01 arcsec. The atmosphere is therefore responsible for an incredible loss in resolution of large telescopes. Adaptive optics has the potential to recover the resolution lost to turbulence.

The basic concept for adaptive optics was first suggested by the astronomer Horrace Babcock in 1953. However, technological limitations prevented the first demonstration on an astronomical telescope until the mid-1970s. Until about 1990, most developments were for military applications. However, there has been a resurgence in adaptive optics for astronomy, especially with the development of large, sensitive arrays of infrared detectors that operate in a spectral region where adaptive optics is easier to implement and provides enormous payoff.

How does adaptive optics work in this application? Figure 1 illustrates the basic concept. Light waves arriving at the top of Earth's atmosphere from a distant object of interest are plane waves—the surface of constant phase is a flat, undistorted, plane wave front. As these planes travel down through the atmosphere, they encounter pockets of air at slightly different temperatures, a condition that results from the atmosphere's turbulent motion. Light travels at different speeds through regions of fluctuating temperature, causing the plane wave to become "bumpy" as one area of the wave front falls behind an adjacent area. The image formed at the normal focal plane of the telescope is severely blurred because different parts of the wave front that left one point on the object are traveling in different directions when they arrive at the telescope.

As Fig. 1 illustrates, the normally distorted image can be corrected by adding adaptive optics to the telescope. The objective of the adaptive optics system is to remove the "bumps" in the wave front before it is focused on the camera. This can be done by inserting a small, thin mirror in front of the camera and deforming its surface under the control of a computer to exactly match the distorted shape of the incoming wave front. After reflection from this deformable mirror, the distorted wave front is restored to a plane wave which can be focused to a diffraction limited point by the final camera lens. Corrected wave fronts arriving from different parts of the object are then faithfully registered in a diffraction limited image of the object on the focal plane of the scientific camera.

The key elements in the adaptive optics system are the deformable mirror and a sensor which uses part of the light from the distant object (or a brighter star near the object—a beacon) to measure the residual distortion after reflection from the deformable mirror. A high speed computer analyzes the data from the sensor and sends electrical control signals to pistonlike actuators that push or pull on the thin surface of the mirror, deforming it to the desired shape. The atmosphere changes every few milliseconds, requiring the wave front sensor, computer, and de-

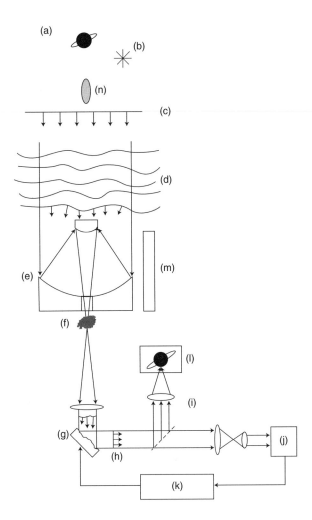

Figure 1 Components of an adaptive optics system: (a) the object of interest, (b) a natural star beacon, (c) undistorted plane wave arriving from the object of interest at the top of the atmosphere, (d) wave fronts distorted by atmospheric turbulence, (e) the imaging telescope, (f) image blurred by atmospheric turbulence at the normal focal point of the telescope, (g) deformable mirror, (h) corrected wave front after reflection from the deformable mirror, (i) a beamsplitter to share light between the science camera and the wave front sensor, (j) the wave front sensor that optically analyzes the distortions in real time, (k) the electronic processor and deformable mirror controller, (l) corrected image formed on the science camera, (m) a laser and projection telescope to create an artificial beacon or guide star in the sky, and (n) the laser guide star for use when a bright natural star is not available near the object of interest.

Figure 2 Images of the planet Saturn. Left panel: Normal image through atmospheric turbulence, no adaptive optics used. Right panel: Image made while correcting atmospheric distortions with laser beacon adaptive optics. This image has much higher resolution, revealing structure and separation in the rings, and bands in the atmosphere. The satellite Titan is clearly resolved. Both images were exposed one second over the wavelength band 0.65–1.0 μm through the 1.5-m telescope at the USAF Phillips Laboratory Starfire Optical Range. The laser beacon was generated by focusing a copper-vapor laser beam in the atmosphere 14 km from the telescope (USAF Phillips Laboratory, Starfire Optical Range).

formable mirror to operate at rates that allow the mirror to change its shape 100 times per second or faster.

One significant limitation of adaptive optics used for astronomy is the lack of bright stars to serve as beacons of light for the wave front sensor. This limitation may be overcome by the use of artificial beacons or laser guide stars created by scattered light from a laser focused in the atmosphere. Useful beacons can be generated by Rayleigh scattering from atmospheric molecules at an altitude of 10–20 km and by resonant scattering from sodium atoms in the mesosphere at an altitude of 95 km.

Figure 2 shows two images of the planet Saturn. The image on the left was made with the adaptive optics turned off and is representative of blurred images caused by atmospheric turbulence. The image on the right was exposed with the adaptive optics in operation. This image was made while using a laser beacon. It is easy to see the increased resolution and detail in the rings and structure in the planet's atmosphere. Also, the satellite Titan is

clearly visible toward the bottom of the image corrected by the adaptive optics.

See also: ABERRATION; ATMOSPHERIC PHYSICS; DIFFRACTION; LIGHT; LIGHT, SPEED OF; REFLECTION; SCATTERING, RAYLEIGH

Bibliography

BABCOCK, H. W. "The Possibility of Compensating Astronomical Seeing." *Publ. Astron. Soc. Pac.* **65,** 229–237 (1953).

BECKERS, J. M. "Adaptive Optics for Astronomy: Principles, Performance, and Applications." *Ann. Rev. Astron. Astrophys.* **31,** 13–62 (1993).

FUGATE, R. Q. "Laser Guide Star Adaptive Optics for Compensated Imaging" in *The Infrared & Electro-Optical Systems Handbook,* Vol. 8: *Emerging Systems and Technologies,* edited by J. S. Accetta and D. L. Shumaker (SPIE Press, Bellingham, WA, 1993).

FUGATE, R. Q. "Laser Beacon Adaptive Optics." *Opt. Photon. News* **4** (6), 14–19 (1993).

HARDY, J. W. "Adaptive Optics." *Sci. Am.* **270** (6), 60–65 (1994).

TYSON, R. K. *Principles of Adaptive Optics* (Academic Press, San Diego, CA, 1991).

ROBERT Q. FUGATE

OPTICS, GEOMETRICAL

Up to the middle of the seventeenth century light was regarded as a stream of particles traveling in straight lines radiating outward from the source. This notion had been reinforced early in the seventeenth century by Johannes Kepler, who introduced the concept that an extended object could be regarded as a multitude of separate points emitting rays in all directions and those entering a lens could be brought to a focus to form an image. This notion that light propagates in straight lines away from a source provides the basis for geometrical optics. Toward the end of the seventeenth century, the wave nature of light was proposed by Christiaan Huygens. He showed that the laws of reflection and refraction could be explained by assuming that light consisted of waves. At first this proposal was rejected because scientists of the time knew waves could be diffracted. The wave concept of light gained favor when it was understood that the diffraction effects are very small because the wavelength of visible light is of the order of 5×10^{-7} m. The wave theory remains an excellent approximation to describe the behavior of light and its ability to form images through the use of lenses and mirrors. The corpuscular theory of light was rejected and remained dormant until the beginning of the twentieth century, when Max Planck and Albert Einstein showed that light must have a particle or quantum nature to explain blackbody radiation and the photoelectric effect.

With the concept that light travels in straight lines firmly in hand, Carl Friedrich Gauss published a classic book on geometrical optics in 1841 and introduced the concept of the focal length for lenses and mirrors. He also developed formulas for calculating the position and size of the images formed. Lenses and mirrors do not form a perfect image. The image is distorted by aberrations such as astigmatism, which is the imaging of a point source off the axis of symmetry of an optical element as a line.

Other forms of distortion are spherical and chromatic aberration. Spherical aberration is the failure of rays from a point source to be focused to a point. Instead they are focused to a spot called the circle of least confusion. Chromatic aberration is caused by the failure of a lens or mirror to focus all colors to the same point. Between 1852 and 1856 the general ideas of aberrations of a lens or mirror were developed, and complex lens systems could be made from the many kinds of glass that were at the disposal of technicians and engineers of the time. Since 1960 computers have been put to work to perform the complex computations for the design of modern multiple lens systems used in binoculars, cameras, and other optical systems.

Pinhole Camera

The ray-like nature of light can be demonstrated through the use of a pinhole camera, which consists of a light-tight box containing a piece of film. Light to form the image enters through a small hole as illustrated in Fig. 1. Rays emanating from the source, the arrows in this case, pass through the pinhole and illuminate the film at the back of the box. Such a camera is limited. To produce a sharp image, a very small hole must be used. If the diameter of the hole becomes comparable to the wavelength of visible light (5×10^{-7} m), then the image is blurred by diffraction. If a large hole is used, multiple rays passing through the large pinhole produce an indistinct image because each object point projects an image of the enlarged pinhole onto the screen, which combines with the image of all the object points to yield a blurred image as is depicted in the lower cartoon in Fig. 1.

It is a simple extension to consider the formation of more perfect images through the use of a system of optical elements consisting for example of lenses, mirrors, or a combination of the two. Image brightness is enhanced and diffraction is minimized by the large size (compared to a wavelength of light) of the optical components. The formation of glass or other transparent media into an image-forming device (i.e., lens), by making one of the surfaces spherical was well known to the ancients. Likewise, mirrors can also be formed into spherical and other image-forming shapes from glass, metal, or other materials. Lenses can also be made from a variety of materials. The first lenses were made from selected window glass. Many satisfactory inexpensive lenses

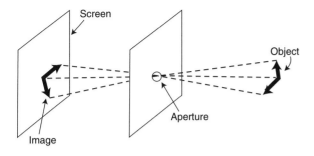

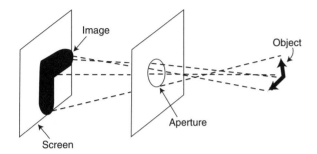

Figure 1 Images produced by pinholes of different sizes.

are now made from plastic by injection molding methods.

In order to have an elementary idea of how to design an optical system, it is sufficient to understand that the image-forming qualities of a lens are controlled by how light passes across a boundary between the lens and some other medium or how it reflects from the surface of a mirror. The light can be reflected from one surface or it can be refracted or bent as it passes across the surface.

Reflection and Refraction

The rules for the behavior of light rays as they pass through different media have been well established by experiment and verified by theory. It has been shown by geometrical considerations from wave propagation and by experiment that upon reflection at a surface, the angle the reflected ray forms with a normal to the surface equals the angle the incidence ray forms with the normal. That is $\angle ABC = \angle CBD$ as shown in Fig. 2.

When a beam of light of a given wavelength passes from one medium to another (i.e., air to water), the wave velocity of propagation changes, and it is said to be refracted. For example in Fig. 2 the light beam is incident on the interface at an $\angle ABC$. According to

wave theory and verified by experiment in about 1621 by Willebrord Snell, the beam propagates in the other medium at a new angle, $\angle A'B'C'$, according to the formula $\sin(A'B'C')/v' = \sin(ABC)/v$. The ratio of the velocity of light in a vacuum, denoted by the letter c, to the velocity of light in a medium, v, is called the index of refraction for the medium and is usually denoted by the letter $n = c/v$. The formula relating the angles ABC to $A'B'C'$ to the indices of refraction of the two media is called Snell's law and is given by $n'\sin(A'B'C') = n\sin(ABC)$.

In a given medium such as water or glass the velocity of light becomes smaller as the wavelength becomes shorter (i.e. more blue). This effect is called dispersion. Thus the refractive index of the medium increases as the wavelength becomes shorter. Angle ABC is called the angle of incidence and $\angle A'B'C'$ is called the angle of refraction. The larger the index of refraction, the greater the ability of the medium to deviate the light. When a refractive medium is formed into a curved shape, like the arc of a circle, to produce a lens, the light rays incident on the surface can be brought to a focus. Such a lens can have convex (double convex, plano convex, or convex meniscus) or concave (double concave, plano concave, or concave meniscus) surfaces.

When light passes from a medium with a high index of refraction, n', to a medium of low index of refraction, n, it is possible that the light becomes completely reflected internally because the ray in the less refractive medium is bent away from the normal. By examining Snell's law one observes there is an $\angle A'B'C'$ less than 90 such that the product of n' and $\sin(A'B'C')$ equals n. This means that the sin $(ABC) = 1$ or $\angle ABC$ equals 90. Because the sin of an angle cannot exceed the value of one, light incident at angles greater than the given $\angle A'B'C'$ must be reflected. For example at a water ($n' = 1.33$)–air ($n = 1.00$) interface, the angle where total internal reflection begins can be computed from the index

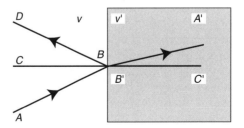

Figure 2 Reflection and refraction.

of refraction of water and air. Substituting these numbers into Snell's law we have

$$1.33 \sin (A'B'C') = 1.00 \sin(90),$$

or

$$A'B'C' = 68°.$$

Therefore, when the angle of incidence of the light ray in the water exceeds 68°, it is reflected internally.

A source of light is an object from which light originates, that is, the rays diverge in all directions from the source. A real image is a source of divergent light. On the other hand, rays can diverge from points representing a virtual image, but these points are not the origin of the rays. A virtual image can be formed by a single lens or mirror or a combination of lenses and mirrors.

The Lens Equation

It is possible to compute the position of the image formed from the focal length, f, the image and the object distances, X_I, and X_O, respectively, using the formula;

$$\frac{1}{X_I} + \frac{1}{X_O} = \frac{1}{f}.$$

This formula works for both mirrors and lenses if the sign convention that follows is used for the various distances. For a lens with light passing from left to right, the object distance is positive if the object is to the left of the lens and negative if the object is to the right of the lens; the image distance is positive if the image is to the right of the lens (real image) and negative if the image is to the left of the lens (virtual image); and the focal length is positive if the lens is convex and negative if the lens is concave. For a mirror, the object distance is positive if the object is in front of the mirror and negative if the object is behind the mirror; the image distance is positive if the image is in front of the mirror (real image) and negative if the image is behind the mirror (virtual image); and the focal length is positive if the mirror is concave and negative if the mirror is convex. The position of the image can be obtained by the simple ray-tracing technique shown in Fig. 3.

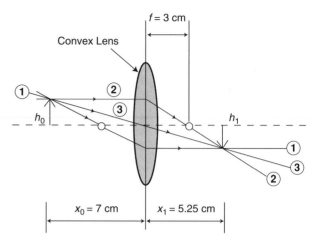

Figure 3   Image position for a convex lens.

In Fig. 3 we use the idea that a parallel ray passes through the focus of an ideal lens, and a ray passing through the center of the lens is undeviated. From our rule, ray 1 passing through the focus emerges parallel from the lens. Ray 2 traveling parallel emerges from the lens and passes through the focal point of the lens, and finally ray 3 passes undeviated through the center of the lens. From Fig. 3 all three rays pass through a point located at a distance $X_I = 5.25$ cm. We can check our construction by the formula with $X_O = 7$ cm, $f = 3$ cm, and $X_I = 5.25$ cm. The lens formula can be derived from the three principle rays by using simple geometry (hence the name geometrical optics). In this example the image is real. The ratio of the image height to the object height is called the magnification of the optical system. A similar construction can be made for a concave mirror, which forms a real or virtual image.

These simple methods may also be applied to charged particles such as electrons and protons. Many common devices, such as television cameras and receivers and electron microscopes, use lenses that depend on electric or magnetic fields to focus the beam of particles. Optical designs used in modern instruments and appliances have been refined so that complex optical systems containing many elements can be designed by ray-tracing methods that use computers. Modern optical devices include photographic and video cameras, telescopes, binoculars, and microscopes, all of which have myriad commercial and consumer uses.

See also: ABERRATION; ABERRATION, CHROMATIC; ABERRATION, SPHERICAL; EINSTEIN, ALBERT; IMAGE, OPTICAL;

IMAGE, VIRTUAL; LENS; LENS, COMPOUND; LIGHT; HUYGENS, CHRISTIAAN; KEPLER, JOHANNES; PLANCK, MAX KARL ERNST LUDWIG; REFLECTION; REFRACTION; SNELL'S LAW

Bibliography

COBB, V., and COBB, J. *Light Action: Amazing Experiments with Optics* (HarperCollins, New York, 1993).

CUTNELL, J. D., and JOHNSON, K. W. *Physics,* 2nd ed. (Wiley, New York, 1992).

GIANCOLI, D. C. *Physics,* 3rd ed. (Prentice Hall, Englewood Cliffs, NJ, 1991).

HECHT, J. *Optics: Light for a New Age* (Scribners, New York, 1987).

DAVID L. EDERER

OPTICS, NONLINEAR

Nonlinear optics describes those optical responses of gases, liquids, solids, or plasmas which are not linearly proportional to the intensity of the light, such as harmonic-, sum-, or difference-frequency generation, multiphoton absorption, Raman scattering, nonlinear refraction and absorption, and phase conjugation.

In optical materials, absorption, refraction, and scattering of light result from the light-induced polarization of the electron clouds surrounding the ion cores. The strength of this dielectric polarization P is related to the time-varying electric field $E = E_0\cos(\omega t)$ of a light wave with angular frequency ω by a complex-valued response function or susceptibility χ:

$$P = \chi E. \tag{1}$$

The susceptibility χ is an intrinsic optical property of any gaseous, liquid, or solid medium and depends on the frequency ω of the illuminating beam. The refractive index and absorption coefficient of the medium are proportional to the real and imaginary parts of $\chi(\omega)$, respectively. $\chi(\omega)$ is also a function of the electric field:

$$\chi = \chi^1 + \chi^2 E + \chi^3 E^2 + \cdots. \tag{2}$$

The linear susceptibility χ^1 produces much larger effects than the nonlinear terms involving χ^2 and χ^3, unless large electric fields from a laser are applied.

An example of the richness and complexity of nonlinear optics is the nonlinear polarization produced by the second-order susceptibility:

$$P^2 = [\chi^2 E_1]E_2, \tag{3}$$

where one wave (E_1) alters the polarization of, or "writes," the medium and a second wave (E_2) "reads" the polarized medium. If the "writing" and "reading" waves have the same frequency ω, the second-order nonlinear polarization is

$$P^2 = [\chi^2 E_1]E_2 = \chi^2 E_{10}\cos(\omega t) \cdot E_{20}\cos(\omega t)$$

$$= \frac{1}{2}\chi^2 E_{10}E_{20} \cdot [\cos(2\omega t) + 1]. \tag{4}$$

Thus the nonlinear medium mixes two photons of frequency ω to produce photons of frequency 2ω or zero frequency, conserving energy and momentum in the process. In general, if the waves E_1 and E_2 oscillate at different frequencies ω_1 and ω_2, optical fields at the sum and difference frequencies, $\omega_1 \pm \omega_2$, are generated. Second-order nonlinearity is exhibited by birefringent materials (e.g., crystalline quartz), deuterated potassium phosphate (KDP), and the ferroelectrics lithium and potassium niobate ($LiNbO_3$ and $KNbO_3$), as well as many others.

The third-order polarization involving χ^3 produces such complex phenomena as nonlinear refraction, nonlinear absorption, and phase conjugation. For example, in some materials the index of refraction depends on the laser intensity I according to the relation $n(I) = n_0 + n_2 I$, where n_2 is called the nonlinear index of refraction. When an intense laser beam strikes a material with $n_2 > 0$, the center of the beam is refracted more than the halo of less intense light, leading to self-focusing. In many materials, the nonlinear index also increases with temperature, leading to catastrophic self-focusing: The greater the distance traversed by the beam in the material, the more tightly it is focused, and the more heat is generated by absorbed light, until finally the electric fields exceed the yield stress of the material.

In optical phase conjugation, the phase of an input probe wave is reversed by reflection from an electron-density grating written in a nonlinear medium by constructive interference of two powerful pump beams. A phase-conjugate reflector reverses both the phase and direction of an incident wave, erasing beam distortions. Phase conjugation is sometimes called "real-time holography," because the phase-conjugate image, like a hologram, is produced by reflection from a grating. In the conceptually related photorefractive effect, light moves electrons from normal atomic sites to traps; the stored electrons which constitute the optical data are recovered by reversing the process.

Multiphoton absorption occurs when an atom, molecule, or ion absorbs two or more photons almost simultaneously to make a transition from one quantum state to another. During the two-photon absorption process, the first photon excites the atom from the ground state (energy ε_g) to a virtual excited state $\varepsilon_{\mathrm{int}}$; this intermediate state, in turn, is excited by a second photon to the metastable state (energy ε_f). This process occurs only if the second photon arrives before the atom "forgets" that it is in the virtual excited state. According to the uncertainty principle, this time must be very short indeed, so that the arrival rate of photons must be very large. Many nanosecond and picosecond pulsed lasers generate laser intensities high enough for multiphoton excitation.

Applications

Nonlinear optical techniques are widely used to create tunable coherent light. Efficient conversion of infrared light from the Nd:YAG at 1.06 μm to the green second harmonic (0.532 μm) or blue third harmonic (0.355 μm) makes it possible to pump dye lasers to obtain tunable laser light in the visible part of the spectrum. This visible laser light can itself be frequency doubled or tripled, yielding tunable ultraviolet coherent light. By mixing different frequencies of laser light in nonlinear crystals, it is also possible to generate tunable infrared radiation by sum- and difference-frequency mixing and by optical parametric techniques.

Nonlinear optical techniques have revolutionized optical spectroscopy. Multiphoton excitation and ionization are widely used to obtain spectral signatures for high-sensitivity detection of atoms and molecules in gases, liquids, solids, in flames and near surfaces. In fact, multiphoton ionization has been used to detect single atoms of one species in a gas of other atoms at atmospheric pressure. Coherent Raman scattering, once limited by weak signals obtained from high-pressure arc lamps, is routinely used for high-sensitivity studies of molecules under exotic conditions (e.g., in flames) or for very small sample sizes (as in thin films).

Phase conjugation is used in optical data storage systems to recover distorted optical images and to produce virtually distortion-free high-power laser beams for both laboratory science and such applications as laser fusion.

Historical Development

Raman scattering was observed in 1928, well before the advent of the laser; the laser, however, made this spectroscopy generally useful through coherent Raman scattering and its many variants. The theory of multiphoton absorption was first described by Maria Goeppert Mayer in 1935, but experimental confirmation of the theory had to await the development of intense pulsed lasers in the 1960s. A major contributor to the new field was Nikolaas Bloembergen, who worked out the theory of nonlinear effects and demonstrated many of the experimental effects on nonlinear optics in optical and semiconductor materials. Sven Hartman reduced to practice the idea of the photon echo, analogous to the spin echo of nuclear magnetic resonance. Boris Y. Zel'dovich developed many practical techniques of four-wave mixing and phase conjugation. Y. R. Shen has pioneered the use of nonlinear spectroscopy to characterize atoms and molecules on surfaces.

See also: HOLOGRAPHY; LASER; MAYER, MARIA GOEPPERT; NONLINEAR PHYSICS; POLARIZATION; THIN FILM

Bibliography

BOYD, R. W. *Nonlinear Optics* (Academic Press, New York, 1992).

LEVENSON, M. D., and KANO, S. S. *Introduction to Nonlinear Laser Spectroscopy*, rev. ed. (Academic Press, New York, 1988).

PEPPER, D. M. "Applications of Optical Phase Conjugation." *Sci. Am.* **260** (Jan.), 74–81, (1986).

SHKUNOV, V. V., and ZEL'DOVICH, B. Y. "Optical Phase Conjugation." *Sci. Am.* **259** (Dec.), 54–59, (1985).

RICHARD F. HAGLUND JR.

OPTICS, PHYSICAL

Physical optics is concerned with the classical wave interaction of light with matter, as demonstrated in interference, diffraction, reflection, refraction, polarization, scattering, and absorption. The simplest materials in their natural states (opaque screens, simple transparent substances, liquids, good conductors) initially dominated physical optics. However, with the development of solid-state electronics, lasers, and the general advance of technology, manipulation of optical materials came to the fore in the last half of the twentieth century. While visible light (wavelength 0.0004 mm to 0.0007 mm) displays both particle and wave properties, physical optics tries to maximize the use of classical concepts and to minimize reference to the underlying quantum mechanics of light and materials.

Although in the 1600s light as a wave was conceived of by René Descartes and mathematically described by Christiaan Huygens, it was not until the early 1800s that definitive experiments by Thomas Young and Augustin Fresnel established without doubt the wave nature of light. In 1864 James Clerk Maxwell published the definitive classical theory that demonstrated ultraviolet, infrared, and visible light (as well as heat radiation) to be electromagnetic waves traveling through space at 300,000 km/s (186,000 miles/s). In the late nineteenth and early twentieth centuries the discovery of radio, x rays, and gamma rays extended this list. Our physical optics survey will be confined to visible light, but the principles apply to all of the above. Wave features essential to physical optics are (a) direct simple addition of multiple waves at a space point ("linear superposition") and (b) direction of travel perpendicular to the radiation vectors $\mathbf{E}$ and $\mathbf{B}$. Essential for striking effects but not fundamental to Maxwell's wave theory are (c) identical wavelengths ("monochromatic" waves) for the multiple waves and (d) an unvarying relative "phase" between any pair of the multiple waves ("coherence"). The "phase difference" $\Delta\phi$ is the angular expression of the difference in time between the peak values of the sine waves relative to the period of the wave, so $\Delta\phi = 2\pi\Delta t/T = 2\pi\Delta L/\lambda$. Coherence is achieved by allowing only waves originating from a common source to reach the observer. The source may have spatial extent (i.e., be larger than a point) and will generally include a large number of wave-emitting atomic electrons. Interference and diffraction are both superpositions of coherent monochromatic waves. If a finite number of waves following different paths are added the effect is termed "interference." It is termed "diffraction" when the summed waves have followed paths that vary continuously from one wave to another and this requires the methods of calculus.

The prototypical interference geometries are the double-slit geometry (Fig. 1) and the thin film (Fig. 2). The phase difference due to path lengths produces addition or cancellation of amplitudes. (There is current interest in the double-slit geometry for *particles* to test the fundamental meaning of quantum mechanics.) In the case of a film of thickness t, the two interfering waves from the first and second surfaces have a path length difference $\Delta L = 2t$ for near-normal incidence. However, the phase difference must be reckoned in terms of the wave length in the film, $\lambda_{film} = \lambda_{air}/n_{film}$; therefore, $\Delta\phi = 4\pi t n_{film}/\lambda_{air}$. Additional π phase shifts occur if $n_1 < n_2$ or if $n_2 < n_3$, which means that the incident wave speed c/n is faster than the speed in the material. Important applications of thin-film interference are found in "invisible" cover panels for meters and in high-quality ("coated") lenses. Newton's rings occur when a spherical glass surface of large radius is in contact

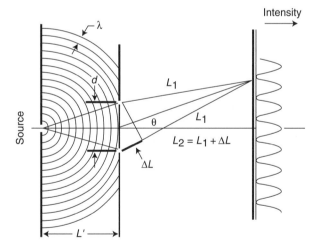

Figure 1 Double-slit interference. The two slits are very long but have a width that is considerably smaller than the wavelength λ. The common source-to-slit pathlength means the slits emit in-phase. The difference in path between slits 1 and 2 and a point on the screen gives a relative phase difference that produces a cosine-squared intensity pattern. The paths shown differ by $\Delta L = 2\lambda$. If the screen is far from the slits then $\Delta L = d \sin \theta$.

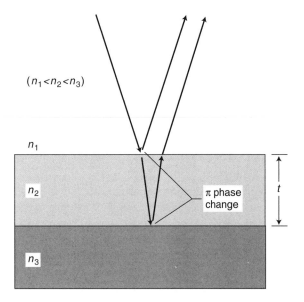

$(n_1 < n_2 < n_3)$

n_1

n_2

n_3

π phase change

t

Figure 2 Thin-film interference. The incident light is partially reflected at both interfaces (1–2 and 2–3). The "optical" path difference between the first and second reflected approaches $2tn_2$ as the incident light approaches normal incidence. For $n_1 < n_2 < n_3$, both reflections are phase reversed (180° or "π" reflection) and the relative phase difference is $\Delta\phi = 4\pi n_2 t/\lambda n_1$. Then the minimum thickness for destructive interference is $n_1\lambda/4n_2 = t$.

with a flat glass surface. Modern process control allows multilayer dielectric filters, which stack many precise thin films, to selectively transmit or block a narrow wavelength range $\Delta\lambda$ of a few Angstroms (1 Å = 1/10,000 of a micrometer) with efficiency close to 100 percent. Albert Michelson and Ernst Mach were important figures in the development of two-wave (or "two-beam") interferometry for precise distance measurement.

Diffraction arises when the size of a coherent source (or aperture) is not small compared to a wavelength. Now pathlength variation across the source must also be included in the total superposition of wave amplitudes. Interesting examples are the single slit and the diffraction grating. The simplest case of the single slit (Fraunhofer diffraction) is characterized by very long paths (or by using lenses focused on infinity). If these distances are not long it is called Fresnel diffraction. Diffraction produces a complex light pattern close to the edge of a shadow, which develops into the pattern of a single slit as the separation between two edges of an aper-

ture approach each other. For a circular aperture of diameter $2r$ the superposition is more complicated, and the first zero of intensity is at $r\sin\theta/\lambda = 0.61$, as compared with $w\sin\theta/\lambda = 1$ for a single slit (Fig. 3). An astronomical telescope with lens diameter $2r$ is limited in performance by diffraction since point sources (i.e., stars) will give diffraction images. A diffraction grating consists of a large number of parallel transmitting open slits in a plate ("transmission grating") or of reflective grooves in a metal surface ("reflection" grating). These devices are "diffraction" gratings because, to get more light intensity, the ratio of slit width to slit spacing w/d is not far from unity. For N slits the phase difference $\Delta\phi = 2\pi w/\lambda$ for a single slit accumulates across the grating to $\Delta\phi = 2\pi Nw/\lambda$. This results in narrow peaks at the double-slit constructive interference locations, where $\Delta\phi = 2\pi dm/\lambda$ ($m = 0, 1, 2,...$). Two close wavelengths λ and $\lambda + \Delta\lambda$ are said to be resolved if $\lambda/\Delta\lambda \leq R$, where $R \equiv mN$ is called the resolving power. A Fresnel zone plate increases the on-axis intensity by blocking those regions of the incident wave with phases relative to the center that would decrease the total amplitude at the corresponding focal point of the zone plate. A more elab-

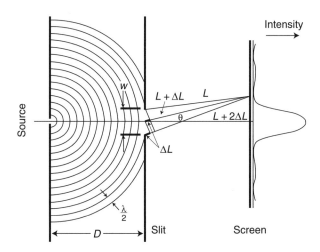

Intensity

Source

w

$L + \Delta L$ L

θ $L + 2\Delta L$

ΔL

$\frac{\lambda}{2}$

D Slit   Screen

Figure 3 Single-slit diffraction. The slit is comparable to, or larger than, a wavelength λ. The path difference between a narrow strip at slit center and a narrow strip at the slit upper edge is $\Delta L = w\sin\theta/2$ when the screen is far away. The whole slit area can be constructed from such pairs. For the angle θ shown, the path difference is $\Delta L = \lambda/2$ (first interference minimum). The width increases as the slit width w is narrowed.

orate scheme (first implemented by Lord Rayleigh) uses thin dielectric rings to reverse those phases instead of merely blocking them, thereby achieving even greater intensity.

Scattering, polarization, and double refraction are described on a classical level as the driving of bound electrons by the incident radiation electric field. Light traveling in materials at apparent speed c/n is actually coherent scattering by atomic electron oscillators. The combined effect of path lengths and the electron oscillator mechanical resonance phase shift produces the slowing lengths and the electron oscillator phase shift produces the slowing effect we call index of refraction n. In fact, for a very thin layer, where the number of scatterers is relatively small, a small "precursor" can be detected that has traversed the film at speed c. The variation of phase and amplitude of the oscillators with the wavelength of the light explains dispersion, and oscillator damping accounts for absorption of light. Crystals that have different binding forces along their various axes have correspondingly different phase shifts and therefore show differing refractive indices along these axes. This will split unpolarized light into two completely orthogonally polarized beams. Double refraction can also be induced by external stress so that transparent models or real structures reveal their stresses through polarization analysis (photoelasticity).

Optical activity in solutions was discovered in 1815 by Jean-Baptiste Biot. This rotation of linearly polarized light is not different in principle from double refraction in normal crystals. It arises from the difference in radial and tangential electron binding in long molecules that guide the polarization by their twist. A pure collection of these molecules, even though randomly oriented, still gives a net polarization rotation since the handedness of the twist is independent of orientation. Many electro-optic phenomena are based on induced double refraction by strong electric fields. The basic physical process is partial alignment of permanent or induced electric dipole moments. When the molecules are aligned by an applied electric field, light can be preferentially blocked or transmitted by a polarizer. In the Pockels effect a strong transverse electric field applied to certain crystals produces a polarization rotation that is linearly proportional to E. The Kerr effect produces rotation proportional to E^2 and is observed in many liquids and gases. Both the Kerr effect and the Pockels effect find application as very fast switches for polarized light.

Magneto-optics was pioneered by Michael Faraday, who felt there had to be a connection between electricity, magnetism, and light and searched diligently for it. In the Faraday effect (where rotation of linearly polarized light is produced by a magnetic field parallel to the beam) the direction of rotation is clockwise or counterclockwise depending on the material and direction of the field and is proportional to the field and length of material. It is understandable in terms of precession of electrons in the magnetic field.

If the intensity of incident light is sufficiently high (i.e., from a powerful laser), the electrons can be driven beyond the linear restoring force range and the second (or even higher) harmonic of the incident frequency appears. Analogously, two intense laser beams can generate light at the sum or difference of their frequencies.

See also: ABERRATION; DIFFRACTION; DIFFRACTION, FRAUNHOFER; DIFFRACTION, FRESNEL; DOUBLE-SLIT EXPERIMENT; FARADAY, MICHAEL; FARADAY EFFECT; FRESNEL, AUGUSTIN-JEAN; HUYGENS, CHRISTIAAN; LIGHT; MACH, ERNST; MAXWELL, JAMES CLERK; MICHELSON, ALBERT ABRAHAM; NEWTON, ISAAC; OPTICS, GEOMETRICAL; OPTICS, NONLINEAR; POLARIZED LIGHT; REFLECTION; REFRACTION; SCATTERING, LIGHT; THIN FILM; YOUNG, THOMAS; ZEEMAN EFFECT

Bibliography

BALDWIN, G. C. *An Introduction to Nonlinear Optics* (Plenum, New York, 1969).

HECHT, E. *Optics* (Addison-Wesley, Reading, MA, 1987).

JENKINS, F. A., and WHITE H. E. *Fundamentals of Physical Optic,* 4th ed. (McGraw-Hill, New York, 1976).

MORGAN, J. *Introduction to Geometrical and Physical Optics* (McGraw-Hill, New York, 1953).

PEDROTTI, F. L., and PEDROTTI, L. S. *Introduction to Optics* (Prentice Hall, Englewood Cliffs, NJ, 1987).

STRONG, J. *Concepts of Classical Optics* (W. H. Freeman, San Francisco, CA, 1958).

R. A. KENEFICK

OPTICS, QUANTUM

See QUANTUM OPTICS

ORDER AND DISORDER

Order is related to "predictability," disorder to the lack of predictability. The Voyager II mission to Neptune is an example of a highly predictable process: The spacecraft followed the desired path very precisely and, as a result, obtained a stunning series of photographs. Systems with a few objects, such as the Voyager–Neptune–Sun–Neptunian–Moon system, are very predictable (but only up to a point). Yet, as the number of objects increases, a point is reached where even the best computers are unable to solve the dynamical equations to predict the motion. For example, the molecules in a roomful of air are so numerous that one never expects to be able to follow the motion of individual molecules, no matter how fast computers become. This is statistical unpredictability.

For molecules and other microscopic systems (atoms, nuclei, elementary particles, etc.) quantum unpredictability is unavoidable. Here, the problem is not one of insufficient computing power. Even very simple systems have a fundamental randomness, built into nature itself, that prevents one from making precise predictions of experimental results, *even in principle.*

A third kind of unpredictability can occur when the number of objects is not large and the objects are not microscopic, for example, several planets in orbit around a star. The orbits of the planets are accurately predictable for awhile, but eventually, no matter how good the computer, the actual orbits will deviate from the calculated ones. This chaotic unpredictability arises because the interactions among the objects are nonlinear. A small change in a planet's motion initially will lead to a large, unpredictable change sometime later; the system is said to show a sensitive dependence on initial conditions. This is frequently called the butterfly effect, referring to the hypothetical possibility that a butterfly flapping its wings in Brazil could cause a hurricane in Miami.

Statistical Disorder

The result of flipping an unbiased coin is completely unpredictable. Yet, the fraction of "heads" that results from flipping many coins is highly predictable: it is 50 percent with only a small uncertainty. Thus, statistical certainty can arise in systems for which the basic process is completely random

and unpredictable. After shaking fifty coins in a box, the chances of getting fifty heads is very small, $(\frac{1}{2})^{50} \approx 10^{-16}$, which is zero for practical purposes. A powerful principle of statistical physics says that *each possible state of a system is equally likely.* There is only one state with fifty heads (HHHHH...H), but there are $\Omega_{50} = 50!/25!25! \approx 1.2 \times 10^{14}$ states with a 50:50 distribution of twenty-five heads and twenty-five tails. Each "state" corresponds to a specific set of twenty-five coins showing "heads." Thus, the probability of getting a 50:50 distribution is 1.2×10^{14} times greater than for "all heads."

The 50:50 distribution is more likely than any other distribution, as shown in Fig. 1. As the number of coins increases, the distribution becomes even more sharply peaked at 50 percent.

A system consisting of "all heads" has a high degree of order. After shaking, the system is very unlikely to remain 100 percent heads: it will almost certainly be close to a 50:50 distribution—almost certainly between thirty-five and sixty-five heads, which are states of considerable disorder. This is a general principle: Systems tend to move from order to disorder.

Entropy, defined as $S = k \ln \Omega$, is a measure of disorder; the larger the disorder, the greater the entropy. From its definition, S is largest where Ω peaks (see Fig. 1). The general principle it that isolated systems tend toward maximum entropy (the second law of thermodynamics). The graph shows that systems move toward the maximum, no matter whether they start out with a majority of heads or of tails.

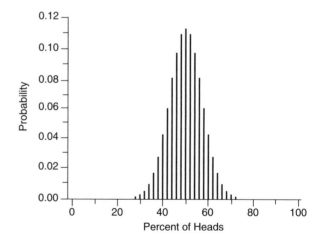

Figure 1 Probability distribution for heads in a coin toss.

A system in equilibrium is at maximum entropy. Thus, as a system moves toward equilibrium, it moves from order to disorder: equilibrium represents the maximum-disorder situation.

Entropy is an extensive quantity. If there are two systems, each with N particles and Ω states, the combined system has $2N$ particles and $\Omega \times \Omega = \Omega^2$ states. Thus, S changes from $k \ln \Omega$ to $k \ln \Omega^2 = 2 (k \ln \Omega)$; that is, the entropy doubles when the number of particles does. Like energy and mass, entropy is an extensive property, meaning that it is additive: If two systems have entropies of S_1 and S_2, the entropy of the combined system is $S_1 + S_2$.

Quantum Unpredictability

At the level of atoms and molecules, nature is fundamentally random. When a spacecraft is sent to the Moon, its impact point is predictable with an accuracy that is limited only by our knowledge of its starting position and velocity. When an electron is fired at a wall, one can predict its impact point only in terms of probabilities, *even with the best possible information about its starting conditions*. Its motion is represented by a spread-out wave, and the probability of impacting a particular point on the screen is determined by the square of the wave's amplitude at that point. If one tries to avoid the spread-out wave packet by creating a "point wave," which is zero everywhere except at one point, the Heisenberg uncertainty principle requires that its momentum be completely unknown. Therefore, its direction of flight is unknown and its impact point will be correspondingly unknown. The impact point will be even less predictable than before. At the microscopic level, nature is inescapably random.

Classical Disorder

A small change in the starting point of a linear system produces a change sometime later that is proportional to the size of the initial perturbation. In a nonlinear system, even a very slight difference in the starting conditions can produce a wildly different, unpredictable result. Figure 2 shows the linear case, where a series of different starting points produces a proportional variation in later positions. For the nonlinear system, the later points are scattered about in a random, unpredictable way: a small initial change produces a large, unpredictable change sometime later.

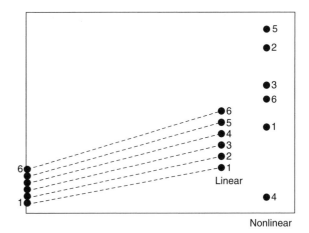

Figure 2 Proportional variation in a linear system.

The orbits of the planets in our solar system are quite stable and predictable over millions of years, but not over billions of years. The interactions are nonlinear because a small shift in position of one planet will change the orbits of the other planets, and these changes will affect the orbit of the first planet. This nonlinearity leads to chaotic orbits over the long term. Chaotic behavior of the atmosphere makes it impossible to predict the weather for more than a few days ahead, and even then there is a large uncertainty.

See also: CHAOS; ENTROPY; EQUILIBRIUM; PROBABILITY; UNCERTAINTY PRINCIPLE

Bibliography

BAKER, G. L., and GOLLUB, J. P. *Chaotic Dynamics: An Introduction* (Cambridge University Press, Cambridge, Eng., 1990).

GLEICK, J. *Chaos: Making a New Science* (Viking, New York, 1987).

PRIGOGINE, I., and STENGERS, I. *Order Out of Chaos* (Bantam, Toronto, 1984).

LAWRENCE A. COLEMAN

OSCILLATION

An oscillation is a repetitive movement from one extreme position, or state, to another. A swinging pendulum, for example, repeats its spatial position and is thus oscillating. The amplitude of an electromagnetic wave returning to an identical value is an example of an oscillatory wave.

Three terms define an oscillation. The time for the oscillation to complete one complete cycle is called the period of oscillation. The inverse of the period is called the frequency ν. Sometimes the angular frequency, ω, is used to designate the motion and is equal to 2π times the frequency ν. That is, $\omega = 2\pi\nu$. The greatest displacement from the equilibrium position is called the oscillation amplitude.

Oscillations can occur in many varieties—mechanical, electrical, optical, and so on. Oscillators are devices that produce periodic signals and are used extensively. Among their many uses, oscillators are used to measure time, to detect signals, and to transmit energy.

Probably the earliest observation of an oscillation was ancient human recognition that the phases of the Moon repeated themselves with an approximately twenty-eight-day period, and that the seasons, too, repeated with regularity. From these observations came the desire to measure time in units smaller than the day.

Most of the early mechanical oscillators were attempts to measure time more precisely. A great impetus for accurate clocks was determining the position of ships at sea. To ascertain the position of his ship, the navigator needed to know the time at some reference seaport when the Sun at the ship's position was directly overhead. So valuable was accurate knowledge of time that sovereigns offered enormous rewards for the most accurate clock.

Little success was achieved in making a practical mechanical clock until a reliable, mechanical oscillator could be developed. An oscillator using a recoil mechanism, the verge and foliot, came upon the scene in the early Renaissance period. Although this was a great improvement over the water clock and other clocks of the time, it still suffered from a lack of precision.

Great improvements were made when Christiaan Huygens developed the pendulum clock following the earlier observations of Galileo. Galileo noted that the period of a clock was regular if the arc of the pendulum's swing was kept small. Galileo, however, was unsuccessful in his attempts to make a practical pendulum clock. Huygens developed the so-called escapement mechanism by which the pendulum's swing could be used to move gears. Huygens is also sometimes credited with the other great advance in

mechanical timekeeping, the hairspring-driven balance wheel. However, Robert Hooke had noted earlier the relationship between a spring's displacement and the force required for the displacement, for example, F = kx, where F is the force, k is the spring constant, and x is the displacement from equilibrium. These studies noted that springs also underwent regular oscillations. In fact, Hooke's notes contained a design of a spring-regulated clock. It is not clear, however, whether he ever built such a clock.

Until Isaac Newton and Gottfried Wilhelm von Leibniz independently invented the calculus, a full mathematical treatment of oscillatory motion could not be developed. In the simple case of a mass m connected to a spring, the calculus makes it easy to show that the frequency of oscillation is given by

$$\nu = \frac{1}{2\pi}\sqrt{\frac{k}{m}}.$$

In contrast, the frequency of a pendulum's motion does not depend upon the mass of the bob. Nonetheless, the mathematical treatment for the pendulum's motion is the same, giving a relationship between the frequency ν, the pendulum's length l, and the acceleration due to gravity g.

$$\nu = \frac{1}{2\pi}\sqrt{\frac{g}{l}}.$$

A rich, mathematical literature developed around the solution of differential equations representing oscillatory motion.

The most accurate modern clocks rely upon electrical oscillators for their operation. Most people are familiar with the quartz-regulated timepiece, which relies upon the natural frequency of oscillation of a quartz crystal. Quartz watches typically are accurate to within a few seconds per month. By far the most accurate clocks are those which use the frequency of atomic transitions as the measure of time, such as the cesium atomic-vapor clock.

The most common oscillators in use are electrical oscillators. The transmission of radio and television signals relies upon the generation of oscillating electromagnetic fields. The earliest radio transmitters used simple electric circuits consisting of inductors and capacitors and produced a radio signal by discharging the energy in the circuit through an arc. After the invention of the triode vacuum tube by Lee DeForest, much more sophisticated oscillators appeared, such as the Hartley oscillator. In modern transmitters the signal to be transmitted is mixed with the carrier frequency, which is produced by an oscillator. For example, transmission of radio and television signals relies upon mixing the audio or video signals with the carrier signal, all of which are produced by oscillators.

The reception of the transmitted signal in modern receivers also involves oscillators. In the superheterodyne type of receiver, the received signal is mixed with a constant-amplitude oscillating signal produced by a local oscillator. The local oscillator's frequency is changed at the same rate as the receiver's tuner so that the resulting beat frequency, called the intermediate frequency, remains at a constant value. The intermediate frequency signal is sent through circuits tuned to this frequency, resulting in superior selectivity and sensitivity.

Oscillators find extensive use in other aspects of electronics. Besides the most common sinusoidal types, oscillators can also produce sawtooth, square-wave, and pulse wave forms. Sawtooth oscillators are used extensively in the drive circuits of display devices, most notably in television receivers, where they sweep the electron beam across the face of the cathode ray tube. Square-wave oscillators are widely used in timing circuits, especially in microprocessor-based applications.

Microwaves are also produced in oscillators. Until World War II, the inductance and capacitance of electric circuits limited the upper range of frequencies that could be produced. The invention of the magnetron by the British led to the first successful radar units and opened the door to the generation of extremely high frequencies. Since then, microwaves have found use in communications, where the higher bandwidth allows much more information to be transmitted per unit time. A more common example, the microwave oven, employs a cavity resonator to excite the natural oscillations in the water molecule, thereby generating heat.

See also: ATOMIC CLOCK; ELECTROMAGNETIC WAVE: GALILEI, GALILEO; HUYGENS, CHRISTIAAN; MOON, PHASES OF; NEWTON, ISAAC; OSCILLATOR; OSCILLATOR, ANHARMONIC; OSCILLATOR, FORCED; OSCILLATOR, HARMONIC; PENDULUM; SEASONS

Bibliography

HALLIDAY, D.; RESNICK, R.; and WALKER, J. *Fundamentals of Physics* (Wiley, New York, 1993).

MAURICE, K., and MAYR, O., eds. *Clockwork Universe* (Neale Watson, New York, 1980).

SHRADER, R. L. *Electronic Communications*, 5th ed. (McGraw-Hill, New York, 1985).

ROBERT N. COMPTON

JOHN E. MATHIS

OSCILLATOR

An oscillator is any mechanical or electrical system that vibrates. Vibrational motion can arise whenever a system is displaced away from a stable equilibrium position. By definition, the equilibrium position is the situation where the system would remain stationary. Stable equilibrium requires that the forces are such as to return the system to the equilibrium position when it is displaced. If, when the system arrives at the equilibrium position, not all the energy has been dissipated, it will continue on past the equilibrium point. The process is repeated, and thus the system will vibrate and be an oscillator.

The simplest mechanical oscillator is a mass attached to a spring. The simplest electrical oscillator is a capacitor connected to an inductor. For both these cases (with an ideal spring and ideal capacitors and inductors), the restoring force will be proportional to the displacement away from equilibrium, and the oscillator will be a harmonic oscillator. For a restoring force proportional to displacement, the situation for the harmonic oscillator, the solution to the equation of motion will be simple sine and cosine functions of time. Oscillators that are not harmonic oscillators are anharmonic. If there is friction in the mechanical oscillator or resistance in the electrical oscillator, the motion will decay with time. This is a damped (or overdamped) oscillator. If there is a source of energy driving the oscillator such that the amplitude of the motion is increasing with time, the situation is underdamped. An example of the later situation would be a child in a swing where, in the beginning, someone is putting energy into the system by pushing the swing, and the amplitude of the oscillations increases.

The frequency (f) of an oscillator is the number of oscillations that occur in a unit of time. The period (T) is the time it takes to undergo one full oscillation, and hence, $T = 1/f$. The amplitude of the oscillator is the maximum distance the system moves away from equilibrium, or, simply stated, the size of the oscillations.

Electrical oscillators are common whenever a control frequency is necessary without an external signal to provide it. A wall clock that plugs into a wall socket is using the sixty cycles per second oscillations of the current provided by the electric company, whereas a battery-driven clock must provide its own oscillator, either mechanical, electrical, or combined. A reference oscillator is an electrical oscillator designed to provide a precise and known frequency. The output of such circuits is periodic but need not be of the simple sine wave form. Other common forms are square waves, triangular waves, sawtooth waves, or trapezoidal waves.

An interesting example of an oscillator that is both mechanical and electrical is the quartz crystal in a watch. The crystal is a mechanical oscillator. All of the atoms vibrate in a simple coherent motion. The frequency of these collective oscillations of the atoms is very precisely determined, thus making the crystal useful as a clock. Through the piezoelectric effect, the energy in the crystal converts between electrical and mechanical. This is useful for two purposes. Electrical energy from a battery can be used to compensate for the loss of energy caused by friction that would otherwise dampen away the oscillations, and the oscillating electrical currents that accompany the mechanical oscillation can conveniently be monitored by attached circuitry. This circuitry simply counts the number of oscillations and uses this to drive the clock.

See also: BATTERY; CAPACITOR; DAMPING; ENERGY, KINETIC; ENERGY, POTENTIAL; EQUILIBRIUM; INDUCTOR; OSCILLATION; OSCILLATOR, ANHARMONIC; OSCILLATOR, FORCED; OSCILLATOR, HARMONIC; PIEZOELECTRIC EFFECT; WAVE MOTION

BIBLIOGRAPHY

GOLDSTEIN, H. *Classical Mechanics,* 2nd ed. (Addison-Wesley, Reading, MA, 1980).

HALLIDAY, D.; RESNICK, R.; and WALKER, J. *Fundamentals of Physics,* 4th ed. (Wiley, New York, 1993).

DAVID J. ERNST

OSCILLATOR, ANHARMONIC

The anharmonic oscillator is any mechanical or electrical system that vibrates or undergoes periodic motion but is not precisely the harmonic oscillator. The harmonic oscillator results from a restoring force that is exactly proportional to the displacement away from equilibrium (the state of the system where it would remain at rest). Any deviation from an exact linear dependence of the restoring force results in an anharmonic oscillator.

A simple deviation from the harmonic oscillator results from friction in a mechanical system or resistance in an electrical system. This deviation is better described as a damped oscillator rather than an anharmonic oscillator.

The harmonic oscillator occurs quite frequently because for any small displacement from a stable equilibrium the force can to some degree of accuracy be approximated by a linear function of the distance. To the same order that there is a correction to the linear dependence of the force there will be anharmonic corrections to the motion. There are several signatures that indicate the motion is not purely harmonic. Since the frequency of the oscillations is independent of the amplitude (the maximum distance away from equilibrium) for the harmonic oscillator, any dependence of the frequency on the amplitude is an indication of anharmonic motion. Any deviation of the time dependence away from the simple sine or cosine function means the motion is anharmonic. A particular feature might be that the motion is not symmetric about the equilibrium point. In this case, the average position of the oscillating particle (or magnitude of the current for a circuit) will change as the amplitude of the oscillations changes. If an anharmonic oscillator is driven by an external force, it will respond not with just the frequency of the driving force, as does the harmonic oscillator, but also with overtones of the driving frequency.

A simple example of an anharmonic oscillator is the pendulum. The motion of a pendulum is approximated by the harmonic oscillator only to the degree that the sine of the angle made with the vertical can be approximated by the angles $\sin \theta \approx \theta$. For small enough angles, this is a reasonable approximation. For larger angles, such as is typical for a child on a swing, the approximation is poor; the frequency of the oscillations will depend on the amplitude, and the motion will not be well approximated by a sine or cosine function of the time.

An interesting physical consequence of anharmonic motion is the expansion of solids as they are heated. If the forces between the molecules were exactly harmonic, then adding heat would cause the amplitude of the oscillations between molecules to increase but the average distance between the molecules would not change and the material would not expand. Because the forces are anharmonic, the increase in the amplitude of the oscillations with an increase in temperature also produces an increase in the average distance between the particles and, hence, expansion of the solid.

See also: MOTION, PERIODIC; OSCILLATOR; OSCILLATOR, HARMONIC; PENDULUM

BIBLIOGRAPHY

HALLIDAY, D.; RESNICK, R.; and WALKER, J. *Fundamentals of Physics,* 4th ed. (Wiley, New York, 1993).

NAYFEH, A. H., and MOOK, D. T. *Nonlinear Oscillations* (Wiley, New York, 1979).

DAVID J. ERNST

OSCILLATOR, FORCED

Many physical systems have one or more natural vibration frequencies and one or more natural modes of energy dissipation (damping) that characterize its behavior when stimulated by a transient impulse. These damped oscillations will eventually stop unless energy is supplied to sustain them. If driven by a series of external impulses at some arbitrary frequency, the system will necessarily vibrate at the driver frequency, but with an amplitude and phase that depend not only on the driver amplitude and frequency, but also on the natural frequencies and damping constants of the system. If the frequency of the external stimulus is varied, the amplitude of the response at or near one of the natural frequencies can exceed the amplitude of the driver, a condition known as resonance. The amount of damping controls the size of the resonant response, the sharpness of the resonance as a function of frequency, and the

phase shift between the stimulus and the response. For small damping the amplitude is maximum when the frequencies of the driver and the oscillator coincide, but for large damping there is a frequency shift. If the damping time of the oscillator is large, the resonance width is small, and vice versa. Thus the product of the frequency width and the damping time is of order unity. This property is known as the uncertainty principle, and it expresses the impossibility of measurements that are arbitrarily precise in both time and frequency.

If the driver and response frequencies are equal but 180 degrees out of phase, the stimulus tends to suppress the response oscillations (e.g., the driven tuned mass dampers in tall buildings that are used to decrease their sway during high winds). If the driver exactly compensates for the energy dissipated by damping, it will sustain the motion (e.g., the escapement in a pendulum clock). If the driver supplies more energy than is dissipated by damping, the amplitude of the system will increase dramatically, and perhaps catastrophically. Many examples of resonance exist, such as the shattering of a wine glass by sound, the use of a microwave oven to drive electrons in water molecules at their natural frequencies, the use of a tuned electrical circuit to filter electromagnetic waves detected in a radio receiver, and the bridge collapses that have occurred when marching troops failed to break cadence. In some cases the external driving force can be initiated by the presence of the responding oscillator itself. For example, alternating vortex swirls can be formed when a fluid flows past an object, which can exert a periodic force on the object that forms them. When the resonance condition is met, huge amplitudes have been observed (e.g., the collapse of the Tacoma Narrows bridge in 1940 and the failure of aircraft wings in the 1960s).

Another application of driven oscillation is given by resonance fluorescence, which is used as a tool in the study of atomic structure. In one application of this technique, observation of the absorption and reemission of laser light of known but variable frequency by an atomic sample permits the determination of the natural frequencies (that characterize the energy level separations) and damping constants (that characterize the level lifetimes) of the atoms.

See also: DAMPING; FLUORESCENCE; FREQUENCY, NATURAL; PENDULUM; PHASE; RESONANCE

Bibliography

FEYNMAN, R. P.; LEIGHTON, R. B.; and SANDS, M. *The Feynman Lectures on Physics,* Vol. 1 (Addison-Wesley, Reading, MA, 1963).

FISHBANE, P. M.; GASIOROWICZ, S.; and THORTON, S. T. *Physics for Scientists and Engineers* (Prentice Hall, Englewood Cliffs, NJ, 1993).

MARION, J. B., and THORTON, S. T. *Classical Dynamics of Particles and Systems,* 4th ed. (Saunders, Fort Worth, TX, 1995).

LORENZO J. CURTIS

OSCILLATOR, HARMONIC

A harmonic oscillator is a physical system that oscillates back and forth around its stable equilibrium position r_e due to a restoring "force" proportional linearly to the "displacement" from r_e. If the displacement is sufficiently small, its motion is called simple harmonic motion (SHM). The harmonic oscillator is not necessarily limited to a mechanical system, such as the spring-mass system and the diatomic molecule discussed under simple harmonic motion. For example, the energy transfer between electric and magnetic fields in a circuit containing an inductance L and a capacitance C (i.e., an L–C circuit) can also be described by an SHM with an angular frequency $\omega = (LC)^{-1/2}$.

Pendulum

In general, a pendulum (or, a physical pendulum) is a rigid body of mass m with an arbitrary shape, pivoted about a fixed horizontal frictionless axis through a point P, at a distance L, from its center of mass O, as shown in Fig. 1. Its moment of inertia about the pivot point P is I. When the pendulum is displaced from the vertical by an angle θ under the gravity, it is subject to a restoring torque

$$\Gamma = -mgL \sin \theta. \tag{1}$$

When θ is small, $\sin \theta \approx \theta$. Together with the relation $\Gamma = I d^2\theta/dt^2$, the motion of the pendulum is described by

$$\frac{d^2\theta}{dt^2} + \frac{mgL}{I}\theta = 0, \tag{2}$$

or, an SHM with an angular frequency $\omega = (mgL/I)^{1/2}$. For a simple pendulum, the mass is concentrated at the end of a weightless rigid rod with $I = mL^2$ and the system oscillates with an angular frequency $\omega = (g/L)^{1/2}$.

Floating Objects

An object with a density that is less than the density of a liquid may float partially submerged in stable equilibrium at the surface of the liquid. According to the Archimedes' principle, the weight of the object, for example, Mg, is balanced by a buoyant force that equals the weight of the liquid replaced by the floating object. If the floating object is slightly pushed into the liquid from its stable equilibrium position by a small vertical displacement y, it will experience a restoring force that equals the increase in the weight of the additional liquid that is replaced by the floating object. Assuming a constant cross section A of the floating object near the floating level, the restoring force is given by $-\rho gAy$, where ρ is the density of the liquid and g is the acceleration due to gravity. The motion is described by a SHM according to the equation

$$\frac{d^2y}{dt^2} + \left(\frac{\rho gA}{M}\right)y = 0 \tag{3}$$

with an angular frequency $\omega = (\rho gA/M)^{1/2}$.

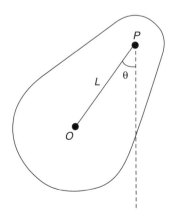

Figure 1 A physical pendulum.

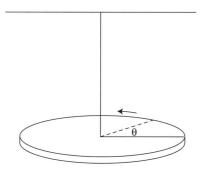

Figure 2 A torsional pendulum.

Torsional Pendulum

A uniform disk, suspended by a weightless rod attached to its center as shown in Fig. 2, is one example of a torsional pendulum. When twisted by a small angular displacement θ, it experiences a restoring torque $\Gamma = -\kappa\theta$, where κ is the torsion constant. The rotational motion is described by an SHM according to the equation

$$\frac{d^2\theta}{dt^2} + \left(\frac{\kappa}{I}\right)\theta = 0 \qquad (4)$$

with an angular frequency $\omega = (\kappa/I)^{1/2}$, where I is the moment of inertia around the vertical axis through the center of mass.

See also: ARCHIMEDES' PRINCIPLE; CENTER OF MASS; MOTION, HARMONIC; MOTION, ROTATIONAL; PENDULUM

Bibliography

FRENCH, A. P. *Vibrations and Waves* (W. W. Norton, New York, 1971).

TIPLER, P. A. *Physics for Scientists and Engineers* (Worth, New York, 1990).

TU-NAN CHANG

OSCILLOSCOPE

The oscilloscope is an electronic instrument designed to record the variation of two or more variables, typically a voltage signal with time. The heart

of an oscilloscope is the cathode-ray tube (CRT) as shown in Fig. 1.

Radiative heat from the filament raises the temperature of the cathode to the point that electrons are emitted and then accelerated toward the fluorescent screen of the CRT, which is at a higher positive potential than the cathode. Adjusting the voltages on the anode allows one to focus the electron beam to a small spot on the screen. A sawtooth voltage (linear voltage increase with time from zero up to a maximum voltage followed by an abrupt return or "fly-back" to zero) is often used to sweep the electron spot from one side of the fluorescent screen to the other. The voltage signal to be measured is amplified and then applied to the vertical deflection plate. As the electron beam moves across the screen, the vertical deflection provides a time history of the

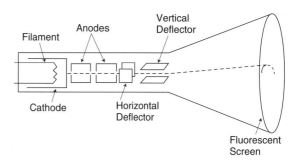

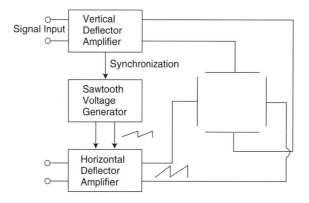

Figure 1 Basic diagram of the cathode-ray tube oscilloscope showing the electronic setup needed to control the amplitude of the input voltage pulses and the horizontal "sweep" voltage ("sawtooth" appearing voltage). The synchronization signal ("sync" control) adjusts the sawtooth voltage to start at the same point on the input voltage, allowing for the repetitive scan of the voltage pulse and therefore visual observation of the voltage pulse.

voltage pulse. A single sweep picture remains on the screen only for a fraction of a second and could be recorded on a photographic film. Such an arrangement would be called a "storage scope." Repetitive sweeps are usually displayed by synchronizing the start of the sawtooth voltage with a common point on the input voltage. It is common to have several input channels for deflection in the vertical direction, allowing the user to compare more than one signal.

Strictly speaking, an oscilloscope is capable of plotting the relationship of any two variables (e.g., voltages). For example, the application of two sine wave (oscillating) voltages to the deflectors, one to the vertical and one to the horizontal deflector, will produce interesting patterns on the CRT called Lissajous figures (ovals, figure eights, etc.). These patterns can provide information such as the amplitude and frequency ratios as well as the phase relationship of the two input signals.

There are basically two types of modern oscilloscopes, the analog and the digital oscilloscope. The analog oscilloscope provides a continuous trace of the voltage pulse as a function of time as described above. The digital oscilloscope converts the analog voltage to a digital signal and stores the magnitude of the voltage at regular time intervals along the wave form. The digital oscilloscope has the great advantage that the signal is stored in a digital form that can be processed in a computer. The computer can then be used to reduce random noise that might occur in repetitive signals, subtract background signals, integrate (i.e., sum) or differentiate the signal. The computer can also perform what is called the Fourier transform of the signal. This is extremely useful in many areas of technology, such as Fourier transform mass spectroscopy and signal processing.

The oscilloscope is one of the most widely used instruments in science and industry. The price of an oscilloscope can vary greatly from a few hundred dollars for a dual-channel analog oscilloscope to tens of thousands of dollars for a four-channel digital oscilloscope.

See also: CATHODE RAY; FOURIER SERIES AND FOURIER TRANSFORM; OSCILLATION

Bibliography

CZECH, J. *Oscilloscope Measuring Technique: Principles and Applications of Modern Cathode Ray Oscilloscopes* (Centrex, Eindhoven, 1965).

LENK, J. D. *Handbook of Oscilloscopes: Theory and Application* (Prentice Hall, Englewood Cliffs, NJ, 1968).

ROBERT N. COMPTON

P

PARALLAX

The general term "parallax" refers to the apparent change in position of an object when viewed from two different locations. In astronomy, parallax is the fundamental method of measuring distances to stars. These can strictly be called trigonometric parallaxes, since astronomers sometimes refer to distance measurements, by whatever technique, as parallax.

The usual parallax measurement procedure is to observe a target star at time intervals of approximately six months, using the orbital motion of Earth around the Sun to change the position of the observatory. The target star's position is commonly measured relative to several reference stars that lie near it in the sky but which are believed to be much more distant than the target (see Fig. 1). (A small correction must be applied to account for the fact that these background stars are not at infinite distance.) Multiple observations every six months for several years are typically required to obtain good precision and to separate the apparent positional change due to parallax from the true change in stellar position known as proper motion. The parallax of a star is defined as half of the maximum apparent positional shift that could be observed from opposite sides of Earth's orbit, or the shift that would be observed if the target were observed from Earth and the Sun. Parallax is expressed as an angle, generally in arcseconds or milliarcseconds. Distances in astronomy are most commonly given in units of parsecs, where 1 pc is defined as the distance at which a star would have a parallax of 1 arcsec (1 pc = 3.26 ly). Stars are so distant, and thus parallaxes so small, that distances can be accurately calculated from the simple formula $d = 1/p$, where d is the distance in parsecs and p is the parallax in arcseconds.

The first parallax measurements were made visually around 1840, but few reliable measurements were made until the introduction of photographic techniques around 1900. Although photography is still used for parallax measurements, nonphotographic techniques have become more important since about 1980. These newest methods include the use of charge-coupled devices (CCDs) instead of photographic plates to record the stellar images, as well as scanning techniques that measure positional changes without actually recording the images.

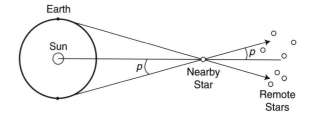

Figure 1 Measuring distance to nearby stars relative to reference stars located near it.

Most pre-1970 parallaxes had large uncertainties, and distances derived from them were often not reliable beyond about 20 pc. Recently, uncertainties as small as 1 milliarcsecond have been obtained, allowing reliable distances to be derived out to a few hundred parsecs. The HIPPARCOS satellite is expected to obtain parallaxes for 100,000 moderately bright stars with uncertainties of 2 milliarcseconds, greatly increasing the number of stars with measured parallaxes.

Geocentric parallax (near-simultaneous observations from two different locations on Earth) was formerly used to derive distances to solar system objects, but this has now been largely superseded by the more precise technique of radar ranging.

See also: STARS AND STELLAR STRUCTURE

Bibliography

KOVALEVSKY, J. "Astrometry from Earth and Space." *Sky and Telescope* **79**, 493 (1990).

LOVI, G. "The Distance Dilemma." *Sky and Telescope* **69**, 45 (1985).

MONET, D. G. "Recent Advances in Optical Astrometry." *Ann. Rev. Astron. Astrophys.* **26**, 413 (1988).

VAN ALTENA, W. F. "Astrometry." *Ann. Rev. Astron. Astrophys.* **21**, 131 (1983).

KYLE CUDWORTH

greatly simplifies this situation. One is only required to compute the moment of inertia about an axis passing through the center of mass (*cm*). Then one can very easily calculate by means of the parallel axis theorem the moment of inertia about any other axis which is parallel to the center-of-mass axis. In symbols, let I_{cm} be the moment of inertia about an axis A_{cm} which passes through the center of mass. Then let A be any other axis of rotation which is parallel to A_{cm}. The moment of inertia I about this axis is given simply by

$$I = I_{cm} + M a^2,$$

where M is the mass of the rigid body and a is the perpendicular distance between the two axes (see Fig. 1). Since the moment of inertia of a point mass about an axis is simply the mass times the square of its distance from the axis, the parallel axis theorem may be put into words as follows: The moment of inertia of a rigid body about any axis equals the moment of inertia about a parallel axis passing through the center of mass plus the moment of inertia of a fictitious point body having the same mass M as the rigid body and located at the center of mass.

It is worth noting that the axes in question, like the center of mass, do not have to be located in a material portion of the body. A simple example to the contrary would be a torus (doughnut) for which

PARALLEL AXIS THEOREM

In describing the rotational properties of a rigid body, the most significant attribute of that body, playing a role comparable to that played by mass in describing linear motion, is its moment of inertia about the axis of rotation. This is a measure of how the mass is distributed about the axis, and takes into account how far individual mass elements are from the axis.

In principle, the axis of rotation may pass through the body at any given position, and the moment of inertia about that axis would have to be laboriously calculated in each particular case. The parallel axis theorem, discovered by the Swiss mathematician Jakob Steiner in the nineteenth century (and hence sometimes called Steiner's theorem),

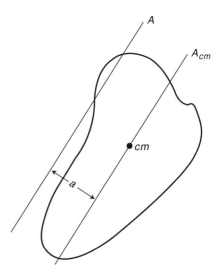

Figure 1   The axes A and A_{cm}.

the center of mass is located at the center of the hole; the axis may pass through this point without ever touching the body. We do not consider mechanical questions of how the torus may be made to rotate about such an axis.

As an illustration of Steiner's theorem, consider a uniform thin rod of mass M and length L. Elementary calculus arguments show that the moment of inertia of the rod about an axis perpendicular to the rod and passing through its center of mass, its midpoint, is

$$I_{cm} = M L^2/12.$$

Then, by the parallel axis theorem, its moment of inertia about an axis perpendicular to the rod but a distance a from its midpoint will be

$$I = M(L^2/12 + a^2).$$

In particular, for an axis at the end of the rod (as in Fig. 2), where $a = L/2$, we obtain

$$I_{end} = M L^2/3.$$

As a second example, elementary arguments show that for a homogeneous solid sphere of mass M and radius R, we have for any axis through the center

$$I_{cm} = 2 M R^2/5.$$

Then, for an axis just touching the edge of the sphere (i.e., a tangent, as in Fig. 3), $a = R$, and Steiner's theorem gives the result

$$I_{tan} = 7 M R^2/5.$$

There is a simple but important corollary to the parallel axis theorem: Consider all possible axes of rotation of a body pointing in a given direction. Then the moment of inertia is smallest about that axis which passes through the center of mass. The proof is immediate, for the moment of inertia about any other axis is, by the parallel axis theorem, greater by the amount Ma^2.

In more advanced treatments of rigid body dynamics, the single quantity, the moment of inertia, is replaced by six quantities, the inertia tensor. Con-

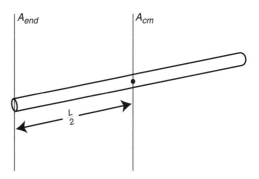

Figure 2 Application of the parallel axis theorem to a uniform rod of mass M and length L. The axes A_{end} and A_{cm} are perpendicular to the length of the rod, parallel to each other, and pass through the end and center of the rod respectively.

sider a point P in the body and a set of xyz coordinate axes having P as the origin of coordinates, as in Fig. 4. Three of the quantities forming the inertia tensor are the moments of inertia about the x, y, and z axes, respectively. The other three quantities, which may be positive or negative, are known as products of inertia. The relation to the simpler moment of inertia concept used previously is that for an arbitrary axis of rotation through P, say A_P, the moment of inertia about axis A_P may be simply calculated from the six elements of the inertia tensor and the direction cosines of A_P. Then the parallel axis theorem can be generalized as follows. Let

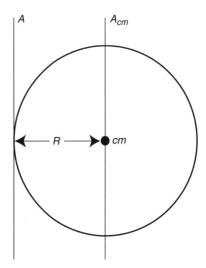

Figure 3 Application of the parallel axis theorem to a uniform solid sphere of mass M and radius R.

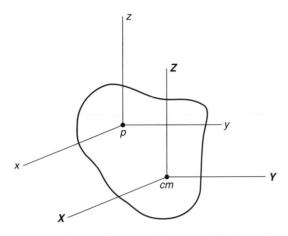

Figure 4 Generalization of the parallel axis theorem to the full inertia tensor. *XYZ* are a set of coordinate axes with origin at the center of mass of the body, while *xyz* are a set of axes with origin at the point *P* but parallel to the set *XYZ*.

there be established at the center of mass, coordinate axes *XYZ* having the center of mass as their origin but parallel to the *xyz* axes previously established at *P*. Then each of the six elements of the inertia tensor relative to the *xyz* axes is equal to the corresponding element relative to the center-of-mass *XYZ* axes, plus the same element calculated for a fictitious point body of mass *M* located at the center of mass.

See also: CENTER OF MASS; CENTER-OF-MASS SYSTEM; INERTIA, MOMENT OF

Bibliography

MARION, J. B., and THORNTON, S. T. *Classical Dynamics of Particles and Systems,* 3rd ed. (Harcourt Brace Jovanovich, Orlando, FL, 1988).

MICHAEL LIEBER

PARAMAGNETISM

A substance (gas, liquid, or solid) is said to be paramagnetic if an applied magnetic field *H* causes the substance to be weakly and reversibly magnetized in the direction of the magnetic field. The magnetic susceptibility is positive, and the magnetization is found to be proportional to the applied magnetic field except at very low temperatures or very large fields. The magnetization of the substance results from the partial alignment of individual magnetic moments associated with the atoms making up the substance, or with the magnetic moments of the conduction electrons in metals (called Pauli paramagnetism).

Individual atoms or ions in the paramagnetic substance may have a magnetic moment, due either to unpaired electrons in the electron shells or to nuclear magnetic moments. Paramagnetism can also arise from the spin magnetic moments of the conduction electrons in metals. This is usually counteracted partially by a weak diamagnetism that is the result of Faraday's law of induction as applied to electron orbital motion. Magnetic moments of electronic origin are usually on the order of a Bohr magneton (9.27×10^{-24} J/T), while those of nuclear origin are typically about 10^3 times smaller.

Consider first the case where an individual atom in a compound has a magnetic moment μ. Such a magnetic moment will tend to be aligned in the direction of an applied magnetic field *H*, and the potential energy of the moment can be written as $V = -\mu H \cos \varphi$, where φ is the angle between the directions of the magnetic moment and the magnetic field. This energy, for reasonable magnetic fields and temperatures, is quite small compared with random thermal energies (on the order of $k_B T$, where k_B is Boltzmann's constant and *T* is the absolute temperature). Under these conditions, the magnetic moment can rotate quite freely, with only a small tendency to be aligned with the magnetic field. If a magnetic field is applied to a substance with *N* atoms per unit volume, each with magnetic moment μ, the moments will be partially aligned and the total magnetization *M* of the system can be written as

$$M \cong \frac{N\mu^2 H}{3k_B T}. \qquad (1)$$

Equation (1) is valid only when the magnetic field is not too large and the temperature is not too small. Equation (1), called the Curie law, results from the competition between the magnetic potential energy and the random thermal energies, and it indicates that the individual atomic moments are on the average more aligned for large fields *H* or small temperatures *T*.

Another assumption in deriving Eq. (1) is that the individual atomic moments do not interact with one another. The interactions between the individual atomic moments may be ferromagnetic (the moments tend to have parallel alignment) or antiferromagnetic (antiparallel alignment). If such interactions are important, then Eq. (1) is modified to

$$M \cong \frac{N\mu^2 H}{3k_B(T - \Theta_P)}.$$
(2)

Equation (2) is called the Curie–Weiss law, and Θ_P is the Curie temperature. It is valid over a range of temperatures T greater than Θ_P. If the interactions between magnetic moments are very strong, then the substance may become ferromagnetic or antiferromagnetic as the temperature is lowered. Thus ferromagnetic and antiferromagnetic substances are paramagnetic above their phase transition temperatures.

Paramagnetism can also arise from alignment of the moments associated with conduction electrons in metals, in which case it is called Pauli paramagnetism. For Pauli paramagnetism, the magnetization is usually only weakly dependent on temperature, as the effective number [N in Eqs. (1) and (2)] of electronic moments that can be aligned with the magnetic field decreases with decreasing temperature.

Some paramagnetic substances include O_2, $KCr(SO_4)_2 \cdot 12H_2O$, $FeSO_4$, and $MnCl_2$. The magnetizations of the first two compounds obey the Curie law, whereas $FeSO_4$, and $MnCl_2$ obey the Curie–Weiss law with negative and positive values of Θ_P, respectively. Note that a positive value of Θ_P indicates that the atomic moments interact ferromagnetically, and a negative value indicates that they interact antiferromagnetically.

See also: ANTIFERROMAGNETISM; BOLTZMANN CONSTANT; CURIE, MARIE SKLODOWSKA; CURIE TEMPERATURE; ELECTRON; FERROMAGNETISM; MAGNETIC MATERIAL; MAGNETIC MOMENT; MAGNETIC SUSCEPTIBILITY; MAGNETIZATION; PAULI, WOLFGANG; SOLID; TEMPERATURE

Bibliography

CULLITY, B. D. *Introduction to Magnetic Materials* (Addison-Wesley, Reading, MA, 1972).

KITTEL, C. *Introduction to Solid-State Physics*, 6th ed. (Wiley, New York, 1986).

ROGER D. KIRBY

PARAMETERS

In physics a parameter is a number that represents a physical property of a system and appears as a constant in the equations describing the behavior of that system. The values for these constants must be determined before the equations can be used to make predictions about the behavior of the system. The more parameters there are to determine, the more measurements must be made before one can either test the theory embodied in the equations or predict future behavior of the system in detail.

There are two basic types of parameters: those that describe a particular system or material, such as the coefficient of friction between two surfaces, and those that are global or universal, such as Newton's constant of gravitation G or the charge of the electron e. These are the same for all matter, anywhere in the universe.

Fundamental Constants

It is interesting to note that in physics (and in all of science) when we ask why a certain effect occurs or why a parameter takes a certain value, the answer lies in a mechanism at some smaller scale. Once we get to a fundamental underlying theory with a few universal parameters, we can no longer ask the question "Why?" about any feature of this theory. The only thing we can say is that the theory provides a description of a large number of effects in terms of a few parameters whose values are given by measurement. If there is no smaller scale, then there is no way we can ever "explain" these parameters. They are simply the fundamental constants of nature.

How do we know when we have reached this point? Is what we now call the standard model of the fundamental interactions truly fundamental? Is there some theory that operates on even smaller scales that provides a mechanism to derive some or all of the twenty or so parameters that describe matter in the standard model from an even smaller set? In the history of science, this type of inquiry has always led to new understanding. Most particle physicists feel that at least some of the parameters of the standard model do indeed have relationships that depend on underlying properties yet to be discerned. Much of the effort of particle physics today is devoted to seeking clues to the nature of such relationships.

Two other fundamental and universal constants of a somewhat different type play a key role in twentieth-century physics. These are Planck's constant h, which is the size of the quantum unit of action (or angular momentum), and the velocity of light c. Each of these represents a fundamental and universal property of nature. They determine the scales on which quantum behavior and relativistic mechanics respectively begin to play a role in physics. While the numerical value of these constants is an artifact of the units we use, the physical values are anything but unimportant. Their fundamental importance in nature is shown by the revolutions in our view of the physical world that occurred early in the twentieth century when we began to understand effects sensitive to quantum behavior and special relativity.

When we study the behavior of many atom systems, another fundamental constant appears, along with the concept of temperature. Boltzmann's constant k_B is defined by the amount of energy needed to cause a specified change of temperature for a specified amount of ideal monatomic gas under specified conditions of temperature and pressure. Clearly, the numerical value of k_B depends on the specified conditions and on the units of temperature and energy so related. Unlike the cases of Planck's constant and the speed of light, Boltzmann's constant does not define a new scale of physics. Its role is to define the concept of temperature better. Temperature can be understood as a measure of the amount of heat energy stored in a system (i.e., the energy of the random motions of the atoms of that system), and Boltzmann's constant gives the relationship of units of temperature to units of energy.

Bulk Parameters

When we study the physics of large-scale objects, it is convenient to introduce many parameters to summarize the effect of the interactions that take place at smaller scales. Thus, we introduce coefficients of friction, densities, bulk elastic moduli, spring constants, expansion coefficients, conductivity, and so on. Their values are different from one material to the next and even for the same material under different conditions. To make predictions about the behavior of an object, we must first measure the values of all the parameters that appear in our equations under the relevant conditions. The values of these parameters for many materials can be found tabulated in handbooks and are the basic starting point of many engineering calculations.

Physicists continue to work to refine their calculational tools so they can calculate more such properties from the underlying theory. One of the aims of understanding matter is to understand all the parameters that appear in its description. In principle, one would like to be able to calculate these parameters from a knowledge of the internal structure of the material. In practice, we are happy if we can do this in just a few simple cases, enough to convince us that our underlying theory is a good one. Since the underlying theory describes many different materials in terms of the same basic building blocks, it contains many fewer parameters than the theory at the larger scale. At the smallest scales, only fundamental and universal quantities appear.

See also: APPROXIMATION AND IDEALIZATION; SCALE AND STRUCTURE

HELEN R. QUINN

PARITY

In its strict definition used in physics, parity denotes a sign designation, + or − (even or odd), which describes how an elementary object transforms under a space-inversion operation.

Space Inversion

Let x, y, z be the Cartesian coordinates in a three-dimensional space. The transformation $x \to -x$, $y \to -y$, and $z \to -z$ is the space inversion.

Mirror Reflection

The transformation $x \to -x$, $y \to y$, and $z \to z$ represents the reflection with respect to a mirror located at $x = 0$ and parallel to the (y, z) plane. Since a 180° rotation along the x axis will change y to $-y$, z to $-z$, but leave x unchanged, space inversion can be achieved by the combined operations of a mirror reflection and a 180° rotation along the axis perpendicular to the mirror.

Examples of Objects with Definite Parity

The location of a point particle is given by its position vector $\mathbf{r}$ whose components are x, y, and z. For brevity, we write $\mathbf{r}(x, y, z)$. Under the space inversion, $\mathbf{r}$ becomes $-\mathbf{r}$, since each of its components changes sign. We call the parity of the position vector odd; that is, under space inversion it becomes the negative of itself.

Under space inversion, $r^2 \equiv x^2 + y^2 + z^2$ is clearly unchanged. We call the parity of r^2 even, since $r^2 \to +r^2$ under space inversion.

Consider two point particles, 1 and 2, located at position vectors $\mathbf{r}_1(x_1, y_1, z_1)$ and $\mathbf{r}_2(x_2, y_2, z_2)$. The scalar product $\mathbf{r}_1 \cdot \mathbf{r}_2$ is defined to be the sum $x_1 x_2 + y_1 y_2 + z_1 z_2$. Under space inversion, because $\mathbf{r}_1 \to -\mathbf{r}_1$ and $\mathbf{r}_2 \to -\mathbf{r}_2$

$$\mathbf{r}_1 \cdot \mathbf{r}_2 \to +\mathbf{r}_1 \cdot \mathbf{r}_2.$$

We call the parity of $\mathbf{r}_1 \cdot \mathbf{r}_2$ even; that is, under space inversion it remains itself.

The vector product $\mathbf{a} = \mathbf{r}_1 \times \mathbf{r}_2$ is defined to be an object with three components given by

$$a_x = y_1 z_2 - z_1 y_2, \quad a_y = z_1 x_2 - x_1 z_2$$

and

$$a_z = x_1 y_2 - y_1 x_2,$$

all in the same Cartesian coordinates. Under space inversion, we see that $a_x \to +a_x$, $a_y \to +a_y$ and $a_z \to +a_z$; consequently, $\mathbf{a} \to +\mathbf{a}$, and we call the parity of the vector product $\mathbf{a} = \mathbf{r}_1 \times \mathbf{r}_2$ even. Under a rotation, $\mathbf{a}$ transforms in the same way as the position vector $\mathbf{r}$, but under space inversion $\mathbf{a} \to +\mathbf{a}$ whereas $\mathbf{r} \to -\mathbf{r}$. Since we call $\mathbf{r}$ a vector, in order to distinguish the sign difference under space inversion, we call the vector product $\mathbf{a} = \mathbf{r}_1 \times \mathbf{r}_2$ a pseudovector.

Now introduce a third point particle, located at $\mathbf{r}_3(x_3, y_3, z_3)$. Form the scalar product π between $\mathbf{r}_3$ and $\mathbf{a} = \mathbf{r}_1 \times \mathbf{r}_2$; that is,

$$\pi = \mathbf{r}_3 \cdot \mathbf{r}_1 \times \mathbf{r}_2 = z_3 x_1 y_2 - z_3 y_1 x_2 + x_3 y_1 z_2$$
$$- x_3 z_1 y_2 + y_3 z_1 x_2 - y_3 x_1 z_2.$$

Under space inversion, $\mathbf{r}_1 \to -\mathbf{r}_1$, $\mathbf{r}_2 \to -\mathbf{r}_2$, $\mathbf{r}_3 \to -\mathbf{r}_3$, and, therefore, $\pi \to -\pi$. We call the parity of

$\pi = \mathbf{r}_3 \cdot \mathbf{r}_1 \times \mathbf{r}_2$ odd because it becomes the negative of itself under space inversion. Under any pure rotation, the square-magnitude r^2 of any position vector and the scalar product $\mathbf{r}_1 \cdot \mathbf{r}_2$ of two position vectors are both unchanged; likewise, $\pi = \mathbf{r}_3 \cdot \mathbf{r}_1 \times \mathbf{r}_2$ is also unchanged. However, under space inversion $r^2 \to +r^2$ and $\mathbf{r}_1 \cdot \mathbf{r}_2 \to +\mathbf{r}_1 \cdot \mathbf{r}_2$, whereas $\pi \to -\pi$. We call r^2 and $\mathbf{r}_1 \cdot \mathbf{r}_2$ scalars, but $\pi = \mathbf{r}_3 \cdot \mathbf{r}_1 \times \mathbf{r}_2$ a pseudoscalar.

State Vector in Quantum Mechanics

The quantum mechanical description of a system with N (spinless) particles is in terms of a function called a wave function, $\psi = \psi(\mathbf{r}_1, \mathbf{r}_2 \cdots, \mathbf{r}_N)$, where $\mathbf{r}_i$ denotes the position vector of the ith particle, with $i = 1, 2, \cdots, N$.

Parity Operator

Under a space inversion, $\mathbf{r}_i \to -\mathbf{r}_i$, and, therefore, the wave function

$$\psi(\mathbf{r}_1, \mathbf{r}_2, \ldots, \mathbf{r}_N) \to \psi(-\mathbf{r}_1, -\mathbf{r}_2, \ldots, -\mathbf{r}_N).$$

We can designate the right-hand side more compactly as $P\psi = \psi(-\mathbf{r}_1, -\mathbf{r}_2, \cdots, -\mathbf{r}_N)$. Under a space inversion, the above formula can be written as $\psi \to P\psi$, and P is called the parity operator.

If we perform two space inversions in succession, under the first one $\mathbf{r}_i \to -\mathbf{r}_i$ and under the second $-\mathbf{r}_i \to \mathbf{r}_i$; therefore, two space inversions in succession leave both $\mathbf{r}_i$ and $\psi(\mathbf{r}_1, \mathbf{r}_2, \cdots, \mathbf{r}_N)$ unchanged. Symbolically we use P^2 to denote two space inversions in succession. Since $P^2\psi = \psi$, we say the square of the parity operator equals unity; that is, $P^2 = 1$.

Parity Quantum Number

From any quantum mechanical wave function ψ, form the linear combination $\psi_\pm = \psi \pm P\psi$. Under space inversion, we have $P\psi_\pm = P\psi \pm P^2\psi$; consequently, since $P^2 = 1$,

$$P\psi_+ = +\psi_+$$

and

$$P\psi_- = -\psi_-.$$

That is, the parity of ψ_+ is even and the parity of ψ_- is odd. We call $+1$ the parity quantum number of ψ_+ and -1 that of ψ_-.

Elementary Particles

For the so-called elementary particles (quarks, leptons, nucleons, pions, muons, kaons, and so on), physicists quite often do not know whether they are indeed elementary, or composites. Whatever the case, each particle or system of particles can always be represented by a state wave function ψ (without specifying the internal variables that ψ depends on). Under space inversion we say $\psi \to P\psi$. Form the combination $\psi_\pm = \psi \pm P\psi$. Again, since $P^2 = 1$, the parity of ψ_+ is even and the parity of ψ_- is odd. Hence, the state wave function ψ of any elementary particle can always be written as a sum $\psi = \frac{1}{2}(\psi_+ + \psi_-)$, each component ψ_+ or ψ_- having a definite parity.

Parity Conservation

If natural law is symmetrical with respect to space inversion, then, if we begin with any state wave function of a definite parity, as time evolves the parity of that state wave function will remain the same. This is called parity conservation.

As mentioned earlier, space inversion is the combined operation of mirror reflection and a $180°$ rotation. In the following, we shall assume natural law is rotationally symmetric. Therefore, space inversion symmetry becomes the same as mirror reflection symmetry. These symmetries imply parity conservation. The converse is also true.

In the state-wave-function decomposition $\psi = \frac{1}{2}(\psi_+ + \psi_-)$ introduced above (for an elementary particle), if parity were conserved, then these two components ψ_+ and ψ_- would evolve independently in time since they are of opposite parities. Therefore, each should be considered a separate entity. With parity conservation, we can always consider an elementary particle to have a definite parity (i.e., either represented by ψ_+ of even parity or by ψ_- of odd parity).

For example, from experiments using the strong and electromagnetic forces, physicists determined the parity of the pion to be odd; that is, under space inversion a pion at rest becomes the negative of itself. Furthermore, under space rotation a pion at rest remains itself. Therefore we call the pion a pseudoscalar (like π in the previously listed examples of objects with definite parity).

Why Would Physicists Ever Believe in Parity Conservation?

Consider the example of two cars that are made exactly alike, except that one is the mirror image of the other; that is, one car has the driver's seat on the left side (as in the United States) and the other on the right (as in Japan), the first one has the gas pedal for the right foot of the driver and the second for the left foot, and so on. Both cars are filled with the same amount of gasoline. Suppose the driver of the first car starts the motor and steps on the gas pedal with his right foot, causing the car to move forward at a certain speed, say 50 mph. The driver of the mirror image car does exactly the same thing, except that he interchanges right with left; that is, he also starts the car, steps on the gas pedal with his left foot, but keeps the pedal at the same degree of inclination. The question is what will be the motion of the mirror image car?

For many people, common sense will say that obviously both cars should move forward at exactly the same speed. Those individuals, however, are thinking just like the pre-1956 physicists: It appears almost self-evident that two arrangements, identical except that one is the mirror image of the other, should always behave in exactly the same way in all aspects, except for the original right–left difference. This is precisely what was found *not* to be true in nature. However, for most physical events it is almost true, the deviation being small. For a class of subnuclear reactions called weak interactions, such as beta decay of any kind of nuclei, the violation is maximal (beta decay refers to decay into electrons).

Mirror Reflection Asymmetry

Stimulated by the strange behavior of elementary particles (θ–τ puzzle described below), physicists T. D. Lee and C. N. Yang proposed in 1956 that weak interactions violate the mirror reflection symmetry (and, therefore, also space-inversion symmetry), whereas strong and electromagnetic interactions are mirror-reflection symmetric. They pointed out that in spite of half a century of experimental work on beta decay, its parity conservation had never been tested. What was required was to begin with two arrangements of, say, cobalt-60 nuclei placed at the

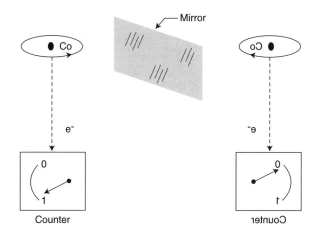

Figure 1   Mirror-image configuration for measuring the parity of cobalt-60 decay.

center of circular electric currents, each being the exact mirror image of the other; that is, in one arrangement the currents circulate one way and in the other they circulate in the mirror-opposite way (Fig. 1). Since cobalt-60 decays produce electrons, one can use two identical counters, again in an exact mirror-image configuration, to measure the number of the decay electrons. If the two counters give different results, then it shows that in weak decays natural law is not mirror-reflection symmetric.

Parity Nonconservation

The experiment suggested by Lee and Yang was completed in 1957 by C. S. Wu, E. Ambler, R. W. Hayward, D. D. Hoppes, and R. P. Hudson. They discovered that beta decay indeed violates mirror-reflection symmetry, and, therefore, parity is not conserved. Soon after, parity nonconservation was also discovered in a large variety of reactions, all involving weak interactions, such as pion decay, muon decay, kaon decay, and so on. As a result, a kaon can decay into an even-parity state of two pions (called the θ mode) as well as an odd-parity state of three pions (called the τ mode). That an "elementary particle," a kaon in this case, can decay into final states of even as well as odd parities was the origin of the so-called θ–τ puzzle that stimulated the whole investigation in 1956.

See also: DECAY, BETA; ELEMENTARY PARTICLES; INTERACTION, ELECTROMAGNETIC; INTERACTION, WEAK; QUANTUM MECHANICS; SYMMETRY; WAVE FUNCTION

T. D. LEE

PARTICLE

Even though the particle notion is one of the most frequently used concepts in physics, a precise and complete definition is surprisingly difficult; it involves a number of subtle, unexpected points. A particle is a discrete separate object that possesses a number of identifiable features. As a discrete object, it can (at least) in principle be localized. The motion of a particle is an expression of the fact that the localization can change in the course of time. Particles (or objects so treated) occur in many sizes—a tennis ball, a rain drop, a planet, a nucleus, an electron. A point particle is an idealization or abstraction, or a limiting case of a particle of finite size. Particles can possess intrinsic qualities that are independent of or unrelated to the physical surroundings. Other features, such as momenta, velocities, and energies, depend on the external circumstances, or a combination of the intrinsic characteristic and outside influences.

The current particle notion is an outgrowth of three distinct trends. One is a philosophical trend starting with the Greeks (Democritus and Lucretius), a second trend stems from physics via Isaac Newton and Daniel Bernoulli, and a third is a chemical trend, especially emphasized by William Prout and John Dalton.

The philosophical discussion started, as is to be expected, from two quite general questions: "What is the world made of?" and "Is matter infinitely divisible?" Democritus conjectured that there was a limit to the divisibility of matter and that matter ultimately consisted of permanent, finite, universal, indivisible entities—atoms. The world was a granular discrete assembly of a (presumably small) number of distinct atoms.

Lucretius showed how increasingly complex structures could be obtained from just a few basic discrete entities (a remarkably modern observation). But the discrete atomic picture was not widely accepted. The atoms appeared too remote from direct sense experiences. This was to be a recurring theme. For rather obscure reasons, atomism—in both the Greek and early Christian cultures—was associated with atheism. Perhaps it was felt that descriptions of the world in terms of fixed atoms limited the options available to divine powers, or even the need for divine intervention.

Newton's fundamental laws of motion used the particle concept explicitly. He states unequivocally that matter is made up of atoms—permanent discrete units. He even hypothesized with surprising foresight that "the smallest particles of matter may cohere by the strongest attraction and comprise bigger particles of 'weaker virtue' and these may cohere again." Bernoulli applied Newton's laws of motion to the molecules that were the conjectured constituents of a gas. In so doing, he created the kinetic theory of gases, which explained a host of phenomena. But the discrete picture of matter, and of nature in general, was not really accepted. Light, in spite of Newton's corpuscular ideas, was a wave in a continuous medium, heat was a continuous material, and electricity was one continuous fluid (possibly two). These conceptions persisted well into the nineteenth century.

The discovery of the chemical elements and the recognition by Prout (ca. 1815) that the specific gravities of atomic species were integral multiples of that of hydrogen suggested some kind of relation between these elements. Furthermore, this order reinforced the idea that each chemical element was made up of a number of identical atoms. The chemical notations developed by Dalton, in formulas such as H_2O, could be a mere bookkeeping device to keep track of the makeup of a compound. But to many practitioners (Dalton, Amedeo Avogadro, André-Marie Ampère), the existence of actual objects—atoms to form physical molecules—was nearly self-evident.

By the second half of the nineteenth century, chemists treated atoms as if they were actual objects with specific properties. They were the real, irreducible building blocks of matter, much the way Democritus had imagined. It is, therefore, not surprising that initially the central problem of chemistry was the understanding of molecules and their properties as a function of their atomic makeup.

Even so, as late as 1890, there were many eminent physicists who had serious misgivings about the reality of atoms and molecules. Most renowned was Ernst Mach, while Max Planck, the originator of quantum ideas, did not convert to the discrete view until late in 1899. The perennial difficulty was that molecules could not be directly observed. To Mach, molecules were "valueless images, or at best, childish fictions." The resulting molecular picture is also extremely unintuitive. A picture where 18 g of water vapor at 100°C contains 60.2×10^{23} entities moving with an average speed of 8,000–10,000 ft/s, colliding 5×10^9 (5 billion) times per second in an apparently random zigzag path, is not easy to visualize or accept.

It was through the observation and analysis of Brownian motion that the reality of molecules was established. Brownian motion is the motion imparted to particles in a colloidal suspension by collisions with the liquid molecules. Other manifestations of the molecular motion are the erratic fluctuations of mirrors in gases. In fact, all fluctuation phenomena are consequences of, or the result of, molecular motions. One of the convincing pieces of evidence was Einstein's analysis of diffusion (another example of an effect of molecular motion), which yielded a numerical value for Avogadro's number (the number of molecules in a gram mol of material), which agreed well with earlier experiments.

By the beginning of the twentieth century, a fairly coherent picture emerged. Molecules exist as real objects. They are the smallest pieces in which a pure material can be subdivided as a matter of principle, while retaining the features of that material. Molecules are definitely particles. They are the discrete identifiable subunits of matter. Molecules are *not* fundamental particles. They can be subdivided further. The components of the molecules—the atoms—are particles too. But the properties of the molecules are quite different from the properties of the atoms, which are the molecular constituents. It is remarkable that while there are many different types of molecules, there are only about 100 distinct types of atoms. It is the task of chemistry to understand the complexity and diversity of molecular species in terms of a relatively few types of atoms. (It was, at one time, believed that atoms were fundamental, and that they had no substructure. If that were true, the atoms would really be Democritus' atoms. But it is not true.)

An absolutely crucial element in this reduction, and of fundamental significance for the understanding of the structure of matter altogether, is the existence of the electron. In the nineteenth century, electricity was generally thought of as one continuous fluid (possibly two). The idea of discrete charges was neither prevalent nor popular. James Clerk Maxwell, in particular, had severe misgivings about the existence of discrete charges (although he strongly believed in discrete molecules). Thus, the discovery of the electron in 1897, having the smallest charge and the smallest mass of any object, came as a major surprise. An electric current is not a continuous flow of electricity but a stream of discrete electrically charged particles. It has never been necessary to ascribe an internal structure to an electron. It behaves as a point-mass and a point-charge. It is stable and truly indivisible: "A perfect particle." Every atom is a composite object, and electrons are universal constituents of every atom.

The mass of an atom is concentrated in a positively charged nucleus. For a neutral atom, containing Z electrons, the nucleus has a charge Z. The mass of the nucleus is usually denoted by A. A typical size of an atom is about 10^{-8} cm ($= 1$ Å). A nucleus is about 100,000 times smaller. A typical nuclear radius is in the order of 10^{-13} cm. It is difficult (it takes major equipment) to affect the nucleus—it is quite easy to remove or add electrons. If $(Z + 1)$ electrons are bound to a nucleus of charge Z, the resulting object is a negative ion. A positive ion with the same nucleus contains $(Z - 1)$ electrons. The structure of an atom is governed by the attractive forces between the electrons and the nucleus. The differences between atoms can, therefore, be traced back to the differences of the respective nuclei. All atoms employ electrons to produce electrically neutral objects.

It was tempting to conjecture that the heavy, small nuclei were truly the ultimate, fundamental objects. But this too turned out to be incorrect. The nucleus of the hydrogen atom has a charge $+1$, and a mass of about 1,820 electron masses (1.7×10^{-24} g). It appears to be indestructible and permanent. Other nuclei (not all) exhibit marked instability. The early discovery of radioactivity by Henri Becquerel, Pierre Curie, and Marie Curie around 1900 demonstrated that certain nuclei spontaneously emit radiation, the so-called alpha, beta, and gamma rays. The beta rays were identified as electrons, and the alpha rays were the (positively charged $+2$) nuclei of helium. The gamma rays are electromagnetic rays. All nuclei (whether stable or not) possess a substructure—they can have different shapes. This is even true for the nucleus of the hydrogen atom, usually called a proton. However, the substructure of the proton was not considered seriously until about 1940, and it was not measured until about 1956.

It was another unexpected surprise when a combination of experiments (the photoelectric effect) and theoretical considerations (Albert Einstein's) showed convincingly that particle-like properties, such as momentum and energy, had to be ascribed to electromagnetic radiation. These particle-like properties were summarized in describing light or electromagnetic radiation as a stream of photons (the same way an electric current was to be visualized as a stream of electrons). For a short period from 1920 to 1930, it appeared that these were just three fundamental particles—electrons, protons,

and photons. This reduction of the number of fundamental particles came to an abrupt halt in the 1930s, when several new particles were found and several more were conjectured. A positron—a particle identical to an electron but with a positive charge—was expected on theoretical grounds and found. A neutron—an object similar to a proton but electrically neutral—was discussed, hoped for, and eventually found (it was not predicted). It actually was a little heavier than a proton. It was recognized early that neutrons were a basic constituent of the nuclei. Wolfgang Pauli conjectured the existence of a massless neutral particle (a neutrino) purely as a means to resolve some puzzles in beta decay. It was found a good deal later. On the basis of an incisive analysis of the strength and range of the proton-neutron force, Hideki Yukawa theorized the existence of mesons—particles intermediate in mass between an electron and a proton. Such a particle was indeed observed in cosmic rays. It was named the μ meson and is now called the muon; it has a mass of about 210 electron masses. Although apparently a verification of Yukawa's prediction, the properties of the μ meson did not match the characteristic of Yukawa's particle. That particle was found later—it is a π meson, and its mass is about 280 electron masses. (The muon is now known to be not a meson but a more massive copy of the electron.)

In the mid-1930s, there were about 8 to 10 particle types. From that time on, the study of matter split into two distinct enterprises. One eventually became nuclear physics. The first problem confronting nuclear physics was the determination of the proton-neutron force. This was immediately followed by the reconstruction of the nuclear properties from the (presumably known) proton and neutron properties. The other endeavor concentrated on the discovery and interpretation of new particles. This became the field of particle physics, made possible by the availability of larger and larger accelerators, and large-scale cosmic ray experiments. These major efforts resulted in the discovery of an enormous number of particles, their masses, charges, lifetimes, and other properties. As a consequence of these investigations, an enormous proliferation of particle properties emerged, so the simple and somewhat simplistic idea of a particle needed substantial extension and modification.

The first new attribute was discovered in connection with atomic spectra. It was observed that electrons possess a quantized, intrinsic, angular momentum ("spin"), which in addition confers magnetic properties on the electron. Electrons, protons, and neutrons all have a spin of 1/2 (in units of $h/2\pi$); a photon has a spin 1. The spin can have values 1/2, 3/2, 5/2 . . . or 0, 1, 2. . . . Quantum mechanics allows no other values. Objects with the 1/2, 3/2 spin values are fermions, and those with integer values are bosons.

Other particle attributes emerge from the type of interactions that are possible for a particle. There are four basic interactions—gravitational, electromagnetic, weak (as in beta decay), and strong (as in the proton-neutron force). But not all particles can experience all interactions. A first classification is the distinction between particles that can participate in strong interactions and those that cannot. Particles that can interact strongly are called hadrons. Protons, neutrons, and π mesons are hadrons; electrons, muons, and neutrinos are not. The hadrons are by far the most numerous—there are at least several hundred hadrons.

An examination of a modern table of particles and particle properties would show the striking fact that there are only six objects of spin 1/2 that do not participate in strong interactions (i.e., the lepton family). Electrons, muons, the even more massive taus, and neutrinos are leptons. There are three distinct types of neutrino, defined by their relationship to the three electron-like leptons. It should be stressed that these additional attributes are introduced to order and systematize large amounts of empirical information. Their utility and significance is determined by the order they can supply to the otherwise unorganized set of data. These attributes do not necessarily possess a transparent geometrical or pictorial interpretation.

Another example of an additional particle attribute is the assignment of baryon numbers. It was observed that the reaction $p \rightarrow e^+ + \gamma$ (a proton decays into a positron and a photon) never occurs. (It is a good thing, too, for the positrons would annihilate with the electrons and the world would rapidly decay into gamma rays.) In fact, in all nuclear reactions, the *number* of heavy, strongly interacting objects (heavier than a proton) is conserved. One assigns heavy particles, protons, and neutrons, a baryon number +1, and electrons, photons, positrons, and π mesons, a baryon number zero. By demanding the rigorous conservation of the number of baryons minus the number of antibaryons, one excludes a number of processes, none of which have been seen.

Another attribute is introduced to describe the grouping, the families of particles, exhibiting very

similar behavior. This attribute, called the isotopic spin, actually counts the number of family members. If T is the isotopic spin, the family has exactly $(2T + 1)$ members. (Clearly, T, like the spin, can be an integer or half-integer). An individual family member is described by T_z, which can assume the values from $-T$ to $+T$. Thus, an individual family member is described by (T_z, T) = (first name, family name). The π mesons, π^+ ($T_z = +1$), π^0 ($T_z = 0$), and π^- ($T_z = -1$) are an isotopic spin triplet ($T = 1$, $2T + 1 = 3$). Isotopic spin applies strictly only to hadrons.

The electrons and the positron are antiparticles. The antiparticle to a particle possesses the identical mass. All its other quantum numbers are the negative of those of the particle. If a particle and antiparticle annihilate, only photons are produced. The lepton family has six members; they include an electron and a negative μ meson. In reactions in which neutrinos occur, an electron is always coupled to its own type neutrino. The μ meson is accompanied by its neutrino. So the leptons, so far, are e^-, ν_0, μ^-, and ν_μ. A third lepton, τ, has been discovered. But the physical properties of e^-, μ^-, and τ^- are so similar that the three pairs, (e^-, ν_e), (μ^-, ν_μ), and (τ^-, ν_τ), appear as heavier copies of the same basic patterns. Furthermore, in the theoretical description, the leptons are the only objects that can rigorously and completely be treated as mathematical points. The leptons appear to have no internal structure. This is very different from the hadrons, which, starting with the proton, were shown to have a substructure (1964). In fact, the scattering of electrons from protons (or neutrons) indicated that the proton acted as if there were three centers within it, where most of the mass is concentrated. These results supported an earlier suggestion by Murray Gell-Mann and George Zweig that hadrons are composite objects with point-like constituents called quarks. Although this may seem like a replay of an old theme, the quarks are remarkable in that they must have fractional charges and fractional baryon numbers—objects never seen.

Furthermore, the extremely strong forces between the quarks would have to be of a new type. Just as the gravitational forces between objects is due to their masses, and the origin of electrical forces is ascribed to the charged state of the bodies, the forces between quarks are a manifestation of a new particle attribute called color. It is an attribute peculiar to quarks. Protons and neutrons do not possess the color attribute. That is why the forces between proton and neutron are not quark-quark forces. If the objects do not possess the color attribute, there are no quark forces, just as if there were no charges, there would be no electrical forces. Most observable particles are a colorless combination of colored quarks. This is not all that different from the circumstance that a neutral atom is composed of charged particles. It takes energy to separate the charges in an atom (the ionization energy). It would take so much energy to split up a colorless combination of quarks, as in a proton, that this energy would create a host of other objects, leaving a combination where no free color is observable.

Both the quarks and leptons were conjectured to be point-like objects. Further, the similarities between quarks and leptons are striking. There are six leptons coming in three groups of two, of which one is neutral and the other has a charge -1. There are six quarks coming in three groups of two, one with charge $2/3$. The other has a charge $-1/3$. The forces between quarks are due to the exchange of bosons, and the forces between leptons are due to the exchange of bosons. In both cases, the objects whose exchange determines the nature of the interaction have been identified, and their characteristics have been determined experimentally. This is one of the great triumphs of particle physics. Quarks and leptons both have spin $1/2$, and quarks and leptons both are permanent and unchanging. Quarks and leptons may appear different from Democritus' atoms, but perhaps they are just that, the fundamental and indivisible constituents of all matter.

See also: ATOM; COSMIC RAY; ELECTRON; ELECTRON, DISCOVERY OF; ELEMENTARY PARTICLES; HADRON; LEPTON; MOLECULE; MOTION, BROWNIAN; MUON; NUCLEAR PHYSICS; PARTICLE MASS; PARTICLE PHYSICS; QUARK

Bibliography

BROWN, L. M.; DRESDEN, M.; and HODDESON, L. *Pions to Quarks: Particle Physics in the 1950s* (Cambridge University Press, Cambridge, Eng., 1989).

GRIFFITHS, D. *Introduction to Particle Physics* (Harper & Row, New York, 1987).

PAIS, A. *Inward Bound* (Clarendon, Oxford, Eng., 1986).

RIORDAN, M. *The Hunting of the Quark* (Touchstone, New York, 1987).

TREFIL, J. S. *From Pions to Quarks* (Scribners, New York, 1980).

MAX DRESDEN

PARTICLE ACCELERATOR

See Accelerator

PARTICLE MASS

The masses of atoms, nuclei, and elementary particles are defined by their acceleration in response to an external force, in accord with Newton's law that force equals mass times acceleration ($F = ma$). To use this definition, we must be careful that the object whose mass we want to measure is not influenced by other forces. Generally speaking, when we measure the mass of a particle, we want to perform our measurements in an environment where the particle is not acted upon significantly by other forces. This is often but not always possible. Some forces are so small they can usually be ignored. For example, all matter is attracted to other matter by the force of gravity, but such forces can generally be neglected in laboratory measurements of particle masses.

The mass of any object is determined in part by the masses of its constituents, but mass is also determined by the energy contained in the object. Because of the equivalence of matter and energy given by Albert Einstein's famous formula $E = mc^2$, particle masses can depend sensitively on their constituents and their interactions. In the laboratory, combining 1 kg of mass with 1 kg of mass gives 2 kg, as measured on laboratory scales, but in particle physics this need not be so. When an electron and a proton bind together to form a hydrogen atom, energy is given off in the form of photons (quantized electromagnetic radiation). The ground state of a hydrogen atom has mass that is very slightly less than the combined mass of its constituent electron and proton. For the hydrogen atom, this difference is about 10^{-8} of the sum of the individual masses.

This effect does not occur for those elementary particles that are not made up of other particles. The electron, for example, appears to be truly elementary; it is not made out of anything else. Nuclei are made up of protons and neutrons, but the mass of a nucleus is not just the sum of the masses of the constituent protons and neutrons; it also receives a contribution from the energy that holds the protons and neutrons together. Similarly, the mass of a proton is the sum of several different effects: the masses of the quarks that make up the proton, the kinetic energy of the quarks' motion inside the neutron, and the potential energy that holds the quarks inside the proton.

In most cases, the potential energy that two particles have with respect to each other goes to a constant as their relative separation goes to infinity. We generally define the potential energy so that infinitely separated objects have zero potential energy relative to each other. This convention defines the mass of constituents as their mass when they are infinitely separated from everything they interact with.

Quarks are different. A free quark, well separated from other quarks, has never been observed. Experiment and theory suggest that quarks are confined inside protons and neutrons by a potential that does not approach a constant for infinite separation, but instead rises with increasing separation. The energy required to isolate a single quark is infinite, and we therefore cannot measure the mass of an isolated quark. We must rely on indirect measurements to determine quark masses. Using a combination of theoretical models and experimental observations, particle physicists have a number of ways to determine the masses of quarks.

In most systems, the energy binding the system together contributes very little to the total mass of the system. However, the masses of protons and neutrons are very large (about 939 MeV/c^2) compared to the masses of their up (u) and down (d) quark constituents, which contribute less than 10 MeV/c^2 to the proton or neutron mass. The rest of the mass has its origin in the kinetic and potential energy of the confined quarks. One approach to understanding this is based on the uncertainty principle. The smaller the uncertainty in the position of a particle, the larger the uncertainty in its momentum. Because quarks are confined inside elementary particles that have small radii, the uncertainty in the quarks' momenta is large. This large uncertainty in the quark momenta leads to large kinetic energies for the confined quarks. This is very different from atoms or nuclei, where the mass of the whole is very close to the sum of the constituents. The forces that confine quarks are well described by the theory called quantum chromodynamics (QCD). Although this theory cannot be solved exactly, computer simulations of QCD indicate that quarks are confined and that most of the mass of protons and neutrons comes from the energy of the quarks rather than their masses.

The masses of many particles can be understood as deriving from more fundamental quantities. For example, the mass of a nucleus is largely determined by the mass of the protons and neutrons that make it up, plus some corrections associated with electromagnetic and nuclear binding energies. Other particles do not appear to be made out of more elementary particles. Electrons and quarks are fundamental particles, and their masses are not predicted from the masses of other particles. In the standard model of elementary particle physics, the masses of quarks and electrons are determined by various parameters of the model, but there is no new prediction made; for each mass, there is a free parameter.

The standard model of particle physics is very successful in many predictions about elementary particles, but its failure to predict the masses of quarks and leptons is seen as a serious shortcoming by many physicists. As of 1996, there was no theory that predicts their masses. The apparent random pattern of these masses has led many particle physicists to suggest that there must be a more complete theory that would explain quark and lepton masses. Such a theory would explain quark and lepton masses in terms of fewer parameters than the standard model.

See also: ELEMENTARY PARTICLES; MASS; MASS, CONSERVATION OF; NUCLEAR BINDING ENERGY; QUANTUM CHROMODYNAMICS; QUARK; QUARK CONFINEMENT

Bibliography

JOHNSON, K. A. "The Bag Model of Quark Confinement." *Sci. Am.* **241,** 112–121 (1979).

QUIGG, C. "Elementary Particles and Forces." *Sci. Am.* 252, 84–95 (1985).

REBBI, C. "The Lattice Theory of Quark Confinement." *Sci. Am.* **248,** 54–65 (1983).

MICHAEL C. OGILVIE

PARTICLE PHYSICS

Particle physics teaches us about what the world is made of and how it works at its most basic levels.

People engaged in this area of science seek to discover basic building blocks, out of which everything in the universe can be constructed, and rules describing the ways these pieces fit together and interact to give us the patterns and diversity we observe throughout nature. Particle physics traces its roots to the ancient Greeks, who introduced the notion of atoms as being indivisible pieces from which everything in the natural world is built. This heritage lead to the very development of the scientific method (with discoveries of the laws of mechanics, planetary motion, electricity, and magnetism along the way) and the revolutionary ideas of relativity and quantum mechanics revealed in the twentieth century.

Based on these historic foundations and illuminating discoveries since the 1970s, particle physics today stands on a historic plateau of understanding, called the standard model, that provides definite answers to the questions of what the world is made of and how it works, at least to the degree that can be tested to date in many different experiments. We know that all observed forms of matter can be built out of subatomic particles belonging to either one of two families called quarks and leptons. Quarks and leptons are known to interact through several distinct forces, with consequences that can be calculated from theoretical formulas. The range of applicability of these ideas stretches from the tiniest things we can measure (roughly 10^{-16} cm in size) to stars and galaxies. There are no generally accepted measurements that contradict the standard model. Yet, its successes raise more fundamental questions about such things as the origin of mass, the character of most of the matter in the universe, the preponderance of matter over antimatter in our world, whether quarks and leptons are built out of even smaller objects, and the ultimate fate of the universe. Testing the limits of validity of the standard model and searching for new physics "beyond" the standard model are the principal activities of today's particle physicists.

Particle physics advances through the interplay of theoretical synthesis and prediction, experimental discovery and confirmation, and developments in technology. For example, some new technical development may permit new experiments that lead to discovery of new phenomena requiring revisions of theoretical ideas, which, in turn, suggest further experimental tests. Purely theoretical ideas, perhaps based on notions of symmetry or mathematical beauty, will suggest new experiments to confirm or refute them. The drive to perform key experiments can push advances in tools and methods that, subsequently, find application in other areas of science and technology, in addition to their utility within particle physics.

Since the beginning of the twentieth century, the most useful experimental approaches in particle physics almost always involve arranging collisions between energetic subatomic particles and detecting the resulting debris. These particles are far too small to be seen in a conventional sense, but examination of the patterns among particles in the debris and rates of collisions reveal internal structures and properties of the forces between the particles. Furthermore, Albert Einstein's relationship between energy and mass, $E = mc^2$, permits the creation and laboratory study of extremely short-lived particles and antiparticles, which have not been plentiful since the earliest moments of the universe, yet which are important if we are to have a complete picture of the particles and forces of nature.

Technology plays key roles in providing us the energetic particles, making them collide, detecting the collision products, and analyzing the results. Particle accelerators, of ever increasing energy, have become the principal tools of particle physics research. An accelerator makes a beam of electrically charged particles—usually protons, the nuclei of hydrogen, or electrons, which can be obtained from a cathode, much like the operation of a television picture tube—and imparts energy to them, raising their speeds to nearly that of light. The accelerated particles can then be used to collide with material targets at rest (fixed-target), or other beams moving in the opposite direction (colliders). Some accelerators are built in a straight line, while others confine the beams in rings of magnets. The choice between linear and circular accelerators is based on technical considerations to optimize performance and minimize cost. Typically, several experiments can be carried out simultaneously on an accelerator or collider. Because of their size and expense, forefront accelerators exist at a small number of sites in the United States, Europe, and Asia, but they are used by scientists, students, and technical staff from universities and laboratories around the world. The largest accelerator in the world today is the Large Electron Positron (LEP) collider at the European Center for Particle Physics (CERN), located near Geneva, Switzerland. It is a ring design, situated in an underground tunnel that is 27 km in circumference. The largest linear accelerator is at the Stanford Linear Accelerator Center (SLAC) in California; it is 3 km long.

Beam energy is the most important parameter in most experiments. Higher energy allows the study of smaller structures within particles and the creation of heavier unstable particles and antiparticles according to the Einstein relation. Today's energy frontier is held by the Tevatron, a circular proton accelerator at Fermilab near Chicago. The Tevatron can accelerate protons and antiprotons to energies of 1 TeV (10^{12} eV). The LEP holds the record—soon to go to 100 GeV (10^{11} eV)—for accelerating electrons and their antiparticles, positrons. Important experiments are being carried out at lower energies, but the opening of new accelerators at higher energies is often accompanied by entirely unexpected discoveries, one of the most exciting events that can be experienced in science.

While particle accelerators provide the energetic particles and collisions used to reveal the structures and forces, particle detectors are our eyes, ears, and noses for sensing what takes place when collisions occur. We want to know what kinds of particles are produced in a collision, in what directions and with what speeds do they move after the collision, and whether new kinds of unstable particles were created. The technical challenges for detectors are many; the particles are too small to see directly, they move at nearly the speed of light, they may decay to other particles in a tiny fraction of a second, and tens or even hundreds may be produced in a single energetic collision. To establish some definitive result may require thousands or even millions of examples of collisions—called events.

The usual way to "see" high-energy particles is through the trail of ionization left behind rapidly moving charged particles as they pass through ordinary matter, in much the same way that a high-flying jet airplane can be detected by its contrail. The ionization path gives the track of the particle. The amount of ionization gives information on the speed. How the ionization varies as particles pass through different kinds of materials, such as gases, plastics, iron, and lead, gives clues to the kind of particle and its energy. Magnetic fields can be used to deflect charged particles; the amount of deflection is related to the particle's momentum. Thus, most experiments employ a succession of magnetic fields and ionization measurements interspersed with different materials to track and identify particles. Elaborate electronics and computer systems record information and analyze it to reconstruct the details of collision events. A modern detector system may weigh a few thousand tons and employ up to a million channels of sensitive electronics, having been built by an international collaboration involving several hundred researchers over a period of years.

To describe what people actually do in particle physics, the work of an imaginary researcher participating on a hypothetical experiment will be followed. Our researcher, let us call her Anne, recently completed a Ph.D. in physics after five years in graduate school at a major midwest university that included a two-year stint at SLAC, where she participated in data-taking on a detector constructed a few years earlier by a collaboration of physicists that included her thesis supervisor. Hers was the tenth Ph.D. dissertation to be completed by students from the six universities participating on the experiment. She accepted a postdoctoral research position at SLAC to work on a new detector being planned for what was to become the world's highest energy electron-positron collider.

Anne joined a group led by a senior physicist holding a joint appointment in the Stanford physics department and at SLAC. There were two other postdoctoral researchers in the group, an assistant professor and a permanent-staff physicist. Two advanced graduate students from Stanford also worked closely with the team. A secretary and two laboratory technicians rounded out the research group.

Being at the host laboratory, Anne's group had the lead in planning the new detector. Their leader was the spokesperson for a new collaboration forming to build the detector. They were joined in this effort by another SLAC group, groups from Berkeley, Harvard, nine other U.S. universities, and contingents from the United Kingdom, Italy, Japan, and the People's Republic of China. The collaboration "courtship" took place over about a year, during which time physics groups with similar research interests, spanning an appropriate range of technical expertise, had joined to propose and, subsequently, to design and build the new apparatus. In total, there were about 200 physicists working in the collaboration, including one Nobel Prize laureate from Italy.

Anne was assigned the task of designing and building the central tracking chamber, a major responsibility. In this effort, she would work closely with one of the other postdoctoral researchers in her group and the Rome University group, who were the world's experts in writing software to track charged particles in magnetic fields of detectors to determine accurately their momenta. The other postdoctoral researcher was interested in the special electronics to detect the ionization trail left behind by particles passing through the special gas in the tracking chamber. Anne worked on the mechanical details of assembling the chamber to the required precision and making sure it fit into the complex set of equipment that would surround it. The chamber was a cylinder 3 m in diameter and 5 m long, which held 100,000 fine wires that would detect the ionization. The central tracking chamber would be at the very heart of the detector, residing inside a large superconducting magnet. Anne worked with technicians and mechanical engineers who were assigned to her research group for this job.

Anne and the technicians built small prototype chambers to test their designs. These were placed in a test beam from the accelerator, where high-speed charged particles were passed through the chambers to measure their performance. Initial tests had not gone well, leading to heated discussions within the group and with the Rome group, who were simulating the tracking system with computer models. The position resolution, the accuracy with which the track's path could be measured, appeared to be poorer than what was needed with the new high energies expected from the new accelerator. These tests were reported to the peer review committee overseeing the experiment and to the laboratory director. Anne discovered that the resolution could be improved by operating the chamber with a different gas, which was then tried in their test-beam setup. The new approach appeared promising, but major engineering changes would be needed to effect the change. Ultimately, the changes were made and the various review committees endorsed Anne's approach.

Next, the chamber was assembled over a ten-month period in one of the large machine shops at SLAC. Anne followed every aspect of this work to ensure its success. She became close friends with the machinists and technicians who, themselves, became deeply attached to the tracking chamber effort. There were continual crises over schedules and engineering details, and although the chamber was completed three weeks behind the original schedule, it was in time to be assembled in the huge detector that was growing on the beam line of the new accelerator. This beam line was a 10-cm diameter pipe through which the accelerator's beams would pass to collide at the center of the detector. The pipe was extremely fragile, being constructed out of a thin, brittle metal to minimize any disturbance to the collision particles that would pass through it as

they headed toward the detector; it maintained nearly a perfect vacuum so the beam particles could pass with relatively small chance of hitting residual gas molecules and creating background events.

After assembling the tracking chamber and other detector elements came the moment of truth: colliding beams from the new accelerator. The detector control room was packed with physicists and technicians as beams were accelerated and prepared for collisions. The process went on day and night for three weeks, until about 3 A.M. one morning, when everything worked and the first collisions were observed. Computer displays around the control room reconstructed pictures of the tracks of the particles detected from Anne's chamber. When the beams were timed to collide at the center of the detector, events began to show up on the computer displays. They looked much like simulated events, with more background tracks and a few dead areas because not all the electronics channels were working, but the experience was exhilarating. Even with the blemishes, Anne and her colleagues in the room felt a sense of awe as they witnessed real collisions, the likes of which had not taken place anywhere in the universe since the first fractions of a second after the big bang.

It would take many more months of running—around the clock, seven days a week—before enough data was accumulated to perform meaningful data analyses. During this time, collaboration members, Anne included, took shifts operating the detector, debugged problems in the parts of the apparatus for which they were responsible, and began to analyze data of interest to them. Anne joined a small subgroup that was interested in measuring the total cross section, the probability that some collision will take place when beam particles pass near each other. This quantity is related to the total number of particles in the quark family, a number of fundamental importance. In her group were two other postdoctoral researchers from U.S. universities, an Oxford professor, and three Japanese graduate students. Anne and the others analyzed the accumulated events, keeping collision events of interest, rejecting background, and searching for experimental biases or instrumental defects. After months of analysis, Anne presented a progress report to the full collaboration at one of its quarterly meetings. Many suggestions and criticisms were made, requiring further analysis. Subsequently, she and her colleagues prepared a draft paper for publication of the total cross section by the collaboration. This was distributed and additional suggestions for improvements were made, but the process was converging toward publication of one of the collaboration's most important early results.

Meanwhile, Anne noticed some very bizarre behavior in a small number of data runs that caused her to hold up submission of the paper. These were probably instrumental effects, but if real, they could herald the existence of entirely new kinds of particles, not built out of quarks or leptons. Anne was about to experience the most exciting and fearful event one can have in science—the opportunity to make a world-class discovery—if the data is correct. It would take courage, toughness, and intellectual honesty to get at the right answer.

Our story, of course, did not actually take place, but it illustrates some of the workings of modern collaborations in particle physics. Indeed, these are really like laboratories within a host accelerator laboratory where people can pursue myriad scientific and technical problems. Particle physics is often described as "big science," but the work mostly takes place in small groups. Young researchers have the opportunity to work with world experts and to assume major responsibilities. By their enthusiasm, energy, and facility with computers, they quickly become leaders. About one-half of the people receiving doctoral degrees in particle physics pursue research and teaching careers; the others, perhaps after a postdoctoral research position, go on to careers in industry and government, where their experience of the broad technical and scientific challenges found in particle physics can serve them well.

See also: ACCELERATOR; ACCELERATOR, HISTORY OF; CYCLOTRON; LEPTON; PARTICLE PHYSICS, DETECTORS FOR; QUARK; SYNCHROTRON

Bibliography

BROWN, L. M., and HODDESON, L. *The Birth of Particle Physics* (Cambridge University Press, Cambridge, Eng., 1983).

CAHN, R. N., and GOLDHABER, G. *The Experimental Foundations of Particle Physics* (Cambridge University Press, Cambridge, Eng., 1989).

PERKINS, D. H. *Introduction to High Energy Physics,* 2nd ed. (Benjamin/Cummings, Menlo Park, CA, 1982).

RIORDAN, M. *The Hunting of the Quark* (Simon & Schuster, New York, 1987).

WEINBERG, S. *Dreams of a Final Theory* (Random House, New York, 1994).

ROY SCHWITTERS

PARTICLE PHYSICS, DETECTORS FOR

Particle physics experiments study collisions of high-energy particles. They use a large detector to reconstruct the tracks of the many particles that emerge from the collision. Since these particles include many short-lived particles and many that are traveling in clusters, almost parallel to each other, at velocities close to the speed of light, detection and reconstruction pose considerable technical challenges.

Most experiments today attempt to study extremely rare processes. They must record millions of collision events in order to collect a significant sample of the rare process under study. The requirement that millions of collisions must be studied means that the detector must be able to collect the data at a very high rate. Analysis of data must be performed by computers. It would take too long to complete the experiment if people had to sort through millions of pictures to find the interesting ones. Detectors such as cloud chambers and bubble chambers, which produce photographic images as their data record, are not used in modern experiments.

This entry will describe a typical modern detector for an experiment in which two oppositely-moving beams of particles collide. When the experiment involves a beam of particles striking a stationary target, the layers of the detector are similar, though the geometry of their arrangement changes.

For colliding beams, the detector is built in cylindrical layers centered around the collision point and surrounding the beam-pipe in which these beams travel. Each layer of the detector is designed to furnish particular information. The detector layers are also repeated in the end caps that enclose the ends of the cylinder so that particles moving nearly parallel to the beam pipe can also be detected. By combining information from all the layers, one can reconstruct the number, directions of travel, and types of particles that emerge from the collision.

There is always some region where the detector layers cannot be placed. For example, there is a cone along the beam directions that cannot be instrumented since space must be left to allow the beams to pass through. Engineering realities also force other regions to be lost. Supports must be placed somewhere to hold up the parts of the detector. Cables to carry out the signals from the inner layers also require some space. The challenge of designing such a detector is to minimize the impact of

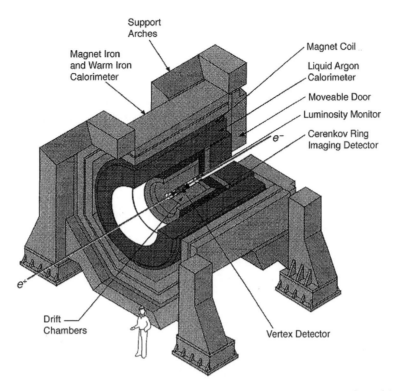

Figure 1 This detector, which is used to detect particles produced in electron-positron collisions at the Stanford Linear Accelerator Center (SLAC), is known as the SLAC Large Detector (SLD). The "person" standing in front of the detector provides an idea of the size of the device.

these practical necessities on the ability to reconstruct the particle tracks.

The inner layers of the detector are typically placed inside the coil of a large toroidal magnet. This magnet produces a strong magnetic field parallel to the beam pipe. The paths of any charged particles moving radially out from the collision are bent by such a magnetic field. The amount of bending is proportional to the radial momentum of the particle. Thus the magnetic field, together with a device to track the particle paths, allows a determination of the momentum for each charged particle.

The tracking device typically has two parts to it. The small inner layer that gives very accurate position information is called a vertex detector. The larger cylinder around this is typically a device called a drift chamber. It too reconstructs charged particle paths. It provides somewhat less precision tracking than the vertex detector but over a much larger volume.

A typical vertex detector is made of layers of narrow silicon strips that alternate in direction so that the overlaps of a pair of layers define tiny regions (called pixels) on the cylindrical surface. When a charged particle passes through a strip, the solid-state circuit attached to the silicon generates a signal to the computer. [Alternatively, the vertex detector pixels might be made from much the same type of charge-coupled devices (CCDs) as one finds in a CCD video camera.] The vertex detector will have three to five layers of pixels, each a slightly larger cylinder than the one before. By a process of "connect the dots," the computer can reconstruct the possible particle paths from the set of pixels that signal the passage of a particle.

The accurate tracks found in the vertex detector can be extrapolated back into the region where there is no detector, toward the collision point. Most of the tracks are seen to come directly from the collision point. However, occasionally one sees a vertex where two or more tracks meet at a point that is a short distance away from the collision point. This shows that a very short-lived particle was produced and then decayed to produce the particles that made these tracks, which travel out from the decay point and not the original collision point. Thus the vertex is the clue to information about particles that are not directly observed. The more accurately the tracks can be defined in the vertex detector, the shorter is the distance from the collision point that can be resolved as a vertex or particle decay point. This in turn allows physicists to determine that extremely short-lived particles were produced and to measure their lifetimes.

The drift chamber reconstructs particle tracks over a much larger volume. It is filled with a readily-ionized gas. It detects the electrons that are knocked out of atoms of the gas by the passage of the high-energy charged particles. Many ions (and corresponding free electrons) are produced along the path of any high-energy charged particle. The chamber is instrumented by a grid of very fine wires that span its length. The pattern of these wires is quite complex, as are the high voltage potentials between them. These are arranged so that any electron formed in a given region of the detector can be pulled toward a particular wire by the electric field caused by the voltage difference between wires. As the electron drifts toward that wire, it gains energy and eventually it too begins to ionize the gas, releasing more electrons that begin to drift with it. Thus, a pulse of electrons arrives at the wire. This pulse is detected and amplified by electronics at the ends of the wire (outside the chamber) and sent to the computer. A computer algorithm reconsructs the paths of all the original high-energy charged particles from the pattern of wires that produce such a pulse and from the times of these pulses compared to the collision time. The curvature of the track due to the magnetic field can be used to determine the sign of the particle charge and to measure its momentum.

The next layer is typically some kind of Cerenkov detector. The aim of this layer is to measure particle speeds, or at least to separate particles into groups on the basis of their speed. Cerenkov noticed the phenomenon that light is emitted in a shock-wave-like cone around any charged particle traveling through some material faster than the speed of light in that material. (In a vacuum, nothing can travel faster than light, but since light is slowed down when traveling through matter, it is possible that a high-energy particle will travel faster than light in a carefully chosen gas or gel.) The angle of the cone of light depends on the particle velocity, so, if this angle can be measured, then the particle velocity can be determined.

Note that once we know both the momentum and the velocity of a particle we know its mass (since momentum p is given by $p = mv/\sqrt{(1 - v^2/c^2)}$, where m is mass, v is velocity, and c is the speed of light). Each type of particle has a distinct mass, so once we know the mass we can restrict the particle type. Thus, for example, a charged pion (mass = $0.14 \text{ GeV}/c^2$) be distinguished from a proton or antiproton (mass = $0.94 \text{ GeV}/c^2$). However, a charged muon has a mass of $0.11 \text{ GeV}/c^2$, which means that it usually cannot be distinguished from a pion given the accuracy of mass determination achieved in this way. Fortunately, muons have quite different interactions with matter than pions, so these particles are distinguishable as explained below.

So far the layers only detect charged particles. The next layers are designed to measure particle energies. These detect energy of neutral particles as well as that of charged particles. They are layers of a dense material such as lead interspersed with layers of plastic scintillator, a material that gives off light when a charged particle passes through it. The scintillator is instrumented with photo-tubes, devices that use the photoelectric effect to turn a light pulse into a current pulse, which can then be recorded by the computer.

Any particle, except a muon or a neutrino, that enters the dense material is slowed down and eventually stopped by a series of interactions in it. These interactions produce additional particles, so one gets a shower of particles that can be seen in the scintillator layers. The depth to which the shower extends and the shape of the shower development can be interpreted by the computer to calculate the total amount of energy deposited in a given shower. The shower shape also distinguishes showers initiated by electrons or photons from those initiated by strongly interacting particles such as protons, neutrons, or pions. To tell a photon from an electron or a proton from a neutron, one simply has to look to see whether or not there is a charged particle track that points to the location where the shower begins. These energy-measuring layers are called the calorimeter (originally the name for a device to measure heat energy or calories).

High-energy muons do not interact very much with dense matter. They go right through the calorimeter layers and the thick steel layer that surrounds the magnet coil (this contains the exterior magnetic fields in a restricted region). An additional scintillator or tracking chamber layer placed outside everything else will detect muons as they leave the detector. All other particles, except neutrinos, are stopped before they reach this layer. Thus, any charged particle track that can be extended out to connect to a track in this outermost layer can be identified as a muon track.

Neutrinos are simply not seen in such a detector; they interact too little. However, the fact that a high-energy neutrino was produced in a collision event can often be inferred because the total energy detected in the detector does not match the collision energy. Furthermore, the total momentum detected is not zero, whereas the initiating collision is known to have zero total momentum. Thus, by conservation of energy and momentum, we know that something escaped detection and, hence, that one or more high energy neutrino's were produced in such an event.

It is a huge endeavor to build and operate the complex multilayered instrument described above. The detector stands on the order of several stories high. An experiment will run for twenty-four hours a day, several months at a stretch, and extend over several years of such runs. Intervening periods are used for repairs, data analysis, and possibly improvements to the apparatus. A typical experimental team today includes more than 100 physicists and engineers. A typical high-energy physics experiment spans more than ten years from conception to completion. Particle physics experiments do not at all match the usual image of one or two scientists working over a laboratory bench.

See also: ACCELERATOR; BUBBLE CHAMBER; CLOUD CHAMBER; CYCLOTRON; PARTICLE PHYSICS

Bibliography

CLOSE, F.; MARTEN, M.; and SUTTON, C. *The Particle Explosion* (Oxford University, New York, 1987).

HELEN R. QUINN

PASCAL'S PRINCIPLE

A hydraulic jack converts a small force into one that is large enough to lift an automobile. The jack works because of Pascal's principle: A pressure applied to any part of a static fluid (gas or liquid) is transmitted unchanged to all parts of the fluid. The principle also explains how we squeeze toothpaste out of the tube;

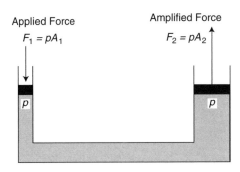

Figure 1 The forces involved in Pascal's principle.

we squeeze one end and the pressure there is transmitted to the far end where it forces the paste out.

When the jack handle is moved, a force F_1 acts on a small piston that converts the force to a pressure p according to the relationship, $F_1 = pA_1$, where A_1 is the area of the piston (Fig. 1). The pressure produced by this first piston is transmitted through the oil in the jack to a second, larger piston. A similar equation relates the force F_2 exerted by this piston to the pressure of the oil: $F_2 = pA_2$, where A_2 is the area of the larger piston. Since the pressure p is the same in both equations, they can be combined to give $F_2 = F_1 (A_2/A_1)$. Thus, the original force is increased by a factor equal to the ratio of the areas of the two pistons. Since this ratio can be quite large, the multiplication of force can be correspondingly large. The drawback is that the small piston must move a lot in order that the large piston may move a little; the ratio of distances moved equals the ratio of areas.

See also: FORCE; PRESSURE

Bibliography

KELLER, F. J.; GETTYS, W. E.; and SKOVE, M. J. *Physics: Classical and Modern*, 2nd ed. (McGraw-Hill, New York, 1993).

LAWRENCE A. COLEMAN

PAULI, WOLFGANG

b. Vienna, Austria, April 25, 1900; *d.* Zurich, Switzerland, December 14, 1958; *relativity theory, atomic and nuclear physics, statistical mechanics, quantum field theory.*

The Austrian-Swiss physicist Wolfgang Ernst Pauli, was born in Vienna, the son of Bertha (Schütz) and Wolfgang Joseph Pauli. His father, originally from Prague, was employed in the medical department of the University of Vienna, becoming a professor of chemistry in 1922; he was one of the founders of the science of colloid chemistry. Pauli's mother was a writer, a member of the artistic and intellectual elite of one of the world's great cultural centers. His sister, Hertha, was also an author as well as an actress. The family was originally of Jewish origin, but his father had become a Catholic, and Pauli was baptized—his godfather was the famous physicist and philosopher Ernst Mach.

In high school, Pauli was an outstanding student (except in languages), with a strong interest in mathematics and astronomy. He read widely in literature and science, paying special attention to Albert Einstein's theory of relativity. In 1918 Pauli enrolled at the University of Munich to study with Arnold Sommerfeld, a famous physicist, an expert on relativity and atomic physics, and the teacher of future Nobel Prize winners, including Werner Heisenberg and Hans Bethe. Pauli completed his Ph.D in only three years, writing a dissertation on the quantum theory of the hydrogen molecule ion. In 1920, while still a student, Pauli wrote a 250-page article on relativity for an encyclopedia at Sommerfeld's request. It was highly praised by Einstein and is still regarded as a major treatise on the subject.

After his Ph.D., Pauli spent a year at another major center of mathematics and physics, the University of Göttingen, with James Franck and Max Born, and then worked for a year in Copenhagen with Niels Bohr, who had originated the quantum theory of the atom. It was at this time that he first took up the problem of the anomalous Zeeman effect, the splitting of atomic energy states, and thus of spectral lines, in the presence of a magnetic field, which was a subject of major interest. It seemed to be an insoluble problem in the Bohr–Sommerfeld quantum picture, which dominated atomic physics. This theory was an extensive elaboration of Bohr's hydrogen model of 1913; thus, it placed the atomic electrons in classical orbits that were restricted by a general set of quantum conditions applied to conserved quantities. For example, angular momenta and their vector components were restricted to being integer multiples of $\hbar = h/2\pi$, where $h = 6.63 \times 10^{-34}$ J·s is Planck's constant.

However, the number of atomic states in a magnetic field was double the number that the theory predicted. In the simple case of sodium, for example, there is one electron outside of a closed shell of electrons (the core) that Bohr–Sommerfeld theory predicts should have zero angular momentum. To account for the extra atomic states, it was proposed that the core should instead have an angular momentum of $\frac{1}{2}\hbar$, an idea that Pauli rejected. His solution was to say that the electron itself has a non-classically describable two-valuedness (*Zweideutigkeit*), so that it was not the core but the external valence electron that was responsible for the doubling of the number of atomic states.

This was a suggestion of the greatest importance, for it played an essential role in explaining the periods in the Mendeleev table of chemical elements. Bohr had already made a start in this direction with his building-up principle, which asserted that in passing from one atom in the table (characterized by the atomic number Z, or number of electrons) to the next one, thus adding another electron with new quantum numbers, the inner electrons kept the same quantum numbers. However, to complete this picture, Pauli's new two-valuedness was needed: Each set of the old quantum numbers labeled not one, but two states. The new form of the principle became known as the Pauli exclusion principle, and it was for this discovery that Pauli was awarded the Nobel Prize for physics in 1945.

Pauli's close friend and collaborator, Heisenberg, in 1925 developed a new quantum mechanics (replacing the older Bohr–Sommerfeld orbit theory), from which most of the modern theory of physics and chemistry has originated. Pauli was the first to apply this to a real physical problem, namely, the hydrogen atom, which he solved completely, taking into account his new quantum number. Meanwhile, also in 1925, two young Dutch physicists, Samuel Goudsmit and George Uhlenbeck, identified Pauli's quantum number as belonging to electron spin. That is, every electron has an intrinsic spin angular momentum of $\frac{1}{2}\hbar$, and thus an intrinsic magnetic moment $e\hbar/mc$, which can take up one of two orientations in a magnetic field. Pauli resisted this rotating electron interpretation for almost a year, but in March 1926 he wrote to Bohr that he would "capitulate completely." He then applied the spin and the exclusion principle to explain the magnetic properties of normal metals (paramagnetism) and thus initiated in 1927 a new research field, the quantum electron theory of metals.

In that same year, the English quantum theoretician Paul Dirac made a quantum theory of the electromagnetic field and a relativistic generalization of the wave function, introduced by the Austrian physicist Erwin Schrödinger in his version of quantum mechanics. Dirac's theory predicted the existence of a positive electron (positron) that could be produced electromagnetically, together with an ordinary negative electron, providing enough energy was available (at least $2mc^2$, with m being the electron mass and c being the velocity of light). Pauli and Heisenberg then wrote two important papers attempting a comprehensive relativistic treatment of the interaction between radiation and matter. They discovered important difficulties in the theory, which had to wait until the 1940s for a satisfactory resolution. Problems of quantum field theory, as it came to be called, occupied Pauli for the rest of his life. As professor at the Swiss Federal Institute of Technology (ETH) from 1928, after serving at the University of Hamburg, Pauli continued his research on wave mechanics. This led in 1933 to another remarkable treatise published as an encyclopedia article.

In December 1930, convinced by reported experiments that a puzzling situation in radioactive beta decay required a "desperate solution," Pauli made such a suggestion in a famous letter to a group of experimental physicists meeting in the German city of Tübingen. Beginning "Dear radioactive ladies and gentlemen," it went on to suggest the existence of a new extremely penetrating neutral particle of very small mass (perhaps zero). This particle was supposed to accompany each beta decay electron emitted and account for what appeared to be energy "missing" from the process. Now called the neutrino, Pauli's particle became an ingredient of a new and successful quantum field theory of beta decay, worked out by the Italian physicist Enrico Fermi at the end of 1933.

During World War II, fearing a German invasion of Switzerland, Pauli moved to the Institute for Advanced Study in Princeton, New Jersey, where Einstein was also in residence. In 1945 he returned to the ETH in Zurich, where he remained until his death in 1958.

In addition to his accomplishments in physics, Pauli was a philosopher and a student of the psychology of creativity, in which he collaborated with the Swiss psychoanalyst Carl Gustav Jung. Because of the profoundly high standards that he brought to his work, Pauli is sometimes referred to as the "conscience of physics."

See also: BOHR, NIELS HENRIK DAVID; BOHR'S ATOMIC THEORY; HEISENBERG, WERNER KARL; PAULI'S EXCLUSION PRINCIPLE; QUANTUM MECHANICS, CREATION OF; ZEEMAN EFFECT

Bibliography

ENZ, C. P. "W. Pauli's Scientific Work" in *The Physicist's Conception of Nature*, edited by J. Mehra (Reidel, Boston, 1973).

FIERZ, M., and WEISSKOPF, V. F. *Theoretical Physics in the Twentieth Century*, memorial volume to Pauli (Wiley Interscience, New York, 1960).

MEHRA, J., and RECHENBERG, H. *The Historical Development of Quantum Mechanics* (Springer, New York, 1982).

PAULI, W. "Remarks on the History of the Exclusion Principle." *Science* **103**, 213–215. (1946).

PAULI, W. *Lectures on Physics* (MIT Press, Cambridge, MA, 1973).

PEIERLS, R. E. "Wolfgang Ernst Pauli." *Biographical Memoirs of Fellows of the Royal Society* **5**, 175–192 (1959).

WEISSKOPF, V. F. "Personal Memories of Pauli." *Phys. Today* **38** (12), 36–41 (1985).

LAURIE M. BROWN

PAULI'S EXCLUSION PRINCIPLE

Pauli's exclusion principle states that no two electrons, or more generally no two identical fermions, can occupy the same quantum state (e.g., be in the same place at the same time). It plays a central role in the understanding of phenomena ranging from elementary particles to stellar structure.

In 1924 Edmund Stoner proposed a model for atoms, consistent with spectroscopic experiments and the Periodic Table, in which each atomic electron has three quantum numbers, corresponding to Niels Bohr's principal and orbital angular momentum quantum numbers n and l, respectively, and an "inner quantum number" $j + 1/2$; the number of electrons in each electron "shell" is $(2j + 1)$, twice the inner quantum number. Wolfgang Pauli (1925) showed that the entire shell structure of electron energy levels could be explained by assigning to each electron a *fourth* quantum number m_j with the $2j + 1$ allowed values $-j, -j + 1, ..., j - 1, j$, provided that a new "exclusion principle" is obeyed: no two electrons can have the same four quantum numbers (n,

l, j, m_j). The fourth quantum number is associated with an intrinsic angular momentum (spin) of the electron, as originally suggested by George Uhlenbeck and Samuel Goudsmit.

Following the development of wave mechanics in 1926, Paul Dirac and independently Werner Heisenberg showed that Pauli's exclusion principle is satisfied automatically if the wave function for systems of electrons is antisymmetric, that is, if it changes sign under the interchange of all the coordinates, including spin, of any two electrons. More generally the wave function for *any* system of identical particles must either remain unchanged when all the coordinates of any two particles are interchanged, in which case the particles are bosons, or it changes sign, in which case the particles are fermions. Bosons have spins that are integral multiples of $\hbar$, whereas fermions have spins that are odd half-integral multiples of $\hbar$ ($\frac{1}{2}\hbar$, $\frac{3}{2}\hbar$, ...). Only fermions obey Pauli's exclusion principle. In nonrelativistic quantum theory this connection between spin and particle statistics is regarded as an empirical fact, whereas in relativistic quantum field theory it follows as a general consequence of causality, as shown by Pauli (1940) in his celebrated "spin-statistics theorem."

The exclusion principle accounts for some of the most basic features of matter in all its forms. Were it not for the exclusion principle all atoms would have essentially the same electronic structure—a shell of electrons about the nucleus. Hydrogen and helium do have a single shell, but the situation is different for lithium, which has three electrons. The first two electrons occupy the same orbital (energy state) with opposite spins but, because of the exclusion principle, the third electron must go into a new orbital, farther on average from the nucleus. Thus, unlike helium, lithium is easily ionized and chemically reactive.

The situation is similar with regard to the shell structure of nuclei. Because the proton and neutron can transform into each other via the weak interaction but are subject to the same nuclear forces, it is useful to regard them as two states of a single "nucleon" differing in an additional intrinsic coordinate or quantum number called isotopic spin. The Pauli principle then requires that no two of the nucleons be in a state defined by the same spatial, spin, *and* isospin quantum numbers. Even more fundamental consequences of the Pauli principle emerge at the subnuclear level. For instance, baryons consist of three quarks, and their experimentally observed levels imply wave functions that are symmetric with respect to the interchange of the spatial, spin, and flavor quantum numbers of any two quarks, in apparent contradiction to the exclusion principle, since quarks are fermions. The firm belief of physicists in the general validity of the exclusion principle led to the hypothesis and subsequent confirmation of a new quark quantum number, color, such that no two quarks can occupy the same spatial, spin, flavor, and color states.

The set of all occupied states in a system of many electrons is called the Fermi sea, and the highest occupied energy level at the absolute zero of temperature is called the Fermi energy. This picture is used in the theory of metals, where Fermi energies are typically on the order of several electron volts, much larger than the average energy $kT = 0.025$ eV for an ideal gas at room temperature. Because the exclusion principle forbids all the electrons from crowding into the lowest-energy states, some of the electrons, even at very low temperatures, have energies near the Fermi energy, that is, energies corresponding to temperatures of several thousand degrees. Heating a metal from $T = 0$ to room temperature therefore has a very slight effect on the energy distribution of the electrons. This explains why the electrons have a negligible effect on the specific heats of metals, and also why metals must typically be glowing hot before electrons can be ejected from them.

The exclusion principle, together with the fact that the electron energy levels of solids occur in distinct bands of energy, is essential to the theory of electrical conductivity and many aspects of modern technology. A solid whose highest occupied band is completely filled in accordance with the exclusion principle is an electrical insulator. Its electrons cannot flow freely when an electric field is applied; loosely speaking, because of the Pauli principle they have nowhere to go. If the highest occupied energy band is only partially filled, on the other hand, the solid is a good conductor of electricity. In semiconductors the gap between a completely filled band and the next allowed, "conduction band" is small, typically about 2 eV or less. At the absolute zero of temperature the conduction band is empty, but the gap is sufficiently small that at room temperature some electrons can cross it and partially populate the conduction band. The electrical conductivity therefore increases with increasing temperature. Electrons can also be promoted into the conduction band of a semiconductor by absorption of radiation, if the energy $h\nu$ of the incident photons exceeds the energy gap; this "photoconductive effect" is the basis for many applications, such as automatic door openers, in which electric currents are controlled by exposure to light. When an electron goes into the conduction band it leaves in the Fermi sea a missing state, or "hole," whose occupation is allowed by the Pauli principle. An electron in the conduction band can make a transition to such a hole and, in so doing, emit light in a process analogous to the emission of a photon when an atomic electron jumps to a state of lower energy. This electron–hole radiative process is the basis of the light-emitting diode (LED) and, when the process is stimulated rather than spontaneous, semiconductor lasers. Similar considerations, based on the creation of conduction electrons by doping a semiconductor, explain the operation of diode junctions and transistors.

See also: DIODE; LASER; NUCLEAR STRUCTURE; PAULI, WOLFGANG; SPIN; SPIN AND STATISTICS; TRANSISTOR

Bibliography

DAVIES, P. C. W. *The Forces of Nature* (Cambridge University Press, Cambridge, Eng., 1979).

FEYNMAN, R. P.; LEIGHTON, R. B.; and SANDS, M. *The Feynman Lectures on Physics,* Vol. 3 (Addison-Wesley, Reading, MA, 1965).

GAMOW, G. "The Exclusion Principle." *Sci. Am.* **201** (July), 74 (1959).

PETER W. MILONNI

HEIDI FEARN

PENDULUM

A pendulum is any object suspended so it is free to rotate about any horizontal axis passing through any point in the body other than its own center of gravity (CG). A simple pendulum consists of a ball of lead (called a bob) or other dense material suspended from a thin flexible string or wire (Fig. 1). Any displacement of the CG from a vertical line passing through the axis and the CG when the body is at rest results in oscillation about the axis after it is released. If the pendulum is heavy and the effects of friction are small, the oscillations can continue for a long time unassisted. Applications of a pendulum

include keeping time and measuring the acceleration due to gravity, knowledge of which is useful in mineral exploration. The famous Foucault pendulum demonstrated the rotation of Earth.

The oscillation of a pendulum closely approximates harmonic motion (that is, sinusoidal) when the angular displacement in radian measure is small compared with one radian (57.29578°) and friction is minimized. Even when the angular displacements are large, the deviation from harmonic motion is small.

A characteristic of harmonic motion is that each oscillation requires the same time, or period, regardless of the amplitude (largest displacement) of the oscillation. Galileo as a student at the University of Pisa, Italy, in 1581 noticed the constancy of the swings of a chandelier in the cathedral and timed them with his pulse. He measured the periods of simple pendulums as a professor at Pisa and at Padua but failed to find any variation with amplitudes as large as ninety degrees. His failure to detect the small decrease in period that occurs with increase in amplitude was due to his lack of a reliable clock.

Christiaan Huygens invented the pendulum clock in 1656 and described it in *Horologium* in 1658.

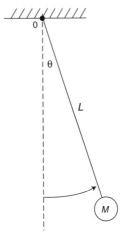

Figure 1 A simple pendulum. It consists of a small heavy ball, called a bob, suspended from a flexible thin string or from a thin rod with gimbals and negligible mass so that the bob can swing freely. The physical characteristics determining the period of oscillation are the mass M of the bob and the distance L from the point O of support of the string to the center of the ball.

(Although Galileo had made an effort to apply the pendulum to clock regulation, he achieved no great success.) In his famous treatise, *Horologium oscillatorium,* published in 1673, Huygens gave the dependence of the period on length and amplitude and described the use of cycloid jaws or cheeks to correct for the slight dependence of the period on the amplitude. This dependence is usually treated extensively in geophysics books.

Isaac Newton extended the proof that cycloid cheeks ensure constancy of period from the case of the uniform gravity field considered by Huygens to the case of force lines converging to a point. This proof appeared, along with Newton's laws of motion and law of gravity, in 1687 in his book *Philosophiae Naturalis Principia Mathematica.*

The formula for the period of an ideal simple pendulum, consisting of a pointlike mass M at the end of a thin, flexible string with length L and negligible mass undergoing small oscillations is

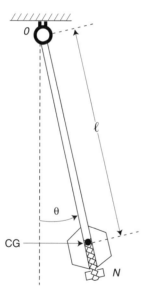

Figure 2 A compound pendulum, also called a physical pendulum. It can have any mass distribution with moment of inertia I and radius of gyration b ($I = M b^2$) about its center of gravity (CG), as long as the distance l between it axis of rotation (O) and the GC is not zero. The compound pendulum found in a pendulum clock can be as simple as a narrow rod with a heavy weight at the bottom end. A nut (N) on the end allows the fine adjustment of the period to coincide with the desired time unit (e.g., a second).

$$T = 2\pi\sqrt{\frac{L}{g}}, \qquad (1)$$

where g is the size of the acceleration due to gravity (≈ 9.80 m/s^2). The ball of a real simple pendulum, however, has finite extent (radius $R > 0$). Including the moment of inertia of the ball about its CG requires adding a term involving its radius of gyration b ($b^2 = 2R^2/5$) about the CG. L in Eq. (1) then becomes the fictitious length

$$L = \ell + \frac{b^2}{\ell} \qquad (2)$$

of an equivalent ideal simple pendulum, where ℓ is the distance from the axis of rotation to the CG of the pendulum. The accuracy of Eq. (1) for the period depends on the accuracy of the approximation that $\sin(\theta)$ is equal to θ. This approximation gets increasingly worse as the angles of oscillation get larger.

In a practical pendulum clock, a compound (or physical) pendulum, usually consisting of a thin rod or bar with one or more heavy weights attached (Fig. 2), operates the clock mechanism. The bar is supported so that oscillations occur in a particular plane. Suspending the pendulum with a short piece of flat spring material can compensate for the amplitude effect since the spring flexes different amounts for different angles.

The moment of inertia and the radius of gyration of a compound pendulum are significantly affected by its mass distribution, while the buoyant and frictional forces of air are affected by its shape and volume. For fixed L, Eq. (2) has two solutions, ℓ_1 and ℓ_2, for l in terms of L such that $L = \ell_1 + \ell_2$. Thus, every rigid pendulum has two axes of rotation (oscillation), separated by a distance L, yielding the same period as a simple pendulum of length L. A pendulum constructed to make use of both axes of rotation is called a reversing pendulum. It is usually a bar or rod with two or more movable weights (Fig. 3) and two sets of knife edges near opposite ends of the bar. The knife edges rest on hard flat surfaces on a support frame and precisely fix the location of the two axes.

According to Norman Feather, Captain Henry Kater used a bar pendulum with knife edges around 1818 to determine the acceleration due to gravity in London. By adjusting the mass carefully he achieved

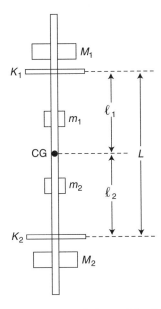

Figure 3 A reversing pendulum. Used in gravity measurements, it can be supported by and oscillate about either set of knife edges (K_1 and K_2) when placed on flat supporting surfaces on its support stand. For the pendulum depicted, masses M_1 and M_2 are large masses having the same shape and volume that are attached to the rod at fixed locations; masses m_1 and m_2, also with equal volumes and shapes, have much smaller masses and are movable for fine-tuning.

equality of periods to within one part in 10^5 upon reversal of the pendulum. Friedrich Wilhelm Bessel worked out a detailed theory in 1826 of Kater's pendulum and suggested a difference technique that required less precision in balancing the pendulum, yet yielded comparable results for gravity measurements. With Bessel's method symmetry of geometrical shape compensates for the effects of air buoyancy and resistance.

Until the 1960s most reliable absolute gravity measurements at reference points around the world were made with reversing pendulums. Relatively straightforward measurements could achieve a precision of a few parts in 10^6. For example, among a number of measurements reported in 1958 by G. P. Wollard for various locations in North America was the value 9.796103 m/s^2 obtained at Denver. In recent decades greater precision and accuracy have been achieved with newer methods, but the instruments are not as portable. Highly precise, portable gravimeters are available but must be calibrated at base stations where accurate absolute measurements have been made.

See also: GALILEI, GALILEO; GEOPHYSICS; HUYGENS, CHRISTIAAN; INERTIA, MOMENT OF; MOTION, HARMONIC; NEWTON, ISAAC; NEWTON'S LAWS; OSCILLATION; PENDULUM, BALLISTIC; PENDULUM, FOUCAULT

Bibliography

FEATHER, N. *An Introduction to the Physics of Mass, Length, and Time* (Edinburgh University Press, Edinburgh, 1962).

GALILEI, G. *Dialogues Concerning Two New Sciences* (Dover, New York, 1954 reprint).

HEISKANEN, W. A., VENING MEINESZ, F. A. *The Earth and Its Gravity Field* (McGraw-Hill, New York, 1958).

NEWTON, I. *Mathematical Principles of Natural Philosophy* (University of California Press, Berkeley, 1962 reprint).

PAGE, C. H., and VIGOUREAUX, P., eds. *The International Bureau of Weights and Measures, 1875–1975*, NBS Special Publication 420 (U.S. Government Printing Office, Washington, DC, 1975).

WOLLARD, G. P. "The Earth's Gravitational Field and Its Exploitation" in *Advances in Geophysics*, edited by H. E. Landsberg (Academic Press, New York, 1952).

CHARLES E. HEAD

PENDULUM, BALLISTIC

The Englishman Benjamin Robbins, author of the book *Principles of Gunnery* (1742), which established the foundation of modern ordnance theory and practice, invented the ballistic pendulum. The ballistic pendulum enabled gunners, for the first time, to measure muzzle velocities accurately. The device is also an instructive example, much used in classroom demonstrations and introductory laboratory experiments, of the application of the laws of conservation of momentum and energy.

The ballistic pendulum (Fig. 1) is a suspended block of wood or other material into which a bullet or any projectile can be shot. The block must be capable of absorbing the projectile without disintegrating. For purposes of the simplest calculation, it is assumed that the block is suspended from a thin, vertical rigid rod with negligible mass that is pivoted

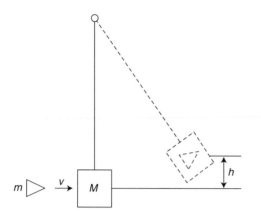

Figure 1 Ballistic pendulum. Projectile enters suspended block whose subsequent rise in height can be used to calculate the initial projectile velocity.

at the top. It is further assumed that there is no significant displacement of this pendulum during the deceleration of the projectile in the block. The motion can be analyzed in two parts, comprising (1) the collision of projectile and block, during which horizontal momentum is conserved, and (2) the rising of the pendulum block (with included projectile) to a height h above its initial vertical position, during which kinetic plus potential energy (mechanical energy) is conserved.

In the first part of the analysis, it can be assumed that momentum is conserved since there are no horizontal external forces acting on the projectile-block system. Because the projectile comes to rest in the block, and both are at rest with respect to their center of mass after the collision, the collision is completely inelastic and there is maximum loss of kinetic energy consistent with conservation of momentum. The lost kinetic energy goes mainly into thermal energy due to the dissipative action of the nonconservative friction force with which the block stops the projectile. Total energy of all forms is conserved. Define the mass and the horizontal component of the velocity of the projectile as m and v, and the mass of the block as M. Let the horizontal component of the velocity of the projectile-block system after the collision be V. Then, from conservation of momentum in the horizontal direction

$$mv = (M + m) V, \qquad (1)$$

which can be solved for v or V.

The kinetic energy lost in the collision is

$$\Delta K = K_i - K_f$$
$$= (1/2)\, mv^2 - (1/2)(M + m)\, V^2, \qquad (2)$$

where K_i is the original kinetic energy and K_f is the kinetic energy after the collision. Substitution of Eq. (1) gives

$$\Delta K = (1/2)\, mv^2 - (1/2)(M + m)\, m^2 v^2 / (M + m)^2$$
$$= (1/2)[mv^2 - m^2 v^2 / (M + m)]$$
$$= (mv^2/2)[1 - m/(M + m)]$$
$$= (mv^2/2)[M/(M + m)]. \qquad (3)$$

The ratio of the kinetic energy lost to the original kinetic energy is

$$\Delta K / K_i = \{(mv^2/2)[M/(M + m)]\}/\{mv^2/2\}$$
$$= M/(M + m). \qquad (4)$$

In the second part of the analysis, the kinetic energy after the collision

$$K_f = (1/2)(M + m)\, V^2 \qquad (5)$$

is converted under the influence of the external force of gravity into potential energy. Note that the external force of the suspending rod on the block is always perpendicular to the direction of motion of the block so that no work is done by this force on the block. Therefore, there is no effect of this force on the conservation of kinetic plus potential energy and

$$(1/2)(M + m)\, V^2 = (M + m)\, gh, \qquad (6)$$

where g is the acceleration due to gravity and h is the height at which the pendulum is instantaneously at rest. Its initial kinetic energy has been converted at this point to gravitational potential energy (Fig. 1). The quantity V is given by

$$V = 2\sqrt{gh}. \qquad (7)$$

The use of Eq. (1) leads to an expression for the pre-collision projectile horizontal velocity component,

$$v = [(M + m)/m]\sqrt{2gh}. \quad (8)$$

Careful measurement of the height to which the center of mass of the pendulum block-projectile rises and the masses of the block and projectile can therefore give a good estimate of the projectile velocity before the collision.

See also: ACCELERATION; CENTER OF MASS; COLLISION; ENERGY, KINETIC; ENERGY, MECHANICAL; ENERGY, POTENTIAL; FORCE; FRICTION; MASS; MOMENTUM; PENDULUM; VELOCITY

Bibliography

GIANCOLI, D. C. *Physics for Scientists and Engineers,* 2nd ed. (Prentice Hall, Englewood Cliffs, NJ, 1988).

SERWAY, R. A. *Physics for Scientists and Engineers,* 4th ed. (Saunders, Philadelphia, 1996).

WEIDNER, R. T., and SELLS, R. L. *Elementary Classical Physics* (Allyn & Bacon, Boston, 1965).

GEORGE E. IOUP

PENDULUM, FOUCAULT

The Foucault pendulum is one of the most interesting and impressive demonstrations of the fact that we live in a rotating noninertial reference frame, though not as impressive as hurricanes, typhoons, and tornadoes. It was described by Jean Foucault (inventor of the gyroscope) in 1851 as an experimental demonstration of Earth's rotation. The basic idea is that a simple pendulum started with no sideways motion will tend to swing in a parallel plane in space in an inertial reference frame (i.e., a frame in which Newton's first law holds). As a thought experiment, consider an observer and a simple pendulum, both on a rotating disk on the surface of a nonrotating earth. Assume the disk has a rotation rate of once per day counterclockwise as observed from above (see Fig. 1). The observer sees the plane in which the pendulum swings rotate clockwise (pre-

cess) through 360° in one day relative to the surface of the disk. Another observer in an inertial frame (i.e., not on the disk) sees the disk rotate under the pendulum while the plane of the pendulum swing remains parallel to its initial plane of oscillation in space.

The description of the Foucault pendulum on the surface of a rotating sphere (the earth) is more complicated than the description on a disk because the plane of the pendulum swing changes in a more complex way as the earth rotates. Only at the North and South poles does the same description as that given above for the rotating disk hold. At the North Pole the pendulum plane appears to rotate through 360° (2π rad) or one revolution per day clockwise as it does on the disk. At the South Pole, the precession relative to the surface of the earth is one revolution per day counterclockwise.

It is instructive to consider a simple pendulum on the surface of the earth at the equator. If the pendulum swings in an east-west equitorial plane, that plane does not change orientation in space as the earth rotates, so the plane of oscillation does not change relative to the surface of the earth. If the pendulum swings in a north-south meridional plane, that plane certainly changes orientation in space as the earth rotates. However, the change does

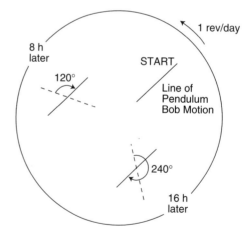

Figure 1 An observer on a rotating disk (one revolution per day counterclockwise) on a nonrotating Earth sees the plane of the pendulum oscillation precess with respect to the surface of the disk at one revolution per day clockwise. The observer not on the disk sees that the plane of the pendulum swing remains parallel to its original direction as the disk rotates.

not rotate the pendulum plane relative to the surface of the earth, so again no rotation of the plane of the pendulum swing is observed on the surface of the earth. Since any plane of swing at the equator can be considered to be a combination of these two perpendicular directions, it is not surprising that no rotation relative to the earth's surface is observed for the pendulum plane of oscillation at the earth's equator.

Characterizing the precession of the plane of oscillation at other latitudes on the surface of the rotating earth is accomplished by solving the equations of motion of the pendulum, subject to certain reasonable simplifying assumptions, such as neglecting the small accelerations associated with the movement of the center of the earth through space and other very small higher order terms. In general, depending on initial conditions, the motion of the pendulum bob is instantaneously an ellipse (if the vertical component of the motion is neglected, appropriate for a long pendulum of small displacement). The important ideas of the Foucault pendulum are illustrated here, however, by looking at only the simplest motion along a straight line, the plane defined by this line and the pendulum support point, and the precession of that plane.

The precession of the plane of the pendulum swing at any latitude is given by (1 rev/day) sin λ, with λ the latitude. The result is consistent with the pole and equator results discussed earlier since at the equator $\lambda = 0°$ and sin $0° = 0$, and at the poles $\lambda = 90°$ and sin $90° = 1$. A sign for the direction of precession can be included or the simple rule, clockwise in the Northern Hemisphere and counterclockwise in the Southern, can be applied. Since sin $30° = \frac{1}{2}$, cities at 30° latitude (New Orleans, Cairo, Pôrto Alegre) have a precession rate of $\frac{1}{2}$ rev/day for a Foucault pendulum.

See also: FRAME OF REFERENCE, INERTIAL; MOTION, ROTATIONAL; OSCILLATION; PENDULUM; PRECESSION

Bibliography

BECKER, R. A. *Introduction to Theoretical Mechanics* (McGraw-Hill, New York, 1954).

McCUSKEY, S. W. *An Introduction to Advanced Dynamics* (Addison-Wesley, Reading, MA, 1959).

GEORGE E. IOUP

PERIODIC TABLE

See ELEMENTS

PERTURBATION THEORY

A perturbation is a disturbance that acts on a physical system, causing an alteration in its motion or its state. Perturbation theory is the study of such disturbances and the calculation of their effects. The equation of motion in the presence of a perturbation is often too complicated to attempt an exact solution, and a variety of special calculational methods have been developed to construct approximate solutions. All the calculational methods of perturbation theory hinge on the assumption that the equation of motion can be separated into two parts, one part describing the unperturbed system, and one part describing the perturbation. The unperturbed part must be simple enough to be solved exactly. Perturbation theory then exploits this exact solution of the unperturbed part to construct an approximate solution when the perturbed part is included.

If the perturbation is small, its effect can be calculated by a method of successive approximations. The essential features of this method are illustrated by the following simple example of the effect of air resistance on a freely falling body, such as a stone or a baseball. For fairly low speeds, the air resistance acting on such a body is proportional to the square of the speed, and the equation of motion is

$$\frac{dv}{dt} = -g + \lambda v^2,$$

where $g = 9.81$ m/s² and λ is a constant that depends on the mass, size, and shape of the body. In this equation of motion, the term λv^2, arising from air resistance, represents the perturbation. The constant λ, which characterizes the magnitude of the perturbation, is called the perturbation parameter. For $\lambda = 0$, the equation of motion reduces to that of free fall, with the exact solution for a body starting with zero velocity at time $t = 0$,

$$v_0 = -gt.$$

This is the unperturbed solution, or the zero-order approximation. To obtain the next approximation, we take the speed from the zero-order approximation, and substitute this into the right side of the equation of motion:

$$\frac{dv}{dt} = -g + \lambda g^2 t^2.$$

Upon integrating this equation with respect to time, we find the first-order approximation:

$$v_1 = -gt + \tfrac{1}{3}\lambda g^2 t^3.$$

To obtain the next approximation, we take the speed from the first-order approximation and again substitute this into the right side of the equation of motion:

$$\frac{dv}{dt} = -g + \lambda(-gt + \tfrac{1}{3}\lambda g^2 t^3)^2$$

$$= -g + \lambda g^2 t^2 - \tfrac{2}{3}\lambda^2 g^3 t^4 + \ldots .$$

We then find the second-order approximation:

$$v_2 = -gt + \tfrac{1}{3}\lambda g^2 t^3 - \tfrac{2}{15}\lambda^2 g^3 t^5.$$

By continuing with this step-by-step procedure, we obtain an approximate solution for v in the form of a series in ascending powers of the perturbation parameter λ. Such series solutions are typical of perturbation theory. If λ is reasonably small, the first few terms of the series suffice to achieve an adequate approximation.

The study of perturbations began in celestial mechanics, where astronomers recognized that the elliptical orbits that the planets describe under the influence of the gravitational attraction of the Sun are disturbed by the extra attractions that the planets exert on each other. Perturbations can be classified as secular or as oscillatory according to their effect on the motion of a system. Secular perturbations involve a continuing ("secular") drift of the perturbed motion away from the unperturbed motion, so the difference between them gradually grows larger and larger (as in a precessing planetary ellipse). Oscillatory perturbations involve a back-and-forth displacement of the perturbed motion relative to the unperturbed motion (as in the motion along a distorted planetary ellipse, which weaves in and out of the unperturbed ellipse).

An elegant and powerful method for dealing with perturbations in the solar system and other frictionless systems is based on the fact that for such a system one can define an energy function, which is known as the Hamiltonian function. This energy function consists of a zero-order energy and a perturbation energy. A general theorem of perturbation theory asserts that the energy change caused in a periodic orbit is equal to the time-average of the perturbation energy over the orbit.

Perturbation theory is widely used in quantum mechanics for calculations of the energies and wave functions of quantum states of atoms or molecules subjected to steady disturbances, such as the disturbances produced by static electric or magnetic fields (time-independent perturbation theory), and for calculations of the probabilities of transitions between quantum states in time-dependent electric and magnetic fields and during collisions (time-dependent perturbation theory, scattering theory). It is also used for calculations of the probabilities of spontaneous transitions between quantum states, which result in the emission of quanta of light by atoms and molecules. Time-independent perturbations are classified as degenerate or nondegenerate. If the system has two or more quantum states of equal energies ("degenerate states"), the perturbation can have a large effect on these states, analogous to the secular case in celestial mechanics. If the system has only states of unequal energies ("nondegenerate states"), the effects remain small, analogous to the oscillatory case in celestial mechanics. For the nondegenerate case, the energy change in first order is equal to the average of the perturbation energy over the wave function of the state.

Perturbation theory plays an essential role in the relativistic quantum theory of systems of many particles, or quantum field theory. The equations of quantum field theory can be solved if the particles do not interact with each other, but the equations become intractable if the particles interact. Perturbation theory is the only available general method for solving the quantum field equations with interactions. The method of successive approximations has been used with excellent results in the quantum field theory of electromagnetic interactions, or quantum electrodynamics. The perturbation parameter is the fine-structure constant ($\alpha = e^2/4\pi\epsilon_0\hbar c \simeq 1/137$), which is fairly small. In the series in ascending powers of α constructed by successive approximations, the first few terms therefore often give adequate accuracy. For instance, the value of the spin magnetic moment of the electron has been calculated to ten significant figures by taking into account terms of up to fourth order in α. The close agreement between this theoretically predicted value and the experimentally measured value is a splendid achievement of modern theoretical physics.

See also: FINE-STRUCTURE CONSTANT; QUANTUM FIELD THEORY; QUANTUM MECHANICS; SCATTERING; WAVE FUNCTION

Bibliography

BORN, M. *The Mechanics of the Atom* (Ungar, New York, 1960).

NAYFEY, A. H. *Introduction to Perturbation Techniques* (Wiley, New York, 1993).

OHANIAN, H. C. *Principles of Quantum Mechanics* (Prentice Hall, Englewood Cliffs, NJ, 1990).

SYMON, K. R. *Mechanics* (Addison-Wesley, Reading, MA, 1971).

HANS C. OHANIAN

PHASE

The phase of a celestial body (e.g., phases of the Moon), of a time- or space-dependent variable (e.g., the phase of an electric alternating current), or of a homogeneous portion of a body (e.g., a liquid phase) generally denotes the momentary appearance, or aspect, or one of the various states of that object. Originating from the archetypal experience of cyclically recurrent appearances of astronomical bodies [from the Greek *phásis*, "appearance (of a star)"], the term "phase" is now used for various purposes in fields like physics, statistics, chemistry, and metallurgy.

Many measurable quantities in physics, such as the movement of a pendulum or the alternating current in an electric circuit, are conveniently expressed by a periodic complex function F in time or space, or a superposition of complex periodic functions, or the real part thereof. The phase of this function is the real argument ϕ in the standard representation of the function by $F = A \exp(i\phi)$, with A the real amplitude. It is often interpreted as a phase angle. Two waves at a given point in space are said to be in phase, if the respective phases are identical up to an even multiple of π. They are in opposite phase if the phases are identical up to an odd multiple of π.

A measurement of a physical observable may depend only on its amplitude, like the measurement of intensities or currents. On the other hand, differences in phase (e.g., between two intersecting light or quantum-mechanical matter waves) manifest themselves through interference patterns in the measured intensity distribution. The relative phase of the electric current in two conducting wires may be studied with an oscilloscope. With a suitable configuration of electromagnets, the phase difference between the currents in the magnets can drive the rotor of an induction motor.

In thermodynamics, a phase is the physically homogeneous portion of a heterogeneous system of matter. Examples include the fundamental solid, liquid, or gaseous phases, which may be further specified as crystalline or amorphous phases. In a pure substance, the different phases can be transformed into each other under specific conditions of temperature and pressure. The quantitative relationship between the phases is expressed in the phase diagram of the substance.

In statistics, the phase of an individual particle is the full set of its spatial and momentum coordinates. The phase of a system of particles then is the

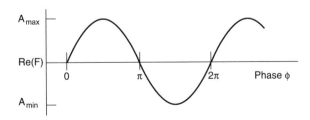

Figure 1 Real part of a periodic function $F = A \exp(i\phi)$.

full set of spatial and momentum coordinates of all its particles. The system is hence said to move in phase space.

See also: CURRENT, ALTERNATING; ELECTROMAGNET; INTERFERENCE; MOON, PHASES OF; OSCILLOSCOPE; PHASE SPACE

WOLFGANG FRITSCH

PHASE, CHANGE OF

Most substances can exist in more than one phase. These phases generally include the gaseous or vapor phase, a liquid phase, and one or more solid phases that are distinguishable from each other by differences in crystal structure (e.g., graphite and diamond are two different solid phases of carbon). These phases are often called states, especially in elementary discussions, although the technical meaning of a state is more restrictive, implying not only the phase but also the values of properties such as temperature, pressure, and volume.

According to the (approximate) generalization known as the law of corresponding states, each substance behaves in much the same way as the temperature and pressure are changed, but on different scales of temperature and pressure. Thus, water goes from liquid to vapor at 100°C, oxygen changes to vapor at −183°C, and iron changes to a vapor at 3,000°C.

For a pure substance (one component), generally only two phases can coexist under arbitrary fixed pressure or temperature, or three phases at a triple point (e.g., the melting point under the equilibrium pressure of pure vapor). Therefore, under fixed pressure (e.g., atmospheric pressure), a phase change of a pure substance occurs at a sharply defined temperature. Two common such temperatures are a melting point (or freezing point) and a boiling point (or condensation point).

Phase changes of mixtures can occur over a range of temperatures. However, the number of phases that can coexist at a given temperature and pressure is restricted by Gibb's phase rule. The number of components present, minus the number of distinct phases, plus the number of uncontrolled external variables (e.g., temperature and pressure) is called the number of degrees of freedom; it is the number of additional conditions that could be imposed on the system. The number of degrees of freedom must be equal to or greater than zero.

The amount Q of thermal energy or heat absorbed by a substance from its surroundings in going from a phase stable at a lower temperature to a phase stable at a higher temperature is commonly called the heat of the transition—for example, the heat of fusion for melting, the heat of vaporization for boiling, and the heat of sublimation for a phase change directly from solid to vapor. Because these are measured under constant pressure (with no work done other than against the atmosphere), the heat of transition is an enthalpy change; $Q = \Delta H$. (Enthalpy H equals the sum of the energy E and the product of pressure and volume PV.) In tables, these heats of transition are often labeled ΔH_f, ΔH_{vap}, and ΔH_{sub}. These values are also known by the archaic term "latent heats" because there is no temperature change when the energy is added.

Each phase has a heat capacity C, which measures the amount Q of thermal energy that must be added for a given increase in temperature ΔT. Because a phase change occurs at a single temperature ($\Delta T = 0$), but requires thermal energy input Q, the heat capacity ($Q/\Delta T$) becomes infinite during a phase transition.

Some changes in phase involve change in a property with no significant change in energy, so $Q = 0$ and the transition may occur over a range of temperatures; these are called second-order phase transitions. Examples include a change from normal helium liquid to superfluid helium, a change of mercury and many other metals from a normal electrical conductor to a superconductor, and a change in a gas at high temperatures to a conducting plasma.

Phase transitions from disordered states (gases or liquids, in contrast to crystals) are quite improbable from the molecular viewpoint, so it is usually necessary to go beyond the transition temperature (to superheated liquid or supercooled vapor or liquid) before bubble, droplet, or crystal formation begins. The absorption or release of thermal energy associated with the phase transition rapidly brings the local temperature back to the equilibrium transition temperature. Ice formation from water droplets in clouds typically occurs at temperatures below −20°C. Some complex compounds, such as sugars, may require months or years to begin crystallization from solution. Honey is a common example.

Phase changes (as well as chemical reactions) between solid phases may occur very slowly, if at all, at temperatures below the normal transition temperature. Therefore, materials such as steel and diamond, which are only metastable, may appear to be stable.

For pure substances, the melting point and freezing point are identical, as are the boiling point (or vaporization point) and condensation point. Usually an impurity is more soluble in one phase of a solvent than in another. Thus salt and sugar are soluble in water (liquid) but not in ice (solid) or water vapor. The presence of the impurity changes the temperature of the phase transition and converts the sharp transition temperature into a range of temperatures. If the impurity is more soluble in the high-temperature phase, the transition temperature is lowered; if it is more soluble in the low-temperature phase, the transition temperature is raised. For example, if salt water is frozen, the first crystals form below 0°C and complete freezing occurs only at lower temperatures, down to −18°C (0°F); salt and ice may begin to melt at low temperatures, completing the process closer to 0°C.

The boiling point is sensitive to pressure. For small changes in temperature, the dependence on pressure is well described by the Clausius–Clapeyron equation, which relates the natural logarithm of the ratio of pressures to the temperature as follows:

$$\ln \frac{p_2}{p_1} = \frac{\Delta H \Delta T}{R T_1 T_2},$$

where p_2 and p_1 are the equilibrium pressures or vapor pressures at T_2 and T_1, ΔH is the heat of vaporization, $\Delta T = T_2 - T_1$, and R is the universal gas constant ($8.314 \ \mathrm{J \cdot mol^{-1} \cdot K^{-1}}$). For typical liquids, the boiling point is proportional to the heat of vaporization. At atmospheric pressure, $\Delta H / T$ is approximately $92 \ \mathrm{J \cdot mol^{-1} \cdot K^{-1}}$, called the Trouton constant.

If T_1 is the freezing point, boiling point, or sublimation point, the change in this temperature, $\Delta T = T_1 - T_2$, caused by small amounts of impurities, may be calculated from an equation similar to the Clausius–Clapeyron equation

$$\Delta X_2 = \frac{(H_1^0 H_1) \Delta T}{R T_1 T_2}.$$

Here ΔX_2 is the difference in impurity levels between the two phases (solid and melt, solution and vapor, or solid and vapor) expressed in mole fraction (number of impurity particles, or moles, divided by the total number of particles, or moles), and H_1^0 and H_1 are the respective molar enthalpy values of the (relatively) pure and impure phases (so $H_1^0 - H_1$ is approximately the heat of fusion, heat of vaporization, or heat of sublimation per mole), and R is the universal gas constant.

Typically, the solubility of one substance in another, called the miscibility, increases as the temperature is raised (sugar is more soluble in hot water than in cold water). However, if the impurity is more miscible with the high-temperature phase, solubility may decrease with rising temperature. Thus gases are less soluble in warm water than in cold water.

Although the qualitative behavior of phase changes with temperature, pressure, or composition is often predictable, and limiting rates of change can be calculated for small changes, the characteristics of real mixtures must generally be found by experiment. A convenient means of presenting such experimental results is a phase diagram. For a single component, temperature may be plotted against pressure. For two components, either the pressure or temperature is fixed and the other is plotted against composition. In each case, when there is more than one degree of freedom, so that temperature and pressure, or temperature and composition, for example, can be changed independently, there is a corresponding area on the phase diagram. Where two phases are present, and, hence, there is only one degree of freedom, a line (or lines) separates the areas of the diagram. A condition of no degree of freedom, as at a triple point, is represented by a point (or points).

An example of a phase diagram for a single substance is the phase diagram for water, shown schematically in Fig. 1. The three areas (three single phases) are vapor, at high temperature and low pressure, solid at low temperature and high pressure, and liquid at intermediate temperature and pressure. The slope of the line along which liquid and vapor are in equilibrium can be calculated from the Clausius–Clapeyron equation. The slope of the line along which solid and liquid are in equilibrium can be calculated from the Clapeyron equation

$$\frac{dP}{dT} = \frac{\Delta H}{T \Delta V},$$

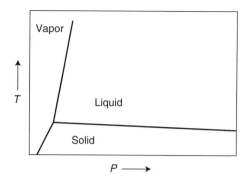

Figure 1 Phase diagram for water (not to scale).

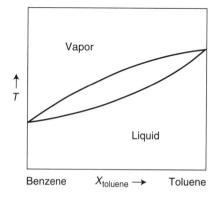

Figure 2 Liquid-vapor phase diagram for benzene and toluene. Temperature is plotted against mole fraction of toluene (at constant pressure).

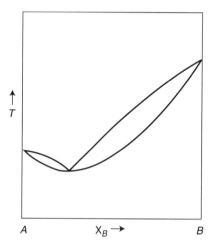

Figure 3 Liquid-vapor diagram of a minimum-boiling azeotrope. The diagram for ethanol (ethyl alcohol) and water is similar, except that the azeotrope is at 95.5 percent ethanol and only 0.3°C below the boiling point (78.4°C) of ethanol.

in which ΔH is the heat of fusion and ΔV is the change in volume upon melting. For water, the slope dT/dP is negative (sloping downward to the right) because ice (solid) is less dense than water (liquid); for almost all other substances, the slope is positive. The point at which the liquid-vapor line terminates, at high temperature, is called the critical point. At high pressures, additional solid phases exist.

Phase diagrams for two substances are constructed by fixing one variable, usually pressure (but sometimes temperature). Then each single phase is again an area on the diagram, lines show equilibrium between two phases, and points indicate no degree of freedom.

If two substances are miscible with each other, a phase diagram like Fig. 2 (the liquid-vapor diagram for benzene and toluene) is obtained. At high temperatures, the vapors form a single phase; at low temperatures, the liquids form a single phase. Liquid and vapor phases (of different composition) are in equilibrium along the two lines separating the liquid and vapor areas. There is no phase present with temperature and composition falling in the region between the two liquid-vapor lines, a region known as the forbidden region. Historically, it has also been known as a two-phase region because the system separates into the two phases, to the left and to the right.

Separation of miscible liquids requires fractional distillation. From any point in the liquid, along the equilibrium line, the material that is first to evaporate is richer in the more volatile component (e.g., benzene in a benzene-toluene mixture). If this material is condensed (at constant composition, and therefore along an imaginary vertical line) farther along the distillation column, again material that evaporates first (isothermally) from it is richer in the more volatile component. After many stages (termed "plates") of separation and condensation, quite pure material may be obtained as the distillate.

Certain miscible mixtures, such as alcohol and water, have phase diagrams with a minimum (or maximum), as shown in Fig. 3. The composition at the minimum (or maximum) is called an azeotrope. At fixed pressure, an azeotrope boils (or melts) as if it were a pure substance, but a change of pressure changes its composition. To separate such materials by fractional distillation, it is usually necessary to add a third substance to "break" the azeotrope. The

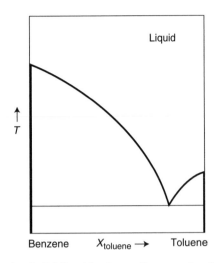

Figure 4 Solid-liquid phase diagram for benzene and toluene. The pure solid phase regions are represented by heavier vertical lines.

third substance can be removed by a second fractional distillation.

When two substances are immiscible in the low-temperature phase and miscible in the high-temperature phase, the phase diagram takes on an appearance like that in Fig. 4 (which shows the solid–liquid equilibrium for benzene and toluene). At high temperatures, the two liquids are miscible. At low temperatures, the two solids are immiscible. The pure solids are very narrow areas (represented in the drawing as heavy single lines) along the left and right, below the melting points. The lowest point at which the liquid phase can exist is called a eutectic point. A horizontal line is added through the eutectic point as a marker. The area between the pure phases (above *and* below the horizontal line) is a forbidden area.

Solids are typically less miscible than liquids, so any impurity lowers the melting point of the host solid. This is the basis of zone refining of solids, such as silicon. A narrow section of molten silicon collects the impurities and can be swept to the end of a rod.

Most phase diagrams can be recognized as composed of those discussed above. When three or more substances are present, phase diagrams must represent cross sections (e.g., at constant temperature) or be represented as three-dimensional figures.

See also: DEGREE OF FREEDOM; ENTHALPY; FREEZING POINT; HEAT CAPACITY; PHASE; PHASE RULE; PHASE TRANSITION; SPECIFIC HEAT; TRIPLE POINT

Bibliography

BAUMAN, R. P. *Modern Thermodynamics with Statistical Mechanics* (Macmillan, New York, 1992).

FINDLAY, A.; CAMPBELL, A. N.; and SMITH, N. O. *Phase Rule,* 9th ed. (Dover, New York, 1951).

ROBERT P. BAUMAN

PHASE RULE

The familiar phases of matter are solid, liquid, and vapor. Usually when more than one phase is present there is a transformation from one into the other: water evaporates to steam from a heated pot, ice melts to liquid when salt is poured on it, and so on. If the phases are in equilibrium there is no net transformation. This is possible only when particular relations among the temperature, pressure, and composition are obeyed. The possibility to change each of these is a thermodynamic degree of freedom. The phase rule of Josiah Willard Gibbs specifies the number of degrees of freedom, or variables that can be changed independently, consistent with equilibrium.

Water provides a first example. The liquid and vapor phases can be in equilibrium over a range of temperatures. The pressure of the vapor is not independent of the temperature since fixing one fixes the other. There is only one degree of freedom. If the water begins to freeze, there is exactly one temperature and one pressure at which all three phases can coexist stably; this is called the triple point. Equilibrium for three phases of one material allows zero degrees of freedom. The values of temperature and pressure depend on the material, but the counting scheme is more general. It does not matter what or how big the system is, or what the proportions of the phases are as long as they are all present.

To derive the phase rule we should consider equilibrium carefully. A difference in temperature (T) induces a flow of heat, as a difference in pressure (p) causes a flow of fluid. Such flows do not exist at equilibrium, so temperature and pressure must be uniform. In a similar way, chemical potential (μ) describes the incentive for material to move across boundaries, such as the boundaries between distinct phases. In equilibrium, it too must be uniform among all the phases present. The key is to count the number of equations which describe equality of all the chemical potentials, without double counting. For a single material the number of equalities is just the number of phases minus one. When there are r chemical components and M phases, there are $r(M-1)$ independent equalities. By contrast, accounting for the composition of a mixture with r components and M phases takes $(r-1)M$ parameters. Adding two for temperature and pressure, the total number of variables is $(r-1)M+2$. The difference between the number of variables and the number of equations gives the number of degrees of freedom: $f = [(r-1)M + 2] - r(M-1)$, or $f = r - M + 2$. This is the Gibbs phase rule.

Mixed materials show the power of the phase rule. Gold and mercury, for example, can mix in all three phases. Two components ($r = 2$) and three phases ($M = 3$) leave one degree of freedom. Unlike pure water, the three phases of this mixture can coexist in equilibrium over a range of temperatures. The same is true of water and salt, even though they mix only in the liquid phase. This allows the use of salt to melt ice. Since there is one degree of freedom, the concentration of salt in the liquid is fixed at each temperature. The ice will melt to dilute the salt to that concentration.

Considered more generally, a phase is a region of uniform physical and chemical properties. The phase rule can be applied to microscopic crystalline forms, for example, when several coexist with boundaries between them. Each distinct molecular structure is a separate phase. The phase rule is indispensable in mineralogy and metallurgy to keep track of the complicated situations that arise in the fields.

See also: CHEMICAL POTENTIAL; DEGREE OF FREEDOM; EQUILIBRIUM; GIBBS, JOSIAH WILLARD; PHASE; PHASE, CHANGE OF; PHASE TRANSITION; THERMODYNAMICS; TRIPLE POINT

Bibliography

CALLEN, H. B. *Thermodynamics and an Introduction to Thermostatistics,* 2nd ed. (Wiley, New York, 1985).
FERMI, E. *Thermodynamics* (Dover, New York, 1956).
ZEMANSKY, M. W., and DITTMAN, R. H. *Heat and Thermodynamics: An Intermediate Textbook,* 6th ed. (McGraw-Hill, New York, 1981).

MICHAEL ELBAUM

PHASE SPACE

Hamilton's equations of motion govern the dynamics of conservative classical systems. For a system with N degrees of freedom, there are N coordinates and N conjugate momenta. Hamilton's equations are $2N$, coupled, first order in time, ordinary differential equations that describe the time evolution of the coordinates and the momenta in a $2N$-dimensional space called phase space. The state of the system at any instant of time is represented in phase space by a $2N$-dimensional point. The time evolution of the system is represented by a trajectory of such points. Two fundamental properties characterize these trajectories:

1. Trajectories in phase space do not intersect at a single instant of time. Since the Hamiltonian dynamics is first order in time, the values of each of the coordinates and momenta at an instant of time uniquely determine the subsequent phase space trajectory. Trajectory crossings would contradict this uniqueness. This also implies that a continuous ensemble of initial phase space points at time t with boundary B evolves into an ensemble with the ensemble boundary B' at time t'. Thus, it is sufficient for many purposes to follow the trajectories of the boundary points of the ensemble instead of the entire ensemble. For large N, this can be an enormous savings in computational work.
2. The $2N$-dimensional volume of a continuous ensemble of phase space points is an invariant of the time evolution. This feature is sometimes referred to as the incompressibility of the phase space flow. It is the essence of Liouville's theorem regarding the invariant time evolution of the ensemble's $2N$-dimensional volume.

The concept of phase space also arises outside of the context of conservative Hamiltonian dynamics. Any system of N coupled, first order in time, ordinary differential equations describes the time evolution of an N-dimensional point in an N-dimensional space of the N-dependent variables. This space is also called phase space. In contrast to the Hamiltonian context in which the phase space has even dimension, it is possible to have an odd-dimensional phase space in this non-Hamiltonian context. Nevertheless, property 1 still holds sway, whereas property

2 is solely a consequence of the sympletic character of conservative Hamiltonian dynamics, which requires a space of even dimension.

An especially important instance of non-Hamiltonian dynamics is dissipative dynamics. In this case, an ensemble of initial points evolves in time towards a phase space attractor, during the course of which the phase space volume changes. The attractor may be a single phase space point or a limit cycle, in which cases property 1 is no longer strictly true since many trajectories can end up at the same phase space point asymptotically in time (trajectories do not actually cross but they do merge at infinite time). Periodically driven dissipative systems can produce chaotic phase space attractors that even have fractal geometry.

Phase space is a fundamental concept in classical mechanics, in statistical mechanics, and for the theory of nonlinear differential equations.

See also: DEGREE OF FREEDOM; STATISTICAL MECHANICS; THERMODYNAMICS

Bibliography

LICHTENBERG, A. J., and LIEBERMAN, M. A. *Regular and Stochastic Motion* (Springer-Verlag, New York, 1983).

RONALD F. FOX

PHASE TRANSITION

When thermodynamic parameters such as temperature T and pressure P change, matter often undergoes a phase transition. This transition involves changes in structure, such as a liquid changing to a solid, or changes in more subtle properties, such as an ordering of the magnetic moments in a paramagnetic solid upon cooling through the Curie temperature, or the vanishing of the electrical resistivity in a superconducting transition.

Phase Equilibrium

Under certain conditions, different phases coexist. The liquid-solid mixture of water at 0°C and standard pressure is a familiar example. The number of possible phases in equilibrium at a given T and P can be determined from the Gibbs phase rule, which is named for Josiah W. Gibbs. Each phase can be characterized by its Gibbs free energy:

$$G = U - TS + PV = H - TS, \qquad (1)$$

where U is the internal energy of the phase, S is the entropy, V is the volume, and H is the enthalpy. For equilibrium between simple phases one and two, such as water and ice, $G_1 = G_2$. For multicomponent phases, the Gibbs free energy of each component must be the same in both phases, $G_{1i} = G_{2i}$.

Order of Phase Transition

Phase transitions are typically characterized by their order. Following a criterion first suggested by Paul Ehrenfest, a phase transition is said to be of nth order if the first $n - 1$ derivatives of the Gibbs free energy are continuous but the nth is not. For all first-order phase transitions, the enthalpy, $H = [\partial(G/T)/\partial(1/T)]_P$, entropy, $S = -[\partial G/\partial T]_p$, and volume, $V = [\partial G/\partial p]_T$, are discontinuous. The discontinuity of H results in heat evolution or consumption accompanying the transition; this is called the latent heat. Freezing, melting, and boiling are examples of first-order phase transitions. Superconducting, magnetic, and chemical order/disorder transitions are examples of second-order phase transitions. It is often not easy to distinguish the order of a phase transition. Weakly first-order phase transitions, for example, have a vanishingly small enthalpy. The existence of an enthalpy is not a good measure either, since the abnormal heat capacity of a second-order phase transition can mimic an enthalpy of transformation if it occurs in a relatively narrow temperature range.

First-Order Phase Transitions: Nucleation and Growth

Most first-order phase transitions proceed by a process of nucleation and growth, involving a large fluctuation of the order parameter within a spatially small region. During freezing, for example, small clusters of the solid phase, called nuclei, form in the liquid. If the clusters appear randomly in space and time, they are homogeneous nuclei; if they form preferentially on dirt or imperfections on the walls of the container, they are heterogeneous nuclei. Small particles, nucleating agents, are often deliberately introduced into a liquid or gas to increase the heterogeneous nucleation rate. Cloud seeding is one example of this.

Gibbs formulated the thermodynamic basis of homogeneous nucleation. Assuming a sharp interface between the nucleus and the original phase, the work of formation W_n for a cluster containing n atoms is

$$W_n = n\Delta G' + \eta \sigma n^{2/3}, \qquad (2)$$

where $\Delta G'$ is the free energy of the initial phase less that of the final phase (negative below the transition temperature), σ is the interfacial energy between the two phases, and η is a geometric constant that depends on the nucleus shape. For spherical clusters, $\eta = (36\pi)^{1/3}\bar{v}^{2/3}$, where $\bar{v}$ is the molecular volume. Below the transition temperature, W_n increases for small clusters, due to the large surface-to-volume ratio, but decreases for large clusters, where the volume-free energy is dominant. A maximum for W_n is then obtained,

$$W_{n*} = \frac{16\pi \bar{v}^2}{3} \frac{\sigma^3}{(\Delta G')^2} \qquad (3)$$

corresponding to a critical cluster size,

$$n* = \frac{32\pi \bar{v}^2}{3} \frac{\sigma^3}{|\Delta G'|^3}. \qquad (4)$$

On average, clusters larger than $n*$ will grow to macroscopic size, while those smaller than $n*$ tend to shrink. Nucleation, then, is the production of clusters larger than $n*$. Assuming that clusters grow by the addition of one atom or molecule at a time, the rate at which nuclei are produced, the nucleation rate, has the form

$$I^S = A \exp\left(-\frac{W_{n*}}{k_B T}\right), \qquad (5)$$

where k_B is the Boltzmann constant, and A contains factors corresponding to the rate at which an atom is attached to the interface of the nucleus. The prediction of an energetic barrier to the formation of a new phase, W_{n*}, explains why it is frequently possible to cool a phase far below its equilibrium transition temperature.

Second-Order Phase Transitions: Chemical Order/Disorder

Second-order phase transitions are characterized by large fluctuations in the order parameter as the transition temperature is approached, and cannot, therefore, be understood by the thermodynamic approach for first-order phase transitions. While the quantitative description is mathematically involved, a simple mean-field analysis can yield qualitative information in some cases. Chemical ordering of different atoms on crystal-lattice sites in alloys and compounds provides a useful example. Consider the simple case of two atoms, A and B, forming an AB alloy that orders by a second-order phase transition. At high temperatures, A and B will be randomly distributed on the lattice due to the dominance of the entropic term TS in the free energy. As the temperature is decreased, the enthalpy H becomes more important, ordering the atoms into specific periodic arrangements, with the exact nature of the ordering dependent on the sign and strength of H. Viewing an ordered AB cubic alloy with the CsCl structure as two interpenetrating simple cubic lattices, with A atoms located on lattice α and B atoms on lattice β, a long-range-order parameter ξ is defined. The number of correct sites (A atoms on the α lattice or B atoms on the β lattice) is $N/4(1 + \xi)$, and those on incorrect sites (A on β and B on α) are $N/4(1 - \xi)$. For complete ordering, $\xi = \pm1$, and for complete disorder, $\xi = 0$. Assuming nearest neighbor interactions and assigning h_{AA}, h_{BB}, and h_{AB} for the strengths of the bonds between like and unlike atoms, the critical temperature, T_c, is

$$T_c = \frac{2[2h_{AB} - (h_{AA} + h_{BB})]}{k_B}. \qquad (6)$$

At other temperatures,

$$\ln \frac{(1 + \xi)}{(1 - \xi)} = \frac{2\xi T_c}{T}. \qquad (7)$$

Based on this discussion, the qualitative behavior of the order parameter for the different types of phase

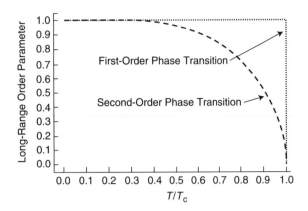

Figure 1 Qualitative behavior of the order parameter for the different types of phase transitions.

transitions can be contrasted. Ideally, ξ is discontinuous at T_c for a first-order phase transition, while for a second-order transition, it decreases continuously as T_c is approached (see Fig. 1).

See also: ENERGY, FREE; ENTHALPY; GIBBS, JOSIAH WILLARD; PHASE; PHASE, CHANGE OF; PHASE RULE

Bibliography

GIBBS, J. W. *Collected Works,* Vol. 1 (Longmans, New York, 1931).

GOODSTEIN, D. L. *States of Matter* (Prentice Hall, Englewood Cliffs, NJ, 1975).

KELTON, K. F. "Crystal Nucleation in Liquids and Glasses" in *Solid-State Physics: Advances in Research and Applications,* Vol. 45, edited by H. Ehrenreich and D. Turnbull (Academic Press, Boston, 1991).

TURNBULL, D. "Phase Changes" in *Solid-State Physics: Advances in Research and Applications,* Vol. 3, edited by F. Seitz and D. Turnbull (Academic Press, Boston, 1956).

KENNETH F. KELTON

PHASE VELOCITY

The definition of phase velocity

$$V = \frac{\omega}{k}, \qquad (1)$$

is so strongly interrelated with the solutions of the wave equation that in order to understand its full significance, we must introduce the wave equation first. The meaning of the terms frequency (ω) and wave number (k) is made clear later in this entry. The wave equation is one of the truly fundamental equations of mathematical physics. For the sake of brevity, we confine our discussion to one space dimension, in which case the wave equation, in its simplest form, governs the propagation of a quantity $\Psi(x, t)$ according to

$$\frac{\partial^2 \Psi}{\partial t^2} - v^2 \frac{\partial^2 \Psi}{\partial x^2} = 0, \qquad (2)$$

where v is the velocity of propagation of the wave. The quantity Ψ could be the transverse displacement of a vibrating string, the pressure in a sound wave, or the electric field strength in an electromagnetic wave. For the three-dimensional examples, the medium must be the same at every point on a given plane, and the properties of the wave motion are then functions only of the distance along a line normal to that plane. Such a wave in an extended medium is called a plane wave.

Assuming v is a constant, the most general expression for describing a traveling (or propagating) wave is

$$\Psi(x,t) = f(x - vt) + g(x + vt), \qquad (3)$$

where f and g are arbitrary functions of the variables $x - vt$ and $x + vt$, respectively, which are not necessarily of a periodic nature, although, of course, they may be. Equation (3) is known as the d'Alembert solution of the wave equation. As time increases, the value of x must also increase in order to maintain a constant value of $(x - vt)$, or decrease to keep $(x + vt)$ constant. Therefore, the function f will retain its original shape and amplitude as time increases if we shift our viewpoint along the x direction, in a positive sense, with a velocity v (and in a negative sense for g). Thus, the functions f and g must represent disturbances that move with velocity v to the right or left, respectively (Fig. 1).

In 1747 Jean le Rond d'Alembert delivered a paper to the Berlin Academy on the problem of the vibrating violin string. In that paper he showed that infinitely many curves could be modes of vibration. His general solution is given by Eq. (3). Most of

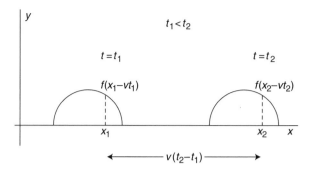

Figure 1 Wave of arbitrary form traveling with velocity v in the positive x direction.

d'Alembert's results were also given by Leonhard Euler in his 1748 paper "On the Vibrations of Strings," which was written after he had read d'Alembert's earlier papers on the subject. An alternative method of solution of the wave equation, developed by Daniel Bernoulli in 1753, produces, even in the one-dimensional case, a complicated system of exponential factors; for a vibrating string the motion is a superposition of its characteristic frequencies. The results of d'Alembert and Euler, and the results of Bernoulli, taken together, indicated that an arbitrary function could be described by a superposition of trigonometric functions. Euler could not believe this, and so he, as well as Joseph-Louis Lagrange, rejected Bernoulli's superposition principle. The French mathematician Alexis Claude Clairaut gave a proof in an obscure paper in 1754 that the results of Euler and d'Alembert, and Bernoulli, were actually consistent. The question was finally settled when Jean Baptiste Joseph Fourier gave his famous proof in 1807 that every periodic function of x satisfying certain very general conditions can be represented as a trigonometric series.

Phase Velocity and Periodic Waves

Because of Fourier's work, we can think of a wave as a sum of Fourier components. One component can be written in complex exponential form (using $i = \sqrt{-1}$) as

$$\Psi(x,t) = A \exp\left[i\frac{\omega}{v}(x-vt)\right] = A \exp[i(kx-\omega t)],$$

$$= A[\cos(kx-\omega t) + i\sin(kx-\omega t)], \qquad (4)$$

where the argument $(kx-\omega t)$ is more convenient than the $(x-vt)$ of Eq. (3) and A is a constant amplitude.

The function $\Psi(x,t)$ describes an infinite harmonic plane wave propagating in the positive x direction, with velocity $v = \omega/k$. The wave possesses a well-defined frequency $\omega = 2\pi\nu$ and the quantity k is called the propagation constant or wave number (i.e., the number of wavelengths per 2π unit length) and has dimensions $[L^{-1}]$. The wavelength λ is the distance required for one complete vibration of the wave

$$\lambda = \frac{v}{\nu} = \frac{2\pi v}{\omega}, \qquad (5)$$

and hence the relation between k and λ is

$$k = \frac{2\pi}{\lambda}. \qquad (6)$$

Certain physical situations can be quite adequately approximated by a wave function of this type, for example, the propagation of a monochromatic light wave in space, the propagation of a sinusoidal wave along an infinite string, or the motion, in quantum mechanics, of a free particle moving in the x direction.

The phase φ of the wave described by Eq. (4) is defined to be the argument of the real part of the wave function

$$\varphi \equiv kx - \omega t. \qquad (7)$$

A fixed value of the phase (as time increases the value of x increases) defines a point on the wave that moves along the x direction in a positive sense (Fig. 2). The condition that the phase remains constant in time defines a phase velocity V. To ensure $\varphi =$ constant, we set

$$d\varphi = 0, \qquad (8)$$

or

$$kdx = \omega dt, \qquad (9)$$

from which

$$V = \frac{dx}{dt} = \frac{\omega}{k} = v, \qquad (10)$$

so that the phase velocity V in this case is just the quantity originally introduced as the velocity. In Fig. 2 we plot the real part of the wave function $\Psi(x, t)$.

It is important to note that it is possible to speak of a phase velocity only when the wave function has the same form throughout its length. This condition is necessary in order that we be able to measure the wavelength by taking the distance between any two successive corresponding points on the wave (two points whose phases differ by 2π) (Fig. 2). If the wave form were to change as a function of time or of distance along the wave, these measurements would not always yield the same results.

The fundamental definition of the phase velocity is based upon the requirement of the constancy of the phase of the primary oscillation and not upon the ratio ω/k. Thus, in general, the phase velocity V and the so-called wave velocity $v = \omega/k$ are distinct quantities. For media in which energy dissipation cannot be neglected, the motion of waves can be described by making the wave number $k = \kappa + i\beta$ complex and the frequency ω real. The wave is then damped or attenuated with distance. To see this, write the wave as

$$\Psi(x,t) = A \exp[i(kx - \omega t)]$$
$$= A \exp(-\beta x) \exp[i(\kappa x - \omega t)]. \qquad (11)$$

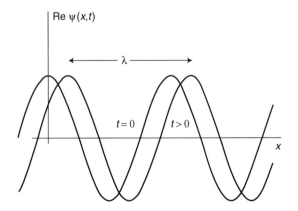

Figure 2 A fixed value of the phase moving in the positive x direction as time t increases.

We note that the wave velocity $v = \omega/k = \omega/(\kappa + i\beta)$ is complex. On the other hand, the phase velocity $V = \omega/\kappa = \omega\lambda/2\pi$ is necessarily always a real quantity because it arises from the requirement that the phase of the primary oscillation $\varphi = \kappa x - \omega t =$ constant. Of course if k is real, then $V = v = \omega/k$.

Group Velocity

If different Fourier components of a wave have different velocities, then combining components leads to interesting new effects.

Another velocity can be defined if we consider the propagation of a varying amplitude that modulates the wave function, the so-called modulation impressed on a carrier. The modulation results in the building up of some groups of large amplitude that move along with a group velocity U. We obtain a simple combination of groups when two wave functions Ψ_1 and Ψ_2 are added. Here, we are making use of the principle that the superposition of various solutions of a linear differential equation is still a solution to the equation. Let us assume, therefore, that we have two almost equal solutions to the wave equation represented by the wave functions Ψ_1 and Ψ_2, each of which has the same amplitude,

$$\Psi_1(x,t) = A \exp[i(k_1 x - \omega_1 t)],$$
$$\Psi_2(x,t) = A \exp[i(k_2 x - \omega_2 t)], \qquad (12)$$

but whose frequencies and wave numbers differ by only small amounts:

$$\omega_1 = \omega + \Delta\omega, \qquad k_1 = k + \Delta k$$
$$\omega_2 = \omega - \Delta\omega, \qquad k_2 = k - \Delta k. \qquad (13)$$

Forming the solution that consists of the sum of Ψ_1 and Ψ_2, we have

$$\Psi(x,t) = \Psi_1 + \Psi_2 = A\{\exp\{i[(k+\Delta k)x - (\omega+\Delta\omega)t]\}$$
$$+ \exp\{i[(k-\Delta k)x - (\omega-\Delta\omega)t]\}\}. \qquad (14)$$

Simplifying this expression via standard trigonometric identities we obtain

$$\Psi(x,t) = 2A \cos[(\Delta k)x - (\Delta\omega)t]\exp[i(kx - \omega t)]. \qquad (15)$$

This represents a carrier with frequency ω and a modulation with frequency $\Delta\omega$. The slowly varying amplitude, corresponding to the term

$$2A\cos[(\Delta k)x - (\Delta\omega)t], \qquad (16)$$

modulates the wave function. The primary oscillation (carrier) takes place at a frequency ω. The situation is represented in Fig. 3 where we plot the real part of $\Psi(x, t)$. In that graph we see a succession of wavelets (ω, k) with varying amplitude ($\Delta\omega$, Δk). The varying amplitude gives rise to beats. The velocity U, called the group velocity, with which the modulations (or group of waves) propagates is given by the requirement that the phase

$$\alpha \equiv (\Delta k)x - (\Delta\omega)t \qquad (17)$$

of the amplitude term be constant. Thus,

$$U = \frac{dx}{dt} = \frac{\Delta\omega}{\Delta k} \rightarrow \frac{d\omega}{dk} \qquad (18)$$

for $\Delta k \rightarrow 0$.

If we do not pay attention to the detailed motion and observe only the average amplitude distribution, we verify that the amplitude curve moves forward with the group velocity U. Looking more carefully at the detailed vibrations, we may see the wavelets moving within the envelope with their own velocity V. The possible situations are summarized in Table 1.

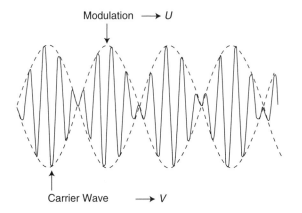

Figure 3 A succession of wavelets with varying amplitudes giving rise to beats.

Table 1 Possible Wavelet Motion Situations

Condition	Interpretation
$U = V$	The group and the wavelets travel along together with no change in aspect.
$U < V$	Both the group and the wavelets move, but the wavelets appear to originate at the rear of the group, progress through the group, and disappear at the front.
$U > V$	The wavelets appear to originate at the front of the group, progress through the group, and disappear at the rear.

Examples of Phase Velocity

The phase velocity V is the relevant quantity in all phenomena involving interference and stationary waves. A very simple experiment provides an excellent example of phase velocity. Just throw a stone in a shallow pond, and look at the rings produced on the surface. They are composed of a small number of short ripples, moving under surface tension. The system as a whole propagates with the group velocity U, but each individual ripple moves with the phase velocity V. In this case $U > V$, and so the ripples build up along the outside ring, move more slowly than the ring, and disappear on the inside of the ring.

The phase velocity is used in the definition of the index of refraction

$$n = \frac{c}{V}, \qquad (19)$$

where c is the velocity of light in a vacuum and enters into the laws of refraction at the interface between two bodies.

When the phase velocity V is a function of k (or λ) for a given medium, that medium is said to be dispersive. The best-known example of this phenomenon is the simple optical prism. The index of refraction of the prism $n = c/V(\lambda)$ is dependent on the wavelength of the incident light (i.e., the prism is a dispersive medium for optical light); and, upon passing through the prism, the light is separated into a spectrum of wavelengths. If the medium is dispersive, the phase velocity V differs from the group velocity U. To show this explicitly, we write

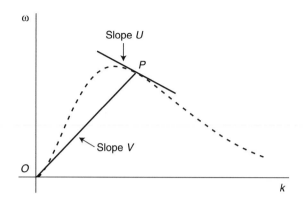

Figure 4 Phase velocity and group velocity graphically represented by a plot of ω as a function of k.

$$U = \frac{d\omega}{dk}. \qquad (20)$$

Since $\omega = kV$, we have (using the chain rule)

$$U = \frac{d\omega}{dk} = V + k\frac{dV}{dk}. \qquad (21)$$

A very useful graphical representation of the phase and group velocities is to plot ω as a function of k (Fig. 4). The slope of the cord $O\,P$ gives the phase velocity V, while the slope of the tangent at point P yields the group velocity U.

If the medium is nondispersive, then $V = $ constant, so that $dV/dk = 0$; hence $U = V$. Vacuum is nondispersive for light ($V = U = c$), but all material media are dispersive. The velocity of sound is approximately constant for long wavelengths, but it depends strongly on the wavelength at short wavelengths, especially when the wavelength is of the order of the distance between molecules. The phase velocity V for sound is then equal to the group velocity U only for long wavelengths.

See also: BEATS; FOURIER SERIES AND FOURIER TRANSFORM; GROUP VELOCITY; WAVELENGTH; WAVE MOTION

Bibliography

BRILLOUIN, L. *Wave Propagation and Group Velocity* (Academic Press, New York, 1964).

HOLLINGDALE, S. *Makers of Mathematics* (Penguin Books, New York, 1989).

MARION, J. B. *Classical Dynamics of Particles and Systems* (Academic Press, New York, 1965).

M. C. CHIDICHIMO

M. STASTNA

PHILOSOPHY

See PHYSICS, PHILOSOPHY OF

PHONON

A phonon is an elementary quantum of vibrational excitation in a lattice of atoms that make up a portion of condensed matter. The atoms in a crystalline solid phase of condensed matter interact in such a way that each of them is more or less located within a fixed region of space that contains their equilibrium positions. The higher the temperature, the greater is the amplitude of vibration of the atoms about their equilibrium positions. When the atoms become slightly displaced from their equilibrium positions, their potential energy increases essentially as the square of the respective displacement distances. When this potential energy is added to the kinetic energy of the atoms' motion, the total collective energy of the excited lattice of atoms is formally equivalent to that of a set of harmonic oscillators. If the displacement of successive neighboring atoms away from their equilibrium lattice positions follows a sinusoidal function with a definite wavelength along some spatial direction, then the resulting lattice wave pattern may propagate through the crystal with a definite frequency. When this collective wave motion of the atoms is quantized according to the rules of quantum mechanics, the energy of excitation becomes represented by the number of elementary lattice wave quanta, or phonons, which may be present for any given wavelength. The vibrational energy carried by the phonons is the quantized excitation energy of the collective motion of the interacting atoms vibrating about their equilibrium positions.

The phonons have definite energies, given by the product of Planck's constant times their respective frequencies. Since phonons also correspond, respectively, to definite wavelengths, they also possess definite wave numbers that define wave vectors once the direction of wave propagation is specified. The phonons' wave vectors play the role of quasi-momentum vectors when multiplied by Planck's constant divided by 2π. The definite phonon quasi-momenta and their corresponding energies that span a spectrum form the basis for a particle-like description of these phonon excitations that make up the vibrational energy levels of the crystalline lattice.

The particle-like properties of the phonon excitations readily lend themselves to explanations of the thermodynamic properties of solids, and also the solid's capacities to conduct heat and electricity. The contribution to the heat capacity of a crystalline solid from the lattice vibrations can be viewed as arising from the thermal excitation of particle-like phonons of different energies. Since lattice imperfections will always be present in a crystal to some extent, the propagation of phonons may be interrupted and they can then be scattered either with some loss of energy or alteration in direction of propagation, or both. Such scattering of phonons as they travel from a hotter region of the crystal, where they are more numerous, to a cooler region, where fewer are excited, serves to limit the rate of transport of phonon energy and therefore the thermal conductivity of the lattice. In a crystal that contains conduction electrons, the phonons may scatter the electrons and limit the ability of an applied electric field to cause the conduction of electricity.

See also: CONDENSED MATTER PHYSICS; OSCILLATOR, HARMONIC

Bibliography

CHAIKIN, P. M., and LUBENSKY, T. C. *Principles of Condensed Matter Physics* (Cambridge University Press, New York, 1995).

KITTEL, C. *Quantum Theory of Solids* (Wiley, New York, 1963).

MICHAEL J. HARRISON

PHOSPHORESCENCE

Some materials luminesce—that is, emit light—when they are subjected to an exciting source (photons, electrons, etc.). If the material continues to luminesce after the exciting source is removed, it is exhibiting either phosphorescence or fluorescence. If the time it takes the "afterglow" to dim depends on the temperature of the material, the material is phosphorescent. In general, the decay time decreases with increasing temperature. If the afterglow decay time is independent of temperature, the material is fluorescent.

Fluorescence and phosphorescence are quantum effects. An atom that is excited by a photon into an elevated energy state may decay to the lower state via some intermediate state (see Fig. 1). The photons emitted in this decay will have an energy lower than that of the incident photon.

Another significant difference between phosphorescence and fluorescence is their respective decay times. An excited atom in a fluorescent material decays to its ground state in about 10^{-8} second, whereas an excited atom of a phosphorescent material may remain in an excited metastable state for several hours.

A common mechanism of phosphorescence occurs when electrons or holes are set free by the excitation and trapped by lattice defects. The electrons or holes are then released from these traps by thermal vibrations in the lattice and recombine with oppositely charged carriers and emit a photon. The intermediate state shown in Fig. 1 represents the state of the system with the electron or hole trapped.

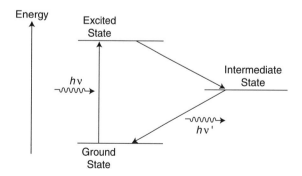

Figure 1

The different decay times for fluorescent and phosphorescent atoms is explained through quantum mechanics. A fluorescent material decays to the lower state via the "allowable" dipole transitions. However, a phosphorescent material is elevated to an excited state in which the decay path to the lower state includes a "forbidden" transition, one that is not a dipole transition. This forbidden decay path, which is another mechanism of phosphorescence, accounts for the long afterglow of such materials.

Phosphorescent atoms in a metastable state can lose energy in forms other than photon emission. For instance, the metastable state could be deexcited, without the emission of a photon, when the atom collides with the wall of its container or with another atom. The extra energy is either absorbed by the wall or transformed into kinetic energy of the other atom. This explains the temperature dependence of the afterglow decay time. If the material is at a high temperature, the excited atoms are more likely to collide with something and give up their extra energies in forms other than photons.

In practice, phosphorescence of atoms is rarely observed because the metastable state is deexcited via some other process. However, a process completely analogous to phosphorescence is commonly observed in nuclei.

Phosphorescent and fluorescent materials are frequently used to convert visually undetectable light into visible light. One common application is a phosphorescent lacquer coating that is applied to optical film to convert x rays into detectable light. Paints made from phosphorescent materials are used to coat watches and fire alarms, and to outline the doors and stairwells of large buildings to reveal the paths to exits in the case of power failure. A common example of phosphorescence occurring in nature is the glow of decaying logs in a damp forest.

See also: EXCITED STATE; FLUORESCENCE; GROUND STATE; LUMINESCENCE

Bibliography

EISBERG, R., and RESNICK, R. *Quantum Physics of Atoms, Molecules, Solids, Nuclei, and Particles,* 2nd ed. (Wiley, New York, 1985).

SERWAY, R. *Physics for Scientists and Engineers,* 3rd ed. (Saunders, Philadelphia, 1990).

SHANE STADLER

PHOTOCONDUCTIVITY

If a periodic potential due to ion cores is included in the quantum-mechanical description of electronic transport, gaps develop between allowed energy bands at the brillouin zone boundaries. These gaps provide a convenient way to explain the electronic behavior of metals, semiconductors, and insulators. In this band theory, the only way electrical current can flow is if at least one band is not completely full or completely empty. The valence band in an insulator is completely full and the conduction band is empty, thus no current can flow in an applied electric field. Semiconductors are insulators at absolute zero, but can conduct electricity at finite temperatures due to thermal excitation of electrons (negative charge carriers) across the band gap, leaving holes (positive charge carriers) in the valence band. Another way to increase the conductivity of a semiconductor (or insulator) is to expose it to photons with energies larger than the band gap. This effect is called intrinsic photoconductivity. When the light is turned off, the electron-hole pair may recombine by the electron returning directly to the valence band, or they may recombine indirectly, through localized defects with energies that lie within the band gap. These defects can also act as sources for electron-hole pairs, so that photons with energies less than the band gap can create electron-hole pairs. This is called extrinsic photoconductivity.

In general, conductivity σ is defined as

$$\sigma = nq\mu,$$

where n is the carrier concentration, q is the charge of the carrier, and μ is the mobility of the carriers, defined as the average drift speed per unit applied electric field. Any changes in conductivity $\Delta\sigma$ can be due either to changes in carrier concentration Δn or carrier mobility $\Delta\mu$. Although both can happen in the same material, photoconductivity in single-crystal materials is most often due to changes in the carrier concentration. If there are f excitations per volume per unit time, and τ is the lifetime of the photoexcited state, then

$$\Delta n = f\tau.$$

In polycrystalline materials there are potential barriers in the intergrain regions that limit electron transport, and photoconductivity can be dominated by photoinduced changes in mobility in the interface regions.

Applications

A photodetector is usually a slab of photoconductor with a constant voltage V applied along one direction. Changes in conductivity are then detected by monitoring the current in the circuit. If the electrodes attached to the photoconductor are a distance L apart, the gain G of the photodetector is given by

$$G = \frac{(\tau_e \mu_e + \tau_h \mu_h) V}{L^2},$$

where the subscripts e and h refer to electrons and holes, respectively. The gain is a measure of the number of charge carriers that flow through the circuit for each photoexcited charge carrier.

Some applications of photoconductivity include infrared detectors (e.g., "night-vision" cameras), video cameras (vidicons), light meters for photography, and electrophotographic reproduction (xerography). Materials are chosen for specific applications based on the size of the photoconductive effect (usually measured by the product $\tau \mu$), spectral response, and speed of response. Typical photoconductive materials include CdS, Ge, InSb, PbS, Si, and ZnS.

See also: CONDUCTION; ELECTRICAL CONDUCTIVITY; ELECTRON; ELECTRON, CONDUCTION; ELECTRON, DRIFT SPEED OF; METAL; QUANTUM MECHANICS; SEMICONDUCTOR; SOLID; SOLID, HOLES IN; TRANSPORT PROPERTIES

Bibliography

CHRISTMAN, J. R. *Fundamentals of Solid-State Physics* (Wiley, New York, 1988).

KEER, H. V. *Principles of the Solid State* (Wiley Eastern Limited, New Delhi, India, 1993).

ROSE, A. *Concepts in Photoconductivity and Allied Problems* (Wiley-Interscience, New York, 1963).

KEVIN AYLESWORTH

PHOTOELECTRIC EFFECT

The photoelectric effect is the process by which electromagnetic energy is absorbed and electrons are emitted when light is incident on a surface of material or when x and gamma rays propagate through a material. Specifically, the entire energy of each photon can be absorbed by an atomic electron, and the electron's final energy is equal to the photon's initial energy minus the electron's binding energy to the atom, molecule, or surface, called the work function. The equation relating photon energy (E_γ), final electron kinetic energy (E_e), and work function or binding energy (W) is

$$E_e = E_\gamma - W.$$

The photon energy is related to the frequency of the light in an important equation first expressed by Albert Einstein, who stated that monochromatic light can interact as individual quanta of energy:

$$E_\gamma = h\nu = \frac{hc}{\lambda},$$

where h is the constant discovered by Max Planck in his theory of blackbody radiation. The value of Planck's constant is $h = 6.626 \times 10^{-34}$ J·s. This formula defines the fundamental nature of photons as individual quanta of energy that make up light of a specific frequency, thus expressing the dual nature of light as particle and wave. The linear momentum of a photon has a magnitude of

$$|\boldsymbol{p}_\gamma| = \frac{h}{\lambda},$$

and conservation of energy and momentum for interactions involving photons is confirmed with the formulas for E_γ and $\boldsymbol{p}_\gamma$.

Light made up of a spectrum of different frequencies is equivalently made up of photons with an identical distribution of energies. The intensity of the light, power per unit area incident on a surface, and the intensity of photons, number per unit time per unit area incident on the surface, are related by

$$\frac{|S|}{h\nu} = \frac{n_\gamma}{A}.$$

The first observations of electrical effects caused by light are attributed to Edmond Becquerel in 1839, who observed the generation of current when sunlight illuminated a cell with a pair of electrodes immersed in an electrolyte. In 1873, Willoughby Smith measured the change in resistance of bars of selenium due to light. Photoemission, first reported by Heinrich Hertz in 1887, was observed during his research into the nature of electromagnetic waves. Subsequent experiments by Wilhelm Hallwachs, Julius Elster and Hans Geitel, Phillip Lenard, and J. J. Thomson led to the compilation of a series of observations consistent with the photoelectric equations. The crucial observations are as follows: (1) The emitted particles are negative and were shown to be electrons. (2) Different types of metal displaying the photoelectric effect differ in their sensitivity to the wavelength of incident light and the more electropositive metals are sensitive to longer wavelengths. (3) The effect produces a photocurrent that is proportional to the intensity of the light. (3) The effect produces a photocurrent that is proportional to the intensity of the light. (4) The emitted electrons have kinetic energy that is independent of the intensity of the light (in contrast to the prediction of the description of light as a purely wave-like phenomenon) but depends on the wavelength of the light. (5) Each material displaying the effect has a distinct threshold wavelength above which no electrons are emitted. (6) The delay between the time light is turned on and electrons are emitted is immeasurably small, and low intensity light produces electrons impulsively at random intervals with the average rate proportional to the light intensity.

As an illustration of the photoelectric effect, consider the setup illustrated in Fig. 1. Ultraviolet light from a mercury arc lamp of wavelength $\lambda = 253.6$ nm, frequency $\nu = c/\lambda = 1.18 \times 10^{15}$ Hz, and photon energy $h\nu = 7.82 \times 10^{-19}$ J $= 4.89$ eV is incident on the aluminum metal surface, and the photocurrent is measured. The work function of aluminum $W = 4.28$ eV $= 6.85 \times 10^{-19}$ J is less than $h\nu$. Thus photoelectrons will be emitted with final kinetic energy of $E_e = h\nu - W = 0.61$ eV $= 0.98 \times 10^{-19}$ J. For intensity such that the total power incident on the surface is $P = 10^{-3}$ W, the number of photons absorbed and electrons emitted per second is $n = P/(h\nu) = 1.28 \times 10^{15}$ Hz, and the negative photocurrent will be 2.0×10^{-4} A or 0.2 mA.

The photoelectric effect is present in many applications. Materials science studies of gases, liquids, and solids all make use of photoelectron spectroscopy or photoemission spectroscopy. Photoemission spectroscopy is a standard method in which instruments measure the final electron energies, intensities, and angular distributions to reveal the electron band structure and other properties of the material. Photoemission spectroscopy is important in the study of materials including high transition temperature (High T_c) superconductors. The atomic structure of heavy atoms and molecular structure are also studied with the photoelectric effect.

Photoelectric cells are made of materials such as selenium with work functions less than the energy of photons of visible light, and electrons are photoproduced with a few electron volts of final kinetic energy. In a photodiode, the photoelectric effect produces electron-hole pairs and measurable current proportional to the absorbed light power. Since the energy required to produce an electron-hole pair is small, photodiodes are sensitive to infrared light. Photodiodes are generally useful for high frequency and low noise applications. Photomultiplier tubes are used to detect weak intensities and even single photons of ultraviolet light. The photons incident on the negatively charged photocathode produce electrons, and the electrons are accelerated to hundreds of electron volts by a high voltage gap. When colliding with the metal dynode, many secondary electrons are produced. A tube with many dynode stages produces many electrons and a significant charge for each photon incident on the photocathode.

The photoelectric effect is also present when photons of many thousand electron volts (keV) to

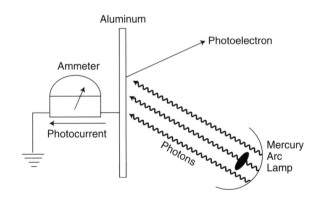

Figure 1 An illustration of the photoelectric effect.

millions of electron volts (MeV) propagate through matter. The absorption of the photon energy can result in the production of photoelectrons, which can be detected as they lose kinetic energy in the material. Gamma rays and x rays interact with matter in three ways: through the photoelectric effect, the Compton effect (scattering and transfer of energy from photon to electron resulting in a lower energy photon and energetic electron), and the production of electron-positron pairs. In gamma-ray and x-ray detectors, these interactions, combined with a way to detect the energy transferred to charged electrons, provide a signal that can be used to detect the photon and in many cases to measure its energy. The probability that the gamma ray or x ray will interact is represented by the photoelectric effect cross section, which depends on the binding energy of the emitted electron. In matter, electrons are bound in shells and for K-shell electrons

$$\sigma_{PE}^{K} = \frac{8}{3}\left(\frac{ke^2}{m_e c^2}\right)^2 \left(\frac{1.5\, m_e c^2}{E_\gamma}\right) a^4 Z^5,$$

where $\alpha \approx 1/137$ is the fine-structure constant, e and m_e are the charge and mass of the electron, and k is Coulomb's constant. Materials with the highest Z are

therefore most effective for absorbing energy. Figure 2 shows total cross sections for the interaction in lead of a photon by the photoelectric effect, Compton effect, and pair production, along with the total interaction cross section versus photon energy. The photoelectric effect is most important in the energy range of 0.1 to 1.0 MeV, below the pair production threshold of $2m_e c^2 = 1.022$ MeV. In Fig. 3, the distribution of energies of scintillation light produced by photons of energy 1.08 MeV in a sodium iodide scintillation detector is shown; 1.08 MeV is below the pair production threshold. The photoelectric effect is evidenced by the peak at the highest energy (labeled the photopeak). This peak corresponds to the final electron kinetic energy given by the full energy of the incident gamma rays minus the work function of a few electron volts. The Compton effect results in an electron of energy lower than the photopeak energy and a lower energy photon with the energies depending on the angle of the change in the photon direction. The maximum photon energy loss, and therefore maximum Compton electron kinetic energy, occurs for 180° (photon backscattering), resulting in the Compton edge labeled in the figure.

Albert Einstein introduced the hypothesis of light as quanta in a paper in 1905, the first of his famous series of four published that year (other papers covered statistical mechanics, Brownian motion, and special relativity). At the time, he was considering blackbody radiation and drew an analogy between the radiation and an ideal gas. This followed Planck's work in 1900 that produced an equation

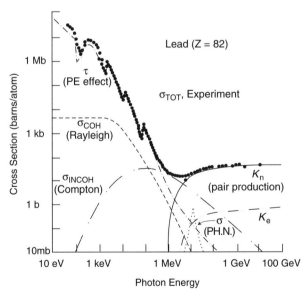

Figure 2 Total cross sections for the interaction in lead of a photon by the photoelectric effect, Compton, and pair production, along with the total interaction cross section versus photon energy.

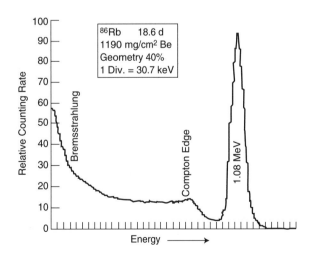

Figure 3 The distribution of energies of scintillation light.

describing emission of radiation from a blackbody. Planck was working to derive from first principles an equation describing all the data down to very low temperatures and long wavelengths. He was successful only after he introduced the unprecedented notion that the blackbody consisted of distinguishable material oscillators each with discrete energy $\epsilon = h\nu$, and that the oscillators' energies are distributed according to laws of thermodynamics and statistical mechanics. Einstein's ideal gas analogy led to consideration of the dependence of entropy on volume in the limit of low temperature, as described by the parametrization of blackbody radiation by Wilhelm Wien. This dependence of entropy on volume was just that for a collection of discrete energy quanta of energy $E = h\nu$, that is, photons. Einstein went on to extend the hypothesis to the interaction of light with material, including the photoelectric effect. This is the work for which Einstein was awarded the Nobel Prize in 1922.

The photoelectric effect played a major role in the emergence of the quantum theory. Einstein's relation of energy, a particle property, to frequency, a wave property, was the first presentation of the wave–particle duality that also underlies interpretations of quantum mechanics. How electromagnetic radiation is manifested depends on the nature of the particular observation or experiment. Wave properties such as interference are readily observable in experiments with electromagnetic radiation from the longest wavelength radio waves to the visible to far into the ultraviolet. Particle properties are revealed by the interaction of radiation with matter in which the energy, momentum, or angular momentum of photons is transferred to electrons and other elementary particles. This requires photons of a wavelength sufficiently short that the photon energy exceeds the electron binding energy or the work function of the material. However, the particular observations of blackbody radiation that led to the quantum hypotheses of Planck and Einstein are the features at the longest wavelengths or lowest energy.

The wave–particle duality applies to matter as well as radiation. The quantum nature of material particles reveal wave properties such as interference and diffraction. The quantum wavelength or de Broglie wavelength of a particle is related to its momentum by the identical relationship to that of a photon.

See also: BROGLIE WAVELENGTH, DE; COMPTON EFFECT; EINSTEIN, ALBERT; PHOTON; PLANCK, MAX KARL ERNST LUDWIG; RADIATION, BLACKBODY; WAVE–PARTICLE DUALITY; WORK FUNCTION

Bibliography

HUGHES, A. L., and DUBRIDGE, L. A. *Photoelectric Phenomena* (McGraw-Hill, New York, 1932).

KNOLL, G. F. *Radiation Detection and Measurement,* 2nd ed. (Wiley, New York, 1989).

ZWORYKIN, V. K., and RAMBERG, E. G. *Photoelectricity and Its Application* (Wiley, New York, 1949).

TIMOTHY E. CHUPP

PHOTOMETRY

A simple definition of photometry is that it is the measurement of light. In this definition, it must be recognized that light is that segment of the electromagnetic spectrum to which the human eye is sensitive. That is, photometry assigns special significance to the response of the human eye as a radiation detector. However, photometry is distinct from radiometry, which similarly measures radiation but assigns no special place to human perception. As a quick example, a green light-emitting diode (LED) emitting the same absolute power (in watts) as a red LED would appear as much as two orders of magnitude brighter, since the eyes' response drops drastically as the wavelength of light moves away from green light. This topic, the eyes' response to different wavelengths of light, is discussed further below.

Photometry, as a quantitative science, began in 1729 when Pierre Bouguer described a device to be used to compare the apparent brightness of two different light sources. It was a simple device consisting of three panels arranged in a T, with eye holes (covered with translucent paper) on either side of the leg of the T. The leg of the T served to direct the light from each of the two sources into each of the observer's eyes. Later developments of photometers had as their aim to increase the accuracy with which a human observer could compare the brightness of two sources. Comparing two light sources is generally the method of choice when using human observers as the detector (called visual photometry), since human observers are rather hopeless at making absolute judgments of brightness but are extremely adept at making comparisons. An experienced observer with well-designed equipment can accurately compare two sources within less than 1 percent. Inasmuch as human observers were the detectors throughout the eighteenth and most of the nineteenth centuries, refinements in photometry require that scientists consider not only such issues as the method of delivering the light to the observer's eye and the means of varying one light source to make the two appear identical (called balancing), but also such seemingly mundane matters as the observer's comfort (accuracy is compromised if the observer is fatigued).

As the accuracy with which brightness comparisons could be made increased, dissatisfaction with the candle, the early light standard, grew. Attempts to obtain a reproducible candle by particularly specifying parameters associated with it found little success. At first, various oil and gas lamps were used to supplant the candle standard. The first, devised by Bertrand Guillaume Carcel in 1800, was the Carcel lamp, in which a pump supplied colza oil (a plant related to cabbage) to the flame at a specified rate. Gas lamps, burning one of a number of hydrocarbons, were proposed and used throughout the nineteenth century. However, even with these new light standards, accuracy in photometry was still limited by the reproducibility of these standard light sources, so other standards were sought. Throughout the first half of the twentieth century (1908 through 1948), the light standard was a bank of carbon filament lamps maintained by the National Bureau of Standards, with similar lamps kept in Britain and France. In 1948, after a delay due to World War II, a standard based on the radiation of a blackbody was adopted. This standard was proposed in 1880 by Jules Louis Gabriel Violle, and adopted briefly from 1889 to 1909, although it was subsequently supplanted by the aforementioned bank of carbon filament lamps. The blackbody radiation of material maintained at the melting point of platinum is the current standard of light emission, although this primary standard is tedious and expensive to use, and current standards are maintained using a secondary standard, tungsten lamps calibrated from the primary standard.

Another product of technological advancement was physical photometry, in which the human observer is replaced by a photodetector of some type. One difficulty with human observers is that, even if all other variables can be controlled (variability of light source, etc.), no two human observers have exactly the same optical response to identical sources, particularly when the sources may be of different colors. In order to deal with this, the notion of a standard observer was introduced, and the visual response as a function of wavelength was tabulated and adopted in 1924 by the *Commission Internationale de l'Eclairage* (International Commission on Illumination). This is referred to as the relative spectral luminous efficiency of radiant energy, or (for obvious reasons) more simply as the visibility function. This is a peaked function with a maximum at 555 nm and a width of 100 nm (that is, the visibility function falls to about half its maximum value when the wavelength deviates by 50 nm from the wavelength of the maximum). However, this describes only vision in bright light, the so-called photopic vision, in which the response of the cones of the eye dominates. As the illumination decreases, the eye makes a gradual transition to scotopic vision, in which the rods of the eye dominate. This transition is truly gradual, in that it requires about five orders of magnitude change in illumination of the retina to complete the transition from purely photopic to purely scotopic vision. The visibility curve for scotopic vision is similar to that for photopic vision, but it is shifted about 50 nm toward the red. That is, in low light levels, things red appear brighter and things blue appear dimmer under the same illumination. This is the so-called Purkinje effect and is one of many difficulties associated with the use of human observers as detectors.

The response to these difficulties was to introduce, primarily in the twentieth century, various physical detectors, to the extent that visual photometry has been almost completely replaced by physical photometry. These physical detectors can be classified by the physical principle by which they convert incident radiation to another signal (typically a voltage or electrical current), those classifications being principally photoelectric, thermoelectric, and photochemical. Photoelectric devices include various devices in which an impinging optical signal is converted into a current or voltage. This could be a photovoltaic or photoemissive cell (the most important of which is the photomultiplier tube), or any of a number of semiconductor devices, such as photodiodes or phototransistors. Thermoelectric devices depend on the incident radiation causing a temperature elevation and include the thermocouple, which depends on the thermoelectric effect, the thermopile (simply a series of thermocouples designed to enhance a single thermocouple's very weak signal), and pyroelectric detectors, in which the response of a material's permanent electric polarization to changes in temperature serves to indicate changes in incident radiation. Photochemical detectors are those in which incident radiation causes a chemical change of some type. The most common of these detectors is photographic film, but photochemical detectors also include photochromic materials. Clearly, none of these detectors mimics the response of the human eye, and so, in order to use them in photometry (as opposed to radiometry), some correction must be imposed, either by introducing an optical filter before the detector so that the combination of the filter's transmission and the detector's response mimics the visibility function, or by resolving the detector's response to a signal as a function of wavelength and then multiplying that signal by the visibility function.

The fundamental unit in photometry is the luminous flux, which is analogous to the radiant flux, or power, emitted by a source. The conversion from power emitted to luminous flux is specified by the equation

$$F = K \int E(\lambda) V(\lambda) d\lambda.$$

In this equation, which William David Wright claimed summarized all of photometry, F is the luminous flux (whose units are lumens), $E(\lambda)$ is the power (in watts) emitted by the source as a function of wavelength, $V(\lambda)$ is the visibility function, and K, whose units are lumens per watt, is the radiation luminous efficacy at the peak of the visibility function. Derived units are the luminous intensity, defined as the luminous flux per unit solid angle, whose unit is the lumen per steradian ($lm \cdot sr^{-1}$), which is also called the candela; illuminance, which is the luminous flux per unit area ($lm \cdot m^{-2}$), or lux; and luminance, which is the luminous flux per unit area and per unit solid angle ($lm \cdot sr^{-1} \cdot m^{-2}$), or the candela per square meter ($cd \cdot m^{-2}$). The value of K, 683 $lm \cdot W^{-1}$ for photopic vision and 1,725 $lm \cdot W^{-1}$ for scotopic vision, is obtained from the definition that the primary standard, a blackbody radiator at the melting temperature of platinum has a luminance of 60 $cd \cdot cm^{-2}$, or 6×10^{-3} $cd \cdot m^{-2}$. For comparison, the Moon or a candle have a luminance of about 0.5 $cd \cdot cm^{-2}$, a 100 W tungsten lamp has a luminance of about 20 $cd \cdot cm^{-2}$, and a sodium vapor lamp has a luminance of about 600 $cd \cdot cm^{-2}$.

A great many other derived units occur in the field of photometry. They are related to the geometry of the eye, its accommodation to light, and so forth.

See also: ENERGY, RADIANT; LIGHT; RADIATION, BLACKBODY; SI UNITS; THERMOELECTRIC EFFECT; VISION

Bibliography

BARROWS, W. E. *Light, Photometry, and Illuminating Engineering* (McGraw-Hill, New York, 1951).

McCLUNEY, W. R. *Introduction to Radiometry and Photometry* (Artech House, Boston, 1994).

PALAZ, A. *A Treatise on Industrial Photometry with Special Application to Electric Lighting*, trans. by G. W. Patterson and M. R. Patterson (New York, Van Nostrand, 1896).

STIMSON, A. *Photometry and Radiometry for Engineers* (Wiley, New York, 1974).

STINE, W. M. *Photometrical Measurements and Manual for the General Practice of Photometry* (Macmillan, New York, 1900).

WALSH, J. W. T. *Photometry* (Constable and Company, London, 1953).

WRIGHT, W. D. *Photometry and the Eye* (Hatton Press, London, 1949).

DAVID NORWOOD

PHOTON

Photons are particles of light (or, more generally, of electromagnetic radiation). They represent the twentieth-century resolution to an ancient debate about the nature of light: Is it a stream of particles ("corpuscles") as Isaac Newton believed, or is it a wave phenomenon, as Christiaan Huygens and others argued? Thomas Young's double-slit interference experiment (1801) appeared to settle the question in favor of the wave theory. In the 1860s James Clerk Maxwell derived all of classical optics from the laws of electricity and magnetism, providing rigorous theoretical justification for Michael Faraday's speculation that light is an electromagnetic wave, and in 1888 experiments by Heinrich Hertz confirmed Maxwell's theory. By the end of the nineteenth century, then, the wave nature of light was firmly established.

But in 1900 Max Planck discovered that he could account for the spectrum of blackbody radiation (the electromagnetic radiation emitted by a hot object) if he assumed that it came in tiny squirts (or quanta), with energy

$$E = h\nu, \tag{1}$$

where ν is the frequency of the radiation, and

$$h = 6.63 - 10^{-34}\,\mathrm{J\cdot s} \tag{2}$$

is Planck's constant (the numerical value of which he chose to fit the experimental data). Planck did not claim to know *why* thermal radiation is quantized—he assumed it was a peculiarity of the mechanism by which the atoms generate the radiation. But in 1905 Albert Einstein suggested a far more radical interpretation: It is in the nature of light itself that it comes in little packages. According to Einstein, the quantization law [Eq. (1)] has nothing to do with the emission process and is not limited to blackbody radiation—rather, it is a universal feature of all electromagnetic radiation.

On this assumption, Einstein produced a beautifully simple theory of the photoelectric effect. When light (or other electromagnetic radiation) strikes a piece of metal, electrons are emitted. In Einstein's analysis, the energy of one light quantum is absorbed by an electron in the metal, which therefore emerges with energy

$$E \leq h\nu - w, \tag{3}$$

where w (the so-called work function of the material) is the amount of energy the electron loses in breaking through the surface. (Because of collisions, the electron may lose some energy before reaching the surface—hence the inequality.) Einstein's formula carries the astonishing implication that the maximum electron energy depends only on the frequency of the light, and not at all on its intensity (a more intense beam would generate more electrons, but their energy would be the same). Einstein's theory came perilously close to reviving the old discredited model of light as a stream of particles, and it was greeted with disbelief and even ridicule. During the following decade, Robert A. Millikan conducted an exhaustive series of experiments to test Eq. (3), at the end of which he was obliged to report that "Einstein's photoelectric equation . . . appears in every case to predict exactly

the observed results. . . . Yet the semi-corpuscular theory by which Einstein arrived at his equation seems . . . wholly untenable" (Pais, 1982, p. 357).

Meanwhile, Niels Bohr's model of the hydrogen atom (1913) proposed that light is emitted when an electron undergoes a "quantum jump" from one allowed energy level (E_i) to another (E_f). According to Eq. (1), the frequency of the radiation is

$$\nu = (E_i - E_f)/h. \tag{4}$$

Bohr's theory lent implicit support to Planck's notion that the quantization of light is attributable to the emission mechanism, and for the next ten years Bohr joined Planck, Millikan, and many others, in disparaging Einstein's view.

It was not until Arthur Compton's experiments in 1923 that the quantum of light was taken seriously by most physicists. In the Compton effect, light scatters off a charged particle and emerges traveling in a new direction with a new frequency. Compton analyzed the problem as an ordinary collision process in which the particle of light carries energy given by Planck's formula, and momentum $p = E/c$ (as required by special relativity for a particle with zero rest mass). Conservation of energy and momentum yields a formula relating the change in wavelength to the scattering angle:

$$\lambda' = \lambda + (h/mc)(1 - \cos\theta), \tag{5}$$

where λ is the incident wavelength, λ' is the scattered wavelength, m is the (rest) mass of the target particle, θ is the scattering angle, and c is the speed of light. Compton's formula perfectly fit the data in his experiments, and the fact that it is based on treating light as an ordinary particle finally convinced the scientific community that Einstein had been right all along. The term "photon" for the Planck/Einstein light quantum was introduced by Gilbert Lewis in 1926.

How are we to reconcile the photon picture with the well-established wave nature of light? It is often said that light sometimes behaves as a wave, and sometimes as a particle. This wave–particle duality, which Bohr elevated to the status of a cosmic principle (complementarity), makes light sound like an unpredictable adolescent who sometimes acts like an adult, and at other times, for no particular reason, like a child. Nothing so arbitrary or mysterious is involved here. When the interaction of light with individual particles is in question (as in the photoelectric effect or the Compton effect) its own particle nature cannot be ignored. But when we are concerned with the collective effect of enormous numbers of photons (as in Young's experiment), quantization is ordinarily irrelevant, and the classical wave theory is perfectly adequate. Of course, you *can* analyse even the double-slit experiment using the photon picture—indeed, it is possible to set things up with an incident beam so weak that only a single photon is passing down the line at any given moment. But the spot on the screen where a particular photon will land cannot be predicted, except in a statistical sense, and it is only after many photons have left their marks that the characteristic interference pattern begins to emerge.

Although Planck's formula [Eq. (1)] was the first step in the tortuous path to quantum mechanics, the light quantum/photon belongs more appropriately to quantum field theory. The photon has zero rest mass and it travels (of course) at the speed of light; there is no such thing as a nonrelativistic photon, and it is awkward, at best, to incorporate photons into the nonrelativistic quantum theory introduced by Erwin Schrödinger and Werner Heisenberg in the mid-1920s. The relativistic quantum theory of photons began with P. A. M. Dirac's equation in 1928, but it turned out to be an extraordinarily difficult problem, which was not completed until the late 1940s with the work of Richard Feynman, Julian Schwinger, and Sin-itiro Tomonaga.

Newton's law of universal gravitation contemplates action-at-a-distance: The earth exerts a force on the Moon, but there is absolutely nothing in between to transmit the force from one to the other— no invisible "chain" connecting the two. Newton was uneasy with the notion of action-at-a-distance and would no doubt have preferred Faraday's field formulation (Faraday was thinking about electric and magnetic forces, not gravitational ones, but the idea is the same): We imagine that the earth sets up a gravitational "field" in its vicinity (its gravitational "odor," if you will), and it is this *field* that exerts a force on the Moon. The field is the "mediator," communicating the influence from one object to another. In Maxwell's electrodynamics, the electric and magnetic fields became dynamical entities in their own right, even though we cannot see or touch them. With the advent of quantum mechanics, the fields must be quantized, and the quantum of the electromagnetic field is precisely the photon. In

quantum electrodynamics (QED), then, the photon is the mediator of electromagnetic forces. This interpretation is the basis for Feynman's diagrammatic formulation of QED.

The modern theory of elementary particles (the standard model) self-consciously mimics QED. There are four kinds of forces: strong, electromagnetic, weak, and gravitational, and each is mediated by gauge bosons: gluons (there are eight of them) for the strong interactions, the photon for electromagnetic interactions, the W^+, W^-, and Z^0 for weak interactions, and the hypothetical graviton for gravity. If Planck's quantum was the father of the photon, the gluon, W, Z, and graviton are its offspring.

See also: ACTION-AT-A-DISTANCE; COMPLEMENTARITY PRINCIPLE; COMPTON EFFECT; DOUBLE-SLIT EXPERIMENT; ELECTROMAGNETIC WAVE; ELEMENTARY PARTICLES; FEYNMAN DIAGRAM; INTERACTION, ELECTROMAGNETIC; PHOTOELECTRIC EFFECT; PLANCK, MAX KARL ERNST LUDWIG; QUANTUM ELECTRODYNAMICS; RADIATION, BLACKBODY; WAVE–PARTICLE DUALITY, HISTORY OF; WORK FUNCTION

Bibliography

BAIERLEIN, R. *Newton to Einstein: The Trail of Light* (Cambridge University Press, Cambridge, Eng., 1992).

FEYNMAN, R. P. *QED: The Strange Theory of Light and Matter* (Princeton University Press, Princeton, NJ, 1985).

PAIS, A. *Subtle Is the Lord* (Clarendon Press, Oxford, Eng., 1982).

DAVID GRIFFITHS

PHOTOSPHERE

Stars are composed of gas (both ionized and neutral) and have no solid surfaces. What appears to be a surface on the Sun is actually a region hundreds of kilometers thick called the photosphere. In the interior of a star like the Sun, the gas is sufficiently opaque that light emitted at one location is reabsorbed at another, typically only a centimeter away. As the outer regions of the star are approached, the material density drops and the gas becomes less opaque. Eventually, the gas becomes sufficiently transparent for visible light to escape, resulting in an apparent surface. Depending on the type of star, the thickness of this photospheric region ranges from hundreds to thousands of kilometers. In an individual star, there is approximately a 20 percent temperature range from the top of the photosphere to the bottom. In the coolest stars, the photospheric temperatures are near 3,000 K, while hot stars may have temperatures of more than 30,000 K.

The gas that forms the photosphere does not stop simply because it becomes more transparent to visible light. It reaches outward for some distance, smoothly transitioning to a cooler region referred to as the stellar atmosphere. In accordance with Kirchoff's laws, as the temperature decreases (going outward through the photosphere and atmosphere) absorption lines form in the cooler material. The observed spectrum of a star arises from the absorbtion and re-emission process occurring through these regions. While the overall spectrum (from the far ultraviolet to the infrared) of the star resembles a blackbody, in localized wavelength regions these processes cause the emergent spectrum to differ considerably from a Planckian.

The important absorption and emission mechanisms in the photosphere include bound-free transitions from ionized atoms and from the negative hydrogen ion (H^-). These mechanisms provide the continuous opacity sources that make the photosphere. Bound-bound transitions from atoms and molecules arise in the cooler gases above the photosphere, and provide more information on the star itself.

The most basic information provided by the spectral absorption lines is evidence for the stellar composition. Early spectroscopists attempted to classify stars based on the strength of the Balmer lines of hydrogen (which lie in the visible part of the spectrum and arise from the first excited state of atomic hydrogen). Stars with the strongest Balmer lines were assigned an A classification, B stars were the weaker, and so on. Some of these classifications (O, B, A, F, G, K, M) survive today as a temperature sequence (O is the hottest). In cool M stars, most of the hydrogen is in molecules, and the temperatures are too low for much of the remaining atomic hydrogen to be in the necessary excited state. In warmer stars, more of the hydrogen is in the excited state, resulting in stronger absorbtion lines. Above the 10,000 K temperature of A-type stars, hydrogen begins to ionize, and again the fraction of hydrogen atoms in the correct excited state drops.

With the work of Celia Payne (Gaposchkin) in the 1930s, stars were found to consist principally of hydrogen and helium, while containing trace amounts of most of the other elements. Like hydrogen, the appearance of these elements depend on their ionization and excitation states. That, in turn, depends on the temperature and density of the photosphere, as well as the atomic structure of each element. Analyses of the spectral absorbtion lines, using the equation of radiation transport (an equation relating changes in the intensity to the emission, absorption, and scattering processes) permit the determination of the abundance of each ionization and excitation state of observed elements. Combining this information with the Saha (ionization equilibrium) and Boltzman (excitation equilibrium) equations permits the determination of elemental abundances, as well as a self-consistent temperature structure of a star's photosphere and atmosphere.

In addition to the composition, the pressure and the temperature structure of the stellar atmosphere, measurements of the Doppler shifts of the absorbtion lines provide astronomers with information on a star's velocity. Stellar velocities provide important evidence on the dynamics of the Milky Way. Velocity measurements of binary stars (in which two stars orbit each other) also allow the determination of the actual masses of individual stars. Determinations of the actual masses, temperatures, compositions, and luminosities of stars provided the basic data on which the theory of stellar evolution is based.

See also: DOPPLER EFFECT; MILKY WAY; STARS and STELLAR STRUCTURE; STELLAR EVOLUTION

Bibliography

ALLER, L. H. *Astrophysics: The Atmospheres of the Sun and Stars,* 2nd ed. (Ronald Press, New York, 1963).

MIHALAS, D. *Stellar Atmospheres,* 2nd ed. (W. H. Freeman, San Francisco, 1978).

DAVID S. P. DEARBORN

PHYSICS

See ASTROPHYSICS; ATMOSPHERIC PHYSICS; ATOMIC PHYSICS; BASIC, APPLIED, AND INDUSTRIAL PHYSICS; BIO-

PHYSICS; CHEMICAL PHYSICS; COMPUTATIONAL PHYSICS; CONDENSED MATTER PHYSICS; EXPERIMENTAL PHYSICS; GEOPHYSICS; HEALTH PHYSICS; MEDICAL PHYSICS; MOLECULAR PHYSICS; NONLINEAR PHYSICS; NUCLEAR PHYSICS; PARTICLE PHYSICS; PLASMA PHYSICS; RADIATION PHYSICS; THEORETICAL PHYSICS

PHYSICS, HISTORY OF

As any other science, physics is a social and intellectual activity that has evolved over time, so it has a history. The history of physics—in the sense of historical studies of the past of physics—has been an integral part of physics since its very beginning, but in different ways. For a long time "history" was considered a natural part of physics, but in a different sense than the term was later understood. For example, the eighteenth-century chemist and physicist Joseph Priestley wrote elaborate "histories" of electricity and optics that were stocktakings of the progress that had taken place in these areas and intended to serve the needs of contemporary scientists. It was only in the nineteenth century that a different conception of the sciences originated, one in which science was seen as a historical phenomenon with a past that is interesting and instructive in its own right. Ernst Mach, the physicist and philosopher, wrote important works on the developments of mechanics, optics, and the science of heat that were critical reviews of the foundations of these sciences. Mach's philosophically oriented history of physics is a genre that has continued to attract interest among historians, philosophers, and physicists. For example, Einstein considered Mach's works the ideal of history of science. History of physics before 1920 was almost exclusively written by physicists, who described the development of their science as a steady progress toward truth, with little attention to the complex social and political conditions surrounding its development. In this view, scientific progress was the result of a small number of great scientists' brilliant thoughts and experiments.

During the period between the two world wars, scholars such as George Sarton and Alexandre Koyré formulated a new program for the history of science as a branch of intellectual history that studied the development of scientific ideas in the context of their own times. This program only became a reality in the 1950s, when history of science received new interest and began to develop into a professional body of scholarship separate from both the sciences and general history. This development started in the United States and was only slowly and incompletely followed in Europe and elsewhere. Since then, history of science has become established as a relatively strong academic discipline, often with its own department or program. Today, history of science is cultivated on an international scale as an autonomous field, although often associated with other fields such as sociology of science, history of technology, and science studies. In this entire development, history of physics has played an important, indeed, often dominant role. Although the histories of the non-physical sciences (biology, geology, etc.) have become increasingly popular, physics has still a central position in modern history of science. Whereas history of physics is part of the more comprehensive history of science, basically a humanistic discipline, it is also related to physics and its institutions. Many national and international physical societies have divisions of history of physics, and in some cases research in history of physics is done within physics departments rather than in departments of history or interdisciplinary studies. Additionally, the American Institute of Physics Center for History of Physics has served as a model for other scientific fields by sponsoring a broad program aimed at preserving correspondence, photographs, books, and other unpublished and published materials and fostering their use by historians, physicists, educators, journalists, and other scholarly and professional groups.

Why is history of physics considered an important study? To many professional historians of science, it is important because it is part of a larger historical process, the development of science and its interactions with social, economic, and political forces. From this perspective, *history* of physics is emphasized, not history of *physics*. Physics, being a science that addresses fundamental scientific questions and is of central importance for the world picture, is part of modern culture; insofar as this culture is a historical product (which it is), the only way to appreciate physics as a bearer and expression of culture is through its historical development. To give an example, quantum mechanical indeterminacy is often discussed in areas outside physics and conceived to have fundamental philosophical implications. To obtain a full understanding of the meaning of this principle, its emergence and development have to be studied historically. But physics is also, and even more visibly, important in its function as producer of knowledge of potential technological significance. This function was particularly important during the last half of the twentieth century, as witnessed by the atomic bomb, the laser, the transistor, and satellite television. How were these devices invented? How were they related to advances in physical knowledge? What were their economic and political implications? Again, such questions can only be answered satisfactorily by a historical study of physics, although not of physics alone. In other words, because of the great importance physics has exerted on modern society, both culturally and economically, history of physics is a key to a better understanding of our present situation.

History of physics is used in a variety of other, less abstract ways. It is an important element in policy-making, where new science and technology projects are often justified—or criticized—by reference to historical experiences. In general, the subject plays a significant, if mostly indirect, role in arguments for funding of basic physical science. Does history not show that important technologies may arise from pure physics of no apparent practical importance? Would radio and television not be inconceivable without theoretical physicists' fundamental studies of electrodynamics in the nineteenth century? In serious history of physics, the exact (but complex) relationship between physical knowledge and technological advances is a prime area, and so is the physicists' use of historical arguments for their own purposes.

Some have argued that the sciences would benefit directly from historical knowledge, that the historically knowledgeable scientist would be in a better position to make new discoveries. Although this belief may be questioned, history of physics does play an important role in the development of physics. It has for a long time been recognized that the historical approach offers certain advantages in the teaching of physics. In traditional teaching, physics is presented in a highly technical and abstract sense, as a timeless collection of factual knowledge and recipes needed to solve problems. By including a historical perspective, a truer picture of physics emerges: a science in flux, the current state of which is the historical result of the struggle of individuals and groups to unravel the secrets of nature. In addition to this human dimension, the historical approach also provides an opportunity to

deal in a more authentic and exciting way with many of the classical results of physics and to incorporate into the curriculum wrong theories and failed experiments. This approach has especially been advocated for nonscience majors and as a means to make physics more relevant and interesting to women and other groups that have been traditionally underrepresented in the subject. By portraying physical research as a human activity, it can be more appealing to these groups.

History of physics also plays a role in physicists' general understanding of their science, including its philosophical foundations. The traditional "logical" approach to philosophy of science has been joined by a more "historical" one that builds directly on the history of science. For most modern philosophers of science, history of science is an indispensable resource. Because of the fundamental nature of physical knowledge, the history of physics is particularly important in this respect.

Historians of physics generally write for an audience of their peers, although their works are also read by physicists and the general public. Some scholarly works have succeeded in winning a large audience and thus have had a much greater effect than usual. Biographies are important in this respect because they often have a readership far beyond narrow academic circles. The success of Abraham Pais's biography of Albert Einstein, *Subtle is the Lord* (1982), shows that there is an audience for seriously researched history of physics. In a different way, Thomas Kuhn's enormously popular *The Structure of Scientific Revolutions* (1962) demonstrates the same thing. Historians of physics also have served as editors of the collected papers of Isaac Newton, Joseph Henry, Albert Einstein, Niels Bohr, and other distinguished physicists, and in this way they have laid the groundwork for much future historical research and writing.

The modern historiography of physics has resulted in a very large literature, which covers almost any conceivable aspect of the development of physics. Yet there are some areas that are more popular than others. The Scientific Revolution in the seventeenth century, with giants such as Johannes Kepler, Christiaan Huygens, and Newton, has traditionally been a central area and continues to attract much attention. The same is the case for the great achievements in the nineteenth century by Michael Faraday, Nicolas Carnot, Lord Kelvin, James Clerk Maxwell, Hermann L. F. von Helmholtz, and others. Until about 1970, twentieth-century physics was an underdeveloped area, but this situation has changed drastically, and today we have a fairly detailed knowledge of how the theories of relativity and quantum mechanics, both pure and applied, have developed. At the same time as there has been a general shift toward more modern physics, there has been a broadening of perspectives and an increased interest in the social and institutional aspects of physics. It has long been recognized that these aspects were of crucial importance during the Scientific Revolution. They have become even more important in modern times when some areas of physics—notably high-energy physics and space science—have become big science. As a response to this development, there has been a strong interest in the interplay between physics, politics, the economy, and the military. The areas of physics traditionally cultivated by historians have been the great theories such as mechanics, electrodynamics, thermodynamics, and relativity. More recently, experimental and laboratory physics have attracted increasing attention. Physical science in a broader sense, including cross-disciplinary fields such as chemical physics, astrophysics, geophysics, and medical physics, has received much less historical attention and is in need of further research.

Who has the "right" to write the history of physics, professional historians of physics or historically interested physicists? There can be a certain tension between the two groups, which may have different conceptions of how history should be written and to whom it should be addressed. Some historians maintain that physicists' accounts of their own history are too internalistic and lack a broad historical perspective. Although examples of such histories can be found, the criticism has no general validity. In fact, much of the best history of physics has been written by physicists with little or no historical training or by historians who originally were professional physicists. History of physics does not necessarily require a training in history, and neither does it necessarily require an advanced degree in modern physics. The literature proves that good history of physics can be written from many different perspectives and thus "belongs" to neither historians nor physicists. Cooperation between the two professional groups has aided both in the past and will be of benefit to both in the future.

See also: CARNOT, NICOLAS-LÉONARD-SADI; EINSTEIN, ALBERT; FARADAY, MICHAEL; HELMHOLTZ, HERMANN L. F. VON; HUYGENS, CHRISTIAAN; KELVIN, LORD; MACH, ERNST; MAXWELL, JAMES CLERK; NEWTON, ISAAC; PHYSICS, PHILOSOPHY OF; RÖNTGEN, WILHELM CONRAD; RUTHERFORD, ERNEST; SOCIETY, PHYSICS AND

Bibliography

BRUSH, S. G. *The History of Modern Science: A Guide to the Second Scientific Revolution, 1800–1950,* (Iowa State University Press, Ames, 1988).

BRUSH, S. G., ed. *History of Physics: Selected Reprints* (American Association of Physics Teachers, College Park, MD, 1988).

COHEN, I. B. *Birth of a New Physics* (W. W. Norton, New York, 1985).

JUNGNICKEL, C., and McCORMMACH, R. *The Intellectual Mastery of Nature: Theoretical Physics from Ohm to Einstein* (University of Chicago Press, Chicago, 1986).

KRAGH, H. *An Introduction to the Historiography of Science* (Cambridge University Press, Cambridge, Eng., 1987).

KUHN, T. *The Structure of Scientific Revolutions* (University of Chicago Press, Chicago, 1962).

PAIS, A. *Subtle Is the Lord* (Clarendon Press, Oxford, Eng., 1982).

SEGRÉ, E. *From X-Rays to Quarks: Modern Physicists and Their Discoveries* (W. H. Freeman, New York, 1980).

WHITAKER, M. "History or Quasi-History in Physics Education." *Physics Education* **14,** 108–112 (1979).

HELGE KRAGH

PHYSICS, PHILOSOPHY OF

Most people today might see physics and philosophy as being about as far apart as two subjects could possibly get. However, until fairly recently, those we now call scientists were known as natural philosophers. Furthermore, many of the generally recognized great physicists—such as Isaac Newton, Max Planck, Albert Einstein, Niels Bohr, Max Born, Erwin Schrödinger, and Werner Heisenberg—were consciously and explicitly interested in philosophical issues related to their creative and ground-breaking work in physics. For example, in the beginning of Book III of his *Mathematical Principles of Natural Philosophy* (1687), the work that provided the foundation for the modern science of classical mechanics, Newton set forth quite explicitly his rules for reasoning in natural philosophy (that is, in science). It is only in modern times that the traditional natural sciences, like physics, chemistry, and biology, have become separated by a great divide from liberal-arts subjects, like philosophy, theology, history, and literature.

In Western tradition, we find some of the earliest systematic discussion of the nature of scientific knowledge among the works of Plato and of Aristotle. While *philosophy* denoted the disciplined study of reality and of human nature, Plato distinguished between mere opinion and science. For him, *science* was the ideal of human knowledge and was seen as necessarily true and unchanging. This can be paraphrased as absolute truth, pure and simple. Axiomatic geometry (of the type typically studied in high school today) provides an example of such a system of knowledge. For Plato, true knowledge had to be real, stable, and unchanging. Belief had to do with mere appearances. Knowledge was infallible, while belief might be true or false. Plato posited a reality far removed from immediate sense experience (realism of natures), while his pupil Aristotle assigned primary reality to immediate sense experience (realism of things). What we are most immediately aware of, Aristotle held, are our sense impressions caused by the phenomena of nature, and these are confused and do not appear to be organized in any obvious fashion. What the mind desires, however, is a comprehension of those basic principles (e.g., laws or theories) that bring order to this myriad of facts and, in some sense, explain them. According to Aristotle, all certain knowledge is gained through logical demonstration proceeding from such true and necessary first principles, whereas these first principles are abstracted from sense experience through observation.

Hence, the goal of science, even from these earliest times, has been to discover those general laws or principles in terms of which one can understand the physical phenomena that we perceive with our senses. Just as Plato and Aristotle differed over the relative emphasis to be placed on abstractions (first principles) versus sense perceptions (observations), so scientists and philosophers throughout the ages have located themselves at various positions between these extremes. This has been one of the great tensions or recurrent themes that has been a focus of philosophical reflections on the nature of scientific knowledge. Thus, rationalists, such as the great French mathematician and philosopher René Descartes, relied mainly upon reason as the source of knowledge, while empiricists relied more upon experience, through observation and experiment, as the basis of knowledge. For instance, Francis Bacon,

a famed advisor and essayist in Elizabethan England, placed more stress on careful induction, which is the operation of discovering and proving general propositions from specific observations, than did Aristotle. That is, he advocated a slow upward advance from many particular instances or observations to small generalizations with repeated checks against new phenomena or situations until one could arrive at very broad generalizations. In Bacon's opinion, Aristotle and his followers had too often gone from scant observations to great generalizations. Although he is often cited as having laid the foundations of modern science with his emphasis on induction to arrive at the laws of nature, the successors of Bacon were not the scientists of the seventeenth and later centuries—such as Galileo and Newton—but the inductive logicians like David Hume and John Stuart Mill, men more concerned with the philosophical problems associated with induction than with constructing specific scientific theories.

Hume, an eighteenth-century British philosopher, held that philosophy is basically an inductive, experimental science of human nature. He was particularly concerned with the notion of causality, the validity of induction, and the necessity of sense data preceding ideas. In analyzing the concepts of cause and effect, Hume pointed out that what we actually observe is only one event *following* another, rather than a *necessary connection* between them. From the occurrence of such constant conjunction of events, we *infer*, as a habit of mind, a necessary connection in terms of cause and effect. In essence, Hume asked what reason anyone had to suppose that future observations would resemble past observations. He challenged the general validity of universal, predictive laws induced from observed facts. As a well-known yet instructive example, from the fact that the Sun has risen every day in the past since the beginning of recorded history, we might induce the "law" that the Sun necessarily rises periodically *without exception*. Can we, however, be *absolutely certain* that the Sun will, in the future, rise as scheduled? No. There is no way around the provisional character of physical laws gained by induction. Hume believed in the necessity of induction, but felt its validity to be *unprovable*.

Mill, a later British philosopher, also wrote extensively on the question of induction. He took *the main problem of the science of logic* to be a justification of the process of induction. Involved in all inductions is the assumption that the course of nature is uniform or that the universe is governed by general laws. This uniformity of the course of nature is a premise (or first principle) that can be used to justify induction. Neither Mill nor anyone else has proven the correctness of this premise so that it remains an assumption, albeit one that few would dispute and without which we cannot do science.

Intertwined historically with, even though conceptually distinct from, this theme of rationalism versus empiricism is another long-standing debate: instrumentalism versus scientific realism. The question here is whether the goal of science is to discover (or, perhaps, to create) laws and theories that allow us to calculate the values of quantities that can be observed (i.e., theories as mere instruments for calculation) or whether successful scientific theories must also give us a realistic and reliable representation of the actual working of the physical universe (i.e., a true picture of the world). This was a central issue in the struggle between the old Earth-centered model of the universe and the Sun-centered one of Copernicus. Galileo defended the Copernican model as a true picture of the heavens. He, like a majority of working scientists still today, was a realist. On the other hand, Hume and Mill fall into the instrumentalist camp. In a similar spirit, the German physicist and philosopher Ernst Mach sought to analyze the logical structure of classical mechanics in order to eliminate unobservable quantities from the laws of nature. This became a hallmark of the program of the logical positivists in the early part of the twentieth century. For these scientists and philosophers of science, no concepts or entities were to be admitted into a theory unless they could be defined in terms of a procedure for measuring them (i.e., physically observable quantities only).

A major figure in laying the foundation for what would become the history and philosophy of science was the Cambridge scholar William Whewell, who wrote monumental treatises on the history and philosophy of the inductive sciences. As he demonstrated in these works, a sound foundation in the history of actual scientific practice is necessary for doing useful philosophy of science. It is for this reason that the history and philosophy of science can give a student a holistic perspective on science as a human creation, as opposed to being seen as some monolithic enterprise that seeks and unerringly grinds out truth. In a tradition similar to Whewell's, other scientists, such as the French mathematician Henri Poincaré and Einstein realized that there is an essential and ineliminable element of convention or of human creativity, and hence some degree of arbitrariness, in all general theories formulated by scientists. Einstein also appreciated that, although theoretical physics seeks general laws from which the phenomena of nature can be deduced, there is no direct and logically compelling path from the phenomena to these laws so that a creative act of insight and a choice of basic concepts and language (akin to Poincaré's conventionalism) are required to produce general theories.

Formal philosophy of science, as a separate branch of inquiry, had taken hold by the early part of the twentieth century. An early influential school, the logical positivists, attempted to demarcate what made scientific knowledge essentially different from knowledge generated in other fields of human intellectual activity. They sought this difference in the *method* that science uses to arrive at its certain, or at least highly reliable, knowledge. Even though it seems fairly evident that science *is* different from, say philosophy or art, the more closely one looks at the creative process of fashioning fundamental laws, the more difficult it becomes to state precisely in what this difference consists. Not only is there no general way to do this, it is unclear that one can even specify consistently some quantitative measure of confidence in the approximate correctness of purported laws of science. The Austrian-born British philosopher of science Karl Popper used this fact to characterize scientific knowledge as *falsifiable*. That is, truly scientific theories, unlike systems of religious or social dogma, make specific predictions about natural phenomena (as does Einstein's general theory of relativity, to use a favorite example of Popper's) that are often unexpected and that *might* be refuted (or falsified) by new observations. Successful scientific theories survive many such rigorous tests, but we still do not *know* that they are true—only that they have not yet been shown to be false. In other words, the goal of truth, or even of approximate truth, is an unattainable one for science. More recently, the sociological component of scientific practice has become another focus of attention since the appearance in 1962 of the influential book *The Structure of Scientific Revolutions* by the American philosopher of science Thomas Kuhn.

So far, only those rather broad philosophical issues that are concerned with a theory of knowledge have been discussed. However, much modern philosophy of science is also devoted to the critical examination of specific theories, such as relativity and quantum mechanics. (There are, of course, other branches of the philosophy of science centered on different sciences such as biology.) Much of this type of analysis of scientific theories is concerned with the implications that our successful and accepted scientific theories have for our view of the physical world. For example, quantum mechanics, with the essential role that it seems to assign to the act of observation, is often interpreted as posing a serious problem for scientific realism. Similarly, it has become clear from an analysis of the experimentally confirmed predictions of quantum mechanics that the actual physical world in which we live cannot have an objective reality that is independent of the act of observing it *and* that is also free from instantaneous action-at-a-distance. The huge literature surrounding what is generally referred to as Bell's theorem is largely the work of philosophers of science and represents a remarkable insight into the quite surprising and unexpected nature of physical reality.

See also: ACTION-AT-A-DISTANCE; BELL'S THEOREM; BOHR, NIELS HENRIK DAVID; BORN, MAX; COPERNICUS, NICOLAUS; EINSTEIN, ALBERT; GALILEI, GALILEO; HEISENBERG, WERNER KARL; MACH, ERNST; NEWTON, ISAAC; PHYSICS, HISTORY OF; PLANCK, MAX KARL ERNST LUDWIG; SCHRÖDINGER, ERWIN

Bibliography

BOHR, N. *Atomic Physics and Human Knowledge* (Cambridge University Press, Cambridge, Eng., 1934).

BORN, M. *Natural Philosophy of Cause and Chance* (Oxford University Press, Oxford, Eng., 1949).

CUSHING, J. T., and McMULLIN, E., eds. *Philosophical Consequences of Quantum Theory: Reflections on Bell's Theorem* (University of Notre Dame Press, Notre Dame, IN, 1989).

EINSTEIN, A. *Ideas and Opinions* (Dell, New York, 1973).

FINE, A. *The Shaky Game: Einstein, Realism and the Quantum Theory* (University of Chicago Press, Chicago, 1986).

FRANK, P. *Philosophy of Science: The Link Between Science and Philosophy* (Prentice Hall, Englewood Cliffs, NJ, 1957).

GILLISPIE, C. C. *The Edge of Objectivity: An Essay in the History of Scientific Ideas* (Princeton University Press, Princeton, NJ, 1960).

HEISENBERG, W. *Physics and Philosophy* (Harper & Row, New York, 1958).

KOURANY, J. A. *Scientific Knowledge: Basic Issues in the Philosophy of Science* (Wadsworth, Belmont, CA, 1987).

KUHN, T. S. *The Structure of Scientific Revolutions*, 2nd ed. (University of Chicago Press, Chicago, 1970).

PLANCK, M. *Scientific Autobiography and Other Papers* (Philosophical Library, New York, 1949).

SCHRÖDINGER, E. *What is Life? & Mind and Matter* (Cambridge University Press, Cambridge, Eng., 1967).

VAN FRAASSEN, B. C. *The Scientific Image* (Oxford University Press, Oxford, Eng., 1980).

JAMES T. CUSHING

PIEZOELECTRIC EFFECT

The piezoelectric effect refers to the ability of some materials to convert mechanical energy to electrical energy and electrical energy to mechanical energy. This effect was discovered by the brothers Pierre and Jacques Curie in the 1880s. Materials which exhibit this phenomena are called piezoelectric materials. When a piezoelectric material is subjected to a mechanical strain (expansion or contraction) charge appears on its surface. This charge results in an electric field and a corresponding voltage. Conversely, when an electric field is applied, a mechanical strain occurs. The former effect is called the direct effect and the latter is called the converse effect. The orientation (polarity) and magnitude of the charge and voltage produced by the direct effect depend on the direction and magnitude of the applied force relative to certain crystallographic directions in the material. The charge disappears when the mechanical force is removed, and the charge polarity reverses when the strain direction reverses; therefore, an oscillating voltage occurs in response to an oscillating strain. Likewise, the direction and magnitude of the strain produced by the converse effect depend on the direction and magnitude of the applied electrical field.

The strength of both the direct and converse effects in a material is determined by its piezoelectric constant d. Another indication of the strength of this effect for a particular material is the electromechanical coupling constant k. The square of this constant equals the fraction of mechanical energy that can be converted to electrical energy, or conversely electrical to mechanical energy.

The direct effect is used for actuators and the indirect effect is used for sensors. For example, piezoelectric materials are used to generate and detect acoustic waves in air (speakers and microphones) or in water. Sonar, fish finders, and depth sounders use the time delay between generating an acoustic pulse and receiving a reflected signal to measure distance to an object. This technique is also used in medical imaging and nondestructive evaluation of materials for cracks or internal faults with high frequency, ultrasonic (> 20 kHz) waves. Because of the low attenuation of ultrasonic waves in most solids and liquids, they can be used to probe deeply into many materials. Ultrasonics are also used for cleaning and abrasive machining. The resonance of vibrating piezoelectric crystals is used for accurate frequency control in radios and watches. Surface acoustic waves (SAW) in piezoelectric materials are used for analog signal processing devices such as bandpass filters, and pulse compression filters. Piezoelectric materials are also used in accelerometers, positioning devices, and pulse generators for stove igniters.

The piezoelectric effect is related to the molecular structure of materials. When the center of positive charge in a material is slightly separated from the center of negative charge, a dipole results. This occurs in materials whose crystal structure is asymmetric. In some materials there is a permanent dipole moment due to an inherent asymmetry in the crystalline structure, whereas in other materials a mechanical strain is required to produce a dipole moment. Of the thirty-two crystal classes, twenty-one lack a center of symmetry. Twenty of these exhibit piezoelectricity. Ten of these twenty exhibit a dipole moment in the unstrained state; the other ten require mechanical strain to produce a dipole moment. When the separation between positive and negative charges shifts in response to mechanical strain, the electric field due to the dipole changes and the charge on the electrodes changes. The separation can also be changed by applying an electric field, resulting in a mechanical strain.

Materials that have a permanent dipole moment also exhibit the pyroelectric (an electric charge in response to uniform heating) and ferroelectric (reversal of the direction of the dipole in response to an electric field) effects. Since the permanent dipole moment can occur in at least two directions and internal stresses can be minimized by combining these different orientations, domains (regions in which dipoles all point in a particular direction) with different dipole orientations will generally form. Materials with a permanent dipole moment usually undergo a transition at some temperature to a higher symmetry structure that lacks a permanent dipole moment; this temperature is called the Curie point. As the temperature increases toward the Curie point, the strength of the piezoelectric effect decreases.

In materials that are single crystals, the dipole moments, which are related to crystalline directions, are constrained to certain directions. In materials that consist of multiple grains (polycrystalline), the crystallographic axes of different grains are aligned in random directions so that the dipoles will cancel each other out unless the material is poled by the application of an electric field to align the dipoles. Poling is also accomplished by applying an electric field at a temperature above the Curie point and cooling through the Curie point. The electric field causes the dipoles to align along a preferred direction.

The piezoelectric effect occurs in many types of materials including single crystals, ceramics, polymers, and composites. Quartz is one of the most common single crystal piezoelectric materials and has excellent temperature stability. It has a piezoelectric constant d, of 2.3×10^{-12} m/V and a coupling constant k, of 0.1. An industrial process for fabricating quartz crystals was established in 1958. Piezoelectric activity in polycrystalline refractory oxides was first discovered in barium titanate (BT) in the early 1940s. Although its d and k of 190×10^{-12} m/V and 0.5, respectively, are significantly higher than quartz because of the relatively low Curie point (115°C), barium titanate exhibits poor temperature stability. In the 1950s piezoelectric effects were discovered in the lead zirconate titanate (PZT) solid solutions. Because of their high Curie points (above 300°C) and large piezoelectric coefficients ($d = 290$ to 370×10^{-12} m/V, $k = 0.5$ to 0.7), these materials have virtually replaced all other materials except for quartz. Both BT and PZT have similar crystal structures and belong to the general class of materials known as perovskites.

The charge q, per unit area A, which is equivalent to the dielectric displacement D, created by the direct effect is describe by $q/A = D = d\sigma$ (where σ is the stress and d is expressed in coulombs/newton). The strain ε, produced by the converse effect, is given by $\varepsilon = dE$, where d is expressed in meters/volt and E is the electric field (volts/meter). The stress and strain are related by the elastic modulus Y, by $Y = \sigma/\varepsilon$. Since the elastic, dielectric, and piezoelectric constants are a function of direction within the material, they must be expressed in tensor form.

All dielectrics also undergo a mechanical deformation in response to an electric field through the electrostrictive effect. This effect is distinguished from the converse piezoelectric effect, because reversal of the electric field does not result in reversal of the electrostrictive deformation. Furthermore, in the electrostrictive effect the deformation is proportional to the square of the field.

See also: ACOUSTICS; CHARGE; CHARGE, ELECTRONIC; CRYSTAL; CRYSTALLOGRAPHY; CRYSTAL STRUCTURE; CURIE TEMPERATURE; DIPOLE MOMENT; ENERGY; ENERGY, MECHANICAL; FERROELECTRICITY; FIELD, ELECTRIC; FINE-STRUCTURE CONSTANT; FREQUENCY; LIQUID CRYSTAL; POLARITY; SOLID; STRAIN; STRESS; SYMMETRY; TEMPERATURE; ULTRASONICS

Bibliography

CADY, W. G. *Piezoelectricity* (Dover, New York, 1964).

JAFFE, B.; COOK, W. R.; and JAFFE, H. *Piezoelectric Ceramics* (Academic Press, New York, 1971).

POHANKA, R. C., and SMITH, P. L. "Recent Advances in Piezoelectric Ceramics" in *Electronic Ceramics*, edited by L. M. Levinson (Marcel Dekker, New York, 1988).

TAYLOR, G. W. ed. *Piezoelectricity* (Gordon and Breach, New York, 1985).

STEVEN D. BERNSTEIN

PIXEL

An image can be reduced to an array of dots, points, or picture elements called pixels or pels. These images can be stored as a pattern of binary numbers in the (digital) memory of an interactive computer graphics display. Such a display has three parts: frame buffer, video monitor, and video controller. The frame buffer, also called the refresh buffer or raster or bitmap, stores the displayed image as pixels with each location in the frame buffer encoding a pixel. The video monitor consists of a cathode-ray tube (CRT) or a liquid crystal display. Its circuitry adjusts the color intensity for each pixel on a line. For definiteness a CRT will be assumed in what follows. The video controller converts the frame buffer's bit map of memory into a video signal that controls a fast moving electron beam inside the picture tube of the video monitor. It collects and organizes one or more bytes of video information and channels them to the video monitor along with synchronizing signals.

Pictures are painted on the CRT from the frame buffer one line at a time from top to bottom. Each horizontal line of pixels is called a scan line. The entire image is scanned dozens of times a second by the video controller to maintain a steady picture when viewed by the human eye.

For the simplest black-and-white image, one can assign 0s to the black pixels and 1s to the white pixels in the frame buffer. An array of 16×16 black-and-white pixels could then be represented by binary values stored in thirty-two 8-bit bytes, containing sixteen scan lines. Each successive byte of data from the frame buffer is read by the video controller and converted into the corresponding video signal. The signal is received by the CRT, which produces a black-and-white pattern on the screen by modulating the intensity of the electron beam. The displayed picture is changed by modifying the contents of the frame buffer to represent a new pattern of pixels.

The memory capacity of a frame buffer is determined by the requirements on the number of scan lines, the number of pixels per scan line, and the number of bits per pixel. The number of bits representing each pixel can be one or more, going to 24 bits and in some cases beyond 64 bits. A raster system with, say, 24 bits per pixel and having a screen resolution of 1,024 by 1,024 pixels requires a frame buffer storage capacity of 3 megabytes (8 bits/byte). Each pixel's position on the screen is implicitly determined by the location(address) of the bits in memory, with 10 or more bits of x and y coordinate precision. Each pixel's intensity can be represented by one or more bits. One bit is sufficient for text and simple graphics; 2 and 4 bits are useful for displaying solid areas of grey or color; and 8 or more bits are required for high quality shaded displays.

The simplest method to encode colored pictures is to define the color components for each pixel. The bits representing the pixel can be divided into three groups of bits each indicating the intensity of one of three primary color components. In an 8-bit byte, 3 bits are normally allotted to red, 3 to green, and 2 to blue. The three components are then fed into the guns of a color CRT.

To enlarge the color range and to provide accurate control over the colors displayed one must use a color map. The values stored in an 8 bit per pixel frame buffer can address a 256-color table. Color options can expand without an increase in raster size with lookup tables. For example, a 9 bit per pixel frame buffer and a color lookup table with 24 bits for each entry enables 512 (i.e., 2^9) colors to be used, with more than 16 million (i.e., 2^{24}) color choices for the lookup table entries.

See also: CATHODE RAY; COLOR; ELECTROLUMINES-
CENCE; LUMINESCENCE; VISION

Bibliography

HEARN, D., and BAKER, M. P. *Computer Graphics* (Prentice
Hall, Englewood Cliffs, NJ, 1986).

KAY, D. C., and LEVINE, J. R. *Graphics File Formats* (McGraw-
Hill, New York, 1992).

MANO, M. M. *Computer Logic Design* (Prentice Hall, Englewood Cliffs, NJ, 1972).

NEWMAN, W. M., and SPROULL, R. F. *Principles of Interactive
Computer Graphics* (McGraw-Hill, New York, 1979).

CAROL ZWICK ROSEN

PLANCK, MAX KARL ERNST LUDWIG

b. Kiel, Germany, April 23, 1858; *d.* Göttingen,
Germany, October 4, 1947; *quantum theory.*

Planck, whose father was a distinguished professor
of civil law at the University of Kiel, was the youngest
of six children. In 1867 the family moved to Munich,
where the children benefitted from their father's
high position in Bavarian society, learning the cultural graces as well as their academic subjects. A program of music, mountain climbing, and religious
training instilled in the Planck children a devotion
to God, family, and country as well as respect for discipline and hard work.

Planck attended secondary school at the Maximil-
ian Gymnasium in Munich from 1867 until his graduation in 1874. Although a shy student, he excelled
in mathematics, astronomy, languages, and physics.
He would later fondly recall his favorite teacher,
Hermann Müller, as a master in the art of making
his pupils visualize and understand the meaning of
the laws governing the physical world. Planck then
attended the University of Munich (1874–1877),
where he decided to study physics and devoted himself to finding the most fundamental physical laws
and constants.

In 1877 he left Munich for a year of study at the
University of Berlin where he came under the tutelage of Hermann von Helmholtz and Gustav Kirchhoff, two pioneers in the field of thermodynamics.
Although not impressed with their teaching styles,
Planck readily took their subject as his own. Inspired
by the writings of Rudolf Clausius, Planck worked on
sharpening the distinction between the first and
second laws of thermodynamics. His doctoral dissertation asserted that all "natural processes" were irreversible and that the sum of the entropies of all
bodies in such processes would always increase. He
was awarded his Ph.D. from the University of Munich in July 1879.

In 1885 Planck was offered the position of associate professor of theoretical physics at the University
of Kiel. Soon thereafter, he married his fiancée,
Marie Merck. The couple would eventually have
four children: a son, Karl, twin daughters, Margrete
and Emma, and a younger son, Erwin. Planck called
this period, "one of the happiest of my life."

In addition to his teaching duties, Planck published several important papers on the thermodynamics of physical and chemical reactions. He
continued to focus his research on the concept of
entropy "since its maximum value indicates a state
of equilibrium [and] all the laws of physical and
chemical equilibrium follow from a knowledge of
[it]." By 1889 he had earned a strong enough reputation to be offered the position of associate professor of theoretical physics at the University of Berlin,
a position that was vacant owing to the death of
Kirchhoff. Planck became an active member of the
Berlin (later German) Physical Society, was promoted to full professor in 1892, and was elected to
the Prussian Academy of Sciences in 1894.

This slow, steady ascent up the professional ladder suited Planck's conservative temperament, but
in 1894 a series of events accelerated his progres-
sion. Heinrich Hertz, the young physicist who had
experimentally proven the existence of electromagnetic waves, died prematurely in the winter of 1894.
Several months later the deaths of August Kundt,
the Director of the Berlin Physics Institute, and
Helmholtz, the leading authority in German physics,
propelled Planck into a position of leadership at the
age of thirty-six. Characteristically, though deeply
saddened by the loss of three of his closest friends
and colleagues, Planck accepted as his duty the
many new demands thrust upon him.

In the mid-1890s many of the Berlin physicists, including Planck, were investigating the light spectrum emitted by a perfect radiator of light, which
was termed a blackbody since it was also a perfect absorber of light. The problem was of interest because
at thermal equilibrium the emitted radiation was
known to depend only on the temperature of the

emitting object, not its composition. Planck theorized that minute, charged resonators comprising the walls of the blackbody could absorb and emit light by oscillating at the same frequency as the light waves. This resonance would result in the body proceeding to an equilibrium temperature.

On October 19, 1900, Planck announced that he had found a formula for the radiant energy emission for the blackbody spectrum that exactly matched the experimental data. He immediately devoted himself "to the task of investing [his radiation formula] with a true physical meaning . . . [which] led me to study the interrelation between entropy and probability." He theorized that the total energy was divided into small, quantized units and then determined the probability W of how these energy quanta could be distributed among the resonators. Applying Boltzmann's equation, $S = k \ln W$ (k is Boltzmann's constant) to calculate the entropy S led Planck to the correct radiation formula. The energy of each quantum was found by multiplying the frequency ν of the oscillating resonator by a new constant h, which Planck called "the quantum of action" but which is now universally referred to as Planck's constant. He reported these findings to the Berlin Physical Society on December 14, 1900.

Others, including Albert Einstein and Niels Bohr, would later use energy quanta to prove the atomic nature of matter, but to Planck, the radiation constants themselves were of fundamental significance. He calculated numerical values for both k and h from the experimental data on blackbody radiation. The value of k allowed him to calculate Avogadro's number and the value of the elementary electric charge. In addition, by using the constants h and k along with the gravitation constant G and the speed of light c, he devised a system of units for the fundamental quantities of mass, length, time, and temperature that were independent of arbitrary standards and so would "retain their significance for all times."

In the years that followed, Planck was an early supporter of Einstein's theory of relativity. Later, as Planck's supervisory role at the university increased, his direct participation in theoretical research lessened. He maintained a full schedule of teaching throughout his career, wrote textbooks on most of the major fields of physics, and played an instrumental role in revising the university science curriculum. As a firm believer in a balanced life, Planck also retained his passion for piano playing and hiking, often in the company of family and friends. He corresponded with the other leading scientific figures throughout Europe and lectured widely on topics ranging from the role of science in society to the relationship between science and religion. As a man of faith, Planck argued that the two were not mutually exclusive and that seeking an understanding of both were required for a complete world view.

Just as Planck began to acquire widespread professional acclaim, his personal life was marred by tragedy. In 1909 his wife died. In 1916 his elder son was killed in action during World War I. The twin girls both died while giving birth to daughters, Margrete in 1917 and Emma in 1919. Although he eventually remarried, Planck wrote mournfully that, "no man is born with a legal claim to happiness, success, and prosperity in life" and that one had no alternative but to "fight bravely in the battle of life, and to bow in silent surrender to the will of a higher power which rules over him."

After the war, most German scientists were ostracized by international organizations, but Planck's reputation for honesty and integrity remained intact. He was awarded the 1918 Nobel Prize in physics for "his contribution to the development of physics by his discovery of the element of action." He was the leading German authority on scientific matters and was instrumental in the reacceptance of German scientists by the international scientific community after the war. He was elected as a member to most of the leading scientific organizations in the world.

Planck retired from the University of Berlin in 1927 after a career spanning more than forty years. He remained active in the Kaiser-Wilhelm Society, which was created to support German scientific research, and to establish and fund new scientific institutes. In 1930 Planck was elected as its President (it would later be renamed the Max Planck Society). He guided and protected the Society through the rise of the Nazi regime, and on one occasion protested personally to Adolf Hitler about the Nazi's anti-Semitic policies.

Even retirement could not shield Planck from more tragedy. His son Erwin, the remaining child from his first marriage, was implicated in the 1944 assassination attempt on Hitler and was brutally executed by the Gestapo in 1945. Planck's home in a Berlin suburb was leveled in an allied air raid. Although he survived the raid, all of his personal belongings, including his library, were destroyed. At the end of the war an American officer brought

Planck and his second wife to the city of Göttingen. Although in failing health, he continued to travel and lecture for two more years in the hope of encouraging "people struggling for truth and knowledge, especially young people." Having enjoyed the best of life with humility and accepted the worst with dignity, Max Planck died on October 4, 1947, at the age of eighty-nine.

See also: PLANCK CONSTANT; PLANCK LENGTH; PLANCK MASS; PLANCK TIME; QUANTUM THEORY, ORIGINS OF; RADIATION, BLACKBODY

Bibliography

GAMOW, G. *Thirty Years That Shook Physics: The Story of Quantum Theory* (Doubleday, New York, 1966).

HEILBRON, J. L. *The Dilemmas of an Upright Man: Max Planck as Spokesman for German Science* (University of California Press, Berkeley, 1986).

KUHN, T. *Blackbody Theory and the Quantum Discontinuity, 1894–1912* (Oxford University Press, New York, 1978).

PLANCK, M. *Scientific Autobiography* (Philosophical Library, New York, 1949).

PLANCK, M. *A Survey of Physical Theory* (Methuen, London 1925).

MARK S. BRUNO

PLANCK CONSTANT

The Planck constant, $h = 6.626 \times 10^{-34}$ J·s, is a fundamental constant of nature, like the velocity of light. It plays a central role in any proper description of atomic and subatomic phenomena. More often one encounters the symbol $\hbar = h/2\pi$, pronounced h-bar. The constant was first introduced by Max Planck in 1900. Upon examining thermal equilibrium between matter and radiation in a cavity, he was driven to the conclusion that in order to explain the observed spectrum of radiation, one had to postulate that at each frequency f, matter and radiation could exchange energy only in multiples of a basic unit or quantum of value $E = hf = \hbar\omega$, where $\omega = 2\pi f$ is the angular frequency. While this postulate explained the data, it was totally mysterious and ad hoc.

Albert Einstein followed this up in 1905 and postulated the reality of these quanta of energy, called photons, as actual particles traveling at the speed of light. This allowed him to explain the photoelectric effect, which was the following. It was known that the electrons in a metal were bound to it, and it took a minimum energy W to liberate them. One way to do this is by the photoelectric effect—by shining light on the metal. It was found that if the frequency was below a threshold, no electrons were emitted, no matter how intense the light. This was very hard to understand in classical terms, since the intensity of light is a measure of the electric force acting on the electrons. Einstein explained this as follows. The incident radiation is composed of photons, each of energy hf. If $hf < W$, the photons will fail to liberate the electrons. Increasing the intensity will not help, since that only increases the rate of photons striking the metal, not the energy carried by each. On the other hand, if $hf > W$, the electrons must come out with residual kinetic energy $hf - W$, as was observed. (Nowadays we can observe multi-photon processes in which an electron is liberated by more than one photon, each with energy less than W, but these are extremely rare.)

Now, if photons are particles that travel at the speed of light, they must obey the relativistic equation $E^2 = c^2 p^2$, and each photon must have momentum $p = E/c = hf/c = h/\lambda$, where λ is the wavelength. It was verified by Arthur Holly Compton, who studied their collisions with electrons, that photons indeed behave this way; that is, they obey the law of conservation of energy and momentum as applied to particles in a collision. Louis de Broglie then argued for wave–particle duality: If light has such particle-like attributes, then particles must exhibit characteristics of a wave with $\lambda = h/p$. It was then verified that electrons impinging on a crystal indeed produce a diffraction pattern corresponding to waves of $\lambda = h/p$.

Note that $\hbar$ has the dimensions of angular momentum. It turns out that in quantum systems the angular momentum is always an integral multiple of $\hbar/2$. Consider for example the single electron in the hydrogen atom. It has intrinsic angular momentum, called spin, equal to $\hbar/2$ and orbital angular momentum equal to an integral multiple of $\hbar$. Thus, $\hbar$ also enters the granularity of angular momentum.

Planck's constant also enters in the Heisenberg uncertainty principle. Whereas in classical mechanics one can say with certainty that if any dynamical variable (like position or momentum) is measured, such and such a value will definitely be measured, in quantum theory we merely assign probabilities for various possible outcomes when these variables are measured. While it is possible to set up situations where any one variable (e.g., position) has nonzero probability for all but one value, when certain pairs of variables (e.g., the position x and momentum p) are measured, there will always be spreads in the probability distributions (denoted by Δx and Δp) that obey the inequality $\Delta x \cdot \Delta p \geq \hbar/2$. Thus, one cannot create a particle with no uncertainty in both position and momentum. The more accurately one measures the position, the less one can know about the momentum (and vice versa).

Consider next a system like a harmonic oscillator of frequency ω. Whereas in Newtonian mechanics it can have any energy, in quantum theory its energy is restricted to $E = (n + \frac{1}{2})\hbar\omega$, where $n = 0, 1, 2, 3. \ldots$ Therefore $\hbar$ also sets the scale for the discreteness of energy of mechanical systems. Thus, the discrete nature of radiation we spoke of earlier follows from the fact that matter, which is its source, can itself exist only in discrete energy levels.

It took so long to discover the existence of h, and the granularity of various dynamical variables, because of its smallness in terms of macroscopic quantities. For example, if you look at a 100-W bulb from a distance of 1 m, you are not likely to realize that energy is arriving in quanta, because roughly 10^{14} photons enter your eyes per second. Likewise, if you consider a billiard ball whose uncertainty in position $\Delta x \simeq 10^{-15}$ m (which is the size of a single nucleus), its uncertainty in velocity is $\Delta p/m \simeq 10^{-18}$ m/s. This means that over a year, its trajectory will be indeterminate by the size of one atom 10^{-10} m. In other words, it will behave like a classical particle with a well-defined trajectory. These are examples of the correspondence principle, which states that the world, explored at scales much bigger than that set by $\hbar$, will appear classical. However, we now understand that the laws of physics at small scales are governed by quantum mechanics and that Planck's constant sets the scale of the quantum nature of matter and energy.

See also: PLANCK, MAX KARL ERNST LUDWIG; QUANTIZATION; QUANTUM MECHANICS; QUANTUM MECHANICS, CREATION OF; UNCERTAINTY PRINCIPLE; WAVE–PARTICLE DUALITY; WAVE–PARTICLE DUALITY, HISTORY OF

R. SHANKAR

PLANCK LENGTH

The Planck length L_p is a fundamental constant with dimensions of length constructed from a combination of the gravitational constant G, the speed of light c, and Planck's constant $\hbar$. Its value is

$$L_p \equiv \sqrt{\frac{\hbar G}{c^3}} = (1.61605 \pm 0.00010) \times 10^{-35} \text{ m}.$$

The fact that a universal length scale could be defined from the three fundamental constants, G, c, and $\hbar$ was noted by Max Planck when he discovered the constant $\hbar$. It is generally believed that if the universe was viewed through a Planckian microscope that could probe distances of order L_P, it would appear radically different. Although such Planckian microscopes are far beyond current technology, the belief is based on the following reasoning. In Albert Einstein's classical theory of general relativity, the distance between any two points is a dynamical variable, determined by solving Einstein's field equation. The laws of quantum mechanics imply that all dynamical variables are subject to the Heisenberg uncertainty principle. The problem of reconciling quantum mechanics with general relativity, that is, constructing a quantum theory of gravity, has not yet been solved. However, the uncertainty principle together with the dynamical character of distances in general relativity implies that distances cannot be defined or measured with arbitrary precision, but are subject to an inherent quantum mechanical uncertainty. A typical size for this uncertainty should involve the constants G, c, and $\hbar$ that govern the theory. L_p is the only length which can be constructed from these constants, and is therefore presumed to characterize the size of the uncertainty. For two points separated by a distance of order L_p, the uncertainty in the distance is of the order of the distance itself. Hence, when viewed under a Planckian microscope, points in spacetime lose their classical characteristic of having precisely defined separations.

The Planck length may be thought of as a minimum distance beyond which our classical notion of spacetime breaks down.

See also: EINSTEIN, ALBERT; GRAVITATIONAL CONSTANT; HEISENBERG, WERNER KARL; LIGHT, SPEED OF; PLANCK, MAX KARL ERNST LUDWIG; PLANCK CONSTANT; PLANCK

Mass; Planck Time; Quantum Field Theory; Quantum Gravity; Quantum Mechanics; Relativity, General Theory of; Uncertainty Principle

Andrew Strominger

PLANCK MASS

The Planck mass, denoted M_p, is a fundamental constant with dimensions of mass constructed from a combination of the gravitational constant G, the speed of light c, and Planck's constant $\hbar$: Its value is

$$M_p = \sqrt{\frac{\hbar c}{G}} = (2.17671 \pm 0.00014) \times 10^{-5}\,\text{g}. \quad (1)$$

An object with mass M_p has rest energy

$$E_p = M_p c^2 = (1.22103 \pm 0.00008)\,10^{28}\,\text{eV}. \quad (2)$$

E_p is known as the Planck energy. Gravitational corrections, to quantum elementary particle processes, whose precise nature is not understood, are expected to become significant when the mass of the particle is of order M_p or the interaction energy is of order E_p. In quantum mechanics an elementary particle of mass m has a minimum quantum mechanical uncertainty in its position given by its Compton wavelength $\lambda_c = 2\pi\hbar/mc$. In general relativity, a total mass m cannot be localized in a region smaller than the Schwarzchild radius $R_s = 2Gm/c^2$ without collapsing into a black hole. For an object of mass $m = M_p$, the Schwarzchild radius and the Compton wavelength are of the same order $\lambda_c = \pi R_s$. Hence gravitational interaction cannot be neglected in describing an elementary particle of mass M_p. It is not presently understood how one incorporates gravitational effects into a quantum theory of elementary particles. M_p thus represents an upper boundary on the mass of a quantum object that can be described as an ordinary elementary particle. A heavier object will collapse into a black hole.

See also: Black Hole, Schwarzchild; Quantum Mechanics; Relativity, General Theory of

Andrew Strominger

PLANCK TIME

Planck time, denoted T_p, is a fundamental constant with dimensions of time constructed from a combination of the gravitational constant G, the speed of light c, and Planck's constant $\hbar$. It is given by the equation

$$T_p = \sqrt{\frac{G\hbar}{c^5}}$$

and has the value

$$T_p = (5.39056 \pm 0.00034) \times 10^{-44}\,\text{s}.$$

A light ray traverses one Planck length in a time T_p. Quantum gravity is expected to be important for any measurement with a time resolution less than or of order the Planck time, for the same reasons that are important for measurements that probe distances shorter than the Planck length L_p. The first T_p seconds of the history of our universe are presumably dominated by quantum gravity effects.

See also: Planck Length; Quantum Gravity

Andrew Strominger

PLANETARY MAGNETISM

Earth is a "rosetta stone" for interpretation of the magnetic fields observed on other planets. About 80 percent of the earth's magnetic field observed at the surface of the planet may be represented as a dipole of magnitude 8×10^{15} T·m³ inclined at about 11° to the planet's rotational axis; this corresponds to a typical surface intensity of about 0.6 G. The 20 percent of the surface field that is not represented by the dipole consists of continental-size regions of field that drift westward at 0.2° longitude per year. Individual patches grow and decay randomly with apparent periods of about 1,000 years. The strength of the dipole also varies; over the most recent 100 years of observation it has decreased by several percent.

Over longer periods, the behavior of the magnetic field is recorded in the magnetization produced in igneous and sedimentary rocks when they form. For example, the igneous rocks spreading away from midocean ridges provide a fairly continuous record of the field over long periods of time. The geologic evidence shows that although the long-term average of the magnitude of the terrestrial field has been fairly constant through time, there have been large short-term fluctuations; in particular, the polarity of the dipole field has changed signs in a random fashion at intervals ranging from 10^4 to 10^6 years. The dipole component of the field vanishes for a brief time interval, leaving only the nondipolar component, before regenerating with interchanged poles.

The magnetic field of Earth cannot be a result of permanent magnetism. The temperature in the interior of the earth is well above the Curie point, and the observed changes in the field require a mechanism involving macroscopic electrical currents in the earth's interior. This in turn requires a conducting medium, provided by the metallic core of the planet. A perfectly conducting medium that is in motion relative to the source of a magnetic field must carry the field with it because it cannot sustain the electric field that would be induced by such relative motion. Maxwell's equations may be used to show that the time behavior of a magnetic field in a real conducting medium of conductivity σ is given by

$$\frac{d\mathbf{B}}{dt} = D\mathbf{\nabla}^2\mathbf{B},$$

where the total time derivative is the rate of change in a parcel moving with the medium and the coefficient D is given by $D = c^2/4\pi\sigma$ ($1/\mu_0\sigma$ in MKS units). In the infinite conductivity limit $D = 0$ and $d\mathbf{B}/dt = 0$, that is, $\mathbf{B}$ is carried with the fluid motion as noted above. The term on the right represents the decay of the current and diffusion of the field out of the medium; in the case of Earth's core, the decay time is about 30,000 years. This, combined with the geologic evidence on reversals, shows that the mechanism must be capable of self-regeneration.

The important factors in the creation of the magnetohydrodynamic dynamo are convection in the fluid portion of the metallic core and Coriolis forces that operate on the convective cells. The convection could result from latent heat associated with phase changes at the boundary between solid and liquid portions of the core or from buoyancy of light ele-

ments, such as oxygen, dissolved in the liquid core and released as the metallic component condenses on the solid, inner core. The Coriolis forces cause differential rotation within the fluid portion of the core, which converts the *poloidal* field (field lines in meridional planes, e.g., dipole) to *toroidal* field (field lines in planes orthogonal to the axis) by winding them as they rise through the differentially rotating fluid. The Coriolis forces also impart cyclonic rotation to the converging fluid elements at the bottom of the convecting cells, which is only partially canceled by the diverging fluid at the top of the cell. This resulting helical flow can convert the toroidal component into poloidal field, thus regenerating the observed external component and preventing its decay. Recent work suggests that an induced field in the solid inner core stabilizes the field generated in the liquid core against short-term variability.

Though the external terrestrial field is mostly dipole, this is not true of the field in the core. The surface moments, when extrapolated to the interior, are essentially all equal in the generation region. The motions of the nondipolar field components at the surface of Earth provide clues concerning the magnitudes of the fluid motions within the liquid core. The westward drift is produced by a differential rotation of outer regions of the core relative to the adjacent mantle with relative velocity of about 0.3 mm/s. The decay period of the continental-size patches of flux is linked to the convection rate within the core and implies a convection velocity of roughly 0.1 mm/s.

Magnetic fields have been observed for several of the other planets in the solar system; Mercury, Jupiter, and Saturn have largely dipole fields similar to that of Earth. Mercury has an intrinsic dipole field consistent with the planet's large iron core but not with its slow rotation and resulting weak Coriolis forces. The metallic hydrogen cores of Jupiter and Saturn provide a fluid conducting medium for field generation, and adiabatic temperature gradients and convection are maintained by the internal energy sources possessed by both planets. Uranus and Neptune have predominantly dipole fields at their surfaces, but the dipoles are displaced by large amounts from the planet centers and are tilted at large angles relative to their rotation axes. These fields may be generated by "ice mantle dynamos" in the mantles of these planets, which are composed of a mixture of water, ammonia, and methane that is partially ionized due to pressure, providing the nec-

Table 1 Fields and Dipole Moments of the Planets

Planet	Field at Equator ($T \cdot 10^5$)	Magnetic Moment ($T \cdot m^8$)	Quadrupole/ Dipole
Mercury	0.04	6×10^{12}	?
Venus	—	0	—
Earth	3.1	8×10^{15}	0.14
Mars	—	0	—
Jupiter	41.1	1.5×10^{20}	0.24
Saturn	2.2	4.8×10^{18}	0.08
Uranus	2.3	3.8×10^{17}	0.7
Neptune	1.3	2×10^{17}	?

essary conductivity. Measurements have provided only upper limits for the magnetic fields for Venus and Mars. Table 1 presents the dipole moments, the surface field strengths of the dipole at the magnetic equator, and the fraction of the surface field that is the dipole term for the various planets.

See also: CORIOLIS FORCE; DIPOLE MOMENT; MAGNETO-HYDRODYNAMICS; MAXWELL'S EQUATIONS

Bibliography

ENCRENAZ, T., and BIBRING, J.-P. *The Solar System* (Springer-Verlag, Berlin, 1990).

GARLAND, G. D. *Introduction to Geophysics* (Saunders, Philadelphia, 1971).

LEVY, E. H. "The Generation of Magnetic Fields in Planets" in *The Solar System: Observations and Interpretations,* edited by M. Kivelson (Prentice Hall, Englewood Cliffs, NJ, 1986).

RUSSELL, C. T. "Solar and Planetary Magnetic Fields" in *The Solar System: Observations and Interpretations,* edited by M. Kivelson (Prentice Hall, Englewood Cliffs, NJ, 1986).

PHILIP B. JAMES

PLANETARY SYSTEMS

The first planets discovered outside our own solar system were found in a most unexpected place—in orbit around the pulsar designated PSR B1257+12. The discovery of planets with roughly the mass of Earth in the violent neighborhood of a pulsar suggests that planet formation may be a relatively common occurrence in the universe. Searches for planets around nearby stars with the mass of the Sun have more recently been successful, first around the star 51 Pegasus, but they have so far only been sensitive to giant planets with masses approximately that of Jupiter. Nearby young stars have also been observed to have disks of dust and gas around them that appear to be in the first stages of the process of planet formation as we understand it.

Searches for extrasolar planets address the fundamental question of the frequency of planet formation as an accompaniment to star formation. This is one of the unknown factors in estimating the chance of finding extraterrestrial life. The types of planets formed must also be considered. Giant planets similar to Jupiter may be necessary to life because they clear newly formed planetary systems of large numbers of asteroids and comets that would otherwise steadily bombard smaller planets similar to Earth.

Pulsar Planets

The environs of a pulsar are an unlikely place to find planets because pulsars are spinning neutron stars formed by the explosion of supernovas. The pulsar in question has a rotation period of only 6.2 msec, indicating that it was spun up after its formation by accreting gas from a normal star in a close binary orbit.

The transfer of mass from the normal star to the pulsar may have produced the conditions for planet formation to occur. Most of the gas falling towards the neutron star initially has sufficient angular momentum to go into orbit around it in an accretion disk, rather than falling straight onto the star, forming a low-mass binary x-ray source. Condensation of dust grains from the elements heavier than helium in the accreted gas is expected. What is more surprising is that the dust could accumulate into planetesimals. Once planetesimals exist, computational models show that they can form planets through repeated inelastic collisions. The existence of the planets is evidence that planetesimal formation did occur, however, suggesting that, despite our limited understanding of this process, it proceeds fairly straightforwardly.

The planets were detected by sensitive measurements of the motion of the central pulsar using the 305-m Arecibo radio telescope. The pulsar emits a pulse of radio emission every time it rotates. Because of its high density its rotation rate is extraordinarily regular. Small shifts in its position can be detected by the greater or lesser time it takes the emitted pulses to cross the greater or lesser distance, since the pulses travel at the finite speed of light. The tug of the orbiting planets causes changes of pulse arrival times of as much as 1.5 ms.

Three planets have been confirmed in this system so far, with orbital periods of 25, 67, and 98 days. Their distances from the pulsar are 0.19, 0.36, and 0.47 times the distance from Earth to the Sun. Their masses can only be determined by the motion they cause in the neutron star along the line of sight. The size of that motion depends on the inclination of their orbits to the line of sight. The closer to edge-on that the system is viewed, the smaller the masses required to produce the observed motions. The minimum masses for the three planets are 0.015, 3.4, and 2.8 times the mass of Earth. The innermost planet has a minimum mass similar to that of Earth's moon. No gas giant planets with masses approaching that of Jupiter have been detected to date in this system.

Planets Around Solar-Type Stars

Planetary systems around nearby normal stars can be discovered either by direct imaging, or by detection of radial velocity variations in the central stars caused by the orbiting planets.

Direct imaging requires extraordinarily high resolution and dynamic range to resolve faint planets from the central star. In the absence of any atmospheric turbulence, the resolving power of a telescope, measured in radians, is $\theta = 1.22\lambda/D$, where λ is the wavelength and D is the diameter of the telescope. For example, the Hubble Space Telescope has a diameter of 2.4 m. At a visible wavelength of

500 nm, it can theoretically resolve objects 2.5×10^{-7} radians apart, allowing the resolution of a planet in a Jovian orbit of radius 7.5×10^{13} cm at distances up to 300 light-years. Practically, such direct searches have, to date, suffered from the difficulty of picking out a faint planet from the glare of the bright central star. Optical interferometry may offer sufficiently high resolution to cleanly separate planets and directly image them, but such techniques are still under development.

Radial velocity searches currently offer the most practical method. The radial motion of the star is determined by measuring the wavelengths of absorption lines in the stellar atmosphere whose wavelengths have been shifted due to the Doppler effect. Jupiter's orbit around the Sun causes the Sun's velocity to change by 12.5 m·s^{-1}. Current radial velocity searches can detect changes of 10 m·s^{-1}, and changes as small as 1 m·s^{-1} should be detectable using 10-m class telescopes such as the Keck Observatory, allowing the detection of most gas giant planets.

The first planet detected by such a radial velocity search was nearly as extraordinary as the planets around pulsars. It is a gas giant with mass half that of Jupiter, but orbiting around its central star, 51 Pegasus, with a period of only four days and an orbital radius only 5 percent that of Earth. A number of other Jovian-sized planets have also been discovered. It currently appears that about one of every hundred solar-type stars in the solar neighborhood has a Jovian-sized planet orbiting it.

See also: DENSITY; DOPPLER EFFECT; GAS; HUBBLE SPACE TELESCOPE; IMAGE, OPTICAL; LIGHT, SPEED OF; MASS; MOTION; MOTION, ROTATIONAL; NEUTRON STAR; PLANET FORMATION; PULSAR; PULSAR, BINARY; RESOLVING POWER; SOLAR SYSTEM; STARS and STELLAR STRUCTURE; SUPERNOVA; SUN; UNIVERSE

Bibliography

LEVY, E. H., and LUNINE, J. I., eds. *Protostars and Planets III* (University of Arizona Press, Tucson, AZ, 1993).

MAYOR, M., and QUELOZ, D. "A Jupiter-Mass Companion to a Solar-Type Star." *Nature* **378**, 355–359 (1995).

PHILLIPS, J. A.; THORSETT, S. E.; and KULKARNI, S. R., eds. *Planets Around Pulsars* (Astronomical Society of the Pacific, San Francisco, CA, 1993).

WOLSZCZAN, A. "Confirmation of Earth-Mass Planets Orbiting the Millisecond Pulsar PSR B1257+12." *Science* **264**, 538–542 (1994).

MORDECAI-MARK MAC LOW

PLANET FORMATION

Planets appear to form as an occasional, but not inevitable, accompaniment to the formation of stars from interstellar gas and dust. The study of planet formation suffers from the small number of known planets, but observations of star-forming regions and newly formed stars can help us understand the conditions under which planets form.

Any theory of planet formation must, of course, also explain the properties of the modern solar system. All the planets orbit the sun in the same direction as it rotates, in nearly circular orbits whose separation increases with distance from the Sun. They tend to rotate in the same direction, as well, though Venus, Uranus, and Pluto are exceptions. Their elemental abundances vary, but isotopic ratios of each element remain nearly constant. The total mass of the planets is only 0.2 percent that of the Sun, and the total mass of the minor objects (asteroids and comets) is only 1 percent that of Mercury, the smallest terrestrial planet, but the planets carry 98 percent of the total angular momentum of the solar system. Radioactive dating of the oldest meteorites gives an age of 4.5 billion years, while the oldest terrestrial rocks are 4.1 billion years old.

As long ago as 1746, the nature of planetary orbits led to the proposal by Immanuel Kant that the planets must have formed from a disk of gas and dust rotating around the Sun. The origin of this disk now can be explained as a consequence of the formation of the central star.

Star Formation

On a dark night, the Milky Way can be clearly seen to have dark patches and streaks against a bright background of distant stars lying far away in the disk of our galaxy. These dark patches are clouds of hydrogen gas and dust many light-years across, with masses as much as a million times that of the Sun. Though massive, they are extraordinarily tenuous, with densities of at most a million atoms per cubic centimeter—lower than the best laboratory vacuum. The dust grains in these clouds have diameters of around a micron, and consist of graphite or silicate, with mantles of water, methane, ammonia, and carbon-dioxide ices. The mass of dust is nearly 1 percent of the total mass of gas. Though these dust grains are widely spaced, the distance through the cloud is so great that they provide suffi-cient opacity to block the passage of light through the clouds. They contain most of the mass of elements heavier than helium, and so assume a central importance in planet formation.

These clouds collapse under the strength of their own gravity if not supported by thermal pressure or magnetic fields. In star-forming regions, the density becomes high enough to overwhelm these sources of support and begin the process of collapse. Initially the collapse is controlled by the resistance of magnetic fields threading the partially ionized gas, until the density increases enough for the gas to fully recombine and no longer interact with the fields. It can then collapse until its thermal pressure increases enough to support it. This pressure eventually increases sufficiently high to initiate nuclear fusion in the core of the protostar, beginning the process of stellar evolution.

Accretion Disks

The gas in these clouds is in constant motion, stirred by exploding supernovas, stellar winds (similar to the solar wind, though often stronger), and collisions with other clouds. As a result, a collapsing protostellar region, that might originally span a light year, will have large amounts of angular momentum. Each parcel of gas in the region will only collapse until its angular momentum is sufficient to maintain it in an orbit around the central mass.

A spherical collapse involves gas moving in all directions that will go into orbits at every angle. Gas on these different orbits will collide where the orbits intersect, forcing all the gas into a single plane so that it has nonintersecting orbits. As further gas collapses into the region, it will fall onto this disk of gas, known as an accretion disk. Similar structures also occur in active galactic nuclei.

Much of the gas in the disk loses angular momentum to the magnetic fields threading the disk and eventually falls onto the central star. The remaining gas and dust provide the raw materials for planet formation. Observations of young stars at infrared and millimeter wavelengths can directly detect thermal emission from the cool material in their accretion disks. Young stars with masses similar to the Sun appear to have accretion disks with masses no greater than the total mass of the planets in our solar system.

More than a third of all stars form in binary or multiple systems. The accretion disks in these systems will be disrupted quite quickly by the varying gravitational forces of the members of the system. Stable orbits for planets only exist if the stars are extremely well separated, so that planets can orbit just one of the stars, or if they are extremely close, so that planets can orbit the entire system. Most such systems probably do not have planets.

Planetesimals

Planets could form from a protostellar accretion disk in two distinct ways: either from collisions of solid objects followed by accretion of gas from the surrounding disk for the gas giants, or by gravitational collapse of dense regions of the gas disk followed by loss of most of the gas from the terrestrial planets. The first of these scenarios looks most likely for the following three reasons. First, none of the planets have as high an abundance of hydrogen and helium as the Sun, though the gas giants have sufficiently strong gravity to prevent the escape of these light elements. Second, separation of the heavier elements into a solid core surrounded by a hydrogen and helium atmosphere would not occur in a gravitationally bound, gaseous protoplanet. Finally, only the first scenario can explain the presence of asteroids, comets, moons, and other small bodies.

Solid objects form from the collisions of dust grains that either came from the interstellar cloud, or condensed in the accretion disk. The dominant forces on dust grains are stellar gravity and gas drag. The dust grains sink to the mid-plane of the disk in Keplerian orbits, while the gas forms a disk stratified both vertically and radially. The radial stratification allows the gas to orbit at slightly less than Keplerian velocities, so that solid bodies orbiting at Keplerian velocities will feel a drag dependent on their size. The disk will most likely also have strong winds driven by local magnetohydrodynamic turbulence. Small dust grains will be blown along with the gas, colliding with each other in the denser midplane of the disk or in the centers of turbulent vortices to form centimeter-sized bodies.

The transition from centimeter-sized bodies to kilometer-sized bodies, called planetesimals, remains the least understood portion of this scenario. Planetesimals are sufficiently large that gas drag no longer strongly influences them, and so they evolve thereafter under the influence only of gravity and collisions with other planetesimals or planets. However, smaller bodies can be slowed down by the gas quite efficiently, so that most would be accreted by the central star in the absence of other complications. It appears most likely that the collective action of these small particles tends to accelerate the disk gas close to the midplane, producing regions where planetesimals can form.

Collisions of planetesimals form the last act of planet formation. The probability of two planetesimals hitting each other and forming a larger object is at first dependent only on their relative speed and physical size. Higher speeds make mutual destruction more likely than formation of a larger object, while larger sizes naturally make collisions more likely. As the largest planetesimals grow, however, their own gravity begins to attract nearby objects, increasing the chance of interactions further. The end result of this process is runaway accretion, where the largest object in each region of the disk sweeps up nearly all the other objects within several thousand years.

Properties of Planets

The observed properties of planets in the modern solar system depend strongly on the violent final stages of runaway accretion when planet-sized objects crash into each other. This epoch can be directly simulated on computers if the assumption is made that nearly all collisions result in the formation of larger objects rather than fragmentation.

Bode's law. Numerical simulations show that the formation of three or four inner planets is a natural consequence of the sweeping up of planetesimals, but that the exact orbital positions and numbers of planets is essentially random, depending only on exactly which objects collide at what time. This shows the rule known as Bode's law to be due entirely to coincidence. This rule gives the orbital radius of a planet in terms of the distance from Earth to Sun as $R_n = 0.4 + 0.3(2^n)$, with $n = -\infty$ for Mercury, 0 for Venus, 1 for Earth, 2 for Mars, 3 for the asteroids, and so on. It is not particularly accurate, with major errors for three out of the nine planets. What is clear is that the planets and planetesimals continued to collide until they were separated by sufficiently large distances to not interact appreciably during the life of the solar system, where we find them today.

Atmospheres. The atmospheres of the terrestrial planets are negligible fractions of their total mass, and probably were originally accreted as ices in solid planetesimals, and then melted and outgassed. The giant planets, however, have hydrogen and helium atmospheres that form significant fractions of their total mass (virtually all in the case of Jupiter). It appears that during their formation they grew large enough to sweep up much of the gas along their orbits through the accretion disk, ceasing to grow only when either they had swept a large gap in the disk, or when the disk was dispersed by the central star. Gaps in protostellar accretion disks may have been observed around young stars, suggesting the possibility of other planetary systems. However, models show that conditions in the protostellar disk must be rather finely tuned for a planet to accrete as much mass as Jupiter, suggesting that such massive planets may not be common.

Satellites and minor bodies. The moons of the gas giant planets formed in two ways. The larger ones probably formed during the accretion of the gas giant atmospheres in small accretion disks by processes similar to those acting in the protosolar disk to form the planets. The smaller moons probably represent planetesimals captured by close interactions with the planet that survived without actually hitting the planet.

The origin of Earth's moon can, surprisingly, best be explained by the impact of a Mars sized planet during the final stages of planet formation. This appears to be the only scenario for lunar formation that accounts for the composition of the Moon, which consists mostly of silicate rocks, with neither a large iron core, nor many ice-forming elements. The collision was so violent that it stripped the silicate crusts almost entirely off the two colliding planets, throwing the material into orbit, where it reaccreted into the Moon.

Comets and asteroids are distinguished from each other primarily by the presence of various ices in comets and their absence in asteroids. Both appear to be examples of planetesimals preserved more or less pristinely from the period of planet formation due to interactions with the gas giants. In the case of the asteroid belt, Jupiter probably swept up so many of the planetesimals in this region that runaway accretion was prevented, and a few planetesimals remained. They have continued to collide with each other, so that modern asteroids have the size distribution expected for objects fragmenting due to collisions. As their total mass is now low, planet formation in this region was effectively prevented by Jupiter.

Comets now reside in two regions: (1) the *Kuiper belt,* lying beyond the orbit of Neptune, that also includes Pluto and a number of 100-km objects discovered within the last few years, and (2) the *Oort cloud,* lying up to 1,000 times the distance from Earth to the Sun. These regions were populated by planetesimals that were ejected from the region close to the gas giants by slingshot encounters with the planets.

See also: ACTIVE GALACTIC NUCLEUS; PLANETARY MAGNETISM; PLANETARY SYSTEMS; SOLAR SYSTEM; SOLAR WIND; STELLAR EVOLUTION; SUN; SUPERNOVA

Bibliography

LEVY, E. H., and LUNINE, J. I., eds. *Protostars and Planets III* (University of Arizona Press, Tucson, AZ, 1993).

LISSAUER, J. J. "Planet Formation." *Annu. Rev. Astron. Astrophys.* **31,** 129–174 (1993).

MORDECAI-MARK MAC LOW

PLASMA

A plasma is a collection of electrons, ions, and neutral atoms in which the net Coulomb force between the charged particles dominates the behavior of the group. The long-range nature of the Coulomb force causes collective modes of motion to be present in plasmas that are not in an un-ionized gas. A plasma can be made from an ordinary gas by raising the temperature sufficiently high, typically above 10,000 K, so that many electrons are freed from bound atomic orbitals. A high temperature is also necessary to reduce the strength of nearby (binary) collisions compared to the distant collective interactions. For the same reason, a plasma must be tenuous.

Several important parameters characterize a plasma: the Debye length (λ_D) is the distance in a plasma over which static electric fields are attenuated to $1/e$ of their initial strength; the plasma frequency (ω_{pe}) is the rate at which electrons collectively oscillate back and forth when disturbed from their equilibrium positions; and the plasma parameter (Λ) is the number of charged particles in a sphere one Debye length in radius. The criteria that an ionized gas of size R behaves like a plasma are $\Lambda > 1$ and $R > \lambda_D$ (see Fig. 1).

Because a plasma is composed of unbound electrically charged particles, plasmas frequently have

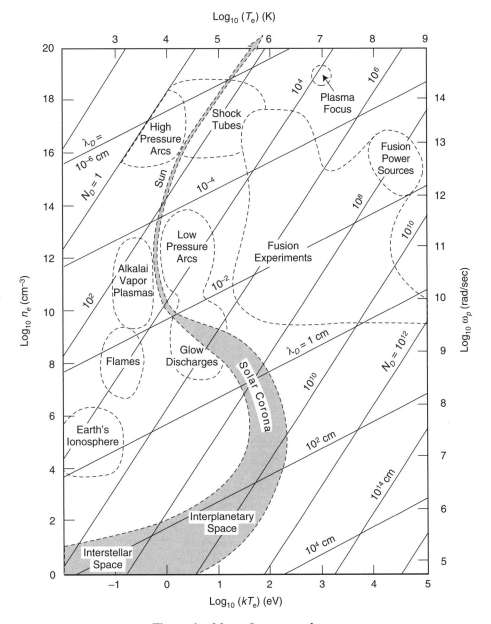

Figure 1 Map of gaseous plasmas.

electrical and thermal conductivities that exceed those of pure metals by factors greater than 1,000. A high electrical conductivity makes it difficult for the densities of the positive and negative charges to significantly differ from each other. Thus, most plasmas are quasineutral. Also, because of its free electrical charges, a plasma's behavior is greatly altered by the presence of magnetic fields. The circular (cyclotron) motion of charged particles around magnetic field lines is in such a sense that the total

magnetic field intensity within the plasma is lowered, making the plasma diamagnetic. A magnetic field also reduces electrical currents perpendicular to its field lines, thus, allowing electric fields, particularly oscillatory ones, to penetrate further than a Debye length, giving plasmas a high dielectric constant. The presence of magnetic fields in plasmas allows additional collective modes, meaning new waves and instabilities. One particularly prominent example is the Alfven wave.

To strip electrons from atoms (ionization) requires the addition of energy, which can be accomplished by bombardment with energetic photons or particles (typically electrons), as well as the aforementioned thermal heating. Plasmas themselves are copious emitters of electromagnetic radiation, primarily by the processes of electron impact excitation of atoms and ions and by bremsstrahlung.

History

Although the special nature of ionized media, specifically flames, was recognized in classical times (Empedocles, ca. 490–430 B.C.E.), it was not until the late nineteenth century, with the work on ionized gases in discharge tubes by scientists such as William Crooke and Johann Hittorf, that the properties of plasmas were first studied systematically. It was Crooke who named (1879) ionized gases the fourth state of matter, though no distinct phase transition separates it from the gaseous, solid, or liquid states of matter. The next landmark in the study of plasmas was the work of Lord Rayleigh (1906) who explained the plasma frequency in terms of the Thomson model of the atom. In the late 1920s, Irving Langmuir made detailed studies of plasma oscillations and plasma diagnostic techniques (which still bear his name). He gave the name "plasma" to this ionized state, to describe its jelly-like behavior.

Interest in plasma physics heightened in the following decades. One reason was the impact of atmospheric plasmas on the propagation of radio waves and the related discoveries of electromagnetic waves emitted from objects of astrophysical significance. A second was the realization that nuclear fusion reactions responsible for the Sun's energy release might be sustained on Earth in plasmas. The commercial value of plasmas has greatly increased, as uses in fluorescent lighting, high-current switches, arc furnaces, semiconductor fabrication, lasers, microwave generation, and so on have been developed.

Occurrences

Plasmas comprise more than 99 percent of the visible universe. The interiors of stars are plasmas because they are so hot. Most of the surface of the Sun is not a plasma because it is too cool for its density. However, the majestic giant prominences seen during eclipses or with x-ray telescopes are plasmas because they contain hotter gases, expelled from regions deeper within the Sun. Extremely energetic plasmas surround black holes and neutron stars. Most of the extensive interstellar clouds are also plasmas, in spite of their relatively cool temperatures, because they are so tenuous and are bathed by photons from nearby stars. It is these photons that maintain the high degree of ionization. As shown in Fig. 1, space plasmas span more than twenty orders-of-magnitude in density and ten in temperature.

Many space plasmas do not emit in the visible range of wavelengths. To see these it is often necessary to use radio antennas. These have also been useful probes of Earth's own plasma layers, the magnetosphere and the ionosphere.

In contrast, plasmas are rarely found on Earth. Naturally occurring terrestrial plasmas include lightning, St. Elmo's fire, the aurora, and flames. Man-made plasmas are ubiquitous, primarily because of a lighting industry based on the greater efficiency of plasmas, compared to incandescent wires, for converting electrical power into usable light. Such plasmas, though hot (ca. 10,000 K) by ordinary standards, can be contained within simple glass tubes because they are tenuous and, hence, transmit little heat to the surrounding glass walls. Plasmas being developed for fusion applications are much larger (ca. 5 m) and hotter (ca. 400,000,000 K), and they must be insulated from the material walls by intense magnetic fields or dense gas layers. Miniature (sub-millimeter) plasmas are a developing technology that have found a niche in flat panel displays.

See also: ALFVEN WAVE; AURORA; BREMSSTRAHLUNG; COULOMB'S LAW; CYCLOTRON RESONANCE; DIAMAGNETISM; DIELECTRIC CONSTANT; FIELD, MAGNETIC; FUSION; IONIZATION; IONOSPHERE; LIGHTNING; MAGNETOSPHERE; PLASMA PHYSICS

Bibliography

BOULOS, M. I.; FAUCHAIS, P.; and PFENDER, E. *Thermal Plasmas: Fundamentals and Applications* (Plenum, New York, 1993).

CHEN, F. F. *Introduction to Plasma Physics,* 2nd ed. (Plenum, New York, 1990).

OCIELLO, O., and FLAMM, D. L. *Plasma Diagnostics* (Academic Press, New York, 1989).

TELLER, E. *Fusion* (Academic Press, New York, 1981).

SAMUEL A. COHEN

PLASMA PHYSICS

Plasma physics is the study of the physical properties of matter in the plasma state. Just as sufficient increases in temperature can change matter from solid to liquid and liquid to gas, so will a further increase of temperature bring matter into the plasma state. In the plasma state, an appreciable fraction of the molecules and atoms are ionized, that is, dissociated into ions and electrons. The gross properties of plasma are therefore significantly different from those of an un-ionized gas. A plasma characteristically emits electromagnetic radiation such as light, bremsstrahlung, or even x rays, and it is an excellent conductor for both electricity and heat. As a conductor, a plasma can be strongly affected by magnetic fields. For example, a plasma can be confined or accelerated by magnetic forces, phenomena that are witnessed, for example, in the long lifetime of the Van Allen belts around Earth and in the motion of arcs in the Sun's corona. Also, in a magnetic field, plasma properties may be strongly anisotropic; the plasma provides very different responses depending on the orientation of the specific effect relative to the direction of the magnetic field. Moreover, plasmas may display a variety of unusual dielectric or refractive properties for intermediate frequency radiation and can easily exhibit nonlocal effects wherein the response to a local stimulus may appear at a later time and often at a remote location. Under special circumstances, plasmas may also exhibit memory phenomena. It is these and other differences from ordinary gases that make the study of plasma physics a fascinating and extraordinarily rich field. Depending on the specific topic, the study may invoke a variety of disciplines within the field of physics, including classical mechanics and chaos theory, statistical mechanics, electricity and magnetism, kinetic theory of gases, and hydrodynamics, together with applied and computational mathematics. These disciplines are often combined to create new areas of science and understanding.

Examples of matter in the plasma state are abundant. Stellar and interstellar matter is mostly in the plasma state, as is matter in the magnetosphere, the ionosphere, in flames, in chemical and nuclear explosions, and in electrical discharges. In addition, a number of high-technology areas involve matter in the plasma state, including gas lasers, free-electron lasers, certain microwave amplifiers, and plasma deposition and etching operations that are basic to the manufacture of computer chips. Also, very significantly, the working fuel in a controlled thermonuclear or fusion reactor would be in the plasma state. Fusion power would use, in a socially and ecologically safe way, the almost limitless nuclear energy that is potentially extractable from the world's abundant resources of light elements, particularly heavy hydrogen (deuterium) and lithium. The kindling temperature for a fusion reactor is around $100,000,000°C$, many times hotter than the center of the Sun, a fact that explains why the reacting fuel would be in the plasma state. It has been very largely the intense study and development of the fusion reactor concept for large-scale electric power generation that has led to our extensive current knowledge of plasmas.

From one point of view, plasma physics becomes both simpler and more interesting at higher and higher temperatures. The reason is simple. Movement of the charged particles that constitute the plasma, the ions and electrons, is dominated by electric and magnetic forces. Unlike an un-ionized gas, collisions with neutral particles are usually unimportant, either because charged-particle collisions almost always constitute a stronger effect or because virtually all the neutrals have been ionized in the high-temperature environment. Moreover, collisions between pairs of charged particles, ion-ion, electron-electron, and ion-electron, also become unimportant at high temperatures because the cross section for such collisions (i.e., the Coulomb cross section) depends inversely on the fourth power of the relative particle velocity. The hotter the plasma, the faster the particles and the smaller the Coulomb cross section. What remains is that the motion of the individual ions and electrons in a plasma depends almost solely on the average electric and magnetic fields in the region surrounding the instantaneous location of the particle. (This region should be large enough to contain many, many ions and electrons, comparable in size to a Debye sphere, as defined below.) These local average fields are, in turn, produced by external sources, if they are present, and are also induced by the net space charge and space currents of the plasma ions and electrons. The entire picture is one of self-consistency: Each charged particle obeys an equation of motion in which its acceleration depends on the local electric and magnetic fields. The electromagnetic fields are, in turn, produced or induced possibly by external sources but also by the space charge and space currents of the plasma particles themselves, now considered collectively.

To proceed further in a discussion of the concepts that underlie plasma physics, it is helpful to introduce some basic lengths and frequencies. Most important is the Debye length, λ_D, which is the characteristic distance over which the plasma is able to neutralize any electrostatic imbalance. For example, the electric field due to any ion is, on average, attenuated by a redistribution in the locations of all the electrons in the "cloud" of electrons surrounding that ion. (Position shifts of neighboring ions also contribute to this attenuation, but the electrons are much more mobile.) As a result, the average electric field around a bare ion of charge Ze is no longer represented by the Coulomb potential Ze/r, but rather by the shielded Coulomb potential $(Ze/r)\exp(-r/\lambda_D)$. The more dense the electron cloud, the shorter the Debye length. Similarly, the hotter and faster-moving the electrons, the less effectively can they perform this shielding. Quantitatively, $\lambda_D^2 = k_BT/(4\pi n_e e^2)$, where k_B is the Boltzmann constant, n_e is the electron density, and e is the charge on an individual electron.

It is frequently said that plasmas are quasineutral, meaning that electrostatic imbalance cannot extend beyond a few Debye lengths. An example is the formation of a Debye sheath around a metallic probe stuck into a plasma.

A common measure of the "quality" of a plasma is the number of electrons in a Debye sphere (i.e., in a sphere of radius λ_D). The larger this number, the less important the effect of two-particle collisions, and the more the plasma acts as an "ideal" collision-free medium. In both the magnetosphere and in fusion-quality plasmas, this number might be as large as a million, or even larger. In fusion experiments, the charged particles move in more-or-less confined orbits determined by powerful magnetic fields, and the mean-free-path for these particles against two-particle collisions can be kilometers long, much, much longer than the dimensions of the laboratory apparatus.

A characteristic frequency for plasmas is the plasma frequency. Consider a neutral plasma formed of equal densities of ions and electrons. Now compress a thin slab of electrons without moving the nearby ions. The local ion and electron space charges no longer neutralize each other, the compressed electrons at first repel each other, are accelerated, overshoot their original positions so that they are now attracted back toward their starting points, overshoot again, and so on. The angular frequency for this oscillation is the plasma frequency, ω_p. Quantitatively, $\omega_p^2 = 4\pi n_e e^2/m_e$, where m_e is the electron mass. In a second interpretation, one easily sees that the time required for an electron moving with thermal speed $(k_BT/m_e)^{1/2}$ to travel to a Debye length is simply $1/\omega_p$.

Another set of characteristic frequencies is the rate at which the particles gyrate in the background magnetic field, if one is present. For particles of charge Ze and mass m, the angular gyrofrequency in cgs-Gaussian units is $\Omega = ZeB/mc$, where B is the strength of the magnetic field and c is the velocity of light. Individual charged particles in a magnetic field move in helical orbits, the axis of the helix being parallel or anti-parallel to the direction of the magnetic field. The projection of the orbit perpendicular to the magnetic field is a circle, and the radii of these circles introduce new characteristic lengths. The length is called the gyroradius, $\rho = v_\perp/\Omega$, where $v_\perp$ is the speed of the component of the particle velocity that is perpendicular to the magnetic field. Often the rms value of $v_\perp$ is used to characterize the gyroradii for a group of particles.

In an ordinary gas, the dominant interaction is the rapid occurrence of two-particle collisions with the result that the local velocity distribution of the gas particles is always very close to a Maxwellian distribution. In a sound wave, for example, the gas in a high-pressure region is compressed, and a gas molecule experiences an increased rate of collisions by molecules coming from the side of higher density, urging the original molecule in the direction of lower density and lower pressure. On the other hand, in a hot plasma, two-particle collisions are a subdominant effect. The charged particles in the plasma respond to the average electromagnetic fields. But even though the intermediate mechanisms are vastly different, the final effects for a gas and for a plasma may be similar. In a plasma, the electric field due to a cloud of compressed ions will attract electrons and repel neighboring ions; quasineutrality will keep the local electron and ion densities almost exactly equal. The net result is a sound wave driven by the combined ion and electron pressures.

Speaking more generally, it is possible to consider plasmas as highly conducting fluids and to describe the plasma dynamics and transport by sets of fluid equations that correspond to conservation of number, of momentum, of energy, and so on for the electrons and for each species of ions present in the plasma. One needs an additional equation, such as an ad hoc adiabatic or isothermal law, for closure, plus Maxwell's equations for the electromagnetic fields. The two sets are coupled in that the source terms in Maxwell's equations, that is, the densities of electric charge and current, are easily expressed in terms of the macroscopic densities and velocities of the charged plasma particles. A closely related set of equations, and one that offers a very useful description of low-frequency long-wavelength plasma phenomena, is the magnetohydrodynamic set.

What gives this field its fascination, however, is the way in which plasma phenomena differ from gas physics. A very interesting example appears even in the fluid equations, namely, the occurrence of anisotropic pressure. In a hot plasma, thermalization of the particle distributions may be very slow. In the absence of collisions, the free-streaming along the field lines of a slightly nonuniform magnetic field will quickly symmetrize the distribution of velocities in the plane normal to the direction of the magnetic field, but this process will not tend to equalize the "parallel" and "perpendicular" pressures. A frequently appropriate pressure tensor is then $\mathbf{P} = \mathbf{1}p_\perp + \hat{\mathbf{b}}\hat{\mathbf{b}}(p_\parallel - p_\perp)$, in which $\mathbf{1}$ is the unit dyadic, $\hat{\mathbf{b}}$ is the unit vector along $\mathbf{B}(\mathbf{r})$, $p_\parallel = nk_BT_\parallel$, $p_\perp = nk_BT_\perp$, and $T_\parallel$ and $T_\perp$ are the temperatures for motions parallel and perpendicular, respectively, to $\mathbf{B}$. Separate laws apply to $p_\parallel$ and to $p_\perp$, and the appropriate fluid relations then are the double-adiabatic Chew–Goldberger–Low equations. Moreover, the velocity distributions even within the parallel and perpendicular projections may remain non-Maxwellian for long times, so that characterizing such distributions with macroscopic descriptives such as temperature or fluid velocity may be inadequate and perhaps misleading. A better description is given by the full velocity distribution function $f(\mathbf{r},\mathbf{v},t)$, which is the density of particles as a function of time in 6-dimensional space, the three space dimensions and the three velocity dimensions. An analysis of plasma phenomena from this standpoint falls into the discipline termed kinetic theory. The equation that describes the evolution of $f(\mathbf{r},\mathbf{v},t)$ is the Boltzmann equation; for hot plasmas, the term in this equation that represents the effects of collisions is often dropped, in which case the equation is referred to as the Vlasov equation. The sequence of fluid equations, mentioned in the previous paragraph, are just the successive velocity moments of the Boltzmann equation.

It was mentioned in the first paragraph that plasmas may be confined or accelerated by magnetic fields. Concerning force directed perpendicular to the magnetic field, the interaction is through the familiar "motor force," $\mathbf{j} \times \mathbf{B}/c$, $\mathbf{j}$ in this case being the density of current flow in the plasma. But magnetic force can also be exerted parallel to the magnetic field. When charged particles move in a static magnetic field that is increasing in strength along the particle's helical path, some of their "parallel energy" is converted into "perpendicular energy." [In single-particle motion, the quantity $mv_\perp^2/B$ is an adiabatic invariant, that is, it is approximately constant in situations where $B(\mathbf{r})$, in the particle's frame of reference, does not change too quickly with time. If B increases along the path, $v_\perp^2$ must then also increase.] Since the total particle energy will be constant, the particle will be reflected at the point where its "parallel energy" has dropped to zero. From a fluid-equation point of view, this "magnetic mirror" force appears in the divergence of the stress tensor $\mathbf{P}$. Use is made of this principle in the "magnetic mirror" geometry for plasma confinement, which confines plasma in a weak-field region located between two strong-field regions. Leakage of plasma particles through the mirrors may be very slow, dependent on pitch-angle scattering in velocity space due to two-particle collisions.

A host of novel plasma phenomena are revealed in the study of plasma waves. In an unmagnetized plasma, in addition to the sound waves and plasma oscillations mentioned above, there is a rich spectrum of Van Kampen modes and their nonlinear counterpart, Bernstein–Greene–Kruskal waves. A sampling of possible modes for a magnetized plasma includes Alfven waves, magnetosonic waves, drift waves, ion and electron cyclotron waves, lower and upper hybrid modes, and whistlers. An especially interesting and important effect is the collision-free absorption of wave energy by those particles that, because of the Doppler effect, "see" the wave at zero frequency, or at their gyrofrequency, or at a harmonic thereof (i.e., those particles with parallel velocity $v_\parallel$ that satisfies $\omega - k_\parallel v_\parallel - n\Omega = 0$, where ω is the angular frequency for the wave, $k_\parallel$ the parallel wave number, and where n may be any integer, positive, negative, or zero). The processes are called, respectively, Landau damping ($n = 0$), or cyclotron ($n = \pm 1$), or cyclotron-harmonic damping ($|n| \geq 2$). Other novel phenomena associated with plasma waves include finite gyroradius effects, wave trapping of particles, wave echoes, waves whose group velocity may be at a large angle with respect to the phase velocity, spontaneous conversion from one mode to another during propagation through inhomogeneous plasmas, and many other effects.

Closely related to the study of waves is the study of plasma instabilities. For small amplitude waves, the relation is simple: An instability appears as a wave for which the frequency is a complex or imaginary number. Almost any deviation from a thermal distribution can be a source of free energy that may drive a plasma instability. For fluid instabilities, the driving force comes from a nonthermal arrangement of the plasma and the magnetic field in configuration space. They go by such picturesque names as the interchange, ballooning, kink, tearing, or rippling modes. Under the interchange mode, for instance, which is a magnetized-plasma counterpart to the well-known Rayleigh–Taylor instability, adjacent magnetic lines of force pass by each other like the extended fingers on two hands. In the kink mode, the bending of a current-carrying column is aggravated by the increased magnetic pressure of lines squeezed together on the concave side of the bend. In this vein, it is worth noting that dissipative processes such as resistivity and viscosity, generally considered mechanisms for wave damping, may actually enhance magnetohydrodynamic instability by eliminating perfect-conductivity restraints on plasma motion. Totally different in nature are the microinstabilities, in which the free energy comes from a non-Maxwellian distribution in velocity space. In both cases, unchecked instability leads to plasma turbulence and enhanced transport rates for material and energy. Now, the design of a successful controlled fusion reactor based on magnetic confinement will require that the rate of plasma transport across the magnetic field be close to its theoretical minimum. Therefore, the subjects of plasma transport and instability have been ones for intense scrutiny by fusion-plasma experimentalists, with highly sophisticated diagnostic techniques used on large experimental devices such as tokamaks and stellarators and by plasma theorists using both rigorous analysis and the most powerful computers. In a similar vein, scientists looking toward successful controlled fusion via inertial confinement (i.e., laser- or particle-beam-driven implosion of micropellets of frozen deuterium and tritium) also seek, by both experimental and theoretical means, to achieve dynamic behavior in the imploding pellet that is close to a theoretical ideal. Keenly interested as well in the topics of plasma instability and turbulence are the many space scientists and astrophysicists who seek to explain a great number of highly diverse phenomena, including the Sun's coronal arcs, the solar wind, Earth's bow shock, Earth's magnetotail, and Earth's relation to auroral phenomena, as well as phenomena associated with pulsars and with supernova explosions.

The relatively new field of plasma physics has been found to span a large number of disciplines in classical physics. Moreover, the joining together of concepts from electricity and magnetism—classical and statistical mechanics, for instance—has led to totally new areas of scientific knowledge and has given us understanding of fascinating phenomena observed both in our laboratories and in our astrophysical surroundings. The field has proved to be incredibly fertile for both experimental and theoretical research, and it still poses great challenges that await solution by tomorrow's scientists.

See also: ALFVEN WAVE; FIELD, MAGNETIC; FUSION; FUSION POWER; INTERSTELLAR AND INTERGALACTIC MEDIUM; MAGNETOHYDRODYNAMICS; MAGNETOSPHERE; PLASMA; VAN ALLEN BELTS

Bibliography

CHEN, F. F. *Introduction to Plasma Physics and Controlled Fusion,* 2nd ed. (Plenum, New York, 1984).

KRALL, N. A., and TRIVELPIECE, A. W. *Principles of Plasma Physics* (McGraw-Hill, New York, 1973).

NICHOLSON, D. R. *Introduction to Plasma Theory* (Wiley, New York, 1983).

SCHMIDT, G. *Physics of High-Temperature Plasmas,* 2nd ed. (Academic, New York, 1979).

STIX, T. H. *Waves in Plasmas* (American Institute of Physics, New York, 1992).

THOMAS HOWARD STIX

PLASMON

Because of the long range of the Coulomb interaction, the density fluctuations of electrons in a plasma—that is, electrons moving in a uniform background of positive charge—display collective behavior at long wavelengths that is characterized by oscillations at the plasma frequency $\omega_p = (4\pi n e^2 / m)^{1/2}$, where e is the electron charge, m is its mass, and n is the electron density. As was emphasized by David Bohm and David Pines, who, during the period 1949–1953, were the first to study this phenomenon under the quantum conditions found in metals and other solids, the physical origin of this collective mode is the influence on a single electron of the average self-consistent field of the other electrons. A long-wavelength plasmon, the basic quantum of plasma oscillation, thus possesses the energy $\hbar\omega_p$, where $\hbar$ is Planck's constant.

For plasmas at metallic densities—$(10^{22}/cm^3) < n \lesssim (10^{23}/cm^3)$—the energy of a plasmon ranges from some 6 eV to 25 eV. Since this is greater than the energy that the most energetic single electron in the metal possesses, no internal electron in the metal can excite a plasmon. Plasmon excitation occurs when a charged particle with an energy large compared with that of a plasmon passes through the metal. It was first observed, but not identified, in experiments on discrete energy losses in metals carried out by Georg Ruthemann, who found that the energy loss spectrum for kilovolt electrons passing through thin metallic foils of beryllium (Be) and aluminum (Al) consisted in several comparatively narrow lines, separated by 19 eV (for Be) and 15 eV (for Al). Bohm and Pines showed that these energy transfers were exactly those to be expected for plasmon excitation, since the plasmon energies for these materials, calculated on the assumption that the outermost valence electrons were free, were 19 eV and 16 eV, respectively.

Plasmon dispersion reflects plasmons' coupling to the single-particle excitations in the valence band; for a nearly free electron-like metal, plasmons at long wavelengths (small wave vectors k) follow the Bohm–Pines dispersion relation, $\omega(k) \cong \omega_p + (3/10)\, k^2 v_F^2$, where v_F is the electron Fermi velocity, whereas for wave vectors $\sim k_c = \omega_p/v_F$, plasmons can no longer be regarded as an independent entity, since valence electrons act to damp the plasma oscillations. This is the quantum analog of the Landau damping of classical plasma oscillations by electrons moving in phase with, and hence absorbing energy from, the plasma waves. Plasmon dispersion is determined by measuring the angular distribution of electrons scattered by thin films. Such experiments also provide valuable information on the critical wave vector k_c beyond which plasmons cease to be a well-defined excitation.

At the 1954 Solvay Congress, Nevill Mott showed quite generally that for any solid (metal, insulator, or semiconductor) in which the valence electrons are weakly bound and the core electrons are strongly bound, the interband excitations characteristic of the solid-state environment would not greatly influence the plasmon energy or the critical wave vector for their existence, k_c. He thus explained why plasma oscillations at very nearly the free-electron plasma frequency are quite a general property of solids, as many subsequent electron energy loss spectroscopic (EELS) experiments demonstrated.

Surface plasmons, a second kind of plasma excitation associated with the collective motion of electrons in the immediate vicinity of a vacuum–solid interface, were proposed by Rufus Ritchie in 1957; their existence was soon established experimentally. Experimental verification of the existence of a third member of the plasmon family, the acoustic plasmons with a linear dispersion relation, which can arise in the out-of-phase motion of electrons and holes of quite different masses in some semiconductors, took rather longer. Proposed by Philippe Monthoux and David Pines in 1956, they were detected thirty years later.

Today, plasmons have joined phonons and quasiparticles (individual electrons with their associated screening clouds) as elementary excitations found in nearly all solids. Thanks to the increasing sophistication of both EELS and vacuum techniques, they are often used as diagnostic tools to study the electronic structure and possible surface contamination of new materials.

See also: EXCITATION, COLLECTIVE; PLASMA; PLASMA PHYSICS

Bibliography

PINES, D., and NOZIÈRES, P. *The Theory of Quantum Liquids* (Addison-Wesley, Reading, MA, 1966).

DAVID PINES

POLARITY

Electric and magnetic forces, unlike gravitational forces, can be attractive or repulsive. Whether two electrical charges attract or repel each other depends on their polarity. Charges with the same polarity repel each other and charges with opposite polarity attract each other. Coulomb's law states that the force between two charges is proportional to the product of the charges and inversely proportional to the square of the distance separating them.

At the subatomic level, many elementary particles are electrically charged. The polarity assigned to the charge on an electron is negative and to the charge on a proton is positive. An object becomes positively charged by losing electrons and negatively charged by gaining electrons. A simple experiment illustrates the electrical forces that can occur between objects. Rubbing a glass rod with a piece of silk cloth transfers electrons from the rod to the silk leaving the rod with a positive polarity. The charged rod will attract bits of paper placed near it. Although the bits of paper are electrically neutral, the rod causes electrons in the paper to move toward it. This makes the side of the paper nearest to the rod negatively charged and the side away from the rod positively charged. Since electrical forces decrease with distance, the attractive force between the positive rod and the electrons on the near side of a piece of paper wins out over the repulsive force between the rod and the positively charged far side of the paper. Thus, the bits of paper move toward the rod. Once the bits of paper touch the rod, however, electrons move from the paper to the rod. The bits of paper become positively charged, but the number of electrons transferred is insufficient to affect the polarity of the rod. The rod remains positively charged and repels the now positively charged bits of paper. The same thing happens with a hard rubber rod rubbed with a piece of fur. However, in this instance, the rubber rod acquires electrons from the fur and becomes negatively charged.

Magnetic forces are somewhat different because magnetic poles do not exist in isolation. Nevertheless, a bar magnet behaves as if it had magnetic charges of opposite polarity near its ends. If one end of a bar magnet attracts one end of another magnet, it will also repel the other end of that magnet. The earth has magnetic properties that liken it to a bar magnet with one end near the North Pole and the other near the South Pole. The polarity of a magnet is defined by which of the earth's poles attracts the magnet. The end of a compass that points toward the magnetic North Pole of the earth has a north or positive polarity, while the end that points toward the magnetic South Pole has a south or negative polarity. Thus, the earth itself has negative polarity at the North Magnetic Pole and positive polarity at the South Magnetic Pole.

See also: CHARGE; CHARGE, ELECTRONIC; COULOMB'S LAW; MAGNET; MAGNETIC BEHAVIOR; MAGNETIC MATERIAL; MAGNETIC MONOPOLE; MAGNETIC POLE

Bibliography

FISHBANE, P. M.; GASIOROWICZ, S.; and THORTON, S. T. *Physics for Scientists and Engineers,* 2nd ed. (Prentice Hall, Upper Saddle River, NJ, 1996).

HOWARD SAMS EDITORIAL STAFF. *Basic Electricity and an Introduction to Electronics,* 3rd ed. (Howard Sams, Indianapolis, IN, 1975).

GIULIO VENEZIAN

POLARIZABILITY

Atoms, which normally have spherically symmetric electron charge distributions centered on their nuclei, become elongated in the presence of an electric field **E**. This stretching is caused by the opposite directions of the forces due to the field felt by the oppositely charged nucleus and electron cloud. It results in an atomic dipole moment, $\mathbf{p} = Ze d\hat{\mathbf{z}}$, where Z is the number of electrons in the atom, e is the electron charge, and d is the average separation between the electrons and the nucleus along $\hat{\mathbf{z}}$, the direction of **E**. The polarizability, α, of an atom is defined by the equation $\mathbf{p} = \alpha\mathbf{E}$, and has units of centimeters cubed in the CGS system. An analogous definition holds for the polarizability of molecules, which may also have permanent dipole moments due to their chemical structure (e.g., HF). If **E** oscillates sinusoidally in time with an angular frequency ω, α will depend on ω and is called the dynamic (as opposed to static) polarizability. The two quantities are essentially equal unless ω is comparable to optical transition frequencies.

The size of atomic polarizabilities can be estimated as follows. Electric fields applied to the atom will generally be only a small fraction of fields in the atom's interior; it is difficult to generate laboratory fields much greater than 10^5 V·cm^{-1}, whereas the electric field that the proton exerts on the electron in a hydrogen atom is 5×10^9 V·cm^{-1}. The atom can be expected to stretch by a fraction of its size comparable to the ratio of the applied to the atomic electric fields:

$$\frac{d}{a} \approx \frac{E}{e/a^2},$$

where a is a length characteristic of atomic diameters, typically 1 Å. Thus we find that α is of the order of the size parameter cubed, or about 10^{-24} cm³. Actual atomic values range from 0.2 (in units of 10^{-24} cm³) for He to 60 for Cs. Molecular polarizabilities are also in this range, even for large polyatomic species. On the periodic chart of the elements, α generally increases with row number and decreases with column number, and is closely correlated to the ionization potential.

Two standard techniques for determining polarizabilities are to measure the dielectric constant of a gaseous sample, or to measure the deflection of a beam of the atoms or molecules to be studied in an inhomogeneous electric field. The former method involves the use of a capacitor with the sample introduced between its electrodes as a dielectric medium. The capacitance C of the device can be measured with a bridge circuit with $C = C_o K$, where C_o is the capacitance with no sample and K is the dielectric constant. Then

$$\alpha = \frac{K-1}{4\pi N},$$

where N is the number density of the sample. In an inhomogeneous electric field, an electric dipole feels a force in the direction of the maximum field gradient. Thus the field creates the dipole and its spatial derivative deflects it. If the atomic sample is formed into a narrow beam that is perpendicular to the field and its gradient, the magnitude of the beam's deflection is then a direct measure of α.

See also: ATOM; CAPACITOR; DIELECTRIC CONSTANT; DIPOLE MOMENT; FIELD, ELECTRIC; MOLECULE

Bibliography

MILLER, T. M., and BEDERSON, B. "Atomic and Molecular Polarizabilities: A Review of Recent Advances" in *Advances in Atomic and Molecular Physics,* Vol. 13, edited by D. R. Bates and B. Bederson (Academic Press, New York, 1977).

PURCELL, E. M. *Berkeley Physics Course,* Vol. 2: *Electricity and Magnetism,* 2nd ed. (McGraw-Hill, New York, 1985).

TIMOTHY GAY

POLARIZATION

Polarization is a property of any transverse wave. A transverse wave is a wave in which the direction of vibration is perpendicular to the direction of propagation, such as in the case of light. Light is composed of oscillating electric and magnetic fields. The polarization of a light wave depends on the direction of the electric field vector that lies in a plane perpendicular to the direction of propagation of the light wave.

In general, natural or ordinary light is not polarized. This means that the directions of the electric fields of successive waves are randomly distributed. However, in certain cases, the electric field may point in one direction only, producing linearly polarized light. A linear polarizer is a material that can polarize natural light by only transmitting those waves that have electric fields pointing in one specific direction. If linearly polarized light is passed through another polarizer whose transmission axis is not lined up with the light's polarization axis, only a fraction of the light will be transmitted.

Light can also be circularly polarized. This occurs when the electric field has components along two perpendicular directions that are equal in amplitude but 90° out of phase. The electric field vector then traces out a circle as a function of time. Depending on the sense of rotation of this vector with respect to the direction of propagation, the light can be either left circularly polarized or right circularly polarized. In general, the amplitudes in the two directions do not have to be equal, nor is their phase difference necessarily 90°. The wave can then also be elliptically polarized, where the tip of the electric field vector traces out an ellipse as a function of time.

An ensemble of transverse light waves may contain some waves that are polarized and some that are unpolarized. This is called partially polarized light.

History

The polarization of light was first discovered in 1690 by Christiaan Huygens. He found that a ray of light passing through a crystal became divided into two rays of equal intensity, except when the light passed through the crystal in a direction parallel to the axis of the crystal. He also discovered that when an emergent ray passed through, a second crystal would divide into rays of equal or unequal intensity, depending on the orientation of the second crystal. However, Huygens did not understand the nature of polarization, and over a century passed before his discovery was supplemented by that of scientists such as Louis Étlenne Malus and David Brewster.

Production of Polarized Waves

Huygens's observation was an example of producing polarization through selective absorption processes. When light passes a medium which is anisotropic, meaning that the x and y directions respond differently to the entering electromagnetic wave, the electric field vector in a certain direction is partially or fully absorbed such that the electric field of the emergent wave in that direction is reduced, thus decreasing the intensity of the transmitted light. Polarized light can also be produced by controlling the radiation process, such as through the design of an antenna or the orientation of the radiating atoms.

Polarization can also result from reflection from a surface or the scattering of light off an atom of a molecule. Linearly polarized light passing through a piece of cellophane or clear plastic can be made to become circularly polarized. These materials have the property that the index of refraction is different along different directions. Since the speed of the light in the material (and thus its electric field vector) depends on the index of refraction, this has the effect of pushing the components of linearly polarized light out of phase. By changing the thickness of the material, one can control the phase difference between the perpendicular components of the vector, producing circularly polarized light.

Applications

Polarization can be used to determine the directional properties of a medium. For example, the orientation of an atom or molecule can be determined by examining the dependence of its photoabsorption on incident polarized light. For distant galaxies, the existence of a galactic magnetic field and its direction can be determined from the polarization of the light emitted from the Galaxy.

Atoms or molecules may become polarized through the process of photoabsorption of polarized light. These polarized atoms or molecules may be used to enhance or reduce chemical reaction rates since their reactivity depends on their directional properties. With the availability of lasers that can be made polarized and synchrotron radiation that is generally polarized, the range of scientific studies and applications based on the polarization of light continues to grow.

See also: HUYGENS, CHRISTIAAN; POLARIZABILITY; POLARIZED LIGHT; POLARIZED LIGHT, CIRCULARLY; SCATTERING, LIGHT

Bibliography

WOOD, R. W. *Physical Optics* (Dover, New York, 1967).

CHII-DONG LIN

POLARIZED LIGHT

Light, deduced Thomas Young, is a wave of vibrations that occur in the plane perpendicular (transverse) to the direction of propagation. The light is polarized if those vibrations are not random in direction. One usually takes the polarization direction to be that of the electric field.

Linear, Circular, and Elliptical Polarization

If the electric field everywhere along a wave vibrates back and forth along one direction, the light is said to be linearly polarized in that direction, or plane polarized in the plane containing both that direction and the direction of propagation. Reflected or scattered light tends to be linearly polarized along the direction perpendicular to the plane of incidence or of scattering (that is, the plane containing the propagation vectors both before and after reflection or scattering).

If, in contrast, the electric field at any fixed point is of constant magnitude but its direction rotates in the transverse plane, the light is circularly polarized. If the rotation is counterclockwise as viewed by someone receiving the light, it is said to have positive helicity, meaning that the photon's spin angular momentum is oriented parallel to its linear momentum. If the rotation is clockwise, the light has negative helicity, and the spin and linear momentum are antiparallel.

Intermediate to circular and linear polarization is elliptical polarization, in which the electric field at a fixed position in space rotates while changing magnitude, sweeping out an ellipse. Linearly polarized light is elliptically polarized in the limit that the eccentricity approaches 1, whereas circularly polarized light is elliptically polarized in the limit that the eccentricity approaches 0.

Circularly polarized light is said to be right- or left-handed; unfortunately, authors differ on the use of these terms. According to "natural" electromagnetic theory and particle physics, a wave of positive helicity is right-handed because at a fixed point in space the field turns in the sense of a right-handed screw moving in the direction of propagation. On the other hand, at an instant in time, the field traces out a left-handed spiral in space and, furthermore, the counterclockwise motion in the transverse plane, as seen by someone looking toward the source, may also be described as left-handed. Consequently, most chemists and workers in traditional optics refer to positive-helicity waves as left-handed.

Monochromatic Waves

Most light sources produce waves of many different frequencies. If these waves pass through a prism or diffract off a grating with many fine parallel lines, they are dispersed with components of different frequency traveling at different angles. The portion of the light passing through a thin slit aligned perpendicular to the dispersion plane is restricted to a small range of frequencies or colors. As the slit is made narrower, the light approaches monochromatic radiation, characterized by a single color and a single frequency of oscillation. Light from a single atomic transition—as emitted, for example, by a single-mode laser—is very nearly monochromatic.

The vibrations of the electric field in monochromatic waves are sinusoidal in time and space. According to Maxwell's equations, the electric field in the plane perpendicular to the propagation direction behaves like a pendulum bob suspended on a string from a point. The pendulum can swing back and forth, from side to side, or in some combination of such motions. Its position at any instant can be described by two components of its horizontal displacement. (Variations in height, which are ultimately responsible for the restoring force, need not be considered explicitly because the height is simply related to the horizontal displacement.) Each component of the horizontal displacement oscillates

sinusoidally and is independent of the other component. The pendulum is said to have two degrees of freedom. Both components of the displacement vibrate at the same frequency. One can set the pendulum in motion with any combination of side-to-side and back-and-forth motions, but once started, it continues to vibrate in a fixed pattern, tracing out a closed elliptical orbit that repeats at the pendulum frequency. (Here both the friction that tends gradually to reduce the amplitude of the motion, and the rotation of the earth, which causes a slow precession of the ellipse, are ignored.)

The electric field in a monochromatic light wave is similar. At a given position, the components of the horizontal displacement of the pendulum are replaced by the components of the electric field in the transverse plane. At any point in space, the field traces out a closed ellipse, and the field cycles around the ellipse at the frequency of the wave. There is no randomness to the vibrations of the electric field, and the monochromatic wave is said to be fully polarized. Unpolarized or partially polarized light exists only when waves of different frequency are mixed together incoherently.

The elliptical polarization of any monochromatic wave can be expressed either as a superposition of two plane-polarized waves or as a superposition of two circularly polarized waves rotating in opposite directions. The amplitudes of the two circularly polarized waves must be equal if the resulting superposition is to give linear polarization, and one of the amplitudes must vanish if a circularly polarized wave is to result.

The polarization of a light beam can be represented by a vector in an abstract three-dimensional space. The vector is of unit length if the beam is fully polarized and its direction uniquely gives the type of polarization. For partially polarized light, the length of the vector gives the fraction of polarization in the beam. The vector and a parameter specifying the total intensity of the beam are given by the four Stokes parameters. The polarization can be analyzed by observing the transmission of the light through a quarter-wave plate followed by a linear polarizer as a function of the alignment of the quarter-wave plate.

Polarizing Elements

When electromagnetic waves pass through matter, their fields exert oscillating forces on the charges in the medium. The frequencies of visible light are high enough (up to roughly 10^{15} Hz) that only the

Figure 1 A Nicol prism, used as a linear polarizer, employs internal reflection to eliminate one polarization component of light in a birefringent crystal like calcite.

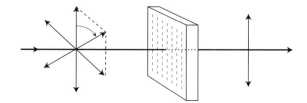

Figure 2 Polaroid film contains a dichroic material that absorbs one plane-polarized component of the incident beam.

electrons move appreciably, with typical amplitudes of a fraction of a nanometer $(10^{-9}$ m). They absorb energy from the wave and then act as a miniature antenna array to reradiate it at the same frequency. The interaction slows the propagation of the wave and thus reduces the speed of light (its wave velocity) in the medium. When the speed of light changes, so, generally, does its direction of propagation, a phenomenon called refraction. The ratio of the speed of light in a vacuum to that in the medium is called the index of refraction of the medium.

When light is incident on the interface between two uniform media with indices of refraction n_1 and n_2, part of the light is transmitted into medium 2 and the rest is reflected back into medium 1. The amount reflected, given by the Fresnel equations, depends on the polarization. When the angle between the incident light and the normal vector to the interface equals the Brewster angle $\theta_B =$ arctan (n_2/n_1), the reflected light is fully polarized with the electric vector perpendicular to the plane of incidence. At this angle, the perpendicular polarization vector in medium 1—that is, the one lying in the plane of incidence—is parallel to the refracted propagation direction in medium 2, and the vibrations generated in medium 2 have no component in this direction. A stack of plates, all oriented at the Brewster angle, can reflect as much as 99 percent of the light polarized perpendicular to the plane of incidence and thus can serve as a simple polarizer of the transmitted light.

The index of refraction generally depends on the frequency of the wave; close to resonant transition frequencies of the medium, it may change rapidly and some of the energy may be absorbed. If the distribution of matter is not homogeneous over distances of roughly a wavelength—for example, in dilute gases or in condensed matter with marked inhomogeneities—some of the reradiated energy will be scattered away from the incident direction.

In most isotropic media, absorption and refraction are independent of light polarization. Crystalline solids, however, are generally anisotropic and frequently display double refraction or birefringence, in which waves of perpendicular linear polarizations have different indices of refraction. Crystalline solids may also be dichroic, which means that light of different linear polarizations experiences differing amounts of absorption. Such anisotropic material in effect splits incident light into components of perpendicular polarizations, and refracts or absorbs the two components independently.

By exploiting the different propagation directions of refracted components in a birefringent material like calcite, and subjecting one of them to total internal reflection at an oblique interface, one can construct linear polarizers known as Nicol prisms (see Fig. 1). Since the work of Edwin H. Land in the late 1920s, large, inexpensive linear polarizers can be made from Polaroid sheets, which are plastic films containing many small, needle-shaped crystals of a dichroic material such as quinine iodosulfate (herapathite), all aligned along a common axis (see Fig. 2).

A layer of birefringent material will shift the relative phases of the components by an amount proportional to its thickness. In a quarter-wave plate, the component polarized along the slow axis is shifted $n + 1/4$ cycles more than the component polarized along the fast axis, where n is a nonnegative integer. Circularly polarized light incident on one side will emerge from the other side linearly polarized along a direction that bisects the fast and slow axes, and vice versa: incident light linearly polarized along a bisector will emerge circularly polarized (see Fig. 3). The combination of a linear polarizer followed by a quarter-wave plate can thus produce a circular polarizer.

Polarizers are often called polarization filters, but it should be understood that because of the way electromagnetic waves superimpose, polarizers can actu-

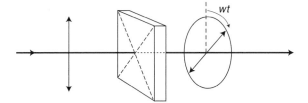

Figure 3 A quarter-wave plate shifts the relative phases of the two polarization components by $n + 1/4$ cycles, where n is an integer. When the incident light is linearly polarized along the direction that bisects the fast and slow axes of the plate, circularly polarized light emerges.

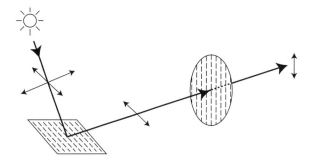

Figure 4 The polarization direction in Polaroid sunglasses is usually vertical in order to reduce glare reflected from horizontal surfaces.

ally change the direction of polarization present. Thus, crossed linear polarizers, with perpendicular polarization directions, block or filter out all light. If one were to think of the polarizers as passive filters acting on individual photons of light, then the crossed polarizers would continue to block all light even if a differently aligned filter were sandwiched in between. However, this is not the case, as a simple experiment will verify: if a linear polarizer with an intermediate polarization axis is inserted between the crossed polarizers, some light is transmitted through the sandwich.

Solutions of molecules, such as sugar, that lack a center of inversion often have different indices of refraction for electromagnetic waves of opposite helicity, a phenomenon referred to as optical activity or circular birefringence. Many simple molecules behave in a similar manner when placed in a magnetic field aligned along the direction of wave propagation. This behavior, called the Faraday effect, is related to the Zeeman effect, which splits absorption lines in magnetic fields. Incident waves are effectively split into waves of opposite helicity, and the relative phase difference of these waves increases as the light propagates through the solution. If plane-polarized light is used, the plane of polarization will be rotated by an amount proportional to the product of the path length and the difference in the indices of refraction for the two helicities. A measurement of the rotation can help identify the solute or its concentration. If the presence of the magnetic field is the cause of the rotation, then the rotation continues to increase when the direction of the light is reversed. In optically active solutions, in contrast, the rotation is reversed with the direction of propagation.

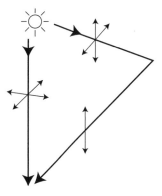

Figure 5 Although direct sunlight is essentially unpolarized, sunlight scattered by the atmosphere is linearly polarized perpendicular to the scattering plane. Some animals can detect this polarization and use it for navigation.

Uses of Polarized Light

A major use of polarizers is to reduce the glare of light reflected off polished surfaces. Polaroid sunglasses usually transmit only light with vertically oscillating electric fields, thus eliminating much of the light reflected from horizontal surfaces, which is predominantly polarized in the horizontal plane (see Fig. 4). Polarizers on camera lenses can be rotated to reduce the glare from surfaces of any orientation or to reduce the intensity of sunlight scattered in the sky (see Fig. 5). Eyeglasses with lenses made of linear polarizers aligned perpendicular to each other can be used to produce three-dimensional effects in movies that project scenes onto the screen from two slightly different perspectives, through corresponding polarizers.

The polarization of spectral lines and of diffuse light often indicates important information about the sources. In atomic collision experiments in which atoms are excited by collisions with photons, electrons, ions, or atoms, it gives information about the atomic multipole moments that can be interpreted in terms of the shape of the excited atom. The polarization data can also assist in the identification of the electronic transitions or emission mechanisms responsible for the radiation, and provide data on the interactions experienced in the source. Polarization thus is a versatile tool that is applied to laboratory plasma diagnostics as well as to spectra from astronomical sources.

Polarimeters that can measure the rotation of the plane of polarization to about 10^{-4} degree are commonly used to analyze bulk properties of crystals and solutions. Recent refinements, driven by the desire to measure electroweak interactions in atoms, have increased the sensitivity of such instruments by more than an order of magnitude. Measurements are often carried out as a function of the frequency of the transmitted light in instruments called spectropolarimeters. Ellipsometers measure the change in polarization upon reflection and are useful in measuring properties of thin films and surfaces.

Crossed polarizers are often used to measure strain in transparent materials such as plastic and glass. The strain is seen as regions of induced birefringence, manifested by contours of dark lines and, if white light is used, by bands of color. Some materials, such as liquid nitrobenzene, become birefringent under the application of a strong electric field. When placed between crossed polarizers, they can form fast optical switches or shutters, turned on and off by the application of an electric field. Liquid crystal display (LCD) panels, common in read-outs for calculators, digital watches, voltmeters, and other instruments, use relatively small applied electric fields to orient rod-shaped nematic molecules and thereby to change the polarization of light and its transmission through a polarization filter.

Although the human eye is not very sensitive to different types of polarization, some animals, such as some ants, bees, and horseshoe crabs, can detect linear polarization, orient themselves, and navigate based on the polarization from scattered sunlight. Many humans are able to see a fleeting image of a yellow bow tie, known as Haidinger's brushes, oriented perpendicular to the electric-field vector of polarized light. It is only a few degrees wide and appears close to the point of fixation when the linear polarizer is rotated in front of a plain white or blue background.

See also: BIREFRINGENCE; BREWSTER'S LAW; DICHROISM; DICHROISM, CIRCULAR; DIFFRACTION; ELECTROMAGNETIC WAVE; FARADAY EFFECT; FIELD, ELECTRIC; FIELD, MAGNETIC; GRATING, DIFFRACTION; LIGHT; OPTICS; POLARIZATION; POLARIZED LIGHT, CIRCULARLY; REFRACTION; SCATTERING, LIGHT; WAVE MOTION; ZEEMAN EFFECT

Bibliography

BAYLIS, W. E.; BONENFANT, J.; DERBYSHIRE, J.; and HUSCHILT, J. "Light Polarization: A Geometric-Algebra Approach." *Am. J. Phys.* **61,** 534–545 (1993).

BORN, M., and WOLF, E. *Principles of Optics,* 3rd ed. (Pergamon Press, Oxford, Eng., 1965).

KLIGER, D. S.; LEWIS, J. W.; and RANDALL, C. E. *Polarized Light in Optics and Spectroscopy* (Academic Press, San Diego, CA, 1990).

WATERMAN, T. H. "Polarized Light and Animal Navigation." *Sci. Am.* **193,** 88–94 (1955).

WILLIAM E. BAYLIS

POLARIZED LIGHT, CIRCULARLY

Electromagnetic radiation is transmitted in the form of a transverse wave in which the electric and magnetic fields at any point in space and time lie in a direction perpendicular to the direction of motion. This is a beam of light. The polarization of the beam is described by the locus of points which the tip of the electric field vector traces out as the wave is propagated. The radiation is circularly polarized when these points follow a spiral centered about the direction of motion. The radius of the spiral is equal to the amplitude of the electric field vector, $|E_0|$. The beam is right circularly polarized when the tip of the vector rotates in the clockwise direction, when looking in the direction of the on-coming radiation, as a function of time and left circularly polarized when the rotation is counterclockwise. This relationship can be expressed mathematically by imagining mutually perpendicular axes, x and y, in a plane at position z_0 along the direction of propagation and writing

$$E_x = E_0 \cos(\omega t - kz_0); \quad E_y = E_0 \cos\left(\omega t - kz_0 + \frac{\pi}{2}\right),$$

$$E_x = E_0 \cos(\omega t - kz_0); \quad E_y = E_0 \cos\left(\omega t - kz_0 - \frac{\pi}{2}\right)$$

for right and left circularly polarized radiation, respectively.

Circularly polarized light can be produced or analyzed on an optical bench through the use of a quarter-wave plate made from a naturally birefringent material such as calcite or quartz cut with the optic axis parallel to the surface. Radiation incident normal to the surface of the crystal and linearly polarized at 45° with respect to the optic axis will be split into two components of equal amplitude, one parallel and one perpendicular to the optic axis. The thickness d of the quarter-wave plate is chosen so that the difference in phase shift acquired between the two components upon passing through the crystal is exactly $\pi/2$, as required by the above set of equations. This is equivalent to a difference in optical path length of $|n_0 - n_e| d = \lambda/4$, where n_0 and n_e are the indices of refraction governing the behavior of the two components. The direction, or handedness, of the circular polarization depends on the choice of the angle, ±45°, which the linearly polarized radiation makes with the optic axis, whether the birefringent crystal is positive, $n_0 < n_e$, or negative, $n_0 > n_e$.

The absorption of circularly polarized light by atoms leads to the creation of an excited state in which the angular momentum is preferentially oriented in a given direction. This property can be used to produce an ensemble of spin-polarized atoms by optical pumping with circularly polarized light. Such a state has a net magnetic moment which is observable through its interaction with an applied magnetic field. The application of a magnetic field to certain materials can also cause the materials to exhibit circular dichroism, the Faraday effect. A series of magnetic domains in a thin film or in a solid surface will also respond differently to right and left circularly polarized radiation. This makes surface photoemission with circularly polarized radiation a sensitive tool for the study of the magnetic structure of materials.

See also: Birefringence; Dichroism, Circular; Electromagnetic Radiation; Faraday Effect; Field, Magnetic; Polarization; Polarized Light

Bibliography

Fowles, G. R. *Introduction to Modern Optics,* 2nd ed. (Holt, Rinehart and Wilson, New York, 1975).

Hecht, E. *Optics,* 2nd ed. (Addison-Wesley, Reading, MA, 1987).

Klein, M. V., and Furtak, T. E. *Optics,* 2nd ed. (Wiley, New York, 1986).

C. Denise Caldwell

POLARON

An electron in the conduction band of a solid interacts with the other electrons in the ion cores. This coupling, since it arises from the interaction of many indistinguishable electrons, can be long range (if the cores are charged as in polar materials like NaCl) or short range (if electrons are excluded from the already occupied orbitals, as in Si).

The periodic part of the core interaction, which comes about when the ion cores are in their equilibrium positions, leads to a band structure for the extra electron. This means that the electron in question has a nontrivial energy $E(\mathbf{k})$ versus crystal wave vector $\mathbf{k}$. In some cases, the free particle energy-momentum relation, that is, $E(\mathbf{k}) = \hbar^2 k^2 / 2m^*$ with an effective mass m^*, is a good approximation ($\hbar =$ Planck's constant).

Deformation of the ideal lattice, which in most cases is accurately described in terms of simple harmonic vibration of the atoms in the crystal (phonons), leads to coupling between the motions of the single conduction band electron and the phonons. The coupling between electrons and phonons is quantitatively characterized by an interaction Hamiltonian (energy operator) of the form,

$$H_{\text{int}} = \sum_{\mathbf{q}} Q_{\mathbf{q}} (b_{\mathbf{q}}^+ - b_{-\mathbf{q}}) e^{i\mathbf{q}\cdot\mathbf{r}}.$$

The operator $b_{\mathbf{q}}^+ (b_{\mathbf{q}})$ creates (destroys) phonons of momentum $\hbar\mathbf{q}$. The ordinary function $Q_{\mathbf{q}}$ characterizes the strength of the coupling, and the vector $\mathbf{r}$ is the coordinate of the electron.

In many crystals there are different kinds of phonons; for example, in an ionic crystal like NaCl there are acoustic phonons and optical phonons.

For the longitudal optical phonons, the case most often considered, the frequency $\omega(\mathbf{q}) \cong \omega$ a constant and

$$Q(\mathbf{q}) \sim \sqrt{\alpha}\,\frac{1}{\mathbf{q}}.$$

For this so-called Frohlich coupling, the interaction strength increases without limit as $\mathbf{q} \to 0$ because of the long-range force between the conduction electron and the charged polarizable, vibrating ions. The dimensionless coupling constant is

$$\alpha = \frac{e^2}{\hbar}\left[\frac{1}{\varepsilon_\infty} - \frac{1}{\varepsilon}\right]\sqrt{\frac{m^*}{\hbar\omega}}.$$

Here $\varepsilon_\infty(\varepsilon)$ are the high (low) frequency dielectric constants of the solid. In many cases (i.e., for real materials) α may be greater than unity. For example $\alpha \cong 5$ for NaCl.

The main effect of the electron phonon coupling is to dress the electron with a cloud of polarization (i.e., an electron in motion is accompanied by crystal lattice distortions surrounding it). This dressed object has been called a polaron. It has a complex internal structure, but it still interacts with the phonons of momentum $\mathbf{q}$ with an effective dressed interaction. This residual coupling scatters the polaron from one crystal momentum $\hbar\mathbf{k}$ to another state $\hbar\mathbf{k}'$ ($\mathbf{q} = \mathbf{k}' - \mathbf{k}$) and also leads to a weak attractive interaction between pairs of polarons.

Interest in the polaron problem, which began in the 1930s, has continued for many reasons. First, it is relevant to the applied physics of semiconductors (i.e., transistors in Si, quantum-well lasers in GaAs, etc.). Second, it provides a simple physically relevant model of a nonrelativistic particle, the electron, interacting strongly ($\alpha > 1$) with a quantum field, the phonons. Many unique and powerful theoretical techniques have been developed to approximately solve this problem and the theoretical results have been subject to experimental tests. In addition, the many-polaron problem has had a significant impact on the BCS theory of superconductivity and to a lesser extent on possible explanations of the new high temperature superconductors.

See also: ELECTRON; ELECTRON, CONDUCTION; PHONON; PLANCK CONSTANT; SEMICONDUCTOR; TRANSISTOR

Bibliography

KUPER, C. G., and WHITFIELD, G. *Polarons and Excitons* (Plenum, New York, 1962).

SUTTON, A. P. *Electronic Structure of Material* (Clarendon, Oxford, Eng., 1993).

P. M. PLATZMAN

POLYMER

A polymer is an extremely large molecule (macromolecule) that is composed of small, molecular subunits that are repeated over and over (hence the name poly, which means many, and mer, which means unit). For example, polypropylene is composed of the following chains:

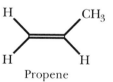
Polypropylene

It therefore may also be represented according to the repeat unit as $[\text{-CH}_2\text{CH}(\text{CH}_3)]_n$, where n equals several thousand. The sources of the repeating units are small molecules called monomers, which are linked together by chemical bonds in what is called a polymerization reaction. For polypropylene, the monomer is propene (also called propylene), which undergoes a chain reaction, converting the double bonds of the monomers to single bonds between monomer units to yield the final polymer:

Propene

Joining monomers end to end in such a fashion produces an addition polymer. This class of polymers is most often produced from alkenes; common examples include polyethylene, polystyrene, polyvinyl chloride, and Teflon. One distinctive trait of addition polymers is that the chemical composition of the polymer is identical to that of the monomer.

Another type of polymer is produced by condensation reactions in which two monomers react with each other forming chemical bonds between themselves and splitting out a small molecule (usually water). Thus, the chemical composition of condensation polymers produced in such a manner is different from that of monomers. Condensation polymers are classified according to the type of chemical linkage formed, and several types exist. Polyesters result from the reaction of a dialcohol with a dicarboxylic acid. The most common polyester solid in the United States is Dacron:

$$HOCH_2CH_2OH + HO_2C-\bigcirc-CO_2H \xrightarrow{-H_2O}$$

$$\left[-O-CH_2CH_2-O-\overset{O}{\overset{\|}{C}}-\bigcirc-\overset{O}{\overset{\|}{C}}-O-\right]_n$$

Polyamides are produced by the reaction of a diamine with a dicarboxylic acid. They constitute a family of polymers called nylon, known for forming very strong fibers. The most common nylon is nylon 66:

$$H_2N(CH_2)_6NH_2 + HO\overset{O}{\overset{\|}{C}}(CH_2)_4\overset{O}{\overset{\|}{C}}OH \xrightarrow{-H_2O}$$

$$\left[-NH(CH_2)_6NH-\overset{O}{\overset{\|}{C}}(CH_2)_4\overset{O}{\overset{\|}{C}}-\right]_n$$

Polycarbonates are clear, heat-stable polymeric materials with high impact resistance. They are produced by a condensation reaction of a dialcohol with phosgene. The most common polycarbonate is produced by the reaction of bisphenol A with phosgene:

$$HO-\bigcirc-\overset{CH_3}{\underset{CH_3}{\overset{|}{\underset{|}{C}}}}-\bigcirc-OH + \overset{O}{\overset{\|}{C}}\overset{}{\underset{Cl}{}}\overset{}{\underset{}{Cl}} \xrightarrow{-HCl}$$

$$\left[-O-\bigcirc-\overset{CH_3}{\underset{CH_3}{\overset{|}{\underset{|}{C}}}}-\bigcirc-O-\overset{O}{\overset{\|}{C}}-\right]_n$$

Polyurethanes are used as coatings when high impact and abrasion resistance is required and in foams for bedding and car upholstery. They are produced by the reaction of a dialcohol with a diisocyanate:

$$O=C=N-R-N=C=O + HO-R-OH \longrightarrow$$

$$\left[\overset{}{\underset{O}{\overset{\|}{C}}}-NH-R-NH-\overset{}{\underset{O}{\overset{\|}{C}}}-O-R-O\sim\right]_n$$

For polymers to be useful, they must be chemically inert under the conditions in which they are used; that is, they must not be attacked by air or water and, in some cases, they must be stable under high temperatures. Polymers may be divided into two broad categories based on their response to heat: thermoplastic materials, which soften and melt at elevated temperatures, and thermosetting polymers, which remain hard and eventually decompose on heating.

The physical properties of polymers make them versatile and valuable. For example, Teflon is very chemically inert and stable at high temperatures, but its slipperiness makes it useful on nonstick cooking ware. Similarly, not only do synthetic fibers resist mildew and moth infestation, but their superior tensile strength often makes them more practical than natural fibers.

Depending on the length of polymer chains and temperature, a polymer may exist either as a partially crystalline solid, a glass, a rubbery solid, or a viscous liquid. In general, the longer the average length of the polymer chains, the less like fluid the polymer is at a particular temperature. Typically, when chain lengths exceed 10,000 bonds, a polymer is solid at normal temperatures and does not display liquid flow. Polymers usually begin to develop mechanical strength after fifty monomer units have been strung together; they become stronger as the chain length increases to 500 units, but beyond this, little increase in strength is observed.

In a solid polymeric material, polymer chains are coiled and intertangled with each other much like a can of worms. If the polymer is stretched, the chains slowly untangle and the polymer appears to flow. In such a case, the material adopts the new, stretched shape. This is typical of plastics, which can usually be stretched between 20 percent and 100 percent of their length. However, if crosslinks are introduced that tie the chains together at average intervals of between one hundred and one thousand bonds, the polymer will still stretch up to ten times its unstretched length, and, when released, it will spring back to its original length. Therefore, such polymers are elastic and are often called elastomers.

An example of a highly useful elastomer is natural rubber, which is composed of cis-1,4-isoprene units. Ordinarily, this polymer is tacky, but, in a process called vulcanization, disulfide (-S-S-) crosslinks are introduced by heating a mixture of the polymer and sulfur. The resulting rubber is not only elastic but also much harder than natural rubber and no longer tacky. As a consequence of the cutoff of U.S. supplies of natural rubber during World War II, several varieties of synthetic rubber were developed. As with natural rubber, the desired degree of elasticity is realized by introducing cross-links through vulcanization. One example of a synthetic rubber is neoprene $[CH_2-CCl=CH-CH_2-]_n$, which is obtained by polymerizing 2-chloro-1,3-butadiene (chloroprene):

Natural (Hevea) Rubber

Almost all organic polymers are electrical insulators, and some are used for covering electrical wires. However, special types of polymers, such as polyacetylenes, polypyrroles, and polythiophenes, when doped with certain inorganic species, are electrical conductors. The common feature of these polymers is conjugated π bonds in the backbone of the polymer; that is, alternating double and single bonds:

Polyacetylene

To this point, only organic polymers have been discussed. There are also polymers that have a molecular backbone that does not consist mainly of carbon. These inorganic polymers tend to have a higher Young's modulus and a lower failure strain than organic molecules. Inorganic polymers are classified according to the composition of the backbone: for example, silicones $(-SiR_2O-)_n$, phosphazenes $(-N=PR_2-)_n$, alumoxanes $(-AlRP-)_n$, and polythiazenes $(-N=S-)_n$. The latter polymer is remarkable in that, like polyacetylene, it has conjugated π bonds. As a result, it displays metallic conductivity and becomes a superconductor at 0.26 K.

The best-known inorganic polymers are silicones. They are an extensive family, and, through variation of the pendant R groups and the chain length, as well as cross-linking, a wide variety of materials may be prepared, including oils, waxes, resins, and elastomers. As a result, silicones have an extremely large number of applications, especially in the area of elastomers, since they outperform most natural and synthetic organic rubbers at both elevated and low temperatures.

One particularly important application of inorganic polymers is as precursors to ceramic fibers. These fibers, which are useful for composite materials, can be prepared by forming fibers of the inorganic polymer and then pyrolyzing it at elevated temperatures. This causes the organics to be burned off, leaving the inorganic elements behind as a ceramic fiber.

See also: ELASTIC MODULI AND CONSTANTS; MOLECULAR PHYSICS; MOLECULE

Bibliography

ALPER, J.; and NELSON, G. L. *Polymeric Materials: Chemistry for the Future* (American Chemical Society, Washington, DC, 1989).

BILLMEYER, F. W., JR. *Textbook of Polymer Science,* 3rd ed. (Wiley, New York, 1984).

BIRLEY, A. W.; HEATH, R. H.; and SCOTT, M. J. *Plastics Materials: Properties and Applications,* 2nd ed. (Chapman and Hall, New York, 1988).

BRYDSON, J. A. *Plastics Materials,* 5th ed. (Butterworths, Boston, 1989).

COWIE, J. M. G. *Polymers: Chemistry and Physics of Modern Materials* (Chapman and Hall, New York, 1991).

MARK, J. E.; ALLCOCK, H. R.; and WEST, R. *Inorganic Polymers* (Prentice Hall, Englewood Cliffs, NJ, 1992).

SEYMOUR, R. B., ed. *Conductive Materials* (Plenum, New York, 1981).

STEVENS, M. P. *Polymer Chemistry: An Introduction,* 2nd ed. (Oxford University Press, New York, 1990).

ALLEN APBLETT

POSITRON

The positron is an elementary particle with the same mass and spin as the electron but with the opposite charge and magnetic moment. The positron is the antiparticle twin of the electron and, like the electron, has a mass of approximately 9.11×10^{-31} kg, or 0.511 MeV/c^2 (about $1/1,836$ that of the proton). Each carries a spin angular momentum of one-half the quantum unit ($\frac{1}{2}\hbar = 0.527 \times 10^{-34}$ J·s). The positron and electron, designated e^+ and e^-, respectively, are classified as leptons and considered point-like particles; they display no measurable size or internal structure down to about 10^{-19} m, which is the smallest scale accessible with current accelerators.

The positron was discovered by Carl D. Anderson in 1932 during a study of cosmic-ray tracks deflected by a magnetic field in a cloud chamber. With this apparatus, Anderson observed tracks of particles that interacted with a lead plate in the same way as high-energy electrons but that curved in the opposite direction, indicating that they were positively charged.

The first hint of the existence of the positron had arisen some four years earlier from the theoretical calculations of P. A. M. Dirac. Dirac combined the ideas of quantum mechanics and special relativity to produce a mathematical description for the behavior of spin one-half particles. This description, encompassed in a differential equation that now bears his name, provided a quantitative description of the spin-related properties of the electron, including an extremely accurate calculation of the magnetic moment of the electron and of the energy levels of the hydrogen atom. This equation has four distinct solutions, two of which describe the properties of the electron and correspond to the two possible orientations of the spin. The other two solutions, which imply the existence of states with negative energy, were initially dismissed by skeptics as unphysical. If such states existed, it was argued, real electrons would promptly fall into them, with catastrophic implications. In 1930 Dirac proposed that these negative-energy states indeed existed but were filled by an infinite sea of electrons throughout the universe. The Pauli exclusion principle prevented any positive-energy electrons from falling into these lower levels. Any unoccupied state, or "hole," in this sea would appear as a particle with positive energy similar to the electron but with a positive charge. Dirac suggested that the proton, which was well known by then, corresponded to this positive particle, even though his theory required that it have the same mass as the electron. This suggestion was refuted by J. Robert Oppenheimer, Hermann Weyl, and others and was abandoned by Dirac when he explicitly predicted the existence of a new positive particle in 1931. The subsequent discovery of the positron was a stunning demonstration of the predictive power of this theory. Richard P. Feynman later developed an alternative view in which positive-energy antiparticles propagating forward in time were equivalent to negative-energy particles propagating backward in time. This interpretation led to the same properties for the positron but did not require the infinite sea of negative-energy electrons and had applicability to other particles as well, including spin-0 particles.

A positron is stable and will not decay spontaneously, but, when passing through matter, it will eventually collide with an electron, resulting in the annihilation of both particles. For collisions at low relative velocity, the annihilation usually produces two photons that are emitted in opposite directions, each with an energy equal to the mass of one of the colliding particles.

When an electron and a positron collide with high relative momentum and annihilate, two or more particles emerge from the collision to carry off the energy. Any combination of particles that satisfies the fundamental conservation laws (such as conservation of total electric charge) may be produced by annihilation. This combination may include particles much more massive than the electron, so long as the center-of-mass energy of the produced particles equals the center-of-mass energy of the initial e^+e^- pair. For example, when an electron and positron collide with a center-of-mass energy near 92 GeV, the most likely outcome is the production of a single massive Z^0 boson, which decays almost instantaneously into a burst of lighter particles.

Beginning in the early 1970s, a series of storage ring accelerator facilities were constructed for accelerating and colliding counter-rotating beams of electrons and positrons at ever-higher energies. In these machines, the beams of electrons and positrons are guided by an array of steering and focusing magnets around a circular or oval path through a chamber, or beam pipe, from which the air has been evacuated. A machine of this kind can typically store beams for hours while only a negligible fraction of the circulating particles are lost in collisions with stray ions or with the vacuum chamber. This technique has proven extremely powerful for the study of elementary particle physics and has led to the discovery of numerous quark-antiquark bound states and to the experimental elucidation of the electroweak force.

Positrons can be produced by either the radioactive decay of certain nuclear isotopes or by pair production, the interaction of a high-energy photon with a strong electromagnetic field, as found near the nucleus of a heavy atom, which converts the photon into an electron and positron. In nuclei that are proton rich (i.e., nuclei that have fewer neutrons than stable isotopes of the same element), a proton may decay to a neutron by the emission of a positron and a neutrino, leaving a nucleus with an atomic number reduced by one and a slightly smaller energy. For many laboratory applications, positrons may be conveniently produced with commercially prepared sources such as ^{22}Na (half-life 2.6 years) or ^{58}Co (half-life 71 days). Low-intensity positron beams (up to about 10^7 e^+/sec) based on radioactive sources have a variety of applications, especially in solid-state surface physics.

Positron sources based on radioactive decay have not been practical in the high-energy accelerators used for particle physics research because they do not provide the high peak-pulse intensity typically needed for such accelerators. Problems associated with the manufacture and handling of short-lived, intensely radioactive isotopes are also an impediment to their use. This approach, however, may prove useful in future continuous-beam accelerators.

Positron beams for high-energy accelerators are produced using the pair-production process, typically by directing a beam of energetic electrons into a plate of tungsten or similar heavy metal. The electrons interact with the tungsten nuclei to produce energetic photons (bremsstrahlung radiation), which in turn generate more electrons and positrons through pair-production interactions in the high electromagnetic fields internal to the tungsten atoms. Both the electrons and positrons produced in this way produce more photons, resulting in a cascade of particles that grows as it passes through the plate. The positrons emerging from the plate are distributed over a broad momentum spectrum and angular distribution but must be captured and accelerated to form a useful beam. Because of practical limitations on the size and performance of accelerator components, only a small fraction of these positrons end up in the final high-energy beam.

Inefficiencies in producing positrons do not usually pose a significant limitation for a storage-ring facility because the positrons may be accumulated gradually, typically over several minutes, as they are produced, accelerated, and injected into the storage ring. Storage rings in use for high-energy physics research typically operate with up to about 10^{12} each of electrons and positrons distributed among a few short bunches and with energies up to several tens of giga-electron-volts (GeV) per particle. More advanced designs under construction in the 1990s might accumulate up to about 10^{14} particles in each beam distributed among a thousand or more short bunches to achieve a high rate of e^+e^- collisions.

To reach center-of-mass energies of a few hundred giga-electron-volts or more, future facilities are likely to use two linear accelerators, one for the electrons and one for the positrons, pointed at each other end to end. Positrons will not circulate endlessly in a linear collider, but rather will be produced, accelerated, and collided once each. Such a machine will require a source of positrons intense enough to produce the desired collision probability on each pulse of the accelerators. A prototype for such a machine is the SLAC Linear Collider at Stanford University. In this machine, short (10^{-11}s) pulses of about 3×10^{10} electrons, each having been accelerated to 33 GeV, are directed into a target to produce positrons. The target is a 21-mm-thick water-cooled disk made of 74 percent tungsten and 26 percent rhenium, an alloy chosen for its high atomic number (to provide a high e^+ yield) and mechanical properties that enable it to withstand the sharp temperature rise and resulting thermal stress induced by each pulse. Each pulse of electrons produces a burst of over 10^{11} positrons, most of which are distributed within an angular radius of about 20 degrees. Positrons with momenta in the range of about 2 to 20 MeV/c are collected and accelerated to 200 MeV/c in a short high-gradient linear accelerator surrounded by strong magnetic focusing fields. The positron bunches emerging from this system are then accelerated to 1.2 GeV and processed though a small storage ring, where the properties of each bunch, including its longitudinal and transverse sizes and momentum spectrum, are transformed to make a tightly focused beam suitable for acceleration in the high-energy linac to the final e^+e^- collision point. A net yield of one positron in the final high-energy beam for each electron incident on the target has been achieved using this technique. Pulses of over 3×10^{10} positrons, repeating 120 times per second, are routinely produced at the SLAC facility for particle physics research applications.

See also: ACCELERATOR; ANTIMATTER; DIRAC, PAUL ADRIEN MAURICE; LEPTON; POSITRONIUM

Bibliography

ANDERSON, C. D. "The Positive Electron." *Phys. Rev.* **43**, 491–494 (1933).

DIRAC, P. A. M. "The Quantum Theory of the Electron." *Proc. R. Soc. London, Ser. A* **117**, 610–624 (1928).

DIRAC, P. A. M. "The Quantum Theory of the Electron, Part II." *Proc. R. Soc. London, Ser. A* **118**, 351–361 (1928).

DIRAC, P. A. M. "A Theory of Electrons and Protons." *Proc. Soc. R. Soc. London, Ser. A* **126**, 360–365 (1930).

DIRAC, P. A. M. "Quantized Singularities in the Electromagnetic Field." *Proc. R. Soc. London, Ser. A* **133**, 60–72 (1931).

SCHULTZ, P. J., and LYNN, K. G. "Interaction of Positron Beams with Surfaces, Thin Films, and Interfaces." *Rev. Mod. Phys.* **60**, 701–779 (1988).

ROGER A. ERICKSON

POSITRONIUM

Positronium (Ps) is a short-lived bound state of an electron and a positron. Positronium can be pictured as a hydrogen atom in which the proton has been replaced by a much lighter positron. This bound state has energy levels similar to those of the hydrogen atom but with half the spacing, because the reduced mass (as used in calculating atomic energy levels) is half the electron mass; that is, the center of mass of the system is midway between the two constituent particles. As a result, the ionization energy of positronium is 6.8 eV, about half that of hydrogen. The magnetic moment of the positron is much larger than that of the proton; therefore, the spin-dependent interaction produces hyperfine splitting in the positronium level structure, which is much larger than that of hydrogen. Unlike the hydrogen atom, positronium is intrinsically unstable, even in its ground state, because the electron and positron can annihilate into photons.

Positronium was first predicted in 1934 by Stjepan Mohorovicic, though it was not detected experimentally until seventeen years later. By that time, calculations of energy levels, decay rates, and other properties had already been published by several other researchers. The first observation was reported in 1951 by Martin Deutsch in an experiment in which positrons emitted from a ^{22}Na source were slowed in Freon gas until they combined with electrons. The formation of positronium was deduced by identifying decay photons with scintillation detectors. Other methods for producing positronium, involving interactions of positrons with solid surfaces or powders, have been developed to support a variety of research applications. One such method involves bombarding a metal surface with slow positrons in a vacuum. Decays of positronium ejected into the vacuum may then be observed without the complication of interactions with a gas or other material.

Positronium provides a means for testing the fundamental interactions described by quantum electrodynamics (QED), a central theory of modern physics. Because the positron and electron are both leptons and have the same mass and the same magnitude of magnetic moment, the positronium system is simple enough that QED can be used to compute observable features with great accuracy and without the ambiguities that arise when applying the theory to ordinary atoms. Theoretical ambiguities arise even with hydrogen, the simplest of ordinary atoms, because of the finite size and hadronic structure of the single-proton nucleus.

QED provides a framework for calculating the probabilities of interactions among leptons and photons and can be used to predict approximate values directly. Accurate predictions of observable physical quantities within this framework, however, also require the evaluation of more complicated processes, generally referred to as radiative corrections, involving the creation and annihilation of virtual lepton pairs and the exchange of virtual photons among the constituent particles. The energy levels and decay rates of positronium can, in principle, be calculated to very high precision with QED; however, the calculations become complex and tedious beyond the first-order radiative correction.

Positronium in its ground-energy level can exist in either of two states: a singlet state, in which the electron and positron spins are antiparallel, called parapositronium and designated $^1S_0$; or a triplet state, in which the spins are parallel, called orthopositronium and designated $^3S_1$. Charge-conjugation symmetry and angular momentum conservation allow the singlet state to decay only into an even number of photons (usually two) and the triplet state to decay only into an odd number of photons (usually three). Decay of isolated positronium into a single photon is impossible because such a decay would violate momentum conservation; however, the single photon decay of the $^3S_1$ state is possible if another particle is present to carry the recoil momentum. The probabilities of direct decays into four or five photons are much smaller by a factor of order α^2 (the fine structure constant squared) and have not yet been observed experimentally. The mean lifetimes of parapositronium and orthopositronium are about 0.1 ns and 140 ns, respectively.

Energy levels above the ground state are produced with much smaller probability, making studies of these levels more difficult. Photons emitted from the $n = 2$ to $n = 1$ transition, corresponding to the Lyman-α spectral line of hydrogen, were reported by Karl F. Canter, Allen P. Mills Jr., and Stephan Berko in 1975, along with a measurement of the fine-structure energy separation between the 2^3S_1 and 2^3P_2 states. In their experiment, positronium in the $n = 2$ state was produced by directing positrons from a ^{58}Co source onto a copper target inside a metal cavity. A radio-frequency field was applied to drive the transition from the 2^3S_1 state to the 2^3P_2 state, which occurred at 8,625 MHz, as predicted. The 2^3P_2 state decayed to the 1^3S_1 state with the emission of the 2,430Å Lyman-α photon, which was detected with a photomultiplier tube, and, finally, the 1^3S_1 state decayed to three gamma rays, which were detected with a scintillation counter. The results of this experiment were fully consistent with the predictions of QED.

Precision experimental work in the late twentieth century sparked renewed interest in positronium. Several experiments have measured decay rates of orthopositronium that are slightly higher, on the order of one part in a thousand, than predicted by theoretical estimates. Experiments of this kind are done with great care to identify and correct for any small effects that could influence the outcome of the measurements. This discrepancy has prompted refined calculations of higher-order corrections and even led to speculations about possible exotic particles, as yet undetected, into which positronium might decay. As experiments are performed with greater accuracy and theoretical calculations are carried out with greater sophistication, this discrepancy may disappear. If the discrepancy persists, it would pose a serious problem for QED.

See also: POSITRON; QUANTUM ELECTRODYNAMICS; SPECTRAL SERIES

Bibliography

CANTER, K. F.; MILLS, A. P., JR.; and BERKO, S. "Observations of Positronium Lyman-α Radiation." *Phys. Rev. Lett.* **34,** 177–180 (1975).

DEUTSCH, M. "Evidence for the Formation of Positronium in Gases." *Phys. Rev.* **82,** 455–456 (1951).

DEUTSCH, M. "Three-Quantum Decay of Positronium." *Phys. Rev.* **83,** 866–867 (1951).

MILLS, A. P., JR.; BERKO, S.; and CANTER, K. F. "Fine-Structure Measurement in the First Excited State of Positronium." *Phys. Rev. Lett.* **34,** 1541–1544 (1975).

RICH, A. "Recent Experimental Advances in Positronium Research." *Rev. Mod. Phys.* **53,** 127–165 (1981).

ROGER A. ERICKSON

POTENTIAL

See CHEMICAL POTENTIAL; ELECTRIC POTENTIAL; ENERGY, POTENTIAL; IONIZATION POTENTIAL

POTENTIAL BARRIER

The term "potential barrier" is used when systems are described in terms of energy flow. For example, consider a simple pendulum as shown in Fig. 1. The pendulum has its greatest kinetic energy, that is, energy by virtue of motion, when it passes through its equilibrium position at $\theta = 0$ (straight down). The pendulum has its greatest potential energy, that is, energy by virtue of location, when it reaches its maximum height at $\theta = \pm\theta_0$. The sum of the kinetic and potential energies is the total energy, which is constant in this case. As the pendulum rises, its potential energy increases, and since the total energy remains constant, its kinetic energy must decrease. Because kinetic energy never can be negative, the pendulum must stop when its total energy becomes all potential. One way of expressing the fact that the potential energy cannot increase any further is to say that the pendulum has "run into a potential barrier." The point at which this occurs is called a turning point since the pendulum reverses its direction of motion there.

Potential barrier acquires a more visual meaning when the energy flow is illustrated graphically, in an energy diagram. The energy diagram representing the motion of a simple pendulum is shown in Fig. 2. The horizontal axis represents the angular displacement from equilibrium, with $\theta = 0$ denoting the equilibrium position (straight down) and $\theta = \pm \pi$ denoting the totally inverted position (straight up). The vertical axis represents energy. The solid curve shows the potential energy at each θ, the line of pluses $(+++)$ shows the total energy at each θ, and the difference between them is the kinetic energy at each θ. The pendulum is represented by a point on the line of constant energy. At $\theta = 0$, the point representing the pendulum is on the line of constant energy above $\theta = 0$, and the energy is all kinetic. As the pendulum moves away from equilibrium in the direction of increasing θ, the point representing it remains on the line of constant energy and also moves in the direction of increasing θ. Since the total energy remains constant while the potential energy increases, the kinetic energy must decrease. Thus, the diagram shows that the pendulum must slow down as it moves away from equilibrium. The farther from equilibrium the pendulum moves, the smaller its kinetic energy becomes until the total energy is all potential at $\theta = \theta_0$, whence the pendulum stops. In the energy diagram, the point representing the system also stops at θ_0 and has the appearance of "running into a potential barrier," that is, into the potential energy curve. After hitting the potential energy curve at θ_0, the system point moves back along the line of constant energy in the direction from which it came until it hits the potential energy curve at $-\theta_0$, where it again stops and reverses its direction of motion. Thus, potential barrier acquires a more picturesque meaning when we watch the system point in an energy diagram bounce off the potential energy curve at each turning point, since its motion resembles that of a real elastic ball bouncing off real barriers at $\pm \theta_0$.

The simple pendulum is an example of a bound system, that is, of a system whose motion is confined to a bounded region. Potential barriers also occur for systems that are unbounded. For example, consider an elastic ball hitting a wall. Describing what happens in terms of forces, we say that the wall exerts a force on the ball, which eventually brings the ball to rest and reverses its direction of motion. Viewing this same process in terms of energy flow, we say that as the ball hits the wall, its kinetic energy decreases while its potential energy increases, since

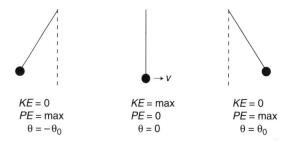

KE = 0
PE = max
$\theta = -\theta_0$

$\rightarrow v$
KE = max
PE = 0
$\theta = 0$

KE = 0
PE = max
$\theta = \theta_0$

Figure 1 A simple pendulum. When the pendulum moves through its equilibrium position, the total energy is all kinetic. At the highest point in its motion, the total energy is all potential, and the pendulum stops since it has run into a potential barrier. (Courtesy of T. Jayaweera)

kinetic energy is being stored in the ball's deformation. When the ball comes to rest against the wall, we say it has "run into a potential barrier" because its total energy is all potential. Since the ball can proceed no further in its original direction of motion, it begins moving back in the direction from which it came. Once the ball moves away from the wall, its motion is unbounded since there is nothing in this direction to impede its travel. Thus, potential barriers also occur in unbound systems such as balls bouncing off walls, alpha particles colliding with heavy nuclei, carts rolling up hills, and so on.

In general, one can use either the force or energy framework for describing and predicting motion.

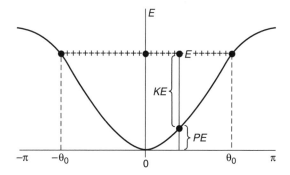

Figure 2 The energy diagram for the simple pendulum of Fig. 1. The point representing the pendulum moves along the line $(+++)$ of constant energy E, and the pendulum's potential energy is denoted by the solid curve. The turning points of the motion, where the pendulum runs into potential barriers, are denoted by $\pm \theta_0$. (Courtesy of T. Jayaweera)

The energy description, however, has a distinct advantage: A great deal of information can be obtained from studying energy flow without actually having to solve any equations of motion. This simplifies many problems and, furthermore, means that studying energy flow provides a lot of information even when the equations of motion are too complicated to solve. Thus, the framework of energy is used throughout physics, and potential barriers and turning points play important roles in understanding many different types of physical systems.

Potential Barriers in Classical Mechanics

Studying the energy flow of a simple pendulum actually enables us to understand a number of different systems since the simple pendulum is just one example of a simple harmonic oscillator. Similarly, studying the potential energy function $V(r) = -\alpha/r$ enables us to understand many different gravitational and electrostatic systems. For example, when $\alpha = GMm$ (with $G = 6.67 \times 10^{-11}$ in SI units), $V(r)$ is the gravitational (Kepler) potential energy of a mass m a distance r from a second mass M fixed at $r = 0$. When $\alpha = -kQq$ (with $k = 1/4\pi\varepsilon_0 = 8.99 \times 10^9$ in SI units and 1 in cgs units), $V(r)$ is the electrostatic (Coulomb) potential energy of a charge q a distance r from a second charge Q fixed at $r = 0$. For motion in three dimensions in such a potential, the total potential energy function depends only on r and is given by

$$U(r) = \frac{L^2}{2mr^2} - \frac{\alpha}{r},$$

where L is the total angular momentum of the particle, which, for our (central) potential $V(r)$, is con-

stant. The extra term $(L^2/2mr^2)$ originates in the rotational kinetic energy, but because r is the only variable it contains, it behaves like a potential energy keeping particles with nonzero angular momentum away from the center of force at $r = 0$; thus, we study the radial motion with this combined potential energy. A graph of the various potential energies is shown in Fig. 3, with the dash-dot-dash curve representing $(-\alpha/r)$, the dashed curve representing $(L^2/2mr^2)$, and the solid curve representing their sum $U(r)$.

As mentioned above, the real power of the energy diagram lies in the amount of information that can be obtained from it without having to solve any equations of motion. For example, looking more closely at Fig. 3, we note that $U(r)$ has a minimum, and by differentiating $U(r)$ and setting the result equal to zero, we find that this minimum occurs at $r = L^2/m\alpha$. Thus, whenever the total energy equals $-m\alpha^2/2L^2$, the particle moves at a fixed distance $r = L^2/m\alpha$ from the center of force; that is, the particle moves in a circular orbit with a constant speed. If the total energy is greater than this value but still less than zero, the particle's radial motion will be bounded between the two turning points $r_0 = r_{\max}$ and $r_0 = r_{\min}$. At these points the total energy will be all potential and will be given by

$$E_0 = \frac{L^2}{2mr_0^2} - \frac{\alpha}{r_0}.$$

Multiplying both sides of this equation by r_0^2 and using the quadratic formula, we find that

$$r_{\max} = -\beta \left[1 + \left(1 + \frac{L^2}{2m\beta^2 E_0} \right)^{1/2} \right]$$

and

$$r_{\min} = -\beta \left[1 - \left(1 + \frac{L^2}{2m\beta^2 E_0} \right)^{1/2} \right],$$

with $\beta = \alpha/2E_0$ (remember, $E_0 < 0$). Since bounded, periodic motion between two fixed radii corresponds to motion in an ellipse, the energy diagram shows us that whenever the total energy is between $-m\alpha^2/2L^2$ and zero, the particle will move in an elliptical orbit. Furthermore, from our knowledge of

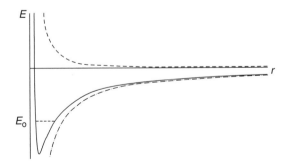

Figure 3 The energy diagram for a bound mass or charge ($E < 0$). (Courtesy of T. Jayaweera)

conic sections, we conclude that the semi-major axis of this ellipse is $a = (r_{max} + r_{min})/2$ and its eccentricity is $e = (r_{max} - r_{min})/(r_{max} + r_{min})$.

When the total energy is greater than zero and $V(r)$ is a repulsive potential ($\alpha < 0$), the particle will run into only one potential barrier, and this turning point will be the closest an unbound particle with angular momentum L can get to the center of force at $r = 0$. This particular potential barrier is called the centrifugal barrier since, for small r, the main contribution to it comes from the term $L^2/2mr^2$ whose gradient is the centrifugal force (a fictional force that occurs when we apply Newtonian mechanics in rotating reference frames).

Thus, a great deal of information about the three-dimensional motion of a particle with potential energy $V(r)$ can be deduced from an energy diagram using relatively simple mathematics. The same type of analysis can be carried out with other important physical problems by defining the relevant potential energy functions, such as the Yukawa potential energy, the general relativistic and perturbation theory corrections to the Kepler potential energy α/r, the potential energy of a spinning, symmetric top, the Morse potential energy function for an atom in a diatomic molecule, the isotropic oscillator potential energy $V(r) = kr^2/2$ for the vibrational motion of a diatomic molecule, the van der Waals potential energy, and so on.

Potential Barriers in Quantum Mechanics

A number of interesting potential barriers occur in (nonrelativistic) quantum mechanics, where energy diagrams are used to understand some of the more surprising features of quantum systems. The most famous example is that of alpha decay, in which a nucleus emits an alpha particle and changes into another nucleus. In 1928 George Gamow and, independently, Edward U. Condon and Ronald W. Gurney showed that alpha decay could be understood as an alpha particle inside a nucleus tunneling through the potential barrier it encounters at the nuclear edge. Figure 4 shows an approximation of the potential energy of such an alpha particle with total energy E_α. Classically, the alpha particle could not move beyond the nuclear radius at $r = a$ since it would run into a potential barrier there. But in quantum mechanics, the alpha particle is described by a wave function ψ, and its behavior at the nuclear edge can display properties not normally associated

with particles. In particular, ψ is nonzero at the classical turning point $r = a$ and decays exponentially just beyond it in the classically forbidden region $a < r < b$. (This region is classically forbidden because to be in it the particle would have to have a potential energy greater than its total energy, or a negative kinetic energy.) Consequently, there is a small but finite probability of finding the alpha particle outside the nucleus. In other words, if the height of the barrier is not too much greater than the alpha particle's energy E_α, and if the barrier's width $(b - a)$ is not too large, then ψ can be nonzero in the classically forbidden region where $V(r) > E$. A nonzero ψ in this region results in ψ being nonzero for $r > b$. Although the probability of finding an alpha particle at points $r > b$ is quite small, nonetheless, since the alpha particle encounters the nuclear barrier quite frequently (about 10^{21} times per second), quantum theory predicts that some experiments will find it outside the nucleus. In fact, treating alpha decay as tunneling through a potential barrier (also called barrier penetration), correctly predicts all the results found experimentally.

Analogs to quantum tunneling occur in a variety of classical wave systems. For example, in classical optics, a light wave traveling through glass can be directed so that it hits a glass-air interface at an angle greater than the critical angle. In this case, all of the wave will be totally internally reflected. However, if a second piece of glass is put close to the first, then a small part of the wave will tunnel through the air

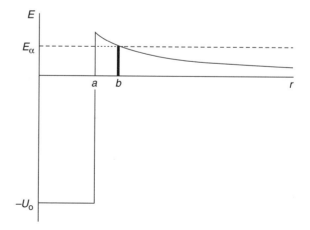

Figure 4 The energy diagram for an alpha particle inside a nucleus of radius a. E_α is the energy of the alpha particle, and $(b - a)$ is the width of the potential barrier. (Courtesy of T. Jayaweera)

barrier between them and appear in the second piece of glass. This phenomenon, called frustrated total internal reflection, was first observed by Isaac Newton around 1700 and now is routinely demonstrated in undergraduate physics labs. Similar phenomena also occur with other types of waves.

Tunneling through potential barriers also provides the basis for our understanding of spontaneous nuclear fission, of how electrons pass through thin oxides and insulators (e.g., tunnel diodes), of the dynamics of the ammonia molecule and the ammonia maser, of the Josephson effect and superconducting quantum interference devices, and of Scanning Tunneling Microscopes.

See also: BARRIER PENETRATION; DECAY, ALPHA; ENERGY, KINETIC; ENERGY, POTENTIAL; JOSEPHSON EFFECT; MASER; MOMENTUM; PENDULUM; SCANNING TUNNELING MICROSCOPE; SUPERCONDUCTING QUANTUM INTERFERENCE DEVICE; VAN DER WAALS FORCE

Bibliography

EISBERG, R., and RESNICK, R. *Quantum Physics* (Wiley, New York, 1974).

KIBBLE, T. W. B. *Classical Mechanics*, 3rd ed. (Longman, New York, 1985).

KLEPPNER, D., and KOLENKOW, R. J. *An Introduction to Mechanics* (McGraw-Hill, New York, 1973).

SPOSITO, G. *An Introduction to Classical Dynamics* (Wiley, New York, 1976).

TAYLOR, J. R., and ZAFIRATOS, C. D. *Modern Physics for Scientists and Engineers* (Prentice Hall, Englewood Cliffs, NJ, 1991).

THORTON, S. T., and REX, A. *Modern Physics for Scientists and Engineers* (Saunders, New York, 1993).

TIPLER, P. *Elementary Modern Physics* (Worth, New York, 1992).

MARK D. SEMON

POTENTIOMETER

A potentiometer is an electrical device consisting of a length of resistance wire (or resistive material) with some resistance R and a tap point that slides on this length of resistance wire, making electrical contact with it. There are three electrical connections to a potentiometer: points A and B at the ends of the resistance wire and the sliding tap point T (see Fig. 1). The volume control of a radio or stereo is a simple, inexpensive potentiometer; in contrast, a precision potentiometer is an expensive instrument that can be used to measure voltage to high precision.

If connections are made only to the tap point T and to point A of a potentiometer, it becomes an adjustable resistor, or rheostat. The resistance between points T and A is indicated by R_1, and this varies from 0 to R as the slider is moved over the length of resistance wire from the end near point A to the end near point B.

The potentiometer gets its name from the fact that it can "meter out" varying amounts of electrical potential, or voltage, from the output between the tap point T and one end of the potentiometer wire (e.g., A). Suppose a battery with emf V is connected between end points A and B as shown in Fig. 1. Let the resistance between B and T be R_2 while the resistance between T and A is R_1. These two resistances constitute a voltage divider so that the voltage between tap T and end A (V_{TA}) is a fraction of the voltage between B and A (V_{BA}). We have that $V_{TA} = [R_1/(R_1 + R_2)] V_{BA} = (R_1/R) V_{BA}$. The resistance ($R_1 + R_2$) is a constant equal to the resistance R of the potentiometer. As the slider is moved along the resistance wire, resistance R_1 varies from 0 to R, and the voltage V_{TA} between the tap and A varies from 0 to V_{BA}. This is the basic way to produce a variable amount of voltage from a fixed voltage.

In the case of a radio volume control, the voltage V_{BA} presented to the potentiometer is an audio-frequency voltage corresponding to the sound wave. The variable amount of voltage from the potentiometer tap (V_{TA}) is sent (through an amplifier) to

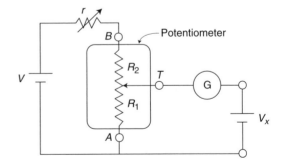

Figure 1 Schematic diagram of a potentiometer. The potentiometer is shown connected with working voltage V, rheostat r, and galvanometer G to measure voltage V_x.

the loud speaker, and the amount of sound that we hear is varied as the slider is moved.

In a precision potentiometer, the ratio of the resistances R_1 and R_2 is adjustable to high precision. A working battery V is connected through rheostat r to the potentiometer, and the rheostat r is adjusted until the voltage V_{BA} is some precise voltage (e.g., 1.6000 V). An unknown voltage V_x is connected through a galvanometer to the slider tap T, and the R_1/R_2 ratio is adjusted until the galvanometer indicates no current flow. Then the voltage V_x is determined to be $(R_1/R) V_{BA}$. To calibrate the voltage V_{BA}, a standard cell with precisely known voltage is connected in place of V_x, the R_1/R_2 ratio is set corresponding to this voltage, and rheostat r is adjusted for no galvanometer current.

Voltages can be measured to five significant digits and down to microvolts by a quality precision potentiometer. However, the measurement process is slow and the instrumentation bulky; most routine precision voltage measurements are now done with precision digital voltmeters. A potentiometer can be used to calibrate a digital voltmeter.

See also: ELECTRICAL RESISTANCE; ELECTRICAL RESISTIVITY; RESISTOR; VOLTMETER

Bibliography

JONES, L. *Electrical and Electronic Measuring Instruments* (Wiley, New York, 1983).

STOUT, M. B. *Basic Electrical Measurements,* 2nd ed. (Prentice Hall, Englewood Cliffs, NJ, 1960).

DENNIS BARNAAL

PRECESSION

Precession is the phenomenon that occurs when the axis of rotation of a spinning body, and hence its angular momentum, changes direction in space with time. For example, under certain conditions, the rotational axis of a spinning gyroscope or a toy top will trace out a conical motion. Figure 1 shows a top that has been tipped from its upright position so that its rotational axis is at some angle θ with respect to the vertical. The weight acting through the center of mass of the top produces a torque that is in a direction perpendicular to the spin axis. This torque causes a change in the direction of the angular momentum or spin axis, which results in the axis sweeping out the indicated conical path. Not all forces exerted on spinning bodies cause precessional motion. The direction and point of application of the force is an important consideration; it must not be parallel or antiparallel to the axis to produce the effect. In other words, the force must produce a torque.

One of the most interesting examples of precessional motion is that of Earth. Like a spinning top in the absence of a torque, Earth tends to keep its axis of rotation pointing in a particular direction; presently in the direction of Polaris, the North Star. If Earth were a perfectly homogeneous sphere the gravitational forces that the Sun and Moon exert on it could not produce a torque and the axis would forever point in this direction. Earth, however, is not homogeneous and it bulges somewhat near the equator. The resulting differential forces acting on Earth produce a torque that tries to align its spin axis perpendicular to the plane of Earth's orbital path around the Sun. However, the result is that the axis precesses, sweeping out a cone. This effect is shown in Fig. 2. The time it takes to complete this precessional cycle is about 26,000 years. Since the seasons depend both on the angle and the direction of Earth's spin axis with respect to its orbital plane, this phenomenon is called the precession of the equinoxes. In about 13,000 years the seasons will

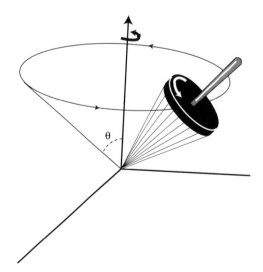

Figure 1   Spinning top.

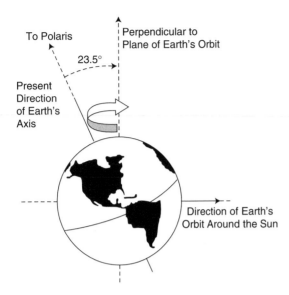

Figure 2   Precession of the equinoxes.

shift by six months so that in the Northern Hemisphere spring will begin in September and winter in June. Approximately 4,000 years ago, Earth's spin axis pointed toward Thuban in the constellation Draco and 14,000 years from now the bright star Vega will be in that position.

Another example of precessional motion is the situation where a spinning electron in an atom interacts with an external magnetic field. The spinning electron acts like a small magnet, that is, it has a magnetic dipole moment. The magnetic field tries to align the magnetic moment along the field direction, producing a torque on it. This torque causes the spin axis of the electron to precess around the direction of the magnetic field. This precession is called Larmor precession and gives rise to the splitting of spectral lines known as the Zeeman effect.

See also: PRECESSION, LARMOR; SEASONS; SPIN, ELECTRON; ZEEMAN EFFECT

Bibliography

CHAISSON, E., and McMILLAN, S. *Astronomy Today* (Prentice Hall, Englewood Cliffs, NJ, 1993).

EISBERG, R., and RESNICK, R. *Quantum Physics of Atoms, Molecules, Solids, Nuclei, and Particles* (Wiley, New York, 1974).

FOWLER, P. *Understanding the Universe* (West, New York, 1990).

SERWAY, R. A. *Physics for Scientists and Engineers,* 3rd ed. (Saunders, Philadelphia, 1990).

THORNTON, S. T., and REX, A. *Modern Physics for Scientists and Engineers* (Sanders College Publishing, Fort Worth, TX, 1993).

S. G. BUCCINO

PRECESSION, LARMOR

The motion of the magnetic moment of a particle, atom, or molecule about the axis of an applied magnetic field is known as Larmor precession. The magnetic moment (or alternatively, the angular momentum vector of the particle) sweeps out a cone about the direction of the magnetic field at a constant angular frequency. For an atom, the angular frequency of precession, known as Larmor's frequency, is numerically equal to $eB/2mc$, where e/m is the charge-to-mass ratio of the electron, c is the speed of light, and B is the magnitude of the applied magnetic field.

The motion described above resembles that of a symmetrical spinning top with one end held fixed, whose axis of rotation sweeps out a cone about the vertical direction. For a top, the precessional motion results from the combined action of the force of gravity and of the contact force which holds the end of the axis fixed. Since these forces do not lie along the same line, the top experiences a net torque. The torque is directed so as to make the top fall, but the inertia of the spinning top (i.e., its state of motion) deflects it instead in a direction perpendicular to the applied forces.

The magnetic moment of an atom is produced by the motions of electrons about the central nucleus. These subatomic particles are electrically charged, and their motion constitutes an electrical current that encircles the atomic nucleus. The current produces a magnetic field, which is another way of saying that the atom acts like a magnet, with its poles defining the direction of its magnetic moment. When the magnetic atom is placed near an external magnet, magnetic forces result in a torque which causes the orbits of the electrons to precess about the axis of the external field. This precessional motion was predicted on general principles by

Joseph Larmor in 1895, long before anyone really knew how an atom is constructed.

Historically, Larmor precession played an important role in the interpretation of the normal Zeeman effect, in which the spectroscopic absorption or emission lines of an atom are split into three lines in the presence of a magnetic field. The frequency shift of the radiation is precisely the Larmor frequency divided by 2π.

See also: FIELD, MAGNETIC; MAGNETIC MOMENT; MOMENTUM; PRECESSION; ZEEMAN EFFECT

Bibliography

FANO, U., and FANO, L. *Physics of Atoms and Molecules* (University of Chicago Press, Chicago, 1972).

HERZBERG, G. *Atomic Spectra and Atomic Structure* (Dover, New York, 1944).

MICHAEL J. CAVAGNERO

PRECISION

See ACCURACY AND PRECISION

PRESSURE

Pressure is the force per unit area exerted by a fluid (a liquid or gas) on the wall of an object in contact with it. The force exerted by the wall on the fluid "contains" the fluid.

Why limit the definition to a fluid? The difficulty with defining a pressure within a solid arises from the possibility of shear stress and from possible anisotropy. A fluid in equilibrium can have compressive or tensile stress, but it flows in response to shear stress. In other words, shear stress results in elastic deformation in a solid, but not in a fluid.

If a definition of pressure inside a solid is desired, one option would be to choose the average of the compressive stresses along three mutually perpendicular directions. If these are all equal, the same definition as for a fluid results. In other words, if compressive (or tensile) stress is isotropic, it can be called a positive (or negative) pressure.

Units

The unit of measure to use in the meter-kilogram-second system when pressure is defined as force per unit area is the newton per square meter. This unit is called the pascal (Pa). However, many of the following pressure units are used in various specialties:

$$1 \text{ atm} = 1 \text{ bar}$$
$$= 1.013 \times 10^5 \text{ Pa} = 1.013 \times 10^4 \text{ dyn/cm}^2$$
$$= 760 \text{ mm of mercury} = 760 \text{ torr}$$
$$= 1{,}033 \text{ cm of water}$$
$$= 14.7 \text{ psi.}$$

In health-related fields, for example, blood pressure is measured with a sphygmomanometer (pressure cuff), usually using mercury in a vertical tube. Using mercury instead of a less dense fluid makes the height of the tube manageable. If the 1,033 cm of water is divided by the equivalent 76 cm of mercury, the resulting value of the specific gravity of mercury is 13.6; that is, the density of mercury is 13.6 times the density of water. Therefore, a much taller instrument would be needed if water were used instead of mercury to measure a typical blood pressure of 120 over 80: 120 torr on the strong part of the heartbeat (systolic pressure) and 80 torr on the weakest part (diastolic pressure).

Air Pressure

Air pressure is measured using a barometer. The classical laboratory instrument is a vertical glass tube, closed at the top, filled with mercury, and dipping into a pool of mercury at the bottom. Even though the tube was filled with liquid mercury when it was horizontal, a Torricellian vacuum forms at the top when it is turned to stand vertical; that is, the mercury falls by the amount of the vapor pressure of mercury. This is only 0.0012 mm—not an important correction to the "reading" of the barometer. Isopropyl alcohol (rubbing alcohol), on the other hand, has a vapor pressure of 42 mm of mercury at room temperature. Because of this difference in vapor pressure, a bottle of alcohol left open will affect the room's air. In a small enough room, 42/760 = 6 percent of the air in such a situation would be alcohol vapor.

Tire pressures are measured using a rather simple spring gauge, giving readings such as "29," meaning 29 psi. However, that is not 2 atm as simple arithmetic would lead one to believe. It is important to remember that the other end of the gauge is out in the open air, exposed to roughly 1 atm. This ambient pressure must be added to the gauge pressure. So although the *gauge* pressure is 2 atm, the *absolute* pressure is 3 atm. The same correction (same warning) applies in scuba diving. If the pressure gauge on the tank reads 100 atm, the absolute pressure in the tank is really 101 atm. Why such high pressures for deep-sea diving? For every 10.3 m down into the water, the pressure increases by 1 atm. A diver who goes down 20 m would want to make sure the tank has a lot more than 3 atm of pressure, because it is the excess pressure over the ambient pressure at the diver's depth that opens the breathing valve.

Gas Laws

The relation between the pressure and volume of an enclosed fluid is called its equation of state. At low pressures, gases obey an equation called the ideal gas law, which says that the density is proportional to the pressure. At higher pressures, gases become less compressible than the ideal gas.

See also: GAS; IDEAL GAS LAW; PRESSURE, ATMOSPHERIC; PRESSURE, SOUND; PRESSURE, VAPOR

Bibliography

MACHLUP, S. *Physics* (Wiley, New York, 1988).
YOUNG, H. D., and FREEDMAN, R. A. *University Physics* (Addison-Wesley, Reading, MA, 1996).

STEFAN MACHLUP

PRESSURE, ATMOSPHERIC

The atmosphere of the earth has a remarkably homogeneous composition of 78 percent nitrogen, 21 percent oxygen, and 1 percent other gases. However, properties such as temperature, pressure, density, and velocity vary considerably with location and altitude. The total mass of the atmosphere is approximately 5.2×10^{18} kg, more than 99 percent of which is within 30 km of the surface. The surface area of the earth is about 5.1×10^{14} m^2. Since pressure is defined as force per unit area, the weight of the atmosphere (mass $\times$ gravity) divided by the area suggests that the atmospheric pressure at the surface must be about 10^5 N/m^2. The vertical variation of atmospheric pressure is governed by the condition of hydrostatic equilibrium. The pressure at any height is determined by the weight of the atmosphere above it, and decreases exponentially with altitude. In the mile-high city of Denver the average pressure is about 80 percent of sea level. On the summit of Mt. Kilimanjaro (5,900 m) the pressure is 50 percent of sea level, and on Mt. Everest (8,850 m) it is less than 30 percent of sea level. Horizontal differences are described as high- and low-pressure weather systems. Although the pressures in these regions differ from the global average by less than 1 percent, they are responsible for the trade winds and the spawning of fronts and storm systems.

Two types of instruments are used to measure atmospheric pressure: the mercury barometer and the aneroid barometer. The former consists of a vertical glass tube filled with mercury, sealed at the top, with its bottom immersed in a reservoir of the liquid. In 1643 scientists noted that the pressure of air on the reservoir would support a column of mercury 76 cm high. The aneroid instrument, invented in 1843, consists of a partially evacuated metal bellows coupled to a dial indicator via levers and springs. As the external pressure changes the expansion or contraction of the bellows turns the dial. This is a portable and sensitive indicator of changes in pressure, but requires careful calibration to provide accurate readings.

The mercury barometer provides the definition of standard atmospheric pressure. One atm is the pressure exerted by a column of mercury exactly 0.76 m high. With a density of $\rho = 13,595$ kg/m^3 and gravity $g = 9.807$ m/s^2, 1 atm $= \rho g h = 1.013 \times 10^5$ N/m^2. In SI units 1 Pa = 1 N/m^2, so 1 atm = 101.3 kPa. In MKS units 1 atm = 1.013 10^6 dyn/cm^2, and in English units = 14.7 lb/in^2. Laboratory measurements are often reported in "torr," where 1 torr = 1 mmHg. Therefore, 1 atm = 760 torr. Meteorologists use a unit called "bar," where 1 bar = 10^6 dyn/cm^2. Thus 1 bar =0.987 atm, and 1 atm = 1,013 mbar. Barometric pressure is usually reported in units of "inches of mercury" in newspaper and television weather reports, where 1 atm = 29.92 in Hg.

Atmospheric pressure has several familiar effects. The temperature at which water boils is 100 °C at sea level, but only 90 °C in Vail, Colorado, at an altitude of 3000 m. Sucking liquid through a soda straw and pumping water from a well both involve lowering the pressure at the receiving end and letting atmospheric pressure push the liquid through the tube. The aerodynamics of airplane wings, sailboats, and curve balls are all governed by the variation of atmospheric pressure over their surfaces.

See also: ATMOSPHERIC PHYSICS

Bibliography

PEIXOTO, J. P., and OORT, A. H. *Physics of Climate* (American Institute of Physics Press, New York, 1992).

DENNIS EBBETS

PRESSURE, SOUND

Sound waves in a fluid are normally longitudinal waves; the molecules move back and forth in the direction of wave propagation, producing alternate regions of compression and rarefaction. This results in small variations in the fluid pressure above and below its equilibrium pressure (atmospheric pressure in the case of sound waves in air).

Although the pressure variations in a sound wave are extremely small, the ear can respond to pressure changes less than one-billionth (10^{-9}) of atmospheric pressure. The threshold of audibility, which varies from person to person, corresponds to a sound pressure amplitude of about $2 \times 10^{-5} \, \text{N/m}^2$ at a frequency of 1,000 Hz. The threshold of pain corresponds to a pressure approximately one-million (10^6) times greater, but still less than 1/1,000 of atmospheric pressure. Pressure variations for sound waves in denser fluids, such as water, are even smaller.

Because of the wide range of pressures encountered in sound fields, it is convenient to measure sound pressures on the logarithmic scale of decibels (dB). We define sound pressure level (L_p or SPL) as $L_p = 20 \log p/p_0$, where p is the root-mean-square

(rms) sound pressure and p_0 is the reference level. (In air, $p_0 = 20 \, \mu\text{Pa}$ or 2×10^{-10} times normal atmospheric pressure, while in water $p_0 = 1 \, \mu\text{Pa}$.)

Sound pressure level is usually measured with a sound level meter, which consists of a microphone, an amplifier, and a meter calibrated to read in decibels. Sound levels have one or more weighting networks to provide the desired frequency response. Generally, three weighting networks are used; they are designated A, B, and C. The C-weighting network has an almost flat frequency response, whereas the A-weighting network introduces a low-frequency roll-off in gain that bears rather close resemblance to the frequency response of the human auditory system. Measurements of environmental noise are usually made using the A-weighting network; such measurements are properly designated as $L_p(A)$ or SPL(A) in dB, although the units dBA and dB(A) are often used to denote A-weighted sound level. Many sound level meters have both fast and slow response, the slow response measuring a short-term average sound level.

To obtain the sound pressure level from multiple sound sources, we add the mean-square pressures (average values of p^2) at a point. Two uncorrelated sound sources, each of which would produce a sound level of 80 dB at a certain point, will together give 83 dB at that point. When two waves of the same frequency reach the same point, however, they may interfere constructively or destructively. If their amplitudes are both equal to A, the resultant amplitude can be anything from zero to 2A.

The sound power level (L_W or PWL), also expressed in decibels, describes the strength of the sound source. It is defined as $L_W = 10 \log W/W_0$, where W is the sound power emitted by the source and the reference power $W_0 = 10^{-12}$ watts. The relationship between sound power level and sound pressure level depends on several factors, including the geometry of the source and the room. If the sound power level is increased by 10 dB, the sound pressure level also increases by 10 dB, provided everything else remains the same. If a source radiates sound equally in all directions and there are no reflecting surfaces nearby (a free field), the sound pressure level decreases by 6 dB each time the distance from the source doubles.

One should not confuse *loudness* or *loudness level* with sound pressure level. Loudness describes a subjective response to sound, and the sensitivity of the ear depends on such parameters as the frequency, the spectrum, and the amplitude envelope of the sound.

See also: SOUND; SOUND ABSORPTION; WAVE MOTION

Bibliography

KINSLER, L. E.; FREY, A. R.; COPPENS, A. B.; and SANDERS, J. V. *Fundamentals of Acoustics,* 3rd ed. (Wiley, New York, 1982).

ROSSING, T. D. *The Science of Sound,* 2nd ed. (Addison-Wesley, Reading, MA, 1990).

ROSSING, T. D., and FLETCHER, N. H. *Principles of Vibration and Sound* (Springer-Verlag, New York, 1994).

THOMAS D. ROSSING

PRESSURE, VAPOR

Vapor pressure is the pressure of a vapor (gas) in equilibrium with a solid phase or liquid phase of a thermodynamic system. The term usually refers to saturated vapor. It can be read from a pressure-temperature plot of a phase boundary. A *P-T* plot of the phase boundaries of a water-ice-steam system is shown in Fig. 1.

As the temperature changes, the pressure of the vapor phase in contact with a condensed phase (i.e., liquid or solid phase) changes. The slope of this change dP/dT is given by the Clapeyron equation:

$$dP/dT = l/T(v_g - v_c),$$

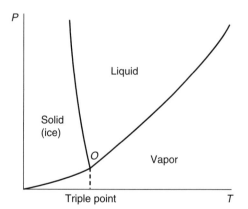

Figure 1 *P-T* plot of the phase boundaries of a water-ice-steam system.

where l is the latent heat involved in crossing the phase boundary and v_g and v_c are molar volume of the vapor and condensed phase, respectively. Molar volume of the liquid and solid states are small compared to that of the gas phase and can be ignored. In addition, we observe that the vapor pressure is sufficiently low in most cases to allow ideal gas approximation. Thus,

$$v_g - v_c \approx v_g \approx RT/P.$$

This approximation facilitates the integration of the Clapeyron equation as follows:

$$l = R(dP/P)/(dT/T^2) = -R\, d \ln P/d(1/T).$$

Since we can assume l/R to be almost over a reasonable temperature range, the above can be integrated to

$$\ln P = -(A/T) + B,$$

where A and B are constants. (These constants, A and B, must be adjusted if one is to use a semi-log plot to record experimental data on vapor pressure.) A somewhat better approximation is known as the Rankine–Dupré formula and is given by

$$\ln P = -(A/T) + B - C \ln T,$$

where B and C are constants involving specific heats of the phases. A version of the expression with common logarithm is given by

$$\log P = -(A'/T) + B' - C' \log T.$$

For a vapor phase in equilibrium with a solid phase, the simplifying assumptions for the molar volume and the gas phase is significantly better than for the vapor-liquid phase equilibrium. One can derive expressions in this case to enable experimental determination of the heat of sublimation at absolute zero and a constant known as the vapor pressure constant.

Measurement of Vapor Pressure

If the vapor pressure to be measured is 10^{-3} torr (1 torr = 1 mm Hg or 1.333 Pa) or larger, one can employ ordinary methods of pressure measurement. One can connect the vessel containing the liquid or solid to a pressure gauge, such as a Bourdon gauge or McLeod manometer.

When the system in equilibrium with the vapor is a solid, especially a refractory metal of very high melting point, the vapor pressure is too low to be measured by the conventional method. In this case, one must use one of the indirect methods to measure the vapor pressure.

Langmuir's method uses a kinetic assumption that the rate of evaporation of a solid is equal to the rate at which the molecules of the vapor strike the phase boundary under equilibrium condition. The method thus consists of measuring the rate of mass loss of the solid at constant temperature and uses the kinetic approximation formula

$$P = Y \sqrt{(2\pi RT/m)},$$

where Y is the rate of mass loss per unit area of the solid and m is the molecular mass of the substance.

Knudsen's method is the variation of the above. Instead of measuring Y, the vapor is extracted through an orifice of known area and then condensed in a cold trap. Rate of mass gain of the cold trap yields the value of Y.

See also: PRESSURE; PVT RELATION

Bibliography

BAUMAN, R. P. *Modern Thermodynamics with Statistical Mechanics* (Macmillan, New York, 1992).
CALLEN, H. B. *Thermodynamics and an Introduction to Thermostatistics* (Wiley, New York, 1985).

CARL T. TOMIZUKA

PRISM

See NICOL PRISM; REFRACTION

PROBABILITY

Probability theory applies when there are a number of possible outcomes of an experiment, and you would like to know how likely each one is to occur.

For example, suppose you toss a coin. It can come down heads, H, or tails, T. If the coin is not weighted, both options are equally likely, so the probability of getting H is $p_H = 1/2$, while the probability of getting T is $p_T = 1/2$ also. Note that $p_H + p_T = 1$. In this simple experiment, the number of outcomes possible is $N = 2$, and the probabilities add up to 1.

In general, if an experiment has N outcomes, the ith possibility has probability p_i, and

$$\sum_{i=1}^{N} p_i = 1. \tag{1}$$

Equation (1) simply states that an experiment must result in one of its possible outcomes; it is fundamental in probability theory.

Look at a different example. Suppose you throw a die. The possible outcomes are the numbers $i = 1$, 2, 3, 4, 5, 6. If the die is not weighted, then $p_i = 1/6$ for each number, and $\sum p_i = 1$. But suppose the die has been weighted, so that 6 is twice as likely to occur as any other number, while 1 through 5 are equally likely. Then $p_1 = p_2 = p_3 = p_4 = p_5$ and $p_6 = 2p_1$. Putting that information into Eq. (1) gives $p_1 = 1/7$ and $p_6 = 2/7$. In this experiment, the chance of throwing a 6 is 2/7. The chance of *not* throwing a 6 is 5/7, the sum of the probabilities of all the other numbers.

Now consider a two-stage experiment in which you first toss an unweighted coin and then throw the weighted die. How many possible outcomes are there? The answer is $2 \times 6 = 12$. Since tossing the coin and throwing the die are independent of each other, probabilities simply multiply. The probability of getting H and 2 is just $(1/2) \times (1/7) = 1/14$, while the probability of getting H and 6 is $(1/2) \times (2/7) = 2/14$. You can easily check that Eq. (1) is satisfied. Note that this simple multiplication of probabilities only applies if the two experiments are independent of each other. If they are not, you must use the more complex Bayes's theorem.

Again, suppose you perform an experiment with N possible outcomes and probabilities p_i, but let us now extend the probability arithmetic to include the concept of a desired or successful outcome. We assume the p_i are constant and do not change if the experiment is performed many times. Equation (1) is always valid.

Let us define the concept of a successful experiment. In the case of the weighted die, a success could be throwing a 6. Then the probability of a success in a single throw is $p = 2/7$. The probability of failing is

$$q = 1 - p. \tag{2}$$

Suppose you throw this die many times, say N times. Each throw is independent of every other throw, so probabilities are simply multiplicative. The chance of throwing N sixes is p^N. The chance of throwing $N - 1$ sixes, and then failing the last time is $p^{N-1} q = p^{N-1} (1 - p)$. The chance of throwing n sixes in a row, and then $(N - n)$ non-sixes, is $p^n q^{N-n}$. Notice that in this last statement the *order* of sixes versus other numbers mattered. Instead, suppose you are only interested in the probability of throwing n sixes out of N tries, when the order does not matter. Then the probability $P(n)$ is given by the binomial distribution:

$$P(n) = \frac{N!}{(N - n)! \, n!} \, p^n \, q^{N-n}, \tag{3}$$

where the combinatoric factor involving factorials is just the number of ways of choosing n successes from N trials.

For example, suppose you throw the weighted die 60 times. What is the probability of seeing 20 sixes? Equation (3) says $P(20) = (60!/40! \, 20!) \, (2/7)^{20} (5/7)^{40} = 0.07869$. The *average* number of sixes in this kind of experiment is

$$\bar{n} = Np, \tag{4}$$

so we would expect to see $60 \times (2/7) = 17.14$ sixes, if the experiment of throwing the die 60 times were repeated many times.

Binomial distributions are peaked about their averages. As the number of trials N increases, they tend to become more sharply peaked, and the average departure from the mean, given by the standard deviation σ, where

$$\sigma = \sqrt{\frac{\sum (n - \bar{n})^2}{N - 1}} = \sqrt{Npq}, \tag{5}$$

decreases. For our die experiment, $\sigma = 3.5$. Notice that for a binomial distribution the ratio

$$\frac{\sigma}{\bar{n}} = \frac{p}{Nq}, \tag{6}$$

showing that the distribution gets more sharply peaked about the mean as N gets large, or the fluctuations get smaller.

Gaussian Distributions

The Gaussian distribution, sometimes called the normal distribution, is an approximation of the binomial distribution when N gets very large. As N increases, the distribution becomes more and more sharply peaked about the average value, and $P(n) \to 0$ when n is far from $\bar{n}$. You can use Stirling's approximation for $N!$, $n!$, and $(N - n)!$ (all large):

$$\ln n! \approx n \, (\ln n) - n + \tfrac{1}{2} \ln(2\pi n). \tag{7}$$

The result is

$$P(n) = \frac{1}{\sqrt{2\pi\sigma^2}} \exp\left[-\frac{(n - \bar{n})^2}{2\sigma^2}\right], \tag{8}$$

where $\bar{n}$ and σ are given by Eqs. (4) and (5).

The Gaussian distribution is shown in the Figs. 1 and 2 for the weighted die experiment, for $N = 200$ and 1,000 trials, respectively. Note that as N gets larger, the distribution becomes more sharply peaked about its mean.

It can be shown that the meaning of the standard deviation σ in this context is that if you do the experiment a large number of times N, the result of the experiment will lie within $\pm\sigma$ of the mean 68.3 percent of the time, within $\pm 2\sigma$ of $\bar{n}$ 95.5 percent of the time, and within $\pm 3\sigma$ of $\bar{n}$ 99.7 percent of the time. Because of this, we call the interval from $\bar{n} - \sigma$ to $\bar{n} + \sigma$ the 68.3 percent confidence interval. The interval from $\bar{n} - 2\sigma$ to $\bar{n} + 2\sigma$ is the 95.5 percent confidence interval, and the interval $\bar{n} - 3\sigma$ to $\bar{n} + 3\sigma$ is the 99.7 percent confidence interval. If you do the experiment one more time, you can say that at the 99.7 percent confidence level you expect to find the result within $\pm 3\sigma$ of the mean. At the 68.3 per-

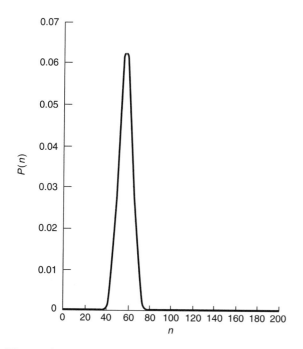

Figure 1 Gaussian distribution for the weighted die experiment, where $N = 200$.

cent confidence level, the result will lie within $\pm\sigma$ of the mean, and so on.

Gaussian distributions occur whenever the outcome of a particular trial is random. Probabilities of specific outcomes may be known, but the result of any one trial is random.

Gaussians are often used to analyze data. Suppose you perform an experiment but random errors occur, causing repeated measurements of a quantity to lie close to the mean or average value, but scattered randomly about it. Then Gaussians are useful, because despite the errors, the true value of the quantity being measured is likely to be close to the average of the measured values.

As a simple example, assume you have a block of glass. It is supposed to be very homogeneous and have constant refractive index ν. You would like to know the refractive index of this material precisely, so you measure it many times, using many different experimental methods. The errors are then probably random, and the average value should be close to the true value of ν:

$$\bar{\nu} = \frac{\sum_{i=1}^{N} \nu_i}{N}. \tag{9}$$

You would also like to know the standard deviation. It is

$$\sigma = \sqrt{\frac{\sum_{i=1}^{N} (\nu_i - \bar{\nu})^2}{N-1}}. \tag{10}$$

You might also want to know how close to the average experimental value the true value really is. In other words, what is the error on the mean, or the standard error. Call this ε:

$$\varepsilon = \frac{\sigma}{\sqrt{N}}. \tag{11}$$

The true value should then lie between $\bar{\nu} - \varepsilon$ and $\bar{\nu} + \varepsilon$.

For example, suppose the refractive index of the block of glass just described is measured 100 times, and the mean value is found to be $\bar{\nu} = 1.45$ with standard deviation $\sigma = 0.20$, then $\varepsilon = 0.02$, and you would say the true value lies in the range $\nu = 1.45 \pm 0.02$, or $1.43 \leq \nu \leq 1.47$.

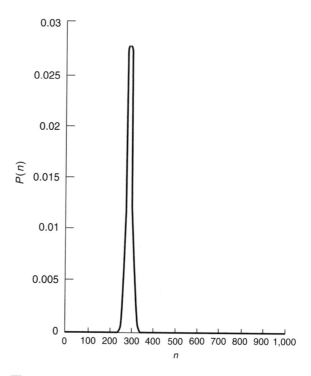

Figure 2 Gaussian distribution for the weighted die experiment, where $N = 1,000$.

Poisson Distribution

We have discussed how the binomial distribution [Eq. (3)] describes the frequency with which a particular number n occurred. Its generalization for large N, the Gaussian distribution, is useful in determining the true value of a quantity being measured, experimental error, and confidence levels. There is another approximation to the binomial distribution, useful for experiments when N is large, but the probability of a success p is very small, that is $p << 1$. This is called the Poisson distribution.

Equation (3) shows the binomial distribution. Note that $P(n) \propto p^n$. Since p is supposed to be very small, $P(n)$ will be approximately zero for all but small n. When p is small, we can approximate: $\ln(1-p) \approx -p$, so

$$q^{N-n} = (1-p)^{N-n} \approx \exp[-(N-n)p]$$

$$\approx \exp[-Np]. \qquad (12)$$

Also,

$$\frac{N!}{(N-n)!} \approx N^n, \qquad (13)$$

which follows from Stirling's approximation [Eq. (7)]. Then Eq. (3) reduces to $P(n) \approx [(Np)^n/n!] e^{-Np}$. However, $\overline{n} = Np$. Call this number $\lambda = \overline{n} = Np$, the mean value. Then

$$P(n) \approx \frac{\lambda^n}{n!} e^{-\lambda}. \qquad (14)$$

This is a form of the Poisson distribution. It is easy to show that Eq. (1) is still obeyed, and in fact the upper limit on the summation in Eq. (1) can be replaced by ∞, since N is large and $P(n) \approx 0$ for all but small n.

A good example of the use of Poisson distributions in physics is in describing radioactive decay of a substance. The rate of decay, that is, the number of disintegrations per second, is usually slow. Suppose we watch the sample for a time interval t. Divide t up into small increments of time $\Delta t = t/N$, where the number of time intervals N is very large. Then the probability p of a disintegration occurring in any particular time interval Δt is very small. The probability $P(n)$ of n atoms having decayed after time t is well approximated by the Poisson distribution.

Statistical Fluctuations

Fluctuations in systems continually occur and are fundamental, or natural, to a system in which the outcome of an experiment is any one of several possible values randomly determined.

Consider again the experiment of tossing an unweighted coin. Suppose a whole experiment consists of tossing the coin 200 times. On the average you expect to see 100 heads. Will you *always* see 100 heads if you toss the coin 200 times repeatedly? Clearly, the answer is "no." The number of heads seen will be distributed about the mean value of 100 and described by Gaussian statistics.

A physical example of this consists of measurements of the temperature in a room. Suppose a room is sealed off completely from its surroundings, and suppose it contains a large number of air molecules (nitrogen, oxygen, carbon dioxide, argon, and so on) in thermal equilibrium at 20°C. Suppose you hold a thermometer in position for a few seconds at a given location and take a reading of the temperature. If you take a lot of readings at different places and average the results, the mean temperature will be seen to be 20°C. But the value of the temperature at any specific location may differ appreciably from 20°C. If you could measure the temperature at many different points in the room at the same time, you would thus see spatial fluctuations in temperature. If you measured the temperature at any single point repeatedly at different times, you would find it is not constant; there are temporal fluctuations. All random classical fluctuations are governed by binomial, or Gaussian, distributions. The quantity used to measure the size of fluctuations is $\sigma/\overline{n}$.

Statistical Error and Systematic Error

For the experiment in which the errors were random, the Gaussian distribution could be used to analyze the data. This kind of error, random in any given measurement, is called a statistical error. It would be nice if all experiments could be designed to have only statistical errors present, because we would know how to analyze them. Unfortunately, most experiments are full of systematic errors.

The simplest type of systematic error is error due to miscalibration, or inadequacies, of an instrument. Suppose, for example, you are trying to measure the acceleration due to gravity at the surface of the earth, g, by measuring the length of time it takes for a certain object to fall a certain vertical distance in an evacuated tube under gravity. It should be easy to calculate g from this data. You expect to find $g = 2H/t^2$, where H is the distance fallen and t is the time interval. You measure the time interval t using a photo-timer. When the object is released, the timer starts counting; when the object passes a point H below its starting point, the timer stops. You take a lot of measurements, find the average value of g, and use Gaussian-based error analysis to calculate g, and the error in g. You may still get an answer that is wildly incorrect, despite your efforts. Suppose the timer does not work perfectly. Suppose it always measures the time of flight to be longer than it really is. Then your experiment will consistently give values of g that are too small. Experiments are rife with systematic errors. Instrument problems are only one source of them. A quantity being measured may depend on parameters (temperature or air pressure, for example) that have not been allowed for and may not be known. Experimentalists strive to overcome systematic errors. It is possible, *in principle*, in classical physics to overcome systematic errors and to allow for statistical ones, although it is rarely easy to do so.

Probability in Classical Versus Quantum Physics

Let us go back to the experiment of tossing an unweighted coin that comes down heads as often as it comes down tails. It does so at random. A classical physicist would argue with the last remark. He or she would say the result, H or T, is not really random at all. Rather, the result depends on the exact details of how the coin was tossed. What was its initial angular velocity? What was its exact orientation relative to the horizontal plane? What was its initial kinetic energy? How, exactly, was each atom or molecule in the coin moving at the instant it was spun? A puritanical classical physicist would say that if *everything* about the coin and its surroundings were known, the outcome of the tossing experiment would be certainly known too. No randomness would be involved. Gaussian distributions, error analysis, fluctuations, and so on would not be needed; it is *only* because we do not know all the details that statistics and probability theory are needed. We need them whenever a very large number of items (coins, atoms, molecules, and so on) are involved, and the details of their state are so complex, so much information about them is required to describe them

completely, that such detailed description is hopeless, and we are forced to resort to statistics. A puritanical classical physicist would say, though, that *in principle*, nothing is random, and every experiment has a completely deterministic outcome. If we know all about the system to begin with, we can calculate exactly what will happen to it in time.

Standard quantum physics has another way of looking at this issue. Consider a very simple small system: a single electron bound by electromagnetic forces to a single proton, a hydrogen atom. In quantum mechanical analysis, we consider the electron as able to exist in certain states, or orbitals, called eigenstates. In the ground state the electron orbits the proton with certain low energy. In excited states the electron orbits the proton with higher energy. Suppose you observe a cloud of hydrogen atoms in thermal equilibrium with one another at some temperature T. Some electrons will be in the ground state, some will be in the first excited state, and some will be in higher states. Focus on one particular hydrogen atom. Standard quantum mechanics says that the state of that atom is described by a superposition of all the eigenstates, weighted by probability factors, which depend on the temperature T of the gas. If you then perform a measurement to determine what the state of that atom is, the possible results of the measurement are the eigenstates. But the measurement will show the atom to be in one and only one of the eigenstates possible, even though a moment before you did the measurement you assumed the atom was in a superposition of all states. Probability theory tells you how likely each eigenstate is to occur when you do the measurement. In the standard approach to quantum mechanics, then, probability theory is essential, being built into the theory from the beginning. Standard quantum mechanics is intrinsically a probability theory itself.

Statistical Uncertainty and Quantum Indeterminacy

Compare two experiments. The first is that in which the refractive index ν of a block of glass was measured. Suppose the errors made in that experiment truly seem random and you apply Gaussian theory. You measure ν many hundreds, even thousands, of times, and the mean value still turns out to be 1.45. You believe this is really the true value. Yet there is still uncertainty. There is always a finite value of the error on the mean, ε, given by Eq. (11), and there is always a nagging doubt that maybe there is some systematic error you have overlooked. However good your statistics, you can never be 100 percent certain of the value of the quantity measured. This is statistical uncertainty.

Quantum indeterminacy is fundamentally different. In this section we discuss only the view from standard quantum theory. Quantum mechanics says there is a basic incompatibility in measuring certain quantities accurately. The idea is summarized in Heisenberg's uncertainty principle. Take a free electron. You would like to measure its position and momentum simultaneously. Quantum mechanics says you cannot do so with infinite precision. Suppose the electron is confined to move in the x direction only. Suppose you measure the position as $x = \bar{x} \pm \Delta x$, where $\bar{x}$ is an average, or most likely value. Somehow, you also measure the electron's momentum simultaneously. You find it to be $p = \bar{p} \pm \Delta p$. Again, $\bar{p}$ is an average value. The uncertainty principle says that Δx and Δp cannot both be zero. Rather,

$$\Delta x \Delta p \gtrsim \hbar. \tag{15}$$

Consequently, if you measure x precisely, so $\Delta x = 0$, then the uncertainty in the measurement of momentum is infinite, and vice versa. The best you can hope for is to measure position, with some small error, and to measure momentum, also with a small error. The best result you can hope to obtain is $\Delta x \Delta p \sim \hbar$.

In standard quantum mechanics, some variables are fundamentally incompatible and cannot be measured with precision however good the experiment is, or however good the statistical data are. The errors invoked are seen as inherent to the quantities being measured. This idea is basically, and totally, different from the hypothesis in classical mechanics that it is theoretically possible to measure any two quantities as accurately as desired simultaneously.

See also: DISTRIBUTION FUNCTION; ERROR, SYSTEMATIC; UNCERTAINTY PRINCIPLE

Bibliography

DAVIS, J. C. *Statistics and Data Analysis in Geology* (Wiley, New York, 1986).
MERZBACHER, E. *Quantum Mechanics* (Wiley, New York, 1970).
REIF, F. *Statistical Physics* (McGraw-Hill, New York, 1967).
SPIEGEL, M. R. *Theory and Problems in Statistics* (McGraw-Hill, New York, 1961).

ROBERT H. DICKERSON

GAYLE COOK

PROPER LENGTH

The special theory of relativity proposed in 1905 by Albert Einstein is based on two postulates: the principle of relativity and the constancy of the speed of light. One of the natural consequences of the postulates is that time must be relative. It is obvious that if time is relative, Newton's concept of absolute space would also have to be questioned. After all, space and time appear together in the definition of speed, and the speed of light is an absolute quantity. If motion affects time as in time dilation, motion must also affect space; otherwise, the speed of light would not be the same to all observers, as required by the principle of the constancy of the speed of light.

Space has three dimensions, meaning that a point in space requires three independent numbers, for example, (x, y, z) in the familiar Cartesian coordinate system. If the relative motion between two observers takes place along the x axis, we can call the x axis the "longitudinal" dimension and the y and z axes the "transverse" dimensions. It turns out that motion affects only the longitudinal dimension, that is, the dimension parallel to the direction of motion.

Specifically, Einstein showed that if an object, say a rod, moves at a velocity v, its length must contract in the direction of motion. This relativistic effect is called length contraction, and is described quantitatively by

$$L = L_0 \sqrt{1 - v^2/c^2}$$

where L_0 is the length of the rod at rest, L is the (longitudinal) length of the rod moving at a velocity v, and c is, as usual, the speed of light. The length of an object at rest is called its proper length.

As in the case of proper time, the deviation of observed length from its proper value is negligibly small in ordinary experience—less than 0.5 percent even at velocities as high as 10 percent of the speed of light. At higher velocities, however, the deviations become considerable; in fact, as the speed of an object approaches the speed of light, its length in the direction of its motion would contract to almost zero.

Applied to interstellar space travel, the idea of length contraction becomes particularly interesting. Suppose an astronaut volunteers to travel to a star 50 ly away; this would be the proper distance between Earth and the star. With such a distant destination, the astronaut's outbound trip, even traveling at near the speed of light, would take nearly 50 years according to the Earth-based clock. With the passage of another 50 years for the inbound trip, a total of 100 years, then, would elapse on Earth during the astronaut's entire trip.

From the astronaut's frame of reference, however, the distance between Earth and the star would contract, and the star would not be as far. If the spaceship could travel at near the speed of light, say at $v = 0.9998c$, the astronaut would observe the distance to contract to a mere 1 ly:

$$L = L_0\sqrt{1 - v^2/c^2} = 50 \times 0.02 = 1 \text{ ly}.$$

At such a speed, the star would take only 1 year to reach the astronaut; hence the astronaut would age only 1 year during the outbound trip. Similarly, during the inbound trip, Earth would take only a 1-year return to the astronaut. Consequently, the astronaut would age only 2 years during the entire trip. On Earth, by contrast, a total of 100 years would have elapsed during the astronaut's absence.

So, in relativity theory, space and time are relative. Even a simple question such as "How far is New York from Tokyo?" would thus lose its precise meaning. To be unambiguous, this question would have to be rephrased, "What is the *proper* distance between New York and Tokyo?"

See also: LIGHT, SPEED OF; PROPER TIME; RELATIVITY, SPECIAL THEORY OF; TIME DILATION

Bibliography

MERMIN, N. D. *Space and Time in Special Relativity* (McGraw-Hill, New York, 1968).

TAYLOR, E. F., and WHEELER, J. A. *Spacetime Physics,* 2nd ed. (W. H. Freeman, New York, 1992).

SUNG KYU KIM

PROPER TIME

The modern view of time is that it is not absolute, as ordinary experience would suggest, but rather it is relative, as required by special relativity. The relativity of time means that time intervals measured by clocks are different to different observers in relative uniform motion. That is, clocks that are in motion with respect to each other keep time at different rates. Consequently, unless we specify the state of motion for a clock, time measurements would be ambiguous and without precise meaning. By contrast, time measured by a clock at rest would be unambiguous and would be the same to all observers. Hence, it is given a special name—proper time. The relationship between proper time and observed time is given by

$$t_{\text{observed}} = \frac{t_{\text{proper}}}{\sqrt{1 - v^2/c^2}},$$

where v denotes the velocity of the object being observed and c is the speed of light.

To illustrate the usefulness of proper time, let us consider the half-life of a radioactive particle—specifically, the muon. The muon has a half-life of 1.53 μs at rest; the muon's proper half-life would, of course, be measured by a muon-based clock. What we mean by the muon's half-life is that by the time 1.53 μs have elapsed on the muon-based clock, half of the muons in our sample would have decayed. Suppose we now observe muons in motion. The relativity of time implies that the half-life of the moving muons would be different. Its exact value would, of course, depend on the velocity of the muons; it could, in fact, assume any value greater than the proper half-life.

Suppose, to be definite, that we observe muons moving at 99 percent of the speed of light (i.e., $v = 0.99c$). Then, according to Albert Einstein's time-dilation equation, the observed half-life would be

$$t_{\text{observed}} = \frac{1.53}{\sqrt{1 - (0.99c)^2/c^2}} \ \mu\text{s} = 10.85 \ \mu\text{s}.$$

The longer half-life means that muons "live" longer when they are in motion than when they are at rest. Just how much longer would, of course, depend on the velocity of the muon. But all observers would agree that the proper half-life of the muon is 1.53 μs.

The relativistic deviation of observed time from proper time is negligible for velocities encountered in ordinary experience; even for velocities as high as 10 percent of the speed of light the deviation is only about 0.5 percent. But there are many phenomena where velocities approach the speed of light. Such phenomena include the creation of new particles in large accelerators and jets of matter associated with some active galaxies. Such high-energy phenomena would, of course, be incomprehensible without Einstein's theory of relativity.

See also: PROPER LENGTH; RELATIVITY, SPECIAL THEORY OF; TIME DILATION

Bibliography

MERMIN, N. D. *Space and Time in Special Relativity* (McGraw-Hill, New York, 1968).

RUCKER, R. V. B. *Geometry, Relativity, and the Fourth Dimension* (Dover, New York, 1977).

SUNG KYU KIM

PROPULSION

Propulsion is the process of causing an object to move forward by exerting a force on it. The person or thing that exerts the force on this object is called the propellant. As a result of the action of this force on the object, the object is propelled forward.

An elementary example of propulsion is provided by a person walking or running. As the person's foot pushes back on the ground, the ground exerts an equal force forward on the foot, thereby accelerating the person forward. The backward force exerted by the foot on the ground and the forward force exerted by the ground on the foot provide an example of Isaac Newton's third law of motion. Newton noted that whenever one object exerts a force on a second object, the second object exerts an equal and opposite force back on the first object. In his book *Principia Mathematica* (1687) Newton stated this principle in the following way: "To every action there is always opposed an equal reaction." The acceleration that results from the forward force on the person's foot is explained by Newton's second law of motion: The acceleration of an object is directly proportional to, and in the same direction as, the net force exerted on it and inversely proportional to its mass.

The explanation provided by Newton's second and third laws of motion for the process of walking or running applies equally well to all other examples of self-propulsion by a living organism, whether the organism moves on land, in the water, or through the air. For example, as a dolphin's flippers exert a backward force on the surrounding water, the water exerts a forward force on the dolphin's flippers, thereby propelling the dolphin forward.

If the net force, acceleration, and mass, are represented by F, a, and m, respectively, then the common form of Newton's second law of motion can be expressed as the equation $a = F/m$, or equivalently, $F = ma$. Mass and force have inverse effects on acceleration. The greater the net force on an object, the greater the object's acceleration. But the more mass an object has, the more resistance it will offer to acceleration. For a speed skater at the start of a race, a greater force exerted by the skate backward against the ice results in a greater forward acceleration. For two speed skaters exerting the same backward force on the ice, the skater of greater mass will have a smaller acceleration.

The same principles apply to the propulsion of mechanical devices such as automobiles, helicopters, and airplanes. In an automobile, the force produced by the engine is conveyed to the drive wheels of the car. As the drive wheels rotate, the tires exert a force on the road, and the road exerts an equal and opposite force on these tires, thereby accelerating the automobile. To propel an automobile forward, the tires must exert a force backward on the road so that the road will exert a forward force on the tires. To change the direction of the automobile so it is propelled backward, the direction of rotation of the drive wheels is reversed so that the tires exert a force forward on the road.

A helicopter is propelled upward by the rotation of blades that are shaped to force air downward. Associated with this downward force on the air is an upward reaction force that the air exerts on the helicopter blades, a force that is called lift. An airplane, whether powered by propellers or jet engines, is also kept aloft by a lift force. As the plane moves forward, a stream of air strikes the wings, which are tilted so that the air is forced downward. The reaction force exerted by the air upward on the wing provides the lift.

For a propeller-powered airplane the forward propulsion of the plane is a result of the propeller blades exerting a backward force on the air, which then exerts a forward force on the propeller blades. In a jet aircraft or a rocket, as a result of the ignition of fuel within the engine, some combustion products are vented out of the rear of the engine as exhaust gases, while combustion products propelled in the opposite direction within the engine exert a forward force on the craft, thereby propelling it forward. While jet and rocket engines are based on the same principle of action-reaction forces, there is an important difference between the two kinds of engines. In a jet engine, fuel burns in the presence of oxygen pulled in from the atmosphere. Since a rocket is designed to travel above the atmosphere, it must contain both its fuel and oxidizer (e.g., liquid oxygen) for combustion.

The same physical principles, Newton's laws of motion, explain the propulsion of a punctured balloon or a toy water rocket. When an inflated balloon is punctured, air vents out of the hole, a consequence of the force exerted on the venting air by the pressurized air still within the balloon. The venting air exerts an oppositely directed reaction force on the air within the balloon, resulting in the forward propulsion of the balloon. A toy water rocket is propelled forward by a similar venting of a pressurized mixture of water and air.

Propulsion may also be understood from an alternative formulation of Newton's second law of motion: The net force on an object is equal to the rate of change of the object's momentum. Propulsion occurs when one body is propelled forward by a force exerted by a second body, while the second body is propelled backward by an equal and opposite force exerted by the first body. For such an action-reaction pair, the change in momentum received by one of the bodies is equal and opposite to the change in momentum received by the other body. The sum of the changes in momenta of the action-reaction pair that generates propulsion is therefore zero. Momentum is said to be conserved. When a jet aircraft is propelled forward, the forward change of momentum of the jet is equal in magnitude to, but in the opposite direction from, the change in the momentum of the exhaust gases. When an automobile is accelerating forward, the change of momentum of the automobile in the forward direction is equal in magnitude to the change in momentum imparted to the road in the backward direction. This principle is likewise illustrated by the small rocket propulsion units that astronauts use during space walks. The combined momentum of the rocket and fuel has the same value just before and just after ignition. For the combined momentum, the forward momentum that the rocket gains from the combustion process exactly cancels the backward momentum of the exhaust gases.

See also: MOMENTUM; MOMENTUM, CONSERVATION OF; NEWTON, ISAAC; NEWTONIAN MECHANICS; NEWTON'S LAWS

Bibliography

ASIMOV, I. *Asimov's Biographical Encyclopedia of Science and Technology,* 2nd ed. (Doubleday, New York, 1982).

MARCH, R. H. *Physics for Poets,* 2nd ed. (McGraw-Hill, New York, 1978).

ROGERS, E. M. *Physics for the Inquiring Mind* (Princeton University Press, Princeton, NJ, 1977).

RUTHERFORD, F. J.; HOLTON, G.; and WATSON, F. G. *The Project Physics Course,* 3rd ed. (Holt, Rhinehart and Winston, New York, 1981).

MARTIN M. MALTEMPO

PROTON

The hydrogen atom consists of a bound state of an electron and a proton. The nuclei of other atomic elements consist of protons and neutrons bound together by the strong interaction. Free protons can be found in cosmic rays and can be produced by particle accelerators. Experiments by Wilhelm Wien in 1898 and later by J. J. Thomson in 1910 identified a positively charged particle with mass about equal to that of a hydrogen atom among the particles found when gas streams were ionized. In 1919 Ernest Rutherford, studying experiments where nitrogen was bombarded with alpha particles, again found such positively charged particles, which he identified as a hydrogen nucleus. By 1920 he concluded this was an elementary particle, which he named the proton, after the Greek "protos" for "the first," to denote its primary position among the nuclei of atomic elements.

The mass of the proton is $m_p = 938.272$ MeV/c^2 $= 1.6726 \times 10^{-27}$ kg. It is 1,836 times heavier than the electron. Many experimental searches have been undertaken to look for the decay of the proton into lighter particles, and thus far none have been observed. Independent of its mode of decay, a mean lifetime limit of $\tau > 10^{25}$ years can be set. Higher limits have been established for some specific decay modes, such as $\tau > 10^{32}$ years for the $p \to e^+ + \pi^o$ decay.

The electric charge of the proton is positive. It is opposite in sign but equal in magnitude to the electric charge of the electron, $q_p = -q_e = -e$. From the observed neutrality of matter, one can deduce the limit $|q_p + q_e|/e < 10^{-21}$. The limit on the electric dipole moment of the proton is about $d_p < 10^{-7}\,e\,\mathrm{fm}$ ($1\,\mathrm{fm} = 10^{-15}\,\mathrm{m}$), and the mean-square charge radius of the proton, obtained from experiments in which electrons are scattered on protons, is about $0.72\,\mathrm{fm}^2$. The proton has an angular momentum of $\hbar/2$, positive parity, and a magnetic moment of $2.792847\,\mu_N$. ($\mu_N = e\hbar/2m_p c = 0.1050\,e\,\mathrm{fm} = 3.152 \times 10^{-14}\,\mathrm{MeV\cdot T}^{-1}$, is the nuclear magneton.) Other properties of the proton's internal structure have been obtained from experiments in which electrons and neutrinos are scattered on protons.

The neutron is a particle whose structure has many similarities to that of the proton. It was discovered by James Chadwick in 1932 as a very penetrating form of radiation emitted in the bombardment of beryllium with alpha particles. Its mass is slightly heavier than that of the proton, $m_n = 939.565$ MeV/c^2. The neutron decays with a half-life of $T_{1/2} = 614.8 \pm 1.4\,\mathrm{s}$ by the decay mode $n \to p + e^- + \bar{\nu}_e$, where $\bar{\nu}_e$ is the electron antineutrino. The neutron is neutral in charge to within the limit $q_n < 10^{-21}\,e$. The limit on the electric dipole moment of the neutron is about $10^{-10}\,e\,\mathrm{fm}$, and the mean-square charge radius obtained from the scattering of neutrons with energies of a few electron volts on atomic electrons is about $-0.12\,\mathrm{fm}^2$, where the minus sign indicates a predominance of negative charge near the surface. The neutron has an angular momentum of $\hbar/2$, positive parity, and a magnetic moment of $-1.913043\,\mu_N$. Properties of the neutron's internal structure have been obtained from electron and neutrino scattering experiments on deuterium, after subtracting the contribution from the proton.

The similarity in the mass of the proton and neutron, along with the fact that they have the same angular momentum (spin) and that the strong interaction between them is nearly identical, leads to the introduction of the isospin concept. In the "isospin" concept an abstract particle called a nucleon is rotating in an abstract isospin space, which is related to their difference in electric charge. The proton and neutron together are collectively referred to as nucleons. The nucleon is the lightest mass particle of a group that has half-integer angular momenta (in units of $\hbar$). This group is collectively referred to as baryons. The proton and neutron have antiparticles called the antiproton and antineutron, respectively. The antiproton was discovered in 1955 by Owen Chamberlain, Emilio Segre, Clyde Weigand, and Thomas Ypsilantis using the bevatron at the Berkeley Radiation Laboratory. Shortly after, the same machine aided in the discovery of the anitneutron.

Atomic nuclei are composed of protons and neutrons (nucleons) held together by the strong interaction. The proton and neutron composition of a given nucleus can be specified in the form $^A Z$, where $A = Z + N$, and Z and N are the number of protons and neutrons, respectively. The number of protons in the nucleus determines the number of atomic electrons and hence the atomic (chemical) properties. In the notation $^A Z$ the Z is often replaced by the chemical symbol. Isotopes are nuclei with a fixed number of protons but with differing numbers of neutrons. For example, the stable isotopes of calcium ($Z = 20$) are ^{40}Ca, ^{42}Ca, ^{44}Ca, ^{46}Ca, and ^{48}Ca. For light elements, the most stable isotopes have $N \approx Z$, due to the stronger proton-neutron interaction compared to the proton-proton and neutron-neutron interactions and also due to the fact that the kinetic energy is minimized for $N = Z$. For heavy elements the effect of the Coulomb repulsion between protons becomes relatively more important and this results in $N > Z$ for the most stable isotope.

The properties of the nucleon are important for establishing and setting limits on the conservation laws. The stability of the proton leads to the concept of baryon conservation. The nucleon and electron are assigned baryon numbers $B_n = 1$ and $B_e = 0$. The rule that baryon number is conserved, together with the fact that the proton is the lightest baryon, prohibits the proton from decaying. However, the grand unified theory (GUT) predicts that there are superheavy gauge bosons whose interactions would allow for baryon nonconservation, and hence, proton decay. The experimental limit on the lifetime of the proton sets important constraints on these models. Time reversal invariance implies that the electric dipole moments must vanish, and the limits on the nucleon electric dipole moments provide one of the best experimental tests of this rule. (The presence of electric dipole moments in atoms and molecules is due to the existence of two nearly degenerate mass states of the same system. This is not the case for the proton or neutron.)

Unlike the electron, the nucleon is not a fundamental particle. This is indicated by the deviation of the magnetic moments from their Dirac values of $\mu_p = \mu_N$ and $\mu_n = 0$ as well as the finite extension of their charge and angular momentum (spin) structure. The spin and isospin properties of the nucleon together with the other baryons and mesons can be classified using the mathematical methods of group theory to the symmetry group $SU(3)$. Study of the group theory related to $SU(3)$ lead Murray Gell-Mann and George Zweig to the hypothesis that these particles could be understood in terms of an underlying structure of "quarks" bound together by the strong interaction. In the simplest quark model, the nucleon consists of three valence quarks, uud for the proton, and udd for the neutron, where u stands for the up quark, which has an electric charge of $q_u = (2/3)e$, and d stands for the down quark with an electric charge of $q_d = -(1/3)e$. The quarks have an angular momentum of $\hbar/2$, an isospin $I = 1/2$, and a baryon number $B = 1/3$. The valence quark model with the quarks taken to be fundamental particles qualitatively explains the observed properties of the proton and neutron. The valence quark model can also be successfully applied to mesons and other baryons. More completely, the nucleon is envisioned as three valence quarks together with "sea" quarks and antiquarks held together by the interactions with gluons under the rules of quantum chromodynamics (QCD). Solutions of the QCD equations are difficult but are being pursued.

Protons have been accelerated up to energies of about 10^6 MeV (1 TeV) and collided with electrons, protons, or nuclei in order to study their internal structure and to produce new particles. Accelerated protons have also been used to destroy cancer cells either directly or via the neutrons produced in a subsequent reaction. Protons make up the primary part of the galactic cosmic rays. Very high energy protons that enter the upper atmosphere eventually collide with a nucleus resulting in a shower of particles that reach Earth and can be detected experimentally.

See also: BARYON NUMBER; DIPOLE MOMENT; ELECTRON; GRAND UNIFIED THEORY; INTERACTION, STRONG; ISOTOPES; NEUTRON; NUCLEON; QUANTUM CHROMODYNAMICS; QUARK; SPIN

Bibliography

CLOSE, F., MARTEN, M., and SUTTON, C. *The Particle Explosion* (Oxford University Press, Oxford, Eng., 1987).

LEDERMAN, L., and SCHRAMM, D. N. *From Quarks to the Cosmos: Tools of Discovery* (W. H. Freeman, New York, 1995).

B. ALEX BROWN

PULSAR

The discovery of pulsars is often considered one of the three most important discoveries in astronomy in the second half of the twentieth century, along with quasars and the microwave background. In 1967, Jocelyn Bell, Anthony Hewish, and their collaborators at Cambridge University constructed a detector to observe quasars in radio frequencies. While looking at one of the charts produced by the array, Bell noticed a bit of scruff that appeared every one and a third seconds. The period of repetition of the signal was so regular that it seemed like an artificial source, but after a careful search, Bell and Hewish ruled this out and concluded that the pulses were astronomical in origin.

There are now nearly 1,000 pulsars known of the type discovered by Bell and Hewish, with periods ranging from approximately 1 ms to nearly 5 s. The periods of almost all such pulsars increase with time (a very small number of pulsars have short decreases of periods called glitches), but the increase is extremely slow, ranging from 10^{-5} s/yr for some young pulsars to 10^{-12} s/yr for the fastest pulsars; these last rival the best atomic clocks for stability. Shortly after pulsars were discovered, Thomas Gold and others realized that their extremely short variability times require a source that is small by astrophysical standards, implying that white dwarfs, neutron stars, or black holes are the only candidates. White dwarfs are too large to rotate, vibrate, or orbit faster than about once a second, and black holes have no physical surface to which a regular emitter can be attached. This leaves neutron stars as the only candidates for pulsars. Furthermore, since neutron stars vibrate a thousand times per second, far too rapidly to explain most pulsars, the period of two neutron stars orbiting each other in a binary would *decrease* with time (due to gravitational wave emission); therefore, pulsars have to be *rotating* neutron stars.

Most pulsars are detected by their emission of radio waves, but since the early 1970s, many pulsars have been found to emit x rays, and a few emit optical light or gamma rays. Initially, pulsars were classified according to the wavelength in which they were discovered (e.g., radio pulsars or x-ray pulsars), but it is now known that a more fundamental physical distinction comes from their energy source.

The majority of pulsars are powered by the decay of their rotations, which is caused by torques applied by their magnetic fields, typically around 10^{12} G. The pulsations probably occur because the emission is beamed and the beam sweeps around, lighthouse-like, as the pulsar rotates. Oddly enough, although Franco Pacini predicted a few months *before* the discovery of pulsars that rotating magnetized neutron stars would emit pulsed radio waves, astrophysicists have still not reached a consensus about the precise cause of the emission. The radiation cannot be thermal because the radio emission is strong enough that the required temperatures are impossibly high, up to around 10^{30} K. The radio emission therefore has to be some kind of coherent radiation, but its exact nature is still a mystery.

There are about thirty known pulsars that derive their energy from accretion instead of rotation, and their properties are different from those of the rotation-powered pulsars originally discovered by Bell and Hewish. Accretion-powered pulsars have relatively long periods, from around 60 ms to more than 1,000 s, and the changes in their periods are both faster and less regular than the changes in rotation-powered pulsars; some of them have periods that decrease with time. Accretion-powered pulsars are in binary systems, typically with a normal star or red giant as a companion. Mass from the companion star flows onto the neutron star, and the matter hitting the neutron star generates energy that is typically manifest in x rays or gamma rays. Despite the differences in how they are powered, the regularity of pulsations seen in both rotation-powered pulsars and accretion-powered pulsars comes from rotation.

The study of pulsars has had many unexpected benefits. Doppler shifts from rotation around a companion are easily detected, and careful observations of binary pulsars, such as PSR 1913 + 16, have confirmed that gravitational radiation occurs just as predicted in the general theory of relativity. For this discovery, Joseph Taylor and Russell Hulse received the 1993 Nobel Prize in physics. Similar analysis of Doppler shifts in the pulsar PSR 1257 + 12 led Alex Wolszczan and Dale Frail to discover the first planets outside our solar system. Pulsars are also our best probe into the properties of very dense matter, and in the future, they may give important clues about the structure of our galaxy and about spacetime ripples created near the beginning of the universe.

See also: BLACK HOLE; COSMIC MICROWAVE BACKGROUND RADIATION; GALAXIES AND GALACTIC STRUCTURE; NEUTRON STAR; PULSAR, BINARY; QUASAR; RED GIANT; STARS AND STELLAR STRUCTURE

Bibliography

GOLD, T. "Rotating Neutron Stars as the Origin of the Pulsating Radio Sources." *Nature* **218,** 731–732 (1968).

HEWISH, A.; BELL, S. J.; PILKINGTON, J. D. H.; SCOTT, P. F.; and COLLINS, R. A. "Observation of a Rapidly Pulsating Radio Source." *Nature* **217,** 709–713 (1968).

MANCHESTER, R. N., and TAYLOR, J. H. *Pulsars* (W. H. Freeman, San Francisco, 1977).

PACINI, F. "Energy Emission from a Neutron Star." *Nature* **216,** 567–568 (1967).

SHAPIRO, S. L., and TEUKOLSKY, S. A. *Black Holes, White Dwarfs, and Neutron Stars* (Wiley, New York, 1983).

WOLSZCZAN, A., and FRAIL, D. A. "A Planetary System Around the Millisecond Pulsar PSR 1257 + 12." *Nature* **355,** 145–147 (1992).

M. COLEMAN MILLER

PULSAR, BINARY

In 1974, during a search at the Arecibo Observatory in Puerto Rico for new pulsars, Russell Hulse and Joseph Taylor found PSR 1913 + 16, a new pulsar that had unusual properties. The periods of most pulsars are very steady, increasing only slightly over time. In contrast, the pulse period of PSR 1913 + 16 changed cyclically over a time of 7 h 45 min. Such changes suggested to Hulse and Taylor that the pulsar is in a binary system, and shortly afterward, they realized that they had discovered a system of two closely orbiting neutron stars. Because pulsars are very reliable clocks, timing of their pulses, combined with application of the general theory of relativity, allowed Hulse and Taylor to determine the mass of the neutron star and its companion, as well as their orbital separation (less than 700,000 km, twice the distance between Earth and the Moon). The masses are consistent with a system of two neutron stars, each with a mass of 1.4 times the mass of the Sun.

The discoverers immediately realized that this system can provide tests of general relativity. In particular, the orbital motion causes the neutron stars to lose energy via gravitational waves and to spiral together over time. Direct detection of gravitational waves from PSR 1913 + 16 is impossible, even with the best instruments, since the waves are extremely weak. However, observation of the timing changes of the pulsar tells us the rate of inspiral, and this can be compared to the predictions of general relativity. General relativity predicts that the change of period due to this effect should be approximately 75×10^{-8} s/yr, and this has been confirmed with great accuracy, although scientists had to wait until 1979 for enough data to be gathered to observe such changes.

Since the discovery of PSR 1913 + 16, four other rotation powered pulsars and about thirty accretion powered pulsars have been found in binaries. Overall, about 1 percent of known pulsars are found in binary systems, much less than the fraction of ordinary stars in binaries. Timing of the inspiral of the other rotation powered pulsars in binaries has yielded additional confirmation of predictions for gravitational radiation. For their discovery of PSR 1913 + 16 and their subsequent analysis of its orbital decay, Hulse and Taylor received the 1993 Nobel Prize in physics.

See also: GRAVITATIONAL WAVE; NEUTRON STAR; PULSAR

Bibliography

HULSE, R. A., and TAYLOR, J. H. "Discovery of a Pulsar in a Binary System." *Astrophys. J.* **195**, L51–L53 (1975).

LYNE, A. G., and GRAHAM-SMITH, F. *Pulsar Astronomy* (Cambridge University Press, Cambridge, Eng., 1990).

TAYLOR, J. H.; FOWLER, L. A.; and McCULLOCH, P. M. "Measurement of General Relativistic Effects in the Binary Pulsar PSR 1913 + 16." *Nature* **277**, 437–440 (1979).

TAYLOR,, J. H., and STINEBRING, D. R. "Recent Progress in the Understanding of Pulsars." *Ann. Rev. Astron. Ap.* **24**, 285–327 (1986).

TOMASZ BULIK

PVT RELATION

When one performs a series of thermodynamic measurements and obtains an empirical plot among P (pressure), V (volume), and T (temperature), the result can hardly be called an equation of state since it lacks an explicit analytical relation among the parameters. In such a case the graphic plot is frequently called the PVT relation.

In a strict sense, the so-called equation of state is defined as an intensive parameter (such as temperature T, pressure P, and chemical potential μ) expressed as a function of extensive parameters such as volume V, entropy S, and mole numbers N_1, N_2, and so on (number of molecules of the first kind measured in the unit of mole, etc.). Thus, for example, one of the equations of state of a single component system ($N_1 = N$, $N_2 = 0$) has the form

$$T = f(S, V, N).$$

In reality the most common equation of state is that of an ideal gas: $PV = NRT$, where R is the universal gas constant. This does not satisfy the above definition of equation of state given in the framework of axiomatic thermodynamics. Thus the term "PVT relation" is sometimes used instead to describe any functional relation among these thermodynamic variables.

The terms "extensive" and "intensive" parameters deserve some explanation. Extensive parameters are state variables such as volume, mole number, or entropy that double their values when the system is doubled. For instance, one has a gas in a box of volume V with mole number N, pressure P, temperature T, entropy S, and chemical potential μ. If one combines it with an identical system, that is, another box of the same gas with the same values of V, T, N, S, P, and μ, the new volume will now be $2V$ and the mole number $2N$. Intensive parameters are the quantities that do not double but remain the same, such as temperature, pressure, and chemical potential. In the above example, the temperature of the gas in each of the boxes is T, and thus the temperature of the combined system would still be T. Pressure P and chemical potential μ would remain the same as well.

In general, intensive parameters of the system are easier to measure experimentally than its extensive parameters. For instance, volume is more difficult to measure since it requires counting the cubic millimeter of all its nooks and crannies as compared to attaching a pressure gauge at the boundary of the system. Since an intensive parameter near the surface (a small subsystem) is identical to the rest of the system, one can measure intensive parameters for the entire system near the system boundary.

See also: STATE, EQUATION OF; THERMODYNAMICS

CARL T. TOMIZUKA

PYROELECTRICITY

Pyroelectricity, among the oldest of reported physical properties, was initially identified more than twenty-three centuries ago by Theophrastus as the power of attraction that tourmaline develops on being heated. Pyroelectric materials characteristically undergo a change in electric polarization in response to a variation in temperature. All pyroelectric materials are polar and hence lack centers of inversion symmetry; they are therefore also piezoelectric, although the converse does not necessarily apply. In addition, some pyroelectrics exhibit ferroelectricity.

The change in spontaneous polarization (P_s) along the ith axis of a pyroelectric crystal (ΔP_i) is proportional to the temperature change (ΔT) in the crystal, with $\Delta P_i = p_i \Delta T$, where p_i is the pyroelectric coefficient. The subscript $i = 1$, 2, or 3 corresponds, respectively, to the a, b, or c crystal axis. A single coefficient p_i suffices in all point groups that possess a unique direction, namely 2, $mm2$, 3, $3m$, 4, $4mm$, 6, and $6mm$. The vector quantities P_s and p_i may lie along any direction in the mirror plane of crystals belonging to point group m and hence the coefficient in such cases is specified by two p_i components; in triclinic crystals with point group 1 the pyroelectric vector may assume any direction within the unit cell and therefore has three p_i components.

In principle, ΔP_i may be measured either with the shape and size of the crystal held constant by the application of constraints, that is, at constant strain, or with the crystal held free and allowed to undergo normal thermal expansion, that is, at constant stress.

The first procedure reveals only the primary (or true) pyroelectricity; in the second procedure, an additional effect arises from a piezoelectrically induced polarization change, namely secondary pyroelectricity. It is simpler to measure the pyroelectric moduli using the latter rather than the former procedure, thereby leading to an experimental sum of the primary and secondary coefficients. Nonuniform heating with its resulting thermal gradients may cause nonuniform stresses and strains and may hence produce tertiary or false pyroelectricity through uncontrolled piezoelectric effects. Such effects can be misleading and are best eliminated experimentally to avoid false results. The magnitude of the total primary and secondary p_i coefficients (i.e., p_i^T) in nonferroelectric pyroelectrics ranges from about 10^{-2} to 10^3 $\mu C \cdot m^{-2} \cdot K^{-1}$, in ferroelectric pyroelectrics p_i^T ranges from 10^0 to 10^4 $\mu C \cdot m^{-2} \cdot K^{-1}$.

Quantitative measurement of the pyroelectric moduli, using either of the above procedures, may be made by static, dynamic, or indirect methods. In a commonly used technique, a radiant light source is used to heat the crystal and a charge integration amplifier collects the resulting pyroelectric charge. The coefficient p_i^T generally exhibits a characteristically nonmonotonic temperature dependence resulting from the thermal dependence exhibited separately by the component primary and secondary coefficients, the total temperature dependence may be of either sign.

The origin of the pyroelectric effect is a combination of thermally dependent ionic and electronic charge distributions, as illustrated by the simple case of ZnO. Each zinc atom in hexagonal crystals of zincite is located at the unit cell origin with one oxygen atom at a distance of 1.98 Å from it along the polar c axis. Symmetry places three related oxygen atoms on the opposite side of and at the same distance from the zinc atom to complete a slightly distorted ZnO_4 tetrahedron. The small charge of about $0.2\ e$ on all atoms, which would give zero P_s if the tetrahedron were regular, gives rise to a net polarization caused by the geometrical distortion. The atomic displacement of oxygen, with respect to the zinc atoms in ZnO, is reported to have a thermal dependence of $\sim 0.3 \times 10^{-4}$ Å$\cdot$K^{-1}, leading to a direct relationship with the primary pyroelectric polarization of $\sim 0.06 \times 10^{-4}$ Å$/\mu C \cdot m^{-2}$.

The polarization change produced in a pyroelectric material on being heated by any radiation within a wide energy range, from the far infrared to beyond the ultraviolet, has led to widespread use of the property in a series of devices including infrared detectors, temperature sensors, motion sensors, calorimeters, thermal imagers, and pyroelectric vidicon tubes.

See also: CALORIMETRY; CRYSTAL; FERROELECTRICITY; PIEZOELECTRIC EFFECT; TEMPERATURE

Bibliography

HELLWEGE, K.-H., and HELLWEGE, A. M., eds. *Landolt-Börnstein: Numerical Data and Functional Relationships in Science and Technology,* Group III, Vol. 18 (Springer-Verlag, Berlin, 1984).

LANG, S. B. *Sourcebook of Pyroelectricity* (Gordon & Breach, New York, 1974).

S. C. ABRAHAMS

Q

QUANTIZATION

The word "quantization" refers to the fact that dynamical variables like momentum, energy, and angular momentum, which are allowed to have a continuum of values in classical mechanics, are typically found to have only certain discrete or quantized values in atomic and subatomic systems. For example, the orbital angular momentum of the electron in the hydrogen atom has to be an integral multiple of $\hbar = 1.05 \times 10^{-34}$ J·s. The quantity $h = 2\pi\hbar$ is known as Planck's constant. Likewise, the total energy of that electron (kinetic plus potential) is restricted to have the values

$$E = -\frac{me^4}{2\hbar^2 n^2} \qquad (n = 1, 2, 3 \ldots),$$

where m and e are the electron mass and charge, and n is the number of quantum units of orbital angular momentum.

Let us start with the rather mysterious emergence of quantization and work our way to the fairly complete understanding we have today. In 1900 Max Planck was analyzing thermal equilibrium between matter and radiation. From very general thermodynamic principles he could calculate the spectrum of radiation in equilibrium with matter; that is, the amount of energy per unit volume at each frequency. Planck found that he could reproduce the observed spectrum by making a simple but totally ad hoc assumption: Energy can be exchanged between matter and the radiation field only in multiples of the unit quantum

$$E = hf,$$

where f is the frequency. This means in particular that as we go to higher frequencies, the basic unit of exchange gets larger and larger and, hence, less and less accessible. While this assumption gave a fit to the measured spectrum, it was a bolt from the blue.

Next, Albert Einstein proposed in 1905 that if energy exchange was taking place in quanta of energy $E = hf$, then these quanta must be treated as genuine particles, called photons, and not as a mathematical device. This allowed him to provide the following very appealing explanation of the photoelectric effect, in which electrons are ejected from metal when one shines light on it. It was found that if the frequency was below a threshold, no electrons are emitted, no matter how intense the light. This was very hard to understand in classical terms, since the intensity of light is a measure of the electric force acting on the electrons. Einstein suggested the incident radiation is composed of photons, each of energy hf. If $hf < W$, the work function, which is the energy needed to free an electrom from the metal, the photons will fail to liberate the electrons. Increasing the

1255

intensity will not help, since this only increases the rate of photons striking the metal, but not the energy carried by each. On the other hand, if $hf > W$, the electrons must come out with residual kinetic energy $hf - W$, as was observed. Thus, in one stroke Einstein provided an explanation of photoelectric effect and established the reality of the photons.

If photons are genuine particles that travel at the speed of light, they must obey the relativistic equation $E^2 = c^2p^2$. Thus each photon must have momentum $p = E/c = hf/c = h/\lambda$, where λ is the wavelength. Arthur Holly Compton verified experimentally in 1923 that in their collisions with electrons, photons do indeed obey the laws of conservation of relativistic energy and momentum with the above mentioned values.

The next logical step comes with the recognition that conservation of energy requires that, if radiation comes with quantized energies, then matter, its source, must also have quantized energy levels. Consider for example the hydrogen atom, which is made up of one electron and one proton. It is found to emit radiation only at some special frequencies. Assuming this radiation is emitted one quantum at a time, if the frequency is f, the atom must lose energy $E = hf$ in the process. The most natural explanation for this is that the atom can have only discrete energy levels and $E = hf$ is the difference between two such levels. (If the allowed energies are continuous, it is hard to see why a continuum of E's and hence a continuum of frequencies are not seen.) But what determines these allowed levels? How does Planck's constant enter, as it must? The answer came from Niels Bohr in 1913. In the Bohr model, one assumes the proton (which weighs roughly 2,000 times the electron) is at rest and the electron goes around it on a circle of radius r. Equating the requisite centripetal force to the Coulomb attraction, we find

$$\frac{mv^2}{r} = \frac{e^2}{r^2}, \qquad v^2 = \frac{e^2}{mr}.$$

The total energy is then

$$E = \tfrac{1}{2}mv^2 - \frac{e^2}{r} = -\frac{e^2}{2r}.$$

So far E is not quantized since r can have any value. Bohr now postulated that the only allowed values of angular momentum were inter multiples of $\hbar$

$$mvr = n\hbar \qquad (n = 1, 2, 3 \ldots).$$

Note that Planck's constant has units of angular momentum.

It is easy to solve these equations and to show that the only allowed values of r are

$$r = n^2 \frac{\hbar^2}{me^2} \equiv n^2 a_0,$$

where a_0 is called the Bohr radius and has a value of about 10^{-8} cm. The allowed energies are then

$$E_n = -\frac{e^2}{2a_0 n^2} = -\frac{me^4}{2\hbar^2 n^2} = -\frac{\text{Rydberg}}{n^2}$$

as quoted in the beginning. The energy has been expressed in terms of a unit called a Rydberg = 13.6 eV. The integer n is an example of a quantum number.

We visualize the energy levels of the atoms as in the Fig. 1. The lowest level corresponds to $n = 1$, the one above it corresponds to $n = 2$, and so on. When the electron makes a transition from one labeled n_1 to another labeled n_2, it emits a photon of energy

$$E(n_1, n_2) = \frac{me^4}{2\hbar^2} \left[\frac{1}{n_2^2} - \frac{1}{n_1^2} \right]$$

and frequency $f = E/h$. This simple formula agrees with the observed spectrum of hydrogen to a fantastic accuracy.

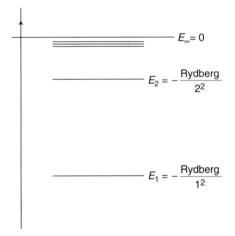

Figure 1 Energy levels of an atom.

While Bohr provided a model of the atom that reproduced the spectra, he did not fully furnish the complete formalism that would tell us what was quantized and in what way in each situation. We now trace that part of the development using the simplest example of a particle moving in one dimension to illustrate the origin of quantization.

We start with Louis de Broglie, who argued in 1924 that if light, which exhibited wave-like properties such as diffraction and interference, can behave like a particle, then particles like electrons must exhibit wave-like particles. An electron of momentum p must, just like the photon, exhibit properties of a wave of wavelength $\lambda = h/p$. This was verified by Clinton Joseph Davisson and Lester H. Germer. What they found is equal to the statement that if you send electrons of momentum p through a double slit and looked at how they hit a screen on the other side, they form an interference pattern characterized by a wave with wavelength $\lambda = h/p$. Mathematically, the electron traveling along the x direction (transverse to the slits, which lie in the yz plane) was described by a plane wave

$$\psi(x) = e^{ipx/\hbar},$$

called its wave function, whose interference pattern in the double-slit geometry was first to be computed using classical wave optics. The corresponding intensity $|\psi|^2$ at any point on the screen gave the likelihood of finding the electron there.

Consider now a particle moving along a one-dimensional ring of circumference $2\pi R$. Let x be the coordinate along this ring measured from some arbitrarily chosen starting point. If the particle has momentum p, we can describe it once again by the wave function

$$\psi(x) = e^{ipx/\hbar}.$$

But now we must insist that since moving a distance $2\pi R$ brings us to the same point, $\psi(x)$ also returns to the same value under this operation. Thus,

$$\psi(x) = \psi(x + 2\pi R)$$
$$e^{ipx/\hbar} = e^{ip(x + 2\pi R)/\hbar},$$

which in turn implies that

$$pR = n\hbar \qquad (n = 0, \pm 1, \pm 2 \ldots),$$

since $e^{2\pi n i} = 1$. Thus, momentum is quantized to these values, depending on $\hbar$ and radius of the ring. From the quantization of momentum, we can deduce the quantization of angular momentum to the values

$$L = pR = n\hbar$$

and thus of the kinetic energy to the values

$$E = \frac{p^2}{2m} = \frac{\hbar^2 n^2}{2mR^2}.$$

This simple case shows us how things work in general: In quantum theory the particles are described by wave functions and placing some restrictions on the wave function (single valuedness in this example) leads to the quantization of variables like momentum, energy, and angular momentum.

A generalization of this idea to the degrees of freedom of the electromagnetic field leads to the result that its energy is also quantized into bundles, these being just the photons. Indeed all fundamental particles like electrons, quarks, and so on may be understood as the quanta of the corresponding fields.

See also: COMPTON, ARTHUR HOLLY; COMPTON EFFECT; DAVISSON–GERMER EXPERIMENT; EINSTEIN, ALBERT; ENERGY LEVELS; MOMENTUM; PHOTON; PLANCK, MAX KARL ERNST LUDWIG; QUANTUM FIELD THEORY; QUANTUM MECHANICS; QUANTUM MECHANICS, CREATION OF; QUANTUM NUMBER; WAVE–PARTICLE DUALITY, HISTORY OF; WORK FUNCTION

Bibliography

SHANKAR, R. *Principles of Quantum Mechanics,* 2nd ed. (Plenum, New York, 1994).

R. SHANKAR

QUANTUM

Quantum is an indivisible, specified amount. The charge of the electron e is a quantum in that all observed charges are integer multiples of this fundamental quantity. *Quantized* and *quantization* signify discrete (as opposed to continuous) values of a physical parameter (e.g., total energy, angular momentum). *Quantum behavior* refers to laws applicable at the level of single particles. These laws may be totally different from what applies to macroscopic many-particle objects, which exhibit *classical behavior*. Quantum behavior is at times counterintuitive, because our everyday experience is based on macroscopic effects. However, careful experimentation with atoms and other few-particle systems shows that the quantum theory fits the data, whereas no classical-type theory can do so.

While the rules for the line spectra of gases (a prime manifestation of the discrete energy values of electrons in the atom) were well established at the beginning of the twentieth century, it was to explain the (continuous) spectrum of a blackbody that the quantum concept was first introduced. In 1900 Max Planck was able to correctly explain the energy density (the energy per unit volume per unit frequency within a blackbody cavity) by assuming that the wall of the cavity consisted of *resonators* (of unknown nature) of frequency ν. The energy E of the resonators could not assume arbitrary values, but was *quantized,* that is, could only be equal to integer multiples of $h\nu$,

$$E = nh\nu \qquad (n = 1, 2, 3,...),$$

where h is a constant. By fitting the energy density formula to the experimental data, Planck determined $h = 6.55 \times 10^{-34}$ J s. The now-accepted value of Planck's constant is $h = 6.626 \times 10^{-34}$ J·s $= 4.136 \times 10^{-15}$ eV·s.

Albert Einstein's explanation of the photoelectric effect (1905) was a milestone in establishing the quantum character of light. The key concept was that the *photon* (quantum of light) carries energy $h\nu$, where ν is the frequency. Thus the energy that could be delivered by a light beam of frequency ν, is a multiple of $h\nu$, which for the case of red light is about 2 eV. This is a very small amount of energy. A 1 mW He-Ne laser delivers over 3×10^{15} photons per second. When light is incident on a metal, an electron can absorb a single photon, and if $h\nu$ is larger than the energy binding the electron to the metal (the work function ϕ) the electron can escape (photoemission of electrons). If $h\nu$ is less than ϕ, photoemission will not occur, even if the intensity of the incident light is increased. This is a quantum effect: The electron can not absorb several photons of energy less than ϕ and overcome the forces binding it to the metal. The photoelectric effect is a manifestation of the quantum-particle characteristics of light, and provides a standard method for measuring h.

In 1924 Louis-Victor-Pierre-Raymond de Broglie proposed that particles exhibit wave characteristics as well, and associated a wavelength

$$\lambda_B = h/p$$

To a particle of linear momentum p $(= mv)$. This concept was equivalent to the basic assumption of Niels Bohr in his theory of the hydrogen atom (1913). Bohr assumed that the angular momentum of the electron orbiting around the nucleus must be a multiple of $h/2\pi$ $(= \hbar)$ or

$$mvr = n\hbar \qquad (n = 1, 2, 3,...),$$

where m is the mass of the electron, v its velocity, and r the radius of the circular orbit. To each orbit corresponds an energy

$$E_n = -\frac{m e^4}{8 \varepsilon_0^2 h^2 n^2} \qquad (n = 1, 2, 3,...),$$

where ε_0 is the permittivity of free space $(= 8.854 \times 10^{-12}$ F/m). The above equation yields $E_1 = -13.6$ eV (the ground state), $E_2 = -3.40$ eV (first excited state), etc., $E_\infty = 0$ (the electron is no longer bound to the nucleus and can assume any positive value of energy). To jump from the $n = 1$ to the $n = 2$ state the electron must absorb one photon of energy $E_2 - E_1 = 10.2$ eV. Conversely, the electron may jump from the $n = 2$ to the ground state by emitting one 10.2 eV photon. Hence, the absorption and emission of radiation are quantized. The theory accounts for the main features of atomic line spectra. Classical physics does not justify Bohr's assumption of quantized angular momentum, and does not even predict stable orbits for the electron (the Ernest Rutherford model). According to classical physics, the circular motion of the electron is an accelerated motion. Thus, the orbiting electron would continuously emit electromagnetic radiation and eventually spiral into the nucleus (the Joseph John Thomson model, which pictured the electrons embedded in the nucleus). Note that Bohr's angular momentum quantization is equivalent to the requirement that the circumference of the orbit ($2\pi r$) is a multiple of λ_B of the electron.

A main feature of waves is that they can be diffracted by obstacles of size comparable to the wavelength. For example, an electron accelerated from rest through a voltage of 150 V has λ_B of about 1 Å, which is in the order of atomic spacings in crystals. Thus electrons should be diffracted by arrays of atoms in the same way that a beam of light is diffracted by a grating. This experiment was done by Clinton J. Davisson and Lester H. Germer (1927), and confirmed de Broglie's hypothesis of wave properties for particles.

The wave character of particles has many significant implications. In a simple *gedanken* or thought experiment, an electron beam is incident normal to an impenetrable barrier with a slit. From our experience with classical particles we expect that a detector on a screen behind the barrier will register a pattern corresponding to the shadow of the barrier. However, if λ_B of the electron is comparable to the width of the slit, the wave character becomes prominent. The pattern spreads like the fringes observed with diffraction of light. With light, intensity maxima on the screen will occur whenever the path traveled by two waves from the edges of the slit differ by a multiple of the wavelength λ, and minima will occur when the path difference is an odd multiple of $\lambda/2$. In the case of electrons this interpretation is unjustifiable. If the incidence rate of electrons is lowered so that one electron crosses the slit at a time, there will not be other electrons to interfere with. Thus we have to accept that a single electron crossing the slit can land anywhere on the screen: the probability of arrival at a given point is given by a distribution similar to the diffraction pattern. In this way, it is impossible to predict exactly where any particular electron will hit the screen. Thus quantum behavior allows us to predict a distribution of outcomes for many electrons passing through the slit but not the outcome of an individual event. The inability to predict the exact outcome of a single event stems from the fact that it is impossible to know simultaneously the momentum and position of a particle with arbitrary precision. The product of the uncertainties in position Δx and momentum (in the x direction) Δp will be

$$\Delta x\, \Delta p \approx h$$

at best.

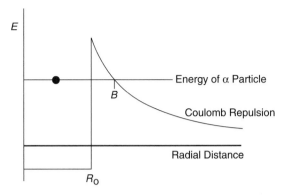

Figure 1 Model for the potential of the α particle as a function of the radial distance from the center of the nucleus. R_0 is the radius of the nucleus. The particle can tunnel through the region between R_0 and B which is forbidden by classical physics.

The probabilistic nature of the physical world is clearly manifested by the spontaneous emission of α particles by heavy nuclei. The α particles are held in the nucleus by a high potential, as shown in Fig. 1. From the classical point of view, if the α particle had kinetic energy larger than the height of the barrier, it would escape (in less than 10^{-21} s), otherwise, it would never escape. Yet in a given sample, the time it takes for half of the heavy nuclei to each emit one α particle (the half-life $\tau_{1/2}$) can range from a fraction of a microsecond to billions of years depending on the nucleus involved. The interpretation (George Gamow, Edward Condon, and Ronald Gurney, 1928) is based on purely quantum effect, namely tunneling. In the quantum theory, a particle has a small, yet finite probability P, of crossing a potential barrier, even if its total energy is less than the height of the barrier. The α particle bounces back and forth between the walls of the barrier about 10^{21} times every second. Thus, the α particle will escape after a time $\tau = 10^{-21}/P$ seconds on the average. This is the mean lifetime which is equal to $\tau_{1/2}/\ln 2$. Note that P is rather small, that is, if $\tau = 1$ s the odds of a particle escaping are 1 out of 10^{21} attempts.

It is worth emphasizing that the quantization is introduced in terms of h. The effects of this fundamental quantity on mechanics of macroscopic objects are too small to detect, thus classical physics is valid as long as h is negligible.

See also: BROGLIE, LOUIS-VICTOR-PIERRE-RAYMOND DE; DAVISSON–GERMER EXPERIMENT; DIFFRACTION; DIF-

FRACTION, ELECTRON; EINSTEIN, ALBERT; GROUND STATE; LIGHT, WAVE THEORY OF; MOMENTUM; PHOTO-ELECTRIC EFFECT; PHOTON; QUANTUM MECHANICS; RADIATION, BLACKBODY; UNCERTAINTY PRINCIPLE; WAVELENGTH; WAVE MECHANICS; WORK FUNCTION

Bibliography

BEISER, A. *Concepts of Modern Physics*, 5th ed. (McGraw-Hill, New York, 1995).

FEYNMAN, R.; LEIGHTON, R.; and SANDS, M. *The Feynman Lectures on Physics*, Vol. 3 (Addison-Wesley, Reading, MA, 1965).

PANOS J. PHOTINOS

QUANTUM CHROMODYNAMICS

Quantum chromodynamics (QCD) is the modern theory of the strong (nuclear) interactions. "Chromo" comes from the Greek for color. Thus, the subject of *chromo*dynamics deals with the forces between objects with a property, explained below, called color. A fruitful contrast is *electro*dynamics, which deals with the forces between objects with electric charge. Chromodynamic interactions take place at distances where quantum mechanics is essential. Therefore, one usually focuses on *quantum* chromodynamics.

In electricity there are positive and negative charges. Combining equal amounts of each yields a neutral system. In QCD, on the other hand, one should think of three "colors." Combining each in equal amounts yields a colorless system. This is much like the primary colors red, blue, and green of human vision; in a color television, for example, red, blue, and green light combine to make white. The analogy carries over to the colors cyan, yellow, and magenta, which are complementary to the respective primary colors. Thus, the combination of all three complementary colors is colorless. Finally, the combination of a primary color and its complement, say green and magenta, is neutral.

The analogy between the color of chromodynamics and the color of vision cannot be taken any further. At a fundamental level, the latter depends on the physiology of the eye and the brain. The colors of QCD summarize features of a kind of symmetry, called $SU(3)$ symmetry. The common feature—but, hence, the origin of the name chromodynamics—is that certain three-way and two-way combinations of colored objects are neutral—or colorless—objects.

Quarks, Gluons, and Hadrons

The objects in nature with $SU(3)$ color are called quarks. Quarks are assigned primary colors red, blue, and green, and their antiparticles—antiquarks—are assigned cyan, yellow, and magenta. Three quarks, one each of red, blue, and green, form a kind of particle called a baryon. (Three antiquarks, one each of cyan, yellow, and magenta, form an antibaryon.) The most familiar examples of baryons are protons and neutrons, which make up atomic nuclei, but there are many others that have been found in a wide variety of experiments. A quark and an antiquark of complementary colors (say, green and magenta) form another kind of colorless particle called a meson. An example of a meson is the pion, which is seen in cosmic rays, but, again, there are many others that have been found. Indeed the quark model was originally introduced without color in 1964 (by Murray Gell-Mann and George Zweig) as a device for keeping track of the plethora of baryons and mesons.

Were the story to end here, the additional property of color would be a minor complication to the quark model. In 1973 Gell-Mann, Harald Fritzsch, and Heinrich Leutwyler proposed new fields, analogous to electric and magnetic fields in electrodynamics, that are sensitive to the quarks' color. In the quantum theory these chromoelectric and chromomagnetic fields exist as quanta called gluons. "Gluon" is a whimsical adaptation of the word glue; in QCD the gluons are responsible for binding—or "gluing"—the quarks into baryons and mesons. [Analogously, the photon of quantum electrodynamics (QED) binds oppositely charged particles into atoms.]

The most remarkable property of gluons is that they are not just sensitive to color, but they carry color themselves. To be precise, a gluon carries both a quark-like color (red, blue, or green) and an antiquark-like color (cyan, yellow, or magenta). This is completely unlike the photon of QED, which senses electric charge but is electrically neutral. The color of the gluon is, ultimately, responsible for the remarkable features of QCD discussed throughout this entry.

Physicists use the term "hadrons" for particles like baryons and mesons. A fair picture of how hadrons emerge in QCD is as follows. Imagine that a gluon is a little bit of string with red, blue, or green at one end and cyan, yellow, or magenta at the other. In QCD (as opposed to a quark model without gluons) the picture of a meson as a quark and antiquark makes sense only if the quark and antiquark are the same point in space. Separating them is possible only if a piece of gluon string intervenes, with colors always compensating each other locally, as in Fig. 1a. Similarly the three-quark picture of a baryon must be generalized to include a three-point "Mercedes Benz" star of gluon string, as in Fig. 1b.

Gluons have color. As a consequence, there should be a third kind of hadron, made of gluons only. To see how these so-called glueballs might arise, connect several bits of string together until the ends have complementary colors, say, red and cyan.

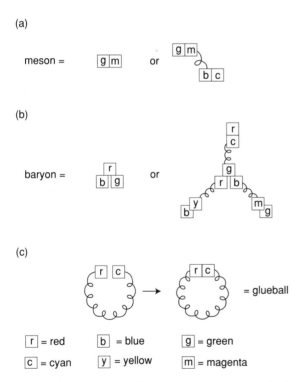

(a)

meson =

(b)

baryon =

(c)

= glueball

$\boxed{r}$ = red $\boxed{b}$ = blue $\boxed{g}$ = green

$\boxed{c}$ = cyan $\boxed{y}$ = yellow $\boxed{m}$ = magenta

Figure 1 Quark and gluon structure of the hadrons. (a) A meson can exist either in a quark-antiquark state or in a quark-antiquark-gluon state. (b) A baryon can exist either in a three-quark state or as three quarks with a three-point gluon configuration. (c) A glueball is a loop of gluon string without any quarks.

Then pull the ends around into a loop; red and cyan compensate each other, leaving the loop colorless as a whole, as in Fig. 1c. Glueballs are an uncontroversial theoretical prediction of QCD, but their existence remains without unambiguous experimental verification.

Confinement and Asymptotic Freedom

QCD and QED belong to a class of mathematical models called gauge theories. In quantum gauge theories, forces between particles arise when certain quanta of fields—the gauge bosons—are exchanged. In QCD and QED the gauge bosons are the gluon and photon, respectively. If only one (massless) gauge boson is exchanged, the force behaves as $1/r^2$, where r is the separating distance, as in Coulomb's law or in Newton's law of gravitation.

Let us write the quark-antiquark force of the strong interaction $F(r) = -\alpha_s(r)/r^2$. The numerator $\alpha_s(r)$, called the strong coupling, is introduced because more than one gluon can take part in the exchange. In the framework of QCD, when a gluon is exchanged to generate the force, the gluon can split into two. This can happen because the gluon carries color. (The photon, on the other hand, carries no electric charge, and, hence, cannot split into two photons.) After the split, the two gluons can recombine into a single gluon again. The split and recombination has a striking effect on the force: The strong coupling $\alpha_s(r)$ becomes a function of distance. It gets stronger at longer distances and weaker at shorter ones.

The consequences are remarkable. Recall the above picture of a meson as a quark and antiquark at the ends of a gluon string. Now imagine pulling the quark and antiquark apart. At a distance of about 1 fm (1 fm = 10^{-15} m) the force becomes constant, and a great deal of energy is stored in the string. (Indeed, roughly 98 percent of the rest energy of a proton is in the gluon strings.) Pull further, and eventually the energy can be converted into a new quark-antiquark pair. The quarks and antiquarks pair up to form two mesons. In other words, it is impossible to separate a quark far from a companion antiquark (or, as in a baryon, from two companion quarks). This is called quark confinement.

At short distances, on the other hand, the strong coupling weakens. Deep inside a hadron, therefore, the quarks and antiquarks can be thought of as nearly noninteracting. Indeed, as the distance between the quarks becomes smaller and smaller, the coupling α_s vanishes. The absence of an interaction in asymptotic limit is called asymptotic freedom.

The mechanisms of quark confinement and asymptotic freedom occur only in a special class of quantum field theories, to which QCD belongs. Both features are in accord with laboratory observations. On the one hand, there is no serious evidence for quarks outside of hadrons; they are confined. On the other hand, at short distances asymptotic freedom is well established. Heisenberg's uncertainty principle provides a profound correspondence between short distances and high momenta (and, thus, high energies). So, by colliding high-energy hadrons in a particle accelerator, one can probe the distance dependence of $\alpha_s(r)$. The data from numerous experiments verify asymptotic freedom quantitatively.

High-Energy Scattering Experiments

QCD describes not only the forces that bind hadrons together, but also the forces exerted when hadrons are scattered off each other, or when other particles collide to produce hadrons. Such scattering experiments provide an impressive body of data that agrees with detailed predictions of the theory of QCD. Although examples abound, we shall discuss only three experiments, in historical order. The first and third together demonstrate the colored quark structure of the hadrons. The second directly verifies the existence of the gluon.

The first experiment is in electron-proton scattering. A possible setup is to use a target of solid frozen hydrogen—this is almost completely made of protons. The target is placed at the front of a detector apparatus. Then an accelerator is used to create a beam of electrons (traveling near the speed of light), which impinges on the target. The objective is to count how often the electron scatters off the proton. As a rule, the proton will be smashed to smithereens, but the details of the smithereens are not measured.

Electron-proton scattering is relatively simple. Electrons have (as far as we know) no structure, and they can interact with the proton only through thoroughly understood electromagnetic interactions. If the proton consists of quarks, then one might anticipate that the probability to scatter an electron off a proton is equal to the density of the quarks inside the proton times the probability to scatter an elec-

tron off a quark. A lengthy mathematical derivation within QCD does indeed substantiate the decomposition into two factors. A derivation with Feynman diagrams provides a formula for the probability to scatter an electron off a quark. Unfortunately, the density of quarks inside a proton cannot be calculated with the theoretical tools available today. Instead, one solves the equation for the density of quarks inside a proton, in effect using the electron-proton scattering experiment to measure the density.

A group led by Jerome Friedman, Henry Kendall, and Richard Taylor were the first to make such measurements at the Stanford Linear Accelerator Laboratory (SLAC) in the late 1960s. In the historical development of QCD, their results played a crucial role. Their data could be explained by the hypothesis that the proton consisted of nearly free constituents trapped inside a confining bag. But these ideas were contrived and had some shortcomings; gradually, it became clear that the proton also contained constituents that do not interact with electrons. With the benefit of hindsight, QCD offers an explanation: At the short distance probed by the electron, the quarks are asymptotically free; at larger distances, the chromodynamic force gets strong, confining the quarks inside the proton; a proton contains not just quarks, but gluons as well.

The second experiment is the production of hadrons in electron-positron collisions. (The positron is the antiparticle to the electron.) Beams of electrons and positrons are directed at each other and meet in the middle of a big detector. When an electron collides with a positron, the two particles disappear. Their energy is stored in a taut bundle of electromagnetic field. This configuration is unstable and, on occasion, decays into a quark and antiquark, which hurtle away from each other with tremendous momentum.

According to QCD, the reaction proceeds as follows. To conserve color locally, a gluon string stretches out between quark and antiquark. By and by, the string fragments into more quarks and antiquarks. Suppose, for illustration, one secondary pair is created; then the secondary quark pairs with the primary antiquark, and vice versa, to form two mesons. But usually the string fragments into dozens of secondary quarks and antiquarks, so the final state contains many mesons, plus, perhaps, a few baryons and antibaryons. A signature of the primary quark and antiquark remains, however; the hadrons emerge in collimated streams, or jets, of particles,

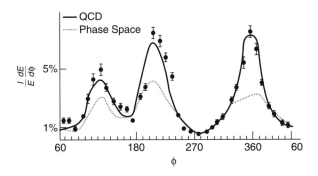

Figure 2 Comparison of energy flow in three-jet events in electron-positron scattering. The points are data from the experimental group MARK J. The solid curve is the prediction of quantum chromodynamics. The dotted curve (phase space) shows what would happen if the third jet were produced randomly (Gottfried and Weisskopf, 1984, with permission of authors).

aligned along the directions of the primary quark and antiquark.

About 10 percent of the time, a quark or antiquark radiates a high-energy gluon, leading to a third jet. A calculation using QCD predicts not only the frequency, but also detailed properties of events with three jets. As enthusiasm for QCD grew in the late 1970s, experiments at SLAC in the United States, the *Deutsches Elektronen-Synchrotron* (DESY) in Germany, and elsewhere searched for and saw three-jet events. An especially satisfying result is reproduced in Fig. 2, from DESY's experiment MARK J. It compares the energy flow of three-jet events with the theoretical prediction from QCD. The agreement verifies not only the presence of a mechanism for producing a third jet, but also that the mechanism is gluon radiation, in accord with QCD.

The third experiment considered here is in proton-antiproton scattering. The experimental setup is similar to electron-positron collisions, but with protons and antiprotons circulating in the accelerator. The proton contains quarks, the antiproton of antiquarks. As in electron-proton scattering, the scattering probability can be written as a product; the probability to scatter an antiproton off a proton is equal to the density of quarks inside a proton times the probability to scatter an antiquark off a quark times the density of antiquarks inside an antiproton. Because of the symmetry between matter and antimatter, the density of antiquarks inside the antipro-

ton is identical to the density of quarks inside the proton. Furthermore, the density of quarks in a proton is a feature that exists independent of a collision; it should be the same as in electron-proton scattering. With the densities measured in the first experiment and a formula, from QCD, for the probability to scatter an antiquark off a quark, one has a clear-cut prediction for the proton-antiproton experiment.

The crucial test is whether the prediction from QCD agrees with observations of proton-antiproton scattering. One must choose a signature of the scattering; a useful one is a single jet in the final state. Studies of single-jet production were carried out first at the European Center for Nuclear Research (CERN) in Geneva and more recently—at higher beam energy—with the Tevatron accelerator at Fermi National Accelerator Laboratory in Batavia, Illinois. The curve in Fig. 3 gives QCD's prediction for the one-jet final state for Fermilab's energy. The energy of the jet ranges over a factor of a hundred; according to QCD, the number of observed events should decrease its increasing jet energy by more than ten orders of magnitude. The symbols in Fig. 3 give results from two different detectors, CDF and

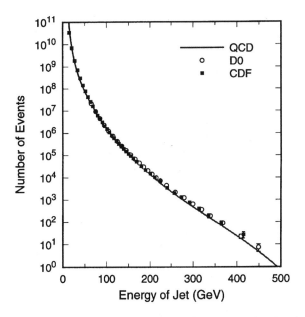

Figure 3 Number of events producing one jet (and anything else) in proton-antiproton scattering. The solid curve is the prediction of quantum chromodynamics. The circles are from the experimental group D0. The squares are from the experimental group CDF.

D0, in the Tevatron. The data agree spectacularly well with the theoretical prediction of QCD.

QCD and Hadron Masses

Quantum chromodynamics is part of the standard model of elementary particles. It occupies a special place in the standard model because it gives a thoroughly consistent mathematical theory of the natural world. Given the quark masses and the strong coupling (at a single, fixed distance), QCD should explain strong interaction phenomena at all distance scales, all energies, all temperatures of the universe, and so on. This is exceptional and awe inspiring; QCD calls out to scientists for tests of the most exacting standards.

At high energies or, equivalently, at short distances, asymptotic freedom makes theoretical analysis of QCD tractable and reliable. As Figs. 2 and 3 attest, the confrontation of theory and experiment is splendid. At longer distances, however, the confining force plays a strong role, and theoretical physicists must turn to other methods to extract information from QCD.

For example, from QCD one should be, and in principle is, able to compute the masses of all hadrons. But the most relevant distance for the hadron masses is the hydronic radius, $r \approx 0.5$ fm, where the strong coupling $\alpha_s(r)$ is large. At these distances one must treat the chromoelectric and chromomagnetic fields fully, instead of one gluon at a time. Moreover, in a quantum field theory, like QCD, many field configurations are important. The high strength of the coupling and the need for many field configurations presents a tremendous challenge. The most promising line of attack, suggested by Kenneth Wilson in 1974, employs large-scale computer calculations. The computer is programmed to generate (numerical representations of) the most important configurations. It is then not too difficult to obtain the masses from various averages over the set of configurations. The difficulty lies instead in the generation of the fields. Physicists have been able to complete the computer calculations only with certain undesirable, but computationally frugal, approximations.

To succeed in calculating the hadron masses from QCD will be a grand achievement, for it will prove that quantum chromodynamics explains the strong interactions from the tiniest distances probed by particle accelerators out to the diameter of the atomic nucleus.

See also: BARYON NUMBER; BOSON, GAUGE; COLOR CHARGE; GAUGE THEORIES; HADRON; QUANTUM ELECTRODYNAMICS; QUARK; QUARK CONFINEMENT

Bibliography

FRITZSCH, H. *Quarks: The Stuff of Matter* (Basic Books, New York, 1983).

GOTTFRIED, K., and WEISSKOPF, V. F. *Concepts of Particle Physics* (Oxford University Press, Oxford, Eng., 1984).

WEINGARTEN, D. H. "Quarks by Computer." *Sci. Am.* **274** (2), 117 (1996).

ANDREAS S. KRONFELD

QUANTUM ELECTRODYNAMICS

Quantum electrodynamics (QED) is the theory that describes light (electromagnetic radiation) and its interaction with matter (i.e., with electrons and other charged particles). It was formulated in the late 1920s by P. A. M. Dirac, Werner Heisenberg, Pascual Jordan, and Wolfgang Pauli, and completed in the early 1950s by Freeman J. Dyson, Richard P. Feynman, Julian Schwinger, and Sin-itiro Tomonaga. Although they worked independently, both Schwinger and Tomonaga constructed the theory in terms of relativistic quantum fields, while Feynman used a diagrammatic approach based on trajectories of particles in spacetime (now called Feynman diagrams). Feynman's method proved to be an incredibly effective computational tool, which Dyson then showed was derivable as a power series expansion known as perturbation theory from the formulations of Schwinger and Tomonaga. Dyson also proved that various divergences (i.e., infinite values for calculated quantities), present in both the field-theoretic and diagrammatic approaches, could be eliminated (to all orders in perturbation theory) by redefining how the mass and charge parameters appearing in the theory were related to the values observed in the lab. This method of handling the divergences (now called renormalization) had been explored earlier by Dirac, Heisenberg, Pauli, Victor Weisskopf, Hendrik A. Kramers, and Sidney Michael Dancoff and has continued to be improved on to the present.

QED developed as the result of an impressive interplay between theory and experiment. It was, in part, the newly developed technology of microwaves that made possible the extremely precise measurements of the spectrum of hydrogen (the Lamb Shift) by Willis E. Lamb and Robert C. Retherford, and the magnetic moment of the electron by Polykarp Kusch and Henry M. Foley. Both of these results, published in 1947, stimulated rapid theoretical advances that in turn stimulated experimenters to develop new techniques for making even more precise measurements. At present, although there is still room for improvement, both theory and experiment agree to a remarkable degree of accuracy and over an incredible range of energies. For example, the agreement between experiment and theory has been confirmed to within parts per million (ppm) in measurements on atomic spectra, which detect energies of order 10^{-6} eV, and to within 1 percent in electron-positron colliding beam experiments, which occur at energies of around 100 GeV. Thus, theory and experiment are in excellent agreement over an energy range of about fifteen orders of magnitude. At energies higher than about 100 GeV, the weak and strong interactions become important and QED must be viewed as part of a more comprehensive theory that includes them and perhaps gravity as well.

Background

At the beginning of the nineteenth century there were two different theories of light: a particle theory and a wave theory. The particle theory fell into disfavor after interference effects were demonstrated in the early 1800s. It was almost totally set aside in the late 1800s when James Clerk Maxwell showed that all electric, magnetic, and optical phenomena could be derived from four equations (Maxwell's equations), and that these equations *predicted* electromagnetic waves that traveled with speed $c = 2.9979 \times 10^8$ m/s. Since this value was close to the speed of light that had been measured earlier, Maxwell hypothesized that light itself was an electromagnetic wave and, by the early 1900s, the electromagnetic wave theory of light was well established. In fact, the visible, infrared, ultraviolet, radio, x ray, gamma ray, and so on all can be viewed as electromagnetic waves of different wavelengths; in what follows, we use the word "light" to refer to the whole electromagnetic spectrum rather than to just that part visible to the human eye.

The first indications that a wave theory alone could not account for the behavior of light appeared in three experiments: blackbody radiation, the photoelectric effect, and the Compton effect. Together, these experiments supported the hypothesis that light was made of particles, now called photons, each with energy $E = hf$ and momentum $p = hf/c$, where f is the frequency of the light involved and h = Planck's constant = 6.63×10^{-34} J·s. At this point, although the properties of photons simply were inferred from these experiments, the success of the photon model motivated physicists to search for a more fundamental theory from which photons and their properties could be derived.

Outline of the Theory

The foundations of the quantum (or photon) theory of light, that is, of QED, were established in the period from 1926 to 1940 by Dirac, Heisenberg, Pauli, Enrico Fermi, and Jordan. What emerged was a theory whose starting point is the classical treatment of electromagnetic fields based on Maxwell's equations and Hamilton's approach to the derivation of such equations of motion from a function expressing the energy density at each point in space, the Hamiltonian density. The quantity $H_{free} = (\mathbf{E}^2 + \mathbf{B}^2)/8\pi$ is the total energy per unit volume (at position $\mathbf{r}$ and time t) of the electromagnetic field in the absence of charges and currents. H_{free} is the Hamiltonian density for the free electromagnetic field, and $\mathbf{E}$ and $\mathbf{B}$ are the electric and magnetic field vectors at the point $(\mathbf{r},t)$. As in Maxwell's treatment, the $\mathbf{E}$ and $\mathbf{B}$ fields are then expressed in terms of the electromagnetic potentials φ and $\mathbf{A}$ since, in the Hamiltonian formalism, the interaction between the electromagnetic field and the current and charge densities $\mathbf{j}$ and ρ is expressed by the interaction energy $H_I = -(\rho\varphi - \mathbf{j} \cdot \mathbf{A}/c)$. It is found that if H_{free} is written in terms of the Fourier transforms of φ and $\mathbf{A}$, then the Fourier coefficient for each frequency obeys the differential equation for a simple harmonic oscillator. Consequently, the Hamiltonian density for the free electromagnetic field is found to be equivalent to that of a system of uncoupled harmonic oscillators. It is at this point that we quantize the fields. That is, we assume that each harmonic oscillation is a quantum oscillator with a discrete set of energies $E = nhf$, which we interpret as a given number n of photons, each with energy hf and momentum hf/c. (In more technical terms, we treat the Fourier coefficients as quantum mechanical operators whose action is to create and destroy photon.) Thus, the original goal of deriving photons from a more fundamental basis is achieved.

Once the free electromagnetic field has been quantized, the next step is to treat its interaction with charged particles. The total Hamiltonian density is written as $H = H_{free} + H_{el} + H_I$, where H_{el} is the Hamiltonian density for the free, quantized electron (or any lepton) field; that is, the field obtained from the Dirac equation in the absence of the electromagnetic interaction. Unfortunately, exact solutions corresponding to H have not been found, so an approximation method is used to obtain numerical predictions that can be compared with experiment. A perturbation series, called the S-matrix expansion, is derived in which each successive term involves higher powers of the electron's charge $e = 1.6 \times 10^{-19}$ C, or more precisely, higher powers of the dimensionless constant $\alpha = 2\pi e^2/hc$ (called the fine-structure constant), which is approximately equal to $1/137$. Because the Coulomb force between two charges is proportional to e^2, both e and α are referred to as coupling constants, since each is a measure of the strength with which the electromagnetic interaction couples one charge to another. Terms in the perturbation series are referred to as being first-order, second-order, and so on, according to the power of α they involve. By evaluating terms in the S-matrix expansion, predictions can be made that are accurate to a given power of α and then compared with experiment. As the experimental values become more precise, the theoretical predictions are made more precise by calculating terms in the series involving higher powers of α. For example, as we discuss below, the best theoretical value of the magnetic moment of the electron is obtained by evaluating all the terms in the perturbation series up to and including α^4. (Evaluating just the terms of order α^4 requires 891 Feynman diagrams, each involving about one hundred thousand terms.)

New Insights into Nature

Every new major theory illuminates aspects of nature that were not recognized previously. For example, special relativity introduced (among other things) the conversion of mass into energy and energy into mass via $E = \gamma mc^2$, and the Dirac equation for the relativistic electron introduced antiparticles. Perhaps the main new feature of QED is the photon

and the way it participates in the electromagnetic interaction as the mediator of the electromagnetic force between charges. Before the mid 1800s, the electromagnetic interaction was viewed as a Newtonian force acting over the distance between charges. Next, in Maxwell's theory, electric and magnetic fields were viewed as existing at each point in space and the force on a charged particle was attributed to the electromagnetic field at the point the charge occupied. In QED, the electromagnetic interaction is viewed as resulting from an exchange of photons between charges; that is, photon exchange replaces the Maxwellian electromagnetic field as the source of the electromagnetic interaction. This picture, of an interaction being the result of an exchange of mediating particles, has been successfully extended to describe the weak and strong interactions and forms the current view of how these interactions take place.

QED also resolves the wave–particle duality of light, that is, the fact that in some experiments light behaves like a wave while in others it behaves like a particle. Since all experiments now are described in terms of the exchange of photons, it appears that this resolution is accomplished through a more comprehensive particle theory. Yet the photon of QED is not a particle in the usual (classical) sense. For example, it does not have a precise spacetime trajectory; it is not a "thing" with spatial extent and nonzero (rest) mass; it does not move with any speed other than the speed of light; and two photons with the same energy, momentum, and polarization are indistinguishable from each other. Basically, a photon is simply one unit of the electromagnetic field possessing a definite energy, momentum, and polarization.

Finally, QED alters our view of the vacuum, which classically was viewed as simply the empty state. Although the expectation values of **E** and **B** in the vacuum state are both zero, the expectation values of $\mathbf{E}^2$ and $\mathbf{B}^2$ are nonzero. This means that fluctuations of the vacuum can occur. Indeed, as we discuss in the next section, fluctuations of the vacuum are verified experimentally in the sense that they contribute a significant amount to effects observed in the lab. Furthermore, in the presence of an external (classical) electromagnetic field, these fluctuations create particle-antiparticle pairs that polarize like the constituents of a dielectric medium. This effect is called polarization of the vacuum, and it has been confirmed experimentally. Thus, our concepts of light and the vacuum both are radically altered by QED.

Experimental Tests of QED

Magnetic moments. Electrons, positrons, protons, muons, and so on all have intrinsic properties called mass, charge, and spin. Although spin had to be introduced in an ad hoc fashion to nonrelativistic quantum theory, Dirac showed that it is an automatic consequence of a quantum theory consistent with special relativity. In addition, Dirac's theory predicted that the spin **S** of an electron is related to its magnetic moment $\boldsymbol{\mu}$ by the relation $\boldsymbol{\mu} = (e/2mc)[\mathbf{L} + g\mathbf{S}]$, where m is the mass of the electron, **L** is its orbital angular momentum, and the constant g (called the electron's gyromagnetic ratio) is exactly two. However, this value of $\boldsymbol{\mu}$ results from treating the electron as a quantum field and the electromagnetic field as a classical field. In 1947 Kusch and Foley found experimental evidence that g actually is slightly larger than two. This result stimulated Hans A. Bethe, Schwinger, Feynman and others to account theoretically for the discrepancy, known as the anomalous magnetic moment of the electron, by using the quantized electromagnetic field and the Hamiltonian density H discussed above. They found that QED predicted a g-factor slightly larger than two, with the precise numerical value dependent upon how many terms in the perturbation series are evaluated. Each term in the series is interpreted as the degree to which the electron interacts with its own electromagnetic field, and the higher the order of the term, the less of a contribution it makes to the numerical value of the electron's magnetic moment (at least in the terms evaluated so far).

The most recent experimental values of the g-factor for the electron, $g(e^-)$, and its antiparticle the positron, $g(e^+)$, were measured by suspending each in a Penning trap:

$$g(e^-) = 2[1.0011596521884(43)]$$

$$g(e^+) = 2[1.0011596521879(43)],$$

where the value in parentheses means that the last two decimal places have an experimental uncertainly of ± 43 (43 parts in 10^{13}).

The best theoretical value of $g(e^-)$ at the present time, obtained through a collaboration of several groups, is accurate through terms of order α^4. Using the best current value of the fine-structure constant, $\alpha^{-1} = 137.0359979\,(32)$, they found

$$g(e^-) = 2[1.00115965214(4)],$$

with the same value predicted for the positron. Most of the uncertainty in the theoretical value results from the current experimental uncertainty in the fine-structure constant. However, the agreement between theory and experiment, to within 0.0001 ppm, is still spectacular. As Feynman notes (1985, p. 118): "This accuracy is equivalent to measuring the distance from Los Angeles to New York, a distance of over 3,000 miles, to within the width of a human hair."

There also is excellent agreement (to within 0.01 ppm) between the measured magnetic moments of positive and negative muons and the theoretical values resulting from evaluating terms in the perturbations series up to and including order α^5.

Lamb shift. Both the Dirac and Schrödinger theories predict that the $2S_{1/2}$ and the $2P_{1/2}$ states in hydrogen will have the same energy. However, in 1947, Lamb and Retherford found experimentally that the energy of the $2S_{1/2}$ state is higher than that of the $2P_{1/2}$ level by about 4.4×10^{-6} eV, which results in light being emitted with a frequency of about 1,060 MHz. To put this number in perspective, recall that the Schrödinger equation predicts the energy levels of an electron in hydrogen to be $(-13.6 \text{ eV})/n^2 = -\alpha^2 mc^2/2n^2$, where n is any nonnegative integer. So atomic transitions that end in the $n = 2$ state correspond to light being emitted with frequencies around 10^8 MHz. There are several corrections to these simple calculations of atomic energy levels when one includes effects of high order in α. The fine-structure correction results from including the first-order relativistic correction to the electron's kinetic energy and the interaction between the electron's spin and the magnetic field it experiences as a result of its motion relative to the proton. The energy levels are shifted by about 4.5×10^{-4} eV when $n = 2$, a shift of order $\alpha^4 mc^2/2n^2$, which causes the frequency of light emitted to be shifted by about 10^4 MHz. The hyperfine correction, which results from including the interaction between the spins of the electron and proton, is about 10^{-7} eV, or of order $\alpha^4(mc^2/2n^2)(m/m_p)$, where m_p is the proton's mass, 1,836 m. Thus, the Lamb shift in the $n = 2$ level is intermediate between the fine and hyperfine corrections and is about 10 percent of the fine-structure correction. (It was from this particular analysis that α received its name, since it provided a convenient measure of the relative scale of the corrections that accounted for the fine structure that already had been observed in spectral lines.)

As with the magnetic moments discussed above, the Lamb shift is accounted for when the quantized electromagnetic field is used in place of the classical one. The best experimental values for the Lamb shift in hydrogen, given in terms of the frequency of light emitted in the transition, are

$$f(2S_{1/2} - 2P_{1/2}) = 1,057.845(9) \text{ MHz}$$

and

$$1,057.851(2) \text{ MHz.}$$

The best theoretical values for the Lamb shift are

$$f(2S_{1/2} - 2P_{1/2}) = 1,057.853(13) \text{ MHz}$$

and

$$1,057.871(13) \text{ MHz,}$$

with the first value resulting when the root-mean-square (rms) radius of the proton is taken to be 0.805 (11) fm, and the second when it is taken to be 0.862 (12) fm. Experiment and theory agree to within 10 ppm if the first (older) value of the proton's radius is used, but they are in less satisfactory agreement (of about 20 ppm) if the second (newer) value is used. The main difficulty is in determining the rms radius of the proton rather in than the theory of QED itself.

The terms of the perturbation series that contribute to the Lamb shift can be described as follows: first, the interaction of the electron with its own electromagnetic field (the self-energy correction), which leads to the modification in the electron's magnetic moment described above; second, the spontaneous production of electron-positron pairs in the neighborhood of the nucleus, which polarize and thus partially screen the electron from the proton's charge (the vacuum polarization correction); third, fluctuations of the vacuum, which alter the motion and kinetic energy of the electron; and fourth, corrections involving the proton's size, charge distribution, mass, and motion.

Lamb shifts also have been measured in other transitions in hydrogen, as well as in deuterium, hydrogen-like atoms, helium, helium-like atoms,

positronium, and muonium. All of these results are in excellent agreement with the theoretical predictions.

Hyperfine structure. As mentioned above, an additional correction to the spectrum of hydrogen results when the interaction between the spins of the electron and proton is taken into account since the electron has one energy if its spin is parallel to the spin of the proton (spin-1 hydrogen) and another if their spins are anti-parallel (spin-0 hydrogen). The frequency of light emitted between these two levels in the ground state of hydrogen is the most accurately measured number in physics:

$$f = 1,420.4057517667(9) \text{ MHz.}$$

The current, most accurate, theoretical value of the this splitting is

$$f = 1,420.45199(14) \text{ MHz,}$$

with the main uncertainties coming from the experimental uncertainty in the fine-structure constant and the lack of precise knowledge about the structure of the proton. Thus, theory and experiment currently agree to about 30 ppm.

The hyperfine splitting between the spin-0 and spin-1 levels in the ground state of muonium also has been measured very accurately and found to be

$$f = 4,463.30288(16) \text{ MHz.}$$

Since muonium is essentially an electrodynamic system, with the weak and strong interactions making almost no contribution, the comparison between theory and experiment provides an especially rigorous test of QED. Currently, the best theoretical value is

$$f = 4,463.303(2) \text{ MHz,}$$

and the agreement with experiment, to within 1 ppm, is excellent. The uncertainty in the theoretical value arises from the uncertainty in the muon's mass, and the uncertainty in the value of the fine-structure constant.

Measurements of hyperfine splitting also have been made in higher energy levels of hydrogen and in positronium, deuterium, and tritium. All are found to be in excellent agreement with the theoretical predictions when the proper nuclear structure and recoil corrections are taken into account.

Other tests. The predictions of QED also have been tested in many other experiments, such as in high energy electron-electron and electron-positron scattering, where they agree to within 1 percent with theoretical calculations made through order α^3, in the scattering of light by light (Delbruck scattering), in electron-positron pair production and pair-annihilation, in the Casimir effect (in which two neutral metal plates experience a slight attraction due to fluctuations of the vacuum), and so on. At present, there is no known discrepancy between experiment and theory, and the most precise theoretical predictions and experimental measurements are in excellent agreement.

Difficulties with the Theory

The main difficulty of the QED theory is that when higher terms in the perturbation series are evaluated, some of the resulting integrals are divergent (that is, infinite). However, these infinities can be isolated and removed (in every order of perturbation theory) by redefining the charge and mass parameters that appear in the theory. In principle, even if the infinities did not occur, it still would be necessary to renormalize the charge and mass parameters because of the way the Hamiltonian density H is divided up in the perturbation approach. More specifically, the Hamiltonian densities for the free electromagnetic field H_{free} and the free electron field H_{el} together form the unperturbed system, and the interaction term H_I is considered as the perturbation. The mass parameter m appears in the unperturbed Hamiltonian, so by virtue of using a perturbation approach, it represents the mass an electron would have in the absence of the electromagnetic interaction. For this reason, it is called the bare mass. The physical (or dressed) mass is then the bare mass plus that part of the mass resulting from the interaction of the electron with its own electromagnetic field. Neither the bare mass nor its interaction correction are observable; the only observable is their sum, the physical mass. Consequently, each can be calculated separately, and even

though each involves integrals that diverge, all that matters is that when they are added together to give the physical mass, the divergent integrals cancel. Thus, the diverging terms disappear in the process of renormalizing the mass.

Exactly the same considerations apply to the electron charge. The parameter e in the theory is the bare charge; that is, the charge the electron would have in the absence of the electromagnetic interaction. This charge is then renormalized so that the physical (or dressed) charge is identified as the sum of the bare charge and the charge arising from the interaction of the electron with its own electromagnetic field. As in the renormalization of mass, the diverging integrals in the bare charge and its correction cancel when the two are added together to obtain the physical charge.

Most physicists now accept the renormalization procedure, since it leads to predictions that are in excellent agreement with experiment and since modern methods have made it more rigorous. However, many continue to think that subtracting infinities is not a valid mathematical procedure, no matter how elegantly it is performed. To these physicists, the renormalization process represents a serious problem in the fundamental structure of the theory. Other physicists think that the success of renormalization in producing numbers verified by experiment implies that it is, or someday will be, justified. Still others think that the divergences might be removed when QED is incorporated into a more comprehensive quantum field theory that includes gravity as well as the strong and weak interactions.

A second difficulty with QED is that no one has been able to show that the perturbation series converges, or that if it converges, it converges to the correct limit. In fact, in 1952 Dyson argued that the series probably diverges and that at best it is an asymptotic series, that is, a series whose first few terms approach a limit but which then diverges as more terms are added. More specifically, he argued that if the coupling constant is a fixed number, then the series may not converge, but if the coupling constant is considered as a variable, then as it tends to zero, increasingly long initial segments of the expansion should converge to the right limit. This would explain why the sum of the first few terms is in such good agreement with experiment. However, Dyson's argument is not a rigorous mathematical proof, so it is not regarded as conclusive (even though he found it so convincing that he left the field to pursue research in other areas).

Conclusion

Because of its spectacular agreement with experiment, QED is considered by many to be one of the most successful theories in physics. Although some difficulties remain, most physicists accept the theory as being basically correct. Furthermore, many features of QED have been incorporated with great success into the recent theories of strong, weak, and electroweak interactions, thus reinforcing the procedures and points of view upon which it is based and solving some of the difficulties of definition that occur for QED in isolation, but not for the combined theory.

See also: DIRAC, PAUL ADRIEN MAURICE; ELECTROMAGNETISM; FERMI, ENRICO; FEYNMAN, RICHARD; FEYNMAN DIAGRAM; HEISENBERG, WERNER KARL; HYPERFINE STRUCTURE; INTERACTION, STRONG; INTERACTION, WEAK; LAMB SHIFT; MAGNETIC MOMENT; MAXWELL'S EQUATIONS; PAULI, WOLFGANG; RADIATION, BLACKBODY; RELATIVITY, SPECIAL THEORY OF; RENORMALIZATION

Bibliography

DEHMELT, H. "Less is More: Experiments with an Individual Atomic Particle at Rest in Free Space." *Am. J. Phys.* **58,** 17–27 (1990).

FEYNMAN, R. P. *QED: The Strange Theory of Light and Matter* (Princeton University Press, Princeton, NJ, 1985).

GRIFFITHS, D. J. *Introduction to Elementary Particles* (Wiley, New York, 1987).

JAUCH, J. M., and ROHRLICH, F. *The Theory of Photons and Electrons,* 2nd expanded ed. (Springer-Verlag, New York, 1976).

KINOSHITA, T., ed. *Quantum Electrodynamics* (World Scientific, Teaneck, NJ, 1990).

MANDL, F., and SHAW, G. *Quantum Field Theory* (Wiley, New York, 1984).

RAYMER, M. G. "Observations of the Modern Photon." *Am. J. Phys.* **58,** 11 (1990).

SAKURAI, J. J. *Advanced Quantum Mechanics* (Addison-Wesley, Reading, MA, 1967).

SCHWEBER, S. S. *QED and the Men Who Made It* (Princeton University Press, Princeton, NJ, 1994).

TELLER, P. *An Interpretive Introduction to Quantum Field Theory* (Princeton University Press, Princeton, NJ, 1995).

WEINBERG, S. *The Quantum Theory of Fields,* Vol. 1: *Foundations* (Cambridge University Press, Cambridge, Eng., 1995).

MARK D. SEMON

QUANTUM FIELD THEORY

Quantum field theory is the correct way to describe systems like the electromagnetic field in accordance with the laws of quantum mechanics (and any other restrictions, such as those due to relativity). It forms the basis of modern elementary particle physics and proves invaluable in other many-body problems like superconductivity and the physics of solids.

Let us begin with a single mass coupled to an anchored spring and vibrating along the *x* direction. This oscillator is a system with one degree of freedom, since only one coordinate, *x* is needed to describe it. A system of *N* masses would have *N* degrees of freedom. A field is a system with infinite number of degrees of freedom. For example, if we had an infinite array of masses in a line, coupled to their neighbors by springs, the displacements of each mass from equilibrium would constitute a field. Another example is the electromagnetic field whose degrees of freedom are the values of the electric and magnetic field at each point in space. Notice that the latter has an infinite number of degrees of freedom within any finite volume, and this causes many problems. The aim of quantum field theory is to describe how these degrees of freedom behave, following the laws of quantum mechanics.

We will work our way to the field by starting with the simplest system: a single mass *m* attached to a spring of force constant *k*. In classical mechanics this mass will vibrate with frequency $\omega = 2\pi f = \sqrt{k/m}$. It can have any finite amplitude x_0, and the corresponding energy will be $E = \frac{1}{2}kx_0^2$.

In the quantum version the oscillator can only have energy $E = \hbar\omega(n + \frac{1}{2})$, where $n = 0, 1, 2 \ldots$ and $\hbar = 1.05 \times 10^{-34}$ J·s is Planck's constant. Note that the lowest energy is not zero but $\frac{1}{2}\hbar\omega$, which is called the zero-point energy. The uncertainty principle in quantum mechanics, which forbids a state of definite location and momentum, does not allow the classical zero energy state in which the mass sits at rest in the equilibrium position. Next, the fact that the levels are equally spaced means that instead of saying the oscillator is in a state labeled by *n*, we can say that there are *n* quanta of energy $\hbar\omega$. This seemingly semantic point proves seminal as we shall see.

Now consider two masses *m* attached to springs of force constant *k* as in the Fig. 1. The coordinates x_1 and x_2 are coupled to each other and the motion is quite complicated. But consider the fol-

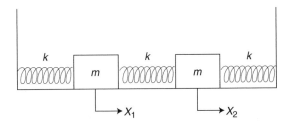

Figure 1 Two masses attached to a string with constant force.

lowing combination of coordinates, called normal coordinates

$$X_1 = \frac{x_1 + x_2}{\sqrt{2}}, \qquad X_2 = \frac{x_1 - x_2}{\sqrt{2}}.$$

These behave like independent oscillators of frequency $\omega = \sqrt{k/m}$, $\sqrt{3k/m}$, respectively. To see this, imagine starting off the masses with equal displacements, so that X_1 is nonzero and $X_2 = 0$. Since the middle spring is undistorted, the masses will begin moving in response to the end springs. Since these are identical, the condition $x_1 = x_2$ (same as $X_2 = 0$) will be preserved for all times and x_1 and x_2 and hence X_1 will vibrate at $\omega = \sqrt{k/m}$. On the other hand, if both masses are given equal and opposite displacements, then $X_1 = 0$ initially. The middle spring is distorted twice as much as the end springs (so the effective force constant felt by the masses is $3k$), and the masses vibrate with equal and opposite displacements at $\omega = \sqrt{3k/m}$. That is $X_1 = 0$ and X_2 vibrates at $\omega = \sqrt{3k/m}$. If we are given some arbitrary initial displacements so that both X_1 and X_2 are nonzero, we compute their future values (which is easy since they behave like independent oscillators) and go back to x_1 and x_2 at the end. We discuss the two mass problem in such detail since the same strategy works for any number of masses.

Imagine now that we have N such masses coupled to their neighbors by N such springs of equilibrium length a arranged around a circle of length $L = Na$. (We shall use units in which $m = k = 1$.) Let $\phi(n)$ be the displacement of mass numbered n, where $1 \le n \le N$. The system can once again be reduced to a set of decoupled oscillators. The normal coordinates are just the Fourier coefficients:

$$\phi(K) = \sqrt{\frac{1}{N}} \sum_n \phi(n) e^{-iKna}.$$

The requirement that the points numbered n and $n + N$ are one and the same implies that, for a consistent definition of $\phi(K)$ we must require

$$K = \frac{2\pi m}{Na} \qquad m = 0, \pm 1, \pm 2, \dots,$$

while the fact that K and $K + 2\pi/a$ are indistinguishable in the exponential factor limits m to the range of size N and K to an interval of width $2\pi/a$, which we choose to be $-\pi/a \le K \le \pi/a$. The system is equivalent to N oscillators with frequencies $\omega(K) = 2(1 - \cos Ka)$. If we let $N \to \infty$, the allowed values of K become continuous in the interval $[-\pi/a, \pi/a]$. The quantity ϕ is a classical field. [Note that once we know $\phi(K)$ for all K, we can calculate $\phi(n)$ from the inverse Fourier transform

$$\phi(n) = \sqrt{\frac{1}{N}} \sum_{K = \frac{2\pi m}{Na}} \phi(K) e^{iKna}.$$

The quantum version will be obtained by treating the decoupled oscillators as per quantum mechanics. Each oscillator of frequency ω can only have energy $E = \hbar\omega(n + \frac{1}{2})$, where $n = 0, 1, 2\dots$. This means that even in the ground state the field has zero-point energy $\frac{1}{2}\hbar\omega$ per oscillator, which can have observable effects. (The zero point energy of the electromagnetic field leads to the spontaneous decay of atoms.) Next, the fact that the levels are equally spaced means that instead of saying the oscillator is in a state labeled by n, we can say that there are n quanta of energy $\hbar\omega$ and momentum $\hbar K$, where the identification of $\hbar K$ with momentum comes from examining the interaction of the system, described in terms of the decoupled oscillators, with any external probe. Thus, the quantum state of the field is specified by saying how many quanta there are at each K. In the present problem, the quanta are called phonons, because in a real crystal, such vibrations of the masses are quantized sound in the medium.

If these methods are applied to the electromagnetic field, which has degrees of freedom at each point in space (i.e., $a = 0$), the allowed K values will go from $-\infty$ to ∞, which has some important consequences that we will soon discuss. The quanta of the field are called photons.

Both quanta above come from bosonic oscillators, for which n is not restricted; any number of quanta are allowed for each K. When a macroscopic number of bosonic quanta, say, photons, are present in a state with some K, we perceive it as a classical (electromagnetic) field at that wave number and the corresponding energy density. To describe fermions like electrons as quanta, we need to quantize a fermionic oscillator that cannot support more than one quantum per state K. There is no classical manifestation of such a field, which is why it is so unfamiliar.

In the cases considered so far, the number of quanta in each oscillator stays fixed, because the potential energy is quadratic in the variables. Upon adding higher order interaction terms, this changes; there can be change in these numbers. Since the quanta correspond to particles, this means either that particles change their energy or momentum values or that new particles are created. Consider quantum electrodynamics (QED), in which a term quadratic in the electron field and linear in the photon field is added to the total energy with a coefficient e, which is the electron's charge. This cubic term describes, for example, a process in which an electron emits or absorbs a photon. This can cause, among other things, two electrons to scatter from their original momentum states to new ones. The results are computed in a perturbation series in e or equivalently $\alpha = e^2/\hbar c \simeq 1/137$, called the fine-structure constant and represented by Feynman diagrams, which reproduce the series. For example, Fig. 2 shows a process second order in e (or first order in α) in which an electron emits a photon and

recoils, while another captures the photon and recoils the other way. Thus, the quanta of the field (photons) also mediate interactions between other quanta (electrons). If we go to higher orders in the expansion, we can have more complicated scattering diagrams, say one wherein one of the electrons emits two photons and then either both are absorbed by the other or one is absorbed by the emitter itself and the other is absorbed by the other. The Feynman diagrams go on to infinite order in number of photons (or equivalently, powers of α). However, one can calculate just the first few terms of the series and obtain excellent numbers since α is so small. In a general theory, there may be no such small parameter.

Consider now a lone electron that emits a photon and reabsorbs it, so that for a while we have an electron and a photon. This can be shown to change the observed mass of the electron from the value m_0 to

$$m = m_0 + \alpha I,$$

where m is the observed mass, calculated to order α in perturbation theory, and I is an integral that sums over the various electron-photon states that can occur between the emission and reabsorption. This sum diverges (i.e., gives an infinite result) due to the infinite range of moments for the quanta. Since the observed mass of the electron is finite this is, at first sight, a disaster. However, physicists found a way to make sense of the calculation by a prescription known as renormalization. In this prescription, all energy and momentum sums are cut off at some large value Λ. Thus $m = m_0 + \alpha I(\Lambda)$, but one now requires that m_0 itself is Λ dependent in such a way that m is finite and Λ independent. Next, one finds that the scattering rate of two electrons comes out infinite also, at higher orders. One now says that the coupling is not given by $\alpha \simeq 1/137$, but by $\alpha_0(\Lambda)$, chosen so that the scattering rate comes out finite and independent of the value of the artificial cutoff Λ. Remarkably enough, once these two parameters are so chosen, no new infinities arise, a feature called renormalizability. A renormalizable theory is one that gives finite and cut-off independent results for all measureable quantities after only a fixed finite number of such redefinitions. Renormalizability has been a guiding principle in arriving at the theory of strong interactions (quantum chromodynamics, or QCD) and in the unified theory of electromagnetic and weak interactions, the Glashow–Weinberg–Salam

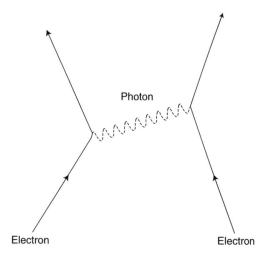

Photon

Electron   Electron

Figure 2   Feynman diagram of a process in which an electron emits a photon and recoils, while another captures the photon and recoils the other way.

(GWS) model. When we say the right theory, we mean the right formula for the energy of the system in terms of its coordinates, since everything follows from this.

Another important guide in searching for the right theory is symmetry. In particle physics the predictions have to be invariant under spacetime symmetries, like translations and Lorentz transformations, or internal symmetries, such as isospin symmetry, in which a proton and neutron are exchanged. In the study of a magnetic system, one requires invariance under rotations of the applied fields so that all directions in space are equivalent. Each symmetry implies a conservation law. For example, translation symmetry leads to conservation of momentum, while rotational symmetry leads to the conservation of angular momentum. An important symmetry is local gauge symmetry, in which the theory is invariant when the fields are multiplied by phase factors that vary from point to point in spacetime. QED, QCD, and the GWS model are theories are of this type.

Typically the ground state or vacuum state is invariant under all the symmetries of the interactions. For example in a theory invariant under a shift of the coordinates, the lowest energy state will look the same when translated. In a magnet, however, the ground state may be magnetized in some direction, chosen randomly from all equivalent ones, and will look different when rotated. This is called a spontaneous breakdown of symmetry.

See also: DEGREE OF FREEDOM; FEYNMAN DIAGRAM; FINE-STRUCTURE CONSTANT; GAUGE THEORIES; PHONON; PHOTON; QUANTUM CHROMODYNAMICS; QUANTUM ELECTRODYNAMICS; QUANTUM MECHANICS; QUANTUM MECHANICS and QUANTUM FIELD THEORY; SYMMETRY BREAKING, SPONTANEOUS

Bibliography

ABRIKOSOV, A. A.; GORKOV, L. P.; and DZYALOSHINSKI, I. E. *Methods of Quantum Field Theory in Statistical Mechanics* (Dover, New York, 1963).

BJORKEN, J. D., and DRELL, S. D. *Relativistic Quantum Mechanics and Relativistic Quantum Fields* (McGraw-Hill, New York, 1980).

ITZYKSON, C., and ZUBER, J. B. *Quantum Field Theory* (McGraw-Hill, New York, 1980).

MAHAN, G. D. *Many Body Physics* (Plenum, New York, 1981).

R. SHANKAR

QUANTUM FLUID

At sufficiently low temperatures, assemblies of freely mobile particles (e.g., gases or liquids) enter a regime where their properties are dominated by quantum mechanics. Such systems are known as quantum fluids. They exhibit a range of extraordinary properties, such as superfluidity.

A criterion for estimating the temperature below which quantal behavior will occur in a given system can be derived from de Broglie's hypothesis that a particle of mass m moving at velocity v has an associated wavelength λ given by

$$\lambda = \frac{h}{mv}, \tag{1}$$

where h is Planck's constant. When the thermal velocity is small enough for λ to exceed the average interparticle separation a, quantum effects are to be anticipated. According to the classical equipartition theorem, which is applicable for temperatures T above the quantal regime, the average thermal kinetic energy per particle in a gas is $\frac{3}{2}k_BT$ where k_B is Boltzmann's constant. Equating this to $\frac{1}{2}mv^2$ yields the thermal velocity $v = (3k_BT/m)^{1/2}$, and thus, using (1), the criterion for quantal behavior can be written as

$$\frac{h}{(3mk_BT)^{1/2}} > a. \tag{2}$$

For fixed values of m and a, this inequality can in principle always be satisfied by making T small enough, provided, of course, that the system does not solidify. The properties of the resultant quantum fluid depend crucially on whether its constituent particles are bosons or fermions. Strictly, the criteria for quantal behavior are that T should be smaller than the Bose–Einstein condensation temperature or much smaller than the Fermi characteristic temperature, respectively.

When an ideal boson fluid (one composed of point particles with no interparticle forces between them) is cooled, the particles can crowd into the lower energy quantum states because there is no restriction on the number of particles permitted to occupy one state. The thermodynamic and transport properties depart slightly from the values for the classical gas and then, very suddenly, at a temperature given approximately by (2), the system undergoes a phase transition to a new state. Further cooling results in Bose–Einstein condensation, corresponding to a significant fraction of the particles being crowded together in the zero-momentum ground state. The properties of the system are then markedly different from the classical gas behavior seen at higher temperatures. For example, the heat capacity varies as $T^{3/2}$, instead of being equal to the constant value of $(3/2)R$.

When an ideal fermion fluid is cooled, the sequence of events is quite different because of the restriction that a given quantum state cannot be occupied by more than one particle. Rather than a sudden transition to a new behavior, the properties of the system change smoothly with further cooling once criterion in (2) has been satisfied. The heat capacity, for example, falls gradually and tends asymptotically toward a linear dependence on T.

Quantum fluids are rare in nature because nearly all materials solidify at temperatures far above those where (2) would be satisfied. The main exceptions are the two stable isotopes of liquid helium ^{3}He (a fermion) and ^{4}He (a boson); the mobile charge carriers (fermions) in metals and other conductors; monatomic hydrogen (bosons) and deuterium (fermions); and the proton and neutron (fermion) fluids that are believed to exist in neutron stars. In the latter case, (2) can be satisfied because of the enormous density, and correspondingly small value of a, even though the temperature T may be 10^8K.

Real quantum fluids are far from being ideal, partly on account of their finite interparticle forces and partly because, in some cases (e.g., liquid helium), the size of the particles is comparable with their separation. This nonideality has important consequences. Not only does it lead to quantitative departures from ideal Bose–Einstein and Fermi–Dirac gas behaviors, but it is also an essential prerequisite for the occurrence of superfluidity in the liquid heliums and superconductivity in certain metals, alloys, and other conductors. Although the onset of superfluidity in liquid ^{4}He is closely associated with Bose–Einstein condensation, the condensate fraction of an ideal boson gas would not in fact be a superfluid. Superfluidity in liquid ^{3}He (and, probably, in neutron stars), and superconductivity in metals, arises through a phase transition in which Cooper pairs of ^{3}He atoms, nucleons, or electrons are formed. Because the Cooper pairs each have even spin, they are bosons, and can immediately undergo Bose–Einstein condensation in much the same way as liquid ^{4}He. The pairing interaction would not occur in an ideal Fermi–Dirac gas, which would not, therefore, be expected to display superfluidity. Superfluidity has been observed and studied in liquid ^{3}He and ^{4}He, it has been sought in monatomic hydrogen, and it has been inferred to occur within neutron stars. Superfluidity of the electron gas in conductors (i.e., superconductivity) has been known since 1908.

Superfluids are characterized by wave functions or order parameters that are macroscopic, filling the entire system and corresponding to highly correlated motion of their constituent particles. They exhibit quantum phenomena on a macroscopic scale, and the very different systems mentioned above share many properties in common. For example, at low velocities, a superfluid flows without dissipation (friction). This feature can be understood in terms of the high degree of correlation, because a dissipative event would require simultaneous changes in the motions of all the particles in the system, which is inherently improbable. As the flow velocity is increased, however, a critical value will be reached where a sudden transition occurs to more conventional dissipative behavior; in superconductors, this would correspond to a critical current density.

Rotation in superfluids is quantized. When a beaker of superfluid ^{4}He is rotated slowly, for example, the liquid does not turn, but remains at rest relative to the fixed stars. For higher angular velocities of the beaker, one or more quanta of rotation may appear in the liquid, taking the form of quantized vortex lines, like little eddies threading the liquid parallel to the axis of rotation, but the liquid does not rotate as a whole. The analog of quantized vortex lines in superconductors is magnetic flux lines. Most of the angular momentum in a rotating neutron star is believed to reside in quantized vortex lines.

Superfluids possess numerous other fascinating characteristic properties in addition to these examples and, in the case of superfluid ^{3}He and perhaps the electron fluids in certain superconductors, they can exhibit magnetism, a high degree of anisotropy, and even liquid crystal-like behavior.

See also: Boltzmann Constant; Condensation, Bose–Einstein; Cooper Pair; Fermions and Bosons; Friction; Liquid Helium; Neutron Star; Superconductivity

Bibliography

Leggett, A. J. "Low-Temperature Physics, Superconductivity and Superfluidity" in *The New Physics*, edited by

P. C. W. Davies (Cambridge University Press, Cambridge, Eng., 1989).

LOUNASMAA, O. V., and PICKETT, G. R. "The ³He Superfluids." *Sci. Am.* **262,** 64–71 (1990).

McCLINTOCK, P. V. E.; MEREDITH, D. J.; and WIGMORE, J. K. *Matter at Low Temperatures* (Wiley, New York, 1984).

MENDELSSOHN, K. *The Quest for Absolute Zero,* 2nd ed. (Taylor and Francis, London, 1977).

P. V. E. McCLINTOCK

QUANTUM GRAVITY

The phrase "quantum gravity" refers to the as-yet-unknown theory that consistently incorporates both general relativity and the uncertainty principle of quantum mechanics. Given the overwhelming experimental evidence for both quantum mechanics and general relativity, it is virtually certain that such a theory must exist. While candidate theories do exist, it is not yet known which, if any, are correct.

In the first half of the twentieth century, the subject of quantum field theory was developed for the purpose of incorporating quantum effects into the classical theory of electromagnetism, resulting in the theory of quantum electrodynamics. The same methods were later applied to the strong and weak interactions with great success. However, they notably fail in their application to classical general relativity. Straightforward attempts lead to a theory plagued by infinities, instabilities, and lack of predictive power.

Rather than being discouraged by the apparent incompatibility of quantum mechanics and general relativity, many physicists regard the problem of reconciling the two theories as an exciting theoretical challenge. One reason for this is illustrated in Fig. 1. Historically, major conceptual advances in physics have arisen from incompatibilities between well-established areas of knowledge. For example, Newtonian gravity is incompatible with special relativity because the gravitational force is transmitted instantaneously. This contradicts special relativity, in which nothing can exceed the speed of light. The need to resolve this conflict led Albert Einstein to his theory of general relativity. It is hoped that resolution of the apparent conflict between general relativity and quantum mechanics will similarly lead to a

fundamental advance in our understanding of the laws of physics.

An enormous obstacle to progress on this important problem is the absence of any direct experimental results to guide the development of quantum gravity. The reason for this absence is that under normal circumstances the effects of quantum gravity are extremely small and, hence, difficult to measure. Quantum gravity comes into play only when both general relativistic and quantum mechanical effects are non-negligible. For example, general relativistic corrections to the gravitational field at a distance D from an object of mass M become large when the mass exceeds

$$M \geq \frac{Dc^2}{G},\qquad(1)$$

where G is Newton's gravitational constant and c is the speed of light. Quantum effects on the other

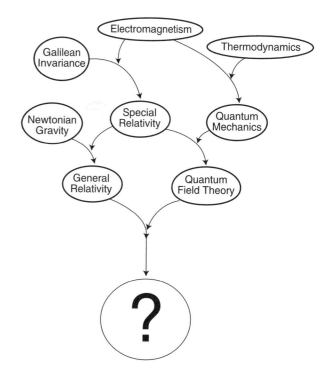

Figure 1 Major advances have in the past arisen from incompatibility of different physical theories. Currently, general relativity and quantum mechanics appear incompatible. The as-yet-unknown theory that reconciles them is referred to as quantum gravity.

hand are large when the mass is small so that its Compton wavelength exceeds D

$$M \leq \frac{\hbar}{Dc}, \tag{2}$$

where $\hbar$ is Planck's constant. If these two inequalities are ever to be simultaneously satisfied, D must obey

$$D \leq L_p, \tag{3}$$

where

$$L_p = \sqrt{\frac{G\hbar}{c^3}} = 1.616 \times 10^{-33} \text{ cm.} \tag{4}$$

L_p is known as the Planck length. Quantum gravitational effects therefore become important at distances less than of order L_p. The most precise known short-distance probes are high-energy particle accelerators, which able to study particle structures on scales of order 10^{-16} cm. Quantum gravitational corrections are generically expected to be suppressed by powers of the ratio of the Planck length to the scale being measured. Thus, quantum gravity corrections at current particle accelerators are at most of order 10^{-17} and, hence, unmeasurable. Direct experimental measurements of quantum gravitational effects are thus unlikely in the foreseeable future.

Nevertheless, candidate theories of quantum gravity are highly constrained by internal self-consistency. Indeed, efforts in the last half of the twentieth century have yielded only one significant candidate for a complete, self-consistent theory of quantum gravity, known as superstring theory. These theories reduce to quantum field theories at large distance scales, but fundamentally new physical phenomena appear when distances of order L_p are probed. The tiny quantum gravity corrections at large-distance scales are calculable and free of the infinities that plague straightforward applications of quantum field theory to general relativity. However, it is not known whether a superstring theory whose predictions match physical reality in every detail exists, and it is difficult to conceive of an experiment that might help settle the issue.

See also: ELECTROMAGNETISM; GRAVITATIONAL CONSTANT; QUANTUM ELECTRODYNAMICS; QUANTUM FIELD THEORY; QUANTUM MECHANICS; RELATIVITY, GENERAL THEORY OF; SUPERSTRING

Bibliography

GREEN, M. B.; SCHWARZ, J. H.; and WITTEN, E. *Superstring Theory* (Cambridge University Press, Cambridge, Eng., 1987).

MISNER, C. W.; THORNE, K. S.; and WHEELER, J. A. *Gravitation* (W. H. Freeman, San Francisco, 1973).

ANDREW STROMINGER

QUANTUM MECHANICAL BEHAVIOR OF MATTER

The fundamental object in nonrelativistic quantum mechanics is the wave function $\psi(x,t)$, a probability amplitude whose absolute square, $|\psi(x,t)|^2$, gives the relative probability that a particle is located in the vicinity of point x at time t. Specifically, if dx is infinitesimal, then the probability of finding the particle between locations x and $x + dx$ at time t is

$$P(x, x + dx) = |\psi(x,t)|^2 \, dx \tag{1a}$$

with

$$\int_{-\infty}^{\infty} |\psi(x,t)|^2 \, dx = 1. \tag{1b}$$

(For simplicity, we shall deal only with one spatial dimension. However, all results can be generalized straightforwardly to the three-dimensional case.)

The way in which the existence of a probability amplitude leads to results that cannot be understood from classical physics can be seen from the example of particle propagation from a point source that is separated from a distant detecting screen by a pair of slits, as shown in Fig. 1. Classically the particle must pass either through slit 1 or through slit 2 leading to a pattern as shown in Fig. 1a. On the other hand, according to quantum mechanics the wave-like nature of matter requires the superposition of amplitudes associated with both slits, yielding a net amplitude

$$\text{Amp} \sim \exp\left(i2\pi\frac{d_1}{\lambda}\right) + \exp\left(i2\pi\frac{d_2}{\lambda}\right) \qquad \text{(2a)}$$

and

$$\text{Prob} = |\text{Amp}|^2 = 4\cos^2\left(\pi\frac{d_1 - d_2}{\lambda}\right), \qquad \text{(2b)}$$

where d_1 and d_2 are the path lengths associated with passage through slits 1 and 2, respectively, and the wavelength is given by de Broglie's relation, $\lambda = h/p$. Since $d_1 - d_2 \approx L\sin\theta$ the intensity pattern found on the screen—Fig. 1b—is of a very different character than that predicted from classical considerations.

The wave function is, of necessity, a complex number whose time development is determined by

$$i\hbar\frac{\partial}{\partial t}\psi(x,t) = \hat{H}\psi(x,t), \qquad \text{(3)}$$

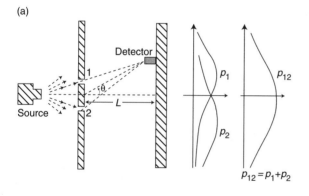

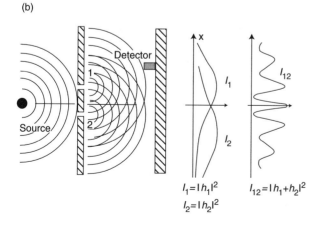

Figure 1 (a) Classical and (b) quantum mechanical behavior of matter passing through a pair of slits.

where $\hbar \equiv h/2\pi$ is Planck's constant. Here, $\hat{H}$ is the Hamiltonian operator, which is generated from the corresponding classical quantity $H(p, x)$ via the substitution $p \rightarrow -i\hbar(\partial/\partial x)$, in which case the time evolution prescription becomes the Schrödinger equation:

$$i\hbar\frac{\partial}{\partial t}\psi(x,t) = \left(-\frac{\hbar^2}{2m}\frac{\partial^2}{\partial x^2} + V(x)\right)\psi(x,t). \qquad \text{(4)}$$

In the absence of a potential—that is, when $V(x) = 0$—the equation possesses plane wave solutions

$$\psi(x,t) = N\exp\left[\frac{i}{\hbar}(px - Et)\right], \qquad \text{(5)}$$

where the energy E and momentum p are related by the classical physics expression

$$E = \frac{p^2}{2m}. \qquad \text{(6)}$$

Comparison with the usual wave form

$$\exp\left[i(kx - \omega t)\right] \qquad \text{(7)}$$

reveals the de Broglie relation between wavelength and momentum p,

$$p = \frac{h}{\lambda}, \qquad \text{(8)}$$

as well as Planck's condition relating the energy E and angular frequency ω,

$$E = \hbar\omega. \qquad \text{(9)}$$

The plane wave Eq. (5) has definite momentum, so that there is no uncertainty—$\Delta p = 0$. On the other hand, since $|\exp[(i/\hbar)(px - Et)]|^2 = 1$ for all x and t, the plane wave probability to be located at position x at time t is independent of x and t and is completely delocalized in position—$\Delta x \sim \infty$. This is unrealistic. It is possible, however, to localize the particle by using the feature that the Schrödinger equation is linear and homogeneous, so that any superposition of plane wave solutions is also a solution. Thus,

$$\psi(x,t) = \int_{-\infty}^{\infty} \frac{dp}{(2\pi\hbar)} \, \phi(p) \exp\left[\frac{i}{\hbar}\left(px - \frac{p^2}{2m}t\right)\right] \quad (10)$$

is a solution for an arbitrary function $\phi(p)$. A classic and instructive example involves a Gaussian shape localized about $p = 0$ with width (uncertainty in momentum) $\Delta p \sim \hbar/\sigma$:

$$\phi(p) = (8\pi\sigma^2)^{1/4} \exp\left(-\frac{\sigma^2 p^2}{\hbar^2}\right), \quad (11)$$

in which case the integration in Eq. (10) can be analytically performed, yielding

$$\psi(x,t) = \left(\frac{\sigma^2}{2\pi}\right)^{1/4} \frac{1}{(\sigma^2 + i\frac{\hbar t}{2m})^{1/2}}$$

$$\exp\left(-\frac{x^2}{4(\sigma^2 + i\frac{\hbar t}{2m})}\right). \quad (12)$$

Since

$$|\psi(x,t)|^2 = (2\pi\sigma^2(t))^{-1/2} \exp\left(-\frac{x^2}{2\sigma^2(t)}\right) \quad (13a)$$

with

$$\sigma^2(t) = \sigma^2 + \frac{\hbar^2 t^2}{4m^2\sigma^2}, \quad (13b)$$

the particle is shown to be localized about the point $x = 0$ with width (uncertainty in position) given by $\sigma(t)$.

Such a superposition of plane wave solutions is called a wave packet. This example indicates two particularly interesting properties: (1) Heisenberg's uncertainty principle,

$$\Delta p \Delta x \geq \frac{\hbar}{2}, \quad (14)$$

and (2) the property of "spreading"—the spatial widening of the wave packet as time evolves; indeed,

$$\Delta x = \sigma(t) = \left(\sigma^2 + \frac{\hbar^2 t^2}{4m^2\sigma^2}\right)^{1/2}. \quad (15)$$

Although exhibited above for the particular example of a Gaussian shape, these two properties are general features of arbitrary wave packets.

The uncertainty principle can be generalized to additional variables by introducing the concept of operators—specific mathematical operations on quantum mechanical wave functions. An example is the momentum operator

$$\hat{p} \rightarrow -i\hbar \frac{\partial}{\partial x} \quad (16)$$

or the position operator

$$\hat{x} \rightarrow x. \quad (17)$$

In general, the value of the classical quantity (observable) $<\mathcal{O}>$ associated with an operator $\hat{\mathcal{O}}$ is given by the weighted average or expectation value

$$<O> = \int_{-\infty}^{\infty} dx \, \psi^*(x,t)\hat{\mathcal{O}}\psi(x,t). \quad (18)$$

If the action of such an operator on a wave function simply reproduces the same wave function times an ordinary number, this number is said to be an eigenvalue, and the wave function is an eigenstate of the operator. When this situation occurs, the value of the classical observable associated with this operator is definite and equal to the eigenvalue. An instructive example is the case of the Hamiltonian operator $\hat{H}$. If

$$\hat{H}\psi_n(x,t) = E_n\psi_n(x,t), \quad (19)$$

then the state ψ_n has definite energy E_n and the wave function can be written as

$$\psi_n(x,t) = \psi_n(x) \exp\left(-\frac{i}{\hbar}E_n t\right). \quad (20)$$

One can expand an arbitrary wave function $\chi(x, t)$ in terms of these definite energy states:

$$\chi(x,t) = \sum_n c_n \psi_n(x) \exp\left(-i\frac{E_n}{\hbar} t\right). \qquad (21)$$

Then, for example,

$$<E> = \sum_n |c_n|^2 E_n \qquad (22a)$$

with

$$\sum_n |c_n|^2 = 1, \qquad (22b)$$

and $|c_n|^2$ represents the probability of having energy E_n.

A wave function can be a simultaneous eigenfunction of two or more such operators if and only if these operators commute—that is, if the order in which they are applied is immaterial. When the order matters, that is, if

$$[\mathbb{O}_1, \mathbb{O}_2] \equiv \mathbb{O}_1\mathbb{O}_2 - \mathbb{O}_2\mathbb{O}_1 \neq 0, \qquad (23)$$

then an uncertainty relation is obtained. Thus, for example,

$$[\hat{p}_i, \hat{x}_j] = \left[-i\hbar\frac{\partial}{\partial x_i}, x_j\right] = -i\hbar\delta_{ij}, \qquad (24)$$

leading to the Heisenberg uncertainty relation

$$\Delta p_i \Delta x_j \geq \frac{\hbar}{2} \delta_{ij}. \qquad (25)$$

Another example is the angular momentum operator

$$\hat{\mathbf{L}} = \hat{\mathbf{x}} \times \hat{\mathbf{p}}, \qquad (26)$$

for which

$$[\hat{L}_i, \hat{L}_j] = i\epsilon_{ijk}\hat{L}_k. \qquad (27)$$

This lack of commutativity is associated with the physical result that rotation, for example, about the x and then about the z direction is not the same as first rotating about z and then about the x direction. It requires that any wave function must have

$$\Delta L_x \Delta L_z \geq \frac{\hbar}{2}. \qquad (28)$$

Finally, since $E \rightarrow i\hbar \, (\partial/\partial t)$

$$\Delta E \Delta t \geq \frac{\hbar}{2}, \qquad (29)$$

which expresses the feature that a state with a finite lifetime Δt must of necessity represent a superposition of energy states with a spread ΔE.

A simple example of a nonfree particle involves motion in the potential well

$$V(x) = \begin{cases} V_0 & |x| > \\ & a \end{cases}. \qquad (30)$$

Then definite energy solutions, which satisfy the time-independent Schrödinger equation

$$\left(-\frac{\hbar^2}{2m}\frac{d^2}{dx^2} + V(x)\right)\psi_n(x) = E_n\psi_n(x), \qquad (31)$$

will have a completely different character depending on whether the energy E is greater than or less than the height of the well V_0. If $E > V_0$, then, just as in the case of classical physics, a solution can be found no matter what the energy. However, if $E < V_0$, then solutions to the Schrödinger equation outside the well, that is, for $|x| > a$, satisfy

$$\frac{d^2}{dx^2}\psi(x) = \kappa^2\psi(x), \qquad (32a)$$

with

$$\kappa^2 = \frac{2m}{\hbar^2}(V_0 - E) > 0, \qquad (32b)$$

and are the exponential forms $\exp(\pm\kappa x)$. The wave function cannot be allowed to become infinite, since this is not consistent with the probability interpretation of this function. Thus, only exponentially decreasing solutions are allowed for large $|x|$, and this can only be satisfied for a discrete (i.e., quantized) set of energies—$E = E_0, E_1, \ldots, E_{n_{max}}$. This quantization is a general feature of such bound state wave functions.

As a final example, consider a potential well having walls with finite width

$$V(x) = \begin{cases} 0, & |x| < a \\ V_0, & a < |x| < b. \\ 0, & b < |x| \end{cases} \quad (33)$$

In this case, if $E > V_0$ then again, just as in classical physics, solutions with *any* energy are allowed. On the other hand, if $0 < E < V_0$, then solutions exist that are nearly bound, in that they have the approximate form of ordinary potential well solutions—Eq. (31)—when $|x| < b$, but $e^{ik|x|}$ when $|x| > b$, corresponding to outgoing waves. The physical interpretation is that while at time $t = 0$ the particle can be confined within the well (i.e., $|x| < a$), at later times there exists a finite probability to have escaped into the region $|x| > b$—the particle is said to have "tunneled" through the barrier. According to classical physics this cannot happen—the tunneling phenomenon is an intrinsic property of quantum mechanics. The lifetime T of such a state is found to be

$$T^{-1} = \frac{\hbar v}{2a} \exp[-2\kappa(b - a)]. \quad (34)$$

Here $v = \sqrt{2E_n/m}$ is the classical velocity within the well, so that the factor $v/2a$ represents the number of collisions per second with the barriers at $x = \pm a$, while the exponential represents a penetration factor with κ^2 defined as in Eq. (32b). Such a tunneling picture provides a very successful description of the phenomenon of alpha decay.

See also: BARRIER PENETRATION; EINGENFUNCTION AND EIGENVALUE; QUANTIZATION; QUANTUM MECHANICS; QUANTUM MECHANICS AND QUANTUM FIELD THEORY; SCHRÖDINGER EQUATION; UNCERTAINTY PRINCIPLE; WAVE FUNCTION; WAVE PACKET

Bibliography

GOSWAMI, A. *Quantum Mechanics* (Wm. C. Brown, Dubuque, IA, 1992).

GRIFFITHS, D. J. *Introduction to Quantum Mechanics* (Prentice Hall, Englewood Cliffs, NJ, 1995).

BARRY R. HOLSTEIN

QUANTUM MECHANICS

Toward the end of the nineteenth century all known physics fell under the heading of what is now called nonrelativistic classical physics. The key entities were particles (billiard balls, planets) and waves (in water, on a string, in vacuum). Particles were localized bundles of energy and momentum and moved according to Newton's laws. Waves were spread out, obeyed some wave equation and exhibited diffraction, interference, and so on.

Subsequent experiments revealed shortcomings of this picture on two fronts. It was found that Newtonian mechanics had to be modified when dealing with motions at velocities comparable to that of light and replaced by Einstein's relativistic mechanics. It was also found that at atomic scales one needed quantum mechanics even in nonrelativistic problems. Problems that involved both large velocities and small scales require relativistic quantum field theory. We are concerned here with just nonrelativistic quantum mechanics.

We now discuss the types of results that led to the birth of quantum mechanics, replacing the actual experiments by schematic ones for pedagogical reasons. Consider a source of light of some color, a screen and a double-slit placed in between, as in Fig. 1. Let us imagine an array of light detectors mounted on the screen. If we open just one of the two slits numbered 1 or 2, the light forms a diffraction peak behind that slit, with intensities I_1 or I_2 as in Fig. 1a. If we open both, we get the interference pattern with intensity I_{12} as in Fig. 1b. In particular, at the minima of the pattern we actually get less light

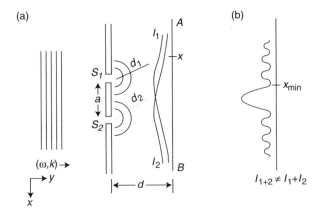

Figure 1 Double-slit experiment.

by opening an extra slit, and this is well explained in wave theory as being the result of destructive interference. If we now reduce the intensity of the light source we begin to see a new effect. It is found that light energy arrives, not continuously but in staccato fashion, with a burst here and a burst there. Each burst carries an energy $E = hf$, where f is the frequency of the light and $h = 6.626 \times 10^{-34}$ J·s is called Planck's constant. Frequently one uses $\hbar = h/2\pi$, called "h-bar," in terms of which $E = \hbar\omega$, where $\omega = 2\pi f$ is the angular frequency. Each burst also carries momentum $p = \hbar k$, where $k = 2\pi/\lambda$ is the wave number. Since $E^2 = p^2 c^2$, these bursts are naturally to be viewed as massless particles, called photons. (Historically, the photon concept was introduced by Albert Einstein to explain the photoelectric effect.) Thus, we find that light is not actually a continuous wave but a discrete barrage of particles, the photons. While classical mechanics can readily accommodate the notion that light, which appears continuous in a large scale is actually discreet and made of photons, it cannot accommodate the dynamics of the photons. The problem is the interference pattern. Why do less photons arrive at some points when we open both slits than when we open just one? The possibility that photons from one slit bounce off the ones from the other to form this pattern are ruled out by the fact that even if we have just one photon at a time in the experimental region, the longtime distribution on the screen still shows interference. If a photon travels like a classical particle, it has to go via one of the slits. To a photon going via slit 1, the fact that the other slit is open or closed, or even exists, would be immaterial and the number arriving with both slits open would equal the sum of the numbers that arrive when just one or the other slit is open (i.e., $I_{12} = I_1 + I_2$). While the interference pattern is very hard to describe in terms of particles, it is very natural in wave theory. It was Max Born who provided the following correct probability interpretation of this situation. Light is made of particles, the photons. Each photon of momentum p approaching the double slit along the y axis has associated with it a plane wave, called its wave function

$$\psi(y) = Ae^{ipy/\hbar}.$$

When this plane wave hits the double slit, it forms an interference pattern on the screen. The intensity $|\psi|^2$ at any point on the screen is proportional to the probability of finding the photon at that point. Note that the wave function is used to describe *each* photon. When a huge number of photons hit the screen over a long period of time from a weak source or in an instant from a bright source, the probability distribution turns into the number distribution.

Louis de Broglie then argued that if light, originally believed to be a wave, could show particle-like properties, then objects like electrons, thought to be particles, must show wave-like properties. In particular, an electron of momentum p must have associated with it a plane wave of de Broglie wavelength

$$\lambda = \frac{2\pi\hbar}{p}.$$

Indeed a double-slit experiment with electrons reveals an interference pattern of the predicted form. (In actuality, the wave nature of electrons was confirmed in a diffraction experiment by Clinton J. Davisson and Lester H. Germer.)

Whereas in classical physics a particle moving along the x axis has a definite position, in quantum theory it has a wavefunction ψ and $|\psi(x)|^2 dx$ gives the probability of finding the particle between x and $x + dx$. But a particle in classical mechanics also has a definite momentum p. Quantum theory again gives odds $|\psi(p)|^2 dp$ for getting a value between p and $p + dp$. However, $\psi(p)$ is not a new function, it is just the Fourier transform of $\psi(x)$:

$$\psi(p) = \int_{-\infty}^{\infty} \psi(x) \exp(-ipx/\hbar)\, dx.$$

One refers to $\psi(x)$ and $\psi(p)$ as the coordinate and momentum space wave functions.

The theory of Fourier transforms tells us that a wave function $\psi(x)$ of width Δx (defined by any reasonable measure) will have a transform $\psi(p)$ of width Δp, such that

$$\Delta x \Delta p \approx \hbar,$$

which is Heisenberg's uncertainty relation, which states that one cannot have states of perfectly well-defined position and momentum. As mentioned earlier, a single plane wave of definite momentum describes a particle of certain momentum p, but such a plane wave has $|\psi(x)|$ a constant; that is, it describes a particle of completely indeterminate position. Likewise, the Fourier transform of a wave function concentrated at one point in space will have $|\psi(p)|^2$ independent of momentum.

Consider now a typical wave function $\psi(x)$. If the position of a particle is measured and found to have a value x_0, immediately after the measurement, its wave function is sharply localized at x_0. Likewise, if its momentum is measured to give a value p, the wave function immediately after the measurement is a plane wave of wavelength $\lambda = \hbar/p$. This change in the wave function, from something with a width in x or p to something with support at just one value of the measured variable, is called the collapse of the wave function.

We now turn to another feature: quantization. Consider a particle of momentum p in a ring of radius R so that the coordinate x runs from 0 to $2\pi R$. The wave function is

$$\psi(x) = \frac{1}{\sqrt{2\pi R}}\, e^{ipx/\hbar},$$

where the prefactor ensures that the total probability adds up to 1:

$$\int_0^{2\pi R} |\psi(x)|^2 dx = 1.$$

We now demand that if we go once around the circle, ψ returns to its original value. This means

$$\frac{2\pi R p}{\hbar} = 2n\pi \qquad (n = 0, \pm 1, \pm 2 \ldots)$$

and

$$L = pR = n\hbar.$$

This is a simple example of quantization: Angular momentum L is quantized into a multiple of $\hbar$. If we look at the energy of the particle, we find that it too is quantized:

$$E = \frac{L^2}{2mR^2} = \frac{n^2\hbar^2}{2mR^2}.$$

One refers to n as the quantum number. In general, variables in quantum mechanics can take only some allowed and quantized values.

So far we have focused on kinematics of ψ and its interpretation. The dynamics of ψ is given by the Schrödinger equation

$$i\hbar\, \frac{\partial\psi}{\partial t} = \left[-\frac{\hbar^2}{2m}\, \frac{\partial^2\psi}{\partial x^2} + V(x,t) \right]\psi,$$

where V is the potential in which the particle finds itself. If V is time-independent, one can find solutions of the equation such that

$$\psi(x,t) = e^{-iEt/\hbar}\psi_E(x),$$

where $\psi_E(x)$ obeys the time-independent Schrödinger's equation

$$\left[-\frac{\hbar^2}{2m}\, \frac{\partial^2\psi}{\partial x^2} + V(x) \right]\psi_E(x) = E\psi_E(x).$$

These are called stationary states, because none of the probabilities depend on time since it appears only in the phase factor and drops out of $|\psi(x, t)|^2$. For example, if we choose $V(x) = \frac{1}{2}m\omega^2$, the harmonic oscillator potential, it will be found that solutions (which do not blow up at $x \to \pm\infty$) exist only for certain quantized values of energy $E = n\hbar\omega$. At each energy, the corresponding $\psi_E(x)$ will not vanish abruptly beyond the classically allowed region, $|x| \leq \sqrt{2E/m\omega^2}$, but instead decay exponentially.

These are the highlights of one particle in one dimension. One particle in three dimensions will be described by a wave function $\psi(x, y, z)$ (or its three-dimensional Fourier transform if we are interested in its momentum distribution), while two particles in three dimensions will be given by $\psi(x_1, y_1, z_1; x_2, y_2, z_2)$, and so on. As in classical mechanics, there exist variables besides just x and p, for example, angular momentum.

Consider now two identical particles. If we release two of them at some points and catch up with them later, it is impossible to tell which is which, since there are no interpolating continuous trajectories as in classical mechanics. Thus, the theory must assign the same probability for finding them at one pair of points as for finding them at another pair with the coordinates interchanged. This in turn means that

$$\psi(x_1, y_1, z_1; x_2, y_2, z_2) = \pm \psi(x_2, y_2, z_2; x_1, y_1, z_1),$$

where ± 1 is the only possible prefactor if we require that two exchanges be equal to no exchange. Particles that choose the $+$ sign are called bosons, and particles that choose the $-$ sign are called fermions. Relativistic quantum field theory tells us that particles with integer spin will be bosons and those with half-integer spin will be fermions. Note that two identical fermions cannot be at the same point, since in this case we would get $\psi = -\psi = 0$. In general, the rule, known as the Pauli exclusion principle, is that two identical fermions cannot be in the same quantum state, which is central to the explanation of atomic structure.

Although few people dispute the correctness of quantum mechanics, no one who has thought about it can fail to notice how strange it is. The probability description is the hardest to swallow. Over the years, people have attempted to come up with some hidden variables, which are not visible to us but control the particles in a deterministic way and manage to reproduce all observed correlations predicted by quantum mechanics and seen in experiments. It was John Bell who brought to focus the point that any such theory would be non-local; that is, it would have the feature that measurements made at one region in space influence those in another that is arbitrarily far away and possibly outside the light cone.

See also: DAVISSON–GERMER EXPERIMENT; PAULI'S EXCLUSION PRINCIPLE; PHOTON; QUANTIZATION; QUANTUM MECHANICAL BEHAVIOR OF MATTER; QUANTUM MECHANICS, CREATION OF; QUANTUM MECHANICS AND QUANTUM FIELD THEORY; QUANTUM NUMBER; SCHRÖDINGER EQUATION; UNCERTAINTY PRINCIPLE; WAVE FUNCTION

Bibliography

COHEN-TANNOUDJI, C.; DIU, B.; and LALOE, F. *Quantum Mechanics* (Wiley, New York, 1977).

FEYNMAN, R. P.; LEIGHTON, R. E.; and SANDS, M. *The Feynman Lectures*, Vol. 3 (Addison Wesley, Reading, MA, 1965).

MERMIN, N. D. "Quantum Mysteries Revisited." *Am. J. Phys.* **58**, 731 (1990).

SHANKAR, R. *Principles of Quantum Mechanics*, 2nd ed. (Plenum, New York, 1994).

R. SHANKAR

QUANTUM MECHANICS, CREATION OF

The period of creating modern atomic theory, which is called quantum mechanics (QM) or wave mechanics (WM), lasted roughly five years, from 1923 to 1928. In this time the theoretical concepts and mathematical schemes were established and a standard physical interpretation was achieved. We discuss here the first part of this period, the events of the years from summer 1923 to summer 1926, when the theory was formulated and immediately applied to solve fundamental problems of atomic and molecular physics.

The so-called old quantum theory (OQT), set up by Max Planck, Albert Einstein, Niels Bohr, Arnold Sommerfeld, and others between 1900 and 1922, had given the first handle to explain features of atomic and molecular structure and its characteristic radiation, including the spectra of hydrogen and the x-ray spectra of all elements, as well as the splitting of the spectral lines in external electric and magnetic fields (the Stark and Zeeman effects). However, when applied in detail, especially to multiple-electron atoms, it revealed serious contradictions. By the end of 1924 the experts recognized that OQT even failed to account for the properties of the simplest atomic system, the hydrogen atom, when exposed to external electric *and* magnetic field, let alone the fundamental unresolved difficulties posed by crucial experiments—notably the Compton effect (1923) and the splitting of atomic beams in inhomogeneous magnetic fields (Stern–Gerlach effect, 1922)—and the new organization of atomic energy terms by Pauli's exclusion principle (1925). Therefore, the most active theoreticians of atomic physics in Copenhagen, Göttingen, and Hamburg called for a "new" quantum theory of the atom or simply "quantum mechanics." Several elements of QM existed already at that time, more or less recognized and accepted: de Broglie's matter waves (1923); the Bose statistics for light quanta and (ideal) gas atoms (1924); and a particular approach to deal with the scattering of light by atoms and related phenomena, then called dispersion theory (1924).

The direct path to the first scheme of QM started in summer 1923, when Max Born, Werner Heisenberg, and Wolfgang Pauli wanted to create a "discrete" physics, in contrast to the continuous physics described by the laws of classical dynamics. They obtained the new scheme by replacing certain differential quotients in the classical dynamical equations by difference quotients. Thus, Hendrik Kramers in spring 1924, and Kramers and Heisenberg later that year, found dispersion-theoretical formulas explaining satisfactorily the experimentally observed scattering of light by atoms. In the spring of 1925 Heisenberg proceeded to the decisive step. The classical theory of multiply-periodic systems described the motion of electrons in atoms with the help of the Fourier series, consisting of harmonic functions (with frequencies ω, 2ω, 3ω, etc.) multiplied by coefficients (whose square characterized the intensity of the emitted radiation). Heisenberg now rewrote the classical Fourier series by what he called the quantum-theoretical reformulation; that is, he replaced the Fourier amplitudes systematically by the quantum theoretical transition amplitudes $a(n, m)$, depending on two quantum numbers n and m, and the harmonic frequencies $n\omega$ by the quantum frequencies $\omega(n, m)$. As in the classical theory, all properties of the atomic system could then be expressed by the quantum-theoretical series and products of them. But when Heisenberg wrote down the multiplication law for these series, he observed that the product of two dynamical variables (like position and momentum) of a given atomic system does not commute in general, unlike in the classical theory. Still the theoretical scheme, which he formulated in his pioneering paper of July 1925, seemed to represent a great number of experimental observations on atoms and their radiation.

Born and Pascual Jordan expressed, still in the summer of 1925, Heisenberg's new quantum variables by infinite Hermitean matrices and rewrote in particular his quantum condition—the reformulated Bohr–Sommerfeld phase integral—as the matrix commutation relation between the momentum p and position q of the atomic particle. Together with Heisenberg, they established in the fall the scheme matrix mechanics (MM), the earliest mathematical form of quantum mechanics containing a, at least in principle, complete dynamical description of (multiply-periodic) atomic systems also under the action of external forces (three-man paper, submitted in November 1925).

MM was immediately applied to the simplest examples of atomic theory. The Göttingen theoreticians calculated several atomic and molecular properties (November 1925), and Pauli in Hamburg succeeded in evaluating the discrete energy states of the (non-relativistic) hydrogen atom (January

1926). A further application to the anomalous Zeeman effect and the relativistic hydrogen atom demanded the use of the electron-spin hypothesis and of extended versions of MM, such as operator mechanics, the theory with differential operators that Born developed with Norbert Wiener in December 1925, or the q-number theory of P. A. M. Dirac. The latter noticed in the summer of 1925, after a visit with Heisenberg, that the results of the Göttingen physicist could be obtained from another quantum-theoretical reformulation of the classical Hamilton–Jacobi theory of multiply-periodic systems: He had just to replace the classical Poisson bracket, an expression containing differential quotients, by a commutator, that is, the difference of commuted products of the dynamical variables involved (November 1925). Dirac approached during the following months an increasing number of atomic problems, from the hydrogen atom to multiple-electron systems, and was able to treat the relativistic scattering problem of x rays from atoms, the Compton effect (April 1926).

Thus, by early 1926 QM existed in three slightly different mathematical forms: the matrix mechanics of Born, Heisenberg, and Jordan; the q-number theory of Dirac; and the operator mechanics of Born and Wiener. All three schemes were based on the mathematics that David Hilbert had developed between 1904 and 1910 for dealing with linear integral equations, although this mathematical basis needed substantial extension, which was achieved only after 1926. With respect to the physical foundation, these schemes took into account the particular discreteness observed in the existence of stationary atomic states and in the emission and absorption of radiation by atoms. Dirac's theory and the one of Born and Wiener had the advantage that they allowed, in principle, the analysis also of scattering problems. However, for its range and easiness of application, a fourth scheme quickly won dominance, the so-called wave mechanics (WM), which was based on quite different ideas and seemed to work with rather different mathematics.

The origin of WM goes back to 1909, when Einstein observed that the radiation in Planck's blackbody law, if analysed statistically, exhibits simultaneously wave and particle properties. From 1923 to 1924 Louis de Broglie generalized this idea; he associated with each atomic particle, notably electrons, a matter wave whose wavelength λ depends on the momentum p of the particle: $\lambda = h/p$, with h denoting Planck's constant. Then, in late 1925, Erwin

Schrödinger of Zurich University wrote down a wave equation for the hydrogen atom and derived from it the (non-relativistic) energy states of the atom. Unlike the Göttingen QM or Dirac's q-number theory, WM worked with standard mathematics (i.e., differential equations), just like classical wave theory. The discrete quantum states of the atom emerged from applying boundary conditions; otherwise, no discreteness seemed to be involved in the new atomic theory. Schrödinger also obtained the continuum states of the hydrogen atom (January 1926).

The Zurich professor quickly succeeded in extending his time-independent wave equation for the electron in the hydrogen atom,

$$\left[\frac{\partial^2\psi}{\partial x^2} + \frac{\partial^2\psi}{\partial y^2} + \frac{\partial^2\psi}{\partial z^2}\right] + \frac{8m\pi}{h^2}\left(E + \frac{e^2}{r}\right)\psi = 0$$

(with E and e denoting the energy and the charge of the electron and r its distance from the nucleus), to the general case of multiple-electron atoms in two ways: either by using a mechanical-optical analogy and invoking a multidimensional Riemannian space for the wave propagation (February 1926), or by applying a new wave-mechanical reformulation of the classical Hamiltonian–Jacobi theory (March 1926). That is, he wrote down a wave equation $H(p, q)\psi(q) = E\psi(q)$ for the time-independent wave amplitude $\psi(q)$, in which $H(p, q)$ denotes the Hamiltonian operator depending on the position variables q and the momenta $p = (h/2\pi i)\partial/\partial q$.

Although WM worked with continuous, in generally complex, functions and differential equations—in contrast to MM, where discrete matrices provided the mathematical tool—it satisfied the commutator conditions of Born, Heisenberg, and Jordan or Dirac, respectively. Schrödinger now demonstrated in his third paper (quoted above) the complete formal equivalence between MM and WM; in particular, he wrote down for each matrix expression or equation the corresponding wave mechanical relation. Independently, Pauli provided a similar equivalence proof in a letter to Jordan (April 12, 1926), and finally Carl Eckart did so in Pasadena, California (June 1926). Actually, Cornelius Lanczos of Frankfurt had noticed previously the connection between the Born–Heisenberg–Jordan discrete matrices and a field-like representation, as provided by a reformulation of the fundamental equations of MM as linear integral equations of the type considered by Hilbert two decades earlier (November 1925).

The wave-mechanical methods offered many older and younger quantum theorists the opportunity to deal with a great number of problems: Schrödinger and Erwin Fues computed the band spectra of diatomic molecules; Schrödinger, Paul Epstein, and Ivar Waller treated the Stark effect; and Gregor Wentzel worked out the intensity of the hydrogen spectrum and of x-ray spectra. It might be added in this context that, like Born, Heisenberg, and Jordan in MM, Schrödinger developed in WM a perturbation theory, by which he was able to evaluate the behavior of atoms and molecules when disturbed by external forces (May 1926). Finally, Heisenberg partly used the wave formalism in his calculation of the energy states of helium, one of the triumphs of the new QM in general (July 1926).

In practice, a very important breakthrough was achieved by Born with the first serious treatment of collision problems. In June and July 1926, he calculated in WM the scattering of electrons or atoms by atomic systems. Not much later, Schrödinger and Walter Gordon discussed the relativistic scattering occurring, for example, in the Compton effect. In August 1926, Dirac gave another fundamental application of WM by demonstrating that to each of the two types of quantum statistics, the Bose–Einstein statistics of light quanta or the Fermi statistics of electrons, there corresponded either symmetrical or antisymmetrical wave functions—that is, the wave functions kept sign or changed it when the coordinates of two microscopic objects were permuted.

From the very beginning of his investigations of WM, Schrödinger had one expectation. It should provide an *anschauliche* interpretation of atomic processes in terms of pictures known from the classical description of wave phenomena: destructive and constructive interferences, beat frequencies, and so on. Intuition and visuality, he hoped, would thus return to atomic physics, although he was aware of the fact that the wave function of multiple-electron atoms lacked a direct interpretation in the real, three-dimensional space. At any rate—about this he was certain—one would be able to completely get rid of the strange quantum jumps, which Heisenberg and his followers claimed to be the central aspect of atomic theory. However, Schrödinger's reality interpretation of the wave function received a decisive blow from Born's collision theory. Born stated that the amplitude of the scattered wave represented only the probability of finding an atomic particle at a certain place: "Schrödinger's quantum mechanics yields a very definite answer on the problem, but one does not obtain a causal relation. From the point of view of our quantum mechanics no quantity exists which in a single collision fixes its effects causally." This probabilistic interpretation of the scattered wave function—indeed, of any wave function—which Schrödinger never accepted fully, led to the later standard interpretation of QM, especially to Heisenberg's uncertainty relations and Bohr's principle of complementarity.

See also: BOHR, NIELS HENRIK DAVID; BORN, MAX; COMPLEMENTARITY PRINCIPLE; DIRAC, PAUL ADRIEN MAURICE; HEISENBERG, WERNER KARL; PAULI, WOLFGANG; PAULI'S EXCLUSION PRINCIPLE; PERTURBATION THEORY; QUANTUM MECHANICS; QUANTUM THEORY, ORIGINS OF; SCHRÖDINGER, ERWIN; STARK EFFECT; STERN–GERLACH EXPERIMENT; UNCERTAINTY PRINCPLE; WAVE–PARTICLE DUALITY, HISTORY OF; ZEEMAN EFFECT

Bibliography

CLINE, B. *The Questioners: Physicists and the Quantum Theory* (Thomas Y. Crowell, New York, 1965).

JAMMER, M. *The Conceptual Development of Quantum Mechanics* (McGraw-Hill, New York, 1966).

KEMBLE, C. *The Fundamental Principles of Quantum Mechanics* (McGraw-Hill, New York, 1937).

MEHRA, J., and RECHENBERG, H. *The Historical Development of Quantum Theory*, 5 vols. (Springer-Verlag, Berlin, 1982–1987).

SEGRÈ, E. *From X-Rays to Quarks: Modern Physicists and Their Discoveries* (W. H. Freeman, San Francisco, 1980).

HELMUT R. RECHENBERG

QUANTUM MECHANICS AND QUANTUM FIELD THEORY

Near the end of the nineteenth century, all of the known physics fell under the heading of what is now called nonrelativistic classical physics. The key entities were particles (billiard balls, planets) and waves (in water, on a string, in vacuum). Particles were localized bundles of energy and momentum and moved according to Newton's laws. Waves were spread out, obeyed some wave equation and exhibited diffraction, interference, and so on.

A classical field was a system with an infinite number of degrees of freedom. Consider, for example, an infinite number of identical masses along a line, connected to their neighbors on both sides by identical springs. One could define a one-dimensional field, the values of which correspond to the displacements of each mass from its equilibrium position. Another example is the electromagnetic field described by the values of the electric and magnetic field strengths at each point in space. While both fields have infinite degrees of freedom, the latter has an infinite number even within a finite volume, a feature that proves very important. The fields obey field equations of motion, for example, Maxwell's equations, in the second example predict the behavior of the fields given their starting configuration. The fields also supported the waves referred to earlier: the compressional waves in the string of masses corresponded to sound, while the waves of the electromagnetic field corresponded to light, radio waves, x rays, and so on (depending on the wavelength).

Subsequent experiments revealed shortcomings of this classical picture on two fronts. It was found that Newtonian mechanics had to be modified when dealing with motions at velocities comparable to that of light, and replaced by Einstein's relativistic mechanics. It was also found that at atomic scales one needed quantum mechanics even in nonrelativistic problems. Problems that involved both large velocities and small scales required relativistic quantum field theory.

To keep the discussion simple, we study physics in one space dimension only and start with just one nonrelativistic particle of mass m. In Newtonian mechanics it is described at any time by its coordinate x and velocity $v = dx/dt = \dot{x}$. (Note we here introduce to notation $\dot{x}$ for rate of change.) Newton's second law prescribes the acceleration and determines the future evolution. Equivalently we can start with the energy function (or Hamiltonian):

$$H(x,p) = \frac{p^2}{2m} + V(x),$$

where $p = mv$ is the momentum and $V(x)$ is the potential energy. Then Hamilton's equations

$$\dot{x} = \frac{\partial H}{\partial p}, \qquad \dot{p} = -\frac{\partial H}{\partial x}$$

can be used to derive equations for the motion with any potential energy function V. All other dynamical variables are functions of x and p. As an example consider a harmonic oscillator, for which $V(x) = 1/2kx^2$. It is easily found that in this case x and p oscillate with angular frequency $\omega = \sqrt{k/m}$. The energy and amplitude of oscillations are arbitrary.

However, the above description proved inadequate. First it was found that light exhibited particle-like properties: a continuous wave of light at frequency f and wavelength λ was found to be made up of photons of energy $E = hf$ and momentum $p = h/\lambda$, where $h = 6.626 \times 10^{-34}$ J·s is called Planck's constant. (Often one uses $\hbar = h/2\pi$.) These photons were *bona fide* particles in that one could apply to them all the laws of relativistic kinematics (energy and momentum conservation) in their collisions, say with electrons. On the other hand, the same wave of light, when passed through a double slit would exhibit an interference pattern expected of a wave of wavelength λ. We now accept Max Born's prescription that each incident photon in the beam (moving along the x axis) is to be described by a wave function

$$\psi(x) = \exp(ipx/\hbar)$$

whose interference pattern at the other side of the double slit is to be computed by standard wave theory, and whose final intensity $|\psi|^2$ at the screen is a measure of the probability of finding the photon at that point. We also know that an electron (universally accepted as a particle) of momentum p, is again described by the plane wave $\exp[ipx/\hbar]$, and that an electron beam would exhibit the same interference pattern. This phenomenon, in which particles are described by wave functions that control their probability of being found at any point, is known as wave particle duality.

If we accept the notion that the classical description is replaced by a probabilistic one based on the wave function, several kinematical questions arise: For example, if a particle of momentum p is described by the plane wave given above, what functions describe a particle with some other definite attribute, say energy? In addition there is the dynamical question: What equation controls the time evolution of ψ?

First note that the plane wave obeys the equation

$$-i\hbar \frac{d}{dx}\psi(x) = p\psi(x).$$

In other words, the wave function of a particle with definite momentum goes into p times itself when we we take $-i\hbar$ times its derivative. This relation between momentum and the derivative turns out to be a general rule. To find the wave function of a particle with definite energy we simply take $H(x,p)$, replace every p there by $-i\hbar(d/dx)$ and require that the wave function $\psi_E(x)$ obey

$$H\psi_E(x) = \left[\frac{1}{2m}\left(-i\hbar\frac{d}{dx}\right)^2 + V(x)\right]\psi_E(x) = E\psi_E(x).$$

This is called the time-independent Schrödinger equation. For the harmonic oscillator

$$\left[-\frac{1}{2m}\hbar^2\frac{d^2}{dx^2} + \frac{1}{2}kx^2\right]\psi_E(x) = E\psi_E(x).$$

To solve such an equation one looks for functions $\psi_E(x)$ that satisfy it. It turns out that for a randomly chosen energy E the solution blows up, that is, its value tends to infinity as x goes to an infinite distance from the origin. Given that $|\psi(x)|^2 dx$ is the probability of finding the particle between x and $x + dx$, and the total probability must add up to 1, this is not acceptable, that is, it is a mathematical solution but cannot represent a physically possible system. However at some special energies given by

$$E = \left(n + \frac{1}{2}\right)\hbar\omega, \qquad n = 0, 1, 2, \ldots$$

the solution vanishes at spatial infinity. These are the only allowed values of energy for this system. This is an example of quantization of a dynamical variable and n is an example of a quantum number. The wave function corresponding to the ground state, or lowest energy ($n = 0$ above) is given by

$$\psi_0(x) = \left(\frac{m\omega}{\pi\hbar}\right)^{1/4}\exp\left(-\frac{m\omega x^2}{2\hbar}\right).$$

Observe that $|\psi_0(x)|^2$ is peaked at the origin, has a width of roughly $\Delta x \approx \sqrt{\hbar/m\omega}$ and is nonzero in the region that would be forbidden in classical mechanics to a particle of this energy. As we move up in n, these nonclassical features disappear. For example, if the energy is a joule, n will be around 10^{34}, the granularity of energy (i.e., the discrete gap between n and n + 1 units at $\hbar\omega$) will be negligible compared to the energy itself, also the leakage of $\psi_n(x)$ to the classically disallowed region will become utterly negligible. This return to classical mechanics for large quantum numbers is an example of the correspondence principle.

So far we have seen that $\psi_n(x)$ describes a particle of energy $E = (n + 1/2)\hbar\omega$. We also know the odds for finding it at any point, namely $|\psi_n(x)|^2$. What if we measure its momentum? What are the possible outcomes and what are the odds? The answer is that

$$\psi(p) = \int_{-\infty}^{\infty} \exp\left(\frac{-ipx}{\hbar}\right)\psi_n(x)\,dx$$

where $|\psi(p)|^2$ gives the probability density for getting a value p. Thus the wave function for p is the Fourier transform of the wave function for x. For the $\psi_0(x)$ above, we find that $\psi(p)$ has the same functional form, but a width $\Delta p \simeq \sqrt{m\omega/\hbar}$. Thus

$$\Delta x \cdot \Delta p \simeq \hbar,$$

which is an example of Heisenberg's uncertainty principle, which asserts that certain pairs of variables cannot simultaneously have arbitrarily small widths in their probability distributions.

Note that the energy levels $E = (n + 1/2)\hbar\omega$ are uniformly spaced. This allows one to interpret $n\hbar\omega$, the excess energy above the ground state, as the total energy of n quanta each of energy $\hbar\omega$. Thus instead of saying the oscillator is in the nth excited state, we say there are n quanta. This seemingly semantic reinterpretation is very seminal. The part $\hbar\omega/2$, which exists in the ground state, is called zero point energy. In quantum theory such an energy is inevitable since the uncertainty principle would not allow for a ground state in which the particle is at rest at the origin where the potential is smallest.

The evolution of ψ with time is given by the time-dependent Schrödinger equation

$$i\hbar\,\frac{\partial\psi}{\partial t} = H\psi,$$

where H is obtained by replacing every p in the classical Hamiltonian by the operator $-i\hbar\,\partial/\partial x$.

Note that if we begin with $\psi(x, 0) = \psi_E(x, 0)$, which obeys the time-independent equation $H\psi_E(x, 0) = E\psi_E(x, 0)$, then the solution is very simple:

$$\psi(x, t) = \exp[-iEt/\hbar]\psi_E(x, 0).$$

This means in particular that $|\psi(x, t)| = |\psi(x, 0)|$, that is, the probability distribution does not change with time. This is called a stationary state. If we apply this idea to the hydrogen atom, where an electron is attracted to the nucleus by the Coulomb potential, we will find that only certain energies are allowed by the time-independent Schrödinger equation. The corresponding wave functions give the time-independent probability distributions. If however the atom is disturbed, it will jump from one allowed level to another emitting or absorbing a photon of frequency corresponding to the energy difference.

Notice that wave functions corresponding to definite energy have definite frequency $\omega = E/\hbar$. However a system that exists for a finite time Δt does not have a well-defined energy, since a well-defined frequency exists only for an infinitely long train. This leads to the time-energy uncertainty relation

$$\Delta E \cdot \Delta t \simeq \hbar.$$

What this means is that in quantum theory energy conservation does not restrict processes over short times. It is not that energy conservation is violated, but rather that energy itself is not well defined over short times, just as the period of a pendulum that has been oscillating for just a few cycles is not well defined.

Having quantized a system with one degree of freedom, we turn to fields. In the case of the string of coupled masses, we can define a field that consists of the Fourier transforms of the displacements of the masses rather than by the displacements of themselves. In classical treatments, it was found that the system described in these variables is simple; they behave like a set of decoupled oscillators with frequencies

$$\omega(K) = 2(1 - \cos Ka),$$

where K labels the Fourier transform variable and a is the equilibrium spacing between the masses. Note that due to the periodicity of the cosine, K is restricted to

$$-\frac{\pi}{a} \leq K \leq \frac{\Pi}{a}.$$

The quantum treatment of the field, that is, the quantization of the field, amounts to treating these oscillators quantum mechanically. The quantum field is then described by how many quanta there are at each K. These quanta are called phonons.

If the electromagnetic field is similarly quantized, that is, in quantum electrodynamics, the quanta are just the photons. The K values are unrestricted in this case and the field is defined in the continuum. What we call an electromagnetic wave of momentum K actually corresponds to a quantum state with a huge number of photons in the corresponding oscillator state.

If one quantizes the field for the electron, one gets the electrons as its quanta, with the important restriction that $n = 0, 1$ are the only possible values. (This is the Pauli exclusion principle at work.) This type field, called a fermionic field is not very familiar because it has no classical macroscopic manifestations. The photon or phonon fields with no restriction on n are bosonic fields.

In the preceding discussion it was assumed that the Hamiltonian on energy function was quadratic in the coordinates and momenta. If this were exactly true, the number of quanta would never change. In practice there are nonquadratic terms, which either describe interactions with the rest of the universe (e.g., the mass spring system is tweaked by an observer) or nonlinear effects within the system (the springs do not exactly obey Hooke's law). These can be treated by perturbation theory, if the nonquadratic terms are small. In other words, if α is a small parameter characterizing the nonquadratic term, the answer to a question like how often two phonons can merge to become a third that carries the total momentum and energy, can be developed in a power series in α.

In quantizing a relativistic theory like quantum electrodynamics, we run into a new problem due to the fact that over short periods of time, the energy is ill defined and according to $E = mc^2$, so is the number of particles. Thus even a system that began with just two electrons can evolve into fifty-four particles over short times. This is what makes relativistic quantum field theory a notoriously difficult subject.

However, the agreement between theory and experiment in quantum electrodynamics is nothing short of phenomenal.

See also: CORRESPONDENCE PRINCIPLE; FERMIONS and BOSONS; GROUND STATE; KINEMATICS; NEWTONIAN MECHANICS; PHOTON; QUANTIZATION; QUANTUM ELECTRODYNAMICS; QUANTUM FIELD THEORY; QUANTUM MECHANICS; QUANTUM MECHANICS, CREATION OF; QUANTUM NUMBER; SCHRÖDINGER EQUATION; UNCERTAINTY PRINCIPLE; WAVE–PARTICLE DUALITY

Bibliography

ITZYKSON, C., and ZUBER, J. B. *Quantum Field Theory* (McGraw-Hill, New York, 1980).

SHANKAR, R. *Principles of Quantum Mechanics*, 2nd ed. (Plenum, New York, 1994).

R. SHANKAR

QUANTUM NUMBER

Systems small in size, for example, atomic and subatomic particles, are correctly described by quantum mechanics. Quantum numbers are integers (n) and half integers ($n + 1/2$) that are necessary to characterize such systems. Under certain situations many physical quantities, such as energy and angular momentum of a quantum system, may acquire only discrete values (i.e., they may not vary in a continuous fashion). Quantum numbers are mathematical expressions of this discreteness.

The idea of quantum numbers, and consequently the field of quantum mechanics, was first introduced by Max Planck in 1900 in his effort to explain the energy spectrum of blackbody radiation. Planck achieved excellent agreement with experimental data by postulating that radiation from a blackbody can be considered to be made up of harmonic oscillators, each of which has a discrete energy value given by $E_n = (n + 1/2)hf$, where h is the Planck's constant, f is the frequency of oscillation of the oscillator, and n is a quantum number that can have values 0, 1, 2, ... This idea of quantization of energy was totally new, because previously (in classical physics) it was thought that a harmonic oscillator could have any amount of energy. It is important to understand that blackbody radiation is made up of radiation of *all* possible frequencies. Consequently the energy from a blackbody may have any arbitrary value, but a constituent oscillator cannot; an individual harmonic oscillator oscillates at a natural frequency f, and it can have only those values of energy allowed by the quantum number n [i.e., $(1/2)hf$ for $n = 0$, $(3/2)hf$ for $n = 1$, etc.].

Quantization of energy and momentum can be inferred directly from the wave nature of matter. Consider a wave in one dimension. If it is bound between two points separated by a distance L, it cannot *travel* away from this region. Such a condition demands that the wave behave as a standing wave; the two endpoints have amplitude zero, and consequently only an integral number of half-wavelengths fit in the length L (see Fig. 1), that is, the wavelength is quantized, $\lambda = 2L/n$ where $n = 1, 2, 3, \ldots$. Since the momentum of a wave/particle is related to its wavelength through the de Broglie expression, $p = h/\lambda$, it must take only discrete values. The energy is quantized through the equation $E = E_1 n^2$, where E_1 is the energy of the lowest state.

The number of quantum numbers necessary to describe a wave/particle is equal to the number of dimensions in which that particle exists. For example, a particle in a one-dimensional box is described by one quantum number. A particle in a three-dimensional box is described by three quantum numbers, one for each of the three dimensions. Quantum numbers are only applicable to bound systems. If the particle is not bound, then the energy and momentum of the wave/particle will not be quantized.

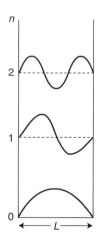

Figure 1 Standing waves.

Quantization is seen even for photons. A laser, for example, is a classic example of the quantization of an electromagnetic field in a cavity. It is interesting to note that if a particle within a laser cavity is in some higher quantum state, it can only relax to a lower state through the release of a photon if the energy and momentum of the photon released satisfies the quantum conditions of the laser cavity. If it does not, the particle will remain in that higher state indefinitely.

The electronic structure of atoms and molecules is based on quantization. In an atom, negatively charged electrons are bound by the electrostatic attraction of a positively charged nucleus. Boundary conditions again demand that the electron behave as a standing wave. The symmetry of the atom suggests that the three dimensions in which the electron exists are best represented in spherical polar coordinates (i.e., radius, azimuthal angle, and polar angle). These three dimensions give rise to three quantum numbers: the principal quantum number n, the azimuthal quantum number l, and the magnetic quantum number m_e, respectively. The first one relates to energy, the second to angular momentum, and the third to the z component of the angular momentum. Relativistic effects allow for a fourth quantum number, the spin quantum number s. The above description is valid for any particle moving under a central force (i.e., a force directed toward a central point), such as the attractive force of the positively charged proton on the negatively charged electron in a hydrogen atom that is directed toward the center of the atom. The range of values the above quantum numbers can acquire are: $n = 1, 2, \ldots$; $l = 0, 1, 2, \ldots, n-1$; $m_l = -l$, $-l + 1, \ldots, 0, 1, 2, \ldots, +l$; $s = 1/2$. The above simply means that energy, angular momentum and its z component, and the spin can take only discrete values; for a given energy, there is an upper limit on the angular momentum; and for a given angular momentum, only certain values of its z component are allowed. Further quantum numbers show up in description of this system, for example, the z component of the spin is quantized ($m_s = \pm 1/2$), the total angular momentum j is quantized ($j = l \pm s = l \pm 1/2$), and the z component of the total angular momentum m_j is also quantized ($m_j = m_l + m_s$).

For systems made up of more than one particle, one uses a vector model in which the angular momenta are added as vectors to give the total angular momentum. Quantum numbers show up here as well, for example, total angular momentum J and its z component M_J turn out to be quantized. Additional quantum numbers also show up in the description of elementary particles. Spin J, isospin I, and its third component I_z, electric charge Q, baryon number B, lepton number L, hypercharge Y, and strangeness S are some of these.

See also: BARYON NUMBER; LEPTON; MOMENTUM; OSCILLATOR, HARMONIC; PHOTON; QUANTIZATION; QUANTUM MECHANICS; RADIATION, BLACKBODY; STANDING WAVE

Bibliography

OHANIAN, H. *Modern Physics*, 2nd ed. (Prentice Hall, Englewood Cliffs, NJ, 1995).

SHAFIQUR RAHMAN

DAVID STATMAN

QUANTUM OPTICS

Quantum optics is a description of electromagnetic radiation as a group of particles known as photons. Their quantum mechanical and statistical properties are important. Electromagnetic radiation can be described either quantum mechanically as a collection of photons, each with energy $E_{ph} = \hbar\omega$, where $\hbar$ is Planck's constant and ω is the photon's angular frequency, or classically as an electromagnetic wave, with specified electric and magnetic fields. The quantum mechanical description is always required when the number of photons is small. The "optics" in quantum optics refers to the electromagnetic radiation while the "quantum" refers to the particle nature of the light.

The photon description of the electromagnetic radiation was postulated by Max Planck and Albert Einstein as an explanation of two experimental observations, the photoelectric effect and the blackbody spectrum.

The quantum mechanical description of photons is similar to that of the harmonic oscillator. A cavity with conducting walls supports radiation modes with discrete energy levels (photon energies). In free space the cavity walls are at infinity. Each mode satisfies the harmonic oscillator equations. There is no limit to the number of photons in each mode.

Coherence properties such as diffraction and interference depend on the fluctuations of the electromagnetic radiation. The quantum mechanical fluctuations are proportional to the square root of the average number of photons in the mode.

When the photon density is large, the electromagnetic radiation can be equivalently described by a photon or field picture. To use the field picture, photons must be in many modes. The propagation is described by Maxwell's equations. The energy density of the wave is equal to the photon density multiplied by the energy per photon and is proportional to the square of the electric field strength.

The interaction of electromagnetic radiation with a two-energy-level atom provides a good approximation to many atomic systems. When the photon energy, $E_{ph} = \hbar\omega$ is close to the energy spacing between the levels, E_{10} (see Fig. 1), population will be transferred between the lower and upper energy levels with a Rabi frequency proportional to the square root of the photon density (laser electric field). Rabi oscillations can be described in either the field or photon picture as stimulated absorption and emission of radiation.

Spontaneous emission, where an electron in an upper energy state makes a transition to a lower energy state, emitting a photon with $\hbar\omega = E_{10}$, must be explained quantum mechanically because it can occur in the absence of external fields. In quantum electrodynamics, this is described by vacuum fluctuations.

Quantum optics provides an important testing ground for quantum mechanical principles, including wave–particle duality. The interference pattern generated by light passing through two slits is a wave-like property. If one slit is closed, the interference pattern is destroyed because one "knows" which slit the photon (particle) passed through.

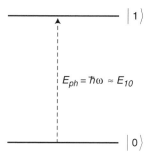

Figure 1 The interaction of electromagnetic radiation with a two-energy-level atom.

G. I. Taylor (1909) showed that the two-slit interference pattern is generated even if there is only one photon at a time passing through the slits. Experiments have also shown that a photon is indivisible.

See also: DIFFRACTION; ELECTROMAGNETIC RADIATION; INTERFERENCE; LASER, DISCOVERY OF; MASER; PHOTOELECTRIC EFFECT; PHOTON; QUANTUM ELECTRODYNAMICS; RADIATION, BLACKBODY; WAVE–PARTICLE DUALITY

Bibliography

KNIGHT, P. L., and ALLEN, L. *Concepts of Quantum Optics* (Pergamon, Oxford, Eng., 1983).

DAVID D. MEYERHOFER

QUANTUM STATISTICS

Where large numbers of a particular species of particle are involved, the description is a matter of statistics. Six examples of species of particle are: strange quarks, 5,000 Å photons, electrons, neutrons, helium atoms, and water molecules. Primarily the concern is with a gas, which means that the particles are noninteracting: aside from elastic collisions each particle feels nothing of the presence of the others. Examples are nitrogen gas and the electron gas that fills a metal.

For such a noninteracting ensemble of any one species of particle, the classical view is that they are identical but nevertheless intrinsically distinguishable. Under this assumption particles obey classical statistics, also called Maxwell–Boltzmann statistics to honor those responsible for its mathematical formulation.

Unless we believe that particles can have attributes inaccessible to experiment, this notion embodies a contradiction: "Identical" means that no experiment can unearth a difference between two of them, but "distinguishable" means that there must be some experiment that can reveal the distinction between them. Hence Maxwell–Boltzmann particles are distinguishable indistinguishable particles. No such particles have been found.

The quantum view is concerned with observables; it is based on observable behavior. Quantum theory recognizes indistinguishability as being the essential property characterizing a particle species. All particles of a given species are indistinguishable from each other.

The concept rests on experiment. Two particles are indistinguishable if exchanging them produces no measurable effect on the universe. Quantum mechanics does not admit attributes beyond experiment. If no experiment exists that can detect their exchange, then the two particles are indistinguishable.

Remarkably enough there are two distinct classes of indistinguishable particles. Each species fits into one class or the other. That there are two classes arises from the mathematics that follows below.

In quantum mechanics the state of a system is governed by amplitudes for events. Consider a group of particles. There is an amplitude for any measurable event to happen concerning them. Choose an event. There's an amplitude for it. Now consider the original group but with two of the particles exchanged. Choose the same event. There is an amplitude, albeit different, for it to happen.

It is not the amplitude itself but rather its absolute value squared that is the physically accessible quantity. This absolute square gives the probability for the event—something that is measurable.

Indistinguishability means that, for all possible events, the absolute squares must be equal for these two amplitudes—the direct and the exchanged one. This is the mathematical statement that no event can reveal the exchange. It says that any following event has a probability independent of whether it proceeded from the original or the exchanged condition. Any possible detection of the exchange is thus forbidden, so the two particles are indistinguishable.

But, for their absolute squares to be equal, there are two ways the two amplitudes can be related; their ratio may be either $+1$ or -1. So there are two classes of indistinguishable particles. Those with an exchanged amplitude ratio of $+1$ are called bosons. Those with -1 are called fermions.

Examples of fermions (fermion species) are strange quarks, electrons, and nitrogen atoms. Examples of bosons are photons, alpha particles, oxygen atoms, and nitrogen molecules (N_2).

The critical feature characterizing fermions is that no two indistinguishable fermions can be in the same state. This attribute, called the Pauli exclusion principle, is a direct mathematical consequence of the -1 feature of fermions. For two electrons with identical labels the exchanged state of the pair must both be equal to, and the negative of, the original state of the pair. This condition is satisfied only by zero. Thus the amplitude for two fermions of the same species to have identical state labels is zero. They are excluded from being in the same single-particle state.

Every member of an ideal gas of indistinguishable particles has the same single-particle spectrum of states. That feature is a mark of their indistinguishability. We can always label these states with some index, ν. Then in the gas some number of particles are in the first ($\nu = 1$) state, some in the second ($\nu = 2$), and so on. The number of particles in state ν is $n(\nu)$. The distribution $n(\nu)$ specifies how many particles there are in each microstate ν. The state of the whole gas—its macrostate—is governed by the function $n(\nu)$.

The total energy of any macrostate is the energy connected with each microstate $\varepsilon(\nu)$ multiplied by the number $n(\nu)$ of particles in that state added together for all the states possible. If we call this macrostate energy E, then

$$E = \sum \nu \, \varepsilon(\nu) \, n(\nu)$$

The three different classes of indistinguishability yield quite different energy spectra. This is readily seen for the case illustrated in the figure of a gas containing a mere two particles with the single-particle spectrum of each member limited to three states, the upper two being of equal energy. So $\varepsilon(1) < \varepsilon(2) = \varepsilon(3)$.

For any particular distribution of bosons $n(\nu)$ the number of classical configurations that achieve it is

$$\frac{[n(1)\,n(2)\,n(3)\ldots]!}{n(1)!\ n(2)!\ n(3)!\ldots},$$

the factorial of the sum divided by the product of factorials. An example is the boson distribution $n(1) = 0$, $n(2) = n(3) = 1$ shown in Fig. 1 (one particle

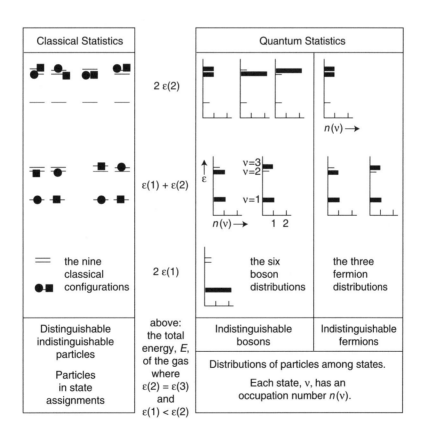

Figure 1 For a two-particle gas constrained to three single-particle states, the correspondence between state spectra in classical statistics and in quantum mechanics is shown. That they differ in structure and thus in thermodynamic behavior expresses their different philosophical underpinnings.

each in the two states $\nu = 2$ and $\nu = 3$) for which there are two classical configurations corresponding to the one boson distribution.

A gas of indistinguishable particles held at a fixed temperature T models the typical macroscopic system accessible to experiment. The temperature constraint gives rise to the renowned thermal average distributions called the Maxwell–Boltzmann, the Bose–Einstein, and the Fermi–Dirac. These arise solely from the degeneracy structure. The degeneracy is the number of gas states possible with any energy E. As can be seen in Fig. 1, the three energy levels of the two-particle gas exhibit degeneracies of 1, 4, and 4 for classical statistics but 1, 2, and 3 for bosons and 0, 2, and 1 for fermion quantum statistics. Hence the thermodynamic behavior in each case will differ discernibly from the others.

See also: FERMIONS AND BOSONS; MAXWELL–BOLTZMANN STATISTICS; PAULI'S EXCLUSION PRINCIPLE

Bibliography

CHESTER, M. *Primer of Quantum Mechanics* (Krieger, Malabar, FL, 1992).

FEYNMAN, R. P.; LEIGHTON, R. B.; and SANDS, M. *The Feynman Lectures on Physics*, Vol. 3 (Addison-Wesley, Reading, MA, 1965).

REIF, F. *Fundamentals of Statistical and Thermal Physics* (McGraw-Hill, New York, 1965).

MARVIN CHESTER

QUANTUM THEORY, ORIGINS OF

Quantum theory has its origin in the study of so-called blackbody radiation, the kind of continuous radiation that is emitted by an ideal blackbody in thermal equilibrium. In 1859 the German physicist Gustav Robert Kirchhoff deduced that the radiation from such blackbodies is characterized by a universal function that depends on the temperature and frequency, but not on the kind of substance. He also showed that the characteristic blackbody radiation is found inside a cavity of constant temperature filled with radiation. For this reason the radiation is also known as cavity radiation. Kirchhoff recognized the

theoretical importance of the universal function, given as, for example, the energy density $u(\nu, T)$ as a function of the frequency ν and the absolute temperature T. The problem of finding this particular distribution function became an important task for physicists in the late nineteenth century, and eventually it resulted in the new quantum theory.

The first insight into Kirchhoff's function did not relate to its spectral shape but to the total energy density of the blackbody radiation, that is, to $u(\nu, T)$ integrated over all frequencies. According to the Stefan–Boltzmann law, established empirically by the Austrian physicist Josef Stefan in 1879 and proved theoretically by his compatriot Ludwig Boltzmann five years later, the total energy is proportional to the temperature raised to the fourth power ($E = \sigma T^4$). In his derivation, Boltzmann relied on thought experiments and results obtained from James Clerk Maxwell's electromagnetic theory of light. Inspired by Boltzmann's reasoning, the German physicist Wilhelm Wien attacked the problem of the spectral distribution and derived his displacement law in 1893. This law states how to deduce the spectrum for a certain temperature from the knowledge of the spectrum for any other temperature. There were several attempts to derive, or guess, the distribution function $u(\nu, T)$, but it was only in 1896, when Wien suggested a law, that a good candidate was found. Wien's radiation law (not to be confused with his displacement law) seemed to give a good account of the experimental data known at the time.

Max Planck used Wien's result in his own research activity, which aimed at understanding the irreversibility of naturally occurring processes in terms of the entropy associated with electromagnetic fields. He hoped to prove that the Maxwell equations, if applied to hypothetical oscillators (or resonators) emitting and absorbing radiation, would lead to a kind of entropy function. This led him to a theoretical study of the blackbody radiation and a more satisfactory derivation of the assumedly correct Wien radiation law. However, around 1899 experimentalists at Berlin's Physikalisch-Technische Reichanstalt showed that Wien's formula was only approximately true; it failed for small values of ν/T (high temperature, low frequency). The improved data, obtained primarily by Otto Lummer, Heinrich Rubens, and Ernst Pringsheim, forced Planck to reconsider his work, and in 1900 he had ready a new formula for the entropy from which a new distribution law followed—Planck's radiation law.

Most textbooks claim that Planck arrived at his law in an attempt to avoid the "ultraviolet catastrophe" to which the laws of classical statistical mechanics (the equipartition theorem) would otherwise lead. If applied to the electromagnetic vibrations of blackbody radiation, the equipartition theorem leads to an expression for $u(\nu,T)$ that corresponds to an infinite total energy density—an absurd result and, hence, catastrophical. The result appeared indirectly in Lord Rayleigh's derivation of $u(\nu,T)$ from 1900—usually referred to as the Rayleigh–Jeans law—but it was of no importance for Planck's route to his radiation law. Planck did not make use of the equipartition theorem, so he was not confronted with any ultraviolet catastrophe.

Planck announced his new distribution law at a meeting of the Berlin Physical Society on October 19, 1900. Although it lacked rigorous theoretical justification, it agreed excellently with the most recent measurements; since it included both Wien's law and the Rayleigh–Jeans law as special cases (for very large and very small values of ν/T, respectively), it seemed to be the right answer. In this first version it rested on a classical basis, but very soon Planck was led to introduce a nonclassical element in his search for a better, more physical explanation. Exploiting statistical-mechanical ideas due to Boltzmann, Planck realized that a more satisfactory derivation required the energy to be divided in discrete parcels: "We consider . . . [the energy] as composed of a finite number of discrete equal parts and employ for this purpose the natural constant $h = 6.55 \times 10^{-27}$ erg sec. This constant multiplied by the common frequency ν of the resonators gives the energy element ε in ergs." This first introduction of the famous Planck constant (h) and Planck's energy quantization law $E = h\nu$ was presented to the German Physical Society on December 14, 1900.

Although later realized to be of historic importance, the quantization hypothesis did not occupy a prominent position in Planck's work, and he seems to have been uncertain about its significance. He had quantized the energy only in order to justify a formula he already knew agreed with experiment; he later described the introduction of energy quanta as "an act of desperation." The basically conservative Planck hesitated to interpret his result as contradicting classical physics and tended to conceive it as a mathematical rather than a physical innovation. At any rate, whereas Planck's radiation formula was quickly accepted, the introduction of energy quantization did not arouse much attention. A revo-

lution had taken place, but for some time it went unnoticed.

Planck's ideas of energy quanta were extended and, at the same time, transformed by 26-year-old Albert Einstein in an important paper of 1905, which Einstein himself characterized as "very revolutionary." In this paper, appearing the same year as the theory of relativity, Einstein proposed to revive the corpuscular theory of light, conceiving electromagnetic radiation as consisting of discrete portions of energy. This was a drastic proposal that went much beyond Planck's more restricted conception of quanta that related to the changes in energy of material oscillators and not to radiation itself; moreover, it went against the wave theory of light for which there was impressive experimental evidence. In spite of this evidence, Einstein boldly suggested that light (and other electromagnetic radiation) consists of discrete energy quanta and that these can be produced and absorbed only as units. To investigate the consequences of this "heuristical view," Einstein turned to the theory of blackbody radiation, which he combined with considerations based on the second law of thermodynamics. However, the radiation law he built upon was not Planck's, but Wien's. From Wien's law Einstein found an expression for the radiation entropy that was analogous to the entropy of a gas, a striking result that indicated a corpuscular nature of radiation. Developing such general arguments, Einstein concluded that "monochromatic radiation ... behaves with respect to thermal phenomena as if it were composed of independent energy quanta of magnitude $(R/N_0)\beta\nu$." Einstein's light quanta were later named photons, a name first suggested in 1926 by the American chemist Gilbert N. Lewis.

It is noteworthy that Einstein did not write the energy of his light quanta as $h\nu$, but used the pre-Planck form $(R/N_0)\beta\nu$, where R is the ideal gas constant, N_0 Avogadro's number, and β is a constant appearing in Wien's radiation law. Although Einstein did refer to Planck in his paper, he used neither Planck's radiation law nor his idea of quantized oscillator energies in his own arguments. Contrary to what is often stated in textbooks, Einstein simply did not build on Planck's work. In fact, when he wrote his paper, he believed Planck's theory not only to be irrelevant to his own, but contradictory to it. It was only a year later, in 1906, that Einstein realized that Planck's theory and his own rested on the same foundation. Not only was Einstein's theory of light quanta very different from Planck's, his approach and ambitions were as well. To Einstein, the existence of light quanta followed from Wien's radiation law, experimentally confirmed for high frequencies, and he welcomed rather than resisted the break with classical physics necessitated by the light quantum hypothesis.

Einstein's work of 1905 is often referred to as his paper on the photoelectric effect, but in fact his analysis of this effect constituted only a minor part of the paper and was not of crucial importance for the light quantum hypothesis. On Einstein's hypothesis that $E = (R/N_0)\beta\nu$, it was reasonable to assume that the photoelectric effect consisted in transfer of energy from the light quanta to the bound electrons and then that the maximum kinetic energy of the photoelectrons would satisfy the relationship $E_k = (R/N_0)\beta\nu - W$, where W is a work function characteristic for the metal exposed to light. Einstein's equation is usually written $E_k = h\nu - W$. As Einstein pointed out, it implies that the maximum energy of photoelectrons depends only on the frequency and not on the intensity of the light. Einstein's photoelectric formula was a genuine prediction; in 1905 there were not as yet any measurements of the (E, ν) relationship. For this reason alone, photoelectric measurements did not, and could not, act as an experimental basis for his inference of light quanta.

Einstein's revolutionary theory of light quanta was ignored or rejected by almost all physicists. After all, there was overwhelming evidence for the wave nature of light (interference and diffraction), and Einstein's prediction of a linear energy-frequency relationship between photoelectrons and light remained unconfirmed for a long time. It was only about 1915 that Robert Millikan provided convincing proof of the relationship, and even then most physicists refrained from concluding from the correctness of Einstein's equation to the correctness of his light quantum hypothesis. Millikan himself dismissed it as a "bold, not to say reckless, hypothesis."

All the same, Einstein's hypothesis marked a new stage in quantum theory and formed the starting point for Einstein's further development of the theory in 1906–1907. He now accepted and transformed Planck's theory into a much extended version that covered not only radiation and interaction between matter and radiation, but also the kinetic theory of heat. This was the first time that the universality of the quantum hypothesis was recognized, and in this sense Einstein's work of 1907 may be said to mark the start of a proper quantum *theory*. After having given a new and physically more lucid derivation of Planck's distribution law, Einstein applied it to a simplified model of solids consisting of atoms vibrating independently and at the same frequency. He arrived at a formula for the molar specific heat capacity (c) that differed in a striking way from the classical result (the Dulong–Petit law), according to which $c = $ constant. Einstein's result approximated this result for large temperatures, but for the absolute temperature approaching zero it predicted a vanishing specific heat. In 1907 there were no reliable low-temperature data, so Einstein's prediction lacked confirmation. However, at that time the German physical chemist Walther Nernst had just announced his heat theorem (or third law of thermodynamics), which implies entropy differences to disappear at absolute zero. The experimental program initiated by Nernst to prove his heat theorem also produced data that agreed well with Einstein's prediction. In 1911 Nernst was convinced about the correctness of Einstein's theory of specific heats, and he also became an early and influential supporter of its background, the quantum theory.

Like most other fundamental theories, quantum theory was not "discovered" at a specific time and place. Retrospectively, Planck's presentation of December 14, 1900, may be celebrated as the "birthday of quantum theory," but at that time Planck would not himself have admitted paternity. It took about ten years until the seeds hidden in Planck's theory were developed into a stage that can reasonably be called a theory of quanta. Had it not been for Einstein's pioneering work, the infancy might have lasted considerably longer.

See also: EINSTEIN, ALBERT; PLANCK, MAX KARL ERNST LUDWIG; QUANTUM MECHANICS, CREATION OF; QUANTUM THEORY OF MEASUREMENT; RADIATION, BLACKBODY; STEFAN–BOLTZMANN LAW

Bibliography

HAAR, D. TER *The Old Quantum Theory* (Pergamon Press, Oxford, Eng., 1967).

HERMANN, A. *The Genesis of Quantum Theory (1899–1913)* (MIT Press, Cambridge, MA, 1971).

JAMMER, M. *The Conceptual Development of Quantum Mechanics* (McGraw-Hill, New York, 1966).

KANGRO, H. *Early History of Planck's Radiation Law* (Crane, Russak, New York, 1976).

KLEIN, M. J. "The Beginnings of the Quantum Theory" in *History of Twentieth Century Physics,* edited by C. Weiner (Academic Press, New York, 1977).

KUHN, T. S. *Black-Body Theory and the Quantum Discontinuity, 1894–1912* (Oxford University Press, New York, 1978).

SEGRÈ, E. *From X-Rays to Quarks: Modern Physicists and Their Discoveries* (W. H. Freeman, San Francisco, 1980).

HELGE KRAGH

QUANTUM THEORY OF MEASUREMENT

The process of measurement consists of the interaction between the system on which the measurement is performed and the apparatus or measuring device. The response of the apparatus depends on the state of the system. There are two purposes to measurement in physics: to obtain from the response of the apparatus information about the state of the system and to evaluate the manifestation of this state, which gives it physical meaning.

In classical physics, measurements on a particle can in principle determine its state to have a position and momentum at a given time, represented by a point in phase space or a tangent vector in space-time. In quantum physics, however, it is not possible to determine simultaneously the position and momentum of a particle. The maximum information is obtained by measuring a complete set of commuting observables, which determines the state as a ray (one-dimensional subspace) of the Hilbert space of state vectors for the system. Nevertheless, quantum measurements raise profound physical and philosophical questions concerning the physical meaning of the quantum state.

Consider, for example, a photon whose state is described in quantum theory by a wave function extended in space. This may be verified by passing a stream of similarly prepared photons one at a time through a diaphragm with two slits on the far side of which is an array of closely packed, highly efficient photomultipliers. Then an interference pattern would be recorded by the array, as the photons accumulate there, consistent with the assumed common wave function for each photon. But as each photon interacts with the array of photomultipliers, which is the apparatus, it triggers only one photomultiplier. Quantum theory cannot describe the process by which the initial extended wave function of the photon is transformed into the localized wave packet observed by this photomultiplier. Only the probability of this transition can be predicted.

According to Niels Bohr, the chief architect of the Copenhagen interpretation of quantum theory, no quantum-mechanical analysis is possible of the process of detection of each photon because the macroscopic apparatus must be described in classical terms. Yet the array of photomultipliers, as with any apparatus, is composed of protons, neutrons, electrons, and photons that are subject to the laws of quantum mechanics, which, in principle, are capable of describing the joint behavior of any number of such microsystems.

The interaction of the single-photon wave function with the array is a quantum mechanical process. Each extended photon wave packet, as it reaches the array, is a superposition of states, each of which corresponds to arrival at one of the photomultipliers. It follows from the linearity of the quantum equation of motion (Schrödinger's equation) that the final state of the array must be a coherent superposition, or linear combination, of states in each of which just one of the photomultipliers registers the photon. But in the laboratory, only one photomultiplier is observed to register the photon.

In *Mathematical Foundations of Quantum Mechanics* (1932), John von Neumann presents a careful, systematic discussion of the measurement process in which both the observed system and the apparatus are treated quantum mechanically. Von Neumann considered the special case where the interaction between the microsystem being measured and the apparatus does not disturb the initial state of the microsystem when the microsystem is an eigen state of the observable being measured, later called a measurement of the first kind. The apparatus's response depends on the eigenvalue of this observable. So, if the microsystem is originally in a linear combination of eigen states, then it follows from the linearity of the time-evolution operator in quantum theory that the composite system of the microsystem and the apparatus evolves into the same linear combination of states, each of which is a product of one of the eigen states and the corresponding state of the apparatus. Such a superposed state, which cannot be factored as a simple product of system state and apparatus state, is called an entangled state. In practice, each of the apparatus states that are entangled with the eigen states of the system has a "pointer" in some definite state of "pointing". In the above example, the "pointer states" are the states of the array in which one of the photomultiplier tubes has detected the photon, which is the microsystem being observed.

Von Neumann, and many after him, felt that what is observed cannot correspond to a coherent superposition of distinct pointer states. He was thus led to postulate two kinds of motion in quantum mechanics: the continuous, unitary deterministic dynamics predicted by Schrödinger's equation, which in the measurement process results in the superposition of pointer states; and the discontinuous, stochastic process that occurs only at the end of a measurement process where the state of the combined microsystem and apparatus suddenly and spontaneously reduces to one of the elements of the superposition, and so to a definite, observed, measurement outcome. This second type of motion is called the collapse of the wave function or the reduction of the wave packet, and its assumed occurrence is called the projection postulate. Which outcome actually occurs can only be predicted probabilistically. This is the only point where probability enters quantum theory.

According to the Copenhagen interpretation of quantum theory, the wave function of a quantum system provides a catalog of probabilities for the possible outcomes when measurements are performed on the system. This view has been aptly expressed in the following statement of John A. Wheeler: "The quantum wave has the same relationship to a particle as a weather prediction does to snow."

The Measurement Problem

Quantum theory, as usually formulated, does not account for the occurrence of events in the apparatus when a measurement is performed on an individual quantum system. Similarly, if the occurrence of such events is accepted as a primitive concept (as in the Copenhagen interpretation or the Feynman path integral formulation of quantum mechanics), no criteria are given by quantum theory for the conditions under which such events occur, apart from the vague stipulation that the apparatus should be macroscopic. This is the measurement problem.

The ad hoc collapse postulate of von Neumann's to explain the outcome of a measurement has several problems. His postulate lies outside the domain of Schrödinger dynamics and apparently only occurs when the interaction between the microsystem and the apparatus constitutes a measurement, which is not clearly defined. But why are measurement interactions singled out, and how large does the apparatus have to be for the interaction to be regarded as a measurement? Also, what Schrödinger dynamics predicts as the final state of the microsystem and apparatus may be decomposed in countless ways into superpositions of elements of sets of other linearly independent states. How does nature know which set (called the preferred basis) is relevant when it comes to applying the collapse mechanism? It is also not clear whether the collapse process is consistent with relativistic quantum theory because a collapse that is instantaneous in one Lorentz frame may not be instantaneous in another.

One suggestion about how to avoid collapse was provided by Werner Heisenberg. His theory involved replacing the initial pure state of the macroapparatus, which is in practice unknowable, by a mixed state in the quantum analysis of the measurement process. But in his 1932 book, von Neumann had already succeeded in showing that this suggestion does not work for measurements of the first kind. This was the first of a series of what were later called insolubility proofs. These proofs, worked out by Eugene Wigner, Abner Shimony, Arthur Fine, and others, established for increasingly general types of measurement interaction—including those considered disturbing or second kind—the failure of Heisenberg's early suggestion.

Another early idea for explaining the preferred basis, which reemerged in various forms later, is associated with the process of de-coherence. As a consequence of the large number of degrees of freedom associated with the microsystem and apparatus, or their coupling to the rest of the universe, the final superposed state is highly entangled with the states of the environment. So, the state of the microsystem and apparatus—or rather of the ensemble of similarly prepared joint systems—can for all practical purposes be replaced by an appropriate mixture of states corresponding to definite pointer positions. Some researchers have tried to make de-coherence do the work of the collapse process without requiring collapse as a separate mechanism. But reflection on the case of each system shows that an extra stochastic process is still involved when the system decides in which element of the mixture it will end up.

Attempts to Solve the Measurement Problem

The measurement problem has generated an enormous literature since the work of von Neumann, and the plethora of attempts to solve it is far greater than can be summarized here. The majority, however, may be characterized as promoting either (1) a change in the orthodox interpretation of quantum theory or (2) a change in the fundamental dynamics. What follows is a brief account of representatives of both approaches.

Promotion of a change in the orthodox interpretation of quantum theory. Von Neumann's analysis presupposed that the representation of the final state (wave function) of the microsystem and apparatus by a coherent superposition of distinct pointer states is incompatible with experience or with definite measurement outcomes occurring in the world. This assumption has been questioned on two grounds.

First, it presupposes that quantum mechanics is a complete theory in the sense of excluding any hidden variables associated with the microsystem and the apparatus that would single out a definite measurement outcome even when the wave function does not. Contrary to von Neumann's expectations, a fully consistent hidden-variables interpretation of quantum mechanics was provided in 1952 by David Bohm, following earlier work of Louis de Broglie. This strictly deterministic but nonlocal theory accounts for definite events occurring in measurement processes as due to particles having well-defined spacetime trajectories that are guided by the quantum wave. Consequently, it does not require a collapse mechanism for the wave function in order to account for these events. A purported solution to the measurement problem based on a stochastic hidden-variables theory, known as the modal interpretation of quantum mechanics, has also been articulated by a number of other researchers.

Second, in a seminal paper published in 1956, Hugh Everett also postulated that the collapse does not occur. The wave function of the entire universe under-goes a linear, unitary evolution as determined by Schrödinger's equation. The price of this evolution is acceptance of the simultaneous reality of the many worlds into which the familiar world splits, owing to the linearity of the evolution equation. In the example of detecting the photon, the world splits into a large number of worlds, in each of which a different photomultiplier detects the photon. The experimenter who observes the triggering of a photomultiplier is in one of these many worlds, in each of which there are duplicates of this experimenter observing different photomultipliers trigger.

Promotion of a change in the fundamental dynamics. Another approach to solving the measurement problem, without appeal to hidden variables or many worlds, involves clearly and systematically providing the details of the putative collapse mechanism. This means replacing the time-dependent Schrödinger equation by a new, universal, nonlinear dynamical equation that contains an explicit stochastic element. No primitive notion of measurement should enter into such an approach, and yet the usual, successful Schrödinger dynamics should reemerge to a good approximation in the treatment of the microworld. In the latter half of the twentieth century, considerable work has been done along these lines, resulting in a series of models of localization of the wave function, which occurs spontaneously or due to gravity.

A very recent development in quantum measurement theory is the "protective observation" of a wave function of a single system. This development involves "protective measurements", allowed by the laws of quantum mechanics, in which the wave function neither becomes entangled with the apparatus nor collapses. The entire process is deterministic. This absence of a stochastic element makes it unnecessary to use an ensemble of similarly prepared systems to observe the wave function as in the usual measurement. A protective measurement may be carried out by means of a suitable interaction that the system undergoes (the protection), compared to which the measurement interaction is weak and adiabatic. By monitoring the response of the apparatus, the expectation value of the observable in the given state of the system may be obtained. By doing such measurements for a sufficient number of observables, the wave function may be reconstructed.

Protective measurement does not fulfill the first goal of measurement, which is to obtain previously unknown information about the state of the system that is represented as a ray in Hilbert space. However, it shows the manifestation of a spatially extended wave function for a single particle, which has not been achieved before. It therefore suggests that the quantum wave is real and thereby questions the purely epistemological meaning given to the wave function by the Copenhagen interpretation. But because protective measurement circumvents the collapse by avoiding the entanglement that is a prerequisite for collapse, its implications for solving the measurement problem are unclear.

See also: ACTION-AT-A-DISTANCE; LOCALITY; SCHRÖDINGER EQUATION; WAVE FUNCTION; WAVE PACKET

Bibliography

AHARONOV, Y.; ANANDAN, J.; and VAIDMAN. L. "Meaning of the Wave Function." *Phys. Rev. A* **47** (6), 4616–4626 (1993).

BUSCH, P.; LAHTI, P. K.; and MITTELSTAEDT P., eds. *The Quantum Theory of Measurement* (Springer-Verlag, Berlin, 1991).

SQUIRES, E. J. *Conscious Mind in the Physical World* (Adam Hilger, Bristol, Eng., 1990).

VON NEUMANN, J. *Mathematical Foundations of Quantum Mechanics,* trans. by R. T. Beyer (Princeton University Press, Princeton, NJ, 1955).

WHEELER, J. A., and ZURCK, W. H., eds. *Quantum Theory and Measurement* (Princeton University Press, Princeton, NJ, 1983).

WIGNER, E. "The Problem of Measurement." *Am. J. Phys.* **31**, 6–15 (1963).

JEEVA ANANDAN

H. R. BROWN

QUARK

Quarks are a class of fundamental particles that are subject to the so-called strong force. The proton and neutron are examples of subatomic particles consisting of quarks. Physicists say quarks have six flavors. This simply means there are six types of quarks. For each type of quark, there is a corresponding antiparticle type.

Approximately 200 types of particles (mesons and baryons) and the patterns of their decays are explained by the six flavors of quarks and their interactions. Since all of the more massive flavors of quarks are unstable and decay via weak interactions, all of the particles that contain them are unstable except for the proton, which is the lightest baryon.

Quarks have electric charge of either $\frac{2}{3}$ (the up, charm, and top flavors) or $-\frac{1}{3}$ (the down, strange, and bottom flavors), compared to the -1 charge of the electron and the $+1$ charge of the proton. The somewhat whimsical names of the various quark types serve to distinguish them from one another but have no deeper meaning.

A peculiar feature of the quark model is that, according to current understanding, quarks cannot be isolated. Quarks are only found confined inside particles called hadrons.

However, their presence inside hadrons has been clearly observed. In the late 1960s, high-energy electron scattering of hydrogen and deuterium was studied by Jerome I. Friedman, Henry W. Kendall, and Richard E. Taylor. In much the same way as the alpha-scattering data of Hans Geiger and Ernest Marsden gave evidence for the nucleus within the atom, the results of these experiments could only be explained by the existence of quarks within the protons and neutrons. Friedman, Kendall, and Taylor received the 1990 Nobel Prize for this important discovery. There are now many other experimental results showing the existence of quarks.

Quarks were first introduced in 1964 by Murray Gell-Mann and George Zweig. They argued that all mesons and baryons are composites of three species of quarks and antiquarks, called *u, d, s* (up, down, strange), of electric charges $(\frac{2}{3}, -\frac{1}{3}, -\frac{1}{3})$. These charges had never been observed, so the introduction of quarks was treated more as a mathematical explanation of flavor patterns of particle masses than as a postulate of actual physical objects. Later theoretical and experimental developments allow us to now regard the quarks as real physical objects, even though they cannot be isolated. Gell-Mann won the Nobel Prize for his contributions to the understanding of strangeness, quarks, and their interactions.

Stimulated by the repeated pattern of leptons (the electron and muon and their corresponding neutrinos), several papers suggested a fourth quark carrying another flavor to give a similar repeated pattern for the quarks. Sheldon L. Glashow and James Bjorken coined the term "charm" (*c*) for the fourth quark. Very few physicists took this suggestion seriously at the time. In fact, the quark model was accepted rather slowly.

The critical importance of a fourth type of quark in the context of the standard model was recognized by Glashow, John Iliopoulos, and Luciano Maiani in 1970. A fourth quark allows a theory that has flavor-conserving Z^0-mediated weak interactions but not flavor-changing ones. The discovery of particles containing the fourth quark type, the $J/\psi = c\bar{c}$ in 1974 and the *D* mesons ($c\bar{d}, c\bar{u}$) and their antiparticles in 1975, was a major step in the establishment of what is now called the standard model of particle physics.

The fifth and sixth quark types form a third repetition of the pattern established by the less massive quark types. They were predicted to exist when a third lepton type (the tau) was discovered in the same experiments that found the J/ψ and the *D*'s.

However, the masses of these particles were not predicted, and physicists were surprised to find that the top quark is very much more massive than any of the others.

The pattern of quark masses and the pattern of their weak decays are among the outstanding puzzles of particle physics. Particles containing the bottom quark were discovered in 1977. Direct evidence for the top quark was not found until 1994–1995. Because it is so massive, the top quark decays so rapidly after it is produced that there is never time for any hadrons that contain it to be formed.

In 1973 a quantum field theory (now part of the standard model) of quarks and gluons and the strong forces was formulated by Harald Fritzsch and Gell-Mann. The standard model theory of these strong forces implies that each quark carries one of three types of strong charge, also called color charge. An antiquark has one of the three complementary color charges (anti-colors). Thus, there are in fact eighteen quark types, three colors for each of the six flavors. However, all quarks of a given flavor have identical masses, so we do not distinguish the different colors of quark by different names.

When a quark emits or absorbs a gluon, its color can be changed. Quarks emit and absorb gluons very frequently within a hadron, therefore there is no way to observe the color of an individual quark—that is why we generally do not even bother to denote the color charges.

"Color" has *nothing* to do with the colors of visible light; it is simply a cute name used because there is a crude analogy to real color in that three different colors can combine to be color neutral. No observed particles made of quarks have a color charge. Instead, all of them are color neutral. There are two ways to make color-neutral combinations of quarks: (1) Opposite colors (i.e., colors and anti-colors) attract. When each color and anti-color is present with equal probability in the quantum state, it is color-neutral. The observed mesons are all color neutral $q\bar{q}$ combinations. (2) Three different colors or three different anti-colors will attract and form a bound system that is color neutral. This yields the baryons and antibaryons made out of qqq or $\bar{q}\bar{q}\bar{q}$ combinations, where each quark (q) or antiquark ($\bar{q}$) has a different color.

See also: COLOR CHARGE; ELEMENTARY PARTICLES; FLAVOR; INTERACTION, STRONG; QUARK, BOTTOM; QUARK, CHARM; QUARK, DOWN; QUARK, STRANGE; QUARK, TOP; QUARK, UP

Bibliography

FRITZSCH, H. *Quarks: The Stuff of Matter* (Basic Books, New York, 1983).

GELL-MANN, M. *The Quark and the Jaguar: Adventures in the Simple and the Complex* (W. H. Freeman, New York, 1994).

GLASHOW, S. L. *The Charm of Physics* (Simon & Schuster, New York, 1991).

LEDERMAN, L. M., and SCHRAMM, D. S. *From Quarks to the Cosmos: Tools of Discovery* (Scientific American Library, New York, 1989).

RIORDAN, M. *The Hunting of the Quark* (Simon & Schuster, New York, 1987).

R. MICHAEL BARNETT

QUARK, BOTTOM

Evidence for the existence of the b quark was first seen in 1977 in data taken at Fermi National Accelerator Laboratory by Leon Lederman and collaborators. They looked at the invariant mass ($m^2 = E^2 - p^2$, where E and p are the energy and momenta of the particles they saw in their detector) of $\mu^+\mu^-$ pairs produced when high energy protons bombarded a nuclear target. Expectation was that the number of μ pairs would decrease with increasing μ pair mass. However, right around masses of 10 GeV/c^2, there was an unexpected excess. This indicated the observation of a new particle.

Initial speculation that this enhancement was evidence for a new short-lived particle made from a new quark (the b quark) bound by the strong interaction to its antiparticle (a $\bar{b}$ quark) was confirmed by data taken at accelerators in Hamburg, Germany and at Cornell University in Ithaca, New York, where several of these $b\bar{b}$ bound states (called Υ resonances) were observed. Also observed were the relatively long-lived B mesons which are formed from b quarks bound to light $\bar{u}$ or $\bar{d}$ antiquarks. The Υ resonances ($b\bar{b}$ bound states) have provided an interesting laboratory for the study of the strong interactions that bind the quarks, and the B mesons have been helpful in developing an understanding of the weak interactions that control their decay.

Quarks are thought to be the most fundamental and indivisible building blocks of nature. The matter we see around us in everyday life is made of up

(u) and down (d) quarks. At high energy accelerators, heavier quarks called strange (s), charm (c), beauty or bottom (b), and truth or top (t) are produced. The quarks are arranged in pairs

$$\begin{pmatrix} u \\ d \end{pmatrix} \begin{pmatrix} c \\ s \end{pmatrix} \begin{pmatrix} t \\ b \end{pmatrix}$$

of increasing mass, where the upper member of the pair has electric charge $+\frac{2}{3}e$ and the lower member has electric charge $-\frac{1}{3}e$ (e is the magnitude of the charge of the electron: $e = 1.6 \times 10^{-19}$ Coulombs). The b or bottom quark is so named because it has charge $-\frac{1}{3}e$, putting it as the lower member of a pair. It is occasionally referred to as the beauty quark in a tribute to its exotic nature.

The b quark was not discovered until 1977 because its large mass of about 5 GeV/c^2 meant that it could only be produced at the new high energy accelerators built during the 1970s. It is quite puzzling to physicists why a fundamental particle like the b quark has such a large mass; the explanation for the quark masses is one of the great unsolved mysteries of modern particle physics. Because of its large mass, the b quark can decay in a variety of interesting ways. Physicists have studied the decays of the b quark ever since it was first discovered.

In order to study b quarks, physicists study the hadrons (composite particles) that contain a b quark. Single isolated quarks do not exist in nature (another unsolved mystery of modern particle physics). Rather, the particles that we see around us or see evidence for at high energy accelerators are composite objects made of either quark–antiquark pairs (mesons) or combinations of three quarks or three antiquarks (baryons). Some of the hadrons that contain b quarks are listed in Table 1. The lowest mass b hadrons, B^0 and B^-, have been produced

Table 1 b Hadrons

	Quark Content	Status
B^0	$\bar{b}d$	Observed
B^-	$b\bar{u}$	Observed
B_s	$\bar{b}s$	Observed
B_c	$\bar{b}c$	No evidence
Υ	$\bar{b}b$	Observed
Λ_b	udb	Unconfirmed

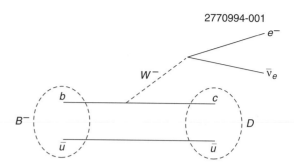

2770994-001

Figure 1 A schematic sketch of how the decay $B \rightarrow Dl\bar{\nu}$ proceeds at the quark level.

in large numbers and studied in many experiments since the early 1980s. The B_s was discovered in 1992, and there is now evidence that the lowest mass b baryon, the Λ_b, has been seen.

In order for a B meson to decay, the b quark must decay into a lower mass quark such as c or u. This means that the b quark must decay via the weak interaction, which is the only interaction that can change quark flavor. Figure 1 illustrates a schematic picture of a B^- meson decaying. The b quark decays into a c quark by emitting a W boson which is the carrier of the weak force. The final state is a D meson made from the c quark and the initial u quark and the decay products of the W boson. In Fig. 1 the W boson is shown decaying into an electron and antineutrino, giving the decay $B \rightarrow De^-\bar{\nu}$. Decays where the b quark turns into a c quark account for about 99 percent of all B meson decays. About 1 percent of the time, the b quark decays into a light u quark.

Because the B meson has such a large mass, there are many allowed decay modes. Only about one quarter of the possible final states in B decay have been observed. The rarest decays are among the most interesting because they test our understanding of the b quark and the weak interaction with the highest sensitivity.

B mesons are currently studied at the Cornell Electron Storage Ring in Ithaca, New York, the Large Electron Positron (LEP) accelerator in Geneva, Switzerland, and the Fermi National Accelerator Laboratory in Chicago, Illinois. Cornell currently has the largest data sample containing about 6 million B mesons. Two new facilities are being built in Stanford, California, and Tsukuba, Japan, which hope to produce 60 million B mesons a year to achieve even greater precision in the study of B mesons.

See also: ACCELERATOR; BOSON, W; HADRON; QUARK; QUARK, CHARM; QUARK, DOWN; QUARK, STRANGE; QUARK, TOP; QUARK, UP

Bibliography

HERB, S. W., et al. "Observation of a Dimuon Resonance at 9.5 GeV in 400-GeV Proton-Nucleus Collisions." *Phys. Rev. Lett.* **39,** 252 (1977).

INNES, W., et al. "Observation of Structure in the Upsilon Region." *Phys. Rev. Lett.* **39,** 1240 (1977).

MISTRY, N. B.; POLING, R. A.; and THORNDIKE, E. H. "Particles with Naked Beauty." *Sci. Am.* **249,** 98 (1983).

MONTANET, L. "Review of Particle Properties." *Phys. Rev. D.* **50,** 1790 (1994).

PERSIS S. DRELL

QUARK, CHARM

The fourth of the six known quark flavors, both in mass and order of discovery, the charmed quark had a particularly dramatic discovery.

Originally conjectured as an extension of the original three quarks, motivated initially by symmetry considerations and later postulated to account for the absence of flavor-changing weak neutral currents, particles containing the charmed quark were discovered in two separate experiments in 1974.

The charmed quark has an electric charge of $+\frac{2}{3}e$, where e is the electron charge, and a mass about 1.5 times that of the proton. The existence of a heavy fourth quark had been postulated to make the repeating pattern of quarks match that known for leptons and to account for the absence of certain flavor-changing weak decays. A fourth, heavier, quark would result in additional higher order decay amplitudes, thereby causing cancellations that accounted for the absence of flavor-changing neutral currents.

The first direct evidence for the charmed quark was the discovery of the J/ψ meson, which is composed of a $c\bar{c}$ quark pair, in November 1974. The J/ψ possesses hidden charm, that is, it is composed of a charmed quark and an anti-charmed quark, and therefore has no net charm quantum number. The unexpected discovery of the J/ψ represented the first window into the world of heavy quarks beyond the three original flavors of the original quark model of Gell–Mann and Zweig. The J/ψ meson was by far the heaviest elementary particle seen up to that time, being more than three times as massive as the proton, and having an extraordinarily narrow decay width, indicating a very long lifetime for the state. Both the mass and the lifetime had been predicted as clear indicators of the existence of new heavy quark flavors. There are a series of mesons with hidden charm, forming a family called charmonium, which form what is termed "the hydrogen atom of quantum chromodynamics" in that they are composed of pairs of charmed quarks and anticharmed quarks bound together by the exchange of gluons in a calculable nonrelativistic system, similar to the binding of proton and electron by photons in a hydrogen atom.

There are two types of elementary particles that have explicit charm quantum numbers: charmed mesons and charmed baryons. Mesons containing a single charmed quark are termed charmed mesons. There are three charmed mesons that decay weakly, the D^+, the D^0, and the D_s^+, composed of a c quark and a d, a u, and an s antiquark, respectively, in a state of zero orbital angular momentum, with their spins anti-aligned. There are also families of excited charmed mesons, D^* and D^{**}, in which the quarks are in higher angular momentum states. These decay via electromagnetic or strong interactions.

There are also charmed quark-containing baryons, having one, two, or three charmed quarks, the lightest being the Λ_c, composed of a udc combination, and the heaviest, the Ω_c, composed of three charmed quarks. For each charmed meson or baryon there is also an antiparticle, found by replacing quarks by antiquarks and vice versa.

Charmed quarks are produced in e^+e^- annihilation or in high-energy hadronic collisions. They can be produced either directly or through the weak decay of heavier b quarks (bottom or beauty quarks). The lifetime of particles with explicit charm flavor is in the range 10^{-12} to 10^{-13} s. The decays are weak interactions. The predominant weak decay of a charm quark produces a strange quark (or anticharm or antistrange) plus virtual W, which subsequently produces either a higher quark plus antiquark or a lepton and antilepton combination.

See also: FLAVOR; LEPTON; QUARK, BOTTOM; QUARK, DOWN; QUARK, STRANGE; QUARK, TOP; QUARK, UP

Bibliography

Fritzsch, H. *Quarks: The Stuff of Matter* (Basic Books, New York, 1983).

Glashow, S. L. *The Charm of Physics* (Simon & Schuster, New York, 1991).

Riordan, M. *The Hunting of the Quark* (Simon & Schuster, New York, 1987).

David Hitlin

QUARK, DOWN

The down quark was one of the three flavors of quark postulated by Murray Gell-Mann in 1963 to explain the various hadrons that were being observed at high-energy accelerators. Direct experimental evidence of the existence of the up and down quarks was obtained in the late 1960s and early 1970s by Richard Taylor and his collaborators in the deep inelastic electron scattering experiments at the Stanford Linear Accelerator Center (SLAC), and from the comparison of these results to similar experiments with initial neutrino beams

The down (d) quark has a mass of about 10 MeV/c^2, which makes it heavier than the up (u) quark. Therefore, the down quark can beta decay as follows:

$$d \rightarrow u + e^- + \nu_e$$

and still conserve baryon number and lepton number. This is what happens in the beta decay of a neutron ($n = udd$: $n \rightarrow p + e^- + \nu_e$. One of the down quarks in the neutron beta decays into an up quark and the neutron changes into a proton ($p = uud$).

The electric charge of the down quark is $Q = -\frac{1}{3}e$, where e is the proton charge ($e = 1.60217733(49) \times 10^{-19}$ C). Its spin is the same as the spin of all the quarks and the charged leptons, $\frac{1}{2}$ in units of $\hbar$ ($\hbar = 6.6260755(40) \times 10^{-34}$ J·s). Thus, three quarks make up the baryons of spin $\frac{1}{2}$ or $\frac{3}{2}$, while a quark-antiquark pair makes a meson of spin 0, 1, or 2. The baryon number of the down quark is $B = +\frac{1}{3}$, the same for all quarks. Because the proton is $p = uud$ and the neutron is $n = udd$, the proton is lighter than the neutron and, hence, is absolutely

stable. Also, since the up quark has electric charge $+\frac{2}{3}$ and the down quark has electric charge $-\frac{1}{3}$, it can be seen that the proton and neutron have electric charge $+1$ and 0, respectively, as they should.

The antiquark $\bar{d}$ also has spin $\frac{1}{2}$, but its other quantum numbers are opposite those of its partner d quark. Thus, $B = -\frac{1}{3}$ and $Q = -\frac{1}{3}e$. For example, the pion (π) is given by

$$\begin{cases} \pi^+ = u\bar{d} \\ \pi^0 = u\bar{u} \text{ and } d\bar{d} \\ \pi^- = d\bar{u}. \end{cases}$$

All of the observed mesons can be represented, with all of their quantum numbers correctly given, as quark-antiquark pairs. Also, all of the observed baryons are correctly described as three-quark states. Thus, all of the hadrons (strongly interacting particles or resonances made of quarks and antiquarks, which included both baryons and mesons) are correctly described by the quark model.

See also: Color Charge; Elementary Particles; Flavor; Interaction, Strong; Lepton; Quark; Quark, Bottom; Quark, Charm; Quark, Strange; Quark, Top; Quark, Up; Spin

Bibliography

Brown, L. M. "Quarkways to Particle Symmetry." *Phys. Today* **19** (Feb.), 44–54 (1966).

Chew, G. F.; Gell-Mann, M.; and Rosenfield, A. H. "Strongly Interacting Particles." *Sci. Am.* **210** (Feb.), 74–93 (1964).

Feynman, R. P. "Structure of the Proton." *Science* **183**, 601–610 (1974).

Glashow, S. "Quarks with Color and Flavor." *Sci. Am.* **232**, 50–62 (1975).

Kokkedee, J. J. J. *The Quark Model* (W. A. Benjamin, New York, 1969).

Mark A. Samuel

QUARK, STRANGE

The strange quark was one of the three flavors of quarks postulated by Murray Gell-Mann in 1963 to explain the various hadrons that were being observed at high-energy accelerators. The three quarks proposed, up (u), down (d), and strange (s), were sufficient to describe all of the known hadrons at the time, both with strangeness ($S \neq 0$) and without strangeness ($S = 0$)

The strange quark has a mass of about 200 MeV/c^2, carries the strangeness quantum number $S = -1$, and is a constituent of all of the strange hadrons (strongly interacting particles or resonances made of quarks and antiquarks, including both baryons and mesons).

The electric charge of the strange quark is $Q = -\frac{1}{3}e$, where e is the proton charge ($e = 1.60217733(49) \times 10^{-19}$ C). Its spin is the same as the spin of all the quarks and the charged leptons, $\frac{1}{2}$ in units of $\hbar$ ($\hbar = 6.260755(40) \times 10^{-34}$ J·s). Thus, three quarks make up the baryons of spin $\frac{1}{2}$ or $\frac{3}{2}$, while a quark-antiquark pair makes a meson of spin 0, 1, or 2. The baryon number of the strange quark is $B = +\frac{1}{3}$, the same for all quarks. The strangeness S of the hadrons is determined by taking the sum of the strangeness of the constituent quarks. Similarly, the baryon number of any hadron is the sum of the baryon number of its constituents, so mesons have baryon number 0.

The antiquark $\bar{s}$ also has spin $\frac{1}{2}$, but its other quantum numbers are opposite those of its partner s quark. Thus, $B = -\frac{1}{3}$, $Q = +\frac{1}{3}e$, and $S = +1$. For example, the K mesons are

$$\begin{cases} K^+ = \bar{s}u \\ K^0 = \bar{s}d \\ \overline{K}^0 = \bar{d}s \\ K^- = \bar{u}s. \end{cases}$$

All of the observed mesons can be represented, with all of their quantum numbers correctly given, as quark-antiquark pairs. Also, all of the observed baryons are correctly described as three-quark states. For example, the Σ baryon is given by

$$\begin{cases} \Sigma^+ = suu \\ \Sigma^0 = sdu \\ \Sigma^- = sdd, \end{cases}$$

with $S = -1$. The Ξ^* baryon is given by

$$\begin{cases} \Xi_0^* = ssu \\ \Xi_-^* = ssd, \end{cases}$$

with $S = -2$.

The Ω^-, which is given by $\Omega^- = sss$, has $S = -3$. The Ω^- was predicted in 1963 by Gell-Mann to complete the set of possible three-quark baryons, and it was discovered experimentally in 1964. It is the only particle known to have strangeness $S = -3$.

See also: COLOR CHARGE; ELEMENTARY PARTICLES; FLAVOR; INTERACTION, STRONG; QUARK; QUARK, BOTTOM; QUARK, CHARM; QUARK, DOWN; QUARK, TOP; QUARK, UP; SPIN

Bibliography

FEYNMAN, R. P. "Structure of the Proton." *Science* **183**, 601–610 (1974).

GLASHOW, S. "Quarks with Color and Flavor." *Sci. Am.* **232**, 50–62 (1975).

KOKKEDEE, J. J. J. *The Quark Model* (W. A. Benjamin, New York, 1969).

MARK A. SAMUEL

QUARK, TOP

The top quark, the heaviest and least stable of the six known quarks, was discovered in 1995 by two independent experiments at the proton-antiproton collider at the Fermi National Accelerator Laboratory outside of Chicago, Illinois. The top quark is a third-generation quark with electric charge $\frac{2}{3}e$ and spin $\frac{1}{2}$. It is the weak decay partner of the bottom quark and carries color charge. The first two generations include the up and down quarks and the charmed and strange quarks. All quarks experience gravitational, electromagnetic, weak, and strong forces.

Top quarks were produced in high-energy proton-antiproton collisions, with 1.8 TeV in the center of mass. The Fermilab collider is the accelerator with the highest center of mass-energy in the world. The top quark decays via the charged weak interaction to an intermediate vector boson and a bottom quark, $t \rightarrow Wb$. The top quark rarely exists outside the collider environment due to its very short lifetime of approximately $\sim 10^{-24}$ s. For example, it decays 10^{12} times faster than its partner, the bottom quark. The corresponding decay rate or width of the top quark is expected to be roughly 1 GeV. Top quarks probably existed during the very early universe. They can only be created in very high-energy particle collisions due to their very large mass. The mass of the top quark has been measured at 176 ±8 (statistical) [±10 (systematic)] GeV by one experiment, the collider detector at Fermilab collaboration of more than 450 physicists, and at 199 ±20 (statistical) [±22 (systematic)] GeV by a second complementary experiment (also at Fermilab) with a similar number of researchers. This mass is extremely large compared to the five lighter quarks, which are roughly $m_u = 5$ MeV, $m_d = 7$ MeV, $m_c = 1.5$ GeV, $m_s = 300$ MeV, and $m_b = 4.5$ GeV. The top quark's mass is comparable to the mass of an entire gold atom, or 188 proton masses. The various masses of the quarks are not understood. They are not predicted by the standard model of the strong, weak, and electromagnetic interactions. The top quark has such a short lifetime that it does not have time to form bound states with other quarks, like $t\bar{b}$, $t\bar{u}$ mesons, before it decays. The six quarks do not behave as free particles. They are confined to exist in bound states of quarks. The top quark is the only quark that decays before it has time to form bound states with other quarks.

Due to the symmetry between quarks and leptons in the standard model, it is thought that the top quark is the last piece in the quark puzzle. This symmetry implies that there are as many quark generations (three) as lepton generations. Experiments in Europe at the Large Electron Positron (LEP) collider and in the United States at the Stanford Linear Accelerator Center (SLAC) have shown that there are only three light neutrinos. Since the known neutrinos are very light, this suggests that there are only three lepton generations and, consequently, three quark generations. The first generation consists of the electron, the electron neutrino, the up quark, and the down quark. The second generation consists of the muon, the muon neutrino, the charm quark, and the strange quark. The third, and possibly final, generation consists of the tau, the tau neutrino, the top quark, and the bottom quark.

See also: LEPTON; LEPTON, TAU; NEUTRINO; QUARK; QUARK, BOTTOM; QUARK, CHARM; QUARK, DOWN; QUARK, STRANGE; QUARK, UP

Bibliography

GLANZ, J. "Collisions Hint That Quarks Might Not Be Indivisible." *Science* **271** (5250), 758 (1996).

MADDOX, J. "The Top Quark Found at Long Last." *Nature* **37** (6518), 113 (1995).

PETERSON, I. "Beyond the Top." *Science News* **148** (1), 10–12 (1995).

STONE, R. "Fermilab Officially Discovers Top Quark." *Science* **267** (5202), 1255 (1995).

MELISSA FRANKLIN

QUARK, UP

The up quark was postulated by Murray Gell-Mann in 1963 as one of the three quark types to explain all of the hadrons that had been observed at high-energy accelerators. Direct experimental evidence of the existence of the up quark as a constituent of the proton and neutron was obtained in the late 1960s and early 1970s in an experiment led by Henry Kendall, Jerome Friedman, and Richard Taylor at the Stanford Linear Accelerator Center (SLAC), in which the deflections of high-energy electrons scattered by a gas target were observed (much as the deflections of alpha particles showed evidence of the nucleus). These data from electron scattering were later supplemented with new data from high-energy neutrino scattering that confirmed the quark model. A comparison of the rates of neutrino-initiated and electron-initiated processes matched the predicted result, which depended explicitly on the $\frac{2}{3}$ and $-\frac{1}{3}$ charges for up and down quarks, respectively. The quark model has now been incorporated into particle-physics theory as the basis of the standard model, which describes all known particles in terms of six quark types and their antiparticles.

The up quark, which has a mass of about 5 MeV/c^2, is the lightest of the six quarks. As the lightest quark, the up quark is absolutely stable in the standard model, since there are no lighter particles into which it can decay without violating baryon-number conservation, which states that the number of baryons (or quarks) minus the number of antibaryons (or antiquarks) is unchanged in all processes. (In modern language, this is called quark-number conservation.) Up quarks (*u*), joined in three-quark bound states with down (d) quarks, are basic to the formation of the proton (uud) and neutron (udd).

The electric charge of the up quark is $Q = +\frac{2}{3}e$, where e is the proton charge ($e = 1.60217733(49) \times 10^{-19}$ C). Its spin is the same as the spin of all the quarks and the charged leptons, namely, $\frac{1}{2}$ in units of $\hbar$ ($\hbar = 6.6260755(40) \times 10^{-34}$ J·s). Three quarks make up the baryons of spin $\frac{1}{2}$ or $\frac{3}{2}$, while a quark-antiquark pair makes a meson of spin 0, 1, or 2. The baryon number of the up quark is $B = +\frac{1}{3}$, the same for all quarks. For example, because the proton is $p = (uud)$ and the neutron is $n = (udd)$, the proton is lighter than the neutron and, hence, in the standard model, is absolutely stable. Also, since the up quark has electric charge $+\frac{2}{3}$ and the down quark has electric charge $-\frac{1}{3}$, the proton and neutron have electric charge $+1$ and 0, respectively.

The antiquark $\overline{u}$ also has spin $\frac{1}{2}$, but its other quantum numbers are opposite those of its partner u quark. Thus, $B = -\frac{1}{3}$, and $Q = -\frac{2}{3}e$. For example, the pion (π) is given by

$$\pi^+ = u\overline{d}$$
$$\pi^0 = u\overline{u} \text{ and } d\overline{d}$$
$$\pi^- = d\overline{u}$$

These pions are unstable and decay as follows:

$$\pi^+ \rightarrow \mu^+ + \nu_\mu$$
(lifetime $T = 2.6030 \pm 0.0024 \times 10^{-8}$ s)

$$\pi^0 \rightarrow \gamma + \gamma$$
(lifetime $T = 8.4 \pm 0.6 \times 10^{-17}$ s)

$$\pi^- \rightarrow \mu^- + \overline{\nu}_\mu$$
(lifetime T $= 2.6030 \pm 0.0024 \times 10^{-8}$ s).

In the case of the π^0, the particle-antiparticle pair $u\overline{u}$ and $d\overline{d}$ annihilate into two photons.

All of the observed mesons can be represented, with their quantum numbers correctly given, as quark-antiquark pairs. Also, all of the observed baryons are correctly described as three-quark states. Thus, all of the hadrons—strongly interacting particles, a class that includes both baryons and mesons—are correctly described by the quark model.

See also: BARYON NUMBER; HADRON; LEPTON; QUARK; QUARK, BOTTOM; QUARK, CHARM; QUARK, DOWN; QUARK, STRANGE; QUARK, TOP

Bibliography

DRELL, S. D. "e^+-e^- Annihilation and the New Particles." *Sci. Am.* **232** (June), 50–62 (1975).

FEYNMAN, R. P. "Structure of the Proton." *Science* **183,** 601–610 (1974).

KENDALL, H. W.; and PANOFSKY, K. H. "The Structure of the Proton and the Neutron." *Sci. Am.* **224** (June), 60–77 (1971).

KOKKEDEE, J. J. J. *The Quark Model* (W. A. Benjamin, New York, 1969).

MARK A. SAMUEL

QUARK CONFINEMENT

According to the quantum chromodynamics (QCD) theory of the strong force, hadrons (baryons and mesons) are composed of quarks. Baryons consist of three quarks and antibaryons of three antiquarks. Mesons are quark-antiquark pairs. The attractive binding force between these quarks is mediated by the exchange of gluons. The theory follows the same pattern as quantum electrodynamics (QED). In this case the forces between charged particles are mediated by the exchange of photons. However, in one respect, electromagnetic and chromodynamics phenomena behave very differently.

Electrons and nuclei bind into atoms. When an atom is sufficiently disturbed, possibly by absorbing a photon or being bumped by another atom, it becomes ionized and constituent electrons are ejected. By contrast, no matter how hard a hadron is struck by another particle, the results of the collisions are always complete hadrons. No single free quarks are ever liberated by the collision. In other words, quarks are permanently confined to the interior of the hadrons that they constitute.

It is not hard to imagine an explanation for this confinement phenomenon. Imagine that the strength of the coulomb forces between electrons and the nucleus were many times larger than it is. Atoms would be far more difficult to ionize. The energy necessary to liberate an electron could be made arbitrarily large. As more and more experiments would fail to discover free electrons, it would be necessary to increase the strength of the coulomb force between nucleus and electron. By analogy, quark confinement can be built into QCD by making the coupling between quarks arbitrarily strong. This can be done by increasing the coupling constant that governs the strength with which a quark will emit gluons when disturbed.

In the imaginary case of the strongly coupled electromagnetic force, the atomic electrons would not only be confined but also get pulled into the nucleus, giving rise to a much smaller atom. Furthermore, it would take huge energies even to excite an atom out of its ground state.

Hadrons, however, while smaller than atoms, have observable finite size. Furthermore, the energy needed to excite a hadron to higher orbital states is not very large. In other words, hadrons behave as though they were composed of relatively weakly interacting quarks, except that free isolated quarks are never liberated.

For quark confinement to coexist with the observed facts about hadrons, a behavior in which the strength of the QCD interaction is variable and depends on the scale of distances between quarks is needed. When quarks are close together, as they are in the interior of a hadron, they interact weakly. When the distances between quarks grow to scales larger than the size of hadrons, they should begin to interact with increasingly strong force.

This interaction can be described in terms of an effective Coulomb force between quarks. The potential energy between a pair of equal but opposite electric charges is given by the coulomb potential:

$$V(r) = -\frac{e^2}{r}. \tag{1}$$

The idea of an effective distance-dependent strength or interaction is equivalent to saying that the charge e of each constituent varies with separation. Thus, e becomes dependent of r:

$$V(r) = -\frac{e^2(r)}{r}. \tag{2}$$

If $e(r)$ is small when r is of order the size of hadrons ($r = 10^{-13}$ cm) or smaller, then hadrons will behave as a collection of weakly interacting quarks. If, however, $e(r)$ increases as r grows, then it requires progressively more energy to eject quarks to great distances. If $e(r)$ grows without bound, then a quark that has been ejected must eventually return to its parent hadron.

The idea of a variable charge for a particle is not without precedent. Electrical charges in electrically polarizable materials, such as H_2O, collect polarization charge around them. As a result, the net charge contained within a radius r varies, and the effective coulomb force is modified, as in Eq. (2). However, the effect is opposite to that desired. At short distances, the unscreened charge is large and becomes smaller as r increases.

Fortunately, QCD is different in its mathematical structure in just the right way. Calculation reveals that as r increases, the effective charge increases so that the forces inevitably grow stronger as quarks separate.

The effect that leads to quark confinement can be understood in an especially graphic way by using the field concept. The electric field surrounding an oppositely charged pair can be represented by field lines. The new effect occurs in QCD, which squeezes the field lines into a narrow bundle or flux tube. The result is a kind of string that connects a quark to its neutralizing partner and keeps it from separating independently.

The theory of quark confinement by electric-flux tubes has interesting implications for the production of particles in high-energy collisions. For example, when a very high-energy electron-positron pair collides and annihilates one another, most often the original energy materializes as a high-energy quark-antiquark pair receding from one another with almost the velocity of light. As they fly apart, a tube of chromoelectric flux materializes between them. Subsequently, one of two things can happen. (1) The pair can simply run out of energy and get pulled back, or, much more likely at high energies, a new quark-antiquark pair is produced in the growing flux tube, thus breaking the tube. (2) The process repeats until all the energy has been converted into outgoing mesons, forming a pair of jets of particles. Such jets are detected in many kinds of particle collisions.

See also: COLOR CHARGE; FIELD LINES; FLAVOR; HADRON; INTERACTION, WEAK; LEPTON; QUANTUM CHROMODYNAMICS; QUANTUM FIELD THEORY; QUARK

Bibliography

FRITZSCH, H. *Quarks: The Stuff of Matter* (Basic Books, New York, 1983).

RIORDAN, M. *The Hunting of the Quark* (Simon & Schuster, New York, 1987).

LEONARD SUSSKIND

QUARKS, DISCOVERY OF

Physicists today regard all matter as composed of fundamental particles called quarks and leptons. Quarks experience the strong force that binds protons and neutrons together inside atomic nuclei, while leptons—which include the familiar electron—do not. Six different kinds of quarks have been identified so far, and there is good evidence that these are the only ones that actually exist.

The quark model emerged in the mid-1960s as a way to explain the great variety of recently discovered subatomic particles. In 1964 Murray Gell-Mann and George Zweig independently proposed that all the known strongly interacting particles could be formed from a set of three fundamental entities (which Gell-Mann dubbed quarks after a line in James Joyce's novel *Finnegans Wake*)—the up (u) quark, the down (d) quark, and the strange (**s**) quark—plus their antiparticles, the antiquarks. Protons and neutrons are composed of three quarks; for example, the proton is a *uud* combination, while the neutron is *udd*. Other strongly interacting particles known as mesons can all be built from a quark plus an antiquark. These quarks had to have electrical charges equal to $\frac{1}{3}$ or $\frac{2}{3}$ that of an electron or proton. Carrying such odd fractional charges, quarks should have been easy to isolate; many experiments in the 1960s searched for quarks, but all failed to turn up any convincing evidence for their existence.

In 1968 a series of experiments at the Stanford Linear Accelerator Center (SLAC) in Menlo Park, California, began to reveal indirect evidence for these quarks. Performed by physicists from SLAC and the Massachusetts Institute of Technology under the general leadership of Jerome Friedman, Henry Kendall, and Richard Taylor, these experiments involved firing intense beams of high-energy electrons into liquid hydrogen targets and observing how the electrons rebounded from the hydrogen nuclei, which consist of only protons. The initial results—that electrons were ricocheting at large angles instead of just plowing straight through— suggested they were striking tiny objects inside the proton.

Could these objects be the quarks? Theoretical work by James Bjorken, Richard Feynman, and others suggested that such might indeed be the case. Further measurements, performed at SLAC from 1968 to 1973 over a wide range of electron energies and scattering angles using both proton and neu-

tron targets, yielded results that were generally consistent with the quark model. Similar experiments at the European Center for Particle Physics (CERN) in Geneva, Switzerland, using high-energy beams of uncharged leptons, known as neutrinos, instead of electrons, confirmed the SLAC results and showed that the constituents of the proton and neutron had just the fractional charges needed for quarks.

But there were a few problems that had to be resolved before the physics community could be completely convinced of the existence of quarks. One was the fact that, although they seemed to be rattling around loosely inside protons and neutrons, individual quarks had never appeared outside in their free state. Another was the requirement that quarks should come not as Gell-Mann and Zweig's original triplet but in pairs, a feature needed by then-emerging unified theories of the fundamental forces of nature.

The first quandary was eventually resolved by the theory of quantum chromodynamics, which stipulates that the force binding two quarks together becomes stronger as the distance between them increases—effectively trapping them in protons, neutrons, and mesons. The other problem was solved by the simultaneous discovery of the J and psi particles in 1974 by groups of physicists led by Samuel Ting of MIT and Burton Richter of SLAC. These massive neutral mesons were found to be combinations of a fourth kind of quark, the charm (c) quark, with its antiquark. Thus, the up and down quarks make one pair, while the charm and strange quarks form a second pair of quarks, as required by theory.

In 1977 physicists at the Fermi National Accelerator Laboratory (Fermilab) near Chicago, Illinois, discovered a fifth quark, the bottom (b) quark. Its very existence meant that there had to be yet another quark, correspondingly named the top (t) quark, in order to complete a third pair of quarks. Searches for this quark were mounted at new particle accelerators in Europe, Japan, and the United States during the ensuing years. In 1995 two Fermilab experiments finally reported the discovery of an extremely heavy top quark with a mass about 190 times that of the proton.

Together with the six leptons, these six quarks provide a complete set of fundamental building blocks from which every kind of matter found so far in nature can be formed. All of them have an intrinsic angular momentum, or spin, equal to Planck's constant h divided by 4π. Although none of these quarks has yet been observed directly in its free state, the indirect evidence for their existence is extensive and compelling. Perhaps the most convincing evidence is the appearance of jets of subatomic particles emerging from particle collisions. Formed by high-energy quarks that materialize almost instantaneously into other, long-lived subatomic particles (which leave observable tracks in particle detectors), these jets are in essence the actual footprints of quarks.

The significance of these discoveries was acknowledged by the Royal Swedish Academy of Science in its award of the Nobel Prize in physics to Richter and Ting in 1976 and to Friedman, Kendall, and Taylor in 1990. What remains to be understood about quarks is why there are three (and only three) pairs of them and why their masses vary so widely—over five orders of magnitude.

See also: COLOR CHARGE; ELEMENTARY PARTICLES; FLAVOR; LEPTON; INTERACTION, STRONG; QUARK; QUARK, BOTTOM; QUARK, CHARM; QUARK, DOWN; QUARK, STRANGE; QUARK, TOP; QUARK, UP

Bibliography

BROWN, L.; HODDESON, L.; RIORDAN, M.; and DRESDEN, M., eds. *The Rise of the Standard Model* (Cambridge University Press, New York, 1996).

CREASE, R. P., and MANN, C. C. *The Second Creation: Makers of the Revolution in 20th-Century Physics* (Macmillan, New York, 1986).

PICKERING, A. *Constructing Quarks: A Sociological History of Particle Physics* (Edinburgh University Press, Edinburgh, 1984).

RIORDAN, M. *The Hunting of the Quark* (Simon & Schuster, New York, 1987).

RIORDAN, M. "The Discovery of Quarks." *Science* **256**, 1287–1293 (1992).

MICHAEL RIORDAN

QUASAR

Quasar is a contraction of the phrase "quasistellar object" and refers to the optical image of the hyperactive nucleus of a distant galaxy that is so bright that it outshines the surrounding stars and appears to be an unresolved point source. Quasars were first recognized in 1963 when an English astronomer, Cyril Hazard, working in Australia used the techniques of radio astronomy to pinpoint the position of a powerful radio source, known as 3C273, with unprecedented accuracy. This enabled a Dutch astronomer, Maarten Schmidt, working in the United States to locate its optical counterpart and understand its spectrum. What Schmidt found was that the spectral lines from 3C273 had wavelengths that were "redshifted" relative to their wavelengths measured in the laboratory, and they appeared to be moving away from Earth with a speed of sixteen percent of the speed of light. Using the Hubble law, which states that the distance of a galaxy is directly proportional to its recession velocity, he was able to deduce that 3C273 was very distant and consequently extremely luminous, more than 100 times more luminous than a typical galaxy, or 10^{39} W.

The total number of quasars has been estimated to be several million, roughly a tenth of a percent of the number of bright galaxies. Most of them are at a great distance from Earth so that the light that we see was emitted when the universe was much younger. The most distant quasar discovered to date has a redshift of $z = 4.9$. This means that its spectral lines have wavelengths that are stretched by a factor of $1 + z = 5.9$. The universe was only about 7 percent of its present age when the light from this quasar was emitted.

Astronomers have learned much more about the properties of quasars since 1963. Quasars have been observed throughout the electromagnetic spectrum from the longest radio waves through the infrared and x rays to the highest energy rays. Most of their power is emitted in the ultraviolet radiation and is believed to originate from the central core of the quasar. Some of this ultraviolet radiation is absorbed by surrounding gas and is re-emitted in the form of spectral lines. Lines emitted by hydrogen atoms and ions of helium, carbon, nitrogen, oxygen, magnesium, and silicon are especially prominent. As most quasars have large redshifts, many of the lines are associated with atomic and ionic transitions in the ultraviolet part of the spectrum but are observed in the visible range. The gas that is responsible for emitting these lines moves with speeds of order 10,000 km/s or 3 percent of the speed of light. This means that the photons that we observe are subject to significantly different Doppler shifts and, consequently, the spectral lines are very broad. This is one of the characteristics by which quasars can be recognized. Much of the infrared radiation is believed to be ultraviolet radiation, reradiated by more distant, dusty gas.

Quasars have also been observed to vary on time scales that can be as short as days, which implies that, in these cases at least, the power derives from a region no larger than the solar system. A minority of quasars, including 3C273, are powerful radio sources and called "radio loud." (The majority of quasars, however, are "radio quiet." Originally the word "quasar" referred to the radio-loud objects, though now it is usually taken to include radio-quiet objects as well.) Radio-loud quasars produce a pair of jets that squirt out of the nucleus and eventually out of the galaxy along antiparallel directions to power two giant radio lobes. The radio-emitting plasma in these jets can move with speeds very close to the speed of light, and when one of these jets is directed toward Earth, we can observe features that move with speeds that appear to be faster than the speed of light. However, this "superluminal" expansion is only an illusion. Some of these jets are also observed at higher frequencies, stretching in one case up to trillion-electron-volt energy x rays.

Quasars are the most spectacular examples of galaxies with active nuclei. The less luminous counterparts of the radio-quiet quasars are called "Seyfert galaxies" and of the radio-loud quasars, "radio galaxies." Many astronomers suspect that most bright galaxies pass briefly through the quasar phase, although it is also possible that, instead, a minority of galaxies remain in this state for most of their lives.

As well as being interesting in their own right, quasars are also valuable probes of material that lies along the line of sight. Most quasars exhibit absorption lines in their spectra, with redshifts smaller than those associated with the quasars themselves. Spectral lines associated with ionic transitions of elements such as carbon and magnesium that have been formed in stars are associated with gas inside galaxies. The large number of lines associated with just hydrogen are identified with primordial gas that has not yet condensed into proper galaxies and stars.

A quite separate effect involving about 1 in 500 quasars is the deflection of rays of light from distant quasars by the gravitational field of intervening galaxies. These gravitational lenses can be so strong that they produce multiple images of a single quasar source. Gravitational lenses can tell us much about the distant galaxies and the size and shape of the universe.

See also: ACTIVE GALACTIC NUCLEUS; ASTROPHYSICS; GALAXIES AND GALACTIC STRUCTURE; GRAVITATIONAL LENSING; REDSHIFT; STARS AND STELLAR STRUCTURE

Bibliography

BLANDFORD, R. D.; NETZER, H.; and WOLTJER, L. *Active Galactic Nuclei* (Springer-Verlag, Berlin, 1990).

FRANK, J.; KING, A.; and RAINE, D. *Accretion Power in Astrophysics*, 2nd ed. (Cambridge University Press, Cambridge, Eng., 1992).

ROGER D. BLANDFORD

R

RABI, ISIDOR ISAAC

b. Rymanov, Austria-Hungary, July 29, 1898; *d.*
New York, January 11, 1988; *atomic and molecular beams.*

Rabi was one of the physicists who helped move the frontier of physics from Europe to the United States during the decade of the 1930s. In his laboratory at Columbia University, Rabi and his illustrious students developed experimental methods to make measurements on individual atoms as they passed in a narrow beam through a long evacuated chamber. The Rabi group was particularly interested in the nucleus of the atom which, with the discovery of the neutron in 1932, became a focus of physical research. By means of magnetic fields located along the path of the beam, they exerted small forces on the nuclei of atoms as they passed through the magnetic fields. This enabled them to measure nuclear properties with unprecedented accuracy. In the process, Rabi and his students developed the magnetic resonance method that today is the basis for magnetic resonance imaging (MRI), a powerful diagnostic tool.

Rabi was born a few miles from the Russian border in far Eastern Europe in 1898. While he was still an infant, the Rabi family moved to the United States, where they settled in New York City. His family was very poor and the young Rabi grew up on the streets of the Lower East Side of the city. He was a good student in the public school system, but it was the discovery of the New York Public Library that set the course for his life. He read a book about astronomy, where he learned about the solar system with the planets revolving around the central Sun. This Sun-centered structure of the solar system was at odds with the religious teachings he learned at home, but for the nine-year-old Rabi, the simple beauty of the Sun with its family of orbiting planets was compelling. He started his education in chemistry, but soon discovered physics, which became his love.

World War II came when Rabi's research was at its most productive. The events in Nazi Germany, however, increasingly preoccupied his thoughts. When a source of microwaves, called the magnetron, was invented in England, it opened the way for the military development of radar. More than one year before the attack on Pearl Harbor, and more than two years before the atomic bomb project began in Los Alamos, the Radiation Laboratory at the Massachusetts Institute of Technology was organized, where physicists began to develop and perfect radar methods. Rabi became an associate director of the Radiation Laboratory in charge of research. Although Rabi was never on the payroll of the Los Alamos National Laboratory, he was a senior advisor to J. Robert Oppenheimer, who directed the development of the atomic bomb. Rabi was proud of his radar work. As he used to say, "The bomb ended the war, but radar won the war."

Rabi won the Nobel Prize in 1944, and after the war, he became active in formulating government policy toward science. As a member of the Science Advisory Committee, he opposed the development of the hydrogen bomb for ethical reasons. Along with the Secretary General of the United Nations, Dag Hammarskjöld, Rabi organized the first International Conference on the Peaceful Uses of Atomic Energy, held in Geneva, Switzerland, during 1955.

See also: MAGNETIC RESONANCE IMAGING; NUCLEAR BOMB, BUILDING OF; OPPENHEIMER, J. ROBERT; RADAR

Bibliography

RIGDEN, J. S. *Rabi: Scientist and Citizen* (Basic Books, New York, 1987).

JOHN S. RIGDEN

RADAR

Radar is an electronic method or system that uses transmitted electromagnetic radio waves to detect, locate, and measure speeds of remote objects. The system exploits two essential physical elements: objects, such as distant aircraft that reflect a small portion of radio waves received from a ground-based antenna back to the transmitter, and electromagnetic waves that travel at the well-known speed of light. Receiving circuits are able to determine the distance of the object from the receiver by measuring the time interval between outgoing and incoming waves. For every 1,000 ft of distance, there is a 2 μs round trip delay, an easily measured quantity.

Furthermore, a moving object exhibits an effect common to all wave phenomena, the Doppler effect, in which the frequency of reflected waves is displaced from the incident frequency in a manner proportional to the speed of the object. This permits measurement of the speed and can be used to discriminate against stationary objects. The object direction is determined by relating the antenna pointing angle to the strength of the reflected wave. For sound waves, the familiar change of pitch as a train whistle passes by is an example of the Doppler effect, while for light, the famous redshift of light emission from stars and galaxies is the Doppler effect applied to the expansion of the universe.

Radar, an acronym of radio detection and ranging, had its origin in experiments in the period from 1932 to 1935 in England by Robert Watson-Watt on behalf of the Royal Air Force. In February 1935, a short-wave radio transmitter of 10 kW was used to illuminate a test fleet of military aircraft some 25 miles away. The return waves were simply displayed on a cathode-ray tube (CRT) screen. This crude arrangement was sufficient to detect and locate the targets, and it led to rapid development of radar in the United Kingdom and United States prior to World War II, which had a profound effect on the outcome of the war.

The wavelength of radar radio waves could be as great as 10 m (with a frequency of 30 MHz) or as short as 3 mm (with a frequency of 100 GHz). Modern radar systems tend to use shorter wavelengths, since the ability to resolve or distinguish distant objects improves with diminishing wavelength. At many airports, a ground control radar uses a rotating antenna to scan 360° and is fed by a microwave oscillator (the klystron tube, producing high-frequency oscillations), while the same antenna is used as a receiver to detect reflected waves. The airport traffic information is displayed on a large CRT with location, speed, altitude, and other identifying markers. The short-wavelength radar allows operators to distinguish one airliner from another at large distances in order to avoid collisions. The radio wavelength must be substantially smaller than the object to be measured for high resolution, and the antenna must be sized appropriately.

Developments in antenna technology have in some cases replaced the rotating antenna with the phased array antenna. This arrangement exploits the fact that a wave incident on a series or array of closely spaced antennas can be electronically combined to produce constructive or destructive wave interference, very much like an optical grating. Introducing electronic phase shifts in each of the antennas is the equivalent of rotating the antenna viewing angle. Such a large array is used as a radio telescope in New Mexico to study the radio emissions of distant galaxies without having to point the antennas physically. Radar has also been used to measure precisely the distance to the planets and Earth satellites, since electromagnetic waves propagate readily through the vacuum of space.

Although most systems in use today employ microwave radio wavelengths, the principle of radar can be applied to any part of the electromagnetic spectrum, including infrared or visible light. Police radars on highways use the Doppler effect to measure violation of highway speed limits, in both the radio and laser versions. In this application, it is necessary to be able to tell which car is speeding, so proper pointing and avoidance of multiple targets is important. Hand-held radar guns are routinely used to measure the speed of pitched baseballs. The Doppler radar is also used by meteorologists to locate rain systems and moving storms. In this case, stationary objects (so-called ground-clutter) are electronically excluded from the detection process, and only those waves that have been frequency shifted by some minimum amount are observed.

It was discovered during early radar research that some radio frequencies were poorly transmitted through rain and fog and thus were to be avoided in practical systems. However, it was later realized that this effect was due to absorption of microwaves by water molecules, which vibrate at characteristic frequencies. This opened a new field of research called microwave spectroscopy, and many molecules were shown to have characteristic radio absorption spectra. The consequence of this research is that radio astronomers are now able to recognize many molecules in space, leading to deeper understanding of the origin of the universe and the chemistry of the solar system. One side effect of water molecule absorption was the invention of the microwave oven, which exploits the water in food as the key substance in direct cooking by microwaves.

Finally, a new application of radar has been invented that produces finely detailed images of large land masses and inaccessible terrain. The scheme is based on the idea that reflected radio waves from a varying topological surface produce continuously variable distance measurements, as in any radar system. This, together with computer storage of return wave information, allows an aircraft or a satellite to overfly a region and a computer to reconstruct a microwave "photograph" in remarkable detail. The shorter the wavelength used, the more fine-grained the resulting detail. This has produced remarkable surface images of the nearby planets, especially of Venus, which has dense clouds and whose surface cannot be observed by ordinary photographic means. On Earth, this new technique has permitted satellite observation of strategically important areas from dusk to dawn, so the cover of darkness is no longer able to mask military activity. Radar technol-ogy has advanced significantly since the early experiments of Watson-Watts.

See also: DOPPLER EFFECT; INTERFERENCE; RABI, ISIDOR ISAAC; RADIO WAVE; SPECTROSCOPY, MICROWAVE

Bibliography

HITZEROTH, D. *Radar: The Silent Detector* (Lucent Books, San Diego, CA, 1990).

SKOLNICK, M. I. *Introduction to Radar Systems* (McGraw-Hill, New York, 1980).

WOOD, D., and DEMPSTER, D. *The Narrow Margin: The Battle of Britain* (Paperback Library, New York, 1969).

ISAAC D. ABELLA

RADIANT ENERGY

See ENERGY, RADIANT

RADIATION, BACKGROUND

See COSMIC MICROWAVE BACKGROUND RADIATION

RADIATION, BLACKBODY

All objects at any temperature radiate energy in the form of electromagnetic waves due to atomic vibrations in their surfaces. Although this thermal radiation can span the entire electromagnetic spectrum, from radio waves to gamma waves, at temperatures under 1,000 K, it is predominantly in the infrared. Most solids and liquids emit a continuous range of wavelengths, while gases usually emit only a finite number of discrete wavelengths.

The amount of radiant energy an object will emit per unit of time per unit of surface area at each wavelength λ is called its spectral radiant intensity, i_λ. Both i_λ and the fraction of incident radiation absorbed at the same wavelength, α_λ, depend on the surface properties of the object. In 1859 the German physicist Gustav Kirchhoff found that objects in thermal equilibrium absorb as much energy as they emit. Kirchhoff also showed that the ratio of i_λ to α_λ is the same for all objects. This ratio is a universal function K_λ and depends only on the temperature of the object and the wavelengths emitted. Thus,

$$K_\lambda = \frac{i_\lambda}{\alpha_\lambda}. \qquad (1)$$

The discovery of this universal function played a key role in the origin of quantum physics. The investigation began by considering what happens when white light, containing all visible wavelengths, is incident on some object's surface. If all the light reflects off the surface, the object appears white. If some wavelengths are absorbed, say all but the blue, the object reflects only the blue and so it appears blue. If all the light is absorbed, the object appears black, and the fraction of light absorbed is 100 percent, making $\alpha_\lambda = 1$. If the black object, or blackbody, is in thermal equilibrium, it is a perfect absorber and a perfect emitter of radiation. According to Eq. (1), i_λ for a blackbody in thermal equilibrium must then be the same as the universal function K_λ. Therefore, finding the function K_λ required an examination of the emission spectrum of blackbody radiation.

Obtaining experimental data for the emission spectrum of blackbodies is hampered by the fact that no actual blackbodies exist on Earth. Approximations can be made by coating an object with soot, lampblack, or charcoal. Still better results are obtained by hollowing out a cavity in a carbon block and blackening the inside of the cavity. The block can be maintained at a constant temperature with an external electric heating coil. Any electromagnetic radiation inside the cavity will continually be absorbed and reemitted until the radiation and the cavity walls reach thermal equilibrium. A tiny hole in the wall of the block will allow a small amount of light to exit the cavity to be analyzed. This cavity radiation is a very close approximation to blackbody radiation.

A graph of the spectrum of blackbody radiation for three different temperatures is shown in Fig. 1. Two important results are apparent from the figure.

First, the total intensity of the emitted radiation, which is the area under each curve, increases dramatically at higher temperatures. Since temperature is a measure of the average kinetic energy of an object's atoms, it is not surprising that objects at higher temperatures radiate more energy than objects at lower temperatures. In 1879 Josef Stefan used such information to derive a relationship between the intensity I (total energy emitted per second per unit of surface area) and the absolute temperature T of an object:

$$I = \sigma T^4, \qquad (2)$$

where σ is a constant with the value of 5.67×10^{-8} W/m²·K⁴. Although Ludwig Boltzmann gave a theoretical basis for this relationship several years later, Eq. (2) only gives the total intensity of the radiation, not the intensity for each wavelength.

Figure 1 also shows that the wavelength for which the maximum intensity occurs is displaced to shorter wavelengths at higher temperatures. This is why, for instance, a hot ember in a fire may glow red or orange, while the flame, which is at a higher temperature, will emit yellow or even blue light. In 1895 Wilhelm Wien used the second law of thermodynamics to show that the wavelength of maximum

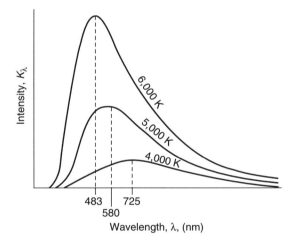

Figure 1 A graph of the intensity of light emitted by a blackbody at three different temperatures. The wavelength of maximum intensity is calculated by Wien's displacement law. The intensity of any individual wavelength can be obtained from Planck's radiation formula.

intensity emitted by an object at some absolute temperature T could be calculated by

$$\lambda_{\max} = \frac{b}{T}, \qquad (3)$$

where b is a constant whose value is 2.9×10^{-3} m·K.

Three years later, Wien also proposed an empirical equation to describe the curves in Fig. 1. He expressed the intensity distribution for a blackbody at various temperatures as

$$K_\lambda = a_1 \lambda^{-5} e^{-a_2 \lambda T} \qquad (4)$$

where a_1 and a_2 are constants. This equation agreed with the experimental evidence at the time, but a series of experiments carried out in 1899 by Wien's former colleagues, Otto Lummer and Ernst Pringsheim, showed that Wien's formula diverged from the data at long wavelengths.

In early 1900, other investigators suggested alternate forms for the radiation formula. The most notable of these was proposed by the English scientist John Strutt, known formally as Lord Rayleigh. From the kinetic theory of heat, Rayleigh derived the equation

$$K_\lambda = 2\pi c \lambda^{-4}(kT), \qquad (5)$$

where c is the speed of light in vacuum, 2.998×10^8 m/s, and k is Boltzmann's constant, 1.38×10^{-23} J/K. Rayleigh's formula agreed with the data at long wavelengths, but, by his own admission, it failed at shorter wavelengths.

The solution was finally found by the German physicist Max Planck, who had worked on the blackbody problem since 1894 and had contributed a theoretical basis for Wien's radiation formula. Planck had begun by assuming that the walls of the blackbody contained individual, submicroscopic oscillators. He found that the energy density of the radiation was related to the average energy of the oscillators, which in turn was related to the entropy of the oscillators.

In early October 1900, the experimentalists Heinrich Rubens and Ferdinand Kurlbaum reported that their most recent data agreed with Lummer and Pringsheim in showing Wien's formula to be significantly in error at long wavelengths. Planck then suggested the following radiation formula for a blackbody as a function of temperature:

$$K_\lambda = \frac{2\pi c^2 \lambda^{-5} h}{e^{hc/\lambda kT} - 1}, \qquad (6)$$

where h is Planck's constant, 6.63×10^{-34} J·s.

Planck's formula matched the data exactly at all wavelengths. It is of interest to note that Planck's formula reduces to Wien's formula for short wavelengths and Rayleigh's formula for long wavelengths. If Eq. (6) is integrated over all wavelengths, the result is the Stefan–Boltzmann equation, Eq. (2). Solving Eq. (6) for its maximum value yields Wien's displacement law, Eq. (3). Planck had succeeded in incorporating all previous work into one extraordinary equation.

Planck's contribution of the correct radiation formula was not, however, his most significant contribution to physics. For his radiation formula to have a firm theoretical basis, an independent means had to be found to determine the entropy, and thereby the energy, of the oscillators.

Using a statistical approach, Planck found that the correct radiation formula could be derived if the energy of each oscillator depended solely on its frequency of vibration. The energy could be equated to the frequency ν by multiplying the frequency by the constant h. The same analysis showed that the energy of each oscillator must also be restricted to some multiple of the product of $h\nu$:

$$E = nh\nu, \qquad (7)$$

where n is an integer. This quantization of the energy was a radical departure from classical physics and was not totally accepted at the time, even by Planck himself. More than twenty-five years passed before the quantized nature of atoms and electromagnetic radiation was fully understood.

Planck's blackbody radiation formula is still used when the emission spectrum of a real heat radiator is studied. The spectral radiant intensity i_λ of any real heat radiator is just the spectral radiant intensity of a blackbody K_λ multiplied by the spectral emissivity of the object ϵ_λ. This experimentally determined coefficient, which ranges from 0 for totally reflective surfaces to 1 for a blackbody, depends on the temperature and surface properties of the object, such as color, roughness, and material composition. The total intensity of light emitted by real heat radiators can also be calculated if Eq. (2) is multiplied by ϵ_λ.

Scientists often find valuable information concerning the thermal properties of materials by measuring their emission spectra and contrasting it with the blackbody spectrum. Many detection devices, like burglar alarms or heat seeking missiles, operate by recording the intensity levels of thermal radiation and comparing the measurements to the expected values calculated from the radiation equations.

Astronomers also use the radiation formulas when studying the light from stars to determine their composition and temperature. For example, the Sun emits a nearly continuous spectrum of light with a wavelength of maximum intensity at around 500 nm. By Eq. (3), the Sun acts like a blackbody with a surface temperature of 5,800 K.

In 1964 a similar measurement led to a startling discovery about our universe. Arno Penzias and Robert Wilson of Bell Labs were trying to use a large radio receiver to monitor an early communication satellite. Persistent interference from every direction in space occurred in the microwave region of the radio spectrum. After eliminating all other possibilities, they concluded that they were detecting electromagnetic radiation corresponding to a blackbody temperature of about 3 K. This microwave background radiation had been predicted to exist if the universe had at some earlier time been hotter and had expanded and cooled. In 1989 the Cosmic Background Explorer (COBE) satellite was launched by NASA to measure this radiation more accurately. The COBE measurements confirmed the existence of electromagnetic radiation corresponding to a blackbody temperature of 2.73 K and, along with more recent satellites, continue to provide much needed data for cosmologists studying the large-scale structure of the universe.

See also: Cosmic Background Explorer Satellite; Cosmic Microwave Background Radiation; Cosmic Microwave Background Radiation, Discovery of; Planck, Max Karl Ernst Ludwig; Radiation, Thermal; Stefan–Boltzmann Law; Thermodynamics

Bibliography

Klein, M. J. "Max Planck and the Beginnings of the Quantum Theory." *Archive for History of Exact Sciences* **1**, 459–479 (1962).

Kuhn, T. *Blackbody Theory and the Quantum Discontinuity, 1894–1912* (Oxford University Press, New York, 1978).

Planck, M. *The Theory of Heat Radiation*, trans. by M. Masius in 1914 (Dover, New York, 1991).

Richtmyer, F.; Kennard, E.; and Cooper, J. *Introduction to Modern Physics*, 6th ed. (McGraw-Hill, New York, 1969).

Segre, E. *From X-Rays to Quarks* (W. H. Freeman, San Francisco, 1980).

Mark S. Bruno

RADIATION, COSMIC MICROWAVE BACKGROUND

See Cosmic Microwave Background Radiation

RADIATION, ELECTROMAGNETIC

See Electromagnetic Radiation

RADIATION, SYNCHROTRON

Synchrotron radiation is light produced by charged particles traveling at nearly the speed of light as they pass through the magnetic field of a dipole bending magnet. Radiation is emitted in an intense fan-like beam tangent to the bending arc. The emission of light by relativistic electrons in circular orbits was predicted before 1945 and was first observed on the General Electric 70-MeV synchrotron in 1947. Research with synchrotron radiation began in 1955 at the 300-MeV Cornell synchrotron. The constant spectral distribution and stable intensity characteristic of storage rings was first provided by the 240-MeV storage ring at the University of Wisconsin. These desirable characteristics have since been developed in electron storage rings all over the world, and the spectral range has been extended into the x-ray regime.

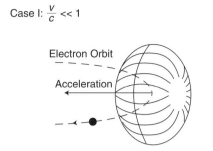

Case I: $\frac{v}{c} \ll 1$

Electron Orbit

Acceleration

Case II: $\frac{v}{c} \approx 1$

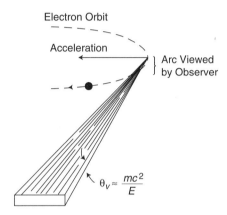

Electron Orbit

Acceleration

} Arc Viewed
by Observer

$\theta_v \approx \dfrac{mc^2}{E}$

Figure 1 Radiation emission pattern by electrons in circular motion.

Synchrotron radiation has many features that make it a very attractive tool for research in a variety of scientific and applied fields of study. The normal radiation of an electron in circular motion at low energy is in a nondirectional radiation pattern (see Fig. 1). However, at relativistic energies, this pattern is folded forward so that the radiation is emitted as a flat "pancake" in the plane of the electron orbit. The vertical opening angle of emission is given approximately by mc^2/E, where mc^2 is the rest mass energy of the electron (0.51 MeV) and E is the total energy. At high electron energy this opening angle is very small, giving rise to very high flux densities on small targets even at distances of 10 to 20 m from the source.

The power is radiated in a smooth continuum (see Fig. 2) up to a critical energy that is determined by the electron energy and the bending radius. Because the spectrum depends inversely on the bending radius, it is possible to shift the spectrum to higher energies by inserting high field magnets in series in machine straight sections. Such magnetic devices are called wigglers or undulators. An array of powerful magnets vibrates the electrons intensely over a short distance. This oscillation, a series of short quick bends in the electron path, causes the beam to emit light. By choosing the proper oscillator period, it is possible to increase the highest energy emission by 100 times in wigglers with 100 oscillations. For 100 oscillations in an undulator,

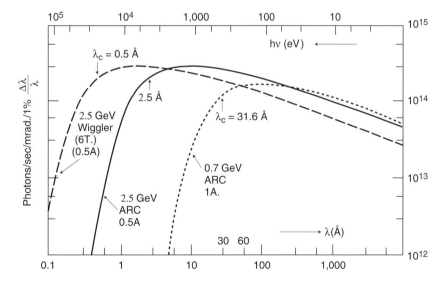

Figure 2 Spectra for the National Synchrotron Light Source at Brookhaven National Laboratory.

1319

which operates on the principle of constructive interferences, it is possible to increase the x-ray brilliance by 10,000 times at a given tunable spectral energy.

Both basic and applied research is carried out with synchrotron radiation. Researchers study the absorption and scattering of light to determine the properties of matter, such as crystal structure, bonding energies of molecules, details of chemical and physical phase transformations, electronic structure, and magnetic properties.

In applied research, surfaces are studied, for example, to improve the performance of catalysts used in petroleum cracking and refining. Synchrotron light is used in x-ray lithography to produce more densely packed computer chips and greater memory capacity. A number of new imaging techniques have been developed, including microtomography (a type of CAT scan capable of nondestructively producing three-dimensional images of microscopic structures), soft x-ray microscopy for the study of *in vivo* biological specimens, and less invasive imaging techniques for medical diagnosis, such as coronary angiography.

See also: CAT SCAN; ELECTROMAGNETIC RADIATION; SYNCHROTRON

KENNETH BATCHELOR

RADIATION, THERMAL

Thermal radiation consists of electromagnetic waves in the infrared to visible light portion of the spectrum. Electromagnetic waves are created by the acceleration of charged particles; thus, the oscillation of the molecules in all objects gives rise to electromagnetic waves that radiate from the object. Electromagnetic waves resulting from the thermal agitation of the molecules in an object are referred to as thermal radiation. The thermal radiation from an object contains a continuous range of wavelengths, with the temperature of the object determining which wavelengths predominate. Infrared light is predominant at lower temperatures, while at higher temperatures an object emits visible light. As an object is heated, one can feel the infrared radiation emitted and, as the temperature increases, the object eventually glows red and then white. An incandescent light bulb is an example of a white-hot object. It emits a full spectrum of visible light, and so it appears white. The bulb also emits in the infrared region, as anyone who has placed their hand near an operating light bulb can attest: the bulb gets very hot. This is in contrast to a fluorescent light, which emits very little infrared light and remains cool. The infrared portion of the spectrum is used by the military and law enforcement agencies to "see" in total darkness. Infrared-sensitive video cameras are used to view the infrared light radiated by all objects and convert the image to visible light on a monitor. This type of device can also be used to determine where energy is being lost from a house or other structure by looking for bright regions on the structure in the infrared region. This type of thermal scanning is called thermography.

The rate at which heat (Q) is radiated by an object (in joules per second, or watts) with surface area A, at temperature T (in kelvins), is described by the Stefan–Boltzmann law, named after its late nineteenth-century discoverers:

$$\frac{\Delta Q}{\Delta t} = \sigma e A T^4.$$

$\sigma = 5.6705 \times 10^{-8}$ W/m^{2}K^4 is a fundamental physical constant called the Stefan–Boltzmann constant. The constant e is the emissivity of the object and varies from 0 to 1. The emissivity is determined by the roughness or smoothness of the surface and by the color of the object. The emissivity describes how well the object radiates energy. Dark, dull surfaces and cavities are the best emitters, while smooth, shiny surfaces with light colors are the worst.

Objects also absorb thermal radiation from their surroundings. Surfaces that are good emitters are also good absorbers and surfaces that are poor emitters are poor absorbers. An object is in equilibrium with its surroundings when it is emitting the same amount of thermal energy that it absorbs. When more is emitted than absorbed, the object cools; if less is emitted than absorbed, the object warms. The net rate of energy (Q) lost or gained by an object at temperature T, with surroundings at temperature T_S, is given by

$$\left(\frac{\Delta Q}{\Delta t}\right)_{\text{net}} = \sigma e A (T^4 - T_S^4).$$

This shows that the direction of the flow of heat depends on the temperature difference between the object and its surroundings, just as in the case of transfer of heat by conduction and convection.

Most shoppers feel that the air temperature in the frozen-foods section of the grocery store is significantly lower than on other aisles. A thermometer would show that there is in fact little difference in temperature. The shoppers are experiencing a situation where their surroundings are radiating much less thermal energy because of the low temperature in the freezers, while the shoppers' bodies continue to radiate heat. The shoppers "feel" the net loss of heat from their bodies.

A practical application pertaining to thermal radiation is the thermos bottle. A thermos is a double-walled bottle that has been evacuated between the walls. The vacuum prevents nearly all heat transfer by conduction and radiation; the only remaining mechanism for heat to enter or exit the bottle is by radiation. To minimize radiative transfers of heat, the walls of the bottle are silvered into a mirror-like surface. Since a smooth shiny surface has a low emissivity, the walls are poor absorbers and emitters of heat. A thermos can keep coffee or soup hot for hours. A similar container called a Dewar flask is used to hold liquefied gases at low temperatures and is designed on the same principles as the thermos.

An object with an emissivity e of one is a perfect absorber and emitter of radiation. Such an object is called a blackbody. The thermal energy radiated from a blackbody does not depend on the material of construction or any other characteristic of the body. This leads to the result that the spectrum of radiation from the blackbody is determined solely by its temperature. The effort to develop a theory explaining blackbody radiation led to the discovery of quantum mechanics by Max Plank in 1900.

The atmosphere of Earth is transparent to the short wavelengths of light from the Sun, allowing sunlight to reach the surface of Earth, where it is largely absorbed. The surface of the earth reradiates part of this energy at longer wavelengths. Earth's atmosphere is opaque to these longer wavelengths, and much of this radiation is absorbed by atmospheric gases (mainly carbon dioxide and water vapor) and re-emitted back to Earth's surface. An equilibrium is reached between the radiation absorbed from the Sun and the radiation Earth emits into space, determining the mean temperature of Earth's surface. Since the burning of fossil fuels releases carbon dioxide into the atmosphere, concerns have been raised that increasing percentages of thermal energy will remain trapped in Earth's atmosphere, resulting in global warming and a higher equilibrium temperature. This possibility has been named the greenhouse effect because of the analogy to how reradiated infrared light is trapped inside glassed-in greenhouses, resulting in a higher temperature in the greenhouse.

A related phenomenon is the trapping of heat by clouds. Cloudy nights are generally warmer than clear nights because some of Earth's radiation is absorbed by clouds and partially reradiated back to the surface. In deserts that are scorching hot in the daytime, the night is often frigid because of the lack of clouds or any appreciable water vapor in the air above the desert.

An interesting application of cooling by thermal radiation is used aboard spacecraft. During the portion of the ship's orbit above the daylight side of Earth, a tremendous amount of thermal energy is absorbed from the Sun. Because the spacecraft is surrounded by a vacuum, the only possible method of cooling is by radiation. Heat is pumped to a radiator on the shaded side of the spacecraft, where it is radiated into coldness of deep space.

See also: CONDUCTION; CONVECTION; ELECTROMAGNETIC WAVE; ENERGY, RADIANT; HEAT; RADIATION, BLACKBODY; STEFAN–BOLTZMANN LAW; THERMAL CONDUCTIVITY; THERMODYNAMICS

Bibliography

HEWITT, P. G. *Conceptual Physics,* 7th ed. (HarperCollins, New York, 1993).

JONES, E. R., and CHILDERS, R. L. *Contemporary College Physics,* 2nd ed. (Addison-Wesley, Reading, MA, 1993).

OHANIAN, H. C. *Principles of Physics* (W. W. Norton, New York, 1994).

YOUNG, H. D. *University Physics,* 8th ed. (Addison-Wesley, Reading, MA, 1992).

ZEMANSKY, M. W., and DITTMAN, R. H. *Heat and Thermodynamics,* 6th ed. (McGraw-Hill, New York, 1981).

JOHN C. RILEY

RADIATION BELT

A radiation belt is a large region that encircles a planet and contains enormous numbers of energetic, electrically charged particles—principally electrons and protons—trapped in the planet's external magnetic field. Such a region has the geometric shape of a doughnut with its axis aligned with the magnetic axis of the planet.

A radiation belt of Earth was discovered in 1958 by James A. Van Allen, a physicist at the University of Iowa, with detectors on the first American satellite, Explorer I, and the third, Explorer III. Later in 1958, a second, outer belt was discovered by U.S. and Soviet investigators. Both inner and outer belts have been studied in detail by many satellites equipped with scientific instruments.

The two principal sources of the particles are the solar wind (the hot ionized gas that streams outward from the sun) and the decay products (electrons and protons) of neutrons moving outward from cosmic-ray-produced reactions in the upper atmosphere. Particles from the solar wind enter Earth's magnetic field and are subsequently accelerated and diffused inward by fluctuating electric and magnetic fields to form Earth's outer radiation belt. A distinguishing feature of the inner radiation belt is the population of neutron-decay protons having energies greater than tens of mega-electron-volts.

The inner belt has a lower boundary at an altitude of about 400 km and extends outward to a radial distance of about 13,000 km. Its structure and composition are relatively stable. The outer belt surrounds the inner belt and extends outward to a radial distance of about 70,000 km. Its composition and structure fluctuate on time scales of minutes, hours, days, and months in response to fluctuations in the flow of the solar wind.

The penetrating radiation in the inner belt adversely affects electronic equipment and restricts prolonged flight of spacecraft carrying human crews to altitudes less than 400 km, though rapid transits through the belt ($\approx$ an hour), as in the Apollo flights to the Moon, are tolerable.

Radiation belts are interior components of a larger region called the magnetosphere, which also contains much greater populations of less-energetic charged particles called plasma. The magnetosphere, the region within which the magnetic field of Earth dominates the motion of charged particles, is bounded by a bow shock on the sunward side of Earth at a radial distance of about 100,000 km and has a long magnetotail extending to a distance of the order of 6,000,000 km on the antisunward side.

Neither the magnetosphere nor the radiation belts are visible in the ordinary sense. The principal geophysical evidences of magnetospheric phenomena are magnetic storms and the luminous polar auroras, the northern and southern lights, produced by the precipitation of energetic electrons (a few kilo-electron-volts) into the upper atmosphere. Nine observable artificial radiation belts of limited lifetimes (days to many months) were produced from 1958 to 1962 by the decay products of fission nuclei from U.S. and Soviet nuclear-bomb bursts at high altitudes. Eventually, particles in both natural and artificial radiation belts are lost into the atmosphere or escape into space.

A huge radiation belt of Jupiter was discovered by radio astronomical observation in 1959 and later studied by this remote technique. It was explored directly and in detail by Pioneer 10 in 1973, Pioneer 11 in 1974, Voyagers 1 and 2 in 1979, and Ulysses in 1992. A radiation belt of Saturn was discovered by Pioneer 11 in 1979 and further studied by Voyagers 1 and 2. Voyager 2 discovered radiation belts of Uranus in 1986 and of Neptune in 1989. Other spacecraft have found that the Moon, Mars, Venus, and Mercury have no radiation belts because of their very weak magnetic fields.

See also: AURORA; MAGNETOSPHERE; SOLAR WIND; VAN ALLEN BELTS

Bibliography

VAN ALLEN, J. A. "Radiation Belts Around the Earth." *Sci. Am.* **200** (3), 39–47 (1959).

VAN ALLEN, J. A. *Origins of Magnetospheric Physics* (Smithsonian Institution Press, Washington, DC, 1983).

JAMES A. VAN ALLEN

RADIATION PHYSICS

Radiation physics can be defined narrowly as the study of the properties of the ionizing radiations of atomic and nuclear physics and the interaction of these radiations with matter. Directly ionizing radiations include electrons, protons, alpha and beta particles, and all the other charged elementary particles and their antiparticles. X rays, gamma rays, and neutrons, as well as other uncharged elementary particles, are indirectly ionizing. The average energy expended per ionization is 33.7 eV for air, a typical value for gases; corresponding values for most solids are approximately a factor of 10 lower. Thus, the field of radiation physics is limited to primary radiation events of energy greater than several electron volts.

Historically, radiation physics is additionally interpreted as encompassing many aspects of health and safety issues arising in the use of radioactive sources and equipment, such as x-ray machines, particle accelerators, and nuclear reactors that produce ionizing radiation.

Sources of Radiation

Radiation physics is an important subdiscipline of physics because of the worldwide use of ionizing radiation sources in science, medicine, and industry. Diagnostic and therapeutic use of x rays in medicine is ubiquitous and has evolved into a highly skilled art. For example, in x-ray therapy, radiation physicists often work with radiologists as part of a team to plan and coordinate the use of filters and multiple x-ray beams in such a way as to maximize the dose to tumors while minimizing doses to surrounding tissues; this may involve the use of CAT scans for localization of critical internal organs and the tumor.

In nuclear medicine, radioactively labeled substances are administered (usually by intravenous injection) to a patient, and the dynamic uptake of the radioactivity into specific organs is followed by a specialized imaging device called an Anger camera. Brachytherapy is a medical treatment mode involving the implantation of sealed radioactive sources into tumors; an optimum dose distribution is achieved by choice of the radioactive isotope used and the distribution of sealed sources within the tumor.

Radioactivity is an unwanted by-product of nuclear reactors used to generate electrical energy or provide power to submarines. A large inventory of radioactivity results from operations surrounding the maintenance of a nuclear arsenal. Unfortunately, none of the five basic human senses can directly detect ionizing radiation, so potentially hazardous uses of ionizing radiation must be very carefully regulated. Health physicists are involved with formulating and interpreting the extensive state and federal regulations for radiation protection and safety, for monitoring radiation levels in the workplace, and for evaluating radiation accidents.

Radiation Dosimetry

The ultimate goal in many applications of radiation physics is to measure absorbed dose or the closely related quantity, dose equivalent. Absorbed dose is defined as the mean energy per unit mass imparted by ionizing radiation at a specified point in an absorber. The SI (*Système Internationale*) unit of absorbed dose is the gray (Gy) where 1 Gy = 1 J/kg, but an older unit, the rad (100 rad = 1 Gy), is still in rather wide use. Absorbed dose is a widely used measure of biologically significant effects such as gene mutations and chromosome breaks.

The starting point for all radiation dose calculations is to characterize the energies, intensities, and types of radiation present in the primary radiation beam or source. Then one must solve a radiation transport problem to determine the absorbed dose of ionizing radiations at target points due to primary radiations emanating from source points. Particularly for indirectly ionizing radiations, this is usually a difficult task requiring considerable skill and judgment of a radiation physicist because of the complexities of the geometry, density, and chemical composition of the source and/or target region (often the human body).

For example, in the cases of x rays and gamma rays, interactions take place by photoelectric absorption, Compton scattering, and, if the energy is more than twice the rest mass energy equivalent of an electron, by production of electron-positron pairs. A primary photon loses energy by a succession of Compton scattering events until the secondary scattered photon is absorbed by a photoelectric event or pair production. Pair production leads to two gamma rays called annihilation quanta, which then further Compton scatter until photoelectric absorption occurs. In one analysis method, these stochastic processes are simulated on a computer and photon histories detailing the kinetic energy delivered to electrons in target regions of interest can be determined. Statistically valid information on absorbed dose is built up by following many such photon histories in a computer intensive procedure generically known as the Monte Carlo method.

Career Path and Working Environment

Radiation physicists will be found anywhere ionizing radiation is used on a large scale. In the medical field, this includes radiology and nuclear medicine departments of hospitals. Some hospitals, especially those associated with medical schools, have separate medical physics departments where basic research on new applications of physics to medicine occurs. For example, the relatively recent positron emission tomography (PET) scanners were developed in this environment.

Health physicists are employed at all nuclear reactor sites and large particle accelerators. Since radiation sources are used in many scientific research projects, nearly all major universities employ health physicists to administer radiation protection and safety programs. The Nuclear Regulatory Commission, the Environmental Protection Agency, and the Department of Energy all have major programs employing the services of radiation physics experts. Interdisciplinary programs of fundamental research on physical and biological aspects of ionizing radiation have been a mainstay of the national laboratories, such as the Oak Ridge National Laboratory, for many years.

Approximately forty universities have advanced degree programs in medical physics. The American Board of Radiology and the American Board of Medical Physics certify medical physicists in therapeutic, diagnostic, and nuclear medicine physics. Board certification is essential for significant involvement in clinical aspects of radiation physics.

A career path in health physics usually begins in one of the approximately sixty universities having undergraduate and graduate training programs designed explicitly for the field of radiation protection. The program of the American Board of Health Physics is directed exclusively toward certification in Health Physics.

See also: CAT SCAN; COMPTON EFFECT; HEALTH PHYSICS; NUCLEAR MEDICINE; RADIOACTIVITY

Bibliography

ATTIX, F. H.; ROESCH, W. C.; and TOCHILIN, E., eds. *Radiation Dosimetry*, 2nd ed. (Academic Press, New York, 1968).

HURST, G. S., and TURNER, J. E. *Elementary Radiation Physics* (Wiley, New York, 1970).

KAHN, F. M., *The Physics of Radiation Therapy*, 2nd ed. (Williams & Wilkins, Baltimore, MD, 1994).

SORENSON, J. A., and PHELPS, M. E. *Physics in Nuclear Medicine;* 2nd ed. (Grune & Stratton, Orlando, FL, 1987).

L. THOMAS DILLMAN

RADIOACTIVE DATING

Radioactive dating uses the radioactive decay of atomic nuclei as a clock to determine the age of materials containing atoms of radioactive nuclides, which are also called radionuclides or radioisotopes. In general, radioactive decay is a reliable clock because it is a property of the atomic nucleus, which is not affected by environmental conditions such as temperature and pressure. Only under extreme conditions of matter, such as the interior of stars, can nuclear properties be affected and the half-lives of certain radionulcides change.

In a typical radioactive-dating experiment, the concentration of a specific radionuclide—that is, the ratio of the number of radioactive atoms to the total number of atoms in a given amount of sample material—is measured. Assuming that one knows the original radionuclide concentration when the material was formed, one can determine the age of the material from this measurement. This is possible because the radioactive decay follows a well-known law: $N_t = N_0\ e^{-\lambda t}$. N_0 is the initial number of radioactive atoms present in the material; N_t is the number of radioactive atoms left after time t has elapsed; e is the basis of the natural logarithm; and λ is the decay constant. A more familiar quantity is the half-life $t_{1/2}$, related to the decay constant by $t_{1/2} = (\ln 2)/\lambda$. The half-life is the time it takes for half of the original radioactive atoms to decay. For a period corresponding to n half-lives, the radionuclide concentration decreases by a factor of 2^{-n}.

To make reliable age determinations, the number of radioactive atoms must change only through radioactive decay and not by any other process such as the migration of radioactive atoms into or out of the material. If the condition of a closed system is not stringently fulfilled, the uncertainty of the age determination increases.

As of 1996, nuclear physics had identified approximately twenty-five hundred nuclides, each characterized by a unique combination of protons and neutrons in the nucleus. Approximately 12 percent of these nuclides are stable. The rest are radioactive, with half-lives ranging from less than 10^{-3} sec for the shortest-lived to 10^{25} years for the longest-lived radionuclide known, ^{128}Te. (The superscript to the left of the elemental symbol is the mass number of the nuclide—the number of protons plus that of neutrons.) These radionuclides provide clocks of vastly different running speeds, some of them much slower than the age of the earth (4.6 billion years) or the age of the universe (10 to 20 billion years). Correspondingly, radionuclides can be used to study almost any time range in the universe.

Accelerator Mass Spectrometry

One method of measuring long-lived radionuclides has revolutionized the field of radioactive dating. The traditional method of detecting radioactive atoms is through the radiation emitted in the decay process. Since the decay rate, dN_t/dt (decays per second), is proportional to the number of radioactive atoms present, and inversely proportional to the half-life, the number of atoms decaying in a given measuring time is

$$t_{\text{meas}} \, (dN_t/dt) = (t_{\text{meas}}/t_{1/2}) N_t \ln 2. \quad (1)$$

For half-lives much longer than the measuring time, very few atoms actually decay. It is therefore much more efficient to detect the radioactive atoms before they decay, which can be done by accelerator mass spectrometry (AMS). In this method, a beam of ions (electrically charged atoms) is formed from the sample material, and the ions are separated according to their mass using electric and magnetic analyzers and a small accelerator. The gain in detection efficiency compared to radioactive decay counting is dramatic: For the most famous radionuclide used for dating, ^{14}C ($t_{1/2} = 5,730$ yr), it is almost a factor of one million, allowing for greatly reduced sample sizes (milligrams instead of grams) and much shorter measuring times (minutes instead of days). AMS allows one to use other cosmogenic radionuclides, with half-lives much longer than ^{14}C, greatly extending the field of radioactive dating.

Dating with Cosmogenic Radionuclides

Most interesting for radioactive dating are long-lived radionuclides naturally produced by nuclear reactions of cosmic-ray primary protons and secondary neutrons with the main nuclides of the earth's atmosphere (^{14}N, ^{16}O, ^{40}Ar). These so-called cosmogenic radionuclides include ^{10}Be ($t_{1/2} = 1.6 \times 10^6$ yr), ^{14}C (5,730 yr), ^{26}Al (7.2×10^5 yr), ^{32}Si (140 yr), ^{36}Cl (3.01×10^5 yr), and ^{39}Ar (269 yr). After pro-

duction in the atmosphere, the radionuclides follow pathways determined by the physical and chemical properties of the element to which they belong. For matter in exchange with the atmosphere on time scales much shorter than the respective half-life, the loss of radionuclides by decay is continuously replenished by the cosmogenic production, and an equilibrium between production and decay is established. As a result, a natural radionuclide concentration exists in this matter. When matter is closed off from this exchange, the radioactive clock starts running, and a dating condition is established.

The production of ^{14}C in the atmosphere proceeds through the nuclear reaction of cosmic-ray secondary neutrons with nitrogen: ^{14}N $+ n \rightarrow p + {}^{14}$C. After production, ^{14}C is first oxidized to ^{14}CO and subsequently to ^{14}CO$_2$. It enters the biosphere through assimilation of ^{14}CO$_2$ by plants and is further distributed by the digestion of plants by animals. As a result, all living organic matter on earth has a minute amount of ^{14}C (1.2×10^{-12}) mixed into stable carbon, consisting of the isotopes ^{12}C and ^{13}C, which means that only one out of 1.2×10^{12} carbon atoms is radioactive. However, 1 mg of carbon still contains 60 million ^{14}C atoms. After death, the radiocarbon clock starts running, and the ^{14}C concentration decreases as time goes on. Under favorable conditions (closed system), radiocarbon dating can be performed on objects as old as ten half-lives (57,300 yr). The original ^{14}C concentration is then reduced by the factor $2^{-10} = 1/1,024$. The equilibrium concentration of ^{14}C in nature is not constant in time because of variation of the earth's magnetic field, which influences cosmic-ray intensity at the surface of the earth and thus the production rate. Fortunately, these natural variations of ^{14}C can be traced precisely back in time for about 10^4 years by measuring the ^{14}C concentration in tree rings, whose dates are established by counting them back continuously from the current time. Beyond 10^4 years, absolute age determination with ^{14}C is less precise because of the lack of a calibration curve. However, it may be possible to extend the calibration by comparing radiocarbon dates with those obtained from the decay of primordial radionuclides, which are not affected by cosmic-ray variations. Applications of radiocarbon dating are numerous. It is used to date archaeological and art objects, fossil bones, groundwater and ocean water, sediments, corals, planktonic organisms, polar ice, and trace gases in the atmosphere such as CO, CO$_2$, CH$_4$.

Dating with Primordial Radionuclides

Primordial radionuclides were present when the solar system formed and have half-lives long enough to survive until today. They are not dependent on cosmic-ray production and are therefore not prone to its variation. Their decay can be used for dating when a measurable change between the abundance of the parent radionuclide and one or more of its decay products (either radioactive or stable ones) can be detected. In order of increasing half-life, the following primordial radionuclides are of interest: ^{235}U ($t_{1/2} = 7.0 \times 10^8$ yr), ^{40}K (1.3×10^9 yr), ^{238}U (4.5×10^9 yr), ^{232}Th (1.4×10^{10} yr), ^{176}Lu (3.5×10^{10} yr), ^{187}He (4.3×10^{10} yr), ^{87}Rb (4.9×10^{10} yr), and ^{147}Sm (1.1×10^{11} yr). The most favorable condition for dating is the absence of the decay product at the start of the time period to be determined.

Potassium-argon dating. This highly successful method for dating certain minerals uses the decay of ^{40}K to the stable nuclide ^{40}Ar, called radiogenic argon. In lava from volcanic eruptions, the radioactive clock is set to zero through the outgassing of argon from the hot minerals. After cooling, the clock starts running, and the ratio of accumulated ^{40}Ar relative to the ^{40}K content in the rock can be used to determine the age provided no argon escapes or intrudes at a later stage. The method allows dating through almost the entire time period of the solar system, that is from 4.6 billion years to about 10^5 years before the present. For example, the appearance of the earliest hominids in Africa several million years ago was established by potassium-argon dating of volcanic tuff associated with fossil bones.

Uranium-series dating. Both ^{235}U and ^{238}U decay through a series of alpha- and beta-decay processes to the stable lead isotopes ^{207}Pb and ^{206}Pb, respectively. On the way to these stable end points, they encounter radionuclides with different chemical properties. Chemical selectivity can lead to a disequilibrium between these products, establishing a dating condition. For example, ^{238}U passes through ^{234}U ($t_{1/2} = 2.5 \times 10^5$ yr) and ^{230}Th ($t_{1/2} = 7.5 \times 10^4$ yr) on its way down to ^{206}Pb. In old minerals, ^{230}Th is present in a radioactive equilibrium between feeding from the decay of ^{234}U and its own decay (to ^{226}Ra). In ocean water, however, uranium stays in solution, whereas thorium precipitates out into the ocean sediments. Therefore, corals growing in ocean water incorporate ^{238}U and ^{234}U but essentially no ^{230}Th. The accumulation of ^{230}Th relative to the ^{234}U and ^{238}U allows a precise age determination. The amount of original thorium incorporated into the corals can be checked by the concentration of long-lived ^{232}Th. Dating from several thousand to about 5×10^5 years is possible. The upper limit requires mass spectrometry rather than alpha-decay counting techniques.

Rubidium-strontium dating. The decay of ^{87}Rb ($t_{1/2} = 49$ billion years) leads to stable ^{87}Sr. Simple dating conditions exist for rocks and minerals where the rubidium content is much larger than that of strontium so that the initial amount of ^{87}Sr is negligible. However, if this condition is not fulfilled, the so-called isochron method can still lead to viable dates. Here, both the abundances of ^{87}Rb and of ^{87}Sr are normalized to the abundance of ^{86}Sr, a stable isotope of strontium with no radiogenic contribution. If one analyzes several minerals of the same rock with different rubidium/strontium ratios, then the higher the ^{87}Rb/^{86}Sr ratio is, the higher the ^{87}Sr/^{86}Sr ratio should be. If the minerals have formed at the same time and have stayed as a closed system for rubidium and strontium, the ratios of ^{87}Rb/^{86}Sr versus ^{87}Sr/^{86}Sr fall on a common slope, which gives the age. This method can be used to determine ages of solar-system materials, including Earth, the Moon, and meteorites in the time range from about 10 million years before the present back to the formation of the solar system (4.6 billion years).

Nucleocosmochronology. One of the most intriguing questions in science is the age of the universe. Different astrophysical estimates exist based on the expansion of the universe (Hubble constant) and the age of the oldest stars (globular clusters). Age estimates range from approximately 10 billion to 20 billion years. Additional age estimates can be obtained from studying the abundance of sufficiently long-lived radionuclides, whose stellar production (nucleosynthesis) in our galaxy is reasonably well understood. These include ^{187}Re, ^{235}U, ^{238}U, and ^{232}Th, which are all produced by rapid-neutron capture (*r*-process). In order to arrive at an age estimate for the universe, abundance ratios of ^{187}Re/^{187}Os (stable), ^{238}U/^{232}Th, and ^{235}U/^{238}U are measured in solar-system material and combined with estimates about their production rates and the chemical evolution of the galaxy. In some cases, subtle effects, such as a shortening of the half-life of ^{18}Re under stellar-burning conditions, have to be taken into account.

Radiation-Effect Dating

The decay of radioactive atoms leads to the emission of radiation (alpha, beta, gamma, fission products), which produces stable physical and chemical changes in exposed materials, particularly in nonconducting solids. The accumulation of these effects—for example, trapping electrons in defects of the crystal lattice and tracks of energetic fission fragments—can be used as a clock, provided one knows the intrinsic or environmental exposure rate of the material and the radiation effects have been set to zero by some natural or artificial process, such as heating.

Thermoluminescence dating. This method is based on the trapping of radiation-liberated electrons in crystal sites from which they can be freed by heating the material. In this process, the electrons release energy in the form of light, which can be detected. The light intensity is a measure of the radiation effects accumulated since the material was last heated and is thus connected to the age of this event. The radiation exposure usually stems from primordial radionuclides, such as ^{40}K, ^{238}U, and ^{232}Th, and can sometimes be established by exposing test materials for a limited time (months). This method is well suited to dating pottery (i.e., the time of firing the ceramics). Also, flint tools and other archaeological objects containing suitable mineral fractions, and exposed to heat in a fireplace, can be dated with this method. The time range of dating with thermoluminescence lies between a few thousand and approximately 3×10^5 years.

Electron-spin-resonance dating. This method is an alternative to thermoluminescence dating. It determines the concentration of radiation-liberated electrons trapped in crystal defects by measuring an electron-spin-resonance (ESR) signal from the unpaired electrons. Unlike thermoluminescence dating, ESR can be performed on opaque materials, such as tooth enamel, and a variety of minerals. The dating range extends from several centuries to many hundred thousand years and possibly beyond 1 million years.

Fission-track dating. This method uses the microscopic damage in certain crystals, such as mica, produced by the highly energetic nuclear fragments emitted in the spontaneous fission of ^{238}U. Spontaneous fission is a rare process relative to the alpha decay of ^{238}U (one fission event out of about 2 million alpha decays), leading to a spontaneous-fission half-life of 8.2×10^{15} years. The fission tracks are approximately 10 μm long and can be enlarged by etching with acid. They can then be viewed and counted with a microscope, and the number of tracks relative to the uranium content gives the age. Under favorable conditions, such as absence of annealing processes, the method is applicable in the time range of about one million to several hundred million years.

See also: DECAY, ALPHA; DECAY, NUCLEAR; ELEMENTS; ELEMENTS, ABUNDANCE OF; ISOTOPES; MASS SPECTROMETER; NUCLEOSYNTHESIS; RADIOACTIVITY; THERMOLUMINESCENCE

Bibliography

GEYH, M. A., and SCHLEICHER, H. *Absolute Age Determination* (Springer-Verlag, Berlin, 1990).

TAYLOR, R. E.; LONG, A.; and KRA, R. S. *Radiocarbon After Four Decades* (Springer-Verlag, New York, 1992).

WALTER KUTSCHERA

RADIOACTIVITY

Radioactivity refers to the spontaneous disintegration of atomic nuclei by the emission of particles or electromagnetic radiation. The main categories of radioactivity include alpha decay (emission of a helium nucleus), beta decay (emission of an energetic electron), and gamma decay (emission of an energetic photon). However, x-ray emission, nuclear fission, and emission of a nucleus of carbon-12 are also observed to occur in spontaneous decay and can be classified as a type of radioactivity.

Discovery of Radioactivity

The first discovery of radioactivity was by Antoine Henri Becquerel in 1896. He found that substances containing uranium would darken a photographic plate even though separated from the plate by glass or dark paper. He also found that these radiations could ionize air, rendering it a conductor of electricity. Hence, an electrically charged body, placed near a sample of uranium, soon lost its charge by conduction through the air. This phenomenon provided a sensitive method for measuring the intensity of radiation. By using this method, Pierre and Marie Curie discovered that thorium is also radioactive. They investigated many uranium compounds and minerals and found that some of the minerals were much more radioactive than pure uranium itself. This suggested that these minerals contained radioactive substances in addition to uranium. Further investigation led to the discovery of the elements polonium (atomic number 84) and radium (atomic number 88) in 1898. Both of these elements are intensely radioactive.

Radioactivity was soon recognized as a more concentrated source of energy than had been previously known. It was Ernest Rutherford, however, who discovered that at least two different types of radiation were emitted: alpha particles, which have a low penetrability through matter; and beta particles, which require roughly 1,000 times as much matter to bring them to rest. In subsequent experiments Rutherford passed radioactive disintegrations through magnetic and electric fields. This demonstrated the presence of gamma rays as a third component.

Alpha Radioactivity

In 1903 Rutherford found that alpha particles are slightly deflected toward the negative pole when passing through an electric field, whereas beta rays are strongly deflected toward the positive pole. Gamma rays, being uncharged, pass through unhindered. From the amount of deflection, the ratio of the charge to mass and the speed of alpha and beta particles could be determined. Particles of mass m, charge q, and velocity v, when placed in a magnetic field of strength B, move in a circular path of radius r_m given by

$$r_m = (mcv/qB),$$

where c is the velocity of light. In an electric field of uniform intensity E around a point source of opposite charge, the radius of curvature r_e is given by equating the electric potential energy Eqr_e to the kinetic energy $\frac{1}{2}mv^2$. This gives

$$r_e = \frac{1}{2}mv^2/Eq.$$

Hence, a measurement of r_m in a magnetic field B determines the ratio mv/q, and a measurement of r_e

in an electric field E gives a value for mv^2/q. By combining these two results, separate values for v and q/m can be obtained. Rutherford found that the velocity of alpha particles emitted by ^{214}Bi was 2.06×10^9 cm/s, and the ratio of q/m was one-half the ratio for a hydrogen atom. The direction of deflection showed that the charge was positive. In order to determine the charge and mass separately, Rutherford and Hans Geiger irradiated a known number of alpha particles per second into a metal plate and measured the current thus generated. The charge was found to be exactly twice the electron charge, but positive instead of negative. The mass must therefore be four times that of hydrogen. This suggested that alpha particles were helium nuclei. This was confirmed in a subsequent experiment by Rutherford and Thomas Royds in 1909 in which alpha particles from the decay of radon gas were allowed to pass through a thin glass tube into an outer chamber. There the alpha particles were stopped and compressed into a discharge tube where the gradual appearance of the characteristic spectrum of helium was observed. This confirmed that alpha particles were indeed helium nuclei.

The theory of alpha decay was first worked out by George Gamow and independently by Ronald Gurney and Edward Condon in 1928. In alpha-decay theory, the alpha particles are viewed as oscillating within the nuclear potential such that they make repeated collisions with the Coulomb barrier of the potential. According to quantum mechanics, there is a certain probability that the alpha particle can tunnel through the barrier and appear outside the nucleus during these collisions. The rate of alpha decay thus relates to the height of the barrier through which the alpha particle must tunnel. For this reason, alpha decay predominantly occurs among elements heavier than lead, although there are a few exceptions among lighter deformed nuclei.

Beta Radioactivity

By 1900, beta particles were shown by Becquerel and others to have qualitative properties similar to those of fast-moving electrons. However, it was not until 1902 that Walter Kaufmann used combined electric and magnetic fields to measure the velocity and charge to mass ratio for beta rays. For slowly moving beta rays, a charge to mass ratio close to that of electrons in a cathode-ray tube was determined. However, the ratio appeared to decrease for faster moving beta particles. The correct explanation for this was found in Albert Einstein's theory of relativity in 1909. In this theory the observed ratio of charge to mass decreases with increasing velocity, precisely as noted by Kaufmann.

Beta decay is a common form of radioactivity. It occurs via the weak interaction in a process whereby a neutron in an atomic nucleus converts to a proton by emitting a fast-moving electron and a neutral neutrino. It can also involve the transformation of a proton into a neutron by the emission of a positively charged antielectron and an antineutrino, or in a related phenomenon referred to as electron capture. In electron capture a proton captures an electron and emits an antineutrino.

The theory of beta decay was first explored by Enrico Fermi in 1934. In this picture the available energy for the decay is divided between the beta particle and the neutrino in all possible combinations. The relative slowness of the beta decay process is due to the fact that the interaction between the nucleons, the beta particle, and the neutrino is so weak.

Gamma Radioactivity

Gamma radiation often accompanies alpha or beta decay in which the daughter nucleus is left in an excited state. The gamma rays are energetic photons that are formed by the de-excitation of a nucleus to a state of lower energy. A gamma ray, like any other electromagnetic particle, represents oscillating electric and magnetic fields. In nuclear decay, gamma rays occur in discrete energies corresponding to the difference between the energies of the initial and final nuclear states. They are typically characterized by the changes in the moments of the nuclear charge distribution when they are emitted. This is referred to as the multipolarity.

A related type of radiation is called internal conversion whereby the energy of the emitting gamma ray is imparted to an electron instead, which then carries the energy away from the system. This phenomenon can give rise to what are called Auger electrons, which are emitted as the atom rearranges to fill the vacant electron orbital left by the emitted electron.

Nuclear Fission

Nuclear fission was discovered by Otto Hahn and Fritz Strassmann in 1938. After the bombardment of uranium with neutrons it was discovered that chemically distinguishable lighter radioactive isotopes are produced. Nuclear fission is a process whereby a nucleus splits into two fragments either spontaneously or induced by incident particles. Spontaneous fission occurs predominantly in the heaviest nuclei that have the largest number of protons concentrated in the nucleus and, therefore, have large Coulomb forces that tend to pull the nucleus apart. Fission typically produces a wide range of products referred to as the yield. Most of the fission yield involves radioactive product nuclei on the neutron rich side of nuclear stability. A great deal of energy is released during the fission process, which mostly appears as the kinetic energy of the fragments. This kinetic energy provides most of the heat energy for nuclear reactors.

Half-Life

In general the rate of decay of an amount of radioactive material is proportional to the number of nuclei present. That is, if there are N nuclei present, then the rate of change of the number of nuclei is

$$\frac{dN}{dt} = \lambda N,$$

where λ is a characteristic decay constant. The solution to this equation is a decaying exponential,

$$N = N_0 e^{-\lambda t},$$

where N_0 is the initial number of nuclei present.

From this, a characteristic time can be determined for a sample of N_0 nuclei to reduce in half due to radioactive decay,

$$\frac{N}{N_0} = \frac{1}{2} = e^{-\lambda t_{1/2}},$$

which gives

$$t_{1/2} = \frac{\ln 2}{\lambda} = \frac{0.693}{\lambda}.$$

The quantity $t_{1/2}$ is referred to as the half-life. It is generally used to characterize the relative decay rates of various radioactive isotopes.

Bibliography

FESHBACH, H. *Theoretical Nuclear Physics: Nuclear Reactions* (Wiley, New York, 1992).

HARVEY, B. G. *Introduction to Nuclear Physics and Chemistry,* 2nd ed. (Prentice Hall, Englewood Cliffs, NJ, 1969).

HEYDE, K. L. G. *Basic Ideas and Concepts in Nuclear Physics: An Introductory Approach* (Institute of Physics, Bristol, Eng., 1994).

PRESTON, M. A., and BHADURI, R. K. *Structure of the Nucleus* (Addison-Wesley, Reading, MA, 1975).

GRANT J. MATHEWS

RADIOACTIVITY, ARTIFICIAL

In contrast to natural radioactivity, as usually encountered in nature, artificial radioactivity is the property of radioactive products obtained by artificial means. The existence of this artificial radioactivity was identified for the first time in January 1934 by Frederic Joliot and his wife Irene Curie, daughter of the renowned Marie Curie.

In about 1928, the Joliot-Curies began a research program to examine the action on matter of alpha rays, emitted by polonium, a natural radioactive element. In 1930, in Germany, Walther Bothe and Herbert Becker found that these rays, sent on beryllium, produced an intense neutral radiation. In 1932 the Joliot-Curies found that this radiation, supposed to be constituted of gamma rays, had the power to eject protons from matter. Unfortunately for them, James Chadwick, in Ernest Rutherford's laboratory in Cambridge, England, showed that this radiation consisted not of gamma rays but of neutrons. A nucleus N was then supposed to be composed of protons and neutrons and could be represented as $_Z N^A$, where Z is the number of protons in the nucleus and A its number of nucleons, that is, its total number of protons and neutrons. For example, alpha (α) rays, or helium nuclei, could be denoted as $_2\alpha^4$.

Chadwick's discovery of the neutron was followed by the discovery of the positive electron e^+ by Carl Anderson in the United States in the study of cosmic rays. The Joliot-Curies then found that, in some cases, alpha rays could produce—apparently in the same reaction and at the same time—a neutron and a positive electron. That was the case for aluminum as a target. In verifying the reality of that curious reaction, they discovered that the irradiated matter continued to emit positrons, even after withdrawal of the alpha ray irradiating source. It was a two-step mechanism. The first reaction was the following:

$$_2\alpha^4 + {}_{13}Al^{27} \rightarrow {}_{15}P^{30} + {}_0 n^1.$$

In a second step, $_{15}P^{30}$, an isotope of phosphorus decayed emitting a positron according to the reaction:

$$_{15}P^{30} \rightarrow {}_{14}Si^{30} + e^+ \ (+ \text{ invisible neutrino}).$$

In their experiment, the Joliot-Curies showed that the irradiated aluminum indeed contained an element having the same chemical properties as those of phosphorus. At that time, it was known (thanks to Rutherford in 1919) that alpha rays can produce nuclear transmutations, but it was believed that the elements thus created were stable elements. The Joliot-Curie's finding therefore was an entirely new discovery.

In 1935 the Joliot-Curies were awarded the Nobel Prize in chemistry for their discovery. At the same time Chadwick received the Nobel Prize in physics for his discovery of the neutron.

After the discovery of the neutron in 1932, Enrico Fermi in Rome began to study the action of neutrons on various elements. He reasoned that transmutations would be easier to induce with neutral particles such as neutrons than with charged particles such as alpha rays. About two months after the Joliot-Curies' discovery, confirmed in many places, Fermi and his team published the results of their neutron experiments. Numerous cases of artificial radioactivity were recorded, and various lifetimes of the products were measured. However, in general, several elements seemed to be created at the same time, and the corresponding reactions were not identified. An element such as uranium, for example, was active, but its fission possibility was identified only in 1938 in Berlin by Otto Hahn and Fritz Strassmann, who found barium in the samples of irradiated uranium.

Practical applications of artificial radioactivity have been many. Very soon after its discovery by Henri Becquerel in 1896, natural radioactivity was found to have important biological effects. Its possible action on lesions or cancer was suspected, and at the beginning of the twentieth century, the action of radium began to be tested in these respects. After World War I, around 1923, the Hungarian Georg de Hevesy (who received the Nobel Prize in chemistry in 1943) invented the technique of radioactive tracers for biological use. He knew that an element and its isotopes have the same chemical properties, and, for studying living matter, he thought to replace an element accepted by an organism with an isotope that could be traced on its way through the body by its recorded radioactivity.

The discovery of artificial radioactivity allowed important progress in these various techniques. For instance, Co^{60} has now replaced radium in the treatment of cancer. Radioactive iodine is used as a remedy for an ill-functioning thyroid gland and also to control its effects. For these two types of uses—therapeutical and biological—a great variety of radioactive elements are used nowadays. They are created mainly in nuclear reactors.

See also: CHADWICK, JAMES; DECAY, ALPHA; DECAY, NUCLEAR; FISSION; NEUTRON, DISCOVERY OF; RADIOACTIVITY; RADIOACTIVITY, DISCOVERY OF; RUTHERFORD, ERNEST

Bibliography

CURIE, I., and JOLIOT, F. "Un Nouveau Type de Radioactivité." ("A New Type of Radioactivity.") *Comptes Rendus Acad. des Sciences* (Paris) **198,** 254–266 (1934).

MURRAY, I. P., and ELL, P. J. *Nuclear Medicine in Clinical Diagnosis and Treatment* (Churchill Livingstone, New York, 1994).

SAMPSON, C. B. *Textbook of Radiopharmacy: Theory and Practice,* 2nd ed. (Gordon and Breach, New York, 1994).

SEGRÉ, E. *From X-Rays to Quarks: Modern Physicists and Their Discoveries* (W. H. Freeman, New York, 1980).

JULES SIX

RADIOACTIVITY, DISCOVERY OF

Surprising as it seems to us today, toward the end of the nineteenth century there was a small but persistent undercurrent of belief that all the great discoveries in science had been made and only a search for greater accuracy in measurements lay ahead. Of course, events provided a different history, especially the remarkable discoveries of x rays in 1895, radioactivity in 1896, the electron in 1897, quantum theory in 1900, and relativity theory in 1905.

Radioactivity's discovery has been called accidental—a lucky observation. But, as Louis Pasteur is said to have remarked, "Chance favors the prepared mind." Henri Becquerel, thus, was not only well trained for his role but intelligent enough to recognize an unexpected result and to investigate it.

Becquerel was third in a line of four notable French physicists who, remarkably, all held the professorship of physics at the Museum of Natural History in Paris. Popular nineteenth-century subjects, such as electricity and magnetism, phosphorescence of solids, and photography, all of which were involved in the discovery of radioactivity, were also family interests. Of special relevance, Becquerel's father was an expert on uranium salts, drawn to them by their exceptionally bright phosphorescence and interesting spectra, and young Henri also investigated this element.

The element uranium was discovered by Martin Klaproth in 1789 and named in honor of the recent discovery of Uranus, the first planet found in recorded history. (The tradition continued with the naming of neptunium and plutonium.) When the periodic table was constructed in 1869, uranium was distinguished as the heaviest element, but it had few commercial uses beyond coloring glass and ceramics.

Photography seems to have entered the laboratory by the mid-nineteenth century, but it found its greatest use after around 1880, by which time convenient, dry, gelatine-emulsion plates were widely available. Light-sensitive emulsions were used to record lightning, spectra, sound waves, objects in motion, and, most important, x rays. Wilhelm Conrad Röntgen was delighted that dry plates were so sensitive to x rays, and the instantaneous, worldwide fame of his discovery was due in large part to the spectacular shadow pictures he made of bones inside living flesh and their obvious application in diagnostic medicine.

X rays were the subject of excited discussion at a meeting, late in January 1896, of the French Académie des Sciences, where Henri Poincaré was convinced that the invisible x rays originated in the glass spot on the vacuum tube that was made fluorescent by the bombardment of cathode rays. Might not other strongly fluorescent bodies, he asked, emit such invisible radiation in addition to visible light? Academician Becquerel, considering himself well qualified for such an investigation, was inspired to test this idea.

Within a few weeks, Becquerel found several phosphorescent materials to be quite active, uranium salts in particular. He wrapped a photographic plate in two sheets of black paper to make it light-tight and placed a lump of uranium salt on it. When developed a few hours later, the region under the crystal showed a smudge of exposure. Later, he placed coins on the paper, covered them with a thin crystalline layer of salt, and was rewarded with silhouettes of the design. Becquerel believed that he had merely confirmed that x rays were produced by luminous sources other than cathode-ray tubes.

The uranium rays were powerful since they could pass through thin sheets of aluminum and copper, although they were somewhat attenuated. In an effort to learn whether the nature of the light that stimulated his crystals to phosphoresce had any bearing, Becquerel tried diffuse sunlight, as well as reflected and refracted sunlight. All of these sources produced exposures on his plates as strong as that from direct sunlight on the crystals.

Because he operated on the theory that the crystals had to phosphoresce in order to emit the invisible rays, Becquerel customarily left his plates and crystals in the sunlight on his window sill. (Their visible glow stopped almost immediately when removed from the sunlight.) At the end of February 1896, he prepared some new experiments but encountered a week of almost totally overcast skies. Awaiting sunnier days, he placed his experiments in a dark drawer. The weather did not cooperate, but Becquerel decided to develop the plates anyway. They had presumably been exposed slightly to diffuse light, and he might be able to report on this "control" experiment at the next day's meeting of the Académy. To his amazement, the plates were very strongly exposed.

This made him wonder if the crystals had to be exposed to light at all. Further tests confirmed that stimulation by sunlight was unnecessary; the crystals could emit invisible radiation long after their visible glow ceased. At this point, he still believed that he was dealing with radiation that was similar to x rays and associated with the phenomenon of phosphorescence.

Becquerel next conducted a series of investigations that strongly resembled those recently performed by Röntgen on x rays; they were, after all, the standard tests of radiation. Since x rays made a gas a conductor of electricity, Becquerel tested the effect of uranium rays on air. They caused the extended gold leaves of a charged electroscope to fall. He (incorrectly) found the rays to be reflected, refracted, and polarized, thereby confirming for him their electromagnetic nature and similarity to x rays.

He was at a loss, however, to explain the intense images on a photographic plate produced by non-phosphorescent uranous sulphate. Pursuing this line of investigation, he melted a crystal of uranium nitrate in a sealed glass tube, which destroyed its ability to phosphoresce. He then allowed it to recrystallize in darkness. All phosphorescence had been destroyed in this process, yet the salt produced results on the photographic plate as strong as crystals exposed to light.

By May 1896 Becquerel reasoned that uranium element was the source of activity. All uranium compounds tested, whether phosphorescent or not, were active. More to the point, uranium metal was far more active than the salts; it was, thus, an atomic phenomenon. Yet Becquerel was not ready to reject all of his previous conjectures, for he argued that this emission from uranium was probably the first example of a metal having a phenomenon such as invisible phosphorescence.

During the first months of 1896, approximately five papers on x rays could be found in each weekly issue of the Académie's journal, the *Comptes Rendus*, while only Becquerel wrote on uranium rays. Worldwide, more than 1,000 papers on x rays were published in 1896 alone. X rays were more popular because they produced sharper photographs faster, and they were useful. Also, vacuum tubes and spark coils needed to generate x rays were more common to physics laboratories than uranium salts. Becquerel himself seems to have lost interest in his discovery; he published seven papers on radioactivity in 1896, two in 1897, and none in 1898. Others, including some very able physicists, contributed only about another dozen papers during the first two years. In a period when a large number of different radiations were under study, Becquerel rays, as they were also called, were not exceptional enough to attract much attention.

In early 1898 Gerhard C. Schmidt of the University of Erlangen reported that he had examined a number of substances and found that thorium emitted invisible rays. Due to confusion in some of the properties claimed by Becquerel, Schmidt could not say that thorium and uranium rays were identical. Shortly thereafter and independently, Marie Curie in Paris also discovered thorium's activity. It is not known why she began her study of this topic, which apparently had been exhausted by Becquerel. Perhaps she saw some interesting questions remaining, or maybe she merely wanted an obscure topic for her doctoral degree research, one on which she would not be "scooped."

Curie brought to her work the careful, exhaustive, and energetic traits for which she became famous. In the belief that Becquerel rays were an atomic property, she planned to examine every element. The intensity of any radiation found was measured with a sensitive electrometer that was designed by her physicist husband, Pierre Curie, and his brother, Jacques. Becquerel's investigation had been partly qualitative (the photographic plates) and partly quantitative (the electroscope). Curie's work was more in the "modern" tradition of science: wholly quantitative, seeking the best numbers possible.

Her working hypothesis was that all of space was constantly traversed by rays analogous to x rays but much more penetrating (and not then—or ever—detected). These rays were absorbed by elements of high atomic weight, such as uranium and thorium, which in turn emitted Becquerel rays as secondaries.

By April 1898 Marie Curie reported that uranium ore was even more powerful that uranium metal; some other active element must be in the mineral. Pierre now joined Marie in this research, which three months later led them to announce the discovery of a new element (which they named polonium after her native Poland), and for the first time to use the term "radioactive," which Marie coined. In the process of isolating polonium, they came across still another activity. By the end of the year, they and their colleague Gustave Bémont revealed the discovery of radium. This new element, the major vehicle for their fame, was more striking than uranium, thorium, polonium, and many other radioelements later discovered, because its radiation was far more intense than that of any substance that could be gathered in tangible amounts. The source of all this energy was of concern, but their working hypothesis, of an all-pervasive and undetected radiation that stimulated radioactivity, sufficed for the moment.

Another physicist entered the field of radioactivity in 1898 and almost immediately became the central figure of this science and of the nuclear physics into which it evolved. This was Ernest Rutherford, then a research student at Cambridge University. He had just completed an investigation with his professor on the ionizing effects of x rays. Now, even before he knew of thorium's activity, Rutherford examined the similar effect of uranium rays on the conductivity of gases. Soon, his attention turned from ionization to the radiations and their emitters.

With the experimental skill that became his trademark—simple experiments that minimized extraneous influences—Rutherford negated Becquerel's belief that uranium rays could be reflected, refracted, and polarized. (Schmidt and Marie Curie also participated in this correction.) Rutherford found uranium radiation was complex and labeled the easily absorbed component "alpha" and the more penetrating part "beta" rays. (Paul Villard in 1900 discovered a third component, named "gamma.")

In 1899 Friedrich Giesel, a commercial chemist in Braunschweig, showed that beta rays could be bent from their paths by a magnetic field. This meant that they were charged particles. The next year, Becquerel, who returned to the subject, determined that betas and electrons were identical. Not until 1903 did Rutherford prove that alphas were positively charged particles. In his first professorship, at McGill University, Rutherford in 1900 showed that thorium gave off a gaseous radioactive product, which he called "emanation," and that in turn it left on exposed surfaces an "active deposit." Working with the Curies in 1899, André Debierne discovered another radioelement, actinium. In the twentieth century, many more radioactive bodies would be found.

Most of these bodies maintained a constant level of activity. But some, such as polonium, thorium emanation, and thorium active deposit, decayed in strength over time. Rutherford plotted such activity changes, finding an exponential curve. This led immediately to the concept of half-life, which was a vital identity badge for each radioelement, especially when their abundance was far too small for normal chemical identification tests.

Rutherford also recognized that the decay curve for thorium emanation and the growth curve for thorium active deposit were identical. This suggested a mother-daughter relationship. With chemist Frederick Soddy, Rutherford in 1902 advanced the transformation theory of radioactivity. The many radioelements were ordered in just a few families, with uranium and thorium at the top of their own series. As they decayed (the time period being so long that their lost material was unmeasurable), daughter products were formed. But they too were radioactive, so their amount would reach an equilibrium value, where the number of atoms formed equalled the number that decayed. At the end of the series was an inactive element, soon to be identified as lead. During the first several years after Becquerel discovered radioactivity, physical study of the radiations predominated. Now, with an atomic theory that claimed the spontaneous transformation of one element into another, radiochemists became increasingly important; one wanted to identify each chemical element.

Rutherford and Soddy placed the source of the radiation's energy in the atom itself. Indeed, the radiation marked the parent atom's transmutation into its daughter. This pointed more than ever to the large store of energy in the atom. In 1903 Pierre Curie and Albert Laborde showed that a radioactive source maintained itself at a temperature higher than its surroundings. This quantitative demonstration sparked a four-decade-long debate over the possibility of harnessing the atom's energy, which ended during World War II with the construction of nuclear reactors and bombs.

A few other applications did not take so long to materialize. From an accidental burn that Becquerel experienced in 1900 by carrying a vial of radium in his pocket, physicians found that the radiation killed cancerous tissue more readily than healthy tissue. By the next decade, hospitals vied to acquire a gram of radium for the only cancer treatment then available, other than the surgeon's knife. One of the techniques of measuring radioactivity was to count the flashes of light the rays caused on scintillation screens. When scintillation crystals were mixed in a paint, and a "dash" of radium added, any object painted (such as a watch dial or an electrical switch) would glow in the dark, a popular use in World War I and later.

Finally, the energy emitted in radioactive decay showed itself as heat, as Pierre Curie and Laborde found. This led Rutherford to suggest a resolution of a decades-long problem. Physicists in the nineteenth century, led by William Thomson (Lord Kelvin), had calculated a very short age of the earth on the basis of the rate of cooling of a once-molten globe. Geologists and biologists felt that they needed more time for the slow changes in their subjects, but they had no independent and valid "clock." With radioactive minerals distributed widely around the earth, providing an additional source of heat, it was possible for Bertram Boltwood, a radio-chemist at Yale University, in 1906 to extend the computed age of the earth—to more than one billion years.

See also: CURIE, MARIE SKLODOWSKA; RADIOACTIVE DATING; RADIOACTIVITY; RADIOACTIVITY, ARTIFICIAL; RÖNTGEN, WILHELM CONRAD; RUTHERFORD, ERNEST; SCATTERING, RUTHERFORD; X RAY; X RAY, DISCOVERY OF

Bibliography

BADASH, L. "Radioactivity Before the Curies." *Am. J. Phys.* **33**, 128–135 (1965).

BADASH, L. "Chance Favors the Prepared Mind: Henri Becquerel and the Discovery of Radioactivity." *Archives Internationales d'Histoire des Sciences* **18**, 55–66 (1965).

BADASH, L. "How the 'New Alchemy' was Received." *Sci. Am.* **215**, 88–95 (1966).

BADASH, L. "Becquerel's 'Unexposed' Photographic Plates." *Isis* **57**, 267–269 (1966).

BADASH, L. "The Completeness of Nineteenth-Century Science." *Isis* **63**, 48–58 (1972).

BADASH, L. *Radioactivity in America: Growth and Decay of a Science* (Johns Hopkins University Press, Baltimore, MD, 1979).

BADASH, L. "The Age-of-the-Earth Debate." *Sci. Am.* **261**, 90–96 (1990).

CURIE, M. *Pierre Curie* (Macmillan, New York, 1923).

ROMER, A. *The Restless Atom* (Doubleday, Garden City, NY, 1960).

ROMER, A. *The Discovery of Radioactivity and Transmutation* (Dover, New York, 1964).

LAWRENCE BADASH

RADIO WAVE

Radio waves are electromagnetic (EM) waves usually produced by oscillating currents in electrical circuits. The existence of these waves, which propagate in space at the speed of light, was predicted by James Clerk Maxwell in 1862 and later demonstrated in the laboratory by Heinrich Hertz in 1888. The full spectrum of electromagnetic waves ranges from low-frequency radio waves; to television waves; to microwaves used in radar; to infrared, visible, and ultraviolet light; then to x rays and gamma rays of extremely high frequency and short wavelength. The universal relation between wavelength, frequency, and the speed of EM waves is that multiplying wavelength and frequency together always gives the speed of light.

The generation of radio waves is closely related to their detection, as an antenna connected to a generating circuit for efficient wave production also can receive waves that induce a current in a receiving circuit. Radio waves travel easily in space, reflect from metal surfaces, and refract in glass and paraffin, but do not penetrate much into salt water due to the conductivity of the fluid.

The use of radio waves as a means of long-distance communication was initially done simply as wireless telegraphy by Guglielmo Marconi in 1895, transmitting information with the long and short pulses of Morse code. The transmission distances increased, and in 1901 transatlantic wireless telegraphy was demonstrated. However, it was not until the invention of the vacuum electron tube "valve" by John Fleming in 1904 that radio waves could be altered continuously to transmit voice and music. This is done by modulating the current in the oscillator circuit; that is, altering the amplitude of the high-frequency oscillation by a slower audio signal. This type of modulation is called amplitude modulation (AM), and is still in use today in the low-frequency radio band 600–1600 kHz.

To receive the desired voice or music information, the modulated wave must then be detected with an antenna circuit in the radio receiver, then demodulated, which is the process of removing the high-frequency carrier wave, leaving the low-frequency audio information to be amplified and passed to sound-reproducing speakers.

One problem with AM transmission is that it is vulnerable to interference by unwanted sources of waves such as lightning flashes, automobile ignition systems, and appliances, which generate amplitude noise at the receiver in the same band as the desired modulation. To overcome this drawback other forms of modulation have been invented. The most important is frequency modulation (FM), in which the desired voice or music information alters the oscillating frequency but not the amplitude. The transmitted wave is then a complicated band of varying frequencies that needs to be demodulated by the receiver. This is done by a special circuit called a "discriminator," whose output response varies according to the frequency of the incoming signal; the resulting amplitude-varying signal is then amplified in the usual way and fed to sound speakers. Frequency modulation signals are not af-

fected by amplitude changes from external noise generators, and tend to be noise-free and of high fidelity, which has made FM the method of choice for music transmission.

See also: ELECTROMAGNETIC SPECTRUM; ELECTROMAGNETIC WAVE; LIGHT, ELECTROMAGNETIC THEORY OF; MAXWELL, JAMES CLERK; OSCILLATION

Bibliography

GLASGOW, S. *From Alchemy to Quarks* (Brooks-Cole, Pacific Grove, CA, 1994).

TERMAN, F. E. *Radio Engineering,* 4th ed. (McGraw-Hill, New York, 1955).

ISAAC D. ABELLA

RAINBOW

The rainbow, the most spectacular display of color in nature, is due to a seemingly insignificant variation with wavelength in the optical properties of water. The index of refraction of water is 1.339 at a wavelength of 400 nm and 1.331 at a wavelength of 700 nm. This small difference means that the shortest visible wavelength (blue-violet) travels through water at a speed that is about 0.6 percent slower than the longest visible wavelength (red). Such a dependence of speed or index of refraction on wavelength is known as dispersion. It means that the different wavelengths of visible light will bend in different directions when entering or leaving any droplet of water in air. The difference in direction ultimately results in the rainbow.

A rudimentary understanding of the rainbow dates back as far as 1637, when the great French scientist and philosopher René Descartes was able to explain why sunlight on rainfall produced an arc of light at all. The explanation of the colors of the rainbow had to wait for the critical discovery of Isaac Newton that white light was really a mixture of all the colors of the rainbow. The explanation originally proposed by Descartes can be reprised by tracing rays through a spherical water droplet in air, as in Fig. 1. The primary rainbow is produced by rays of sunlight like these that reflect once and refract twice at the droplet boundary.

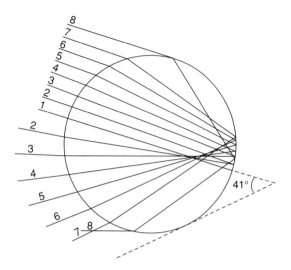

Figure 1 Illustration tracking light rays through a spherical water droplet.

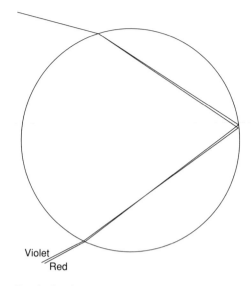

Figure 2 A single ray of white light split into colors.

Figure 1 shows uniformly spaced, parallel rays impinging on half the raindrop. After refraction upon entering, reflection at the back surface, and refraction again upon exiting, the rays are neither parallel nor uniformly spaced. Instead, they are concentrated at an angle of approximately 41° with respect to the incident direction. Such a concentration of rays means that raindrops will appear particularly bright when viewed from this angle, hence the "bow" of light. Of course, Fig. 1 is really symmetric about the axis (ray 1), and rays entering the other half of the droplet similarly concentrate at 41° on the other side. There is really a cone of enhanced brightness coming from each raindrop.

The colors of the rainbow are explained by the fact that water shows dispersion; each wavelength from the white sunlight actually travels through the droplets at slightly different angles. In other words, Fig. 1 is correct for only one wavelength. Because blue light refracts more than red light, the angle of concentration for blue light is about 40°, whereas that for red light is about 42°, with the other colors of the spectrum at intermediate angles. Many arcs or bows of light are formed, each at a slightly different angle and each of a different color. Since sunlight contains a continuous range of wavelengths, the colors of the rainbow blend continuously into each other. Figure 2 shows how a single ray of white light splits into colors; this particular ray produces a difference in exit angle of 1.15° between violet light and red light.

The appearance of the primary rainbow to an observer on the ground is illustrated by Fig. 3. The Sun must be shining on raindrops from a direction opposite the rainbow. For a given observer position, all the raindrops along a line that makes a 42° angle with a line from the Sun through the head of the observer must be sending concentrated red light to the observer. Similarly, all those along a 41° angle must be sending concentrated violet light. The rainbow is not at a fixed *position*, but rather at a fixed *angle* with respect to the Sun-observer line. This is the reason that the rainbow seems to recede before an observer approaching it; when the observer moves in the direction of the bow, new raindrops further in that direction become the ones contributing to the effect, while raindrops that had previously contributed become ineffective for that observer.

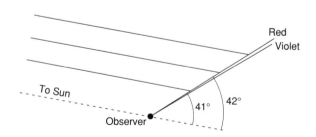

Figure 3 The appearance of the primary rainbow to an observer on the ground.

Although the light leaving a raindrop is concentrated at approximately 41°, Fig. 1 shows that some light comes out at lesser angles; none emerges at greater angles. The light leaving droplets at smaller angles will be visible to the observer of Fig. 2 from raindrops below the rainbow. By contrast, raindrops just above the rainbow should be sending no light toward the observer, since that would be light from angles greater than 41°. The result is that the sky just outside the primary rainbow is noticeably darker, while that inside is noticeably lighter than the sky in general. This effect is easy to see once the observer learns to look for it.

Because the rainbow angle is fixed with respect to the Sun-observer line, the higher the sun, the lower the rainbow, and vice versa; some spectacularly large rainbows are seen when the sun is low in the morning or evening sky. If the observer is viewing sunlight on rain from an elevated position (such as a mountaintop or airplane), so that raindrops are below as well as above the viewing position, then a complete circular rainbow may be seen, with red on the outside and violet on the inside, darker sky outside and lighter sky inside. By contrast, when the observer is on a level plane and the sun is more than 42° above the horizon, no rainbow can be seen because it would be below the horizon, or, in other words, no raindrops above ground are sending concentrated light toward the observer.

The primary rainbow is the brightest one, and often the only one noticed by a casual observer. However, careful observation will usually reveal a larger, dimmer secondary rainbow outside the primary. This secondary rainbow is caused by light rays that make two internal reflections before leaving the droplets. In this case, the light tends to concentrate at an angle of about 51° with respect to the source direction. Again, the different wavelengths of sunlight will concentrate at slightly different angles, giving rise to a colored bow at an angle of 51° to the Sun-observer line and therefore appearing outside the primary rainbow. However, the second reflection reverses the order of the colors from that of the primary bow: red is on the inside and violet on the outside of the secondary bow. Also, the light and dark sky phenomenon is reversed for this rainbow. The net effect is that the sky is lighter both inside the primary and outside the secondary, but darker between the rainbows.

See also: DISPERSION; ELECTROMAGNETIC SPECTRUM; LIGHT; REFLECTION; REFRACTION; REFRACTION, INDEX OF; SKY, COLOR OF

Bibliography

BRAGG, W. *The Universe of Light* (Dover, New York, 1959).

FALK, D.; BRILL, D.; and STORK, D. *Seeing the Light* (Harper & Row, New York, 1986).

GREENLER, R. *Rainbows, Halos, and Glories* (Cambridge University Press, Cambridge, Eng., 1980).

MINNAERT, M. *The Nature of Light and Colour in the Open Air* (Dover, New York, 1954).

WALDMAN, G. *Introduction to Light* (Prentice Hall, Englewood Cliffs, NJ, 1983).

WILLIAMSON, S., and CUMMINS, H. *Light and Color in Nature and Art* (Wiley, New York, 1983).

GARY WALDMAN

RAMAN SCATTERING

See SCATTERING, RAMAN

RAY

See CATHODE RAY; COSMIC RAY; N RAY; X RAY

RAY DIAGRAM

A ray diagram is a geometrical construction that is useful in locating and determining the nature of images formed by optical systems consisting of various types of mirrors and lenses. Ray diagrams use just a few special light rays whose reflections and refractions obey simple rules. Mirrors form images by reflecting light, whereas lenses refract light; thus, the rules are slightly different for the two cases.

A spherical mirror is characterized by two special points: the center of curvature C and the focal point F, both of which are located on the axis of the mirror. These two points are positioned in front of a concave mirror and behind a convex mirror.

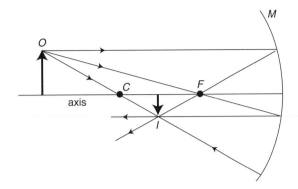

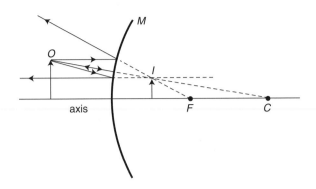

Figure 1 Ray diagram locating real image *I* formed by concave mirror *M* when object *O* is positioned outside center of curvature *C*.

Figure 2 Ray diagram locating virtual image *I* formed by convex mirror *M*.

Rule 1. Any incident ray parallel to the axis is reflected through *F* for a concave mirror, and it is reflected so as to appear to come from *F* for a convex mirror.

Rule 2. Any incident ray that passes through *F* for a concave mirror is reflected parallel to the axis, whereas any incident ray directed toward *F* of a convex mirror is reflected parallel to the axis.

Rule 3. Any incident ray that passes through *C* for a concave mirror or is directed toward *C* for a convex mirror is reflected back on itself.

These rules are applied in Fig. 1 to the construction of a ray diagram for a concave mirror (*M*) that bows away from the object. The three special rays leaving the uppermost point on the object (*O*) converge at a corresponding point on the image (*I*). Once this image point has been determined, it is only necessary to drop a perpendicular to the axis to obtain the entire image. It is observed from the ray diagram that this image is nearer the mirror than the object. Also, the image is inverted with respect to the object, and it is smaller than the object. Since the rays actually pass through the image, this type of image is referred to as a real image. In addition to being visible to the human eye, a real image can be projected on a screen. If the position of the object with respect to the mirror is changed, the position and nature of the image also will change as can be demonstrated by constructing ray diagrams.

Figure 2 is a ray diagram for a convex mirror (*M*) that bulges toward the object. It is observed that the three special rays leaving the uppermost point of the

object (*O*) diverge after reflection. Thus, they must each be extended behind the mirror in order to locate the corresponding image point *I*. An image such as this in which rays appear to emanate from an image behind the mirror is called a virtual image. A virtual image can be seen by the human eye, but cannot be projected on a screen. Also, this image is upright, smaller than the object, and closer to the mirror than the object.

A thin lens with two spherical surfaces is characterized by two focal points *F* and *F'* on either side of the lens along the axis.

Rule 1. Any incident ray parallel to the axis is refracted through the second focal point *F'* for a converging lens, and is refracted so as to appear to come from *F'* for a diverging lens.

Rule 2. Any incident ray that passes through the first focal point *F* of a converging lens emerges parallel to the axis, whereas any incident ray that is directed toward *F* of a diverging lens emerges parallel to the axis.

Rule 3. Any incident ray that passes through the center of either a converging or diverging lens continues undeviated.

A ray diagram for a converging lens (*L*) that is thicker in the center is shown in Fig. 3. For this object position the resulting image is real, inverted, and larger than the object. Also, the image is located at a greater distance from the lens than the object. This type of lens can form a virtual image in front of the lens if the object is placed inside *F*, as is the case when used as a magnifying glass. The ray diagram shown in Fig. 4 for a diverging lens (*L*) that is thin-

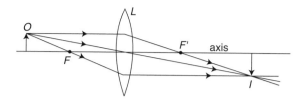

Figure 3 Ray diagram locating real image I formed by converging lens L when object O is positioned outside focal point F.

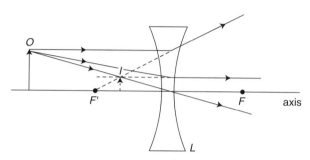

Figure 4 Ray diagram locating virtual image I formed by diverging lens L.

ner in the center shows the resulting image to be virtual, upright, and smaller and closer to the lens than the object.

Many optical instruments, such as telescopes and compound microscopes, contain more than a single mirror or lens. Ray diagrams also can be drawn for these more complex optical systems in order to locate and determine the nature of the final image. The procedure involves treating each optical element in turn; that is, the first element forms an intermediate image that then serves as the object for the second element, and so on until the final image is found.

See also: LENS; LENS, COMPOUND; OPTICS; REFLECTION; REFRACTION

Bibliography

CUTNELL, J. D., and JOHNSON, K. W. *Physics,* 2nd ed. (Wiley, New York, 1989).

FALK, D. S.; BRILL, D. R.; and STORK, D. G. *Seeing the Light* (Wiley, New York, 1986).

HECHT, E. *Optics,* 2nd ed. (Addison-Wesley, Reading, MA, 1987).

TIPLER, P. A. *Physics,* 3rd ed. (Worth, New York, 1991).

ROBERT MORRISS

RAYLEIGH SCATTERING

See SCATTERING, RAYLEIGH

REACTANCE

In alternating current (ac) theory, reactance X refers to the current-limiting property of capacitors and inductors. Like resistance, reactance is a ratio of voltage to current and has the unit ohm (Ω); unlike resistance, it is frequency dependent and does not contribute to energy dissipation. Together, resistance and reactance provide electrical impedance.

Inductive reactance X_L and capacitive reactance X_C are defined by

$$X_L = \omega L,$$
$$X_C = (\omega C)^{-1},$$

where L is the inductance (in henry, H), C the capacitance (in farad, F), and ω the angular frequency of the ac signal [$\omega = 2\pi f$, where f is the ordinary frequency (in hertz, Hz)].

For example, at the frequency 500 Hz, a 2 mH inductance has the reactance

$$X_L = 2\pi(500 \text{ Hz})(2 \times 10^{-3} \text{ H}) = 6.28 \ \Omega.$$

Theory

Inductive reactance. The potential difference V_L across a pure inductor L carrying a current I_L is given by $V_L = L(dI_L/dt)$, where dI_L/dt is the time derivative of the current. Given the ac current, $I_L = I_0 \cos(\omega t)$, where I_0 is the amplitude of the current, and noting that $dI_L/dt = -I_0 \sin(\omega t)$, then

$$V_L = -L\omega I_0 \sin(\omega t) = L\omega I_0 \cos(\omega t + \pi/2).$$

Because of the 90° ($\pi/2$ radian) phase shift between potential difference and current, a point of maximum V_L leads a point of maximum I_L by one quarter of a cycle; or equivalently, I_L *lags* V_L by one quarter

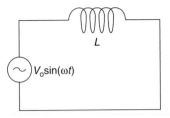

Figure 1 An inductor L driven by a sinusoidal voltage $V_0 \sin(\omega t)$.

cycle. Inductive reactance equals the ratio of voltage amplitude to current amplitude: $X_L = L\omega I_0 / I_0 = \omega L$. For example, consider the sinusoidal potential difference $V = V_0 \sin(\omega t)$ impressed across an inductor L, as shown in Fig. 1. The current amplitude is given by $V_0/\omega L$, and, accounting for the phase lag, $I = (V_0/\omega L)\sin(\omega t - \pi/2)$.

Capacitive reactance. The potential difference V_C across a pure capacitor C carrying an electric charge Q is given by $V_C = Q/C$, where $Q = \int I_C \, dt$ is the integral of the current I_C to the capacitor. If $I_C = I_0 \cos(\omega t)$, $Q = \int I_0 \cos(\omega t) \, dt = \omega^{-1} I_0 \sin(\omega t)$; the constant of integration is set equal to zero assuming that, on average, $Q = 0$. Therefore,

$$V_C = -(\omega C)^{-1} I_0 \sin(\omega t) = (\omega C)^{-1} I_0 \cos(\omega t - \pi/2).$$

Here, a point of maximum capacitor voltage V_C lags a point of maximum current I_C by one quarter cycle. Capacitive reactance equals the ratio of voltage amplitude to current amplitude: $X_C = (\omega C)^{-1} I_0 / I_0 = (\omega C)^{-1}$.

Phasor Representation

Given the ac current $I_0 \cos(\omega t)$, the potential differences across a resistor R, an inductor L, and a capacitor C are, respectively, $V_R = I_0 R \cos(\omega t)$, $V_L = X_L I_0 \cos(\omega t + \pi/2)$, and $V_C = X_C I_0 \cos(\omega t - \pi/2)$. V_R, V_L, and V_C thus can be pictured, as in Fig. 2, as the horizontal projections of three vectors (phasors) of magnitudes $I_0 R$, $I_0 X_L$, and $I_0 X_C$, respectively, rotating with angular frequency ω.

Complex Representation

The phasor representation can be expressed mathematically in terms of complex numbers wherein the current is given by the exponential $I_0 \exp(i\omega t)$, the reactances by the imaginary terms $i\omega L$ and $(i\omega C)^{-1}$, where $i = \sqrt{-1}$; and the resistance R by a real number. Then, the voltage V_R is the real part $\mathrm{Re}[RI_0 \exp(i\omega t)] = RI_0 \, \mathrm{Re}[\exp(i\omega t)] = RI_0 \cos(\omega t)$, using Euler's theorem:

$$
\begin{aligned}
V_L &= \mathrm{Re}[(i\omega L)I_0 \exp(i\omega t)] \\
&= \omega L I_0 \, \mathrm{Re}[\exp(i\pi/2)\exp(i\omega t)] \\
&= \omega L I_0 \cos(\omega t + \pi/2).
\end{aligned}
$$

For example, a real inductor can be modeled as a pure inductor in series with a pure resistance, as shown in Fig. 3; thus resistance and reactance add. The voltage across the combination is given by

$$
\begin{aligned}
V &= \mathrm{Re}[(R + X_L)(I)] \\
&= \mathrm{Re}[(R + i\omega L)(I_0 \exp(i\omega t))];
\end{aligned}
$$

after reduction,

$$V = I_0 (R^2 + (\omega L)^2)^{1/2} \cos(\omega t + \phi),$$

where $\tan \phi = \omega L/R$.

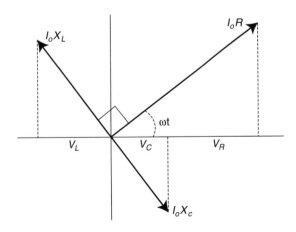

Figure 2 Phasor diagram.

Figure 3 A series RL combination.

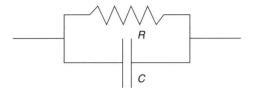

Figure 4 A parallel *RC* combination.

As a second example, a real capacitor can be modeled as a pure capacitor in parallel with a pure resistance, as shown in Fig. 4; thus resistance and reactance add in reciprocal. The voltage across this combination is given by

$$V = \text{Re}[(1/R + 1/X_C)^{-1}(I)]$$
$$= \text{Re}[(1/R + (i\omega C))^{-1}(I_0 \exp(i\omega t))];$$

after reduction,

$$V = I_0 R(1 + (\omega CR)^2)^{-1/2} \cos(\omega t - \psi),$$

where $\tan \psi = \omega CR$.

See also: CAPACITANCE; CAPACITOR; CHARGE; CIRCUIT, AC; CURRENT, ALTERNATING; FREQUENCY, NATURAL; INDUCTANCE; INDUCTANCE, MUTUAL; INDUCTOR; OHMMETER; RESISTOR

Bibliography

DIEFENDERFER, A. J., and HOLTON, B. E. *Principles of Electronic Instrumentation,* 3rd ed. (Saunders, Philadelphia, PA, 1992).

HOROWITZ, P., and HILL, W. *The Art of Electronics,* 2nd ed. (Cambridge University Press, Cambridge, Eng., 1989).

SERWAY, R. A. *Physics For Scientists and Engineers,* 3rd ed. (Saunders, Philadelphia, PA, 1992).

TIPLER, P. A. *Physics For Scientists and Engineers,* 3rd ed. (Worth, New York, 1991).

F. R. YEATTS

REACTION

See CHAIN REACTION; NUCLEAR REACTION

REACTOR, BREEDER

A nuclear reactor that generates more nuclear fuel than it consumes is called a breeder reactor. The term "nuclear fuel," in this context, refers to the fissile isotopes (primarily ^{235}U, ^{239}Pu, and ^{241}Pu), which are the ones that sustain the nuclear chain reaction. New fuel is created when an atom of the fertile isotope ^{238}U absorbs a neutron, becoming ^{239}Pu after one beta decay.

In the early days of nuclear power, the consensus was that the world's supply of economically recoverable uranium was small and that plutonium created in breeder reactors should be substituted for fissile uranium before resources were depleted. Now, uranium reserves are known to be extensive. Considering also the significant quantities of fissile plutonium that have accumulated (approaching a thousand metric tons, worldwide—enough startup fuel for 400 new 1,000-MWe plants), it will be a long time before breeders are necessary.

Net breeding of fuel should not be confused with net production of plutonium. All reactors that have uranium in their fuel contain ^{238}U. Thus, making plutonium does not require a breeder: A typical large power reactor today produces annually about 180 kg more plutonium than it consumes. Essentially all the plutonium in existence was made in thermal reactors—most of it in power reactors, but almost all the plutonium of weapons grade has come from dedicated (thermal) production reactors.

For appreciable net breeding, the energy spectrum of the neutrons in the reactor must be fast (the neutrons must be energetic). Almost all of today's operating reactors are of the thermal variety and cannot be practical breeders because of inherent physical properties. Because a fast reactor is needed to achieve breeding, there is a tendency to use the term "breeder" as a synonym for "fast reactor." That is incorrect. A fast reactor can be configured as either a net producer or net consumer, not only of reactor fuel but also of plutonium.

See also: ENERGY, NUCLEAR; FISSION; REACTOR, FAST; REACTOR, NUCLEAR

Bibliography

AMERICAN NUCLEAR SOCIETY. *Controlled Nuclear Chain Reaction: The First 50 Years* (American Nuclear Society, La Grange Park, IL, 1992).

INTERNATIONAL NUCLEAR SOCIETIES COUNCIL. *A Vision for the Next Fifty Years of Nuclear Energy* (American Nuclear Society, La Grange Park, IL, 1996).

WALTAR, A. E. *America the Powerless: Facing Our Nuclear Energy Dilemma* (Cogito Books, Madison, WI, 1995).

GEORGE S. STANFORD

REACTOR, FAST

A fast reactor is a nuclear reactor in which the neutrons that sustain the chain reaction are fast (have high kinetic energy). This is in contrast with thermal reactors, in which the neutrons are slowed down by a moderator, such as water, before being absorbed in the nuclear fuel. A fast reactor requires no moderator, and for efficient transfer of heat, the coolant usually is a liquid metal (with rare exceptions, sodium), in which case it is also called a liquid-metal reactor (LMR). The fast-reactor power plant, consisting of LMR, steam generator, and electric turbines, is similar in concept to the corresponding conventional pressurized-water reactor (PWR) system.

The concept of a fast reactor has been around since the start of the nuclear age. The first nuclear electric power to be delivered to a commercial grid came in December 1951 from EBR-I, an experimental fast reactor in Idaho. Only recently, however, have technical developments—mainly at Argonne National Laboratory in Illinois and Idaho—demonstrated the full range of what advanced LMRs can do. The two main developments have been (1) a new metallic fuel (rather than an oxide) that has safety and longevity advantages and (2) an economical electrochemical process for recovering and recycling plutonium and other usable materials from spent fuel. Unlike conventional reprocessing, the technology is inherently proliferation-resistant, and there is never pure plutonium (needed for weapons) in any part of the fuel cycle.

Its energetic neutron spectrum gives the fast reactor important capabilities. As part of a fully closed fuel cycle (in which fuel is recycled until it is gone), it can use as fuel all the actinide isotopes (elements with atomic number greater than 89), which means that a fast reactor (1) can extract virtually 100 percent of the energy content of the uranium that was mined to produce the fuel—getting more than 200 times as much energy per kilogram of uranium as

the current U.S. "once-through" thermal-reactor fuel cycle; (2) can consume its own long-lived waste, releasing for disposal only fission products, whose radioactivity decays to negligible levels in less than 500 years; (3) can be configured to completely consume the plutonium and other actinides in the so-called waste from thermal reactors; and (4) could be configured as a breeder reactor if the need arose to supplement existing supplies of fissile fuel.

Thus, fast reactors with an efficient recycle process can be used to manage the inventory of nuclear fuel (plutonium in particular): to consume it (including plutonium from nuclear weapons and the actinides in the spent fuel from today's reactors) when there is too much, and to breed it in small quantities, if decades hence, more than already exists should be needed to start up new reactors. Until all backlogs have been cleared, virtually no plutonium need ever leave a fast-reactor plant (one or more reactors plus a collocated recycling facility)—whatever is brought in or created there is subsequently consumed (except for what is left in the last, spent loading of the last reactor to be shut down when the plant is finally decommissioned). By restricting the world's plutonium supply to what is needed in active inventory, fast reactors can avoid the proliferation concerns that arise in connection with geologic and other long-term storage of material that is of potential interest to would-be makers of nuclear weapons.

See also: ENERGY, NUCLEAR; FISSION; REACTOR, BREEDER; REACTOR, NUCLEAR

Bibliography

AMERICAN NUCLEAR SOCIETY. *Controlled Nuclear Chain Reaction: The First 50 Years* (American Nuclear Society, La Grange Park, IL, 1992).

INTERNATIONAL NUCLEAR SOCIETIES COUNCIL. *A Vision for the Next Fifty Years of Nuclear Energy* (American Nuclear Society, La Grange Park, IL, 1996).

WALTAR, A. E. *America the Powerless: Facing Our Nuclear Energy Dilemma* (Cogito Books, Madison, WI, 1995).

GEORGE S. STANFORD

REACTOR, NUCLEAR

A nuclear reactor is a device in which a controlled nuclear chain reaction takes place. While research is under way on fusion reactors, which would release energy by combining two small atoms to make a larger one, all nuclear reactors today are fission reactors. They release energy through fission, the splitting of large atoms into smaller ones.

Reactors range in size from small research reactors that produce no power to large commercial power plants that produce more than 1,200 MW of electrical power. This entry will center on commercial nuclear power plants because they are the most common reactor type. The fundamental principles and major components are the same for all fission reactors.

Like power plants fueled by gas, oil, or coal, nuclear power plants boil water to create steam. The steam drives a turbine, and the turbine generates electricity. What sets nuclear power plants apart from the others is primarily the use of a nuclear reactor to boil water.

A fission chain reaction takes place in the reactor's core. The major components of the core are fuel rods, control rods, and the moderator.

Fission occurs when a neutron is captured by the nucleus of a fissile atom, one capable of undergoing fission. The most important fissile atom for nuclear power reactors is uranium-235. The isotopes uranium-233, plutonium-239, and plutonium-241 are also fissile, but only uranium-235 is found in sufficient quantity in nature to mine and manufacture into reactor fuel. When the nucleus of a uranium-235 atom is struck by a neutron with an energy of about 0.025 eV, the uranium nucleus splits into smaller nuclei and releases several neutrons. If fissile atoms in the fuel are dense enough, newly released neutrons will be captured by other uranium-235 nuclei, triggering more fission and creating a self-sustaining chain reaction.

The initial fuel in a typical commercial nuclear power plant is about 3 percent uranium-235 and about 97 percent uranium-238, which is non-fissile. As the reactor operates, some uranium-238 nuclei capture neutrons and are transformed through nuclear reactions into fissile plutonium-239. About 40 percent of the nuclear electricity generated in the United States comes from fission of plutonium-239 created in commercial power reactors.

Commercial nuclear fuel takes the form of pellets of uranium oxide, a ceramic. Fuel pellets are sheathed in zirconium alloy cladding to form fuel rods 3 to 4 m long and about 1 cm in diameter. Several hundred fuel rods are placed in a metal framework to form a fuel assembly. The core of a typical commercial nuclear power plant contains several hundred fuel assemblies.

To maintain a perfectly balanced nuclear chain reaction, exactly one neutron from each fission reaction must, on average, be absorbed by exactly one fissile nucleus. If the average is less than one, the chain reaction will shut down and the reactor will stop. If the average is greater than one, the reactor will continually increase power until the core overheats and melts.

Control rods, made of materials that absorb neutrons, are interspersed between fuel assemblies and are moved in and out of the core to control the rate of the fission chain reaction. Inserting all the control rods into the core shuts down the chain reaction. Control rods in commercial reactors are usually made of boron.

Commercial reactors use a moderator to improve the probability that neutrons will be absorbed by fissile atoms. Fission neutrons have average energies of about 2×10^6 eV, about 80 million times higher than the energy of atoms at normal room temperature. By colliding with molecules of the moderator, fission neutrons are slowed to about 0.025 eV, an energy at which they have a high probability of capture by a uranium-235 nucleus. The moderator in most commercial nuclear power plants is light (normal) water circulating between fuel rods and assemblies.

Water also plays two other roles in the reactor. First, it cools the core, keeping it from melting or otherwise deforming and disrupting operations. Second, it carries heat out of the core and into the heat exchanger system, which produces steam to drive the turbines that generate electricity.

Light-water reactors (LWRs), those cooled and moderated by light water, are the world's most common type of commercial nuclear power plant. The two main types of LWR differ in the details of their heat-exchange systems. In the more widely used pressurized-water reactor (PWR), water circulates through the core under high pressure and temperature without boiling. Operating pressure is about 2,250 lb/in.² and operating temperature is about 300°C (about 600°F). The PWR uses a secondary heat exchanger to extract heat from the primary system. Only steam from the secondary loop drives the turbines. In the boiling water reactor (BWR), the other main LWR type, the operating pressure is lower. Water is boiled to create steam, which goes directly to the turbine to generate electricity.

See also: ENERGY, NUCLEAR; FISSION; REACTOR, BREEDER; REACTOR, FAST

Bibliography

AMERICAN NUCLEAR SOCIETY. *Controlled Nuclear Chain Reaction: The First 50 Years* (American Nuclear Society, La Grange Park, IL, 1992).

RHODES, R. *Nuclear Renewal: Common Sense About Energy* (Viking, New York, 1993).

ALAN SCHRIESHEIM

RED GIANT

The ability to determine the true properties of stars began when astronomers were able to measure their parallaxes (and find their distances) in the 1850s. With the distance known, the observed brightness of any star can be converted into an intrinsic, or true, brightness. As the number of measured stars increased, the luminosities were found to have a tremendous range. This range was particularly dramatic among the cooler (red) stars, where most had luminosities between 0.1 percent and 10 percent of that of the Sun ($L_{sun} = 3.89 \times 10^{33}$ erg/sec). However, some red stars were also found to be nearly a million times brighter than the Sun.

The brightness of a star depends on its surface (photospheric) temperature (T) and its radius (R). The temperature is important because the rate at which energy is emitted by each spot on the surface depends on it. According to the Stefan–Boltzmann law, which is obtained by integrating the spectrum produced by a blackbody (Planckian) at a particular temperature,

$$\text{Flux (energy per unit area per unit time)} = \sigma T^4,$$

where σ is the Stefan–Boltzmann constant ($\sigma = 5.6687 \times 10^{-5}$ erg·cm^{-2}·s^{-1}·deg^{-4}).

Therefore, stars with higher temperatures emit more energy from each place on their surface. The star's total luminosity also depends on the surface area:

$$\text{Luminosity (energy per unit time)} = 4\pi R^2 \sigma T^4.$$

Among the red stars where the range of intrinsic luminosities were greatest, the surface temperatures typically ranged from 3,000 K to 4,500 K. This temperature range was too small to explain the tremendous luminosity differences observed; some of them had to be larger—much larger. Stars like Betelgeuse (α Ori) and Antares (α Sco) were found to have radii nearly a thousand times that of the Sun. These radii were confirmed with the advent of speckle interferometry, with which the angular sizes of these large stars could be directly measured.

With the development of the theory of stellar evolution, red giants are now understood to be older stars, in an advanced stage of evolution. Nuclear fusion has converted all of the hydrogen in their cores into helium (or even heavier elements). The Sun is still burning hydrogen, but in about 4 billion years, the hydrogen in the core will become depleted. At that time the structure will begin to change rapidly. The luminosity is expected to grow to a value more than 2,000 times the present brightness, and the radius is expected to increase to more than 100 times the current radius. The red giant stage is relatively short lived, and stars like the Sun will spend only about 10 percent of their lifetimes there. Stellar evolution also shows the eventual fate of red giants. Depending on their masses, they eventually become white dwarfs or explode as supernovas.

See also: STARS AND STELLAR STRUCTURE; STELLAR EVOLUTION; SUN; SUPERNOVA

Bibliography

JASTROW, J. *Red Giants and White Dwarfs,* new ed. (W. W. Norton, New York, 1979).

SCHWARZSCHILD, M. *Structure and Evolution of the Stars* (Princeton University Press, Princeton, NJ, 1958).

DAVID S. P. DEARBORN

REDSHIFT

The spectrum of a star is characterized by distinct spectral lines corresponding to different elements in various ionization states. The radial component of motion of the star results in a redshift or a blueshift, depending on whether the star is receding from or approaching the Sun. The spectrum of a galaxy is a composite of many stars. The light from an elliptical galaxy is dominated by old stars like the Sun, whereas light from a spiral is predominantly from younger, more luminous and hotter stars.

The spectra of galaxies were originally discovered by Vesto Slipher in 1912 to be systematically redshifted. Generally, the fainter the galaxy, the larger the redshift, although some bright galaxies, such as the Andromeda galaxy, were blueshifted. It was not until 1929 that Edwin Hubble obtained distance estimates to the nearby galaxies and found that there was a linear proportionality between redshift and distance out to a recession velocity of about 1,000 km/s. The distance scale was revised and increased by a factor of 5 by 1952, when Walter Baade determined a definitive calibration. By this time, the efforts of Hubble and Milton Humason had expanded tenfold in distance the applicability of Hubble's law, the redshift-distance relation. The modern data, mostly obtained by Alan Sandage and his collaborators between 1960 and 1980, verified that Hubble's law applied out to a recession velocity of more than 100,000 km/s.

The redshifts (and a few blueshifts) measure the motions of galaxies. There is a universal systematic expansion on which is superimposed local to-and-fro motions that are due to the clustering of galaxies. The local self-gravity of a galaxy cluster or group results in an additional component of the measured velocity that may be either toward or away from the observer. This is called peculiar velocity, to distinguish it from the systemic component of recession, and is relatively small. Peculiar velocities average a few hundred kilometers per second in galaxy groups and do not exceed two or three thousand kilometers per second in rich clusters. Consequently, peculiar velocites are important for nearby galaxies, where the recession component is small, and result in occasional blueshifts, as for the Andromeda galaxy, a member of our Local Group of galaxies. Neither galaxies nor groups and clusters of galaxies are themselves expanding; their local self-gravity ensures that they are stable systems. However, they act as beacons, charting the expansion of the universe.

Redshift is defined to be the fractional shift of light at all wavelengths to the red for a receding object. Redshift is interpreted as being due to the Doppler effect, which shifts the wavelength of light according to the amount of relative motion between the source and observer. For velocities small compared to the speed of light, the redshift is equal to radial velocity divided by the speed of light. The record redshift for a distant galaxy is 4.25, which

means that there is a 425 percent shift of its emitted light to the red. This galaxy is receding at 90 percent of the speed of light. As the recession speed approaches the speed of light, the redshift becomes larger and larger. The greatest redshifts measured are for quasars, for which the highest redshift is 4.9.

Redshift is a convenient way of measuring distance, because it denotes a measurable quantity. The actual distance to remote galaxies is uncertain to within a factor of 2, because the conversion from redshift to distance requires knowledge of the Hubble constant. Astronomical determinations of the Hubble constant range from a low value of 40 $km \cdot s^{-1} \cdot Mpc^{-1}$ to a high value of 90 $km \cdot s^{-1} \cdot Mpc^{-1}$. The uncertainty arises because of the use of different distance calibrations. Most of the measurements are indirect, relying on use of objects such as supernovas or planetary nebulas that are calibrated in a few nearby galaxies to which there are several independent measures of distance. Unknown systematic differences between distant and nearby objects may arise because when one looks far away, a greater range of objects is inevitably encountered than one finds in the calibrating galaxies. If the intrinsic dispersion of properties of the distance calibrator is appreciable, this will inevitably lead to uncertainties, since the most distant objects will be those that are intrinsically the brightest. This effect, known as Malmquist bias, is one of the principal reasons that distances determined from redshifts are uncertain by about a factor of 2.

Could the redshift be due to an effect other than the Doppler shift produced by the expansion of space? Another effect that reddens light, extinction by interstellar dust, acts preferentially on shorter, bluer wavelengths and can readily be distinguished from the Doppler shift. In the 1930s, astronomers desperately sought theories of "tired" light, according to which photons, or light quanta, would lose energy and be progressively more redshifted the greater the distance that was traversed by the photons. One motivation was Hubble's alarmingly short distance scale, which led to an expansion age of the universe that was less than that of the earth. In a tired light theory, light of all wavelengths would lose energy by precisely the same fractional amount. Unfortunately, no physical mechanism for tired light was forthcoming.

Astronomers have since measured with considerable precision the same redshift for distant galaxies at wavelengths as diverse as optical, x ray, and radio. Light of such diverse photon energies could not possibly lose energy by the same fractional amount, even if there were a means of achieving this in, say, the optical region of the spectrum. In the meantime, the motivation for seeking a tired light theory largely vanished after Baade's revision of the distance scale in 1952 to imply an expansion age for the universe that comfortably exceeded the age of the solar system.

See also: DOPPLER EFFECT; GALAXIES AND GALACTIC STRUCTURE; HUBBLE, EDWIN POWELL; HUBBLE CONSTANT; UNIVERSE, EXPANSION OF; UNIVERSE, EXPANSION OF, DISCOVERY OF

Bibliography

MADORE, B. F., and TULLY, R. B., eds. *Galaxy Distances and Deviations from Universal Expansion* (Reidel, Dordrecht, 1986).

ROWAN-ROBINSON, M. *The Cosmological Distance Ladder* (W. H. Freeman, New York, 1985).

SILK, J. *A Short History of the Universe* (W. H. Freeman, New York, 1994).

JOSEPH SILK

REFERENCE FRAME

See FRAME OF REFERENCE

REFLECTION

When a wave propagating in one medium encounters a boundary with a second medium, part of the incident wave is returned to the first medium. For sound waves this is called an echo. If the reflecting surface is rough, the waves will undergo diffuse scattering in a variety of directions. If the reflecting surface is smooth, optical light will undergo specular reflection, in which the incident and reflected rays lie in the same plane at opposed equal angles to a line perpendicular to the surface. The reflected intensity depends on the angle and polarization of the light and the nature of the surface. Depending on the media, the phase of the reflected wave can be preserved, inverted, or shifted relative to that of the incident wave.

The phenomenon of reflection has certainly been known since humans first recognized their own images in a surface of water. Aristotle understood the equality of incident and reflected angles. Plato described the properties of a corner reflector, whereby mirrors abutted at a right angle avert the normal left-right reflected inversion when viewed along the abutment, and reverse up and down when viewed perpendicular to it. Corner reflectors also have the property of reflecting all rays back parallel to their incident direction and have applications such as reflective paint on road signs. Corner-cube retroreflector arrays were placed on the Moon by the Apollo astronauts, permitting highly accurate measurements of the Earth–Moon separation by laser ranging.

The intensities of the reflected rays differ for components with polarizations parallel and perpendicular to the plane containing the incident and reflected rays and are given by formulas developed by Augustin Fresnel. Light polarized with its electric field perpendicular to this plane is favored, and at Brewster's angle (where the refracted and reflected rays are perpendicular to each other), the polarization component parallel to this plane vanishes. Polaroid sun glasses are designed to filter out the horizontal component, which dominates specularly reflected light but is only 50 percent of diffusely reflected light.

In moving from a more dense to a less dense medium, there is a critical angle beyond which refraction can no longer occur and the ray is totally reflected (and absorbed). This is the principle by which light can be channeled around corners using fiber optics (which have important communications and medical applications). Channeling of light also occurs in phenomena such as mirages and looming, where differences in air temperature at and near Earth's surface form an optical interface. The rainbow is a result of total internal reflection within rain droplets of rays that are ultimately dispersed into their colors by refraction when they emerge.

Optical reflectivity is related to the presence of conduction electrons as evidenced by the high reflective lustre of metals. Ancient mirrors were made from polished metal, but glass mirrors backed with a tin amalgam were introduced in the seventeenth century, and in 1840 a process was developed for producing silvered glass mirrors. A plane mirror forms a laterally reversed unmagnified virtual image, whereas a convex mirror forms a virtual image with a wider field of view than a flat mirror. Concave spherical mirrors form real inverted images of objects farther away than half the radius of curvature of the mirror (which can fool the eye into seeing a three-dimensional object where none exists), and virtual images of closer objects. Concave mirrors are often used in astronomical telescopes.

See also: BREWSTER'S LAW; FIBER OPTICS; FRESNEL, AUGUSTIN-JEAN; IMAGE, VIRTUAL; MIRROR, PLANE; OPTICS; OPTICS, GEOMETRICAL; POLARIZATION; POLARIZED LIGHT; RAINBOW; REFRACTION

Bibliography

BOYER, C. B. *The Rainbow: From Myth to Mathematics* (Princeton University Press, Princeton, NJ, 1987).

JENKINS, F. A., and WHITE, H. E. *Fundamentals of Optics*, 4th ed. (McGraw-Hill, New York, 1976).

LORENZO J. CURTIS

REFRACTION

Refraction is the bending of a wavefront as it passes across the barrier (interface) from one material to another. Refraction can occur with all types of waves, but it is most commonly discussed in terms of light waves. While Fig. 1 shows the general principle involved with refraction, application of Snell's law will determine the exact amount of bending that occurs for light waves at the interface:

$$n_1 \sin A = n_2 \sin B,$$

where n_1 and n_2 are the refraction indices for the two media, A is the angle of incidence, and B is the angle of refraction.

The index of refraction depends on the materials involved and the wavelength of the light. This variation is responsible for the phenomenon known as dispersion, the splitting of a light ray of mixed wavelengths into its components. For most materials, the index of refraction for blue light is greater than the index of refraction for red light. Therefore, when a beam of white light (a mixture of all colors) is incident upon a prism, each color is bent at a slightly different angle at both the entering and exiting surfaces of the prism, as shown in Fig. 2. This refraction process produces a spectrum of color coming out of the prism with red light bent the least and blue the most. Isaac Newton was one of the first people to study this effect to determine that the colors produced were actually a property of the light, not the prism or any other object. Since that time, it has been well-established that color is related to the wavelength of the light involved.

Refraction at a curved surface, which gives rise to more complicated behavior, is an important consideration in the creation of lenses. The most general solution has been to place two curved surfaces back-to-back, as shown in Fig. 3. By combining curved lenses in this way, the refraction upon entering the lens and refraction upon exiting the lens combine

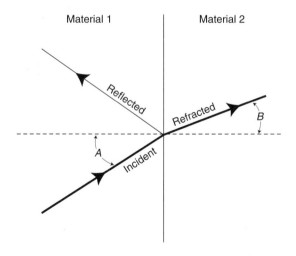

Figure 1 Refraction and reflection at the plane interface between two materials.

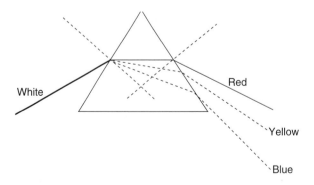

Figure 2 Refraction through a prism to create a color spectrum.

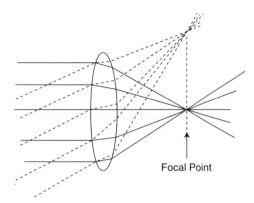

Figure 3 Refraction through back-to-back curved surfaces to bring light rays together at the focal point.

to bring a beam of light into focus at a given focal point, forming an image of the original object. For lensmakers, however, the variation of the index of refraction with color, which can result in chromatic aberration, is an added problem that must be solved. When chromatic aberration occurs, the image is blurred or has colored fringes because the various colors of light are brought to different focal points. The solution to this problem has been to use achromatic lenses, in which the two curved lenses placed back-to-back are made of different materials, such that their dispersions of color neutralize each other but their refractions do not. Because chromatic aberration occurs only with lenses and not mirrors, Newton invented the reflecting telescope, which uses a mirror instead of a lens, to avoid this phenomenon in his studies.

Rainbows are a naturally occurring phenomenon caused by the reflection and refraction of sunlight in raindrops. René Descartes proposed the explana-

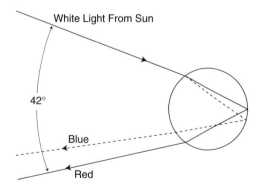

Figure 4 Reflection and refraction in a raindrop to create a rainbow.

tion for the rainbow shown in Fig. 4. White light enters the raindrop and is refracted. The refracted light is then reflected from the back of the raindrop. The reflected light is then refracted again upon leaving the raindrop. The angle between the entering light and the exiting light is about 42°, and Descartes showed that the 42° angle resulted from the spherical raindrop acting as a small lens.

Because the refractive index for light waves varies with the medium involved, scientists have also been able to use refraction as a valuable tool for the study of inhomogeneous materials. Specific types of atoms and molecules each have a unique set of light wavelengths that they will absorb. By using a prism to spread out the spectrum of light passing through a material, scientists could determine which wavelengths had been absorbed, an indication of what atoms or molecules were present. The use of prisms for this purpose has since been replaced by the development of good diffraction gratings.

See also: ABERRATION, CHROMATIC; GRATING, DIFFRACTION; LENS; LENS, COMPOUND; LIGHT; LIGHT, SPEED OF; OPTICS; RAINBOW; REFLECTION; SNELL'S LAW

Bibliography

JENKINS, F. A., and WHITE, H. E. *Fundamentals of Optics*, 4th ed. (McGraw-Hill, New York, 1976).

SMITH, M. A. *Descartes's Theory of Light and Refraction* (American Philosophical Society, Philadelphia, 1987).

RALPH W. ALEXANDER JR.

REFRACTION, INDEX OF

Most science historians credit Willebrord Snell with the discovery of the law that defines the index of refraction. Snell's law states that when a narrow beam of light (a ray) is incident on the boundary (interface) between two transparent media (see Fig. 1), the ratio of the sine of the angle of incidence (B) to the sine of the angle of refraction (C) is constant (i.e., independent of the angel of incidence). This constant, n_r, is the index of refraction of the second medium relative to the first. When the ray is incident from a vacuum, the ratio $n = c/v$, where c is the speed of light in the vacuum and v is the speed of light in the medium, is the index of refraction of the medium entered.

The relative index for any two media is the ratio of the indices of the two media (n_1 and n_2) and is also equal to the inverse ratio of the speeds (v_1 and v_2) of light in the media:

$$n_r = \frac{n_2}{n_1} = \frac{v_1}{v_2}.$$

Therefore, Snell's law can be written as

$$n_1 \sin(B) = n_2 \sin(C).$$

The index of refraction of the upper medium (Fig. 1) can easily be measured by removing the lower medium and increasing the angle of incidence to the critical angle (B_c), at which no refracted ray exits the bottom of medium 1. All the light is then reflected. Snell's law then gives the index relative to air as

$$n_r = \frac{1}{\sin(B_c)}.$$

This equation is helpful in choosing materials for the manufacture of optical fibers used in communications and computer industries, medicine, and other areas. The index also affects the choice of material for and thickness of a thin film used as an optical *coating* to reduce or increase the amount of light reflected from a surface.

Most tabulated indices of refraction were obtained by the method introduced by Isaac Newton in which a ray of light is directed into a prism so that the ray refracts and crosses parallel to the base (see Fig. 2). The angle (D) of deviation between the directions of the incident ray and the final refracted ray is then a minimum (D_m) and the first angle of incidence (B) equals the last angle of refraction (G). Simple geometric and trigonometric proofs establish that the apex angle (A) of the prism and D_m are related to the index of refraction for the prism relative to air by

$$n_r = \frac{\sin\left(\dfrac{A + D_m}{2}\right)}{\sin\left(\dfrac{A}{2}\right)}.$$

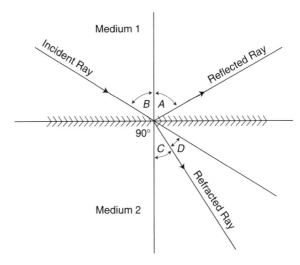

Figure 1 Ray diagram for refraction and reflection, where B is the angle of incidence for the ray of light striking the interface from medium 1, A is the angle of reflection, C is the angle of refraction for the ray in medium 2, and D (which is the difference between B and C) is the angle of deviation for the ray.

Newton discovered in approximately 1666 that the index of refraction varies with the color of light. He found that a prism produces different angles of deviation for different colors and thus separates them. He measured indices for one of his prisms ranging from 1.545 for red light to 1.560 for violet light.

Color separation (dispersion) by prism is the foundation of spectroscopy and is the basis for pow-

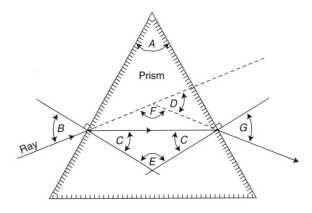

Figure 2 Ray diagram illustrating Newton's prism method for measuring the index of refraction for sunlight.

erful instruments (e.g., spectrographs, spectrometers, and spectrophotometers) that are widely used in medicine, in laboratory and observational sciences, and in engineering. Dispersion in a single lens, however, produces undesired colored halos around images formed on photographic plates and on the retina of the eye. This defect (chromatic aberration) can be minimized by using a compound lens with two or more simple lenses with different indices of refraction.

See also: ABERRATION, CHROMATIC; DISPERSION; LENS; LENS, COMPOUND; LIGHT, SPEED OF; NEWTON, ISAAC; OPTICS; REFLECTION; REFRACTION; SNELL'S LAW; SPECTROSCOPY; THIN FILM

Bibliography

HALLIDAY, D.; RESNICK, R.; and WALKER, J. *Fundamentals of Physics*, 4th ed. (Wiley, New York, 1993).

HECHT, E. *Optics*, 2nd ed. (Addison-Wesley, Reading, MA, 1987).

SERWAY, R. A. *Physics*, 3rd ed. (Saunders, Philadelphia, 1990).

CHARLES E. HEAD

REFRIGERATION

The internal energy of a system is partitioned among modes of behavior called degrees of freedom. These include translations, rotations, and vibrations of molecules, the electronic excitations of atoms, and lattice vibrations. At high temperatures, the energy is equally divided (in what is known as an equipartition of energy) among the available degrees of freedom. The average energy for each degree of freedom is given by $kT/2$, where k is the Boltzmann constant and T is the thermodynamic temperature expressed in kelvins.

To cool a system, energy is extracted from one or more degrees of freedom. The usual way to do this is to place the system in contact with something colder. Refrigerators are devices that provide the required cold environment (cold reservoir).

A refrigerator is a cyclic device within which work is done on a substance called the refrigerant, which extracts heat from a cold reservoir and deposits it into the surrounding environment. Since one would like the maximum extraction of heat for the minimum expenditure of work, a refrigerator is characterized by an efficiency w:

$$w = \frac{\text{heat extracted}}{\text{work done}}.$$

An ideal or Carnot refrigerator, a theoretical rather than actual device, provides the maximum efficiency for given temperatures of the hot (T_h) and cold (T_c) reservoirs:

$$W_{\text{Carnot}} = \frac{T_c}{T_h - T_c}.$$

Joule–Thomson Refrigeration

In most household refrigerators, the refrigerant, usually a substance like freon (CCl_2F_2), extracts heat from a cold reservoir by evaporating while in contact with it (see Fig. 1).

At room temperature and a pressure of about 8 atm, freon is a fully saturated liquid (like carbonated water). In this condition it resides in the condenser of the refrigerator. A compressor pushes the freon from the condenser through a throttling valve that restricts its flow. The refrigerant emerges at a lower pressure, cooler and partially vaporized. This process is known as the Joule–Thomson effect. The cool refrigerant then moves into a much larger volume, the evaporator, in which it evaporates com-

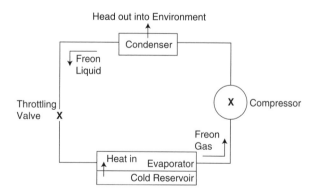

Figure 1 Basic schematic diagram for a common household refrigerator.

pletely, extracting heat from its surroundings. The low-pressure gas then passes through a compressor, becoming a superheated vapor. It then passes into the condenser, giving up its heat to the surroundings and becoming a saturated liquid once again. When applied to refrigerants, such as nitrogen or helium, the Joule–Thomson effect produces temperatures low enough to liquify the refrigerant, 4.2 K in the case of helium.

Cryogenic Refrigeration

The most energetic atoms of a liquid normally evaporate from its surface. If they are removed by pumping them away, they cannot return to the liquid, and what remains is, therefore, cooler. Liquid helium can be reduced to milli-Kelvin temperatures by this process, which is called pumping.

An even more sophisticated refrigeration process, leading to micro-Kelvin temperatures, involves the extraction of energy from a system having magnetic properties. Paramagnetic salts have ions that behave like tiny magnets because of their outermost electrons. These normally unaligned ions acquire energy if they interact with an external magnetic field. When such a salt is placed in contact with liquid helium at 1 K and an external magnetic field is turned on, it aligns the spins and gives them energy. This added energy would normally increase the salt's temperature, but it passes into the surrounding helium, leaving the salt at 1 K. The salt is then insulated from the helium, and the magnetic field is turned off. This process, called adiabatic demagnetization, leaves the spins more aligned than they were initially, which means the system has lower temperature.

Temperatures between 1 K and 10 mK are commonly achieved using the ^{3}He–^{4}He dilution refrigerator. Pumping on either isotope of helium reduces its temperature by efficiently removing the energetic atoms that evaporate. The dilution refrigerator is a more sophisticated and even more efficient way of doing the same thing. When ^{3}He and ^{4}He are mixed, some of the ^{3}He floats on the mixture. Because of rather complicated, quantum mechanical relations between the two isotopes of helium, the ^{3}He actually evaporates into the mixture and is thereby cooled.

See also: ADIABATIC PROCESS; DEGREE OF FREEDOM; EQUIPARTITION THEOREM; HEAT; HEAT PUMP; HEAT TRANSFER; JOULE–THOMSON EFFECT; THERMODYNAMICS

Bibliography

CALLEN, H. *Thermodynamics and an Introduction to Thermostatics*, 2nd ed. (Wiley, New York, 1985).

KITTEL, C., and KROEMER, H. *Thermal Physics*, 2nd ed. (W. H. Freeman, San Francisco, 1980).

REIF, F. *Fundamentals of Statistical and Thermal Physics* (McGraw-Hill, New York, 1965).

ZEMANSKY, M. W. *Heat and Thermodynamics* (McGraw-Hill, New York, 1957).

MORTON A. TAVEL

RELATIVITY, GENERAL THEORY OF

Albert Einstein's theory of general relativity is the currently accepted theory of gravity. It is based on a combination of Einstein's special theory of relativity and the equivalence principle. Experimental tests continue to confirm its predictions of the orbits of bodies ranging from space probes within our solar system to collapsed stars in close orbit around one another.

The Equivalence Principle

Drop an object inside an accelerating rocket, and it will appear to accelerate toward the floor as if acted upon by an invisible downward force. An outside observer will claim that this force is fictitious because it is really the floor of the laboratory that accelerates up toward the object. Einstein's equivalence principle asserts that gravity is actually a fictitious force of this type. There is no way for an observer inside a (sufficiently small) rocket to tell whether the rocket is at rest in Earth's gravity, as in Fig. 1, or accelerating through empty space at 9.8 m·s^{-2}, as in Fig. 2.

In modern treatments of general relativity, the equivalence principle is restricted to experiments done entirely inside "sufficiently small" laboratories. Without the "inside" restriction, an observer can easily tell whether the rocket is at rest on Earth or accelerating in empty space by looking out a window. Similarly, if the rocket is not "sufficiently small," then the direction and strength of the accelerations inside it will vary from place to place when it is on the ground but not when it is in empty space. In a

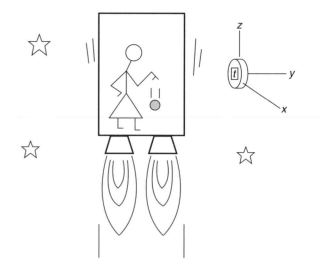

Figure 1 In a rocket at rest on Earth, a dropped object accelerates toward the floor because of Earth's gravitational field. True inertial reference frames accelerate past the rocket on their way to the center of Earth.

more subtle example, the equivalence principle applies to observations of electrically charged particles only if they are shielded inside an electrically neutral laboratory so that they cannot send electromagnetic waves outside the laboratory.

If gravity is a fictitious force, then freely falling objects are force free and, according to Newton's first law of motion, should be unaccelerated in a true inertial frame. In that case, the true inertial frames must themselves be in free fall, as shown in Fig. 1. When you are standing on the ground, the ground presses against your feet, not because it must balance the force of gravity but because it must supply a constant acceleration relative to the true inertial frames that are falling all around you.

Curved Spacetime

To understand the geometrical language of general relativity, begin with a key idea from the special theory of relativity: A world line is a sequence of events in the history of an object. (Think of it as a plot of position versus time.) The world line idea makes it possible to give a simple and purely geometrical form to Newton's first law of motion: An object that is not acted on by external forces always has a straight world line.

What is meant by a straight world line? Within the confines of a sufficiently small, freely falling frame, one can answer this question by using the procedures of special relativity to build Minkowski space and time coordinates x, y, z, t around some point on the world line. A graph of the values of these coordi-

Figure 2 In a rocket accelerating through empty space, a dropped object accelerates toward the floor as if falling in a gravitational field. In a true inertial reference frame, this relative acceleration is explained by the floor accelerating toward the object.

nates along a straight world line will appear straight where it goes through the origin. Away from the origin the line may appear curved because it is outside the "sufficiently small" region where the reference frame is truly inertial. A similar situation can be seen on the curved surface of Earth, where sufficiently small regions can be accurately mapped onto a flat surface but larger regions are distorted.

The small departures from straight-line behavior are important because they correspond to the curvature of the underlying geometry. Consider the two airplanes in Fig. 3. They begin their flights a few miles apart at Earth's equator and head north, perpendicular to the equator. Any local map will show them on parallel initial headings. Both airplanes fly as straight as they can, checking their headings against a series of local maps so that they can be sure that they are turning neither right nor left. When they reach the North Pole, their paths will cross and their headings will obviously be different. Any map large enough to show the whole trip will show their paths curving toward each other. This relative curvature of their paths is a reflection of the curvature of Earth's surface.

Now consider a small laboratory that is falling straight down toward Earth. A cloud of dust particles is initially at rest in the center of the laboratory. The world lines of the dust particles are initially parallel:

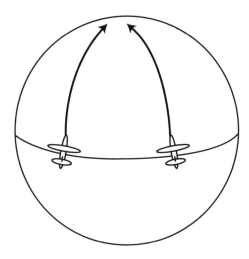

Figure 3 Two airplanes set out on parallel paths, both perpendicular to Earth's equator. Even though the paths are as straight as possible, they curve toward each other and intersect at the North Pole.

They are not moving relative to each other. However, as shown in Fig. 4, the dust particles near the bottom of the cloud are closer to Earth than the others, and they fall with a bit more acceleration. The dust particles near the top of the cloud fall with a bit less acceleration. All of the particles are accelerating toward the center of Earth so that they converge horizontally. Thus the cloud changes shape as shown in Fig. 5, stretching vertically and shrinking horizontally. The initially parallel, locally straight world lines of the dust particles appear to curve relative to each other. Just as the behavior of the airplanes in the previous example was a reflection of the curvature of Earth's surface, the relative curvature of these world lines is a reflection of the curvature of spacetime near Earth.

Einstein's Gravitational Field Equations

Isaac Newton's theory of gravitation has a remarkable property: If the particles in a small cloud are initially at rest relative to one another and no matter is included inside the cloud, then the volume of the cloud will remain exactly constant. In Einstein's theory of gravity, the particles of the cloud trace out straight world lines and the second time derivative of the volume of the cloud is given by a particular combination of spacetime curvature components called the Ricci curvature. When the density of matter is negligible, the local content of Newton's the-

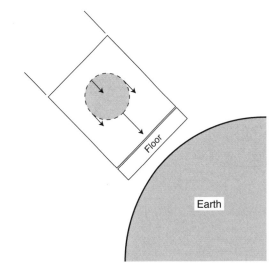

Figure 4 The arrows show the accelerations of different parts of a dust cloud toward the center of Earth. A freely falling laboratory surrounds the dust cloud to provide a frame of reference.

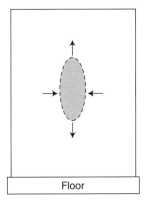

Figure 5 Within a small laboratory that is falling toward Earth, an initially spherical dust cloud stretches into a cigar shape because of the nonuniformity of Earth's gravitational field.

ory can be captured in every local reference frame by Einstein's vacuum field equation:

$$\text{Ricci} = 0.$$

When the density of matter is not negligible, it causes the volume of a cloud of particles to collapse.

In Newton's theory, this collapse is influenced only by the mass density of matter. However, if one attempts to capture this Newtonian result in a statement about the spacetime curvature tensor, the result is not consistent with the conservation of matter and energy. A consistent theory requires a modification of the local content of Newton's theory: Both mass-energy and pressure act as sources of gravitational attraction. With this correction, the stress-energy tensor of matter curves spacetime according to Einstein's gravitational field equation:

Ricci = constant × corrected stress-energy.

For ordinary objects, such as planets and normal stars, the pressure correction to the stress-energy tensor is negligible. However, during the final contraction of a star, it shows itself when increased pressure accelerates the contraction instead of slowing it.

Solving Einstein's Equations

Assign four arbitrary coordinates $x^0(\mathcal{P})$, $x^1(\mathcal{P})$, $x^2(\mathcal{P})$, $x^3(\mathcal{P})$ to each event $\mathcal{P}$ in an extended region of spacetime. For each event $\mathcal{P}$, go into a freely falling laboratory whose history includes that event, and, following the procedures of special relativity, assign Minkowski coordinates $t(\mathcal{P}')$, $x(\mathcal{P}')$, $y(\mathcal{P}')$, $z(\mathcal{P}')$ to each event $\mathcal{P}'$ in the laboratory. For a sufficiently small laboratory, there will be a system of linear relations between changes in the local Minkowski coordinates and changes in the extended coordinates. These relations may be written in the following matrix form:

$$\begin{pmatrix} d't \\ d'x \\ d'y \\ d'z \end{pmatrix} = \begin{pmatrix} f^{(0)}{}_0 & f^{(0)}{}_1 & f^{(0)}{}_2 & f^{(0)}{}_3 \\ f^{(1)}{}_0 & f^{(1)}{}_1 & f^{(1)}{}_2 & f^{(1)}{}_3 \\ f^{(2)}{}_0 & f^{(2)}{}_1 & f^{(2)}{}_2 & f^{(2)}{}_3 \\ f^{(3)}{}_0 & f^{(3)}{}_1 & f^{(3)}{}_2 & f^{(3)}{}_3 \end{pmatrix} \begin{pmatrix} dx^0 \\ dx^1 \\ dx^3 \\ dx^4 \end{pmatrix},$$

where the notation d' is used for infinitesimal changes in local coordinates and the notation d is reserved for changes in coordinates defined over an extended region.

The matrix f specifies what is usually called a moving-frame picture of spacetime geometry. As a summary of the behavior of local Minkowski coor-

dinates in inertial reference frames, it offers direct access to all of the local laws of physics through the principle of equivalence. For example, the line element or metric tensor ds^2 measures proper distances along a space-like curve and may be expressed in terms of the matrix f by expanding the special relativity expression for the line element: $ds^2 = d'x^2 + d'y^2 + d'z^2 - c^2 d't^2$, where c is the speed of light. The Ricci tensor in Einstein's equations may also be expressed in terms of the matrix f and its first and second derivatives with respect to the extended coordinates. Einstein's equations then become a system of partial differential equations for the frame-matrix f as a function of the extended spacetime coordinates.

The solution of Einstein's equations for the spacetime around a spherically symmetric body such as a nonrotating star or planet was found by Karl Schwarzschild in 1916. The spherical symmetry of the spacetime simplifies its description by singling out unique spherical surfaces. Two of the extended coordinates (x^2, x^3) on spacetime can be chosen to be the usual polar angle coordinates θ, φ on these spheres, while the extended radius coordinate $x^1 = r$ can be defined by requiring the areas of the spheres to be given by $4\pi r^2$. If the local Minkowski coordinates are chosen so that the local x axis points in the radial direction and the local y and z axes are tangent to the spheres, then Einstein's vacuum field equations are easily solved for the moving frame coefficients. The relationship between small changes in the extended time coordinate x^0, and small changes in the local Minkowski time coordinate, is found to be

$$d't = \sqrt{1 - (2m/r)}\, dx^0,$$

while the corresponding relationship for radial coordinates is

$$d'x = \frac{dr}{\sqrt{1 - (2m/r)}}.$$

These relationships describe the spacetime outside of a spherical star or planet of total mass m (in appropriate units).

Odd things happen where the extended radial coordinate r has the value $2m$. There, the extended time coordinate x^0 can advance, while the local inertial-frame time coordinate t changes not at all, and small changes in inertial-frame radial coordinate x correspond to no change at all in the extended radial coordinate r. For ordinary stars, the value of $2m$ is just a few kilometers and the peculiar features of the solution can be ignored since this is inside the star where these equations do not apply. For collapsed stars, the events at $r = 2m$ become significant and indicate the presence of a black hole event horizon.

Another important class of solutions describes the spacetime geometry of a universe with all local features averaged out so that space looks the same everywhere and in all directions. These solutions are the basis for the standard model of cosmology. The simplest of these models, the flat, open universe, relates small changes in local Minkowski coordinates to small changes in the extended coordinates by

$$d't = dx^0$$
$$d'x = a(x^0)\,dx^1$$
$$d'y = a(x^0)\,dx^2$$
$$d'z = a(x^0)\,dx^3,$$

where the function a measures the size of the universe and begins with the value zero at the big bang. The form of the function a and thus the connection between the current rate of expansion of the universe and its current age depends on the stress-energy of the assumed matter content of the universe. In this solution, a increases without limit. Other solutions describe universes in which space has the geometry of a closed three-dimensional sphere or the geometry of an open three-dimensional hyperbolic saddle. The closed universe models re-collapse after going through a maximum size, while the hyperbolic models expand without limit.

Small fluctuations in the moving-frame matrix may be described by writing it as a combination $f = e + \varepsilon h$ of a background matrix e and a frame fluctuation matrix h. The limit of small fluctuations is expressed by multiplying h by a smallness parameter ε and expanding Einstein's equations in powers of ε. The first-order term in the expansion requires part of the frame fluctuation matrix h to obey a wave equation in the background spacetime described by the matrix e. These fluctuations, called gravitational waves, travel at the same local speed as light and can carry energy and momentum.

Tests of General Relativity

In Newton's theory of gravitation, an isolated planet should follow a fixed elliptical orbit. After correcting for the attractions of the other planets, the long axis of Mercury's elliptical orbit was found to be shifting by forty-three seconds of arc per century (see Fig. 6). Accounting for this remaining perihelion shift of Mercury was an early success of Einstein's theory of gravity.

Newton, who thought of light as made of particles, could have predicted that the Sun's gravitational field bends light because he would have assumed that the particles of light fall toward the Sun in the same way as anything else. In the language of Einstein's theory, this Newtonian light-bending is due to the curvature of spacetime in the time direction. An additional light-bending effect arises in Einstein's theory because the nonuniform gravitational field of the Sun distorts the geometry of space. For starlight grazing the Sun's photosphere, as shown in Fig. 7, Einstein's theory predicts a deflection of 1.75 seconds of arc, while Newton would have predicted 0.875 seconds of arc.

Detecting this deflection required photographing stars near the Sun during a total eclipse and then, six months later, photographing the same star field with the Sun on the opposite side of Earth. Arthur Eddington's eclipse expedition in 1919 obtained a result compatible with Einstein's prediction and not compatible with Newton's. Since that time, the inconvenience of waiting for total eclipses has

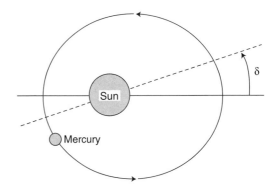

Figure 6 The long axis of Mercury's elliptical orbit changes its direction by an angle $\delta = 43''$ every 100 years because of the effects predicted by Einstein's theory of gravitation.

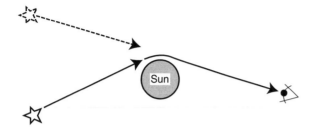

Figure 7 The Sun's gravitational field deflects light passing near the Sun's surface so that the apparent position of a star (dotted lines) differs from its real position.

been circumvented by measuring the deflection (and corresponding phase shifts) of radio waves from quasars and from space probes, as well as radar and laser ranging from the Moon and planets, with results that confirm Einstein's theory.

The equivalence principle—a foundation of Einstein's theory of gravitation—has been tested in a variety of ways. In 1922 Lorand von Eötvös showed that a variety of different substances all fall toward Earth with the same acceleration to within a few parts in a billion. The acceleration of different substances toward the Sun was checked by Robert H. Dicke and his coworkers in 1964 and found to vary by less than one part in a hundred billion. The assertion that inertial frames are actually falling with the same acceleration as freely falling objects was checked in 1960 by Robert Pound and Glen Rebka, who showed that gamma rays are subject to the appropriate gravitational redshift (to within one part in a hundred) when they travel up and down a tall tower.

One of the most striking predictions of general relativity, that a system with a changing mass quadrupole moment will lose energy by emitting gravitational waves, has been confirmed by observations of the binary pulsar discovered by Russell Hulse and Joseph Taylor in 1975. This system consists of a very good clock (the pulsar) in close orbit around another collapsed object. Its motion also confirms the perihelion precession predictions of general relativity to high accuracy.

See also: EINSTEIN, ALBERT; EQUIVALENCE PRINCIPLE; FIELD, GRAVITATIONAL; GRAVITATIONAL WAVE; RELATIVITY, GENERAL THEORY OF, ORIGINS OF; RELATIVITY, SPECIAL THEORY OF; RELATIVITY, SPECIAL THEORY OF, ORIGINS OF; SPACETIME

Bibliography

ABRAMOVICI, A., et. al. "LIGO: The Laser Interferometer Gravitational Wave Observatory." *Science* **256**, 325–333 (1992).

BRUSH, S. G. "How Cosmology Became a Science." *Sci. Am.* **267** (Aug.), 62–71 (1992).

GUTH, A. H., and STEINHARDT, P. J. "The Inflationary Universe." *Sci. Am.* **250** (May), 116–128 (1984).

HORGAN, J. "Universal Truths." *Sci. Am.* **263** (Oct.), 109–117 (1990).

HUCHRA, J. P. "The Hubble Constant." *Science* **256**, 321–325 (1992).

JEFFRIES, A. D.; SAULSON, P. R.; SPERO, R. E.; and ZUCKER, M. E. "Gravitational Wave Observatories." *Sci. Am.* **256** (June), 50–58B (1987).

KRAUSS, L. M. "Dark Matter in the Universe." *Sci. Am.* **255** (Dec.), 58–68 (1986).

MISNER, C. W.; THORNE, K. S.; and WHEELER, J. A. *Gravitation* (W. H. Freeman, San Francisco, 1973).

PARKER, B. "In and Around Black Holes." *Astronomy* **14** (Oct.), 6–15 (1986).

THORNE, K. S. *Black Holes & Time Warps: Einstein's Outrageous Legacy* (W. W. Norton, New York, 1994).

THURSTON, W. P., and WEEKS, J. R. "The Mathematics of Three-Dimensional Manifolds." *Sci. Am.* **251** (July), 108–120 (1984).

WEISBERG, J. M.; TAYLOR, J. H.; and FOWLER, L. A. "Gravitational Waves from an Orbiting Pulsar." *Sci. Am.* **245** (Oct.), 74–82 (1981).

ROBERT H. GOWDY

RELATIVITY, GENERAL THEORY OF, ORIGINS OF

Albert Einstein found his general theory of relativity after eight years of research on two related problems. First was the problem of extending the principle of relativity of the special theory from inertial to accelerated motion. He could make no progress with this problem until he was asked in 1907 to write a comprehensive review article on special relativity. That led him to work on the second problem: Newton's gravitation theory is incompatible with special relativity; it allows instantaneous propagation within the gravitational field. Unlike his contemporaries, Einstein rapidly convinced himself that no straightforward modification of Newton's theory would be adequate. All would violate the exact equality of inertial and gravitational mass, a result whose origins could be traced back to Galileo's observation that all bodies fall alike under gravity.

Einstein's ingenious solution was to construct a new gravitation theory around what he later called "the happiest thought of [his] life," the principle of equivalence. In its original form, the principle asserted that the inertial field of a uniformly accelerated frame of reference in special relativity is a homogeneous gravitational field. Einstein could now use his understanding of acceleration in special relativity to read off the properties of homogeneous gravitational fields, and these properties could then be generalized naturally to other types of gravitational fields. The resulting theory of static gravitational fields was developed from 1907 to 1912. According to it, the speed of light was not constant in a gravitational field so that special relativity held only in the limiting case of no gravitation. Clocks are also slowed by a gravitational field, and light climbing out is redshifted. Light rays traversing a gravitational field are bent, a prediction that Einstein realized in 1911 was open to test through observation during a solar eclipse of the displacement of stars in the star field surrounding the Sun. By March 1912 Einstein had developed a field equation governing all static fields and had treated the effect of these fields on electrodynamics and thermodynamics.

Einstein believed that the principle of equivalence extended the relativity of motion to uniform acceleration. He now sought to complete this extension to all accelerations. He found that, according to his 1912 theory, a test mass experiences a slight acceleration if other masses accelerate past it. This raised his hopes that the inertial forces acting on accelerating bodies may be explained by a hypothesis Einstein attributed to Ernst Mach: The forces are not due to absolute acceleration, which would violate a generalized principle of relativity, but to acceleration with respect to all other masses of the universe.

In the special theory, the principle of relativity is expressed by Lorentz covariance: The theory's laws remain unchanged under Lorentz transformation (between two frames of reference moving at constant velocity with respect to each other). To extend the principle of relativity, Einstein now sought a theory whose laws would remain unchanged under transformation between arbitrary coordinate systems—general covariance. From his student days, Einstein knew of Gauss's theory of curved surfaces, which employed arbitrary coordinate systems. This theory was called to mind by Einstein's 1912 observation that the geometry of a rotating disk ceases to be Euclidean. The decisive step towards general relativity came when Einstein decided that his theory must generalize the flat Minkowski spacetime of special relativity in the same way that Gauss's theory generalized flat, Euclidean space, so that arbitrary gravitational fields came to be associated with a curvature of spacetime.

The theory that resulted from this decisive step was developed in late 1912 and early 1913 by Einstein in collaboration with the mathematician Marcel Grossmann. It contained virtually all the results of the final general theory, including generally covariant laws for free fall, energy-momentum conservation, and the electromagnetic field. But its gravitational field equations were not the celebrated Einstein field equations of the modern theory. In their place, Einstein and Grossmann offered field equations of restricted covariance.

For almost three years Einstein struggled to understand and reconcile himself with the limited covariance of his theory. He even produced an argument—the notorious, fallacious "hole argument"—that purported to show general covariance is physically uninteresting since it would violate determinism. By November 1915 Einstein was convinced that he must return to generally covariant field equations. He chronicled his final steps in a series of four communications in that month to the Prussian Academy. The third demonstrated to his great excitement that his generally covariant theory was able to explain the anomalous 43" of arc per century advance of Mercury's perihelion. The fourth communication (November 25) contained the final generally covariant field equations. Einstein's urgency was heightened by David Hilbert, one of the world's foremost mathematicians, who had turned to the same quest and had communicated the same equations five days earlier to the Göttingen Academy.

Einstein's review of the completed theory in early 1916 concluded with three empirical tests: a redshift in light from massive stars, a deflection of 1.7" arc in starlight grazing the Sun, and the slight advance in the perihelion of Mercury. The third had already been validated; the first proved initially recalcitrant; but the second became the object of celebrated expeditions under Arthur Eddington to Sobral in Brazil and to Principe near Spanish Guinea. Their goal was to measure the gravitational deflection of starlight during the solar eclipse of May 29, 1919. The report of the expeditions' success brought Einstein to the notice of the press and launched him to public prominence. By this time Einstein had already developed the notions of gravitational waves within his theory (1916) and the Einstein universe (1917), which was the first relativistic cosmology. He had also debated whether general covariance expressed a generalized principle of relativity, the first of many debates on whether Einstein's theory had succeeded in extending the principle of relativity to acceleration. He would soon begin publishing his attempts to combine gravitation and electromagnetism geometrically in a unified field theory.

See also: EINSTEIN, ALBERT; EQUIVALENCE PRINCIPLE; FIELD, GRAVITATIONAL; GRAVITATIONAL WAVE; REDSHIFT; RELATIVITY, GENERAL THEORY OF; RELATIVITY, SPECIAL THEORY OF; RELATIVITY, SPECIAL THEORY OF, ORIGINS OF; SPACETIME

Bibliography

EARMAN, J.; JANSSEN, M.; and NORTON, J. D., eds. *The Attraction of Gravitation: New Studies in the History of General Relativity* (Birkhäuser, Boston, 1993).

EINSTEIN, A. *Relativity: The Special and the General Theory,* trans. by R. W. Lawson (Methuen, London, [1917] 1977).

EINSTEIN, A. *The Meaning of Relativity,* 5th ed. (Princeton University Press, Princeton, NJ, 1922).

EINSTEIN, A. "Notes on the Origin of the General Theory of Relativity" in *Ideas and Opinions,* trans. by S. Bargmann (Crown, New York, [1933] 1954).

EINSTEIN, A. *Autobiographical Notes,* trans. by P. A. Schilpp (Open Court, La Salle, IL, [1949] 1979).

LORENTZ, H. A.; EINSTEIN, A.; MINKOWSKI, H.; and WEYL, H. *The Principle of Relativity: A Collection of Original Memoirs on the Special and General Theory of Relativity* (Dover, New York, 1952).

MEHRA, J. *Einstein, Hilbert, and The Theory of Gravitation* (Reidel, Dordrecht, 1974).

NORTON, J. D. "General Covariance and the Foundations of General Relativity: Eight Decades of Dispute." *Rep. Prog. Phys.* **56,** 791–858 (1993).

PAIS, A. *Subtle Is the Lord: The Science and the Life of Albert Einstein* (Clarendon Press, Oxford, Eng., 1982).

JOHN D. NORTON

RELATIVITY, SPECIAL THEORY OF

Special relativity describes the energy and motion of matter and radiation in the absence of gravitational attraction. It was published in its modern form in 1905 by Albert Einstein. The label *special* in special relativity distinguishes it from *general* relativity, which describes the motion of objects and light in the presence of gravitating bodies. Both special and general relativity are so-called *classical* theories; they do not embrace atomic or molecular effects described by quantum mechanics.

Reference Frame and Event

"Relativity theory" is a misleading term, a term that Einstein avoided for years. The relativity of space and time is not the essential thing. Both special and general relativity are based on the central principle that the laws of nature are independent of the viewpoint of the observer.

In relativity, this "viewpoint of the observer" is called a "reference frame." Roughly speaking, a reference frame is a local "container"—such as a spaceship, airplane, automobile, or laboratory room—with respect to which motion is measured using clocks and rulers at rest in that container. Special relativity limits its attention even more narrowly to "inertial reference frames." These are containers with respect to which a free object remains at rest or moves uniformly in a straight line. In an inertial frame a free object "floats in space"—as experienced by an astronaut in an unpowered spaceship, for example. Special relativity compares events as observed in inertial reference frames—unpowered spaceships—moving past one another with constant relative velocity. These are the reference frames discussed in the present article.

Relativity uses reference frames to describe the relation between events. An event is an occurrence determined by both a place and a time of happening. Examples of events are a collision, an explosion, the emission of a light flash, the fleeting touch of a friend's hand. Whether an occurrence is sufficiently localized in space and time to be labeled an event depends on the purpose of examining it. The birth of a baby is an event, important in both time and place to someone who studies family trees. The father and mother, however, experience the birth as a whole series of events from first contraction to delivery, events that can span both time and space.

The Principle of Relativity

Predictions of special relativity seem so bizarre that even today some people try to find alternative theories. Yet none of these alternatives rests on a foundation as utterly simple as does special relativity. Einstein called this foundation the principle of relativity. According to the principle of relativity, *all the laws of physics are the same in every inertial reference frame.* Pull down the shades in our room or vehicle. Then carry out any experiment inside this inertial reference frame. Using many such experiments derive the laws of physics. Someone who carries out the same experiments inside another inertial reference frame will discover the same laws, no matter how fast this second container moves past ours.

The laws of physics contain fundamental physical constants, such as the charge on the electron and the speed of light in a vacuum. According to the principle of relativity, each of these quantities must have the same numerical value as measured in every inertial reference frame. In particular, all observers measure the speed of light in a vacuum to have the value $c = 299,792,458$ m/s. The conclusions of special relativity about space and time spring from the principle of relativity,

including the postulate of the "universal speed" of light.

Time Stretching

Many key ideas of special relativity follow directly from the principle of relativity. Look at the pair of diagrams in Fig. 1. One of the most surprising findings is that the time between two events is typically not the same as measured by two observers moving past one another. This conclusion comes directly from the equality of the speed of light in two reference frames in relative motion: *If light travels a longer path connecting two events in one reference frame than in another, then the time between those events is longer as measured in one frame than in the other.* You can show this result using the diagrams, as explained in the following two paragraphs.

Riding in your transparent unpowered rocket ship, you the reader fire a flash of light upward toward a mirror that you hold 3 m directly above you (left diagram). The flash reflects in the mirror and returns to you. Call the emission of light event A and its reception upon return event B. For you, events A and B occur at the same place. Between these events, the light moves first straight up 3 m then straight down 3 m. *For you, the total time between events A and B is the time that it takes light to travel a total of 6 m.*

The unpowered spaceship in which you carry out these experiments moves from left to right past the

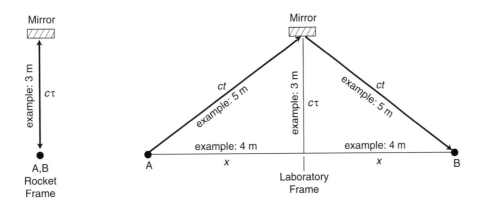

Figure 1 A flash of light emitted at event A is reflected in a mirror and returns to the sender as event B. Events A and B are recorded in both rocket and laboratory frames. Einstein tells us that the measure of the speed of light is the same in each frame. Therefore different path lengths of the light flash in rocket and laboratory frames means different times between events A and B as measured in the two frames. (Symbols *x, ct,* and *cτ* are used later in the article.)

rest of us, who stand in another transparent container called the "laboratory" (right diagram). We also observe the same flash of light emitted at A, reflected at your mirror, and received again at B. But for us in the laboratory your mirror is moving to the right. Thus for us the path of the light slants upward from A along the 5-m-long hypotenuse of the first right triangle (as an example), reflects from the speeding mirror, then slants back down again along the 5-m hypotenuse of the second right triangle to meet you again at event B. *Therefore for us events A and B are separated by the time it takes light to travel a total of 10 m.* For us there is a longer time between events A and B than you measured in your rocket frame.

That's it! The longer a path length for light, the longer a time for light to travel that path at its "universal speed," therefore the longer a time between events as measured in that reference frame. No one has ever found an acceptable way around this simple and powerful result, in spite of the most prodigious efforts. The longer time between two events in one frame than in another is sometimes called "time stretching."

Your light flash and mirror form a kind of clock, call it a "light clock." The principle of relativity assures us that *all kinds of clocks* at rest with respect to one another, once calibrated, must run at the same rate as observed in every frame. Otherwise we could tell which frame we were in by detecting the different rates of these clocks. *All kinds of clocks* includes the "clock" of your body, namely the aging process. Suppose that the mirror was so high above your head that it took six *years* in your rocket for the light to return to you—and ten years by our laboratory clocks. Then you would age six years between these new events A and B; your body "aging clock" and your "light clock" ride together in your rocket frame. In contrast, between events A and B we in the laboratory frame would age ten years.

Time Stretching at All Speeds

These same results follow whatever the relative velocity of the two frames. Prove the general case using symbols in Fig. 1: *x* is *half* the distance between events A and B as measured in our laboratory frame, the length of the horizontal leg of one right triangle in the right diagram; *t* is half the laboratory time between events A and B, the time it takes the light flash to slant upward along one hypotenuse at

the speed *c*. This hypotenuse has (length) = (velocity) times (time) = *ct*.

In your rocket frame the flash moves vertically upward to the mirror in half the time between events A and B. We give your half-time between events the unique Greek letter tau (τ), because in your unique frame A and B occur at the same place. So for you the upward distance covered in time τ is equal to $c\tau$. This vertical span is the same as the vertical leg shared by the right triangles in the diagram at the right. Hence we have expressions for the lengths of all sides of either of these right triangles. Apply the Pythagorean theorem for right triangles: The square of one leg plus the square of the other leg is equal to the square of the hypotenuse, or

$$x^2 + (c\tau)^2 = (ct)^2. \tag{1}$$

Each term has the dimensions of length squared. Put on the right all terms related to the laboratory frame:

$$(c\tau)^2 = (ct)^2 - x^2 \\ \text{(squared timelike interval)}. \tag{2}$$

A wealth of insight comes directly from Eq. (2), such as (1) verification that the times between events A and B as measured in the two frames cannot be the same (discussed below), and (2) we in the laboratory can correctly predict the time you measure between events A and B in your rocket, in spite of the fact that this time is not the same as ours; we simply put our values for *t* and *x* into Eq. (2).

For simplicity in reading the diagram, we have let symbols *x*, *t*, and τ represent *half* the separations in space and time between events A and B in the two frames. But Eq. (2) is still an equality if we double the value of each of these quantities (which gives each squared term four times its original value). Hence Eq. (2) is still correct if we take *x*, *t*, and τ to measure the *full* distance and times between events A and B in the given frames, as we do from now on.

The time τ between two events measured in a frame in which they occur at the same place is called the "proper time" or "wristwatch time." A difference of squares like that on the right of Eq. (2) is called the square of the "timelike interval"—timelike because the magnitude of the time part *ct* is greater than the magnitude of the space part *x*. No matter how fast or slow your rocket, the value of the interval remains the same. We say the interval between two events is "invariant," meaning that it has the same value as measured in all reference frames.

An equation that relates τ and t directly comes from setting $x = vt$, the relation between distance and time as we observe your passing rocket in our laboratory frame. Substitute this into Eq. (2) and divide through by c^2. The result is

$$\tau^2 = t^2 - v^2 t^2 / c^2 = (1 - v^2/c^2) t^2. \qquad (3)$$

Equation (3) encompasses all possible speeds from the very slow to the very fast. The slow speeds v we observe in everyday life are very much smaller than the speed of light c; therefore the value of v/c is very small compared with the value 1. Therefore the expression $(1 - v^2/c^2)$ is approximately equal to 1. Equation (3) tells us that in this case τ and t are essentially equal; the time between events A and B is the same for you in the rocket as it is for us in the laboratory. So at very low relative speeds, special relativity is consistent with our everyday assumption that time is a universal quantity; everyone measures time to have the same value between two events.

At a high relative speed v, however, the outcome is quite different. Imagine that you start from Earth (event A: departure from Earth) and travel to the star Alpha Centauri (event B: arrival at Alpha Centauri). Both events (departure and arrival) occur at the position of your cockpit. Equation (3) tells us that by making v/c closer and closer to the value 1, your trip can take place in shorter and shorter wristwatch time τ *as measured in your spaceship*. By extension of this argument, we arrive at a result that frees the human spirit, if not yet the human body: Given sufficient rocket speed, we can go anywhere in the universe in a single astronaut lifetime!

Equation (3) also gives evidence for the natural speed limit of the universe: the speed of light c. Imagine that we in the laboratory could measure your rocket speed to be greater than the speed of light c. Then v/c (and also v^2/c^2) would have a value greater than 1, and the right-hand side of Eq. (3) would be negative. This would mean that the square τ^2 of your time measurement would also become negative. But this is impossible: No real number, no real time, can be squared to give a negative value. Thus the formula implies that no thing can move at a speed v greater than that of light c. Experiment verifies this: In huge particle accelerators protons or electrons, urged on by electric and magnetic fields

to ever-higher energies, approach the speed of light closely but have never been found to exceed this speed.

Cause and Effect

The analysis this far has omitted a large number of event-pairs, in particular two events that occur at the same time in some reference frame but separated spatially in that frame. Think of two firecracker explosions that occur simultaneously, one in New York City, the other a continent away in San Francisco. In this special "continent frame" the space separation is given the symbol s. We write a new equation similar to Eq. (2) with the two terms on the right interchanged to keep the result positive:

$$s^2 = x^2 - (ct)^2 \qquad (4)$$
$$\text{(squared spacelike interval)}.$$

The distance s between two events, measured in a frame in which the events occur at the same time, is called the "proper distance." The difference of squares on the right side of Eq. (4) is called the square of the "spacelike interval"—spacelike because the magnitude of the space part x is greater than the time part ct.

The right-hand side of Eq. (4) includes the space and time separations between these two events as measured in some other frame that moves past the first. Some consequences can be read from the equation:

- In all frames, x is greater in magnitude than ct, since both sides must remain positive. But ct is the distance that light can travel in the time available between these events. Therefore in all frames nothing, not even a light flash, can move fast enough to travel from one event to the other. Therefore, for events connected by a spacelike interval, one event cannot cause the other event as observed in any frame.

- In some frames, the time t between the events is not zero, they are not simultaneous. *Two events simultaneous in one frame are not necessarily simultaneous in other frames.* As an example, for observers in a rocket streaking across the continent, the two firecracker explosions in New York and San Francisco will *not* occur at the same time. For many people, this is the most difficult concept in special relativity, harder to believe even than the difference in clock rates [Eq. (3)]. The fact that two events simultaneous in one frame are not necessarily simultaneous in another frame is called the "relativity of simultaneity."

Suppose that a flash of light goes *directly* from one event to the other event. Then the distance between them is given by $x = ct$. In this case both the proper distance s and the proper time τ are equal to zero:

$$\tau^2 = s^2 = 0$$
(squared lightlike interval). \hfill (5)

Two events that can be connected by a direct light flash are said to be related by a "null" or "lightlike interval."

Equations (2), (4), and (5) embrace all possible relations between pairs of events that occur along the x direction as described by special relativity. Equation (2) describes two events separated by a timelike interval. A rocket can move between these two events, so that event A can cause event B. This cause-and-effect relation is preserved in every frame. In contrast, not even light can travel between the two events that are separated by a spacelike interval as described in Eq. (4), so that neither of these two events can cause the other. This *lack* of cause-and-effect relation is also preserved in every frame. The boundary between these two cases is provided by Eq. (5), which describes the relation between two events that can be connected by a direct light flash. One event in this pair can cause the other only through the connecting light flash, a connection maintained in every reference frame.

In brief, the threefold categories of timelike, spacelike, and lightlike intervals between pairs of events preserve their possible cause-and-effect relation in all inertial reference frames.

Momentum and Energy

Imagine that a moving particle emits two flashes close together in time on its own wristwatch. We use these two emissions to track the motion of the particle. This is similar to the case that led to Eq. (2), and these two flashes are related by a timelike interval

$$(c\tau)^2 = (ct)^2 - (x)^2 \tag{2}$$

where x and t are measured with respect to the laboratory frame. Starting with this equation we can

say something important (not a derivation!) about momentum and energy of the particle, collapsing into a few lines what took people a generation to devise and verify. Divide both sides of the equation by τ^2 and multiply by m^2c^2, where m is the mass of the particle:

$$(mc^2)^2 = \left(mc^2\frac{t}{\tau}\right)^2 - \left(mc\frac{x}{\tau}\right)^2. \tag{6}$$

On the left appears the famous mc^2, which we will come back to in a minute. The second term on the right contains the fraction x/τ, namely distance x traveled by the particle as measured by our laboratory observer, divided by the time τ taken to move this distance as recorded on the wristwatch carried by the particle. This is a kind of velocity. Mass times velocity is the formula for momentum; call it p. The laboratory observer reckons the momentum to have the value

$$p = m\frac{x}{\tau}. \tag{7}$$

In Newtonian mechanics, the time is thought to be universal, the same in all reference frames. But relativity shows us that the time between the two flashes emitted by the particle has different values as measured in different frames [Eq. (3)]. Equation (7) picks out the proper time τ, the time recorded on the wristwatch carried by the particle, as the time to use in reckoning the particle's momentum.

What about the first term on the right of Eq. (6), the one containing the fraction t/τ? For now, we assert that the first term is the total energy of the particle, call it E. Then, according to Eqs. (6) and (3), energy can be written in two ways:

$$E^2 = \left(mc^2\frac{t}{\tau}\right)^2 = \frac{(mc^2)^2}{1 - v^2/c^2}. \tag{8}$$

Substitute p from Eq. (7) and E from Eq. (8) into the fundamental Eq. (6), which then reads

$$(mc^2)^2 = E^2 - (pc)^2. \tag{9}$$

When the particle is at rest, the momentum p is equal to zero and this equation takes the famous form

$$E_{\text{rest}} = mc^2. \tag{10}$$

Equation (10) reveals that every particle in the universe is a storehouse of energy, useful to us provided we can find ways to transform this rest energy into other forms. Explosion of a nuclear weapon and burning of a star provide spectacular examples of such transformations, but basically every single energy-emitting reaction—down to the lighting of a match—carries with it a conversion of mass to energy.

Now, when the particle is *not* at rest, the momentum p is *not* zero. Then the value of E must be greater than its rest value to keep the right side of Eq. (9) a constant, equal to the left side. The increase in energy E with motion is called the kinetic energy. Equation (8) shows that the total energy—and therefore also the kinetic energy—increases without limit as the particle speed v approaches the speed of light c (as v/c approaches the value 1). And indeed we can add as much kinetic energy as we want to a moving particle, as is done in ever-more powerful and ingeniously designed particle accelerators in order to increase the energy of collisions with other particles. Yet even the highest energy particle never moves faster than light.

Three Space Dimensions

For simplicity, we have so far used one space dimension x in the equations of this article. Of course, there are three space dimensions, the additional two often labeled with the symbols y and z. The more general expressions for the timelike and spacelike intervals, Eqs. (2) and (4), are

$$(c\tau)^2 = (ct)^2 - (x^2 + y^2 + z^2) \tag{11}$$
(squared timelike interval)

$$s^2 = (x^2 + y^2 + z^2) - (ct)^2 \tag{12}$$
(squared spacelike interval).

Lorentz Transformation

Historically, textbooks on relativity have not emphasized those quantities that have the same value for all observers, such as the various kinds of intervals between events described above. Instead, much attention has been paid to what are called the Lorentz transformations, equations that connect the distinct space and time separations between two events as measured in one frame with those separations as measured in another frame moving past the first. We display the Lorentz transformation equations here without deriving them. (The derivation depends on the principle of relativity and some symmetry arguments.) Let unprimed coordinates represent measurements made in the laboratory frame and primed coordinates the corresponding measurements made in a rocket frame that moves past the laboratory with speed v along the positive x direction. Then the Lorentz equations that transform the laboratory space and time separations to rocket space and time separations are

$$t' = \frac{-(vx/c^2) + t}{(1 - v^2/c^2)^{1/2}} \tag{13a}$$

$$x' = \frac{x - vt}{(1 - v^2/c^2)^{1/2}} \tag{13b}$$

$$y' = y \tag{13c}$$

$$z' = z. \tag{13d}$$

In these equations the exponent $1/2$ indicates a square root, and the velocity v has the usual meaning, distance divided by time as measured in the same frame.

Now, "laboratory" is just a label; it could represent simply another unpowered spaceship. Then the only difference between laboratory and rocket frames is the artificial difference we have given them: the rocket moves in the *positive x* direction with respect to the laboratory. So the laboratory moves in the *negative x* direction with respect to the rocket. It follows that the inverse transformation—the one that gives unprimed laboratory space and time separations in terms of primed rocket spacetime separations—derives from Eqs. (13a–d) by interchanging primed and unprimed coordinates and reversing the sign of v, making all v terms positive in the numerators. (Reversing the sign of v does not change the sign of v^2 in the denominators.)

Lorentz Contraction

The Lorentz transformation equations predict a relativistic effect important in the history of the subject, namely, that we as observers will measure an object moving past us at high speed to be shortened—contracted—along its direction of relative motion.

The Lorentz transformation describes the space and time separations between a pair of events. What events can we use to measure the length of a moving object? One answer is to set off two firecrackers, one at each end of the object at the same time ($t = 0$) in our frame, and measure the distance x between these explosions. Set $t = 0$ in Eq. (13b) and multiply through by the square-root quantity. The result is

$$(1 - v^2/c^2)^{1/2}x' = x. \qquad (14)$$

This equation tells us that the length x we measure for the object in the laboratory frame—the distance between simultaneous firecracker explosions—is less than the distance x' between the two ends as measured in the rocket frame in which the object is at rest.

The Lorentz contraction is a curiosity, not much used directly in analyzing experiments. However, it is an example of a deeper principle, the relativity of simultaneity, discussed above following Eq. (4). The two firecrackers may explode at the same time in our frame ($t = 0$), but not in the frame of the rocket ($t' \neq 0$), as you can see by substituting $t = 0$ into Eq. (13a). This does not change the rocket measurement of x' because the object is at rest in the rocket frame; the distance between the explosions at the two ends is the same whether or not these explosions occur at the same time. However, this lack of simultaneity allows observers in the two frames to account for the difference in measured length of the object.

Summary

Special relativity describes the measurements of observers in uniform relative motion based on Einstein's principle of relativity: The laws of physics are the same as observed in every inertial reference frame. Major results are (1) a measure of the separation between events—the interval—that has the same value for all such observers and that preserves cause-and-effect relations between events in all frames, (2) expressions for energy and momentum that correctly describe high-energy particle interactions, including (3) description of transformations between mass and other forms of energy.

See also: EINSTEIN, ALBERT; EVENT; LORENTZ TRANSFORMATION; RELATIVITY, SPECIAL THEORY OF, ORIGINS OF; SPACE; SPACE AND TIME; TIME

Bibliography

EINSTEIN, A. "Zur Elektrodynamik Bewegter Körper." ("On the Electrodynamics of Moving Bodies.") *Annalen der Physik* **17,** 891–921 (1905).

FRENCH, A. P. *Special Relativity* (W. W. Norton, New York, 1968).

MILLER, A. I. *Albert Einstein's Special Theory of Relativity: Emergence (1905) and Early Interpretation (1905–1911)* (Addison-Wesley, Reading, MA, 1981).

TAYLOR, E. F., and WHEELER, J. A. *Spacetime Physics,* 2nd ed. (W. H. Freeman, New York, 1992).

EDWIN F. TAYLOR

RELATIVITY, SPECIAL THEORY OF, ORIGINS OF

The publication of Isaac Newton's *Mathematical Principles of Natural Philosophy* (1687) marked a watershed in the history of western physical science. Its publication and its subsequent influence are important guideposts in deciphering the context in which Albert Einstein created the special theory of relativity more than two centuries later.

Newton not only revolutionized and summarized the explanation of motion, he also synthesized it: Using his three laws of motion and his law of universal gravitation, Newton was able to account for the motion of bodies on or near Earth's surface, as well as the motion of objects, such as planets and moons, distant from Earth's surface. Over the next century, the Newtonian system became the *sine qua non* of natural philosophy, and the Newtonian method became the standard by which one judged a work to be "scientific" or not. By the beginning of the nineteenth century, an explicit understanding had been articulated that *all* phenomena, including mechanical, chemical, biological, and psychical phenomena, would eventually be understood to be direct consequences of Newton's laws of motion. Historians of science have labeled this outlook the mechanical world view.

The Electromagnetic Ether

Between 1865 and 1873 the great British physicist James Clerk Maxwell gave a complete description of electric and magnetic fields and their interactions with only four equations—Maxwell's equations—which, together with the specifications for the force exerted by the fields on electric charges (today known as the Lorentz force equation), provided a synthesis of electric and magnetic effects analogous to the Newtonian synthesis for motion. But long before Maxwell's synthesis, the route to the reduction to Newtonian mechanics of electric, magnetic, and optical phenomena depended on describing, in mechanical terms, the properties of the respective propagating media of ethers. The electromagnetic ether had to be similar to the light ether. It had to be an elastic solid because the disturbances—the varying electric and magnetic fields—were perpendicular to each other and to the direction of propagation of the light wave. And because the speed of light is so high—3×10^8 m/s—the light ether had to be far more elastic than the stiffest steel piano wire, and yet, material objects, including such things as planets and stars, had to move through it as if it were not there. For some people, such seemingly contradictory properties were nothing more than a manifestation of the special nature of the medium. For example, late in the nineteenth century, the British physicist and mystic Oliver Lodge described the ether as "truly the garment of God." Others, for example, Hendrik Antoon Lorentz, the great Dutch theoretical physicist, refused to speculate on the nature of the medium. For them, the ether was nothing more than the equations that described it, that is, Maxwell's equations.

The Newtonian Principle of Relativity

The movement of objects described and predicted in Newtonian mechanics takes place against a backdrop of absolute space and absolute time. Newton and his successors sometimes described this perspective as "the sensorium of God": Only from a perspective outside the physical universe would it be possible to assign absolute values to the spatial and temporal coordinates of events, such as collisions, coincidences, and the other occurrences in the history of the universe.

Mere mortals not being able to command such a perspective, Newton recognized that all spatial and temporal measurements are relative to arbitrarily chosen origins. The three spatial coordinates and the temporal measurements made relative to that origin are termed "a frame of reference." It is a simple matter to show that the *description* of motion from two coordinate systems, or frames of reference, moving at a constant speed relative to each other (inertial reference frames) may not be the same. Observers in different inertial frames of reference observing the same events might disagree on the coordinates and the velocities of objects participating in the events; however, they would all agree that the objects had the same accelerations and masses and, hence, were subject to the same net forces. This result is called the Newtonian, or classical, principle of relativity; it is a result of what are called the Galilean transformation equations—equations first described by Galileo for rationalizing the spatial and temporal coordinates of events observed from different inertial frames of reference.

The Crisis

Since it was possible to formulate the laws of thermodynamics in terms of Newton's laws of motion, thermodynamics must also obey the mechanical principle of relativity as must hydrodynamics and acoustics. If the mechanical principle of relativity was to apply to *all* phenomena, electromagnetic phenomena had to be understood in mechanical terms. One route to such a reduction would be to interpret the propagation of electromagnetic waves through the ether as the propagation of mechanical oscillations through a perfectly elastic solid.

One element of this problem was to understand the manner in which ordinary matter interacted with ether. Since, as far as we know, light comes to us from all parts of the universe, it was generally accepted that the ether filled the universe. It is fair to say that for most nineteenth-century physicists the ether became the benchmark of Newton's absolute space.

There are three possibilities for what happens as matter moves through the ether: At the interface between moving matter and ether, the ether is totally dragged along, partially dragged along, or not dragged along at all. Additionally, if the ether did move, it had to do so perfectly smoothly. There could be no eddies or whirlpools, no discontinuities. If there were, they should manifest themselves in the appearance of distant stars and galaxies in *some* direction as Earth itself moves through the ether.

One way of determining the dynamical relationship between ether and matter would be to compare the speed of light in different directions as Earth moves through the ether. Earth's absolute velocity would be the sum of its velocities as it moved around the Sun, as the Sun moved within the Galaxy, and as the Galaxy moved relative to other galaxies. Doing such experiments in all directions would ensure making some of those measurements when the differences between the outgoing and incoming light beams should be maximized.

The class of experiments that sought to make such a comparison are known as "ether-drift experiments." Many such experiments were done during the nineteenth century. Of all such experiments, undoubtedly the most famous is the Michelson–Morley experiment, first performed in 1881. It was redone with a much more sensitive apparatus in 1887 and has since then been done and redone countless times.

If we only had to rely on the results of the Michelson–Morley experiment, the results would be clear-cut: There is no relative motion between Earth and the ether; the speed of light, whatever its value, appears to be unaffected by the motion of Earth. But, this conclusion flies in the face of the conclusion one is forced to in other ether-drift experiments, where the only reasonable interpretation of the data is that the ether is partially dragged along by Earth. For still other such experiments, the simplest conclusion is that the ether is unaffected by the motion of Earth; the ether is totally at rest. The contradiction in these results underpinned the crisis in late nineteenth-century attempts to incorporate electromagnetic phenomena within the mechanical world view.

Today, with hindsight, it may seem obvious that the core of the problem lay in stubbornly clinging to the concept of an all pervading, perfectly elastic, weightless solid—the ether. Physicists of the day were no less clever than modern-day physicists, however, and the ether was deemed, by almost everyone, as being absolutely necessary if light was, indeed, a wave phenomena. Waves are nothing more than the propagation of physical effects in a medium.

The Electromagnetic World View

Some late nineteenth-century physicists, for example William Thomson (Lord Kelvin), retained an unshaken commitment to the mechanical world view. Most others, following the lead of Lorentz and

Max Abraham, struck out in a different direction. Though Lorentz and Abraham disagreed on the particulars, they both believed that it would be possible to unify the laws of nature by making electromagnetic phenomena fundamental and to explain everything else, including mechanics, in electromagnetic terms.

The Lorentz Theory

To explain the results of the ether-drift experiments as well as to account for various mechanical properties of matter, Lorentz, between 1892 and 1904, by stages, assumed that the totality of matter in the universe was made up of positive and negative electric charges and that interactions between the charges were mediated by the ether. Between 1892 and 1895, his theory assumed that one of the effects of this mediation was to exert a force on extended bodies in motion so as to contract (shrink) the body in the direction in which it moved. He further assumed a new transformation for the time coordinate such that time intervals in inertial frames of reference moving with respect to an observer would appear dilated—the clocks in the moving frame of reference would appear to run slow. For Lorentz this was only an appearance, "an aid to calculation."

By 1904 he had consolidated his theory by assuming a new set of transformation equations—the Lorentz transformation equations—to replace entirely the Galilean transformation equations of Newtonian physics. He further assumed that the fundamental charges that made up all of matter, which by now, after Joseph John Thomson, he called "electrons," were deformable and contracted in the direction of motion as a function of their speed. Having made these assumptions, Lorentz could now account for all ether-drift experiments, could account for recent experimental results that showed the mass of electrons increased as a function of their speed, and could at least provide a sketch of how the mechanical properties of matter might be understood in terms of electromagnetic interactions. Among his conclusions was the fact that regardless of their state of motion, the speed of light would be the same to all inertial observers. Such results followed because observers in different inertial frames of reference would disagree on the time elapsed between two events. For Lorentz such results had little to do with reality—they were aids to calculation. "True" time had to be distinguished from "local" time, which resulted from the calculations made by observers in one frame of reference on the rate at which clocks ran in other inertial frames of reference.

The Abraham Theory

Abraham, a master at manipulating the formal contents of physical theory—the equations—objected to one key provision of Lorentz's theory, the concept of a deformable electron. If the electron were deformable, Abraham argued, then some non-electromagnetic forces would be required to prevent the electron's disintegration because of the repulsive forces between parts of the electron due to the fact that those parts all had the same electrical charge.

Abraham's approach was to stipulate that the electron is an absolutely rigid body, with uniformly distributed charge. He then calculated the inertial forces that would result as the electron moved through its own and any external fields. The result was an electron whose mass increased as a function of its speed not quite to the same degree as Lorentz's deformable electron. Though both Lorentz and Abraham claimed that experiment supported their predictions, the truth of the matter was that the results of experiments were still too coarse to distinguish between the two theories. For all other aspects of electromagnetic theory, Abraham made use of the Lorentz theory, including the Lorentz contraction and the Lorentz transformation equations.

The Contributions of Henri Poincaré

Henri Poincaré, a leading French mathematician, mathematical physicist, and cosmologist, was a close colleague of Lorentz's. Poincaré subscribed to Lorentz's point of view and wrote long penetrating articles for both professional and lay audiences defending the electromagnetic world view. In one of his popular books, *Science and Method* (1906), he confidently told his readers that "beyond electrons and ether, there is nothing."

For Lorentz the ether remained an abstraction whose characteristics were nothing more than the equations that described it. Poincaré asserted that it was the pressure exerted by the ether that held the deformable electron together and provided the additional external force required to contract its diameter along the direction of motion.

Lorentz did not speak of a theory of relativity. Rather, he sometimes referred to what he termed "a theory of corresponding states." Poincaré suggested that using the Lorentz transformation equations rendered the laws of electricity—Maxwell's laws—in the same form in all inertial frames of reference. Observers in different inertial frames of reference might disagree on the magnitude of the electric or magnetic field strength, or the speed and acceleration of various objects participating in events, but they would all agree on the *form* of the laws describing those events. By 1904 Poincaré suggested with more force that, as a matter of empirical generalization, the principle of relativity applied to all physical laws, which, at their root, were electromagnetic in nature.

Einstein's Special Theory of Relativity

At this point, in the spring of 1905, Albert Einstein, then an examiner in the Swiss Patent Office in Berne, wrote and sent in for publication to the prestigious journal *Annalen der Physik (Annals of Physics)* his seminal paper, "On the Electrodynamics of Moving Bodies." Even though he was only twenty-six, Einstein was already operating with a belief that the best physical theories accomplished their tasks with a minimum of axioms. In this paper, which was published in September 1905, after making some initial comments on the meaning of "simultaneity," Einstein began with two axioms or postulates. The first was the principle of relativity—that the laws of physics have the same form in all inertial frames of reference. The second was the invariance of the speed of light—that, regardless of their state of motion, all inertial observers will measure the same value for the speed of any light beam.

With those two postulates in hand, and without any reference to the laws of mechanics, electromagnetism, or any other substantive physical theory, Einstein derived the Lorentz transformation equations. With those transformation equations in hand, he also was able to derive all of the results that Lorentz had derived by assuming the transformation equations. For this reason, many later commentators concluded that Einstein's special theory of relativity ("special" denotes restriction to inertial frames of reference) was a generalization of Lorentz's theory. This is a misconception. Lorentz's theory was a dynamical theory that was premised on the interaction of electrons and ether. Einstein, who said simply that he would proceed without reference to the ether, had constructed a theory of measurement. For Lorentz, the contraction of length was physical; the dilation of time intervals—local time—was only an aid to calculation. For Einstein all the discrepancies were an artifact of measurement, a recognition that judgments of the simultaneity of events were not absolute but a function of the frame of reference from which events are observed.

Behind Einstein's Postulates

In his professional papers, Einstein rarely discussed the details of how he arrived at his formulations of physical theory. But he has provided some tantalizing clues concerning how he came to formulate special relativity based on two seemingly abstract postulates. In an autobiographical sketch he wrote forty years later, Einstein tells us that at the age of sixteen, ten years before he published his first paper on special relativity, he began wondering how a beam of light would appear if one were traveling at the same speed alongside it, for no matter what the circumstance, no one had ever reported light as traveling in empty space at any other speed but the speed of light.

Apparently, the full significance of that observation did not hit him until the spring of 1905. In 1917 Einstein published a popular account of the special theory of relativity that he claimed followed his steps in creating the theory. In that treatment Einstein began by concentrating on the synchronization of clocks not located at the same place. He proposed using light signals for the purpose of synchronization. This is unsatisfactory since there is no guarantee that the speed of light in space is the same in all directions. Furthermore, since nothing can keep up with a light signal, it is not possible to measure the speed of light with but one clock; two clocks placed at different points in space would have to be used. But first, of course, the clocks would have to be synchronized. To break out of this paradox, Einstein stipulated that, in order to come to a meaningful definition of simultaneity in this one inertial frame of reference, the speed of light must be the same in all directions. It is then an easy matter to show that inertial observers will not agree that the clocks in other inertial frames of reference have been properly synchronized. Furthermore, this inertial frame of reference is no different than any other inertial frame of reference. The only way that the stipulation about the speed of light in different directions can

hold in all inertial frames of reference (the principle of relativity) is that it be postulated that the speed of light is an invariant, that is, it must have the same absolute value for all inertial frames of reference (the second postulate). Since all mesurements of length and time intervals in frames of reference other than one's own involve judgements of simultaneity, observers in different inertial frames of reference will disagree on length and time measurements as well as on many other derived physical parameters. However, other key variables will be, like the speed of light, invariant.

To this day, students still struggle to understand that, since nothing can keep up with a light signal, *no experiment can be done to make a one-way measurement of the speed of light with a single clock.* Only if a signal speed higher than the speed of light is discovered will such a measurement be possible. But then, the special theory of relativity will not be invalidated; the symbol for the speed of light in the Lorentz transformation equations will be replaced by the symbol for this new signal speed. Only if we ever should discover a phenomenon for which the signal speed is infinite (a highly unlikely prospect), can we return to the original Galilean transformation equations of Newton that assume, implicitly, that it is possible to transmit information from one place to another instantaneously.

See also: EINSTEIN, ALBERT; ELECTROMAGNETISM, DISCOVERY OF; FRAME OF REFERENCE; FRAME OF REFERENCE, INERTIAL; FRAME OF REFERENCE, ROTATING; GALILEAN TRANSFORMATION; LIGHT; LIGHT, ELECTROMAGNETIC THEORY OF; LIGHT, SPEED OF; LIGHT, WAVE THEORY OF; LORENTZ, HENDRIK ANTOON; LORENTZ TRANSFORMATION; MAXWELL'S EQUATIONS; MICHELSON–MORLEY EXPERIMENT; NEWTON, ISAAC; NEWTON'S LAWS; NEWTONIAN MECHANICS; RELATIVITY, SPECIAL THEORY OF

Bibliography

BORN, M. *Einstein's Theory of Relativity* (Dover, New York, 1962).

EINSTEIN, A. *Ideas and Opinions* (Crown, New York, 1954).

EINSTEIN, A. *Out of My Later Years* (Philosophical Library, New York, 1950).

EINSTEIN, A. *Relativity: The Special and General Theory,* 17th ed. (Crown, New York, 1961).

GOLDBERG, S. *Understanding Relativity: Origins and Impact of a Scientific Revolution* (Berkhaeuser, Boston, 1984).

GOLDBERG S., and STUEWER, R. H., eds. *The Michelson Era in American Science, 1870–1930.* (American Institute of Physics, New York, 1988).

HOFFMAN, B. *Albert Einstein: Creator and Rebel* (Viking, New York, 1972).

HOLTON, G. *Thematic Origins of Scientific Thought: Kepler to Einstein* (Harvard University Press, Cambridge, MA, 1988).

SCHILPP, P. A., ed. *Albert Einstein: Philosopher-Scientist* (Open Court, La Salle, IL, 1970).

STANLEY GOLDBERG

RELAXATION TIME

The earliest use of the term "relaxation time" seems to have been by James Clerk Maxwell in an 1866 treatise on gases. If a stress, for example a sudden change in volume, is applied to a gas, the resulting strain, for example variation in the pressure, will disappear with a characteristic time that he termed the "time of relaxation." The mechanism responsible for the return to equilibrium is dissipative, in this case, viscosity. Although the use of relaxation time has persisted in the description of phenomena in gas, for example, in treatments of atmospheric disturbances, the term has passed into more general use in physics, and characterizes the time for a system to come to equilibrium when new conditions are imposed. As with Clerk Maxwell's original definition, the time is usually taken to be the "*e*-folding time," the time for the transient strain due to a change in the stress on the system to diminish by a factor of $1/e$, where e (equal to 2.71828…) is the base of the natural logarithms. The term is commonly encountered in condensed matter physics as in, for example, "spin-lattice" relaxation time, the characteristic time for the magnetization in a sample to approach a new equilibrium value when the applied field is altered.

Normally, when the stress is altered on a system, the resulting transient strain, S, will "relax," disappear, or dissipate at a rate given by S/T, where T is the relaxation time. Suppose, for example, a stress that has produced an equilibrium strain S_0 is removed from a system. Then,

$$\frac{dS}{dt} = -\frac{S}{T}.$$

The solution to this equation is

$$S = S_0 e^{-t/T},$$

where t is the elapsed time since the stress was removed.

The discharge of a capacitor through a resistor is a simple example of the approach to equilibrium via dissipation of the energy in the strain. Here a capacitor C, given a charge Q_0 (analogous to the stress or displacement from equilibrium) and disconnected from the charging source, is shorted by a resistance R.

The rate of change of the charge on one plate of the capacitor is given by

$$\frac{dQ}{dt} = -\frac{Q}{RC}.$$

Therefore, the charge decreases exponentially with time,

$$Q = Q_0 e^{-t/RC}.$$

In this case, if R is 1 MΩ and C is 1 μF, then RC is 1 s. That is, 1Ω times 1 F is 1 s. The potential on the capacitor, V, is just Q/C and, since C is constant, V will also diminish (relax) exponentially.

Relaxation often can be studied effectively by measuring the response of a system to periodic stress. An example is the dielectric behavior of

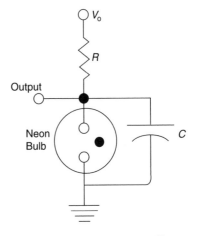

Figure 1 Relaxation oscillator.

water. The permittivity (a measure of the alignment of the electric dipole moment of a water molecule with an applied electric field) of a sample at 20°C decreases significantly when the frequency of the applied field exceeds 17 GHz, the relaxation frequency, that produces maximum energy absorption. The relaxation time is the reciprocal of 2π times the relaxation frequency, in this case 9.3 ps. Microwave ovens take advantage of the energy absorbed by relaxation in this frequency range.

An interesting further example of relaxation is the so-called relaxation oscillator, with the circuit shown in Fig. 1. The relaxation process occurs from the uncharged to the fully charged condition. The potential across the capacitor will rise with the characteristic relaxation time, RC, according to the equation

$$V = V_0(1 - e^{-t/RC}).$$

However, at some potential V_S, typically 100 V, the gas in the neon bulb will ionize and emit a pulse of light. The resistance of the bulb is then very low and the potential across the bulb drops until ionization cannot be sustained and the resistance of the bulb becomes very large. The capacitor begins to charge again and the process is repeated periodically. The wave form of the output voltage is a sawtooth.

Common mechanical examples of relaxation oscillations are the chattering of brakes, the screeching of chalk on a blackboard, and the vibrations of a trumpeter's lips.

See also: STRAIN; STRESS

Bibliography

BROWN, T. B. *The Taylor Manual* (Addison-Wesley, Reading, MA, 1959).

GRANT, E. H.; SHEPARD, R. J.; and SOUTH, G. P. *Dielectric Behaviour of Biological Molecules in Solution* (Clarendon Press, Oxford, 1978).

HALLIDAY, D.; RESNIK, R.; and WALKER, J. *Fundamentals of Physics*, 4th ed. (Wiley, New York, 1993).

HULL, R., and BEAN, J. C. "Dynamic Observations of Relaxation Processes in Semiconductor Heterostructures." *Advanced Materials* **3,** 139 (1991).

KITTEL, C. *An Introduction to Solid-State Physics*, 4th ed. (Wiley, New York, 1971).

PEIXOTO, J. P., and OORT, A. H. *Physics of Climate* (American Institute of Physics, New York, 1992).

H. C. BRYANT

RENORMALIZATION

In quantum field theory calculations, renormalization is a process by which results are expressed solely in terms of measurable parameters, such as the masses and charges of physical particles. This step is necessary because the underlying theories are written in terms of parameters that are not directly measurable.

As a simple example, consider the quantum field theory of a single species of spin-0 particles ϕ that interact through a four-point vertex. The rate of scattering of two of these particles off each other is calculated as the square of a sum of Feynman diagrams:

Each vertex in the diagrams represents a factor of the fundamental coupling constant of the theory, a dimensionless number denoted by g. If this number is sufficiently small, then diagrams with more than a few vertices will be negligible, and the sum of the first few diagrams will give a reasonable approximation of the scattering probability. The crudest approximation is to neglect all but the first diagram; we would then predict that the scattering probability is proportional to g^2. If the rate of scattering is measured in the laboratory, this result can be used to determine the approximate value of g.

To be more precise, however, we should also include diagrams containing two vertices, such as the second and third diagrams shown above. The internal lines in these diagrams represent virtual particles, which we imagine to be constantly appearing and disappearing in the vicinity of the real particles being scattered. We might expect this cloud of virtual particles to affect the strength of the scattering interaction, and indeed, when these diagrams are computed, they are found to have such an effect. Furthermore, the resulting shift in the scattering probability depends in a nontrivial way on the momenta of the incoming and outgoing particles. These effects complicate our experimental determination of g. It is customary to define a measured coupling constant, g_m, as the value of g that one would infer by comparing a measurement of the scattering rate, at some standard values of the momenta, to a calculation of only the first diagram. The underlying parameter g, meanwhile, is called the bare coupling constant, since it is what we would measure if the ϕ particle were stripped of its cloud of virtual particles. To a first approximation, $g_m = g$, but this relation is disturbed by higher-order diagrams. Similarly, since each internal line in a diagram represents an expression that depends on the mass of the ϕ particle, the sum of subdiagrams

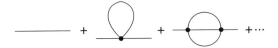

leads to a distinction between the bare mass m and the measured mass m_m. Note, however, that Feynman diagrams such as those shown here allow one to calculate, as accurately as desired, the relations between the bare parameters and the measured parameters.

Renormalization, then, is the process of expressing all predictions of the theory in terms of g_m and m_m rather than g and m. These predictions include the momentum dependence of the $\phi\phi \rightarrow \phi\phi$ scattering probability, as well as the rates of more complicated reactions such as $\phi\phi \rightarrow \phi\phi\phi$. Similarly, in quantum electrodynamics, the Feynman rules are most naturally written in terms of a bare electric charge e and a bare electron mass m, but to compare predictions to experiment, one must eliminate these parameters in favor of the measured values e_m and m_m (both of which are conventionally defined at the limit of zero momentum transfer).

The renormalization process is complicated by the fact that the momentum-independent parts of the shifts from g to g_m and from m to m_m are formally infinite. These shifts arise from diagrams containing loops, for which the Feynman rules specify to integrate over all possible momenta of the virtual particles. These integrals generally diverge, so one must artificially regularize them; that is, make them finite—for instance, by cutting them off at some large momentum scale Λ. One can then "renormalize" and finally take the limit $\Lambda \rightarrow \infty$. For the simple ϕ^4 theory considered here, this procedure is unambiguous and yields well-defined predictions that are independent of the cutoff or other renormalization procedures. The same is true in quantum electrodynamics and many other quantum field theories, so long as the cutoff mechanism does not violate a symmetry of the theory. Such theories are said to be renormalizable.

The divergence of certain parameters in a quantum field theory is troublesome but hardly surprising. Even in classical electrodynamics, the total electrostatic energy in the field around a point charge is infinite, so one must postulate a compensating infinite negative contribution to the electron's observed rest energy $m_m c^2$. Similarly, the divergent integrals in quantum field theory arise from the false assumption that the equations of the theory are valid at arbitrarily small distances, or, by the de Broglie relation $p = h/\lambda$, at arbitrarily large momenta. Cutting off the integrals is merely a way of parametrizing our ignorance of what is going on at these inaccessible scales.

One can determine whether a quantum field theory is renormalizable simply by looking at the dimensions (i.e., units) of its coupling constants. All coupling constants have dimensions of momentum (or energy or mass, if we multiply or divide by c) to some power. If this power is always positive or zero, the theory is renormalizable. However, a coupling constant with negative mass dimension renders a theory nonrenormalizable: Loop integrals then pick up extra factors of Λ in the numerator, and hence diverge as $\Lambda \rightarrow \infty$. The dimension of a coupling constant, in turn, is determined by the number of lines connected to its associated Feynman vertex, with more lines requiring a coupling constant of lower mass dimension. Thus, in renormalizable theories, only the simplest vertices can appear, with no more than four spin-0 particles, or two spin-$\frac{1}{2}$ particles, for instance.

In a nonrenormalizable theory, new infinities arise in the computation of every possible reaction rate. In other words, an infinite number of unobservable parameters would be needed to absorb all the infinities. Such theories would therefore appear to have little predictive power. However, nonrenormalizable interactions still play an important role in realistic models of elementary particles. For example, the effect of Z^0-boson exchange on electron scattering at low momentum (p) can be approximated as a nonrenormalizable four-electron vertex:

The coupling constant for this effective vertex is of order e^2/m^2_Z, where $m_Z = 91$ GeV is the Z^0 mass.

This diagram is therefore suppressed by a factor of roughly $(p/m_Z)^2$ relative to the more familiar photon exchange diagram. Similarly, dimensional analysis suggests that in any effective field theory that describes low-momentum processes, nonrenormalizable interactions will be suppressed by powers of p/M, where M is some large mass scale at which new degrees of freedom become important. If we imagine M to be the scale of grand unification (10^{16} GeV) or quantum gravity (10^{19} GeV), then it is hardly surprising that the present theories of physics below 10^3 GeV are dominated by renormalizable interactions.

See also: CRITICAL PHENOMENA; FEYNMAN DIAGRAM; QUANTUM CHROMODYNAMICS; QUANTUM ELECTRODYNAMICS; QUANTUM FIELD THEORY

Bibliography

FEYNMAN, R. P. *QED: The Strange Theory of Light and Matter* (Princeton University Press, Princeton, NJ, 1985).

GEORGI, H. M. "Effective Quantum Field Theories" in *The New Physics,* edited by P. Davies (Cambridge University Press, Cambridge, Eng., 1989).

PESKIN, M. E., and SCHROEDER, D. V. *An Introduction to Quantum Field Theory* (Addison-Wesley, Reading, MA, 1995).

WEINBERG, S. *Dreams of a Final Theory* (Pantheon, New York, 1992).

DANIEL V. SCHROEDER

REPULSION

See ELECTROSTATIC ATTRACTION AND REPULSION

RESISTANCE

See ELECTRICAL RESISTANCE

RESISTOR

A resistor is an element in an electrical circuit that defines the magnitude of current flow for a particular applied voltage or produces a voltage drop for a particular current. Ohm's law states that the current (in amps) flowing through a resistor is equal to the voltage (in volts) across it divided by its resistance in ohms (symbol Ω). Therefore, for a constant applied voltage, larger resistor values will allow less current to flow. In addition, for a constant current flow, a higher resistance will produce a larger voltage drop across the resistor. If we have a number of resistors in series, since the current flow through all is the same, the voltage drop across each will be in proportion to their value. For example, if two resistors in series have the same value, the voltage across each will be one-half the total applied voltage. Resistors in series therefore act as "voltage dividers." Resistors are also used with capacitors and inductors in oscillator circuits. The resistance determines how the circuit resonates.

Resistors come in a variety of forms. "Discrete" resistors are used in circuits built on printed wiring boards that have components that may be individually replaced. An example of a discrete resistor is shown in Fig. 1. It is composed of a small cylinder of resistive material, typically carbon or a metal oxide, between two metal connections that have wires attached. The colored bands indicate the value of the resistor (in Ω) and the tolerance (in percent). The color code is given in Table 1. For example, a resistor with yellow, purple, red, and silver bands in positions 1, 2, 3, and 4, respectively, has a resistance of 4,700 Ω and a tolerance of ± 10 percent.

"Thick film" resistors are printed on ceramic substrates using a patterned screen and a paste com-

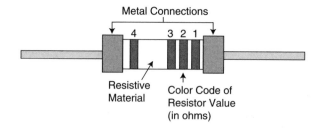

Figure 1 Discrete resistor.

Table 1 Color Code for Resistor Bands

Color	Band 1	Band 2	Band 3	Band 4
Black	0	0	$\times 1$	$\pm 20\%$
Brown	1	1	$\times 10$	
Red	2	2	$\times 100$	
Orange	3	3	$\times 1,000$	
Yellow	4	4	$\times 10,000$	
Green	5	5	$\times 100,000$	
Blue	6	6	$\times 1,000,000$	
Violet	7	7		
Gray	8	8		
White	9	9		
Silver			$\times 0.01$	$\pm 10\%$
Gold			$\times 0.1$	$\pm 5\%$

posed of metal and glass particles. Their resistance may be "trimmed," that is, brought nearer the desired value, by using a laser to remove some of the material. This leads to greater accuracy and consequently provides a tolerance that is considerably better than that of the best discrete resistors.

"Thin film" resistors are made by evaporating a metallic film, such as tantalum or nichrome (a nickel-chromium alloy), onto a substrate and patterning the film to the correct dimensions by selective etching. These resistors may be made very compact, and a high degree of control over the etch process leads to accurate values.

"Integrated" resistors are those found in integrated circuits (ICs). These may be formed by adding specific elements, called dopants, to areas of the semiconductor substrate, which is usually silicon. The dopants change the resistance of the semiconductor; more dopant produces a lower resistance. Alternatively, integrated resistors may be formed from thin layers, such as polycrystalline silicon, which are deposited on the IC and patterned by etching. In either case the resistors may be made extremely small (a few millionths of a meter in size), and are formed at the same time as the other circuit components, such as transistors.

Variable resistors have a sliding contact, the "slider," on a track of resistive material. The length and, therefore, the value of the resistor depends on the position of the slider. An example of their application is in an audio amplifier where they control the volume (by voltage division) and the treble and bass (as part of an oscillator circuit).

See also: DOPING; ELECTRICAL RESISTANCE; OHM'S LAW

Bibliography

CAMPBELL, D. S., and HAYES, J. A. *Capacitive and Resistive Electronic Components*, 2nd ed. (Gordon and Breach, Yverdon, Switzerland, 1994).

SCHWARTZ, S. E., and OLDHAM, W. G. *Electrical Engineering: An Introduction* (Saunders, Orlando, FL, 1993).

MICHAEL N. KOZICKI

RESOLVING POWER

The resolving power R of an optical element or instrument is a quantitative measure of its abiliity to separate spatially two sources of light. In image-forming devices such as microscopes, telescopes, or the human eye, the separation occurs because of the spatial separation of the two sources. For dispersive instruments such as prisms or grating spectrometers, the separation is due to the differing wavelength of the two sources. In both cases, resolving power is ultimately limited by diffraction of the incident light by apertures and/or lenses of the device, but it may also be degraded by physical imperfections.

Light from a point source that has suffered diffraction can no longer be focused to a point, but it has a spatial spread characterized by a central maximum and an attendant regular pattern of ancillary maxima and minima, as shown in Fig. 1a. Such patterns are created, for example, by refracting telescopes as the result of the diffraction of light from a point-like star by the circular aperture of the objective lens. Two images are said to be "resolved" if the angular separation of their respective central maxima is greater than that between one pattern's central maximum and first minimum (Fig. 1b). This is called Rayleigh's criterion. The theory of physical optics provides formulas for the maximum obtainable resolution using this criterion in specific cases.

Telescopes. The value of R is specified as the smallest angular separation between two points of light that can be resolved:

$$R(\text{radians}) = 1.22\frac{\lambda}{d},$$

where d is the diameter of the telescope's objective lens (or primary reflecting mirror) in centimeters and λ is the wavelength of the light in angstroms. For typical visible wavelengths of 5,500 Å, we have

$$R(\text{arc seconds}) = \frac{14.1}{d}.$$

This formula makes the advantage of using large-diameter optical elements apparent.

Microscopes. Resolution is limited by diffraction of the cylindrical beam of light formed by the final lens. Resolving power is defined as the smallest distance between two points on the object that can be resolved:

$$R = \frac{0.61\lambda}{n\sin\theta},$$

where θ is the largest angle between a ray from the object to the first objective lens that is ultimately observed, and n is the index of refraction of the medium between the object and the microscope's objective. (The value of n can be increased by the use of an "oil immersion" objective.) This equation for R can vary somewhat depending on the way in which the object is illuminated. Since $n\sin\theta$ is of the order of one, the ultimate resolution is seen to be comparable to the wavelength of light. The use of shorter wavelengths, obtainable with an electron microscope, may thus be desirable.

Human eye. Diffraction by the eye's pupil limits its angular resolution to about 0.8 arc min (as opposed to 0.1 arc sec for a large telescope), or an object separation of about 0.1 mm at a distance of 25 cm.

Prisms. A prism's "chromatic" resolving power is defined as $\lambda/\Delta\lambda$, where $\Delta\lambda$ is the smallest wavelength difference that can be spatially resoved by the prism's action for spatially coincident incident beams, and λ is the average wavelength of the two. For the case of a triangular prism of base width b that is completely illuminated by the incident light,

$$R = b\frac{dn}{d\lambda},$$

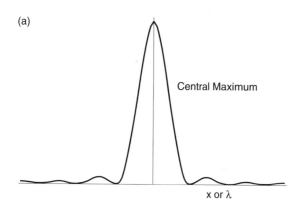

(a)

Central Maximum

x or λ

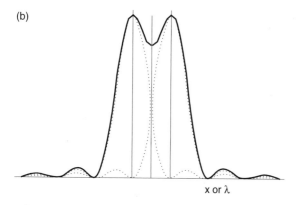

(b)

x or λ

Figure 1 (a) Intensity of a point source after diffraction by an aperture. The abscissa represents position, in the case of an image-forming instrument such as a telescope, or for a dispersive device such as a prism, wavelength. (b) Two intensity patterns that are just resolved using Rayleigh's criterion; the maximum of one pattern coincides with the first minimum of the other. The solid line is the sum of the individual (dotted) patterns.

where $dnd\lambda$ is the derivative of the index of refraction of the prism material with respect to wavelength.

Grating spectrographs. In this case, $R = mN$, where N is the number of grating lines or slits illuminated by the incident light, and m is the order number of the diffraction patterns to be resolved.

See also: DIFFRACTION; GRATING, DIFFRACTION; LENS; LIGHT; OPTICS, PHYSICAL; REFRACTION, INDEX OF; TELESCOPE; VISION; WAVELENGTH

Bibliography

BORN, M., and WOLF, E. *Principles of Optics,* 6th ed. (Pergamon, Oxford, 1980).

JENKINS, F. A., and WHITE, H. E. *Fundamentals of Optics,* 4th ed. (McGraw-Hill, New York, 1976).

SEARS, F. W. *Optics,* 3rd ed. (Addison-Wesley, Cambridge, MA, 1949).

TIMOTHY GAY

RESONANCE

Resonance phenomena pervade nature from the subtle humming of high-tension power lines and the delicate vibration of an irradiated molecule to the catastrophic collapse of a suspension bridge and the synchronized motion of planetary bodies. The basic concept of a resonance centers on mutual or complementary oscillations of two interacting systems. While the effect may range over systems from the subatomic to the galactic, many of the fundamental features display considerable similarities. Three ingredients form the basis of most resonances: a resonator, a driving oscillator, and a coupling mechanism. First, the resonator consists of an object characterized by a specific frequency ω_r of oscillation. A tuning fork, a piece of tubing of a specified length, or a tightly suspended section of thin wire enumerate but a few common examples. Second, the oscillator produces a broad range of frequencies spanning ω_r. Third, some mechanism must exit to transfer the oscillations between the two objects. The traditional example of this configuration consists of a fixed length of wire under tension for the resonator. Such a wire will vibrate and sound at a fixed frequency de-

pending on the composition, size, and tension. The oscillator also contains a fixed length of wire, but with adjustable tension in order to change the vibrational frequency ω. Finally, the sound waves, produced by the oscillator, provide the necessary coupling by traveling the intervening distance and setting the resonator in motion. Plucking the oscillator at an ω much smaller than ω_r, elicits only a very weak response. However, as the frequency approaches ω_r, a stronger vibration results. Very near the resonator frequency, the wire vibrates vigorously, and a tone may become audible. As the frequency increases further, the oscillation dies down. Plotting the response as a function of oscillator frequency gives the behavior displayed in Fig. 1. The position of the maximum occurs near ω_r, while the width of the curve provides information about the nature of the resonator and the coupling. This characteristic response curve appears time and time again for resonances of diverse origins. The classic, if perhaps apocryphal example, depicts the dramatic soprano hitting a high note and shattering a crystal glass (oscillator = prima donna; resonator = glass; coupling = sound waves).

Since phenomena associated with resonances occur in many aspects of daily life, their observation predates any historical or scientific description. The human voice, especially singing, functions from resonating vocal cords. In addition, simple musical instruments such as flutes, lyres, and drums existed in most ancient cultures. Therefore, unlike many physical effects, resonances do not have a single discoverer or a single theoretical explication. In physics, resonances arise in all fields from classical to quantum mechanical. The earliest formal treatments ex-

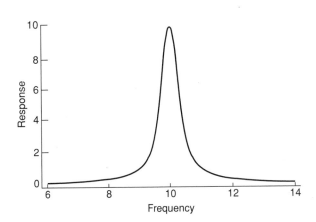

Figure 1   Resonance plotted as a function of oscillator frequency.

tend back to the ancient Greeks with the description by Pythagoras of the harmonics of vibrating strings. Resonances also play important roles in the study of sound, heat, and electromagnetism, and have been addressed by such eminent scientists as Lord Kelvin (William Thomson), Lord Rayleigh (John W. Strutt), James Clerk Maxwell, Heinrich Hertz, and Hendrik Antoon Lorentz. In addition, celestial mechanics focused attention on many forced oscillations as pursued by Pierre Simon Laplace, Karl Friedrich Gauss, and Jules Henri Poincaré. The study of the periodic perturbation of planets led to the earliest investigations into nonlinear effects, which in turn led to the field of chaos. The advent of quantum mechanics opened new realms in which resonances played important roles. Such pioneers in the field as Albert Einstein, Niels Bohr, Erwin Schrödinger, Werner Heisenberg, P. A. M. Dirac, Wolfgang Pauli, and Eugene Wigner examined and elucidated many of these mechanisms. This cast of illustrious characters covers but a few who contributed to our knowledge of this important phenomena.

A brief description of a selected set of resonances will serve as an instructive initiation. As indicated, resonances appear in every branch of physics. A simple mechanical example considers two pendulums connected by a spring. Ideally, the frequency of oscillation depends on the length and mass. Starting one of the pendulums in motion will eventually move the second through the coupling action of the spring. If the frequencies begin poorly matched due, for example, to very different masses of the bobs, very little occurs. On the other hand, when the frequencies match, both oscillate vigorously in phase. On a grander scale, the forces acting among planetary bodies, in special cases, just balance so that the relative positions remain fixed throughout the orbital motions. The Trojan asteroids maintain their position at stable equilateral points relative to the Sun and Jupiter from this delicate balance of gravitational forces. Such resonances also account for gaps and accumulations of matter at particular distances from the Sun and planets. We hear radios and view television largely through electromagnetic resonance effects. An oscillator produces a periodically varying electromagnetic field that broadcasts from an antenna. This field propagates through space until encountering another antenna, which it sets vibrating. An attached oscillator experiences the received modulation of the field. As before, if the two oscillators have similar frequencies, a large response results. Of course, the real transmission consists of many different frequencies at different amplitudes that compose a musical piece or a paid political advertisement.

In classical physics, sound, fluids, and plasma also display such waves resulting from a driving oscillation and the response of the media at certain selective frequencies. The tides of Earth's seas and atmosphere form another example of a resonance. The Moon pulls the atmosphere most effectively when directly above a point on Earth's surface. This raises a tide and a corresponding bulge halfway around the globe. Plotting the response of the atmosphere as a function of time would give a curve similar to Fig. 1 with the maximum near twelve hours. The actual result differs somewhat due to friction and the mismatch of the rotation of Earth and revolution of the Moon. The Moon does not return to a given point above Earth in twenty-four hours since it moves ahead slightly in its orbit during a rotation. Such effects have a dramatic presence during large volcanic explosions, which cause the atmosphere to "ring" for many days or even months.

The microscopic world evinces a variety of effects related to resonances. Quantum mechanics governs the behavior of atoms, molecules, nuclei, and subatomic particles. The bound states of the electrons in the atom come only at discrete values of the energy. Therefore, transitions between these levels, which give rise to absorption and emission of light, occur at specific energies or frequencies. Light also comes in energy packets or quanta, called photons, that have energies proportional to the frequency ($E = \hbar\omega$). An idealized experiment devised to determine the effect of radiation on atoms has several components: a light source, tunable over a broad frequency range; a sample of atoms irradiated by the source; and a detector opposite the source. We select a particular transition between two atomic states of energies E_1 and E_2, corresponding to a transition frequency of $\omega_{12}[= (E_2 - E_2)/\hbar]$. The initial conditions within the sample ensure that all the atoms begin at the lowest level E_1. The source oscillator starts at a frequency ω much smaller than the resonant transition ($\omega \ll \omega_{12}$), and the detector measures the amount of absorption by the media. As expected, the media appears transparent to the light since no transitions occur. However, as the two frequencies approach each other in magnitude, a slight attenuation occurs in the transmission. At ω_{12}, no signal appears as all the light at that frequency goes to exciting the atoms. Increasing the resonance frequency ($\omega > \omega_{12}$) further, again results in transmission. The absorption profile emulates the shape in Fig. 1 rather than just a single narrow line. This width, which arises from many effects such as the natural lifetime, a consequence of the Heisenberg uncertainty principle; thermal motion of the atoms; and collisions, characterizes the lifetime of the state. The observation of such spectral features yields valuable information about the physical conditions within the exposed medium, for example, the temperature and density within a plasma or a distant star.

A very important optical device operates through this type of photon resonance. The absorption of a photon leaves an atom in an excited state. This state, in turn, decays through the emission of a photon by spontaneous or stimulated processes. In most media, a balance ensues between these competing emission and absorption mechanisms. However, if a preponderance of a particular excited state of the atoms were spontaneously excited, then a population inversion would exist. The atoms remove this unstable situation by radiating. Since a large number of atoms radiate in a very narrow frequency range, an intense nearly monochromatic beam emerges. This picture presents a rudimentary version of a laser.

While atomic physics provided the principal examples, molecules, solids, and nuclei also exhibit many important resonance properties since they too have quantized energy levels. Transitions among their states give rise to features that provide valuable diagnostic tools through the observed radiation. Magnetic resonance imaging (MRI) machines that produce intricately detailed pictures of the tissue structure within the body illustrate a composite nuclear-molecular resonance. An external magnetic field causes the intrinsic magnetic moment of the protons within organic molecules to align. This moment flips in response to an externally applied electromagnetic field, resulting in the absorption of radiation within a very narrow (resonant) range. The local magnetic fields of the various molecules modify the external field and produce resonances at different frequencies. Thus, tissues of varying types appear differently.

Collisional processes abound with resonance examples. One of particular interest focuses on the transient formation of an intermediate compound state. The collision between an electron and an atom produces several results. The electron may scatter elastically; transfer energy, provoking a transition between the discrete bound levels; ionize the

atom; or form an excited state of the compound negative ion A^-, schematically represented by

$$e^- + A \rightarrow (A^-)^* \rightarrow e^- + A. \tag{1}$$

Such states, designated by an asterisk, also occur at discrete energy values. Since conservation of energy requires that the incoming electron emerge again in the continuum, this excited state is transient. Therefore, a dramatic effect ensues at an electron energy corresponding to this compound state. The experimentalist measures and the theoretician computes a quantity called the cross section that gives the probability of a particular process occurring at a given energy. A plot of the cross section as a function of the collisional energy displays the same behavior as in Fig. 1 with a maximum near the compound state energy. Also, many complicated chemical collisions between atoms and molecules ($C + AB$) may proceed by similar mechanisms, producing a variety of results such as rotational, vibrational, or electronic excitations (AB^*); dissociation ($A + B$); ionization ($AB^+ + e^-$); and reactions forming new products ($AC + B$). The scattering processes discussed so far fall in the general category of Feshbach resonances. Another type of related resonance occurs for a simpler interaction in which a particle becomes temporarily trapped within a barrier. The cross section again displays the resonance signature with the maximum occurring near the optimal trapping. These shape resonances appear in chemical reactions, in particle collisions with surfaces, in quantum dots, and in nanostructures. Quantum dots have macroscopic dimensions ($\approx \mu m$); however, the electrons within behave according to the dictates of quantum mechanics. The devices consist of alternating layers of crystalline material, which appear to the electrons like a series of narrow barriers. As with a simple barrier, certain configurations favor optimal trapping or tunneling. By this strategy a very effect microswitch emerges, allowing the transmission of electrons (current) only at very precise energies. Both types of resonances also appear in collisions involving nuclei and elementary particles. Such resonances can lead to the production of new types of particles during the scattering events.

Resonances, therefore, exist in all aspects of nature from interactions between the tiniest subatomic particles to the motions of galaxies. This survey has provided but a few representative examples. Each field presents its own special cases that have profound consequences for basic science and everyday life. While diverse, resonance phenomena have a surprising commonality.

See also: COLLISION; CYCLOTRON RESONANCE; ELECTROMAGNETISM; FREQUENCY, NATURAL; HEAT; LIGHT, ELECTROMAGNETIC THEORY OF; MAGNETIC RESONANCE IMAGING; OSCILLATION; PENDULUM; QUANTUM MECHANICS; SOUND; UNCERTAINTY PRINCIPLE

Bibliography

FEYNMAN, R. P.; LEIGHTON, R. B.; and SANDS, M. *The Feynman Lectures on Physics*, Vol. 3 (Addison-Wesley, Reading, MA, 1965).

KASTNER, M. A. "Artificial Atoms." *Phys. Today* **45**, 24–31 (1994).

REED, M. A. "Quantum Dots." *Sci. Am.* **268**, 118–123 (1993).

LEE A. COLLINS

REST MASS

The concept of mass is basic to the study of the motion of matter (mechanics). Mass is a measure of the resistance of a material body to an external force changing the body's motion; this resistance is called inertia. One should not confuse mass and matter. Matter is the stuff that one can feel, and mass is a measure of the resistance to change in velocity of a body made of matter. Mass usually enters into the equations of the motion of bodies independent of which forces alter the motion; in this situation it is called inertial mass.

A quantity called mass also is involved in one of the forces that can alter motion, namely the gravitational force; here it is called gravitational mass. Experiments have shown that these two types of masses are identical to a high degree of accuracy; this identity is the reason that the acceleration of gravity at the surface of the earth does not depend on the mass of the body being accelerated. So, in the following we shall consider only inertial mass, and we shall call it by the simpler word mass.

In classical (pre-twentieth century) mechanics theory, mass is considered to be invariable for an intact body. This theory states that, at all velocities, the mass of the body has a constant value, but in the special theory of relativity the definition of mass is more complex. There is a semantic controversy about the use of the term mass in the special theory of relativity; however, all physicists seem to agree on the concept of rest mass, although some prefer to call it simply mass.

Rest mass is a measure of the resistance of a body *at rest* to its being given a velocity; that is, the resistance to having its velocity changed from zero to some nonzero value. According to the special theory of relativity, when the original velocity is *not* zero, the resistance to changing the velocity depends on the rest mass, but it is larger than for the case when the original velocity *is* zero.

Newton's second law states that the force applied to a body is equal to the rate of change of the momentum with time, where the momentum is the product of the mass times the velocity. If we denote the rate of change with time of a quantity by putting a dot over its symbol, then Newton's second law states $f = \dot{p}$, where f is the force, $p = mv$ is the momentum, m is the mass, and v is the velocity. Therefore, using calculus, $f = \dot{p} = \dot{m}v + m\dot{v}$, allowing for the possibility that the mass can change with time, as it does in the special theory of relativity. When the initial velocity is zero, then $m = \dot{p}/\dot{v}$ is the rest mass, which is often denoted by the symbol m_0. When the initial velocity is not zero and no matter is being added to the body from outside or thrown off from the body to the outside, the special theory of relativity requires that the mass changes with velocity. In this case some physicists call m the relativistic mass, although others prefer not to use that term.

The equation in the special theory of relativity that relates the relativistic mass to the rest mass is $m = m_0/\sqrt{1 - v^2/c^2}$, where c is the speed of light (about 3×10^8 m/s or 190,000 miles/s). Thus, although m_0 does not change as a body's velocity changes, the relativistic mass m does; in fact, it approaches infinity as the velocity approaches the speed of light.

The rest mass of a body can be measured by applying a known constant force f to the body, originally at rest, over a measured short interval of time t. The momentum change of the body is $\dot{p} = f$, and the velocity change (acceleration) is $\dot{v} = v/t$, where v is the final velocity measured at the end of the time interval t. (The measurement of the final velocity involves another even shorter time interval just before the end of the longer short-time interval.) Then one can use $m_0 = \dot{p}/\dot{v}$ to calculate the body's rest mass. For this to be an accurate determination of the rest mass, the final velocity must be very small compared to the speed of light, as it is in most everyday situations.

The special theory of relativity states that a body at rest with rest mass m_0 has a rest energy given by $E_0 = m_0 c^2$. In reactions of two or more bodies, some of the rest energies of the initial bodies can be converted to kinetic energies of the final bodies. Similarly, initial kinetic energies of bodies in a reaction can produce bodies with larger rest masses than the initial reaction bodies had. In fact, these processes are characteristic of chemical, atomic, and nuclear reactions. The rest energy of a composite body is the sum of the rest energies of its component bodies and all internal energies, both potential energies and kinetic energies, involving the components.

See also: ENERGY, KINETIC; ENERGY, POTENTIAL; INERTIAL MASS; MASS; MASS, CONSERVATION OF; MASS-ENERGY; NEWTON'S LAWS; RELATIVITY, SPECIAL THEORY OF

Bibliography

ADLER, C. G. "Does Mass Really Depend on Velocity, Dad?" *Am. J. Phys.* **55**, 739–743 (1987).

BAERLEIN, R. "Teaching $E = mc^2$: An Exploration of Some Issues." *Phys. Teacher* **29**, 170–175 (1991).

JAMMER, M. *Concepts of Mass in Classical and Modern Physics* (Harvard University Press, Cambridge, MA, 1961).

OKUN, L. B. "The Concept of Mass." *Phys. Today* **42**, 31–36 (1989).

SANDIN, T. R. "In Defense of Relativistic Mass." *Am. J. Phys.* **59**, 1032–1036 (1991).

L. DAVID ROPER

REYNOLDS NUMBER

In the 1880s Osborne Reynolds introduced the Reynolds number (Re) from his experimental studies of fluid motion. A fundamental parameter of fluid dynamics, the Reynolds number identifies the nature of the fluid flow. Fluid flow can be either laminar (smooth and steady) or turbulent (irregular, unsteady, and chaotic) after it interacts with the surface of some object. The Reynolds number is related to the macroscopic properties of a system through the fluid density ρ, fluid viscosity η, mean

speed v, and the size of the object D (D depends on the shape and size of the object's surface):

$$\mathrm{Re} = \frac{\rho v D}{\eta}.$$

The units of density ($kg \cdot m^{-3}$), speed ($m \cdot s^{-1}$), D (m), and viscosity ($kg \cdot m^{-1} \cdot s^{-1}$) cancel out in the Reynolds number making it dimensionless (unitless) and therefore independent of many of the specific details of the system.

For a particular Reynolds number, the flow is always the same, regardless of changes in the four parameters (ρ, η, v, and D). The Reynolds number is, therefore, a scaling parameter as it allows many systems or situations to be studied without the expense or inconvenience of a full-sized experiment. This explains the utility of the wind tunnel, where model airplanes and automobiles are tested. As long as the speed, density, viscosity, and object dimensions give the same Reynolds value that will occur in real-life situations, the results of the scaled model accurately portray the actual flow.

The Reynolds number is a ratio of inertia to viscous forces. When Re is small (less than 30), viscosity dominates the motion. When Re is large (greater than 10^5), which typically occurs at large velocities, viscosity does not appreciably affect the mainstream fluid motion.

The Reynolds number is related to the concept of a boundary layer, introduced by Ludwig Prandtl in 1904. The viscous nature of real fluids requires that the relative fluid velocity be zero at a solid surface. The boundary layer is a thin layer of fluid immediately adjacent to a solid surface, in which the velocity rapidly changes from zero to the mainstream fluid velocity. Even if the viscosity is negligible in the mainstream fluid it cannot be neglected in the boundary layer.

For small values of Re (typically less than 1) which occur for low velocity and/or large viscosity the boundary layer remains laminar and can be roughly defined as shown in Fig. 1 (there is no precise edge to the boundary layer).

Notice that the velocity vectors are smaller in the region adjacent to the surface and the fluid is gradually slowed down farther along the surface, increasing the boundary layer thickness farther downstream. For small Re viscosity dominates the motion of the fluid throughout the flow. If a solid travels in the fluid, viscous drag, a frictional force along the direc-

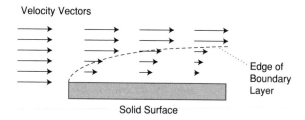

Figure 1 Laminar boundary layer.

tion of motion, impedes the motion. Also, for small Re the Navier–Stokes equation, which describes the dynamics of a viscous fluid, reduces to a simple and commonly known form called Stoke's flow. For example, the viscous drag force on a sphere is given by

$$F = 6\pi\eta r v,$$

where r is the radius of the sphere and v is the speed of the sphere.

The viscosity also introduces vorticity in the boundary layer region. Vorticity is a rotational motion of the fluid particles. For values of Re slightly larger than 1 (approximately 30) vortices form in the region immediately behind the object and the flow remains steady and regular. As Re increases further (to greater than 100) these vortices "peel off" at some point and flow downstream away from the object. When the Reynolds number reaches about 2,000 to 3,000, turbulence develops and the flow is no longer laminar. Turbulence is a chaotic rotational state of the fluid that occurs when the fluid speed and its inertia overcome the viscous forces of the fluid (recall that the Reynolds number is a ratio of these two effects). Fully turbulent behavior occurs

when Re is greater than 10^5. Outside of the turbulent boundary layer and the turbulent wake, the flow may be treated as ideal (nonviscous) since the large Re typically results from either low viscosity or very high speeds (which overcome the viscous effects). Turbulence is related to a nonlinear term in the Navier–Stokes equation meaning that its presence hinders our ability to mathematically predict the fluid motion in the turbulent region. When Re is between 2,000 and 3,000, the flow is unstable and may fluctuate between laminar and turbulent.

Laminar flow offers the least resistance to flow and may be maintained at higher Reynolds numbers by streamlining. As fluid flows over and past an object, as depicted in Fig. 2, the fluid speed increases past point A.

At sufficiently high speeds the boundary layer may separate from the surface (point B), creating a region of low pressure (an adverse pressure gradient) in the viscous, turbulent wake behind the points of separation. This lower pressure creates a drag force on the object, adding to the viscous force already present. Boundary layer separation implies that the vortices, formerly confined to the boundary layer, now penetrate into the mainstream fluid. The location of the separation points (B) depend on boundary layer characteristics, surface roughness, and disturbances in the mainstream flow. With extreme care it is possible to postpone turbulent flow up to Reynolds values of 40,000. Streamlining, as in Fig. 3, greatly reduces the boundary layer separation and minimizes the low pressure region, which in turn reduces the drag. Streamlining requires the leading edge to be rounded, the body to be elongated, and the tail end to be tapered to a point.

Turbulent flow offers more drag on an object than laminar flow, since the presence of the rotational state extracts energy from the fluid. However, experience shows that the largest drag force occurs in the unstable transition region (2,000 < Re < 3,000). Here the total drag force may be four or five

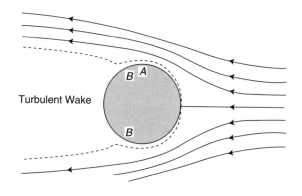

Figure 2 Viscous flow over a solid body.

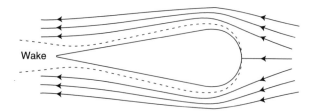

Figure 3 Streamlining to reduce boundary layer separation.

times that in the fully turbulent region, leading to a condition called aerodynamic drag crisis. In many practical cases, where it is not possible to maintain the low-drag laminar flow, it is then beneficial to avoid the drag crisis by insuring that the flow does not occur in the transition region. Roughening a surface, adding dimples (e.g., golf balls), or adding small protuberances (e.g., ailerons on airplane wings) accomplishes the desired effect of increasing the Reynolds number beyond the aerodynamic drag crisis to conditions of fully turbulent flow.

See also: AERODYNAMICS; FLUID DYNAMICS; NAVIER–STOKES EQUATION; TURBULENT FLOW; VISCOSITY

Bibliography

ADAIR, R. K. "The Physics of Baseball." *Physics Today* **48** (5), 26–31 (1995).

BIRKHOFF, G. *Hydrodynamics: A Study in Logic, Fact, and Similitude* (Princeton University Press, Princeton, NJ, 1960).

FISHBANE, P. M.; GASIOROWICZ, S.; and THORNTON, S. T. *Physics for Scientists and Engineers* (Prentice Hall, Englewood Cliffs, NJ, 1993).

FOX, R. W., and McDONALD, A. T. *Introduction to Fluid Mechanics,* 3rd ed. (Wiley, New York, 1985).

HUNT, R. G. "Aerodynamics of Bicycling" in *Physics for Scientists and Engineers,* edited by P. A. Tipler (Worth, New York, 1991).

LIGHTHILL, J. *An Informal Introduction to Theoretical Fluid Mechanics* (St. Edmundsbury Press, Suffolk, Eng., 1989).

KENNETH D. HAHN

RHEOLOGY

Rheology, a branch of condensed-matter physics, takes its name from the Greek root *rei* (to flow) and is defined (by the Society of Rheology) as the "science of deformation and flow." Most of the scientists and engineers who work in this field (rheologists) are concerned with the mechanics of complex, fluid-like substances that exhibit widely different flow behaviors depending on the deformation to which they are subjected. This behavior can range from that of ordinary liquids such as oil or water, whose frictional resistance to flow we call viscosity, to behavior more reminiscent of the rubbery solids we call elastic. One familiar example is the children's play stuff commonly known as Silly Putty which, left to itself flows like a liquid but when deformed rapidly can bounce as a rubber ball. Such "viscoelastic" response is exhibited by a wide range of natural and synthetic materials including high-molecular-weight polymers or macromolecules such as gelatin, unvulcanized rubber, molten plastic, numerous biological fluids such as the synovial fluid that lubricates animal joints, and fluid-particle suspensions, colloidal dispersions and emulsions such as printing ink, pastes, paint, mayonnaise, and ketchup. (Anyone having tried to get the last-mentioned substance out of a bottle may claim amateur status as an experimental rheologist.)

As illustrated by the last example above, many fluids also exhibit other complex flow phenomena such as plasticity (no direct relation to the term "plastic" mentioned below), which implies solid-like behavior below a certain critical load, called yield stress, and fluid-like, continuous flow at higher loads. Fluids that exhibit any of the above effects are generally designated as non-Newtonian, to distinguish them from those obeying Isaac Newton's famous law of viscosity (forces or stresses always directly proportional to rate of deformation).

Many of the interesting rheological properties of non-Newtonian fluids are also essential to various applications in nature and technology. A large number of such fluids are products of modern technology, especially in the manufacture of synthetic high-molecular-weight polymers (or plastics). These industries began a rapid and sustained growth in the years immediately preceding World War II, spurred by an ever increasing demand for new materials such as synthetic rubber for automobile tires, and synthetic fibers and films for all manner of textiles, clothing, and packaging. Many of the trade and scientific names associated with such products (Celluloid, Latex, Nylon, Rayon, Saran-Wrap, polyethylene, Kevlar, and so on) are household words. The need to better understand and describe the flow behavior of such materials in various industrial processes, such as the molding of rubber and plastic, the spinning of artificial fibers, and the blowing of films, has led to a rapid growth of the modern science of rheology.

Apart from the several ideas borrowed from the classical mechanics of viscous fluids and elastic solids, certain basic physical concepts essential to the field of rheology can be traced back to scientists of the nineteenth century, including James

Clerk Maxwell (the founder of the modern theory of electromagnetism), who, in 1868, actually proposed an elementary but still popular mathematical model of viscoelasticity. This so-called Maxwell fluid captures certain essential aspects of viscoelastic behavior and embodies material "memory," in the form of an important parameter now known as the relaxation time. The latter represents the time necessary to "forget" the elastic state resulting from a rapid deformation—the longer a material's relaxation time the more solidlike its behavior, and the shorter the relaxation time the more fluidlike the behavior.

Many of the flow properties of materials such as polymers or macromolecules can be attributed to the long relaxation times associated with relatively sluggish rearrangements of large molecules. Based in part on his conceptual description of the associated, snake-like motions (reptation) of flexible macromolecules, the 1991 Nobel Prize in physics was awarded to the physicist Pierre-Gilles De Gennes who, incidentally, also was recognized for his work on liquid crystals, another class of rheologically interesting materials with useful electro-optical properties. In addition to electrically active polymers and liquid crystals, rheologists also have been recently interested in a class of colloidal or particulate dispersions, called electro-rheological fluids, which develop large yield stress in electric fields.

Early in this century, scientists such as E. C. Bingham, a chemist known for his pioneering work on the yield behavior of paint and other colloidal dispersions (now called Bingham plasticity) had already perceived the need for systematic research aimed at relating the chemical or molecular structure of complex materials to their rheological properties. For that reason, Bingham helped to found in 1929 the (American) Society of Rheology, the oldest and currently largest of about twenty such research societies around the world, which, continues to publish the first scientific periodical devoted to the subject, the *Journal of Rheology*. Apart from this, other professional journals and advanced textbooks, there are interesting educational video cassettes available for novices.

Although difficult to capture in a brief summary, one can cite several rather distinct, on-going activities in contemporary theoretical and applied rheology, including:

1. Rheology of polymers and liquid crystals.

2. Rheology of heterogeneous and particulate systems, such as colloidal and other particulate dispersions (of solids in liquids), as well as emulsions or foams (liquids or gases dispersed in liquids).

3. Development of mathematical models (rheological "constitutive" equations) for flow behavior and their application to laboratory test devices and industrial materials processing.

The last area of activity has been greatly stimulated by the advent of high-speed digital computers that allow one to treat elaborate mathematical equations for the flow of complex fluids through complicated flow geometries.

See also: ELASTICITY; FLUID DYNAMICS; FLUID STATICS; MAXWELL, JAMES CLERK; NEWTON, ISAAC; VISCOSITY

Bibliography

BAFNA, S. "66th Society of Rheology Meeting." *American Laboratory* **27** (1), 23–28 (1995).

BOGER, D., and WALTERS, K. *Rheological Phenomena in Focus* (Elsevier, New York, 1989).

KOKINI, J. L. "Proceedings of the Boston Symposium on Food Rheology, Oct. 17–21, 1993." *J. Rheol.* **39** (6), 1427 (1995).

J. D. GODDARD

RIGHT-HAND RULE

The vector product **A** × **B** (also called the "cross product") of two vectors **A** and **B** also is a vector. This vector is perpendicular to both **A** and **B** (i.e., it is perpendicular to the plane containing **A** and **B**). This is not enough to determine the direction of **A** × **B** because it could point in either of the two opposite directions along the perpendicular line. The right-hand rule picks out the correct direction as follows:

- Imagine rotating the first vector (**A**) into the second vector (**B**).
- Orient your right hand so that the fingers curl in the direction of motion of **A**.
- Then your thumb points in the direction of the vector **A** × **B**.

In Fig. 1, this rule tells us that $\mathbf{A} \times \mathbf{B}$ points up along the line, as shown, rather than down.

According to the right-hand rule, $\mathbf{A} \times \mathbf{B}$ and $\mathbf{B} \times \mathbf{A}$ point in opposite directions. In Fig. 1, your thumb points in opposite directions depending on whether your fingers curl in the direction from $\mathbf{B}$ to $\mathbf{A}$ or from $\mathbf{A}$ to $\mathbf{B}$.

The vector product appears frequently in physics. Right-hand rules have been specifically defined for certain important cases, although they are all examples of the rule given above. All such right-hand rules apply only to three-dimensional space.

The angular momentum $\mathbf{L}$ of a particle at position $\mathbf{r}$ and with momentum $\mathbf{p}$ is defined as $\mathbf{L} = \mathbf{r} \times \mathbf{p}$. The right-hand rule for angular momentum $\mathbf{L}$ says that $\mathbf{L}$ points in the direction of the thumb of the right hand when the fingers curl in the direction from $\mathbf{r}$ to $\mathbf{p}$. Similarly, the right-hand rule for torque $\boldsymbol{\tau}$ says that the torque $\boldsymbol{\tau}$ produced by a force $\mathbf{F}$ acting at $\mathbf{r}$ will point in the direction of the thumb of the right hand when the fingers curl in the direction from $\mathbf{r}$ to $\mathbf{F}$. The torque $\boldsymbol{\tau}$ is defined as $\boldsymbol{\tau} = \mathbf{r} \times \mathbf{F}$, so this right-hand rule follows from the vector product rule.

A straight wire carrying an electrical current produces a magnetic field that circles around the wire. The magnetic field points in the direction of the fingers of the right hand if the thumb points along the wire in the direction of current flow.

The use of right-hand rules instead of left-hand rules is arbitrary and comes from tradition and convention. Defining the direction of the vector prod-

uct by a left-hand rule would reverse the direction of the vector but, if done consistently, would be equally satisfactory. The common three-dimensional rectangular coordinate system is defined in terms of the unit vectors along the positive x, y, and z axes: $\mathbf{i}_x$, $\mathbf{i}_y$, $\mathbf{i}_z$. The usual convention is to define a "right-handed" coordinate system in which the x and y axes are specified, and then the z axis is defined by $\mathbf{i}_z = \mathbf{i}_x \times \mathbf{i}_y$; the right-hand rule determines the direction of $\mathbf{i}_z$. A "left-handed" coordinate system is defined by using the left-hand rule.

See also: Vector; Vector, Unit

Bibliography

Frautschi, S. C.; Olenick, R. P.; Apostol, T. M.; and Goodstein, D. L. *The Mechanical Universe* (Cambridge University Press, New York, 1986).

Lawrence A. Coleman

RIGID BODY

Classical mechanics in its simplest form deals with a "point particle," an idealized object that has mass but no spatial extension. A particle's motion is described by giving its three space coordinates as a function of time, and its motion is determined by the total force, a vector with three components. For mechanics to have wider applicability, it must be able to describe the motion of objects with finite extent, objects such as a planet, a beam, a wheel, and a molecule. For this purpose, the concept of a "rigid body" is introduced, an object that does not change its size or shape, but can move and change its orientation in space. Mathematically, one defines the rigid body to be a collection of mass points such that the distance between any two points remains constant in time. The rigid body is also an idealization (a beam can bend, a molecule can vibrate), but it is a very useful model for many systems.

The rigid body's motion is described in terms of six parameters, three coordinates of a fixed point on the body, and three angles that determine its orientation. The body's motion is determined by the total vector force acting on it and the three components of the total torque acting on it. In the special case of

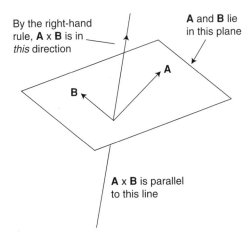

Figure 1 Right-hand rule.

By the right-hand rule, $\mathbf{A} \times \mathbf{B}$ is in *this* direction

$\mathbf{A}$ and $\mathbf{B}$ lie in this plane

A

B

$\mathbf{A} \times \mathbf{B}$ is parallel to this line

a body at rest, the conditions that must hold are that the total force and the total torque both equal zero.

For any system of particles, the force equation is $\mathbf{F} = d\mathbf{P}/dt$, where $\mathbf{F}$ is the total external force and $\mathbf{P}$ is the total momentum, and the torque equation (provided inter-particle forces act along the line connecting the particles) is $\boldsymbol{\tau} = d\mathbf{L}/dt$, where $\boldsymbol{\tau}$ is the total external torque and $\mathbf{L}$ is the total angular momentum. It is convenient to define torques and angular momenta in a coordinate system such that either (a) the origin is fixed in an inertial system, or (b) the origin is located at the system's center of mass. Then if the system is a rigid body, and if the fixed origin is a point of the body, the angular momentum takes the form

$$L_i = \sum_j I_{ij}\omega_j,$$

where i and $j = 1, 2, 3$, corresponding to the three axes. Here ω_i is the ith component of the angular velocity vector, and I_{ij} are elements of a three-by-three symmetric tensor characteristic of the mass distribution of the body. The magnitude of the vector $\boldsymbol{\omega}$ is the angular velocity, and the direction of $\boldsymbol{\omega}$ is the axis about which rotation occurs. A particular body-fixed coordinate system can be found for which the off-diagonal elements of the tensor vanish, in which case the angular momentum is simplified, $L_i = I_i\omega_i$, for each axis (I_{ii} is replaced by I_i). In this coordinate system, the kinetic energy of the system is given by,

$$T = \frac{1}{2}\sum_i I_i\omega_i^2.$$

The I_i are called moments of inertia.

The equations of motion for the components of angular velocity are the Euler equations,

$$I_3\omega_3 - (I_1 - I_2)\omega_1\omega_2 = L_3,$$

and the two similar equations obtained by permuting the three indices.

The spinning top is a rigid body in which a point in the system (viz. the tip on which it spins) is fixed in an inertial frame. The effect of the torque due to the earth's gravity is precession of the axis of the top about the vertical. For an object rotating freely in space, like the earth in its orbit, one considers rotations about its center of mass. Because of Earth's lack of sphericity, it feels a torque due to the gravita-tion of the Sun and Moon, leading to the precession of the equinoxes, with a period of approximately 26,000 years.

See also: CENTER-OF-MASS SYSTEM; FORCE; INERTIA, MOMENT OF; MODELS AND THEORIES; MOMENTUM; PRECESSION; VELOCITY, ANGULAR

Bibliography

MARION, J. B. *Classical Dynamics* (Academic Press, New York, 1965).

SERWAY, R. A. *Physics for Scientists and Engineers,* 3rd ed. (Saunders, Philadelphia, 1990).

SYMON, K. R. *Mechanics,* 2nd ed. (Addison-Wesley, Reading, MA, 1965).

MICHAEL I. SOBEL

RÖNTGEN, WILHELM CONRAD

b. Lennep, Germany, March 27, 1845; *d.* Munich, Germany, February 10, 1923; *x rays.*

Röntgen was the only child of Friedrich Conrad Röntgen, a merchant and cloth manufacturer of Lennep in Germany, and Charlotte Constanze Frowein from Apeldoorn in Holland, to which the family moved in 1848. When Röntgen would not tattle on a classmate, he was expelled from school before graduation. Now ineligible for the Dutch and German universities, he wound up in Switzerland in 1865 to study mechanical engineering at the Federal Polytechnic in Zürich.

Zürich was splendid. He skipped lectures to enjoy the mountains and lake, fell in love with Bertha Ludwig, daughter of a German political refugee, earned his diploma in 1868, and was recruited for research by August Kundt, the professor of physics. In 1869 he qualified for a Ph.D. at the university; then, as Kundt's assistant, he moved to Würzburg and Strassburg, marrying Bertha in 1872. In 1874 he acquired the faculty status that freed him for an independent career and brought him back to Würzburg as a professor of physics in 1888.

He was fifty in 1895 when he first glimpsed the x rays. They were strange, something like light, but of an oddly penetrating kind that would not bend for prisms or focus for lenses. They cast curious shadows of hidden things: brass weights in a wooden box, a steel compass needle in a metal case, the bones of a living hand inside the outline of its flesh. They were impossible to explain, but he could still describe them by recording their behavior, and he stretched his imagination for experiments to reveal every quirk of it.

That description filled seventeen numbered sections in the report that he mailed out as a preprint to announce his discovery on January 1, 1896. With each, he enclosed a set of his shadow photographs, hoping that the pictures might confirm the text and the text might validate the pictures.

Both were effective. Anyone who read the paper could produce x rays; skeletal hands appeared by the dozens, and Röntgen became a reluctant celebrity. He declined the aristocratic title of *von Röntgen,* moved grudgingly when he was called to Munich in 1899, and did not lecture at Stockholm when he received the first Nobel Prize for physics in 1901.

Bertha's health was failing, and Munich was less congenial than Würzburg. They missed their old friends but rejoined them in summer at Pontresina in the Swiss Alps (where Röntgen still enjoyed climbing, as he also enjoyed hunting at their country place in Weilheim). He kept working through the hard days of World War I and the bitter defeat that followed and tried to manage Bertha's bouts of pain with morphine injections. She died on October 1, 1919, and he followed three years later, active to the end.

Was Röntgen a great scientist? In his own way he was. In three brief months he created a new science and gave it to the world. Although the talents by which he established it also kept him from going farther, he opened a new territory for other scientists to prospect and explore for years to come.

See also: RADIOACTIVITY, DISCOVERY OF; X RAY; X RAY, DISCOVERY OF

Bibliography

GLASSER, O. *Wilhelm Conrad Röntgen and the Early History of the Roentgen Rays* (Charles C Thomas, Springfield, IL, 1934).

HEATHCOTE, N. H. DE V. *Nobel Prize Winners in Physics, 1901–1950* (Henry Schuman, New York, 1953).

NITSKE, W. R. *The Life of Wilhelm Conrad Röntgen* (University of Arizona Press, Tucson, 1971).

WATSON, E. C. "The Discovery of X-Rays." *Am. J. Phys.* **13**, 281–291 (1945).

WEBER, R. L. *Pioneers of Science* (Institute of Physics, Bristol, Eng., 1980).

ALFRED ROMER

RUTHERFORD, ERNEST

b. Brightwater near Nelson, New Zealand, August 30, 1871; *d.* Cambridge, England, October 19, 1937; *radioactivity, nuclear physics.*

Rutherford's parents came to New Zealand in the mid-nineteenth century from the British Isles looking for economic opportunities in the distant colony. His father was a small-scale farmer, and his mother, before she married, was a teacher. As one of a dozen children, the young Rutherford learned hard work and frugality, characteristics he carried throughout his life. A scholarship took him to Nelson College, a nearby secondary school of high standards, and a subsequent award in 1889 enabled him to attend one of the colony's few institutions of higher education, Canterbury College, in Christchurch. There he came under the influence of both rigorous, conventional professors (such as Charles Henry Herbert Cook) and a dynamic odd-ball (Alexander William Bickerton). The blend apparently suited Rutherford, for he acquired a solid grounding in mathematics and the sciences while absorbing an unquenchable enthusiasm for their study.

He received his B.A. degree in 1892, stayed an additional year on yet another scholarship to obtain an M.A., and remained one year more for a B.Sc. (bachelor of science degree) in 1894. For the B.Sc., he examined ways to magnetize iron by high-frequency electrical discharges. This was not some obscure research topic; less than a decade earlier, Heinrich Hertz had discovered the radio waves predicted by James Clerk Maxwell. Rutherford was able to construct a detector of these waves. Regarded as a rising star, Rutherford was awarded a scholarship that enabled him to study abroad; he chose the Cavendish Laboratory at Cambridge University, since its director, Joseph John Thomson, was the world authority on electromagnetic phenomena. His entire schooling beyond the elementary level was funded by competitive awards, an indication that social class no longer determined who would be educated.

Cambridge University had just changed its rules to allow the admission of graduates of other universities. Rutherford, thus, in 1895 became the first "research student" in the laboratory. At first, Rutherford continued work on his wireless wave detector, increasing its range. Thomson was so impressed that he soon asked Rutherford to join him in an investigation of the newly discovered x rays.

While Rutherford might have hesitated, for he saw some modest commercial use for his detector and his fiancee was far away in New Zealand, he did not envision the large industry that "wireless" would spawn, nor was he an entrepreneur in the style of Guglielmo Marconi. Besides, it was too much of an honor to be asked by his professor to collaborate on the latest physical phenomenon to be discovered. Thomson had spent much of his career examining the discharge of electricity in gases, and their task now was to look at the effect of x rays on this process. They found that equal numbers of positive and negative ions were formed in the gas. This work led to their joint publication of the theory of ionization, and in 1897 to Thomson's determination of the existence of what soon was called the electron.

The work also primed Rutherford to examine the effect of other forms of radiation, particularly that from uranium. This phenomenon was discovered in 1896 by Henri Becquerel, who almost alone had seemed to exhaust interest in the subject. But attention revived in 1898 when Marie Curie and Gerhard Schmidt independently found thorium also to be "radioactive," and Curie and her colleagues then discovered the new elements polonium and radium. However, even before thorium's activity was noted, Rutherford placed uranium in his ionization chamber. By the use of absorbing foils, he showed that "beta" rays could pass through several thicknesses, while "alpha" rays were stopped quickly.

Universities throughout the British Empire customarily looked to Cambridge to fill their chairs in science. In 1898 Rutherford accepted a full professorship at McGill University, which specifically desired a "research star" to further its already admirable reputation. Once in Montreal, he quickly found a gaseous radioactive product emitted from thorium, which he named "emanation" (called thoron today). Rutherford also noticed that some products lost their activity in time, while others, such as uranium and thorium that had not recently been chemically treated, showed no apparent diminution of activity. Most interesting was recognition that the variable products each had a specific rate of decrease—soon to be called its half-life—which could be used to identify it. This was a valuable insight, since the quantities dealt with in most cases were far too small for chemical identification by normal laboratory means.

With a colleague from the chemistry department, Frederick Soddy, Rutherford saw that freshly purified thorium doubled its activity in the same time that thorium X decayed to half value. In 1902 they theorized a genetic relationship in which the parent decayed into a daughter product, which, if radioactive itself, also decayed, and so on in a series. Their transformation theory placed the energy of radioactivity in the atom itself, with decay marking a change from one element into another. Alchemy had been banished from chemistry for more than a century, but this theory reintroduced it in a modern guise.

Rutherford desired to be closer to the "center of gravity" of the scientific community and in 1907 accepted the physics chair in Manchester, after Cambridge the best in Great Britain. Here too he attracted notable colleagues and built up a strong research school. His laboratory accomplishments brought him numerous awards, including the Nobel Prize in chemistry in 1908.

In Montreal he had shown that alpha rays were positively charged, and thus he took to calling them particles. In Manchester, with his assistant Hans Geiger, he devised a technique of counting the flashes caused by these particles striking a scintillating screen; this valuable quantitative tool was carried further by Geiger in electrical and electronic counters. In an exquisite investigation that furthered his reputation as the finest experimenter since Michael Faraday, he showed with Thomas Royds in 1908 that the alpha was a charged helium atom.

Rutherford's most famous discovery was that of the atom's nucleus. Even at McGill he had noticed a slight "fuzziness" in results when working with alpha particles and wondered if they were scattered. In 1909 he gave to undergraduate Ernest Marsden the task of firing alphas from a radioactive source at a foil target and measuring the number of particles deflected at different angles from the centerline. Most, of course, were bent little, if at all, but some were turned through remarkably large angles—more than 90° in some cases. Rutherford's reaction (which gained in the frequent retelling) is famous: "It was almost as incredible as if you fired a fifteen-inch shell at a piece of tissue paper and it came back and hit you."

By 1911 he claimed to know the explanation for such scattering. The dominant picture of the atom, devised by Thomson, was the "plum pudding" model, with plums of electrons in a pudding of positive electrification. This scheme satisfied some quantitative explanations of small-angle scattering, but it could not cope with the newest data. Now, Rutherford said that the atom was mostly empty space, with its electrons roaming in that domain. Its mass, however, was almost entirely concentrated in a tiny region—the nucleus—which was only about 1/10,000 the radius of the entire atom. Since the atom's charge (other than that on its electrons) also was concentrated on the nucleus, alpha particles coming close enough to the nucleus would be strongly affected by their mutual electrostatic repulsion.

The concept of the nuclear atom was largely ignored at first. But Niels Bohr succeeded in 1913 in using it to explain chemical, radioactive, and spectroscopic data, while applying quantum considerations, a major consolidation of phenomena. Another student from Manchester, Henry Gwyn-Jeffreys Moseley, explained that the regular steps in the x-ray spectra of chemical elements were due to the number of positive charges on the atom's nucleus. Nuclear charge (or atomic number), not atomic weight, was the basis for the periodic table of elements. Yet another Manchester alumnus, Kasimir Fajans, in 1913 explained how alpha and beta decay ordered the numerous radioelements in the periodic table, a rule that made sense only if radioactivity was a nuclear phenomenon.

World War I saw some scientific activity conducted (poison gas, anti-submarine warfare), but most young men with scientific talent found themselves fighting in the trenches. With his laboratory thus depleted of research students, and with war work not taking up all his time, Rutherford found occasion to examine some puzzling results obtained by Marsden. When alphas bombarded hydrogen gas, scintillations were seen on a sceen placed far beyond the range of the alphas. This was easily explained: hydrogen nuclei (soon to be called protons) were bounced forward in the collisions. But when nitrogen was substituted, Rutherford saw similar flashes. These, he explained in 1919, were not simple billiard-ball collisions; instead, a nuclear reaction occurred in which the alphas caused the nitrogen to rupture, and the scintillation-causing protons were explosively released.

In 1919 Rutherford succeeded Thomson as director of the Cavendish, the world's premier physical laboratory. With James Chadwick, who had been his student in Manchester, Rutherford showed in the early 1920s that a number of the lighter elements could be disintegrated by an alpha bombardment, just as had occurred with nitrogen. But the heavier elements, with more protons in their nuclei and thus more repulsive charge against an alpha, resisted this assault. Thus, by the late 1920s, Rutherford encouraged students in the construction of machines to accelerate both electrons and protons in hopes of overcoming the potential barrier of heavier elements. John Cockcroft and Ernest Thomas Sinton Walton succeeded with protons in 1932, disintegrating lithium (not a heavy element) and proving in the process Albert Einstein's famous relationship between energy and mass. Ernest Lawrence's cyclotron, invented in Berkeley, California, soon became the most famous of the "atom smashers," but linear accelerators, such as Cockcroft and Walton's, maintained a place in many laboratories. Rutherford, thus, was associated with atomic disintegration by natural means (the 1902 theory of radioactivity), by semi-natural means (placing a natural alpha emitter near nitrogen in 1919), and by artificial means (using accelerated projectiles).

Another major event in 1932 was the discovery of the neutron by Chadwick. Their existence predicted by Rutherford as early as 1920, neutrons were soon seen as constituents of the nucleus, along with protons, while electrons no longer had a role in models of the nucleus. As an uncharged particle, the neutron entered readily into nuclear reactions, leading, after Rutherford's death, to the discovery of fission, a reaction later used for both reactors and bombs. Fusion, another reaction with bombs in its future, was first achieved in 1934, by Mark Oliphant, Paul Harteck, and Rutherford.

Throughout the first four decades of the twentieth century a debate was conducted by leading scientists about the possibility of using the energy known to be locked in the atom. Because of a remark he made at the 1933 meeting of the British Association for the Advancement of Science (BAAS), calling such hopes unrealistic ("moonshine" was the term he used), Rutherford has been associated with those of little vision. In fact, however, he was precisely correct, for he qualified his opinion: "with the means at present at our disposal and with our present knowledge" (*New York Times*, September 12, 1933, p. 1). It took the discovery of fission and a huge wartime project involving scientists from several nations to make useful atomic energy a reality.

Rutherford was one of the leading scientists of his day and willingly accepted the duties as well as the honors. He was president of the Royal Society from 1925 to 1930, president of the BAAS in 1923, a member of government commissions, and lecturer before innumerable academic and business groups. For his services to science, in 1925 King George V conferred on him the Order of Merit, and in 1931 he was raised to the peerage (as Baron Rutherford of Nelson).

Accustomed to work long and hard, Rutherford regularly took time off for golf and enjoyed dinner conversations at Trinity College's high table (where he was a fellow). He and his wife, Mary, were seasoned travelers to parts of the globe by sea, and around the British Isles and the Continent by auto (they owned one of the first vehicles in Manchester).

Liberal in attitude, Rutherford welcomed women as research students in his three laboratories, although there is little evidence that he then took special pains to encourage them; all his students were expected to be self-starters. He seems to have avoided political matters, except on two known occasions. One occurred in 1934, when his Russian student, Peter Kapitza, was not allowed to return to Cambridge from Moscow. Rutherford, who regarded this as an affront to the concept of the internationality of science, orchestrated an unsuccessful campaign to convince the Soviet government to change its mind. The other event was the expulsion from Nazi Germany, in 1933 and later, of hundreds of Jewish scholars. Rutherford accepted the presidency of the Academic Assistance Council, created to find financial aid and jobs for these refugees.

See also: ACCELERATOR, HISTORY OF; ATOM, RUTHERFORD–BOHR; RADIOACTIVITY; RADIOACTIVITY, ARTIFICIAL; RADIOACTIVITY, DISCOVERY OF; SCATTERING, RUTHERFORD

Bibliography

ANDRADE, E. N. DA C. *Rutherford and the Nature of the Atom* (Doubleday, Garden City, NY, 1964).

BADASH, L. *Kapitza, Rutherford, and the Kremlin* (Yale University Press, New Haven, CT, 1985).

BIRKS, J. B., ed. *Rutherford at Manchester* (Heywood, London, 1962).

BUNGE, M., and SHEA, W. R., eds. *Rutherford and Physics at the Turn of the Century* (Neale Watson Academic Publications, New York, 1979).

CHADWICK, J. ed. *Collected Papers of Lord Rutherford of Nelson,* 3 vols. (George Allen & Unwin, London, 1962–1965).

EVE, A. S. *Rutherford* (Cambridge University Press, Cambridge, Eng., 1939).

FEATHER, N. *Lord Rutherford* (Blackie & Son, London, 1940).

OLIPHANT, M. *Rutherford: Recollections of the Cambridge Days* (Elsevier, Amsterdam, 1972).

RUTHERFORD, E. *Radioactive Substances and Their Radiations* (Cambridge University Press, Cambridge, Eng., 1913).

RUTHERFORD, E. *The Newer Alchemy* (Cambridge University Press, Cambridge, Eng., 1937).

RUTHERFORD, E.; CHADWICK, J.; and ELLIS, C. D. *Radiations from Radioactive Substances* (Cambridge University Press, Cambridge, Eng., 1930).

WILSON, D. *Rutherford, Simple Genius* (MIT Press, Cambridge, MA, 1983).

LAWRENCE BADASH

RUTHERFORD SCATTERING

See SCATTERING, RUTHERFORD

RYDBERG ATOM

See ATOM, RYDBERG

RYDBERG CONSTANT

In 1884, using Anders Angstrom's measured wavelengths for four spectral lines from hydrogen, Johann Balmer created a formula which linked those lines in the simple formula

$$\lambda = \frac{m^2}{m^2 - 2^2}\, h,$$

in which, using the original notation, λ is the wavelength for any transition, h is a constant, and m is a running integer starting at 3. This was the first major step in bringing order into the interpretation of atomic spectra.

In 1890 Johannes Rydberg, finding that there were series of spectroscopic doublets and triplets in various elements, with constant separations of the wave numbers of the transitions involving those groupings within each series, developed a formula which allowed him to calculate the several wave numbers. That formula, again using the original notation, is

$$n = n_0 - \frac{R}{(m + p)^2},$$

in which n is a wave number, n_0 is a constant, m is a running integer, p is a constant characteristic of a given element, and R is a universal constant. R, determined empirically from the spectral experiments, was generally referred to as Rydberg's constant.

That constant assumed special significance with the 1913 publication by Niels Bohr of his papers on the structure of atoms. In the first of those papers, he gave a theoretical derivation of the value of R in the form

$$R = \frac{m_e \mu_0^2 c^3 e^4}{8h^3},$$

where m_e is the rest mass of the electron, μ_0 the permeability of free space, c the speed of light, e the electronic charge, and h is Planck's constant. Since then, R has been designated as "the" Rydberg constant. Its significance lies in the fact that, composed merely of fundamental constants, it holds a critical place in the interpretation of spectral lines and, hence, in establishing the structure of the atoms from which those lines originate.

In terms of the Rydberg constant, Balmer's formula may be written as

$$\frac{1}{\lambda} = R\left(\frac{1}{2^2} - \frac{1}{n^2}\right),$$

in which n is the running integer which characterizes the upper state whose decay to state 2 generates the Balmer lines. The good agreement between Bohr's calculated and the best experimental value for R, in addition to his successful application of a generalized Balmer equation to the Lyman, Balmer,

and Paschen lines in hydrogen, showed that the quantization feature of Bohr's model of the hydrogen atom was basically correct; important corrections were provided in later research, particularly in the evolution of nonrelativistic and relativistic quantum theory, the inclusion of nuclear effects, and the 1947 discovery by Willis Lamb and Robert Retherford of the Lamb shift.

Precision in the measurement of energies of atomic levels is fundamental to atomic physics. For example, definitive measurements of the Lamb shift are needed to confirm or correct quantum electrodynamics. The Rydberg constant is a proportionality factor that appears in expressions for the Lamb shift, transition probabilities, and the energies of atomic states. The desired precision can be achieved by, among other things, improved experiments on spectral lines, on transition probabilities, on the fundamental constants which enter into the Rydberg constant, and on the Rydberg constant itself as an entity.

Much work has been done on the Rydberg constant. For example, in the late 1980s, laser studies of the $2\ ^2S_{1/2}$–$4\ ^2P_{1/2}$ second Balmer transition in hydrogen gave R_∞ as $109,737.31573(3)$ cm^{-1}. In this, R_∞ represents the value for an atomic nucleus of infinite mass, eliminating a center-of-mass correction needed for real nuclei. A limitation on the precision of such measurements is imposed by the relatively short mean life of the decaying state, since a brief lifetime is reflected in a broad spectral width of the emitted line.

In principle, it is preferable to study the $1\ ^2S_{1/2}$–$2\ ^2S_{1/2}$ forbidden transition, since the lifetime of the upper state is 10^6 times longer than that of the $2\ ^2P_{1/2}$ states. These measurements involve the two-photon decay process, which is so improbable that its intensity is weak. Nonetheless, the use of powerful lasers in techniques that avoid the line broadening that normally arises from the Doppler effect makes it possible to do such work. As of this writing, the best value is $R_\infty = 109,737.3156834(24)$ cm^{-1}. The precision in this work was limited by uncertainty in the value of an optical frequency determination for the $2\ ^2S_{1/2}$–$8\ ^2S_{1/2}$ transition in hydrogen.

In addition to establishing R_∞ from studies of the one-electron hydrogen atom, positronium and muonium atoms, also one-electron systems, offer unique advantages, and equally unique difficulties. Both, for example, are free of nuclear effects. However, they are weak sources and have short mean lives. Positronium confronts the investigator with relativistic complications, while an appropriate development of optical spectroscopy of muonium is still in progress. Transitions in heavy ions from which all but one electron have been removed also are of interest because the transitions that are forbidden in hydrogen become dominant for sufficiently high values of the nuclear charge. However, such experiments have yet to be performed with precision approaching that of the work cited above.

See also: ATOM, RYDBERG; BOHR'S ATOMIC THEORY; LAMB SHIFT; POSITRONIUM

Bibliography

CAJORI, F. *A History of Physics* (Dover, New York, 1962).

NEZ, F.; PLIMMER, M. D.; BOURZEIX, S.; JULIEN, L.; BIRABEN, F.; FELDER, R.; MILLERIOUX, Y.; and DE NATALE, P. "First Pure Frequency Measurement of an Optical Transition in Atomic Hydrogen: Better Determination of the Rydberg Constant." *Europhys. Lett.* **24,** 635–640 (1993).

ZHAO, P.; LICHENT, W.; LAYER, H. P.; and BERGQUIST, J. C. "Rydberg Constant and Fundamental Atomic Physics." *Phys. Rev. A* **39,** 2888–2898 (1989).

STANLEY BASHKIN